CSPG MEMOIR 16

CLASTIC TIDAL SEDIMENTOLOGY

EDITED BY

D. G. Smith
Department of Geography
University of Calgary
Calgary, Alberta T2N 1N4

G. E. Reinson
Geological Survey of Canada
3303 - 33rd St. N.W.
Calgary, Alberta T2L 2A7

B. A. Zaitlin
Esso Canada Resources
237 - 4th Ave. S.W.
Calgary, Alberta T2P 0H6

R. A. Rahmani
Sirte Oil Company
P.O. Box 385
Tripoli, Libya S.P.L.A.J.

Canadian Society of Petroleum Geologists

Calgary, Alberta
March 1991

Canadian Society of Petroleum Geologists

ISBN 0-920230-51-2
Calgary, Alberta, Canada

March, 1991

Copies may be obtained from:
Canadian Society of Petroleum Geologists
505, 206 - 7th Avenue S.W.
Calgary, Alberta, Canada T2P 0W7
Phone (403) 264-5610

Printed in Canada
Friesen Printers
Altona, Manitoba, Canada

Foreword

In the summer of 1985, the highly successful *First Research Symposium on Clastic Tidal Deposits* was held in Utrecht, The Netherlands. This conference brought together nearly 200 workers involved in the study of processes and sediments in both modern and ancient tidal environments. The Utrecht Symposium culminated in the 1988 publication "Tidal-Influenced Sedimentary Environments and Facies", edited by P. L. de Boer, A. van Gelder and S. D. Nio. The objectives of the Utrecht Symposium were to provide a forum that would focus on tidal deposits, and to bring together workers engaged in research on a wide range of topics in order to provide a free exchange of field observations, ideas and concepts.

The Second Research Symposium, *Modern and Ancient Clastic Tidal Deposits* held at Calgary, Alberta, Canada, August 22-25, 1989 attempted to build on the objectives of the first symposium. A major reason for hosting the second symposium was an awareness by the organizing committee that much of the Mesozoic stratigraphic succession in the Western Canada Sedimentary Basin had been influenced by tides. By holding the symposium in Calgary, we were able to draw upon the expertise of industry geologists. Their experience with tidal rocks in core and outcrop assured the success of the symposium by bridging modern studies to applications within the stratigraphic record. The Energy Resources Conservation Board (ERCB), Core Research Centre, located in Calgary, was utilized for the core and peel workshop, and proximity to accessible outcrops made the Calgary venue ideal for running local field trips. Four field trips in Alberta were organized by R. A. Rahmani, D. A. Leckie, R. J. Cheel and L. Rosenthal. Field trips to the Bay of Fundy, Nova Scotia, and Willapa Bay, Oregon, were led by R. W. Dalrymple and B. A. Zaitlin, and by H. E. Clifton, R. L. Phillips and R. J. Anima, respectively. A core and peel workshop was organized by G. E. Reinson with the kind cooperation and assistance of the many employees at the ERCB Core Research Centre. Two short courses on "criteria for recognizing tidal rocks" were given by S. D. Nio. The poster session was organized by G. Stewart and D. Keith.

As the Calgary symposium progressed, it became apparent to all participants that several major topics had emerged since the last symposium: 1) criteria for recognizing tidal deposits in ancient rocks, 2) tidal rhythmites in ancient rocks, 3) facies models associated with a tripartite zonation of estuary fills, 4) differentation of subtidal facies in flood- and ebb-tidal deltas and 5) subtidal sedimentation in mixed wave and tidally influenced shelves. The volume of new and exciting research presented at the symposium resulted in a consensus that it would be timely to publish a compendium of papers as a Canadian Society of Petroleum Geologists Memoir.

We feel privileged to have been involved in both the organization of the *Second Research Symposium on Clastic Tidal Deposits* and the editing of this volume. The new ideas, concepts and approaches to tidal sediments and depositional processes that have been developed in the four year period since the last symposium, and the dynamic nature of research presently ongoing, are amply demonstrated in the contents of this volume.

We are pleased to announce that the success of the Utrecht and Calgary meetings has encouraged Burg Flemming of the Senckenberg Institute, Wilhelmshaven, Germany, to organize the *Third International Research Symposium on Modern and Ancient Clastic Tidal Deposits* to be held August 25-28, 1992.

Calgary, March 1991

Derald G. Smith
Gerald E. Reinson

Brian A. Zaitlin
Ray A. Rahmani

Acknowledgments

All the manuscripts were technically evaluated by at least two external reviewers and one or more of the editors, and finally by the production editor. Editing for consistency, grammar and format style, and overseeing the printing of the volume were handled by Jo MacGillivray. The co-editors are grateful to their employers, The University of Calgary, The Geological Survey of Canada, Esso Resources Canada Ltd., and Canadian Hunter Exploration Ltd. for their understanding and assistance from 1987, during the early organization of the symposium, to 1991, and the final production of this book. We wish also to acknowledge the following: Debbie Little, Anne Wilson, Donna Limpert and Joyce Rowden, Department of Geography, University of Calgary, who provided the bulk of secretarial and support services; Eleanor Gordon, Geological Survey of Canada, who provided additional secretarial support as required, and Sylvie Pinard, who translated the abstracts into French.

Last, but not least, a volume such as this would not be possible without input from numerous peers who provided expertise as manuscript reviewers. These reviewers are listed in alphabetical order below:

G. Allen
C. Amos
A. Archer
W. Arnott
G. Ashley
K. Bohacs
J. Boothroyd
J. Bridge
D. Cant
R. Cheel
E. Clifton
R. Dalrymple
R. Davis Jr.
P. de Boer
J. Dionne
R. Dott
W. Duke
A. Embry
K. Ericksson
P. Flach
D. Forbes
M. Gibling
B. Flemming
R. Ginsburg
B. Hayes
M. Hayes
S. Harding
F. Hein
A. Hine
P. Homewood
J. Hopkins
D. James
G. Johnson
J. Knight
R. Kreisa
L. Krystinik
E. Kvale
D. Leckie
M. Lerand
B. Long
J. Luternauer
G. Middleton
T. Moslow
G. Narbonne
M. Nichols
D. Nummedal
G. Plint
P. Putnam
B. Ricketts
B. Rust
R. Schwartz
R. Slingerland
E. Simpson
J. Suter
J. Syvitski
B. Thom
H. Wanless
G. Williams

Table of Contents

Nearshore and Shelf Systems

In Memoriam

GORDON A. STEWART
1934-1991

Gordon Stewart died suddenly on January 12, 1991 at the age of 57 years. At the time of his death he was employed by Saskoil as Technical Advisor and Chief Staff Geologist. Gordon had been a member of the C.S.P.G. since 1955, and throughout the years, served on several organizing committees for conventions, symposia and specific C.S.P.G. activities. Gordon was part of the organizing committee for the Second Research Symposium on Clastic Tidal Deposits upon which this Memoir is based. It is thus fitting that Gordon should be remembered here. This volume is an especially suitable one in which to cite Gordon's contributions in the early recognition of tidal and estuarine facies in the Cretaceous of the Western Canada Sedimentary Basin. In his work (early 1970's) on the McMurray Formation, Gordon was one of the first to propose estuarine, marine-influenced environments for the Athabasca oil sands. At that time such an interpretation was viewed as highly controversial, though it is widely accepted today. Gordon will always be remembered by those who knew him as a consummate professional whose work ethics were guided by honesty and integrity. He will be sorely missed in the Calgary geological community.

D. G. Smith
G. E. Reinson

B. A. Zaitlin
R. A. Rahmani

In: Clastic Tidal Sedimentology, D. G. Smith, G. E. Reinson, B. A. Zaitlin and R. A. Rahmani (eds.). Canadian Society of Petroleum Geologists. Memoir 16, p. ix-xv.

A short historical review of clastic tidal sedimentology

Gerard V. Middleton

Department of Geology
McMaster University
Hamilton, Ontario L8S 4M1

Abstract

The best studied modern tidal environments are tidal flats, tidal inlets and deltas, and subtidal shelf sand banks (including tidal sand ridges and fields of large-scale dunes). At present, much sediment is being deposited in large, tide-dominated river deltas, which remain little investigated and understood. The best criteria for recognition of ancient tidal sediments are: 1) tidal bundles and evidence for other neap-spring rhythms; 2) flaser, wavy and lenticular bedding, particularly where present in 3) fining-upward cycles with supporting evidence for deposition on prograding tidal flats; and 4) crossbedding with evidence for current reversals, such as opposed cross-strata, regularly spaced reactivation surfaces, and mud drapes.

A distinction should be made between morphological elements, such as tidal flats, clearly formed by tidal action, and environmental systems, such as estuaries, composed of a large number of morphological elements, only some of which are tidal.

Résumé

Les environnements de marée les mieux étudiés sont les estrans, passes («tidal inlets») et deltas, ainsi que les bancs de sables sur la plate-forme de la zone infratidale (incluant les crêtes de sables de marée et les champs de dunes de grandes dimensions). Actuellement, beaucoup de sédiments sont déposés dans d'importants deltas de rivière dominés par les marées dont l'étude et la compréhension restent encore limitées. Les meilleurs critères pour reconnaître les sédiments anciens de marée sont: 1) les «faisceaux» («bundles») de marée et les évidences pour d'autres séquences rythmiques mortes-eaux et vives-eaux; 2) les lits de type «flaser», onduleux et lenticulaires, plus particulièrement présents dans les cycles mentionnés ci-après; 3) les cycles à granulométrie décroissante vers le sommet, avec des évidences supportant une sédimentation sur un estran en progradation; et 4) les stratifications obliques avec des évidences de renversements de courants, telles que des couches obliques s'opposant, des surfaces de réactivation régulièrement espacées et des «lentilles de boues pélagiques» («mud drapes»).

Il s'avère important de distinguer les éléments morphologiques, comme par exemple entre les estrans distinctement formés par l'action des marées et les systèmes environnementaux tels que les estuaires qui eux sont composés d'un grand nombre d'éléments morphologiques dont seulement quelques-uns se rapportent aux marées.

Introduction

This paper is a shortened version of the introductory address given at the Second International Research Symposium on Clastic Tidal Deposits. My involvement in research on tidal sedimentation dates from 1968 when George deVries Klein asked me to be a "visiting critic" at a National Science Foundation Seminar he was running in Parrsboro, Nova Scotia. My experiences that summer convinced me that the Bay of Fundy was a magnificent place to study the processes of sediment transport and bedform development. Every twelve and a half hours, Nature runs an experiment in one of the world's largest flumes, and even drains the water out, leaving sedimentologists several hours to sample the sediment, measure the bedforms, and dig trenches.

My main interest in the Bay of Fundy was, and still is, as a place to run natural experiments, but of course the Bay has been studied by those interested specifically in tidal environments, rather than more generally in sediment transport. Studies on the Bay of Fundy began at McMaster University long before we were interested (Swift and McMullen, 1968; Klein, 1970) and have involved workers in several different institutions. Summaries have been given by several authors (*e.g.*, in McCann, 1980; Amos, 1978, 1987; Yeo and Risk, 1981; and Dalrymple *et al.*, 1982, 1990). It can now be claimed that the Bay of Fundy is at least as well understood as any other macrotidal area in the world.

The purpose of this paper is not to describe the Bay of Fundy, but to review the main environments of deposition of modern tidal sediments, and to discuss some problems that geologists have in identifying tidal sediments in the stratigraphic record. I have had the opportunity of seeing modern tidal sediments on the east coast of North America, in the Dutch Wadden Sea, the German North Sea, in Mont St. Michel Bay, the Gironde estuary and Bay of Arcachon areas of France, and in the Severn estuary in England.

The most spectacular ancient examples of tidal sediments I have seen are those exposed in Northern Spain (mostly described in the IAS Guidebook edited by Mila and Rosell, 1985). I thank all those who have shown me their field areas and discussed with me their work on tidal environments. The review that follows does not seek to be comprehensive but selects only a few studies, based on my own experience.

Tidal Sedimentary Environments

Tidal Flats

The first systematic geological studies of tidal environments were of tidal flats. The great German paleontologist Rudolf Richter began these studies in the late 1920s, and set up the Senckenberg Institute at Wilhelmshaven (for an early review in English, see Häntzschel, 1939). At the same time, geologists were much influenced by the work of Van Straaten (1953, 1954). Van Straaten's work had the advantage of being published in English, and it offered a basic explanation of why and how sediment accumulated on mud flats, and where it came from — the sea (Van Straaten and Kuenen, 1957; for later elaboration and refinement of this work see Postma, 1961, 1967; Eisma and Kalf, 1987). The work also provided a model that could be used to identify mud flat sequences in the stratigraphic record. Van Straaten visited Calgary in 1960, and the Alberta Society of Petroleum Geologists (as the Canadian Society of Petroleum Geologists was then called) published his influential review (Van Straaten, 1961). The Van Straaten model essentially consisted of a fining-upward sequence, resulting from the combined action of lateral migration of tidal channels and creeks, and progradation of mud flats and tidal marshes as a result of continued supply of fine sediment from the sea, produced by tidal action. Reineck further developed the tidal creek model (as did Oomkens and Terwindt, 1960). This model was a significant advance in our general understanding of sedimentation in channels, and preceded the work on meandering rivers by Bernard and Major (1963) and Allen (1963). The Dutch and German work exhaustively documented the sedimentary facies typical of the North Sea and other modern tidal flats (for summaries of this work and bibliography see Reineck and Singh, 1980). It was followed by studies of many other tidal flats, in both terrigenous and carbonate environments, and there are now a number of well documented ancient examples, many of which are conveniently compiled in the book edited by Ginsburg (1975).

Out of this work came the first criteria for identifying ancient tidal sediments: 1) fining-upward sequences, produced in some cases by the migration of tidal creeks, but probably more commonly simply by the progradation of tidal marsh and muddy tidal flat over sandy flat and channel deposits; and 2) the interbedding of thin strata of rippled sand and mud that forms the structures Reineck and Wunderlich (1968) called flaser, wavy, and lenticular bedding. Of course, the identification of true tidal flats is complicated by the fact that, where tidal range is low, "meteorological tides" (water level fluctuations produced by changing winds) may be even more important than true tides. Though progradation of tidal flats may be very rapid (LeFournier and Friedman, 1974), sea level change due to eustasy and tectonics may also be very rapid, which makes it hard in practice to use the thickness of the progradational cycle as a measure of tidal range (as proposed by Klein, 1971).

Tidal Shelves

The next major environment described was the sandy tidal shelf environment. Pioneer studies by Van Veen (1935) were followed by the well known studies of Arthur Stride and his associates at the Institute of Oceanographic Sciences in England (Stride, 1963, 1982). These studies concentrated on the continental shelves around the British Isles, and led to a model of shelf sands reworked from older deposits into sand sheets and sand banks during marine transgression. Much of this work made expert use of the technique of Side Scan Sonar to reveal details of shelf bottom morphology in a way that had not been possible earlier (Belderson *et al.*, 1972). The sonographs revealed a heirarchy of large-scale bedforms, with some of the larger ones oblique to the flow, covered by smaller forms nearly transverse to the dominant tidal flow. The larger forms can be seen intertidally in regions of very large tides, such as the Bay of Fundy. Klein (1970) described them and named the larger bedforms "sand waves" and the smaller ones "dunes". The term "sandwaves" (spelled as one word) had been popularized by Stride and his associates, who always maintained that there was a complete spectrum in size and no fundamental difference between large and small forms. Though Klein used the term "dune", most other workers in intertidal areas followed Van Straaten (1953) and used the term "megaripple" to describe the smaller, dune-like forms seen in intertidal areas. At the time, it seemed a good term to use, so long as the term "dune" was considered by most geologists to apply only to subaerial dunes. This is no longer the case, and those working in fluvial environments commonly use the term "dune" for features that are very like many intertidal "megaripples". Though it is certainly useful to continue to distinguish different scales of tidal bedforms, a group of representative research workers, meeting at the 1987 SEPM meeting in Austin, Texas, agreed that the preferred term for all such forms (larger than current ripples) should be "dunes" (Ashley *et al.*, 1990).

Mapping showed that, besides "sandwaves", many sand banks had very large "sand ridges", the crests of which were almost parallel to the dominant tidal flows. Off (1963) pointed out that "sandwaves" and the larger tidal sand ridges were found in many other parts of the world besides the North Sea. Work by Houbolt (1968) and by many others working on modern ridges, has established that such ridges result from flow that is slightly oblique to the ridge axis, and that the flow patterns lead to migration of dunes ("sandwaves") obliquely up the gentle slope of the ridge. At the same time, Klein and Houbolt independently showed that sediment tends to be circulated around the ridge. It follows that the ridges migrate only very slowly compared with the rates at which sediment or even large dunes move, and that most of the outcrop-scale structures in sand ridge deposits are related to dune migration. The same is also true of many large "sandwaves", which have smaller "megaripples" migrating over them (Dalrymple, 1984).

The internal structure of large tidal dunes and ridges remained speculative for many years, and is still incompletely known. Our knowledge is increasing thanks to improved shallow seismic techniques (Berné *et al.*, 1988), theoretical studies by Allen (1980, 1982), box coring (Houbolt, 1968), trenching (Dalrymple, 1984) and outcrop studies of a few probable ancient examples (for a review see Walker, 1985).

Though the model of the European shelves has been widely used by geologists, considerable doubt has arisen about the identification of such deposits in the stratigraphic record. A number of linear sand bodies have been interpreted as ancient tidal ridges, including some from the subsurface of Western Canada (*e.g.*, Evans, 1970). Studies in progress, however, do not confirm a tidal origin for many examples. Few convincing ancient analogues have been described, in spite of the abundance of tidal ridges on modern shelves (for some of the best ancient examples, see Houboult, 1982; Houthuys and Gullentops, 1988a,b).

Tidal Inlets and Deltas

Geologists have long been interested in barrier island systems. Davies (1964) classified coasts into micro-, meso-, and macrotidal. Miles Hayes (1975, 1979) eleborated this classification and showed how the coastal forms varied in importance from one type of coast to another. As wave height and tidal range vary along a coast (Nummedal *et al.*, 1977) so do the morphologies of tidal inlets and the distribution of sand bodies in their tidal deltas. Hayes (1979) used plots of wave height against tidal range to distinguish four basic types of inlets, from wave dominated (Florida) through mixed wave-tidal (U.S. east coast) to tide dominated (Severn estuary or Bay of Fundy). Typical mesotidal inlets have a large ebb-tidal delta. Flow ebbs through a main channel, with elevated channel-margin linear bars, and builds a large, delta-like terminal lobe. The top of this lobe is modified by wave action. Ebb flow tends to persist through the main channel even after the tide has turned, which leads to the development of flood dominated channels on each side of the ebb delta. The flood delta is built into shallower water, where there is much less wave energy. Flood flow constructs a tidal delta with a prominent flood ramp, by which dunes migrate. The highest part of the delta is generally a shallow, sandy tidal flat, and the margin is a relatively steep ebb shield, which tends to deflect the ebb currents around the margins of the delta, where they form ebb spits. Studies of box cores taken by divers, and of morphology monitored over several years, have revealed a lot about the structural sequences characteristic of tidal deltas. Hayes' model has served well for the interpretation of ancient tidal inlets (*e.g.*, Hobday and Horne, 1977), but the number of well described ancient examples is still small. Most investigations carried out in the Paleozoic of eastern North America and in the Mesozoic of western North America have discovered more evidence of wave domination than of tidal deltas. Tidal inlets and deltas are, however, only a small part of the more general barrier island model. For a more complete review than can be given here see Reinson (1984), Boothroyd (1985) and Elliott (1986).

Though Hayes' models were developed mainly for mesotidal deltas, Lambiase (1980) and Dalrymple *et al.* (1990) found that the basic ideas about tidal delta morphology were also quite useful even in strongly macrotidal areas, such as the Bay of Fundy. In particular, the outer Cobequid Bay sandbar complex can be considered (to a first approximation) to be a huge ebb tidal delta.

Estuaries

A substantial part of the literature about estuaries has been concerned with their definition. Though an estuary can be defined in many different ways, geomorphologists and geologists have generally preferred a definition based on geological history (''a drowned valley'') or general ''funnel shaped'' morphology rather than one based on salinity and mixing characteristics (Fairbridge, 1980; Nichols and Biggs, 1985). Because so many different areas have been called estuaries (from Mississippi delta distributaries to Cobequid Bay), and because they include regions of high population, and support various economic activities, they have been extensively studied by many different types of scientists, and the literature is vaste. The volume edited by Lauff (1967) was influential in its day, and was followed by numerous multidisciplinary symposia (Nelson, 1972; Cronin, 1975; and many later volumes sponsored by the Estuarine Research Foundation). Important examples are the Chesapeake Bay and James estuaries (Nichols and Biggs, 1985), the estuaries of the Seine (Avoine *et al.*, 1981) and Gironde (Allen, 1972; Allen *et al.*, 1980) in France, and the Severn estuary in Britain (Allen and Rae, 1988; Harris and Collins, 1985).

An important physical concept for understanding estuaries was derived by Pillsbury (1956) and Langbein (1963): an ''ideal'' estuary should respond to tidal exchange of water by adjusting its shape so that the velocity averaged over the cross-sectional area remains constant in the region where tides are the dominant geomorphological process. This provides an explanation for the ''funnel'' shape of many estuaries (the best example is perhaps the Gironde) and has sedimentological consequences which have so far been only incompletely explored (but see Zaitlin, 1987).

Because most estuaries (as generally understood by geologists) occupy partially drowned river valleys, they show a transition from an upper fluvially dominated section, which is generally being filled with relatively sandy sediment of fluvial origin, a central muddy section, and a section near the mouth dominated by tidal processes, where much of the sediment, including some sand, has been introduced from the coast. This tripartite model for estuaries is well described by Nichols and Biggs (1985) and has been applied to ancient sediments by Rahmani (1988) and by several authors contributing to the core workshop at the Calgary meeting (Reinson, 1989).

An important phenomenon first described from modern subtidal estuarine sediments in Holland (Boersma, 1969) and

Table 1. Delta type for rivers with highest sediment discharge (modified after Milliman and Meade, 1983)

Rank	River	Sediment discharge (million tons/yr.)	Dominance	Tide range (m)
1	Ganges-Brahmaputra	1670	Tidal	3.6
2	Huang Ho (Yellow)	1080	Fluvial	1.1
3	Amazon	900	Tidal	4.9
4	Chang Jiang (Yangtze)	478	Tidal	3.7
5	Irrawaddy	285	Tidal	2.7
6	Magdalena	220	Fluvial/Wave	1.1
7	Mississippi	210	Fluvial	0.4
8	Orinoco	210	(Wave/ Fluvial/Tidal)	1.8
9	Hungho (Red)	160	?Tidal	?4 +
10	Mekong	160	Tide/Wave	2.6
11	Indus	100	Tidal	2.6
12	MacKenzie	100	Fluvial	0.3

popularized by Visser (1980), but soon discovered in many ancient deposits, particularly in Europe (Allen, 1982; Allen and Homewood, 1984; Yang and Nio, 1985), is the rhythmic variation in crossbed thickness typical of some subtidal dune deposits. These deposits are known as "tidal bundles". They are characteristically subtidal, but not necessarily estuarine, and are described in more detail later.

Tide-Dominated River Deltas

Strangely enough, we now associate the term tidal delta mainly with the relatively small deltas found in barrier island systems. These deltas derive their sediment mainly from longshore drift. But most terrigenous sediment is carried down to the oceans by rivers, and much of that by only a few, major rivers. Only a small part of that sediment is extensively redistributed by longshore drift. Most river borne sediment is deposited in river deltas. Milliman and Meade (1983) rank the rivers of the world by their sediment discharge. Table 1 shows their data, modified to include the type of delta (following Galloway, 1975) and the tidal ranges (mainly from Coleman and Wright, 1975). What is most striking about this table is that 8 of the 12 largest deltas in the world are either dominated by tidal processes or show strong tidal influence. It is also remarkable that we know very little about the nature of the sediment in most of these deltas. Perhaps the best documented of these major tide-dominated deltas is the Chang Jiang (Yangtse) — but the available documentation in English (in Ren Mei-e, 1986, p. 264-318; and in Kennedy, 1982, p. 649-704) valuable though it is, is hardly an adequate guide to the identification of ancient analogues.

Summary

It is apparent from this brief review of tidal environments that both tidal deposits and environments occur on many different scales (Table 2). The smallest scale (not included in the table) is the microenvironment in which individual structures formed. Substantial bodies of tidal sediments may be considered to correspond to *morphological elements* in which tidal currents predominated over other processes. Of course, a much more detailed heirarchy of such elements could be recognized than is shown in Table 2. Tidal morphological elements are combined togther with others in which other processes dominate to form *environmental systems*. Thus the two columns of Table 2 only roughly correlate with each other: for example, intertidal flats are found on Barrier Island - Spit Coasts, but are also found fringing estuaries and tidal bays. Though morphological elements are small enough to be relatively clearly defined, there are many variations of major environmental systems and transitions between them: for example, there is a continuous spectrum of systems from tidal bays, through estuaries, to river-mouth tidal deltas, as the discharge of river sediment into the system increases.

Criteria for Recognition of Tidal Sediments

A number of authors have reviewed the problem of recognizing ancient tidal sediments (Van Straaten, 1961; De Raaf and Boersma, 1971; Reineck and Singh, 1980; Bridges, 1982; Clifton, 1982; Weimer *et al.*, 1982; Klein, 1985; Walker, 1985;

Table 2. Distinction between morphological elements and environmental systems (see text for explanation)

Morphological elements	Environmental systems
Intertidal flats (marsh, flats, creeks, etc.)	Barrier island–spit coasts
Tidal inlets and deltas (ebb and flood deltas, channel margin bars, etc.)	Tidal estuaries and bays
Tidal sand banks (ridges, dune fields, sand ribbons, etc.)	Tidal shelves
River-mouth tidal distributaries (poorly known)	Tide-dominated river deltas

Elliot, 1986; Johnson and Baldwin, 1986; Terwindt, 1988), and there is no need to give a comparable review here.

A sceptical historian could recognize several cycles of enthusiasm and disillusionment in the search for ancient tidal deposits: for example, the tidal flats cycle (1950-1975), the offshore sands cycle (1963-1982) and the barrier island/tidal delta cycle (1975-1985). The range of dates assigned is, of course, debatable, and the list is not complete.

A recent period of enthusiasm was set in motion by the recognition of cyclic *bundles* of tidal cross-strata, in Holocene tidal channel deposits exposed in large excavations in the Netherlands (Visser, 1980; the term "bundle" seems to have been first used in this context by Boersma, 1969). The exposures were extraordinary because the deposits clearly displayed evidence of neap-spring semidiurnal tidal cycles, and because the development of the entire deposit took place during historical times, well documented by careful hydrographic surveys of the channel (Van den Berg, 1982; Siegenthaler, 1982; Nio *et al.*, 1983). Within five years, convincing ancient examples had been described from the lower Greensand (Allen, 1982), the Molasse (Homewood and Allen, 1981; Allen, 1984; Allen and Homewood, 1984), and the Roda Sandstone (Yang and Nio, 1985). The hydraulic interpretation of these structures has been studied carefully by several of the authors cited. It seems clear that rather special hydraulic conditions are necessary for the preservation of such clear evidence for astronomical tides: 1) the environment must be subtidal, probably in a laterally migrating tidal channel, with relatively weak, strongly asymmetrical tidal currents; 2) the predominant tide current is able to transport sand, and the subordinate current very little, so that there is little erosion of the cross-strata formed by the predominant tidal phase; and 3) an environment rich in suspended mud (in flocculated or organically pelleted form) is necessary so that each tidal sand layer can be clearly separated from the next. Such conditions are achieved only fortuitously. When they are, the cyclic nature of the deposits (shown by the systematic variation in measured thickness of individual bundles) constitutes very convincing evidence of tidal origin.

Dalrymple and Makino (1989), Williams (1989), and Tessier and Gigot (1989) have described other well documented examples of tidal rhythms, but generally, well preserved cyclicity cannot be documented. Recognition of crossbedding formed by reversing tidal currents is then not such a simple matter, as Allen (1980), Boersma and Terwindt (1981), and Dalrymple (1984) have shown. Klein (1970) suggested that regularly spaced reactivation surfaces within crossbedding sets constituted one line of evidence, and this idea has been further developed and elaborated into the concept of *tidal sigmoids* by Mutti *et al.* (1984; 1985). Such structures have been documented in modern tidal deposits, including the intertidal channel bars of the Bay of Arcachon (Fenies, 1984). Currently we are in a phase of enthusiastic application of this idea (Kreisa and Moiola, 1986; Smith, 1988). The same general type of structure is also produced on a larger scale, by migrating tidal bars (Mutti *et al.*, 1985).

Conclusions

1. The best studied modern tidal environments are tidal flats, tidal inlets and deltas, and subtidal sand banks (including sand ridges and fields of large-scale dunes).

2. The tidal environments in which most modern sediment is being deposited are found in large, tide-dominated river deltas; in comparison with the environments listed under (1) they have been very little studied.

3. Recognition of ancient tidal sediments is still not easy, as can be seen from the number of disputed identifications. The best criteria are: a) evidence for neap-spring (or other tidal) cycles, especially as shown by tidal bundles; b) flaser wavy and lenticular bedding, in fining-upward cycles the features and fauna of which support deposition on prograding tidal flats; and c) crossbedding with evidence of current reversals, including opposed crossbeds, regularly spaced reactivation surfaces, and mud drapes. Ideally, but by no means always in reality, these together produce the phenomenon called tidal sigmoids.

4. It may help to systematize discussion of tidal sediments to distinguish more clearly between *morphological elements*, such as tidal flats, tidal channels, and tidal ridges, which are clearly formed dominantly by tidal action, and *environmental systems*, such as barrier islands, estuaries, and tide-dominated shelves or river deltas, which are complexes generally composed of a wide variety of morphological elements, only some of which are tidal.

Acknowledgments

I thank particularly George deVries Klein for introducing me to the Bay of Fundy; my one-time students (especially Bob Dalrymple, John Knight and Joe Lambiase) who have taught me most of what I know about the Bay of Fundy; George Allen and Hugue Fenies for showing me the Gironde estuary and the Bay of Arcachon, Emiliano Mutti for his incomparable enthusiasm, and Ray Rahmani and Derald Smith for inviting me to participate in the Second International Research Conference on Clastic Tidal Deposits.

References

Allen, G. P. 1972. Etude des Processus Sédimentaires dans l'Estuaire de la Gironde. Université de Bordeaux I, Thèse de Doctorat d'Etat ès Science Naturelles, 314 + xx p.

______ , Salomon, J. C. Bassoullet, P., Du Penhoat, Y. and De Grandpré, C. 1980. Effects of tides on mixing and suspended sediment transport in macrotidal estuaries. Sedimentary Geology, v. 26, p. 69-90.

Allen, J. R. L. 1963. Henry Clifton Sorby and the sedimentary structures of sands and sandstones in relation to flow conditions. Geologie en Mijnbouw, v. 42, p. 223-228.

______ 1980. Sand waves: a model of origin and internal structure. Sedimentary Geology, v. 26, p. 281-328.

______ 1982. Mud drapes in sand-wave deposits: a physical model with application to the Folkestone Beds (Early Cretaceous, southeast England). Philosophical Transactions of the Royal Society, London, v. 306A, p. 291-345.

______ and Rae, J. E. 1988. Vertical salt-marsh accretion since the Roman period in the Severn estuary, southwest Britain. Marine Geology, v. 83, p. 225-235.

Allen, P. A. 1984. Reconstruction of ancient sea conditions with an example from the Swiss Molasse. Marine Geology, v. 60, p. 455-473.

Allen, P. A. and Homewood, P. 1984. Evolution and mechanics of a Miocene tidal sandwave. Sedimentology, v. 31, p. 63-81.

Amos, C. L. 1978. The post-glacial evolution of the Minas Basin, N.S. A sedimentological interpretation. Journal of Sedimentary Petrology, v. 48, p. 965-982.

______ 1987. Fine-grained sediment transport in Chignecto Bay, Bay of Fundy, Canada. Continental Shelf Research, v. 7, p. 1295-1300.

Ashley, G. M., Symposium Chairperson, SEPM Bedforms and Bedding Structures, 1990. Classification of large-scale subaqueous bedforms: a new look at an old problem. Journal of Sedimentary Petrology, v. 60, p. 160-172.

Avoine, J., Allen, G. P., Nichols, M., Salomon, J. C. and Larsonneur, C. 1981. Suspended-sediment transport in the Seine Estuary, France: effect of man-made modifications on estuary-shelf sedimentology. Marine Geology, v. 40, p. 119-137.

Belderson, R. H., Kenyon, N. H., Stride, A. H. and Stubbs, A. R. 1972. Sonographs of the Sea Floor. Amsterdam, Elsevier Publishing Company, 185 p.

Bernard, H. A. and Major, C. F. 1963. Recent meander belt deposits of the Brazos River: an alluvial "sand" model. American Association of Petroleum Geologists, Bulletin, v. 47, p. 350-351 (abstract).

Berné, S., Auffret, J-P. and Walker, P. 1988. Internal structure of subtidal sandwaves revealed by high-resolution seismic reflection. Sedimentology, v. 35, p. 5-20.

Boersma, J. R. 1969. Internal structure of some tidal mega-ripples on a shoal in the Westerschelde Estuary, The Netherlands. Report of a preliminary investigation. Geologie en Mijnbouw, v. 48, p. 409-414.

______ and Terwindt, J. H. J. 1981. Neap-spring tide sequences of intertidal shoal deposits in a mesotidal estuary. Sedimentology, v. 28, p. 151-170.

Boothroyd, J. C. 1985. Tidal inlets and tidal deltas. *In:* Coastal Sedimentary Environments, R. A. Davis Jr. (ed.). New York, Springer-Verlag, Second Edition, p. 445-532.

Bridges, P. H. 1982. Ancient offshore tidal deposits. *In:* Offshore Tidal Sands, Processes and Deposits, A. H. Stride (ed.). London, Chapman and Hall, p. 172-192.

Clifton, H. E. 1982. Estuarine Deposits. *In:* Sandstone Depositional Environments, P. A. Scholle and D. Spearing (eds.). American Association of Petroleum Geologists, Memoir 31, p. 179-189.

Coleman, J. M. and Wright, L. D. 1975. Modern river deltas: variability of process and sand bodies. *In:* Deltas, Models for Exploration, M. L. Broussard (ed.). Houston Geological Society, p. 99-149.

Cronin, L. (editor) 1975. Estuarine Research, Volume II, Geology and Engineering. New York, Academic Press, 587 p.

Dalrymple, R. W. 1984. Morphology and internal structure of sandwaves in the Bay of Fundy. Sedimentology, v. 31, p. 265-382.

______ and Makino, Y. 1989. Description and genesis of tidal bedding in the Cobequid Bay-Salmon River Estuary, Bay of Fundy, Canada. *In:* Sedimentary Facies in the Active Plate Margin, A. Taira and F. Masuda (eds.). Terra Scientific Publishing Company, Tokyo, p. 151-177.

______ , Amos, C. L. and McCann, S. B. 1982. Beach and nearshore depositional environments of the Bay of Fundy and southern Gulf of St. Lawrence. Eleventh International Congress of Sedimentology, McMaster University, Hamilton, Ontario, Guidebook Excursion 6A, 116 p.

______ , Knight, R. J., Zaitlin, B. A. and Middleton, G. V. 1990. Dynamics and facies model of a macrotidal sand bar complex, Cobequid Bay-Salmon River Estuary (Bay of Fundy). Sedimentology, v. 37, p. 577-612.

Davies, J. L. 1964. A morphogenetic approach to world shorelines, Zeitschrift fur Geomorphologie, v. 8 (special number), p. 127-142.

De Raaf, J. F. M. and Boersma, J. R. 1971. Tidal deposits and their sedimentary structures (seven examples from Western Europe). Geologie en Mijnbouw, v. 50, p. 479-504.

Eisma, D. and Kalf, J. 1987. Dispersal, concentration and deposition of suspended matter in the North Sea. Journal of Geological Society of London, v. 144, p. 161-167.

Elliott, T. 1986. Siliciclastic shorelines. *In:* Sedimentary Environments and Facies, Second Ed., H. G. Reading (ed.). Blackwell Scientific Publishing, p. 155-188.

Evans, W. E. 1970. Imbricate linear sandstone bodies of Viking Formation in Dodsland-Hoosier area of southwestern Saskatchewan. American Association of Petroleum Geologists, Bulletin, v. 54, p. 469-486.

Fairbridge, R. W. 1980. The estuary: its definition and geodynamic cycle. *In:* Chemistry and Biogeochemistry of Estuaries, E. Olausson and I. Cato (eds.). New York, John Wiley and Sons, p. 1-35.

Fenies, H. 1984. Faciès séquences et géometrie des depôts de chenaux de marées du Bassin d'Arcachon: une lagune mesotidale. Thèse 3e Cycle, Docteur en Océanologie, Université de Bordeaux I, 278 p.

Galloway, W. E. 1975. Process framework for describing the morphologic and stratigraphic evolution of deltaic depositional systems. *In:* Deltas, Models for Exploration, M. L. Broussard (ed.). Houston Geological Society, p. 87-98.

Ginsburg, R. N. (editor) 1975. Tidal Deposits. A Casebook of Recent Examples and Fossil Couterparts. New York, Springer-Verlag, 428 p.

Häntzschel, W. 1939. Tidal flat deposits (Wattenschlick). *In:* Recent Marine Sediments, P. D. Trask (ed.). Society of Economic Paleontologists and Mineralogists, p. 195-206.

Harris, P. T. and Collins, M. B. 1985. Bedform distribution and sediment transport paths in the Bristol Channel and Severn Estuary, U.K. Marine Geology, v. 62, p. 153-166.

Hayes, M. O. 1975. Morphology of sand accumulation in estuaries: an introduction to the symposium. *In:* Estuarine Research, v. 2, Geology and Engineering, L. E. Cronin (ed.). New York, Academic Press, p. 3-22.

______ 1979. Barrier island morphology as a function of tidal and wave regimes. *In:* Barrier Islands — from the Gulf of St. Lawrence to the Gulf of Mexico, S. P. Leatherman (ed.). New York, Academic Press, p. 1-27.

Hobday, D. K. and Horne, J. C. 1977. Tidally influenced barrier-island and estuarine sedimentation in the Upper Carboniferous of southern West Virginia. Sedimentary Geology, v. 18, p. 97-122.

Homewood, P. and Allen, P. 1981. Wave-, tide- and current controlled sand bodies of Miocene Molasse, western Switzerland. American Association of Petroleum Geologists, Bulletin, v. 65, p. 2534-2545.

Houbolt, J. J. H. C. 1968. Recent sediments in the southern Bight of the North Sea. Geologie en Mijnbouw, v. 47, p. 701-710.

______ 1982. A comparison of recent shallow marine tidal sand ridges with Miocene sand ridges in Belgium. *In:* The Ocean Floor, R. A. Scrutton and M. Talwani (eds.). New York, John Wiley and Sons Ltd., p. 69-80.

Houthuys, R. and Gullentops, F. 1988a. The Vlierzele Sands (Eocene, Belgium): a tidal ridge system. *In:* Tide-Influenced Sedimentary Environments and Facies, P. L. deBoer, A. van Gelder and S. D. Nio (eds.). Boston, D. Reidel Publishing Company, p. 139-152.

______ and ______ 1988b. Tidal transverse bars building up a longitudinal sand body (Middle Eocene, Belgium). *In:* Tide-Influenced Sedimentary Environments and Facies, P. L. deBoer, A. van Gelder and S. D. Nio (eds.). Boston, D. Reidel Publishing Company, p. 153-166.

Johnson, H. D. and Baldwin, C. T. 1986. Shallow siliciclastic seas. *In:* Sedimentary Environments and Facies, Second Ed., H. G. Reading (ed.). Blackwell Scientific Publishing, p. 229-282.

Kennedy, V. S. (editor) 1982. Estuarine Comparisons. New York, Academic Press, 709 p.

Klein, G. deV 1970. Depositional and dispersal dynamics of intertidal sand bars. Journal of Sedimentary Petrology, v. 40, p. 1095-1127.

______ 1971. A sedimentary model for determining paleotidal range. Geological Society of America, Bulletin, v. 82, p. 2585-2592.

______ 1985. Intertidal flats and intertidal sand bodies. *In:* Coastal Sedimentary Environments, R. A. Davis Jr. (ed.). New York, Springer-Verlag, Second Ed., p. 187-224.

Kreisa, R. D. and Moiola, R. J. 1986. Sigmoid tidal bundles and other tide-generated sedimentary structures of the Curtis Formation, Utah. Geological Society of America, Bulletin, v. 97, p. 381-387.

Lambiase, J. J. 1980. Topographic control of sediment distribution on an intertidal sand bar: a case study. Geological Survey of Canada, Paper 80-10, p. 181-187.

Langbein, W. B. 1963. The hydraulic geometry of a shallow estuary. Bulletin of the International Association of Science, Hydrology, v. 8, p. 84-94.

Lauff, G. H. (editor) 1967. Estuaries. Washington, DC, American Association for the Advancement of Science Publication 83, 757 p.

LeFournier, J. and Friedman, G. M. 1974. Rate of lateral migration of adjoining sea-marginal sedimentary environments shown by historical records, Authie Bay, France, Geology, v. 2, p. 497-498.

McCann, S. B. (editor) 1980. The Coastline of Canada: Littoral Processes and Shore Morphology. Geological Survey of Canada, Paper 80-10, 439 p.

Mila, M. D. and Rosell, J. (editor) 1985. Excursion Guidebook. 6th European Meeting, International Association of Sedimentologists, Lleida, Spain, 602 p.

Milliman, J. D. and Meade, R. H. 1983. World-wide delivery of river sediment to the oceans. Journal of Geology, v. 91, p. 1-21.

Mutti, E., Allen, G. P. and Rosell, J. 1984. Sigmoidal cross stratification and sigmoidal bars: depositional features diagnostic of tidal sandstones. Abstracts, 5th European Meeting on Sedimentology, International Association of Sedimentology, Marseille, p. 312-313.
______, Rosell, J., Allen, G. P., Fonnesu, F. and Sgavetti, M. 1985. The Eocene Baronia tide dominated delta-shelf system in the Ager Basin. Excursion Guidebook, 6th European Regional Meeting, International Association of Sedimentologists, Lleida, p. 579-600.
Nelson, B. W. (editor) 1972. Environmental Framework of Coastal Plain Estuaries. Geological Society of America, Memoir 133, 619 p.
Nichols, M. M. and Biggs, R. B. 1985. Estuaries. *In:* Coastal Sedimentary Environments, R. A. Davis Jr. (ed.). New York, Springer-Verlag, Second Ed., p. 77-186.
Nio, S. D., Siegenthaler, C. and Yang, C. S. 1983. Megaripple crossbedding as a tool for the reconstruction of the paleo-hydraulics in a Holocene subtidal environment, southwestern Netherland. Geologie en Mijnbouw, v. 62, p. 499-510.
Nummedal, D., Oertal, G. F., Hubbard, D. K. and Hine, A. C., III, 1977. Tidal inlet variability — Cape Hatteras to Cape Canaveral. Coastal Sediments, American Society of Civil Engineers, p. 543-562.
Off, T. 1963. Rhythmic linear sand bodies caused by tidal currents, American Association of Petroleum Geologists, Bulletin, v. 47, p. 324-341.
Oomkens, R. and Terwindt, J. H. J. 1960. Inshore estuarine sediments in the Haringvliet (Netherlands). Geologie en Mijnbouw, v. 39, p. 701-710.
Pillsbury, G. B. 1956. Tidal Hydraulics. Vicksburg, MS, U.S. Army Corps of Engineers, Revised Edition, 247 p.
Postma, H. 1961. Transport and accumulation of suspended matter in the Dutch Wadden Sea. Netherlands Journal of Sea Research, v. 1, p. 148-190.
______ 1967. Sediment transport and sedimentation in the estuarine environment. *In:* Estuaries, G. A. Lauff (ed.). Washington, DC, American Association for the Advancement of Science, Special Publication 83, p. 158-179.
Rahmani, R. A. 1988. Estuarine tidal channel and nearshore sedimentation of a late Cretaceous epicontinental sea, Drumheller, Alberta, Canada. *In:* Tide-influenced Sedimentary Environments and Facies, P. L. deBoer *et al.* (eds.). Dordecht, D. Reidel Publishing, p. 433-471.
Reineck, H. and Wunderlich, F. 1968. Classification and origin of flaser and lenticular bedding. Sedimentology, v. 11, p. 99-104.
______ and Singh, I. B. 1980. Depositional Sedimentary Environments. New York, Springer-Verlag, Second Ed., 549 p.
Reinson, G. E. 1984. Barrier-island and associated strand-plain systems. *In:* Facies Models, R. G. Walker (ed.). Geological Association of Canada, Reprint Series 1, Second Edition, p. 119-140.
______ (editor) 1989. Modern and Ancient Examples of Clastic Tidal Deposits — A Core and Peel Workshop. Canadian Society of Petroleum Geologists, 126 p.
Ren Mei-e (editor) 1986. Modern Sedimentation in the Coastal and Nearshore Zones in China. New York, Springer-Verlag, 466 p.
Seigenthaler, C. 1982. Tidal cross-strata and the sediment transport rate problem: a geologist's approach. Marine Geology, v. 45, p. 227-240.
Smith, D. G. 1988. Tidal bundles and mud couplets in the McMurray Formation, northeastern Alberta, Canada. Bulletin of Canadian Petroleum Geology, v. 36, p. 216-219.
Stride, A. H. 1963. Current-swept sea floors near the southern half of Great Britain. Quarterly Journal of the Geological Society of London, v. 119, p. 175-199.
______, (editor) 1982. Offshore Tidal Sands: Processes and Deposits. London, Chapman and Hall, 222 p.
Swift, D. J. P. and McMullen, R. M. 1968. Preliminary studies of intertidal sand bodies in the Minas Basin, Bay of Fundy, Nova Scotia. Canadian Journal of Earth Sciences, v. 5, p. 175-183.
Terwindt, J. H. J. 1988. Palaeo-tidal reconstructions of inshore tidal depositional environments. *In:* Tide-Influenced Sedimentary Environments and Facies, P. L. deBoer, A. van Gelder and S. D. Nio (eds.). Boston, D. Reidel Publishing Company, p. 233-263.
Tessier, B. and Gigot, P. 1989. A vertical record of different tidal cyclicities: an example from the Miocene Marine Molasse of Digne (Haute Provence, France). Sedimentology, v. 36, p. 767-776.
Van den Berg, J. H. 1982. Migration of large-scale bedforms and preservation of crossbedded sets in highly accretionary parts of tidal channels in the Oosterschelde, SW Netherlands. Geologie en Mijnbouw, v. 61, p. 253-263.
Van Straaten, L. M. J. U. 1953. Megaripples in the Dutch Wadden Sea and in the Bay of Arcachon (France). Geologie en Mijnbouw, v. 15, p. 1-11.
______ 1954. Composition and structure of recent marine sediments in the Netherlands. Leidse Geologie Mededelser, v. 19, p. 1-110.
______ and Kuenen, Ph. H. 1957. Accumulation of fine grained sediments in the Dutch Wadden Sea. Geologie en Mijnbouw, v. 19, p. 329-354.
______ 1961. Sedimentation in tidal flat areas. Alberta Society of Petroleum Geologists Journal, v. 9, p. 203-226.
Van Veen, J. 1935. Sandwaves in the Southern North Sea. Hydrographic Review, v. 12, p. 21-29.
Visser, M. J. 1980. Neap-spring cycles reflected in Holocene subtidal large scale bedform deposits: a preliminary note. Geology, v. 8, p. 543-546.
Walker, R. G. 1985. Ancient examples of tidal sand bodies formed in open, shallow seas. *In:* Shelf Sands and Sandstone Reservoirs, R. W. Tillman, D. J. P. Swift and R. G. Walker (eds.). SEPM Short Course Notes No. 13, p. 303-341.
Weimer, R. J., Howard, J. D. and Lindsay, D. R. 1982. Tidal flats and associated tidal channels. *In:* Sandstone Depositional Environments, P. A. Scholle and D. Spearing (eds.). American Association of Petroleum Geologists, Memoir 31, p. 191-245.
Williams, G. E. 1989. Late Precambrian tidal rhythmites in South Australia and the history of the Earth's rotation. Journal of the Geological Society of London, v. 146, p. 97-111.
Yang, C. and Nio, S. D. 1985. The estimation of paleohydrodynamical processes from subtidal deposits using time series analysis methods. Sedimentology, v. 32, p. 41-57.
Yeo, R. K. and Risk, M. J. 1981. The sedimentology, stratigraphy, and preservation of intertidal deposits in the Minas Basin system, Bay of Fundy. Journal of Sedimentary Petrology, v. 51, p. 245-260.
Zaitlin, B. A. 1987. Sedimentology of the Cobequid Bay-Salmon River estuary, Bay of Fundy, Canada. Unpublished Ph.D. Thesis, Queen's University, Kingston, Ontario, Canada, 391 p.

KEY WORDS: CLASTIC TIDAL SEDIMENTOLOGY, HISTORICAL REVIEW, TIDAL ENVIRONMENTS, TIDAL FLUVIAL DELTAS

Recognition Criteria
and
Facies Models

In: Clastic Tidal Sedimentology, D. G. Smith, G. E. Reinson, B. A. Zaitlin and R. A. Rahmani (eds.). Canadian Society of Petroleum Geologists. Memoir 16, p. 3-28.

Diagnostic attributes of clastic tidal deposits: a review

SWIE-DJIN NIO AND CHANG-SHU YANG

International Geoservices
Reaal 5-Q, 2353 TK Leiderdorp, The Netherlands

ABSTRACT

The characteristic cyclicities displayed by clastic tidal deposits are unique criteria for recognizing tidal dominance in shallow marine clastic deposits.

Cyclicities detected in megaripple crossbedding structures that formed in tide-dominated environments include: 1) the bundling of sandy foresets bounded by mud couplets and/or a mud drape and a reactivation surface indicating flood-ebb cycles; 2) the thick-thin alternation of successive bundles reflecting the diurnal tidal inequality; 3) the lateral bundle thickness variation and bottomset thickness variation related to semi-lunar cycles of neap-spring tides; and 4) the variation in spring bundle thickness related to the lunar cycles of high-spring/low-spring tides. Reactivation surfaces can be related to the flood-ebb cycle (bidirectional reactivation surface) as well as the neap-spring tidal variation (unidirectional reactivation surface with a regular wavy erosional pattern).

Cyclicities observed in delicate, vertically-stacked, thinly-laminated tidal rhythmites include: 1) the sand/mud couplets related to peak flow/stillstand cycles, which can be correlated to flood-ebb cycles if evidence for other tidal cyclicities can also be found; 2) the vertical couplet thickness variation related to changes in sand transport and mud concentration during diurnal and neap-spring tidal cycles; 3) the variation in the thickness of spring couplets indicating the lunar cycles of high- and low-spring tides; and 4) the vertical thickness variation of rhythmites, reflecting longer cycles of tidal variations or seasonal changes in sediment concentration.

These unique criteria have been established based on comparative studies of modern tidal processes and associated preserved structures and sequences of tidal deposits. Numerous applications of these diagnostic criteria have shown their usefulness in sedimentary facies interpretation of outcrop and subsurface sequences. A clear recognition of tide-dominated sequences can be of importance to sequence stratigraphy and basin analysis (*e.g.*, spatial facies development versus sea level fluctuations).

RÉSUMÉ

Les cyclicités caractéristiques enregistrées dans les sédiments clastiques de marée sont des critères uniques pour reconnaître l'effet dominant des marées dans les sédiments clastiques marins d'eau peu profonde.

Les cyclicités décelées dans les structures de mégarides à stratifications obliques, formées dans les environnements dominés par les marées, comprennent: 1) une série de «structures frontales de pente abrupte» («foresets») de sable, délimitée par des couples de boue et/ou une «lentille de boue pélagique» («mud drape»), et une surface de réactivation indiquant les cycles flux-reflux; 2) une alternance de «faisceaux» («bundles») minces et épais, reflétant l'inégalité diurne de la marée; 3) la variation latérale de l'épaisseur d'un «faisceau» («bundle») et d'un «bottomset», reliée aux cycles semi-lunaires des marées de mortes-eaux et de vives-eaux; 4) la variation d'un «faisceau» («bundle») de marée de vive-eau reliée aux cycles lunaires de hautes et basses marées de vives-eaux. Les surfaces de réactivation peuvent aussi bien être reliées au cycle flux-reflux ou flot-jusant (surface de réactivation bidirectionnelle) qu'à la variation de marée morte-eau et vive-eau (surface de réactivation unidirectionnelle avec un patron onduleux et régulier d'érosion.

Les cyclicités observées dans une succession verticale de séquences rythmiques de marée finement laminées, contiennent: 1) des couples sable-boue reliés au point maximum des cycles écoulement/immobilité qui peuvent être corréler aux cycles flux-reflux si des évidences pour d'autres cyclicités de marée peuvent être identifiées; 2) la variation de l'épaisseur du couple, reliée aux changements dans le transport du sable et de la concentration de la boue durant les cycles diurnes de marées d'eaux-vives et de mortes-eaux; 3) la variation de l'épaisseur des couples de marée de vive-eau qui indique les cycles lunaires des hautes et basses marées de vives-eaux; et 4) la variation de l'épaisseur des séquences rythmiques qui reflète soit des cycles plus longs des variations de marée, soit des changements saisonniers de la concentration du sédiment.

Ces critères particuliers ont été établis sur des études comparatives entre les processus modernes de marée, les structures associées, et les séquences de dépôts de marée. L'emploi de ces caractères diagnostiques s'est avéré utile pour interpréter les faciès sédimentaires de séquences en affleurement et en sous-surface. Une claire identification des séquences dominées par les marées peut être importante pour la stratigraphie séquentielle et l'analyse de bassin (par exemple, relation spatiale des faciès versus fluctuations du niveau de la mer).

INTRODUCTION

Modern intertidal sediments have been studied extensively whereas less attention has been paid to subtidal sediments, because the intertidal zone is more accessible than the subtidal zone. In the geological record, tidal deposits occurred in the systems tracts characterized by high rates of relative sea level rise; that is, in the transgressive systems tract, in the late stage of the shelf margin systems tract and the lowstand wedge systems tract, and in the early stage of the highstand systems tract. The preservation potential of subtidal sediments, therefore, is generally higher than their intertidal counterparts. This means that most of the tide-dominated features recognized in the geological record are of subtidal origin.

A whole series of criteria previously developed for the recognition of tidal deposits (for complete references, see Ginsburg, 1975) based on observations within the intertidal zone, will therefore only have restricted geological applications. It is after the detailed studies of the structural and sequential organization of subtidal deposits that new and geologically more relevant diagnostic criteria for recognizing tide-dominated deposits have been developed (Visser, 1980; Allen, 1981; Siegenthaler, 1982; van den Berg, 1982; Nio and Yang, 1983; Nio *et al.*, 1983; de Mowbray and Visser, 1984; Allen and Homewood, 1984; Amos and Zaitlin, 1984; Yang and Nio, 1985). The features used for defining the criteria are mainly formed in the subtidal zone.

The unique exposures investigated in several large construction pits along the southwest coast of the Netherlands provided an excellent opportunity for detailed study of the structural and sequential organization of subtidal megaripple crossbedded deposits (Fig. 1). These sediments are approximately 400 years old and were formed in a setting similar to that of the present day (Yang and Nio, 1989). It is therefore possible to make comparative studies between present tidal processes and bedforms within the existing tidal channels, and the preserved structures and sequences of subtidal deposits. Similar research has been carried out in other subtidal megaripple crossbedded sequences and in vertically-stacked, thinly-laminated tidal rhythmite sequences. Such comparative studies of process-response models have led to the establishment of distinct diagnostic criteria, which can be correlated with a wide variety of tidal characteristics.

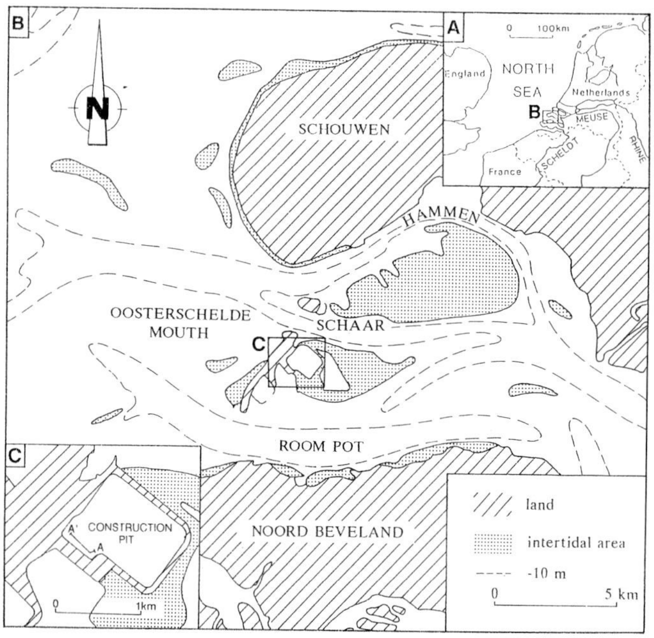

Figure 1. Location map of the construction pit in the Oosterschelde tidal basin, southwest Netherlands. Exposure location is indicated by A-A′ in Fig. 1.

Numerous geological applications of the diagnostic criteria have proved their usefulness in sedimentary facies interpretation and basin analysis (*e.g.*, Homewood and Allen, 1981; Allen, 1981, 1982; Allen and Homewood, 1984; de Boer *et al.*, 1988; Rahmani, 1988; Yang and Nio, 1989; Homewood *et al.*, 1989; Nio and Yang (*in press*). In addition, some of the criteria can also be applied to the interpretation of sedimentary structures in cores (*e.g.*, Reinson, 1989; Banerjee, 1989; Wood and Hopkins, 1989). Tide-dominance is closely related to the basin geometry. Preservation of tidally-influenced or -dominated facies in a sequence stratigraphic framework is largely controlled by a relative sea level rise. Recognition of tidally-dominated facies is becoming an important tool in sequence stratigraphy and basin modeling.

TIDAL CYCLICITIES DETECTED IN MEGARIPPLE CROSSBEDDING

Based on the sedimentary structures of unidirectional, large-scale bedform deposits, two types of criteria have been established for recognizing tide-dominated deposits. Single criteria are unique features that can only be produced by tidal processes. They include: 1) mud couplets or mud drapes; 2) lateral bundle/bottomset thickness variation; 3) diurnal bundle thickness variation; and 4) reactivation surfaces. Additional criteria, although produced by tidal processes, can also be ascribed to other depositional processes. As we will show later, some of the preserved small-scale bedform deposits in geology are also related to their large-scale counterparts.

TIDAL BUNDLES AND BUNDLE SEQUENCES

Tidal channels in estuaries or tidal embayments are usually divided into either flood-dominated or ebb-dominated zones, and tidal currents are commonly characterized by a strongly asymmetrical flood-ebb flow magnitude. The dominance of either the flood or ebb current produces megaripples with lee-slopes dipping in the direction of dominant currents (*e.g.*, Terwindt, 1971; Klein and Whaley, 1972; Dalrymple *et al.*, 1978). Although such megaripples maintain the asymmetry during subordinate flow, they show a regular pattern of change over a flood-ebb cycle. A flood-ebb cycle consists of four distinct stages (Fig. 2): the dominant current stage (from t_1 to t_4); the slack-water stage

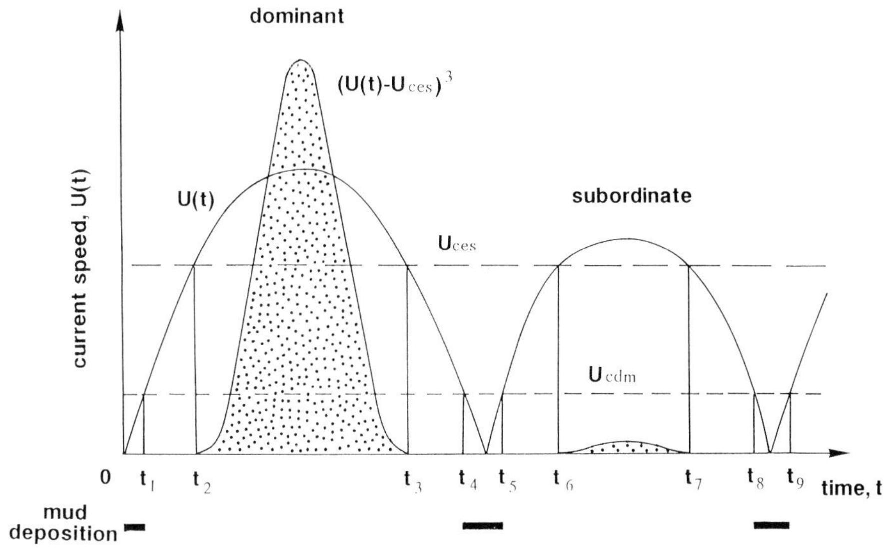

Figure 2. Sand and mud transport/deposition during one cycle of a strongly asymmetrical tide. The dominant tidal current is so strong that its velocity is well above the threshold velocity of sand movement, U_{ces}, causing considerable sand transport. Total sand transport is proportional to the stippled area under the curve $(U_{(t)} - U_{ces})^3$. During the relatively weak subordinate current, however, the velocity is just U_{ces}. The resultant sand transport is negligible. During two slack-water stages, the current velocities drop below the threshold velocity of mud deposition, U_{cdm}, resulting in two periods of mud deposition (modified after Allen, 1982).

after the dominant current (from t_4 to t_5); the subordinate current stage (from t_5 to t_8); and the slack-water stage after the subordinate current (from t_8 to t_9). Each stage is reflected in the deposits of the megaripple (Figs. 3, 4; after Visser, 1980). In the following discussion we assume a generalized case in which the megaripple deposits consist of two populations (fine sand and fine silt); two schematic threshold velocities (that of sand movement U_{ces}, and that of mud deposition, U_{cdm}) are used. The actual values of U_{ces} and U_{cdm} depend on the grain sizes of the sediments involved in a particular case.

Dominant Current Stage

During the dominant current stage, active megaripple migration occurs when the current velocity exceeds U_{ces}, resulting in the deposition of sandy foreset and bottomset laminae. Sometimes peat fragments, shell debris and small mud pebbles may occur within the associated toesets and bottomsets. In some cases co-flow and back-flow ripples (Jopling, 1961; Boersma *et al.*, 1968) can also be found.

Slack-Water Stage After The Dominant Current

During the slack-water stage after the dominant current a mud drape of a few millimetres thickness is deposited over the sandy foreset and bottomset.

Subordinate Current Stage

During the subordinate stage a thin sand layer may be deposited by weak subordinate currents (Fig. 3).

With subordinate currents slightly stronger than these mentioned above, however, not only the mud layer but also a part of the sand deposited by the previous dominant current may be eroded if the current velocity exceeds U_{ces}. This erosional process generally takes place at the upper lee-

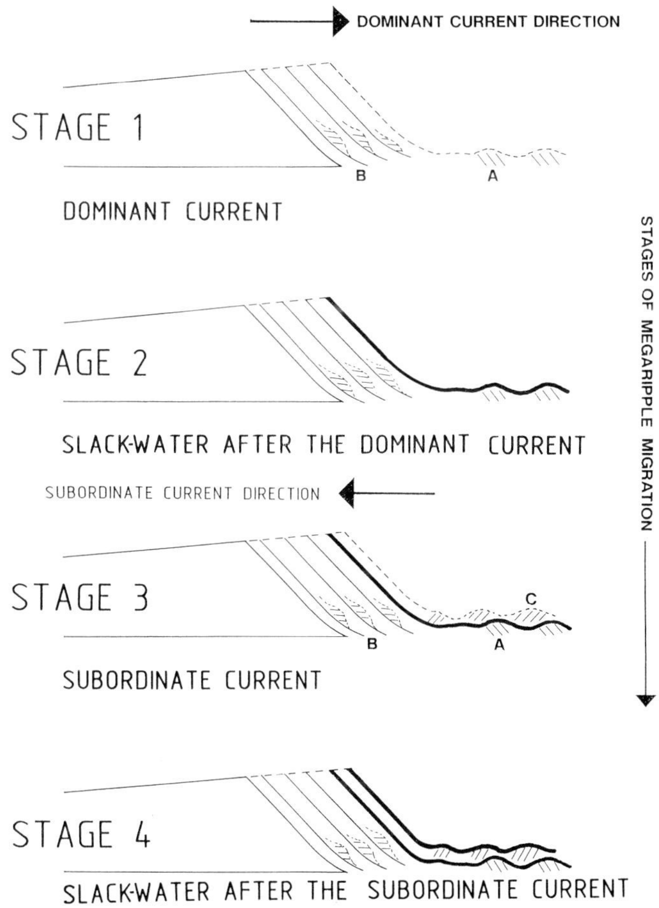

Figure 3. Schematic drawing of megaripple behaviour during a single flood-ebb cycle with weak subordinate currents. **A.** co-flow ripple crosslaminae formed during dominant current stage; **B.** back-flow ripple crosslaminae formed during dominant current stage; **C.** small ripple crosslaminae produced by subordinate current (modified after Visser 1980).

slope of the megaripple (Fig. 4; see also the discussion of dominant and subordinate flow reactivation surfaces). In the troughs of the megaripples no erosion will take place and a thin sand layer a fraction of a millimetre to a few centimetres in thickness may be deposited. If crosslaminae are formed, they are oriented in the direction of the subordinate current.

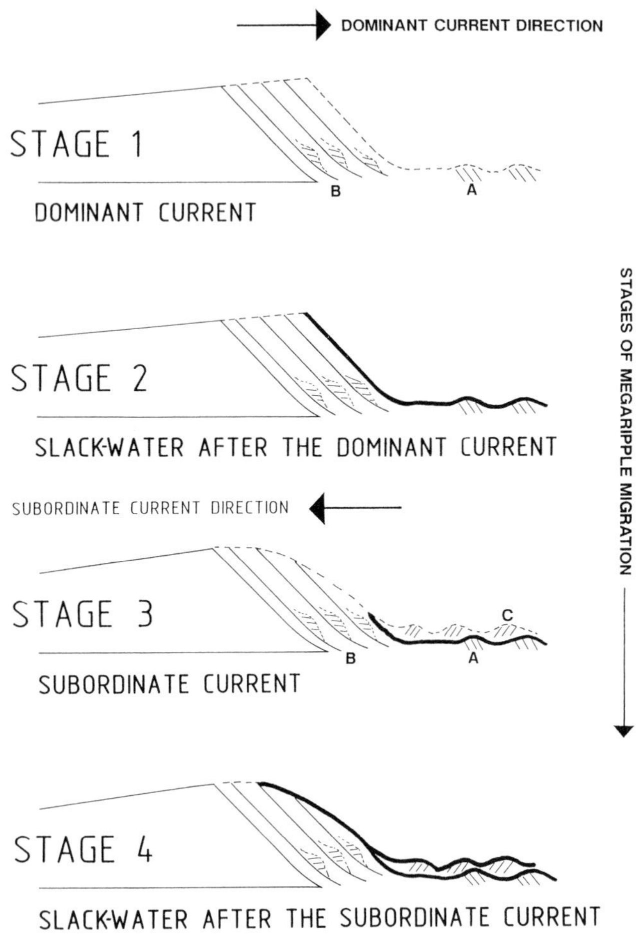

Figure 4. Schematic drawing of megaripple behaviour during a single flood-ebb cycle with slightly stronger subordinate currents. **A.** co-flow ripple crosslaminae formed during dominant current stage; **B.** back-flow ripple crosslaminae formed during dominant current stage; **C.** small ripple crosslaminae produced by subordinate current (modified after Visser, 1980).

Slack-Water Stage After The Subordinate Current

A second mud drape is deposited during this stage. Usually this mud drape can be traced to the top of the preserved lee side.

The above processes may result in a characteristic lateral arrangement of groups of large-scale foreset beds, which are separated by double mud drapes (mud couplets) or single mud drapes at intervals of centimetres to decimetres (Fig. 5). Each foreset group, representing sandy foresets deposited during a single dominant current stage, makes up a bundle (Visser, 1980; Siegenthaler, 1982; Nio *et al.*, 1983; Yang and Nio, 1985). Each bundle is bounded by slack-water mud drapes or, in the case of a slightly stronger subordinate current, by a mud drape and an erosional reactivation surface (de Mowbray and Visser, 1984; see also the discussion of dominant and subordinate flow reactivation surfaces). A series of bundles laid down by successive dominant currents is described as a tidal bundle sequence (Fig. 5).

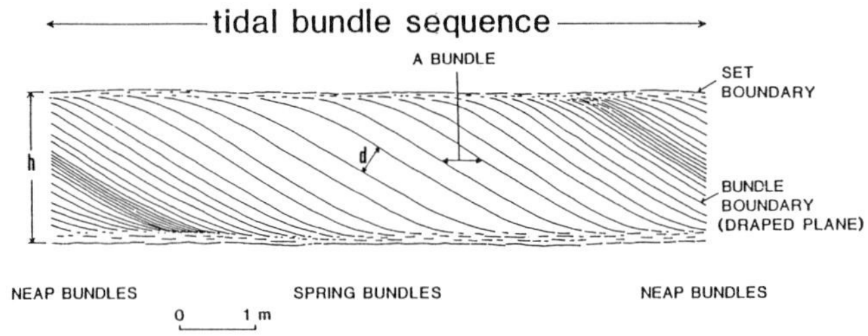

Figure 5. Schematic drawing of a tidal bundle sequence. The bundle boundaries have thin mud drapes and/or erosional reactivation surfaces (after Yang and Nio, 1985).

The Ebb-Flood Cyclicity: Mud Drapes or Mud Couplets

Figure 6 shows typical mud drapes observed in the Oosterschelde construction pits (Fig. 1), which separate a megaripple crossbedded sequence into a number of bundles. The mud layers appear to be coupled in the bottomsets. The two mud layers of the couplet enclose a thin sand layer with crosslaminae oriented in the direction of the subordinate current. The mud couplets are separated from each other by the thicker sand layer deposited by the dominant current. These features may be clearly recognized in the field and in cores. A mud couplet indicates the presence of two slack-water stages during a flood-ebb cycle. This feature is characteristic of the subtidal zone. At the upper lee slope of the megaripple, which is exposed to erosion by subordinate currents, only the mud drape formed during the slack-water period after the subordinate current will be preserved (Fig. 6).

The deposition of sand and mud during the flood-ebb

Figure 6. Megaripple crossbedding, Oosterschelde construction pits, showing the preservation of ebb-flood cycles. **a.** sandy foresets deposited during the dominant current stage; **b.** mud drape deposited during the slack-water stage after the dominant current; **c.** thin sandy layer deposited during the subordinate current stage; **d.** mud drape deposited during the slack-water stage after the subordinate current. As a result of the erosion of the upper part of the megaripple lee slope by subordinate currents, **b** and **c** are preserved mainly in toesets and bottomsets.

cycle of a strongly asymmetrical tide is illustrated in Figure 2. The sand trapped in the lee sides of megaripples comprises two different parts. Usually only a small part moves very close to the bed as tractive load and a larger part is carried in intermittent suspension (Middleton and Southard, 1984). The fact that slack-water deposits commonly consist mainly of mud (Visser, 1980) or silt (Homewood and Allen, 1981) suggests that all the sands transported by the dominant tidal current (even in suspension) must have been deposited during the same stage, before the following slack water. The deposition of suspended sand may be related to the flow separation downstream of a megaripple crest during the dominant tidal current, and will depend on the character of the local boundary layer in the lee of the megaripple crest (for further discussions, see Allen, 1982; Yang, 1986). Substantial amounts of suspended sand may be caught in this separation zone and deposited on the lee slope in the form of foreset beds.

The thickness of a mud drape is related to a number of factors, such as the suspended sediment concentration C_0, the mass settling rate W_s, the excess bed shear stress, and the time of deposition. Field observations show that the thickness of a mud drape will largely depend on the concentration of sediment in suspension, C_0. In typical tidal settings mud laminae will not form below concentrations of 200 mg/L (Amos, pers. comm.). Observations in the Oosterschelde tidal basin show that because of the increasing erosional power of the tidal currents during spring tide, more mud is eroded and transported in suspension into the tidal channel, resulting in the deposition of a relatively thick mud drape.

The deposition of relatively large amounts of mud during the short slack-water stage has been observed in Savannah Harbor, where measurements show a rapid decrease of the near-bed suspended sediment concentration as the flood current wanes (from 1500 to less than 100 mg/L in 2.5 hours at the level 1 ft. above the bed, Krone, 1972). The rather high settling rate in such a case is caused by the flocculation of mud, including physicochemical flocculation processes (Allen, 1982), and more importantly, biological processes, when mud is deposited in the form of faecal pellets (Drake, 1976). Faecal pellets produced by bottom deposit- and filter-feeders in tidal environments usually settle at rates of 0.4 to 2 cm/s, typical of coarse silt and fine sand grains (Haven and Morales-Alamo, 1972). Recent studies by Rust and Nanson (1989) show that mud can be transported in the form of sand-size aggregates. These aggregates, however, are likely to be deposited later than sands during a waning current because they are porous and less dense than quartz or other mineral grains, and therefore have a lower threshold velocity of deposition. This may explain the fact that although the mud can be deposited during a relatively short slack-water period, there is a sharp boundary between the sandy dominant current deposits and the muddy slack-water deposits. In the geological record mud aggregate texture is rapidly obscured by compaction.

After deposition, the mud flocs cohere, resulting in a significant strengthing of the mud drape even after a period of one hour or so (Allen, 1982). This is usually due to diatom polysaccharide production (Amos, pers. comm.). This coherence favours the preservation of mud drapes during the subsequent tidal currents.

An additional feature can be recognized within the lower parts of the foresets. Small-scale distortions in the mud drapes produce "flame structures" (Fig. 7; see also details

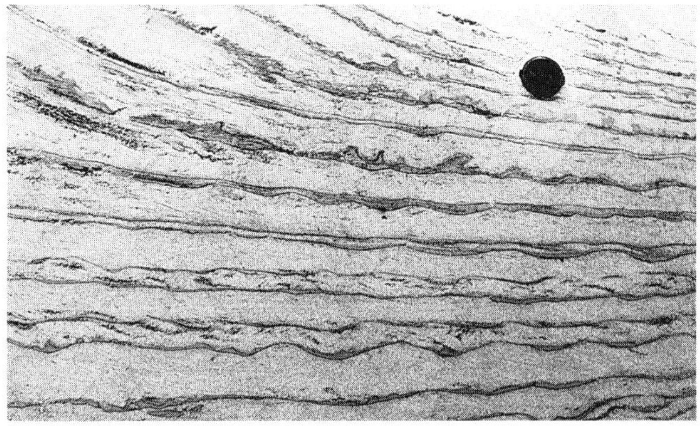

Figure 7. Flame structures in the mud drapes deposited during the slack-water period after the subordinate current stage. Example is from Holocene subtidal deposits, Oosterschelde, southwest Netherlands.

of mud drapes in the lower parts of the foresets in Fig. 6), which can be observed only in the mud drapes deposited during the slack-water stage after the subordinate current. This is caused by a sliding down of the mud drape during the following dominant current stage together with part of the newly deposited sand. The mechanism for the formation of these flame structures can be explained as follows: 1) the angle of repose is produced and controlled by the grain size and shape of sandy deposits transported during the dominant current. This angle is too steep for newly deposited mud deposits, and causes instability of the mud layer; 2) the flame structures can be explained as plastic flow phenomena associated with dewatering and loading by superimposed sand. The resulting pre-pressure buildup causes an upward flow of mud (Allen, 1982; Amos, pers. comm.).

The Neap-Spring Cyclicity: Lateral Bundle/Bottomset Thickness Variation

Lateral Bundle Thickness Variation

Distinct lateral thickness variations of the sand bundles can be recognized in the bundle sequences observed in the outcrop of the construction pit at the mouth of the Oosterschelde tidal basin (see the large megaripple crossbedded sets in Fig. 8 and the lower part of Fig. 9).

From the thickness measurements of the 131 adjacent bundles in a bundle sequence shown in Figure 10, a clear and consistent, cyclic, sinusoidal pattern emerges, with 26 to 28 bundles within a cycle. This number clearly correlates with the number of dominant tidal currents (*i.e.,* 28 or

Figure 8. Photographic mosaic of part of the Holocene exposure, at the Oosterschelde construction pit, showing ebb-oriented sets with well developed bundle sequences. Arrows in the middle set indicate the position of the neap tide periods. Note the thick bottomsets during the neap tides. The lower set also shows bundle thickness variations, but the bottomsets are not well exposed.

29) within a neap-spring cycle (14.7 days) in a semidiurnal tidal regime. Thin bundles correspond to weak dominant currents during neap tide, and the thicker bundles with strong dominant currents of the spring tide. Since the sand transport rate as bedload depends on the 3rd or 4th power of the current velocity, a relatively minor change in tidal current velocity can bring about a major change in sediment transport rate. Changes in dominant current velocity during a neap-spring cycle are therefore recorded in the thickness variations of successive bundles (provided the velocities remain high enough to allow active migration of the megaripples).

A large number of fossil bundle sequences have been recognized in the past few years, such as the Holocene Oosterschelde tidal basin, southwest Netherlands (Visser, 1980; van den Berg, 1982; Siegenthaler, 1982; Nio *et al.*, 1983); the Miocene Swiss Molasse (Homewood and Allen, 1981; Allen and Homewood, 1984); the Eocene southern Pyrenean foreland basin, northern Spain (Fig. 11; Yang and Nio, 1985, 1989); the Eocene of the Californian Borderland, U.S.A. (Clifton, 1979); the Upper Cretaceous basal Horeshoe Canyon Formation in Drumheller of Alberta, Canada (Rahmani, 1988); the Upper Cretaceous Ajali Sandstone of the Anambra Basin, southeast Nigeria (Ladipo, 1988); the Lower Cretaceous Folkestone Beds, England (Allen, 1981, 1982); the Gallup Sandstone in Cretaceous Seaway in New Mexico, U.S.A. (Tillman, 1985); the Cretaceous Cliff House Formation in New Mexico, U.S.A. (Donselaar, 1989); the Late Jurassic Curtis Formation in Utah, U.S.A. (Kreisa and Moiola, 1986); the Cambrian Kanmantoo Fold Belt in Kangaroo Island, Southern Australia; and the Precambrian Port Askaig Tillite in Isle of Islay, Argyll, Scotland (Kessler and Gollop, 1988).

The type of tide (semidiurnal or diurnal) can be established directly from field evidence. Semidiurnal tides will have 26 to 28 bundles within a neap-spring cycle (provided the dominant currents remain strong enough to allow active migration of the megaripples), while diurnal or mixed semidiurnal tides will have fewer bundles within a neap-spring cycle.

Lateral Bottomset Thickness Variation

In addition to the lateral thickness variations of the bundles, thickness variations of the bottomsets can also be observed. The thick bottomsets are related to thinner bundles; this means that well developed bottomsets are formed during neap tides. On the other hand, thin and less well developed bottomsets are related to the thick bundles formed during spring tide (Fig. 8).

The process-related interpretation of the thickening and thinning of bottomsets during neap-spring cycles can be summarized as follows: 1) During spring tide, the relatively strong currents of the dominant tide will have a higher erosional power in the trough. This will produce a deeper trough in front of the megaripple with poorly-developed bottomsets. 2) During neap tide, however, weaker currents will decrease the erosional power of the vortex and favour the formation of thick and well-developed bottomsets.

Because of the erosion (during spring tides) and infilling (during neap tides) in megaripple troughs, a change in megaripple height will occur, where larger megaripples will

Figure 9. Bundle sequence from Holocene deposits (approx. 400 years old), which is exposed at 15 m below MSL in the Oosterschelde construction pit in southwest Netherlands. The thin bundles near the ruler with the half metre scale correspond to the tides during neap tide, and the thick bundles in the middle to the spring tide.

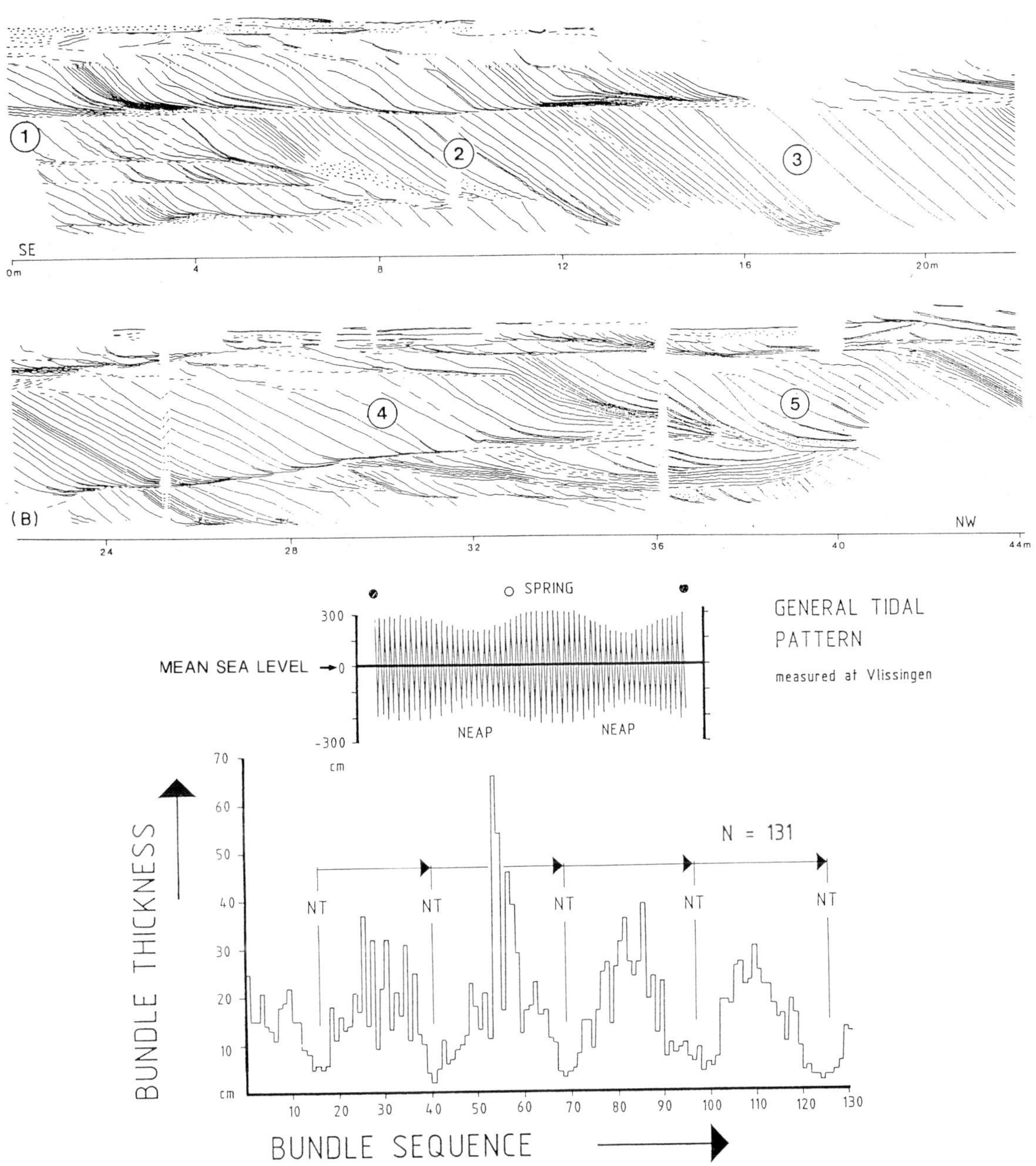

Figure 10. (upper) Drawing of a section through large ebb-oriented bundle bed sets exposed at the Oosterschelde construction pit. Only the mud drapes are drawn to show the bundle structure during neap and spring tides. Numbers refer to neap-spring tide periods recognized in the main set. Field drawing after M. J. Visser (Yang and Nio, 1985). **(lower)** Thicknesses of 131 adjacent bundles in one set revealing a cyclic sinusoidal pattern. The thicknesses are measured from the bundle sequence shown in the upper part. For comparison, a graph of the general tidal pattern at Vissingen is included (modified after Visser, 1980).

be formed during the stronger currents of spring tide, and smaller megaripples during the weaker currents of neap tide. Because of the very high preservation potential of bottomsets, the recognition of the relationship between neap-spring tides and the thickness variations of the bottomsets is an important tool in recognizing tide dominance in ancient clastic sediments.

Diurnal Bundle Thickness Variation

It is well known that in mixed tidal systems, apart from the neap-spring-neap variation, the water levels reached by successive high and/or successive low waters are commonly of unequal height. This diurnal inequality occurs only in semidiurnal or mixed semidiurnal tides. In regions where the inequality is large, the strength of successive flood and/or ebb currents fluctuates (Fig. 12). This will result in the diurnal variation of bundle thickness superimposed on the neap-spring-heap bundle thickness variation. A thick-thin alternation can be detected from the bundle thickness data in the Holocene Oosterschelde (Fig. 13). Similar thick-thin alternations can also be observed in the bundle thickness data from the Miocene of the Swiss molasse and the Cretaceous

Figure 11. Details of the tidal channel sequence in the Lower Eocene Roda Sandstone, southern Pyrenees, Spain (after Yang and Nio, 1985). Note well-developed tidal bundle sequence. Arrows indicate the position of neap bundles.

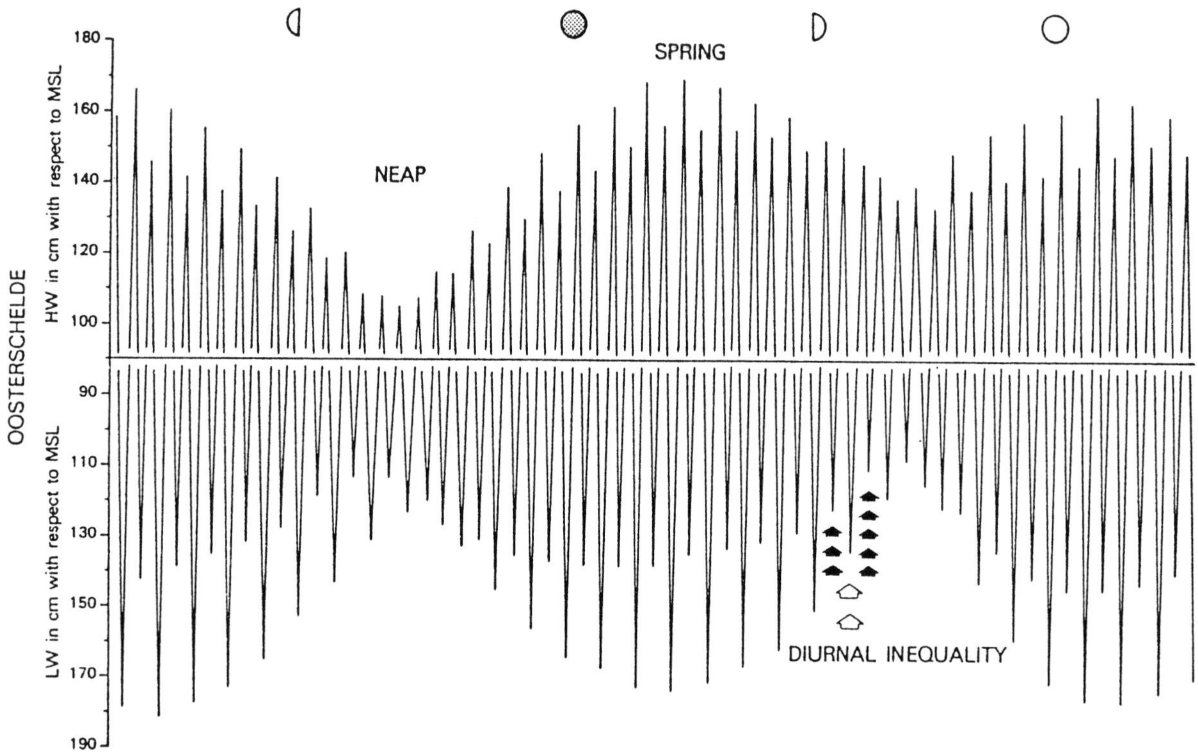

Figure 12. A graph of the general tidal pattern at Oosterschelde, southwest Netherlands, showing neap-spring cycles and the influence of diurnal inequality.

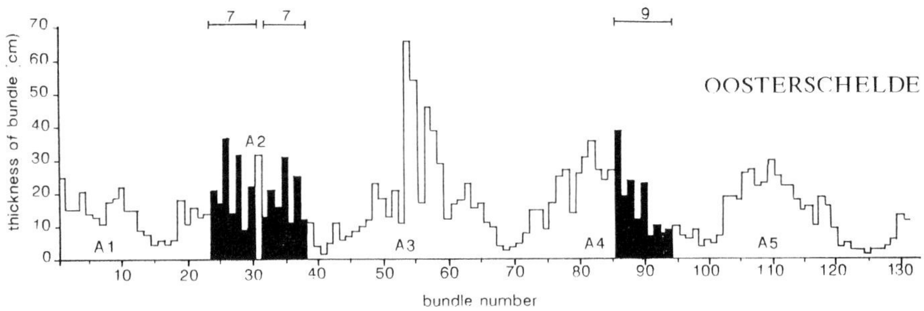

Figure 13. Thick-thin alternations of successive bundles as a result of diurnal inequality (see black coloured bundles). The number of bundles over which the diurnal variation can be observed is indicated above the diagram. (Holocene deposits in Oosterschelde, southwest Netherlands).

Lower Greensand in England. The presence of a diurnal inequality is obvious in outcrops from the Oosterschelde (Fig. 14). In other cases, however, its presence may be more difficult to establish.

The application of a statistical method may help in detecting the presence of the diurnal inequality in the bundle thickness distribution. The thick-thin alternations are superimposed on a simple harmonic signal representing the neap-spring period. A test showing the presence of alternating positive and negative deviations from the general trend suggests the presence of a diurnal inequality in the bundle thickness distribution (Fig. 15). The amounts of the bundle thickness deviations are indicated by the vertical bars and are derived from the following formula:

Figure 14. Thick-thin alternations in a bundle sequence indicating the presence of a diurnal inequality. Numbers indicate the sequence of deposition by ebb- and flood-tidal currents. (Holocene deposits in Oosterschelde, southwest Netherlands; handle is 20 cm long).

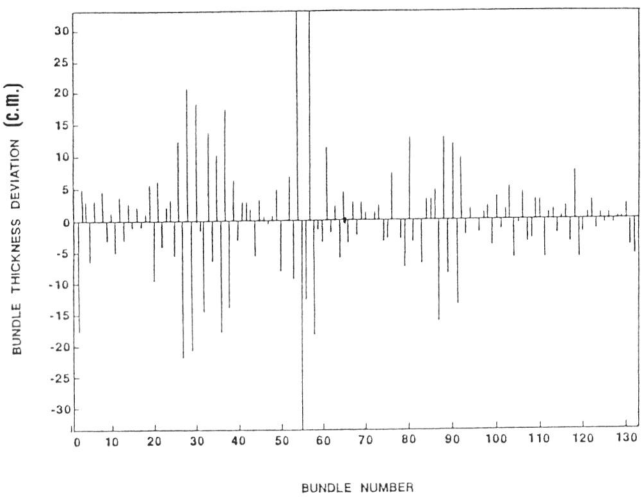

Figure 15. Bundle thickness deviation diagram of the Holocene of the Oosterschelde, southwest Netherlands. See text for explanation.

$$\text{bundle thickness deviation (cm)} = \frac{d_{i-1}+d_{i+1}}{2} - d_i$$

where d_i = thickness of the ith bundle (cm) in a bundle sequence. A positive bundle thickness deviation indicates a thick-thin-thick alternation. A negative bundle thickness deviation, however, indicates a thin-thick-thin succession.

Current velocity measurements (*e.g.*, in the Oosterschelde tidal basin) show that tidal currents within a semidiurnal or mixed tidal regime may drop to such a low magnitude during neap tide that bedform migration ceases. Consequently, the number of bundles within a neap-spring-neap cycle may be less than 28, and the depositional regime can be incorrectly interpreted as a "diurnal tidal regime". In this case, however, the reflection of the diurnal inequality of the tide in the bundle thickness variations will reveal that bedform migration took place during semidiurnal tides and that stillstand periods in the bedform migration during neap tides caused a reduction in the number of depositional events per neap-spring-neap cycle.

In the case of the Holocene deposits of the Oosterschelde it can be concluded that the characteristics of the bundle thickness sequences are in complete accordance with a semi-diurnal tidal regime. Both the neap-spring cycle as well as the diurnal inequality can be detected in the bundle sequence. It is interesting that the relatively small diurnal differences in the tidal ranges in the Oosterschelde can lead to such clear thick-thin alternations of the bundles. This is because the amount of sediment transported is in general related to the third or fourth power of the flow velocity, which again is dependent on the tidal range. Small variations in tidal range will therefore produce a considerable change in the amount of sediment being transported and ultimately deposited on the megaripple lee side.

Reactivation Surfaces

Tidal processes are strongly influenced by the fluctuation of flow velocities. One of the consequences of such unsteadiness is the presence of reactivation surfaces (erosional surfaces), developed within cross-stratified sets of tidal deposits.

Reactivation surfaces can be produced by dominant currents as well as by dominant and subordinate currents (de Mowbray and Visser, 1984). Reactivation surfaces related to dominant flow are formed during the dominant-current phase(s) of the tidal cycle. They are similar to these formed in any unidirectional flow system, such as a fluvial channel. However, occurrences of dominant flow reactivation surfaces with those formed by dominant and subordinate flows are unique features for tidal environments.

Reactivation Surfaces Formed by Dominant Flows (mainly after de Mowbray and Visser, 1984)

Reactivation surfaces of this type are formed by vortex of the dominant tidal current in the trough between two migrating megaripples. The strong erosional power of the vortex during spring tides produces a deep erosional trough, while the weak vortex during neap tides results in a shallow erosional surface. Such a wavy pattern in the dominant flow reactivation surface is characteristic of tidal influences. An example of a dominant flow reactivation surface produced during spring tide in the Cambrian Kanmantoo Fold Belt of Southern Australia is shown in Figure 16.

An interesting case of dominant flow reactivation surface occurs when one megaripple overtakes another (Fig. 17). Even under steady, uniform, and unidirectional flow conditions, bedforms interact with the flow, producing local velocity fluctuations. As a result, there are always differences in migration velocity between individual megaripples (Allen, 1973). Given this variation in migration velocity, it is not surprising that a situation commonly arises in which a faster-moving megaripple starts to overtake a slower-moving one in front (down current transport). Where there is net vertical accretion, the faster-moving megaripple does not destroy the one in front, rather, it climbs up the stoss side. As it approaches the crest of the now underlying megaripple, its lee side vortex starts to erode this crest area (McCabe and Jones, 1977). The resulting erosion surface or reactivation surface is then rapidly buried by avalanching as the upper megaripple overtakes and advances further. The eventual form of the reactivation surface depends on the relative size of the two megaripples concerned and on the flow conditions under which the upper megaripple migrates (Fig. 18).

Figure 16. Dominant flow reactivation surface produced during spring tide. Cambrian Kanmantoo Fold Belt, Kangaroo Island, Southern Australia. Scale is 15 cm.

Reactivation Surfaces Formed by Dominant and Subordinate Flows

This type of reactivation surface is a product of dominant and subordinate flows and is specifically related to tide-dominated environments.

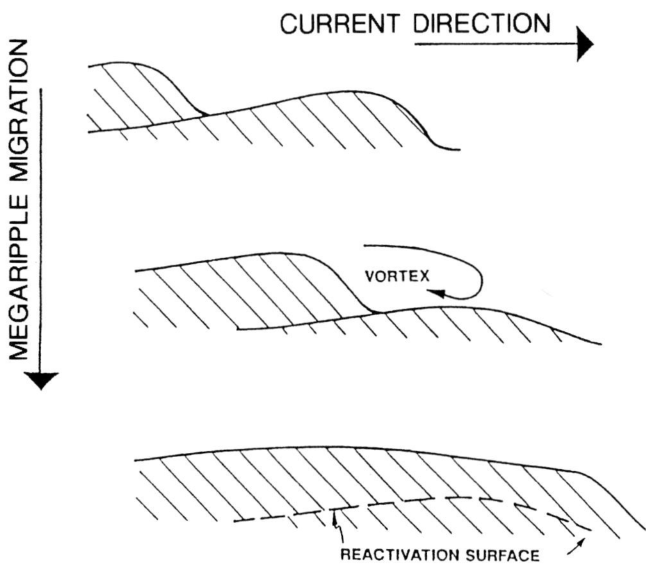

Figure 17. Reactivation surfaces formed by one megaripple overtaking another under a unidirectional flow. See text for explanation.

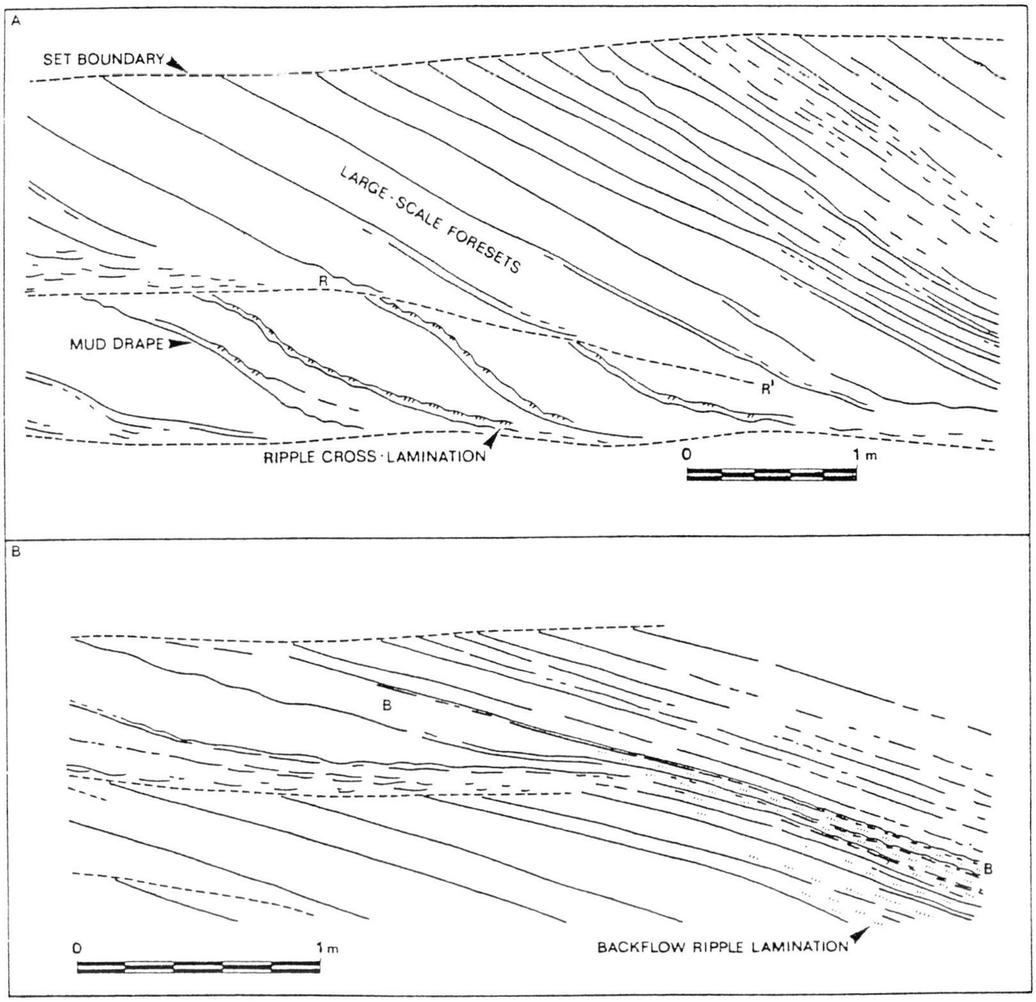

Figure 18. Structures produced by megaripple overtaking. **A.** Overtaking during spring tides has produced a well-developed reactivation surface, R-R′, extending almost to the base of the set. Overtaking took place during spring tides as is reflected in the bundle sequence of the upper set. **B.** Overtaking during neap tides. There was no erosion associated with the overtaking process — the lee side vortex of the upper megaripple was clearly relatively weak. The vortex was, however, capable of producing back-flow ripples on the top of the lee side slope of the lower megaripple. The overtaking process was completed during deposition of bundle **B**.

The form and geometry of this type of reactivation surface is related to the strength of the subordinate current. With very weak subordinate currents there will be no erosion and deposition of only a thin sand layer. The sand in this case is probably derived from other parts of the tidal channel, where subordinate currents are relatively strong. With a stronger subordinate current, however, an increase in the rate of erosion will take place and different forms and geometries of reactivation surfaces will be produced.

In most cases, the subordinate current will erode the bundle deposited during the dominant current stage (Klein, 1970). Based on field observations, five different types of reactivation surface can be distinguished (de Mowbray and Visser, 1984; Fig. 19). They are interpreted as indicating an increasing strength of the subordinate currents. In all the examples mentioned below, the dominant current is always stronger than the subordinate current. With a very pronounced dominant current, a clear asymmetrical tidal pattern will emerge. With increasing stronger subordinate currents, however, this tidal asymmetry will decrease.

Type A: Very Strong Asymmetrical Tide (Fig. 19A)

The subordinate current is very weak compared to the dominant current, and is incapable of eroding the first mud drape deposited during the slack-water period after the dominant flow. In this case the double mud drape or mud couplet is fully preserved. Crosslaminae of the subordinate current are well developed, especially within the bottomsets. Back-flow ripple laminae are commonly present within the lower part of the large-scale foresets of the bundle.

Type B: Strong Asymmetrical Tide (Figs. 19B, 20)

With slightly stronger subordinate currents a 'mild' erosion will occur. In this case only the top of the first mud drape is eroded away and a very small part of the underlying bundle. Generally, the dip of the reactivation surface or

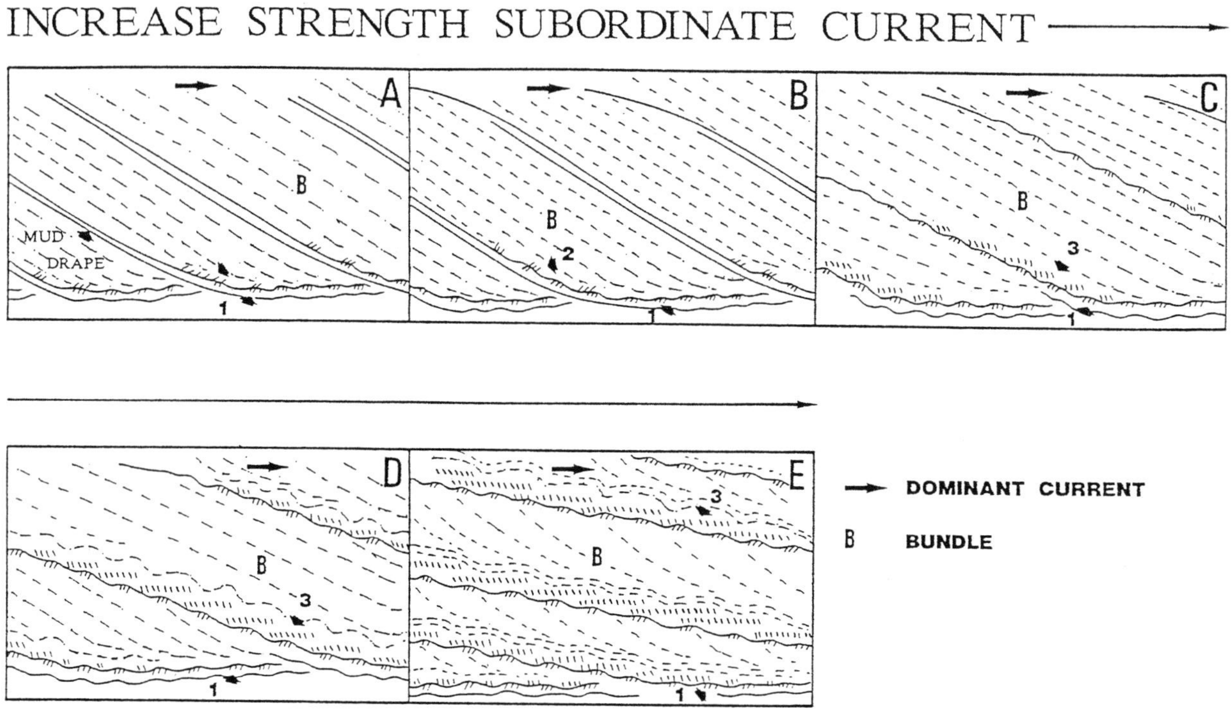

Figure 19. Types of reactivation surfaces in a bidirectional flow system resulting from subordinate current erosion. **A** to **E** show the effects of an increase of the subordinate current with respect to the dominant current; that is, a decrease of the tidal asymmetry (modified after de Mowbray and Visser, 1984).

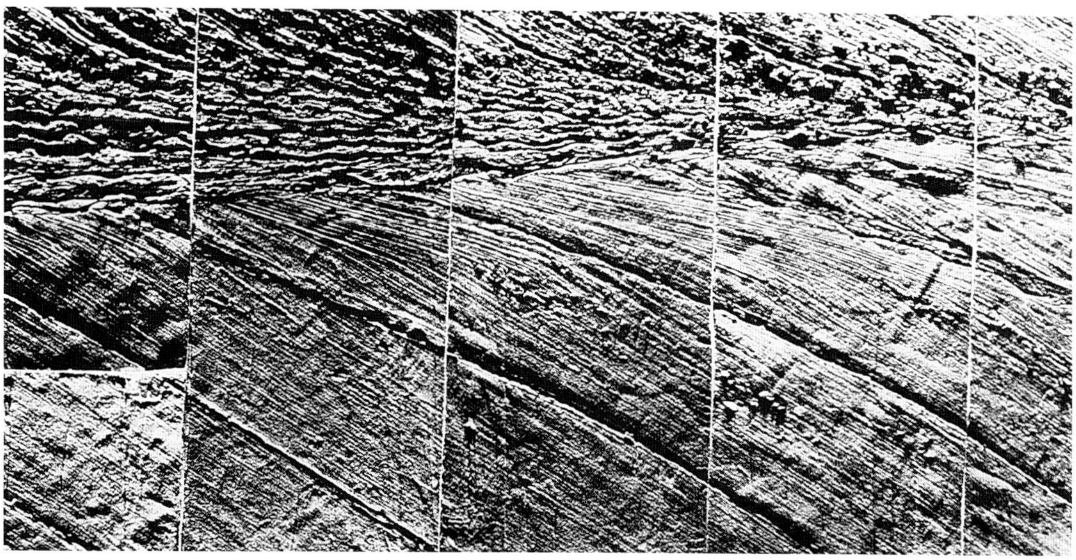

Figure 20. Type B reactivation surface resulting from weak erosion by the subordinate current. Note the small erosional discordance between the upper part of the foreset and the second mud drape. Photograph from a lacquer peel taken from the Oosterschelde construction pit. Bundle thickness is 15 to 20 cm.

'erosional pause planes' (after Terwindt, 1981) is similar to that of the laminae within the bundle.

Type C: Asymmetrical Tide (Figs. 19C, 21, 22)

With a stronger subordinate current, the first mud drape will be stripped off from the entire foreset height. Part of the underlying bundle will also be affected by erosion; the upper parts of the bundle foresets will be more eroded than the lower parts. The erosional discordance therefore is more obvious in the upper parts and will gradually change into a more concordant superposition.

The dip of the reactivation surface is about 25°, relatively high when compared with the average foreset dip of 30°. After the erosive phase, the waning, subordinate current produces crosslaminae on the reactivation surface. The

Figure 21. Type C reactivation surface resulting from intermediate erosion by the subordinate current. Note that the erosion took place over more than 75 per cent of the foreset height. The lower part of the foreset and the bottomsets generally display well developed mud couplets. Arrows indicate small ripple crosslaminae formed by a waning subordinate current. Photograph from Holocene subtidal deposits, Oosterschelde construction pit. Bundle thickness 25 cm.

Figure 22. Same as Figure 21. Arrow indicates small ripple crosslaminae formed by waning subordinate current. Example from the Eocene of the Californian Borderland, U.S.A. Bundle thickness 15 cm.

second mud layer, which is deposited after the subordinate current stage, drapes the ripple forms (see Figs. 21, 22).

Type D: Weak Asymmetrical Tide (Fig. 19D)

A further increase in strength of the subordinate current will result in major erosion of the bundle. As a result, the reactivation surfaces produced by erosion of these stronger subordinate currents will show considerably lower dips (as low as 15°).

Type E: Very Weak Asymmetrical Tide (Figs. 19E, 23)

In systems with a very weak asymmetrical tide (near symmetrical tide), the subordinate currents will be strong enough to erode major parts of the bundle. Such strong subordinate current erosions will produce dips as low as 10°, considerably less than the dips of the foresets. An interesting feature related to this is the occurrence of "flaser forks" (see Fig. 23). The flaser forks occur mainly within the lower parts of the foresets. The direction of the dominant current can be

Figure 23. Type E reactivation surfaces showing severe erosion by the subordinate current. Note the low dips of the reactivation surface (approx. 10°). Megaripple migration, that is, the dominant current, is from left to right. The bundle sequence shows a transition from spring tide to neap tide. Severe erosion by the subordinate currents took place during spring tide. Note the well developed flaser forks during this period. Holocene subtidal deposits, Oosterschelde construction pit (length of the stick is 0.5 m).

deduced from the orientation of the "handles" of the flaser forks.

Variability of Reactivation Surfaces in a Neap-Spring-Neap Cycle and in a Vertical Time Sequence

The formation of different types of reactivation surface is strongly related to different phases in a neap-spring-neap cycle or in a vertical time sequence. It is therefore equally important to be able to recognize typical associations of reactivation surfaces and, if possible, place them within a specific tidal cycle.

Reactivation Surface Associations in a Neap-Spring-Neap Cycle (Fig. 24)

Figure 24 illustrates an association of reactivation surfaces produced during different phases of the neap-spring-neap cycle. This example shows two sets with a dominant current from left to right. The upper set shows a thickening of the bundles, reaching a maximum during spring tide, followed by a thinning indicating a transition toward neap tide. The stronger dominant currents during spring tide produced a deeper erosional surface causing an irregular form in the set boundary. This type of dominant flow reactivation can be found regularly in tide-dominated deposits (see also Fig. 16).

In addition, a secondary dominant flow reactivation surface (R-R' in Fig. 24) can be observed in the upper set. This is related to a megaripple overtaking process during spring tide which took place within 3 flood-ebb cycles (3 bundles). This reactivation surface is very irregular as a result of the large peat fragments in the foresets.

The lower set shows a lateral sequence from thick

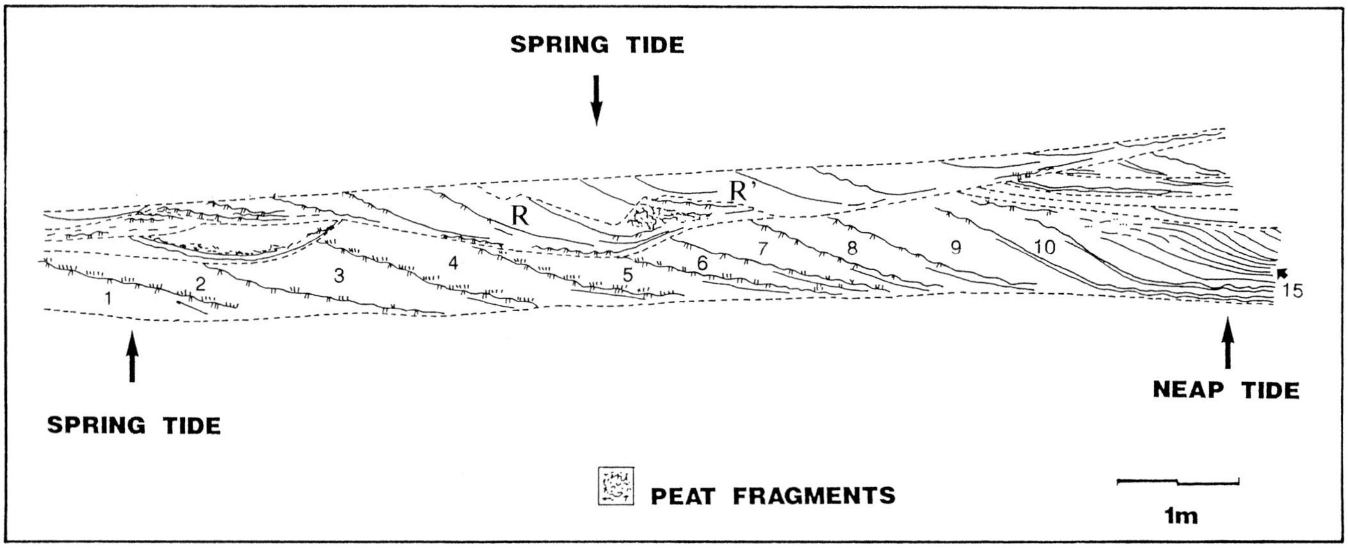

Figure 24. Association of reactivation surfaces during a neap-spring-neap cycle. See text for further explanation (modified after de Mowbray and Visser, 1984).

bundles (spring tide) to thin bundles (neap tide). The spring bundles (bundle nos. 1-4 in Fig. 24) show strong erosion by subordinate currents resulting in type C reactivation surfaces. The erosion by the subordinate currents decreased toward neap tide and produced reactivation surfaces of type B followed by type A at the neap bundles (bundle nos. 11-18 in Fig. 24).

The lateral bundle thickness variation was measured in the lower set and showed a decrease from spring to neap (Fig. 25). In addition, the variation of the dip of the second mud drape has been included in the graph. It shows an increase of the dip toward the neap bundles, indicating a decrease in the rate of erosion by the subordinate currents. During spring tide, more severe erosion by the subordinate currents occurs and results in a decrease of the dip of the second mud drape.

Reactivation Surface Associations in a Vertical Time Sequence (Fig. 26)

Figure 26 shows a vertical sequence in the construction

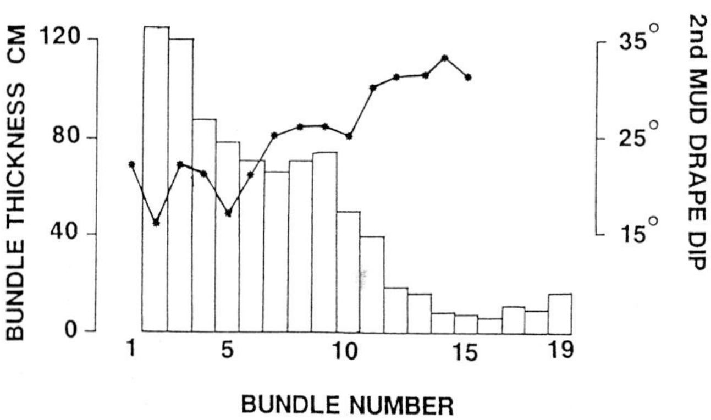

Figure 25. Histogram of lateral bundle thickness measured in the lower set of Fig. 24. See text for explanation (modified after de Mowbray and Visser, 1984).

pit of the Oosterschelde. Most of the sequence consists of ebb-dominated cosets (interval A-C in Fig. 26) and a smaller part consists of flood-dominated cosets (interval D in Fig. 26). It shows that these subtidal channel deposits (approx. 400 years old) were initially formed by ebb-dominated currents and subsequently were influenced by flood-dominated currents. This change is also obvious in the vertical succession of the structural associations of the deposits.

In interval A, which consists of three distinct, ebb-dominated sets (A1-A3 in Fig. 26), the lateral bundle sequence is well developed. Spring as well as neap bundles can easily be recognized together with the different associations of reactivation surfaces. The reactivation surfaces of the spring bundles, however, show a vertical change from the lowermost set (A1) which displays reactivation surfaces of type B, to the uppermost set (A3) which shows reactivation surfaces of type E (X in Fig. 26). It is evident that the subordinate currents were becoming stronger with time. During deposition of set A1 the tidal currents were distinctly asymmetrical, while during deposition of set A3 a weak asymmetry was present.

Interval B does not show a clear lateral bundle sequence, but consists mainly of type E reactivation surfaces. The subordinate currents were presumably very strong and eroded large parts of the bundles, preventing the preservation of a lateral bundle sequence. This means a more consistent increase in the strength of the subordinate currents; that is, the flood currents.

Interval C shows a return to an ebb-dominance with the display of ebb-dominated cosets. The set boundaries show a very irregular pattern (trough crossbedding), indicating the presence of 3-D or type II megaripples. Finally, flood dominance conditions prevailed during the deposition of Interval D.

To summarize, a change in the tidal dominance took

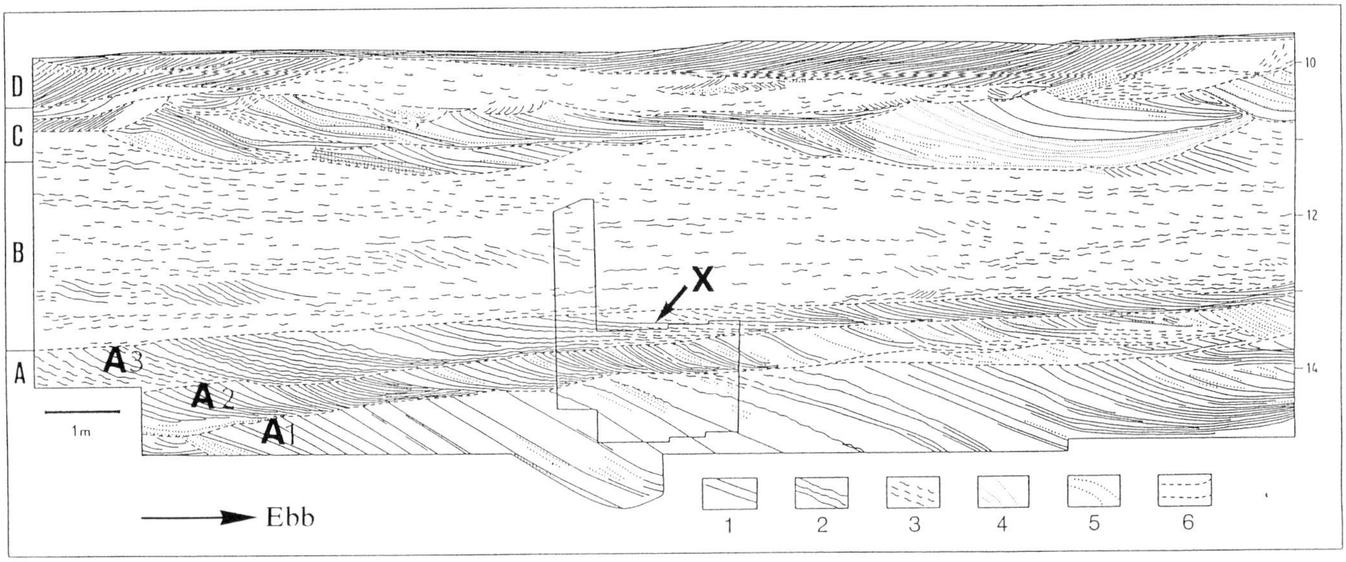

Figure 26. Sequential overview of a Holocene subtidal channel, Oosterschelde construction pit. See text for further explanation. 1) Mud drapes, reactivation surfaces type A to C; 2) Reactivation surface type D; 3) Flaser forks, reactivation surfaces type E; 4) Sand with mud pebbles; 5) Sand with peat debris; 6) Set boundaries. Field drawing by Siegenthaler.

place during deposition within the described tidal channel. The initial ebb-dominance changed to flood-dominance. This is related to a change in the distribution of ebb- and flood-dominated zones in the channel, which was probably caused by a lateral shifting of the channel thalweg. The change from ebb- to flood-dominance is reflected in the deposition of the bundles and the related reactivation surfaces.

Time-Series Analysis of Bundle Thickness Sequence

Time-series analysis methods can be applied to the thickness data of continuous tidal bundle sequences in an attempt to reconstruct the paleotidal regime (Yang and Nio, 1985). This technique involves: 1) Fourier analysis to test the cyclicity of the bundle-thickness sequence and to estimate the periods and the phases of the most important cyclic components; 2) filtering analysis to resolve the bundle-thickness sequence into different components (*i.e.*, diurnal components, random variations, neap-spring components and longer period variations) and also to estimate the relative importance (amplitude) of each component.

From these analyses, useful information can be derived about the paleotidal regime (whether semidiurnal or mixed tides) and also about the possible influence of non-tidal processes, such as storms (their strength, duration, frequency and direction). Time-series analyses of the Holocene tidal bundle sequence in the Oosterschelde tidal basin of southwest Netherlands suggest a number of cyclic variations (diurnal, neap-spring, and longer period) as well as random variations (Fig. 27). Similar analyses have been made for the Eocene tidal bundle sequence of the Roda Sandstone in the southern Pyrenean Foreland Basin of northern Spain. Results from both examples revealed a strong tidal dominance as well as a possible storm influence in these shallow marine environments (Yang and Nio, 1985).

Additional Criteria for Recognizing Tide-Dominated Clastic Deposits

Quite a few of the diagnostic features of clastic tidal deposits were described in earlier publications (for complete references, see Ginsburg, 1975). Most of them, however, cannot be ascribed exclusively to tidal depositional processes, but can also be explained by totally different processes. Therefore, they cannot be used as unique single criteria for the recognition of tidal deposits. The association of a few of them, however, strongly indicates the presence of tidal processes. Therefore the recognition of these additional criteria in the field or in the subsurface should only be considered as a preliminary indication, and a further search should be carried out to find one of the single criteria described earlier.

The most important additional criteria can be summarized as follows: Sigmoidal bedding as it has been first described by Mutti *et al.* (1985) and Herringbone cross-bedding.

Sigmoidal Bedding

Recently, sigmoidal bedding has been used as an important tool for recognizing clastic tidal deposits in the field (Mutti *et al.*, 1985; Kreisa and Moiola, 1986). The sigmoidal shape of the bedding, however, is a result of the weathering of the original bedding. Inhomogeneties in the lithological composition and a particular structural organization of the beds are the main factors in the formation of sigmoidal bedding. Tidal processes produce a specific association of lithotype and structures. During neap tides, a higher mud accumulation can be found in the relatively thick bottomsets. A higher sand accumulation and thicker bundles will occur during spring tides. Weathering of the fine grained parts will eventually result in sigmoidal bedding.

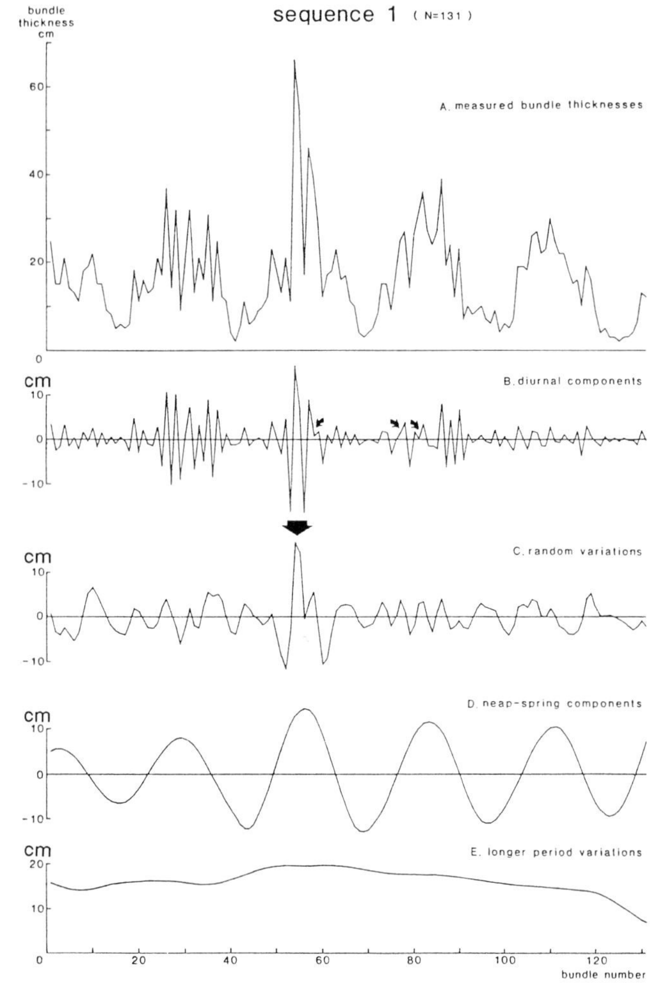

Figure 27. Cyclic variations (diurnal, neap-spring, and longer period) as well as random variations revealed by time-series analyses of the Holocene tidal bundle sequence in the Oosterschelde tidal basin of southwest Netherlands. The sequence consists of 131 bundles (see Fig. 10). The arrows in curve B indicate the departures of the diurnal variation from the regular pattern. Arrowed deviation in curve C is likely to have been caused by a storm event (after Yang and Nio, 1985).

A computer simulation of megaripple crossbedding formed during a neap-spring-neap cycle reveals the following features (Fig. 28): 1) a lateral bundle sequence with thick spring bundles and thin neap bundles. A denser succession of mud layers will occur during the deposition of the neap bundles; 2) thin bottomsets with relatively low amounts of mud will develop during the spring tide, and thick, mud-rich bottomsets during the neap tide. The geometrical interrelationship between the lateral bundle thickness variation and this bottomset thickness variation is well defined in the diagram; 3) reactivation surfaces formed by dominant and subordinate flows indicate strong erosion during the spring tide. Erosion during the deposition of the spring bundles, therefore, will take place along the larger part of the foreset. Toward neap tide, erosion will be weak and limited to the uppermost parts of the foresets.

The interaction of these above-mentioned depositional processes and their products will give rise to a specific geometrical shape. Selective weathering of the fines will result in sigmoidal bedding. It is assumed now that the sigmoidal bedding represents half a neap-spring-neap cycle (see Fig. 28).

Herringbone Crossbedding

Herringbone crossbedding was, up to now, the most common feature ascribed to tidal action. It seems quite logical that within a bidirectional current system herringbone crossbedding will be formed. A tidal system, however, consists of a dominant and subordinate current; that is, a tidally-dominated system is mainly characterized by an asymmetrical tidal pattern. Herringbone crossbedding may be formed within a symmetrical tidal pattern, which is very uncommon within a tidally-dominated environment.

Small-scale pseudo-herringbone crosslaminae can occasionally be recognized in very weak asymmetrical tidal settings where reactivation surfaces of type E are formed (see Fig. 19) or within the bottomsets of tidal bundles (Fig. 29). In the latter case, pseudo-herringbone laminae were produced by the superposition of small-scale ripple sets of the dominant as well as of the subordinate flow and back-flow ripple sets.

Summary

Megaripple crossbedded sets are common in the channelized inshore parts of estuaries or tidal embayments. They may also occur in offshore areas dominated by bidirectional tidal currents or unidirectional baroclinic-driven currents. The bidirectional tidal currents are usually asymmetrical as a result of the separation of flood-dominated and ebb-dominated zones, or the superposition of a unidirectional current on a tidal current. Consequently, megaripples are commonly formed and migrate in the direction of dominant tidal currents. The behaviour of these megaripples, however, shows a regular pattern in response to cyclic tidal movements, producing a number of diagnostic criteria for recognizing tide dominance: 1) The sandy foreset bundle bounded by mud couplets or a mud drape and a reactivation surface reflects the flood-ebb cycle. In offshore areas mud drapes

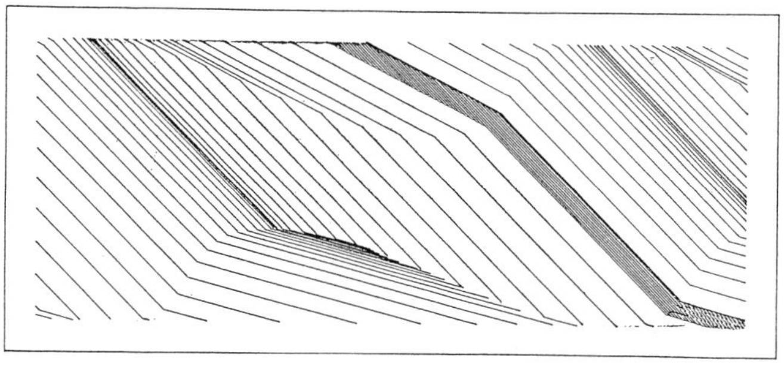

Figure 28. Computer simulated graph revealing the internal structural organization produced by depositional processes during a neap-spring-neap cycle. Shaded area is the bundle deposited during spring tide. A sigmoidal geometry can be recognized between the neap and spring bundle. See text for further explanation.

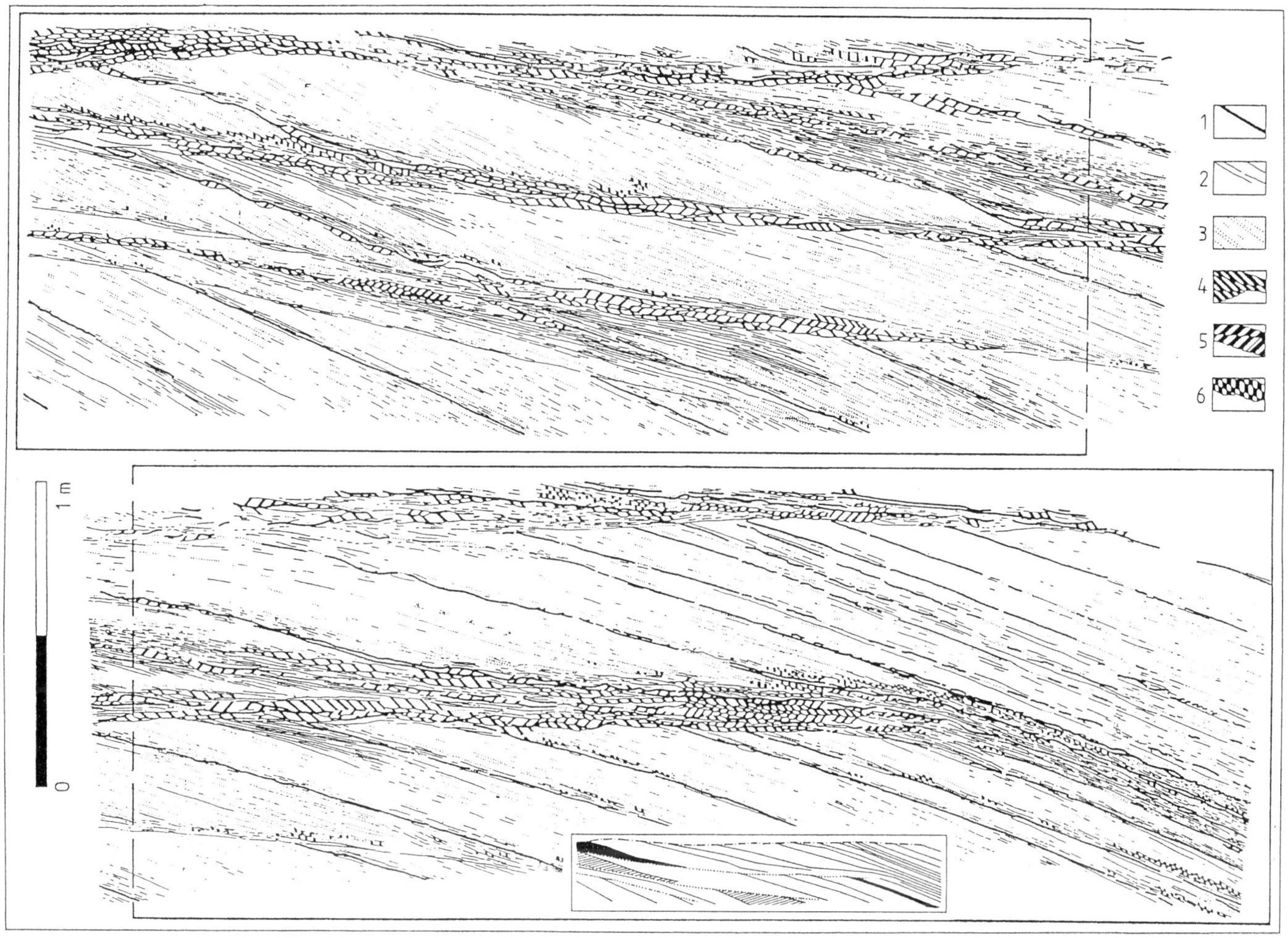

Figure 29. Drawing from a lacquer peel showing details of set boundaries and small-scale structures of the bottomsets. 1) mud drapes; 2) laminae 1 to 2 mm thick consisting of silt or very fine grained sands; 3) laminae 1 to 2 mm thick consisting of coarser sand; 4) small-scale crosslaminae related to the dominant current; 5) small-scale crosslaminae related to the subordinate current; 6) back-flow crosslaminae. Inset: general overview of bundle organization. Full line, bundle boundaries; stippled line, reactivation formed by megaripple overtaking; hatched area, bundles deposited during the same dominant current cycle (modified after Siegenthaler, 1982).

are less common because of the low suspended sediment concentrations. 2) The alternation of successive thick and thin bundles represents the diurnal tidal variation. 3) The lateral bundle/bottomset thickness variation records the neap-spring tidal variation. 4) Reactivation surfaces produced by dominant and subordinate flows are related to the interaction between the dominant and subordinate currents during a flood-ebb cycle, while a reactivation surface produced by dominant flow with a regular wavy erosional pattern indicates a neap-spring tidal variation. All these features are single criteria, which are unique features related only to tidal processes. Single criteria have been applied in the interpretation of sedimentary structures in cores, such as mud drapes and diurnal bundle variations (*e.g.*, Reinson, 1989; Banerjee, 1989; Wood and Hopkins, 1989). There are also additional criteria, such as sigmoidal bedding, and herringbone crossbedding, that may be produced by tidal processes, though they can also be ascribed to other depositional processes. The occurrence of these additional criteria may strongly indicate the presence of tidal processes and a further search should be carried out to find single criteria.

Tidal Cyclicities Recorded in Vertically-Stacked Thinly-Laminated Tidal Rhythmite

Tidal cyclicities are not only preserved in megaripple crossbedding structures, but also in vertically-stacked, thinly-interlaminated sequences of sand and silt-mud laminae. Such sequences are described here as tidal rhythmites. Tidal rhythmite sequences have been recognized in ancient and recent examples. They provide detailed records of the tidal cyclicites at the time of deposition.

Tidal Rhythmites of Upper Ordovician Wenchang Formation in Tonglu of Zhejiang, China

Tidal rhythmite sequences have been recognized widely in Paleozoic sediments of China. Figure 30 shows an example from the Upper Ordovician Wenchang Formation in the Tonglu Area of Zhejiang, China.

The Wenchang Formation was deposited in a shallow marine and nearshore environment. The uppermost part of the Wenchang Formation consists of a set of tidal-flat deposits overlying shallow marine sandstones. These tidal-flat deposits are characterized by rhythmic interlamination of sand and mud laminae (Fig. 30). Three orders of cyclicities can be recognized in this succession:

1. The sand-mud couplets consist of a sand lamina and a mud lamina. The sand lamina of the couplet commonly shows well preserved, small-scale current ripples, indicating deposition by currents. The mud lamina of the couplet commonly drapes the troughs as well as the crests of the sandy ripples, reflecting deposition from suspension. The deposition of the mud lamina in a single stillstand period was presumably caused by flocculation of mud.

2. The vertical couplet thickness variation shows an obvious cyclical pattern. Each cycle involves a progressive increase of couplet thickness, followed by a progressive decrease of couplet thickness. The thick couplet consists of a sand lamina 2 to 5 cm thick and a mud lamina 0.5 to 1.5 cm thick. The thin couplet, however, is characterized by sand laminae only a fraction of 1 mm to a few mm thick, alternating with mud laminae 0.1 to 1 cm thick. Each cycle is about 20 cm thick and contains 16 to 18 sand-mud couplets.

3. A further cyclicity can be detected from unequal thicknesses of the "thick couplets" in successive cycles of vertical couplet-thickness variation. A cycle with less thick "thick couplets" is followed by a cycle with thicker "thick couplets" (Fig. 31).

The three orders of cyclicities can be well correlated to tidal cyclicities. The sand-mud couplets indicate flood-ebb cycles. The vertical couplet thickness variations reflect the semilunar cycle of neap-spring tides. The alternation of less thick "thick couplets" and thicker "thick couplets" in successive cycles of vertical couplet thickness variations may represent the lunar cycle of low-spring/high-spring tides.

It is interesting to note that some thick couplets show a reversal of ripple migration direction (Fig. 30). This may indicate that during spring tides the subordinate current became strong enough to move sand as small migrating ripples. The presence of dominant and subordinate current deposits with two mud drapes may indicate a subtidal environment, which was covered during the two slack-water periods (the high water and low water) in a flood-ebb cycle. The average thickness of a cycle of vertical couplet thickness variations suggests a depositional rate of about 40 cm per month.

Tidal Rhythmites of Upper Cretaceous Basal Horseshoe Canyon Formation in Drumheller, Alberta, Canada (Rahmani, 1988)

The delta-estuarine succession of the Upper Cretaceous basal Horseshoe Canyon Formation can be divided into three facies associations: the delta facies association, the estuary channel facies association, and the coastal plain facies association. The tidal channel point bar in the middle estuary of the estuary channel facies association is characterized by gently inclined lateral accretion bedding (Fig. 31). This facies may be 300 m wide in the dip direction, and 13 m thick in its deepest part. The lateral accretion units consist of a sandstone-shale sequence with a lower sandstone interval and an upper shale interval (Fig. 32).

The sandstone-shale sequence of the lateral accretion unit is characterized by a tidal rhythmite, which shows a fining-upward trend with sandy bundles in the lower sandstone interval and delicate laminations of sandstone-shale couplets in the shale interval above the sandstone (Fig. 32). Figure 32 also shows scour surfaces at the base and the top of the sandstone-shale sequence. This sequence consists of 245 preserved tidal couplets. These sandstones-shale couplets appear to be stacked up vertically in a cyclical pattern (Figs. 33, 34). The regular alternation of thick and thin couplets indicates a diurnal inequality. The alternation of groups of sandy couplets and groups of muddy couplets at intervals of 20 to 30 cm can be related to spring-neap tidal variations, suggesting a rather high deposition rate. It has been interpreted that the entire sandstone-shale sequence was deposited within a half year, when upstream river flooding overprinted the tidal processes in mid-estuary. The scour base was formed on the point bar surface during flood peak. The sandy tidal bundles represent deposition during the high stage of the river, and the sandstone-shale couplets were deposited during the low stage (Rahmani, 1989).

Tidal Rhythmites in the Miocene Digne Basin and the Modern Mont Saint-Michel Bay of France (Tessier *et al.*, 1988)

Typical tidal cyclicities can be recognized in recent deposits of the Mont Saint-Michel Bay and the Miocene deposits of the Digne Basin of France (Tessier *et al.*, 1988; Fig 35). Both sequences show a vertical stacking of thin couplets. Each couplet consists of a sand and a silt-mud lamina, which have been interpreted as representing deposition during the dominant current and the slack-water period, respectively, of a flood-ebb cycle.

In vertical sequences the couplet thicknesses show a cyclical gradual increase of the couplet thickness followed by a gradual decrease, which can be related to the semilunar cycle of spring-neap tides. An average thickness of 10 to 15 cm per semilunar cycle in the Bay of Mont Saint-Michel indicates a depositional rate of 20 to 30 cm per month, while in the Digne Basin the couplet sequence of semilunar cycle is thicker (25-50 cm) and the deposition rate is higher (50-100 cm per month). In the Digne Basin sequence each semilunar cycle contains 20 to 25 couplets. This suggests that the sequence of the Digne Basin was deposited in a subtidal to lower intertidal environment, which was covered by tides during all or most of the neap-spring cycle (Tessier *et al.*, 1988). The sequence in Mont-Saint Michel Bay, however, was

Figure 30. Tidal rhythmite sequence in Upper Ordovician tidal-flat deposits in Tonglu Area of Zhejiang, China. The rhythmic lamination of sand-mud couplets reflects the flood-ebb tidal cycle and the neap-spring tidal variation. Note the reversal of ripple migration directions, which are indicated by arrows.

Figure 31. Lateral accretion of the estuarine point bar (I), which underlies the coal seam (II) and erosionally (dashed line) overlies prodelta sediments (III), Upper Cretaceous basal Horseshoe Canyon Formation, Drumheller, Alberta, Canada. Maximum thickness of the point bar at this location is 13 m. Note multiple internal erosional surfaces (short arrows). Accretion is to the right. **A** and **B** indicate locations of Figures 32 to 34. (Photograph kindly provided by Rahmani).

Figure 32. A sandstone-shale sequence (approximately 2.5 m thick) in a lateral accretion unit of the point bar, showing a fining-upward trend with sandy bundles in the lower sandstone interval and delicate laminations of sandstone-shale couplets in the shale interval above the sandstone. Dashed lines indicate scour surfaces at the base and the top. The sequence consists of 245 preserved tidal couplets, interpreted to have been a result of a seasonal flood overprinting the tidal processes in the mid-estuary. Bar scale is 15 cm. (See A in Fig. 31 for location; photograph kindly provided by Rahmani).

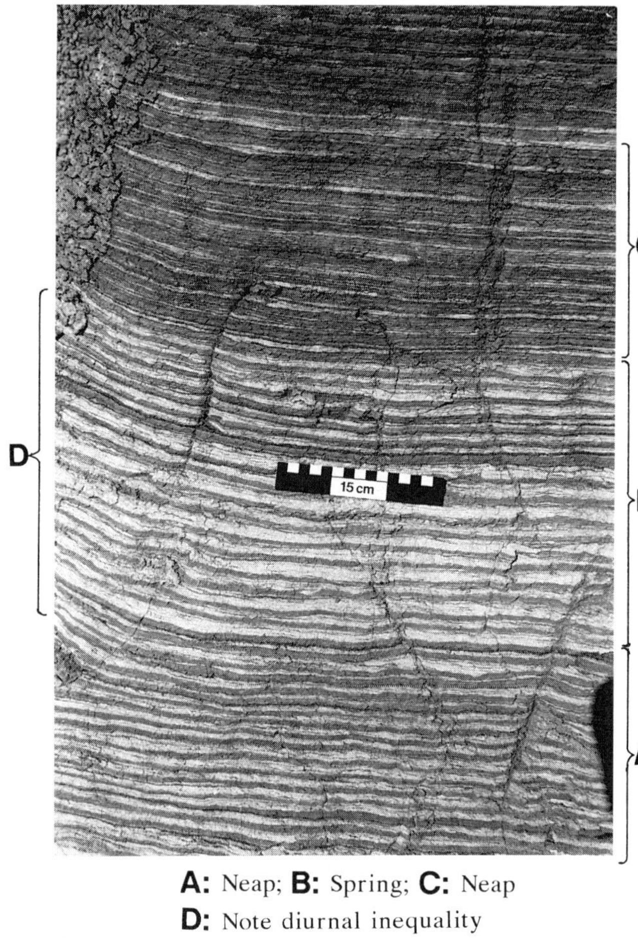

Figure 33. A close-up of the middle part of the sandstone-shale sequence of Figure 32, showing the delicate and rhythmic lamination of sandstone-shale couplets, which appear to be stacked up vertically. Note the alternation of groups of sandy couplets (spring tides) and groups of muddy couplets (neap tides). Note also the regular alternation of thick and thin couplets (diurnal inequality). Bar scale is 15 cm. (See Interval A in Fig. 31 for location; photograph kindly provided by Rahmani).

deposited in an upper intertidal zone characterized by an alternation of sand and silt-mud lamina, which was covered by tides around spring tidal periods. Consequently, each semilunar cycle contains only 10 to 12 couplets.

The successive semilunar cycles of vertical couplet thickness variations show unequal thicknesses of the thick "spring couplets", which may reflect the lunar cycle of lower- and high-spring tides.

Tidal Rhythmites in the Modern Qiantangjiang and Changjiang Estuaries of China

Tidal rhythmite sequences have been recognized in tidal point bar and tidal-flat deposits in macrotidal Qiantangjiang Estuary, mesotidal Changjian Estuary, and other areas along the coast of East China. Both areas are characterized by semidiurnal tides.

An example from a tidal point bar in the Changjiang Estuary displays a vertical sequence of sand-mud couplets (Huang and Wang, 1987). Each couplet consists of a sand and a mud lamina and represents a peak flow/stillstand cycle, which can be correlated to a flood-ebb cycle if clear evidence for other cyclicities (*e.g.*, semilunar and lunar cycles) can also be found. In a vertical sequence successive couplets show a cyclic variation of a gradual thickening followed by a gradual thinning, which can be correlated to spring-neap tidal cycles. Each cycle contains 25 to 26 couplets, which is very close to the number of flood-ebb cycles during a neap-spring period. An average thickness of 12 cm per neap-spring cycle indicates a depositional rate of 24 cm per month. This estimate has been confirmed by field observations that a 50 cm succession was deposited during a period of less than two months (Huang and Wang, 1987).

A: Strong tidal current B: Weak tidal current

Figure 34. A close-up of the middle part of the same sandstone-shale sequence as Figure 32. Diurnal inequality is very clear here. Bar scale is 15 cm. (See B in Figure 31 for location; photograph kindly provided by Rahmani).

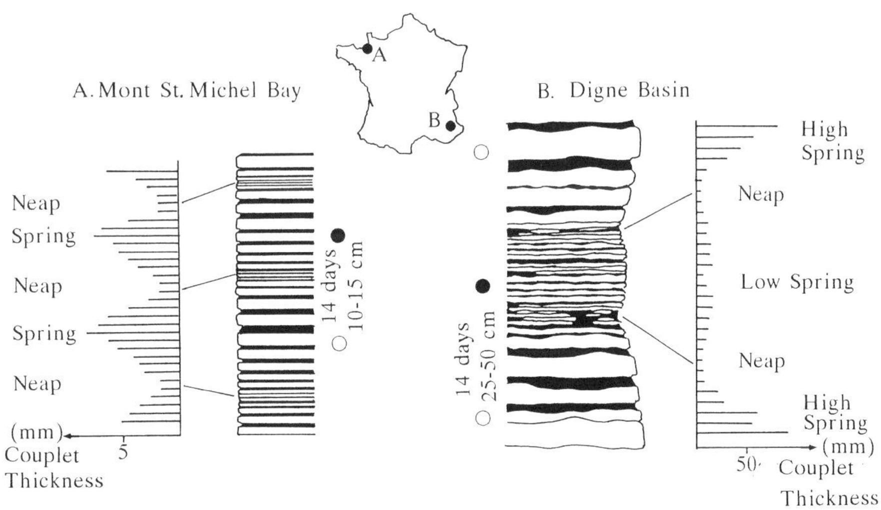

Figure 35. Vertical couplet thickness variation observed in recent deposits of the Mont Saint-Michel Bay and Miocene deposits of the Digne Basin, France (after Tessier *et al.*, 1988).

Precambrian Tidal Rhythmites in a Distal Ebb-Tidal Delta Setting in Southern Australia

Rhythmites of laminated siltstone and fine sandstone from Upper Proterozoic (c.650-800 Ma) glaciogenic formations in South Australia (Elatina Formation, Reynella Siltstone, and Chambers Bluff Tillite) have been interpreted as distal ebb-tidal delta deposits containing a full spectrum of paleotidal cycles (Williams, 1989a, b).

The rhythmites consist of "lamina-cycles". Each lamina-cycle contains up to 16 "laminae". A close observation shows two types of lamina. The "simple lamina" includes only one lamina. The "composite lamina" comprise pairs of "semi-laminae". Each simple lamina or semilamina is a graded layer 0.2 to 3 mm thick with a lower layer of fine grained sand passing upward into a thin band of silty clay. These graded layers can be traced laterally for hundreds of metres. These fine grained deposits are interpreted to have been carried offshore in suspension by turbid ebb-tidal currents and deposited below wave base in a distal ebb-tidal delta, or shelf setting. Therefore only one graded layer of sand-silt was formed during the waning ebb current stage of a flood-ebb cycle. Cyclical changes in the thicknesses of semidiurnal and diurnal laminae (Figs. 36, 37) may reflect variations in the amount of fine clastic material carried offshore because of cyclic variations of paleotidal current velocities.

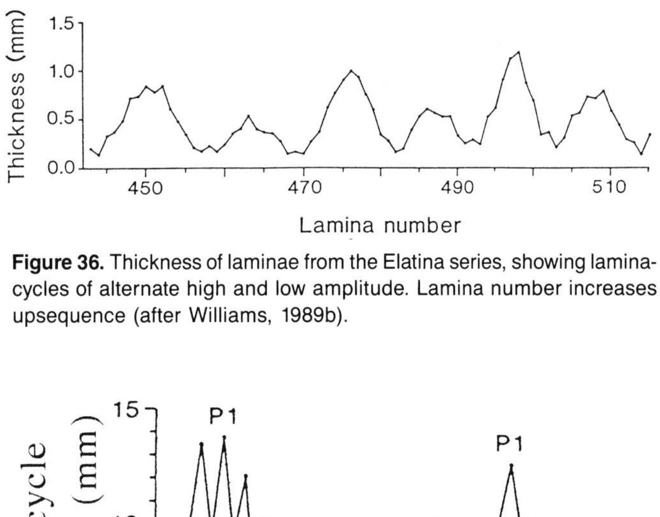

Figure 36. Thickness of laminae from the Elatina series, showing lamina-cycles of alternate high and low amplitude. Lamina number increases upsequence (after Williams, 1989b).

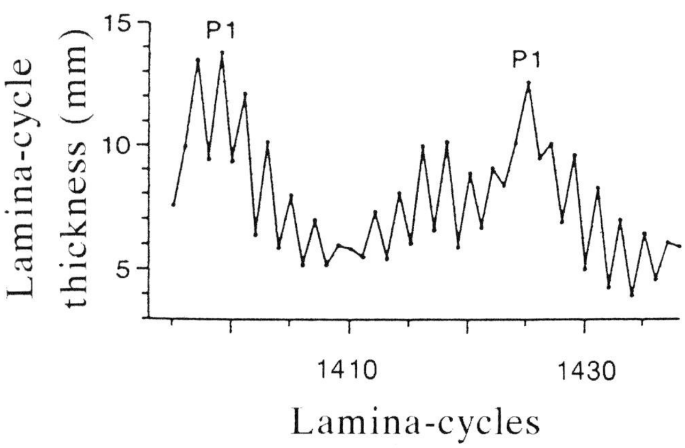

Figure 37. Thickness of lamina-cycles from the Elatina series (lamina-cycle number increases upsequence). The first-order peaks (P1) define the "Elatina Cycle", which represents a Late Precambrian year. The sawtooth pattern reflects the alternation of thick and thin lamina-cycles (after Williams, 1989b).

The rhythmites exhibit the time series records much longer than the time series from tidal bundle sequences or tidal point bar/tidal-flat sequences (*e.g.*, the Elatina lamina-cycle series is about 60 years long), they permit the detection of a full spectrum of palaeotidal cycles ranging from diurnal to the lunar nodal cycle.

1. The simple lamina or the semilamina in a composite lamina was formed during the waning ebb current of a flood-ebb cycle.

2. The paired semilaminae in a composite lamina commonly show unequal thicknesses, thus indicating the diurnal tidal inequality superimposed on semidiurnal tidal cycles. The simple lamina, on the other hand, indicates a diurnal increment in a diurnal tidal regime.

3. The lamina-cycle is bounded by darker, clayey bands where thinner laminae crowd together. It consists of up to 16 laminae and shows a gradual thickening of laminae followed by a gradual thinning, which can be correlated to the fortnightly spring-neap cycle (Figs. 36, 37).

4. Longer periods can be detected from cyclic variations in lamina-cycle thickness. Spectra of Fast Fourier Transform (FFT) of the sequence of lamina-cycle thickness show a number of peaks at about 2 lamina-cycles (representing the alternation of thick and thin lamina-cycles related to the monthly cycle of high and low spring tides), 13.1 lamina-cycles (representing a half year cycle), and 26.1 lamina-cycles (representing the Late Precambrian year) (Fig. 38).

It is difficult to estimate the exact number of days in the Late Precambrian lunar month. Because successive lamina-cycles commonly are of unequal duration, the greatest number of laminae for two successive lamina-cycles in the Elatina series is 29, which provides a minimum value for the number of days per lunar month. A maximum estimate is twice the observed maximum number of laminae per lamina-

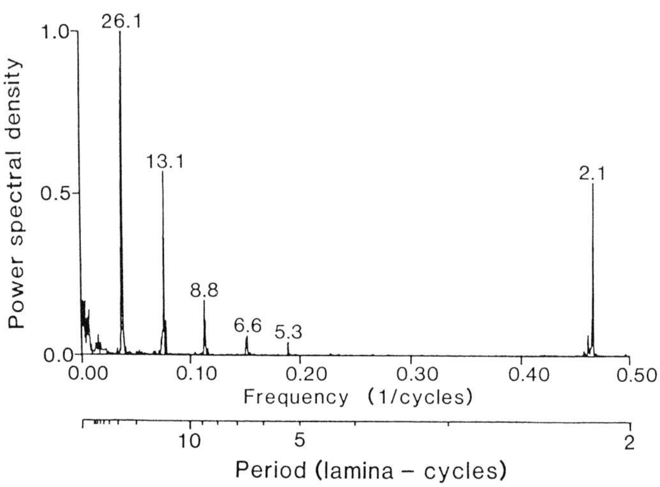

Figure 38. Fast Fourier transform smoothed spectrum for the Elatina sequence of 1580 lamina-cycle thickness measurements (about 60-year-long), normalized to unity for the strongest peaks and with linear frequency scales. The strong period of 26.1 lamina-cycles represents the first harmonic of the Elatina Cycle, with additional harmonics at 13.1, 8.8, 6.6 and 5.3 lamina-cycles. The peak near 2 lamina-cycles reflects the alternation of thick and thin cycles (after Williams, 1989b).

cycle, that is, 16 × 2 = 32. Taking an average of 30.5 (±1.5) days per Late Precambrian lunar month and 13.1 (±0.5) lunar months per year, the Late Precambrian (c.650 Ma) year contained 400 (±20) days. The periods of the lunar apsides and lunar nodal cycles were 9.7 (±0.3) and 19.5 (±0.5) years, respectively (Williams, 1989a, b).

Flaser and Linsen Bedding

Flaser and linsen bedding is commonly observed in vertical sequences of interlaminated sand and mud laminae. It was first described by Reineck and Wunderlich (1968). Significant is that these features were first described in the tidal flats of the German Bight. Within this setting, the flaser and linsen bedding was formed by tidal processes and in a minor way influenced by wave action. The specific depositional processes of flaser and linsen bedding reflect irregular fluctuating conditions. These features are common within a tidally-dominated environment, but also in a wave-dominated setting. Therefore, flaser and linsen bedding alone is not an unique feature in tidally-dominated deposits. The vertical sequence of interlaminated sand and mud laminae with flaser and linsen bedding may show typical tidal cyclicities as described in this section, which are a characteristic feature of tidal-dominated deposits.

Summary

Sequences of vertically-stacked thinly-interlaminated sand and silt-mud laminae with characteristic tidal cyclicities can be described as tidal rhythmites. Tidal rhythmite sequences have been recognized in tidal flats, tidal point bars, and distal ebb-tidal deltas. They provide good records of tidal cyclicities. Different orders of tidal cyclicities can be detected in such tidal rhythmite sequences: 1) the couplet of sand and silt-mud laminae represents a peak flow/stillstand cycle, which can be correlated to a flood-ebb cycle if evidence for other tidal cyclicities can also be found; 2) the alternation of thick and thin couplets indicates a diurnal inequality; 3) a gradual thickening of the couplets in a vertical sequence followed by a gradual thinning reflects a neap-spring cycle; 4) the unequal thicknesses of the "spring couplets" in successive semilunar cycles indicates the lunar cycle of low and high spring tides; and 5) the thickness variation of couplets is related to longer cycles of tidal variations. Although rhythmite sequences can also be formed in other environments, the presence of different orders of cyclicities and their correlation with different orders of tidal cyclicities are undoubtedly sufficient evidence of tidal dominance in such deposits.

In a semidiurnal tidal regime, a neap-spring couplet sequence should contain 27 to 28 couplets. This number, however, may be reduced significantly. Three factors may be responsible for the reduction: 1) in a diurnal tidal regime a neap-spring cycle includes only about 14 couplets; 2) during neap tides, the tidal currents may be too weak to move sand, and mud drapes deposited during successive slack-water periods may merge together; and 3) the upper intertidal zone is covered by tides only during spring tidal period.

Conclusions

The most characteristic feature of clastic tidal deposits is their cyclicity.

Typical cyclicities detected in megaripple crossbedding structures include: 1) the bundling of sandy foresets bounded by mud couplets or a mud drape and a reactivation surface indicating flood-ebb cycles; 2) the thick-thin alternation of successive bundles reflecting the diurnal tidal inequality; 3) the lateral bundle thickness variation and bottomset thickness variation related to semilunar cycles of neap-spring tides; and 4) the variation in spring bundle thickness related to the lunar cycles of high and low spring tides. Reactivation surfaces can be related to the flood-ebb cycle (dominant and subordinate flow reactivation surface) as well as the neap-spring tidal variation (dominant flow reactivation surface with a regular wavy erosional pattern).

Characteristic cyclicities observed in delicate, vertically stacked, thinly laminated tidal rhythmites include: 1) the sand/mud couplets related to peak flow/stillstand cycles, which can be correlated with flood-ebb cycles if evidence for other tidal cyclicities can also be found; 2) the vertical couplet thickness variations related to diurnal and neap-spring tidal cycles; 3) the variation in the thickness of spring couplets indicating the lunar cycles of high and low spring tides; and 4) the vertical thickness variation of rhythmites, reflecting longer cycles of tidal variations.

Although cyclic sequences can also be formed in other environments, the presence of different orders of cyclicities and their correlation with different orders of tidal cyclicities are undoubtedly unique criteria for recognizing tidal dominance in shallow marine clastic deposits.

There are also additional criteria, such as sigmoidal bedding, flaser and linsen bedding, and herringbone crossbedding. These features can be ascribed to other depositional processes as well as tidal. The occurrence of these additional criteria may strongly indicate the presence of tidal processes and a further search should be carried out to find single criteria.

Acknowledgments

Most of the data have been contributed by members from the Comparative Sedimentology Division of the University of Utrecht. We are thankful to the members of the Mase-Oosterschelde work group with whom we spend numerous pleasant and stimulating days in the field. Special thanks are due to C. Siegenthaler who joined our team right from the beginning. We are indebted to R. Visser, T. de Mowbray, R. Donselaar, J. van den Berg, A. van Gelder and M. Cuevas for their enthusiastic participation in the field and in discussions, to R. Rahmani for fruitful discussions on the tidal rhythmites in the basal Horseshoe Canyon Formation and for the original photos that he kindly provided, and to G.

V. Middleton and C. L. Amos for kindly reviewing the manuscript and for their in-depth discussions.

The authors are grateful to International Geoservices for their support in the preparation of this manuscript. Special thanks are due to T. Wong for helping with the drawings and the text, and to H. J. F. van der Veen for preparing the drawings.

References

Allen, J. R. L. 1973. Features of cross-stratified units due to random and other changes in bedforms. Sedimentology, v. 20, p. 189-202.

______ 1981. Lower Cretaceous tides revealed by cross-bedded sets with mud drapes. Nature, v. 289, p. 579-581.

______ 1982. Mud drapes in sand wave deposits: a physical model with application to the Folkestone Beds (early Cretaceous, southeast England). Philosophical Transactions of the Royal Society of London, A 306, p. 291-345.

Allen, P. A. and Homewood, P. 1984. Evolution and mechanics of a Miocene tidal sand wave. Sedimentology, v. 31, p. 63-81.

Amos, C. L. and Zaitlin, B. A. 1984. The effect of changes in tidal range on a sublittoral macrotidal sequence, Bay of Fundy, Canada. Geo-Marine Letters, v. 4, p. 161-169.

Banerjee, I. 1989. Tidal structures in the Glauconitic Sandstone, Countess Field, southern Alberta, Canada. *In:* Modern and Ancient Examples of Clastic Tidal Deposits — a Core and Peel Workshop, G. E. Reinson (ed.). Canadian Society of Petroleum Geologists, Second International Research Symposium on Clastic Tidal Deposits, Calgary, Alberta, p. 89-97.

Berg, J. H. van den 1982. Migration of large-scale bedforms and preservation of cross-bedded sets in highly accretional parts of tidal channels in the Oosterschelde, SW Netherlands. Geologie en Mijnbouw, v. 61, p. 253-263.

Boersma, J. R., Meene, E. A. van de and Tjalsma, R. C. 1968. Intricated cross-stratification due to interaction of a megaripple with its lee-side system of backflow ripples. Sedimentology, v. 11, p. 147-162.

Boer, P. L. de, Gelder, A. van and Nio, S. D. (editors) 1988. Tide-influenced Sedimentary Environments and Facies. Reidel Publishing Company, 530 p.

Clifton, H. E. 1979. Tidal channel deposits of Middle Eocene age, Torrey Pines State Reserve, California. *In:* Eocene Depositional Sytems San Diego, Calfornia, P. L. Abbott, (ed.). Pacific Section, Society of Economic Paleontologists and Mineralogists, Log Angeles, p. 35-41.

Dalrymple, R. W., Knight, R. J. and Lambiase, J. J. 1978. Bedforms and their hydraulic stability relationships in a tidal environment, Bay of Fundy, Canada. Nature, v. 275, p. 100-104.

Donselaar, M. E. 1989. The Cliff House Sandstone, San Juan Basin, New Mexico: model for the stacking of transgressive barrier complexes. Journal of Sedimentary Petrology, v. 59, p. 13-27.

Drake, D. E. 1976. Suspended sediment transport and mud deposition on continental shelves. *In:* Marine Sediment Transport and Environmental Management, D. J. Stanmley and D. J. P. Swift (eds.). John Wiley & Sons, p. 127-158.

Ginsburg, R. N. 1975. Tidal Deposits: a Casebook of Recent Examples and Fossil Counterparts. Springer-Verlag, Berlin, 428 p.

Haven, D. S. and Morales-Alamo, R. 1972. Biodeposition as a factor in sedimentation of fine suspended solids in estuaries. *In:* Environmental Framework of Coastal Plain Estuaries, B. W. Nelson (ed.). Geological Society of America, Memoir 133, p. 121-130.

Homewood, P. and Allen, P. A. 1981. Wave-, tide- and current-controlled sandbodies of Miocene Molasse, western Switzerland. Bulletin of the American Association of Petroleum Geologists, v. 65, p. 2534-2545.

______ , Keller, B., Schoepfer, P. and Yang, C. S. 1989. Facies, sedimentary processes and reconstruction of paleomarine conditions in the Swiss Upper Marine Molasse. Bulletin Société Géologique de France, v. 8 (5), p. 1015-1027.

Huang Naihe and Wang Genfa, 1987. The new criteria of ancient tidal sediments-tidal cycle sequences. Acta Sedimentologica Sinica, v. 5 (2), p. 39-44.

Jopling, A. V. 1961. Origin of regressive ripples explained in terms of fluid-mechanic processes. U.S. Geological Survey Professional Paper 424-D, p. D15-D17.

Kessler, L. G. and Gollop, I. G. 1988. Inner shelf — shoreface — intertidal transition, Upper Precambrian, Port Askaig Tillite, Isle of Islay, Argyll, Scotland. *In:* Tide-influenced sedimentary environments and facies, P. L. de Boer, A. van Gelder and S. D. Nio (eds.). Reidel Publishing Company, p. 341-358.

Klein, G. de V. 1970. Depositional and dispersal dynamics of intertidal sand bars. Journal of Sedimentary Petrology, v. 40, p. 1095-1127.

______ and Whaley, M. L. 1972. Hydraulic parameters controlling bedform migration on an intertidal sand body. Geological Society of America, Bulletin, v. 83, p. 3465-3470.

Kreisa, R. D. and Moiola, R. J. 1986. Sigmoidal tidal bundles and other tide-generated sedimentary structures of the Curtis Formation, Utah. Geological Society of America, Bulletin, v. 97, p. 381-387.

Krone, R. B. 1972. A field study of flocculation as a factor in estuarine shoaling processes. U.S. Army Corps Engineers, Committee on tidal hydraulics, Technical Bulletin 19, 62 p.

Ladipo, K. O. 1988. Example of tidal current periodicities from an Upper Cretaceous sandstone succession (Anambra basin, SE Nigeria). *In:* Tide-influenced Sedimentary Environments and Facies, P. L. de Boer, A. van Gelder and S. D. Nio (eds.). Reidel Publishing Company, p. 333-340.

McCabe, P. J. and Jones, C. M. 1977. Formation of reactivation surfaces within superimposed deltas and bedforms. Journal of Sedimentary Petrology, v. 47, p. 707-715.

Middleton, G. V. and Southard, J. B. 1984. Mechanics of sediment movement. Society of Economic Paleontologists and Mineralogists, Short Course 3, 2nd ed. Tulsa, Oklahoma, 401 p.

Mowbary, T. de and Visser, M. J. 1984. Reactivation surfaces in subtidal channel deposits, Oosterschelde, southwest Netherlands. Journal of Sedimentary Petrology, v. 54, p. 811-824.

Mutti, E., Rosell, J., Allen, G., Fonnesu, F. and Sgavetti, M. 1985. The Eocene Baronia tide dominated delta-shelf system in the Ager Basin. *In:* Excursion guidebook: 6th I.A.S. European Regional Meeting, M. D. Mila and J. Rosell (eds.). Lleida, Spain, p. 577-600.

Nio, S. D., Siegenthaler, C. and Yang, C. S. 1983. Megaripple cross-bedding as a tool for the reconstruction of the palaeohydraulics in a Holocene subtidal environment, SW Netherlands. Geologie en Mijnbouw, v. 62, p. 499-510.

______ and Yang, C. S. 1983. Dynamics, geometry and sequential upbuilding of large subtidal bedforms. *In:* Sedimentation on the continental shelf, with special reference to the East China Sea, Acta Oceanologica Sinica (ed.). Proceedings of the International Symposium, China Ocean Press, Beijing, China, p. 20-36.

______ , and ______ (*in press*). Sea-level fluctuations and the geometric variability of tide-dominated sandbodies. Sedimentary Geology, v. 70.

Rahmani, R. A. 1988. Estuarine tidal channel and near shore sedimentation of a Late Cretaceous epicontinental sea, Drumheller, Alberta, Canada. *In:* Tide-influenced Sedimentary Environments and Facies, P. L. de Boer, A. van Gelder and S. D. Nio (eds.). Reidel Publishing Company, p. 433-474.

______ 1989. Field trip guidebook to Cretaceous tidal estuarine and deltaic deposits, Drumheller, Alberta: Second International Research Symposium on Clastic Tidal Deposits, Calgary, Alberta, Canada, 55 p.

Reineck, H. E. and Wunderlich, F. 1968. Classification and origin of flaser and lenticular bedding. Sedimentology, v. 11, p. 99-104.

Reinson, G. E. 1989. Tide-influenced channel deposits in the Lower Cretaceous Glauconitic Member, southern Alberta. *In:* Modern and Ancient Examples of Clastic Tidal Deposits — a Core and Peel Workshop, G. E. Reinson (ed.). Canadian Society of Petroleum Geologists, Second International Research Symposium on Clastic Tidal Deposits, Calgary, Alberta, p. 98-104.

Rust, B. R. and Nanson, G. C. 1989. Bedload transport of mud as pedogenic aggregates in modern and ancient rivers. Sedimentology, v. 36, p. 291-306.

Siegenthaler, C. 1982. Tidal cross-strata and the sediment transport rate problem: a geologist's approach. Marine Geology, v. 45, p. 227-240.

Terwindt, J. H. J. 1971. Lithofacies of inshore estuarine and tidal inlet deposits. Geologie en Mijnbouw, v. 50, p. 515-526.

Tessier, B., Monfort, Y., Gigot, P. and Larsonneur, C. 1988. Enregistrement vertical des cyclicités tidales; adaptation d'un outil de traitement mathématique. exemples en Baie du Mont Saint-Michel et dans la molasse marine Miocène de Digne. Journée L. Dangeard, Dynamique des Milieux Tidaux, Société Géologique de France, p. 69-70.

Tillman, R. W. 1985. A spectrum of shelf sands and sandstones. *In:* Shelf Sands and Sandstone Reservoirs, R. W. Tillman, D. J. P. Swift and R. G. Walker (eds.). Society of Economic Paleontologists and Mineralogists, Short Course 13, p. 1-46.

Visser, M. J. 1980. Neap-spring cycles reflected in Holocene subtidal large-scale bedform deposits: a preliminary note. Geology, v. 8, p. 543-546.

Williams, G. E. 1989a. Late Precambrian tidal rhythmites in South Australia and the history of the Earth's rotation. Journal of the Geological Society, London, v. 146, p. 97-111.

______ 1989b. Precambrian tidal sedimentary cycles and Earth's palaeorotation: EOS, Transactions, American Geophysical Union, v. 70 (3), p. 33 and 40-41.

Wood, J. M. and Hopkins, J. C. 1989. Reservoir sandstone bodies in estuarine valley fill: Lower Cretaceous Glauconitic Member, Little Bow Field, Alberta, Canada. American Association of Petroleum Geology, Bulletin, v. 73, p. 1361-1382.

Yang, C. S. 1986. On Bagnold's sediment transport equation in tidal marine environments and the practical definition of bedload. Sedimentology, v. 33, p. 465-486.

______ and Nio, S. D. 1985. The estimation of palaeohydrodynamic processes from subtidal deposits using time series analysis methods. Sedimentology, v. 32, p. 41-57.

______ and ______ 1989. An ebb-tide delta depositional model — a comparison between the modern Eastern Scheldt tidal basin (southwest Netherlands) and the Lower Eocene Roda Sandstone in the southern Pyrenees (Spain). Sedimentary Geology, v. 64, p. 175-196.

KEY WORDS: DIAGNOSTIC TIDAL CRITERIA, TIDAL CYCLICITIES, TIDAL BUNDLE SEQUENCE, TIDAL RHYTHMITE

In: Clastic Tidal Sedimentology, D. G. Smith, G. E. Reinson, B. A. Zaitlin and R. A. Rahmani (eds.). Canadian Society of Petroleum Geologists. Memoir 16, p. 29-40.

Sedimentary processes and facies in the Gironde estuary: a recent model for macrotidal estuarine systems

George P. Allen

Total Exploration Laboratory
Avenue de Versailles
78470 Saint-Remy-Les-Chevreuse
France

Abstract

The Gironde estuary was formed by the Holocene drowning of a Pleistocene incised fluvial valley and is presently filling with a regressive wedge of fluvially-derived sediment, organized in a longitudinal facies pattern reflecting the downstream evolution from fluvial, to tidal, to mixed tidal-wave dominant environments. The estuary exhibits a tripartite geomorphological zonation comprising: 1) a mixed fluvial and tidal upper estuary with meandering channels and tidal point bars; 2) a tide-dominated, funnel-shaped lower estuary with nonerosional muddy channels and elongate sandy tidal bars; and 3) a wave and tide-affected inlet with sandy coastal barriers and tidal delta shoals.

Tide amplitudes range from 2.5 to 6 m, and tidal current reversal can occur as far as 130 km from the estuary mouth. A large fluvial sediment load is introduced into the estuary where it is reworked by a combination of density circulation and tidal currents, and deposited in the upper estuary channels and funnel. Suspended sediment is trapped within a turbidity maximum that migrates longitudinally in the estuary, extending landward up to the limit of tidal currents during low river flow. The coarse fluvial sands and gravels are blocked from entering the estuary by tide current asymmetry.

These processes result in a distinct longitudinal facies zonation that parallels the morphological pattern of the estuary. The fluvial-tidal facies transition occurs in the vicinity of the bayline, where the fluvial profile reaches tidal base level, and is marked by an evolution from coarse sand and gravel point bars, to sand and mud estuarine point bars. These are characteristically devoid of alluvial levee and crevasse splay deposits, since the large tidal prism in the estuary damps out alluvial floods. Farther downstream, the limit between the upper estuarine channel and the funnel is marked by a transition from point bars to upward coarsening tidal bars that prograde seaward over estuarine mud. The estuary inlet contains coarse sand with tide and wave structures. These sands are of marine origin and introduced into the estuary from the coast by longshore drift and flood tide currents. They form the transgressive Holocene substratum over which accumulates the present regressive estuarine wedge.

These facies patterns and limits are easily recognizable in cores and outcrops and furnish criteria to delimit paleogeographic boundaries. They also form stratigraphic markers to identify and correlate the different phases of an estuarine valley fill.

Résumé

L'estuaire de la Gironde s'est formé à l'Holocène suite à l'inondation d'une vallée fluviale entaillée au Pléistocène. Le remplissage de cette vallée comprend un prisme de sédiments d'origine fluviatile, accumulé lors d'une régression, et dont la répartition longitudinale des faciès reflète la transition entre un environnement fluvial, un environnement prédominé par l'action des marées et un environnement dit mixte dans lequel vague et marée étaient en interaction. La zonation géomorphologique de l'estuaire est tripartite et se compose: 1) d'un estuaire supérieur où se combinent l'influence fluviale et celle de la marée, avec des chenaux méandriques et des barres d'accrétion latérale de marée («tidal point bars»); 2) d'un estuaire inférieur en forme d'entonnoir, dominé par l'action des marées, avec des chenaux boueux non érosifs et des barres sableuses allongées; et 3) d'un petit bras de mer affecté par la dynamique des marées et des vagues, avec des barrières sableuses côtières et des «hauts-fonds deltaïques de marée» («tidal delta shoals»).

L'amplitude des marées varie de 2.5 à 6 m, et les courants de jusant peuvent être decelés à 130 km en amont de l'embouchure de l'estuaire. La charge de sédiments fluviatiles introduite dans l'estuaire est remaniée à la fois par des courants de densité et par des courants de marée, pour être ensuite redéposée dans les chenaux et dépressions coniques de l'estuaire. Les sédiments en suspension sont piégés dans une zone de turbidité maximale dont la migration longitudinale dans l'estuaire atteint, en amont, la limite des courants de marée associée à un faible débit de la rivière. Les courants de marée asymétrique empêchent les sables grossiers et graviers fluviatiles de pénétrer dans l'estuaire.

Ces processus résultent en une zonation longitudinale des faciès distincte, parallèle à la structure morphologique de l'estuaire. La transition entre les faciès fluvial et de marée avoisine la ligne de rivage où le profil fluvial atteint le niveau de base de la marée. Cette zone transitionnelle est marquée par une évolution de barres d'accrétion latérale («point bars») à sables grossiers et graviers, à des barres estuariennes de sables et de boues. Ces faciès sont caractérisés par l'absence de levées alluviales et de dépôts de «crevasses de débordement» («crevasse splay») puisque l'imposant prisme de sédiments de marée dans l'estuaire réduit l'inondation des plaines alluviales. La limite entre le chenal de l'estuaire supérieur et l'estuaire inférieur se distingue par la transition entre des barres d'accrétion latérale et des barres de marée, à granulométrie plus grossière vers le sommet, s'avançant progressivement vers la mer au-dessus des boues estuariennes. Le bras de mer se compose de sables grossiers avec des structures reliées aux vagues et marées. Ces sables d'origine marine sont introduits dans l'estuaire par des courants de flot et par la dérive littorale. Ils forment le substratum holocène, résultant d'une transgression, sur lequel s'accumule les séquences estuariennes régressives actuelles.

Ces relations et limites de faciès sont facilement identifiables dans les carottes et sur les affleurements, et elles fournissent un outil pour délimiter les limites paleogéographiques. Elles formes également des marqueurs stratigraphiques pour identifier et corréler les différentes phases de remplissage d'une vallée estuaire.

Introduction

Recent advances in the understanding of tidal sedimentation and the increasing application of sequence stratigraphic concepts (Posamentier and Vail, 1988), have spurred a renewed interest in recent estuarine deposits, which furnish analogues for ancient incised valley systems (Reinson *et al.*, 1988; Leckie, 1988; Rahmani, 1988; Wood and Hopkins, 1989). Macrotidal estuaries in particular exhibit marked sediment patterns and facies characteristics, and numerous papers have been published on sediment transport in these estuaries (see Nichols and Biggs, 1985). However, only a few of these relate the estuarine hydrological and sedimentary processes to the regional facies and stratigraphic patterns as they would be preserved in the rock record.

This paper presents a synthesis of the sedimentary processes, facies patterns, and stratigraphic relationships in the modern Gironde estuary, a large macrotidal coastal plain estuary at the mouth of a major fluvial drainage basin in southwestern France (Fig. 1). The major focus will be on the formulation of a general facies model applicable to ancient deposits.

Previous Studies

As in all large estuaries that are the sites of important economic activities, numerous sedimentological studies have been carried out in the Gironde since the end of the 19th century. During the 1930's, the pioneering work of L. Glangeaud led to the discovery and description of the estuarine turbidity maximum (Glangeaud, 1939; Glangeaud and Bonnichon, 1939). More recently, during the past 20 years, the hydrology and sedimentary processes in the estuary have been studied in detail by a number of workers (Bonnefille, 1971; Allen, 1972; Allen *et al.*, 1977; Castaing, 1981). Based on the study of more than 100 vibracores in the estuary, Allen *et al.* (1970), Feral (1970), and Assor (1972) have described the stratigraphic evolution of the estuary since the last glacial lowstand. These studies illustrate the facies patterns of the transgressive-regressive Holocene sediment wedge in the Gironde, and have formed the basis for a recent sequence stratigraphic analysis of the estuary valley fill by Allen and Truilhé (1988).

Estuarine Morphology

The Gironde forms a large, landward-tapering indentation at the mouth of the Garonne and Dordogne rivers, which extends more than 100 km landward from the coast (Fig. 1). A few kilometres upstream from the inlet, the present width is 15 km decreasing to less than 1 km at the city of Bordeaux, located 100 km upstream. The estuary was formed by the drowning of an incised fluvial valley during the Holocene transgression. In the last 5000 years, the rate of sea level rise has slowed and the estuary has been gradually filling with fluvial sediment (Feral, 1970). The resulting estuary fill, described in detail by Allen *et al.* (1970), forms a transgressive-regressive sediment wedge up to 50 m thick beneath the present inlet. More recently, that is, during the last 2000 years, extensive tidal flats and marshes have developed, gradually constricting the estuary and smoothing out

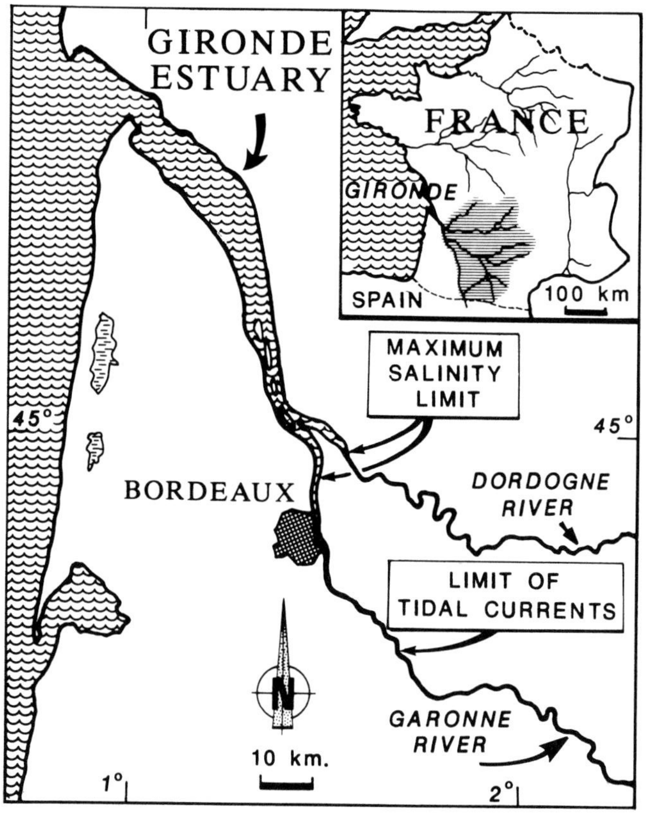

Figure 1. Location of the Gironde estuary and drainage basin of the Dordogne and Garonne rivers. The maximum tide and salinity limits indicated correspond to spring tides during low river discharge.

the irregular pre-Holocene topography, (Dutrait, 1895; Migniot, 1971).

The present morphology of the estuary exhibits a distinct, tripartite, longitudinal zonation (Allen *et al.*, 1973) comprising: 1) meandering, upper estuary channels with sandy and muddy point bars; 2) a smooth, funnel-shaped channel containing linear tidal sand bars and vegetated islands; and 3) a constricted inlet with a deeply scoured tidal channel abuting seaward into a large, arcuate sand shoal (Fig. 2). This morphological pattern is common to all tidal estuaries, although smooth exponential funnel shapes appear to be more characteristic of macrotidal systems (Wright, 1977). Each of these zones exhibits characteristic sediment types and facies reflecting a downstream evolution from fluvial, to tidal, and mixed tidal-wave energy regimes.

An important geomorphological boundary forms the landward limit of the estuary, where the tidal deposits at base level intersect the upward-sloping alluvial plain. This limit constitutes the bayline, as defined by Posamentier and Vail (1988), and is located approximately 100 km landward from the estuary inlet (Fig. 3). Although during low river discharge tidal currents can affect the river upstream from this limit, the bayline coincides with the landward extremity of tidal estuarine facies.

Analysis of historical records dating from Roman times and bathymetric charts since the 17th century indicate that the estuary is filling with sediment, particularly in the intertidal zone (Dutrait, 1895; Castaing, 1981) and the muddy channels of the lower estuary funnel (Allen *et al.*, 1977). Sedimentation rates in these zones can be as much as several centimetres per year (Migniot, 1971). At the same time that the estuary is filling with sediment, the Atlantic coastline adjacent to the estuary is being eroded at an average yearly rate of 1 to 10 m per year (Penin, 1980; Howa, 1987). This erosion is caused by the high energy wave regime and the present lack of sand supply to the coast.

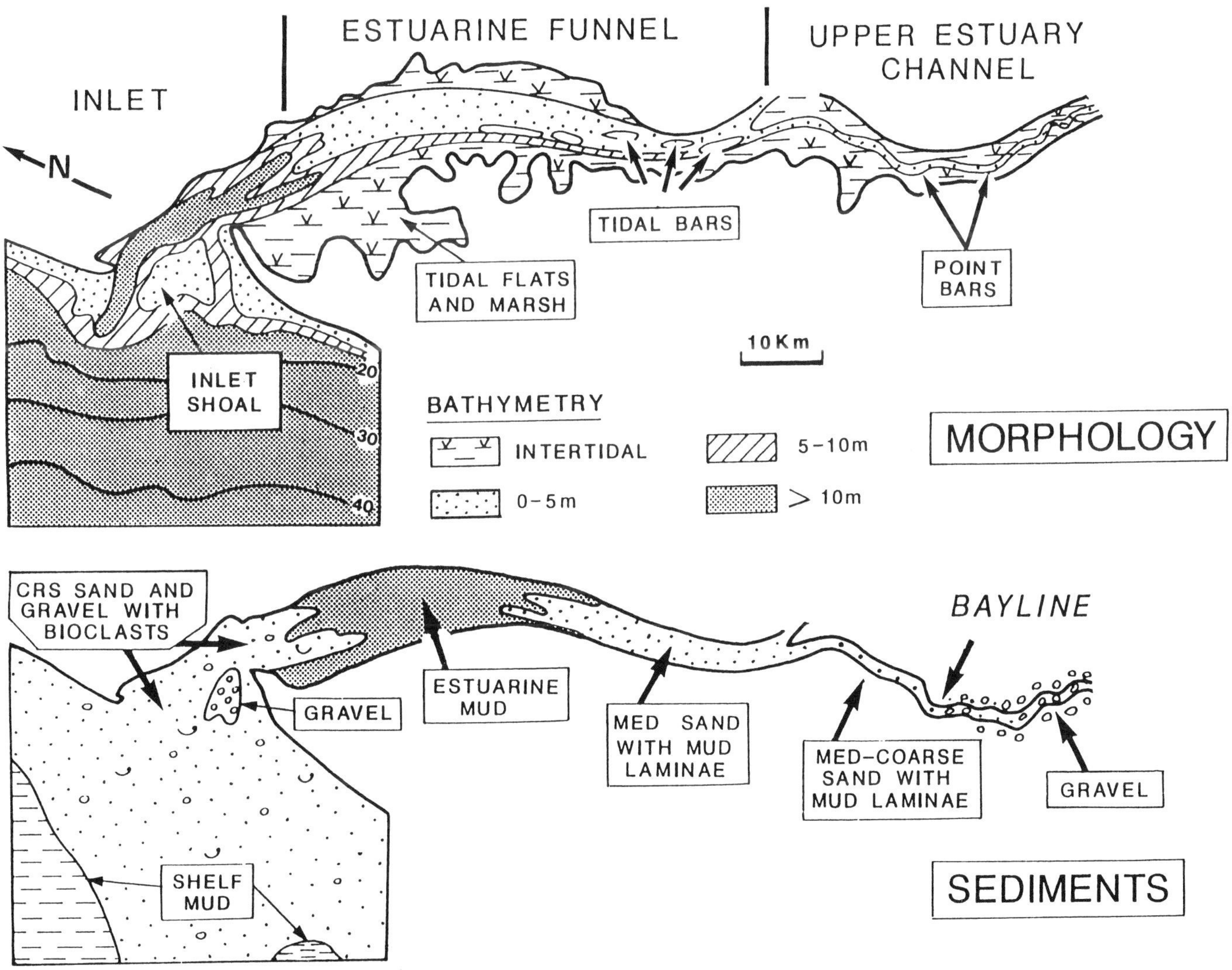

Figure 2. Top: Morphology of the Gironde estuary, illustrating the tripartite geomorphological zonation and major depositional environments. The bathymetry is with respect to the lowest tide level. **Bottom:** Sediment distribution patterns in the estuary. The bayline represents the junction between the fluvial and coastal plain, and coincides with the fluvial-tidal facies transition (see text).

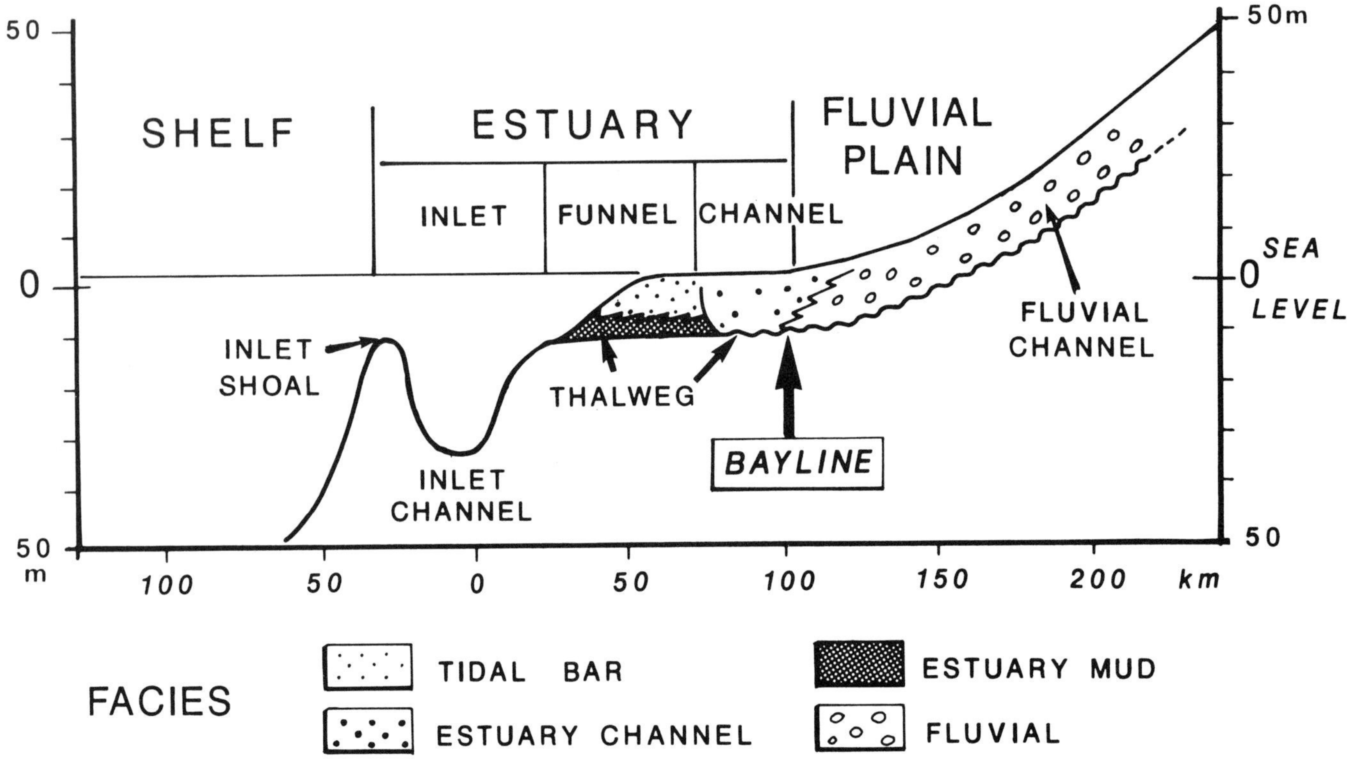

Figure 3. Longitudinal cross-section along the thalweg of the Garonne River and Gironde estuary, illustrating the different morphological zones and sediment types. The depth of the thalweg is relatively constant within the estuarine channels and funnel, but deepens considerably as a result of tidal scour in the constricted estuarine inlet.

Environmental Factors

Tides

Tides in the Gironde are semidiurnal, and in the inlet range from 2.5 m during neap tides, to more than 5 m during spring tides. Friction deforms the tidal wave, causing it to amplify and become asymmetrical as it propagates landward in the upper estuary. This effect is very pronounced in macrotidal estuaries (Coleman and Wright, 1978), and in the Gironde results in higher tide ranges (up to 6 m) and a longer lasting ebb tide in the upper estuary channels. The resulting ebb-flood tidal current asymmetry (Fig. 4) has an important sedimentological effect, as will be discussed later. During low river discharge, spring tide current reversal occurs in the upper estuary channels up to 130 km from the inlet. During periods of high river discharge this limit is shifted downstream about 30 km, and coincides with the bayline.

Waves

Violent storms occur during winter months in the Bay of Biscay, creating a very high energy wave regime, with storm waves exceeding heights of 8 m. The strongest waves are from the west and west-northwest (Dugas and Glangeaud, 1969) and produce a regional littoral drift to the south along the Aquitaine coast (Dugas, 1966; Gensous, 1971). Waves also affect the intertidal zones of the inlet and lower estuary funnel as far as 10 to 15 km upstream from the mouth (Allen, 1971).

Hydrology and Sediment Sources

The average combined discharge from the Garonne and Dordogne rivers varies between 500 and 1000 m^3s^{-1} (Migniot, 1971; Castaing, 1981); however, pronounced seasonal flow variations occur, with floods exceeding 5000 m^3s^{-1} during the winter months. The hydrological mixing regime in the estuary varies seasonally with river discharge. During low river flow, the estuary is well mixed, and the salinity intrusion extends landward up to the junction between the upper estuary channels and the funnel (Fig. 5). During high river discharge, the salt intrusion is pushed back to the mid funnel zone, 40 to 50 km from the inlet, and the estuary is stratified (Allen, 1972). The maximum limit of tidal currents is located about 50 km landward of the salt intrusion. The upper estuary channels are practically never affected by brackish water, and cannot sustain any marine or brackish fauna, whereas the brackish water of the funnel zone maintains foraminifera (Pujos, 1971) and ostracodes (Carbonel *et al.*, 1971).

The sediment budget of the estuary and adjacent coast is illustrated in Figure 6. The total fluvial sediment input averages about 3×10^6 tons per year and comprises about 75 per cent suspended silt and clay, and 25 per cent sand (Migniot, 1971; Castaing, 1981). Several studies (Allen *et al.*, 1977; Castaing and Allen, 1981; Castaing, 1981) have indicated that a significant fraction of the fluvial suspended load (about 60%) is evacuated seaward to the shelf. The entire fluvial

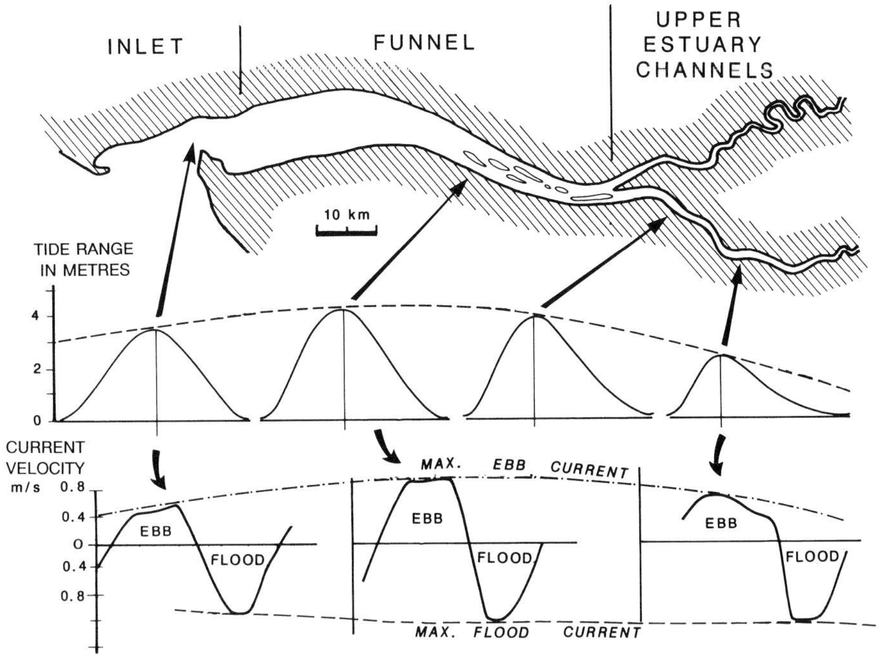

Figure 4. Landward evolution of tides and tidal currents for an average tide during low river discharge. The tide range increases by 1 m from the inlet to the upper estuary channels. Note the landward-increasing, time-velocity asymmetry, which results in longer but weaker ebb currents, and shorter and stronger flood tides in the upper estuary channels.

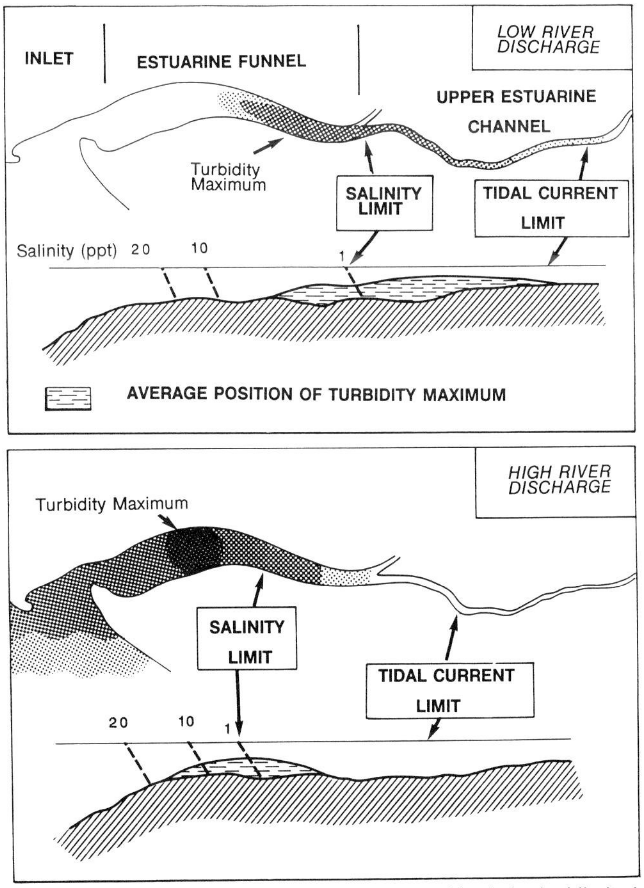

Figure 5. Location of the turbidity maximum and hydrological limits in the estuary during low and high river discharge. Note that during low river flow, the maximum extends a considerable distance landward from the salinity intrusion, and attains the limit of tidal currents. During high river discharge, the maximum is pushed to the distal extremity of the estuary and straddles the limit of the salinity intrusion.

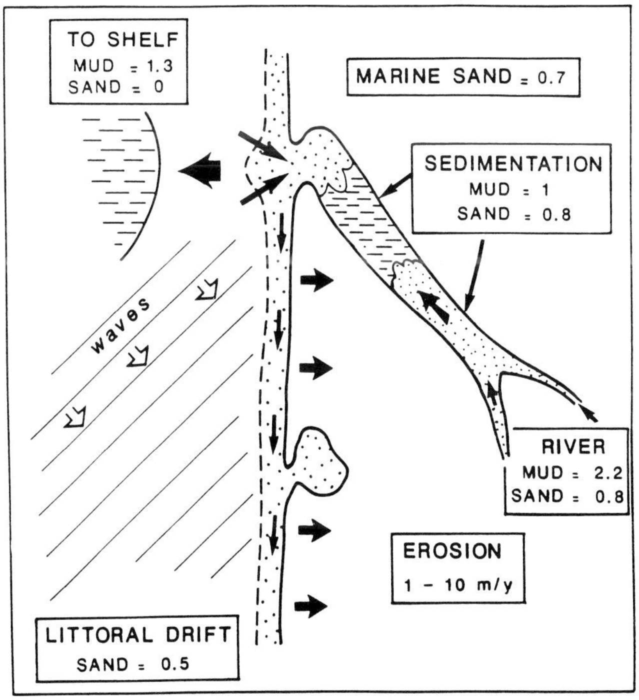

Figure 6. Schematic illustration of the average sediment budget furnished to and out of the estuary, as compiled from various authors. The numbers refer to millions of tons per year.

bedload, however, is trapped within the estuary, and at present no fluvial sand reaches the inlet and shelf, (Allen 1972). Another important source of sediment results from coastal and shoreface erosion, which supplies sand into the estuary inlet by longshore drift and flood tides (about 0.7×10^6 tons/year, Howa, 1987). This reworked marine sand contains gravel eroded from Pleistocene alluvial terraces outcropping in the inlet and inner shelf, and accumulates in the form of ebb and flood tidal shoals (Fig. 2).

Sedimentation Processes and Patterns

Suspended Sediment

The major feature controlling suspended sediment accumulation in the estuary is the turbidity maximum, first observed and described in 1939 by Glangeaud and Bonnichon in the Gironde. Subsequently, turbidity maxima have been described in numerous estuaries and deltas (Postma, 1967; Nichols and Biggs, 1985). The turbidity maximum consists of a zone of high concentrations of suspended sediment extending over a distance of several tens of kilometres (Fig. 5). The Gironde maximum is particularly dense compared to many other estuaries, and exhibits a range of suspended sediment concentrations varying from several tens of mg l^{-1} to several tens of g l^{-1}. The maximum acts as a source of suspended sediment that settles out during tidal slacks.

The turbidity maximum originates from the trapping effect of both density currents and tide current asymmetry. Estuarine density currents result from vertical and longitudinal salinity gradients, which give rise to a net landward flow of bottom water in the salinity intrusion (Pritchard, 1955). This landward flow converges with the net seaward flow of water landward of the salinity intrusion. This convergence, or null zone, coincides with the landward limit of the salinity intrusion, and constitutes an efficient sediment trap for near-bottom suspended sediment (Harleman and Ippen, 1969). The density circulation is well developed when the salinity stratification is pronounced, that is, during high river discharge. During periods of low river discharge however, tide current asymmetry plays the major role in trapping suspended sediment and maintaining the turbidity maximum in the upper estuarine channels (Allen *et al.*, 1980). As shown in Figure 5, during low river discharge, the maximum occurs upstream of the salinity intrusion and extends landward as far as the limit of tidal currents.

In addition to this seasonal cycle, the turbidity maximum is also affected by a neap-spring sedimentation cycle (Fig. 7). During spring tides, the strong currents erode and resuspended muddy sediment, and the maximum expands in volume and sediment concentrations. During neap tides, the low current velocities and prolonged slacks favor net sedimentation during the tidal cycle, and the maximum depletes. This results in a rhythmic alternation of sand and mud laminae, characteristic of tidal estuarine deposits. During periods of very high fluvial discharge, the downstream extremity of the maximum is extruded out of the estuary and furnishes sediment to a mid-shelf mud field, developing at average storm wave base (50-70 m, Fig. 2), (Castaing and Allen, 1981; Lesueur *et al.*, 1989).

Bedload Sediment

The sand in the upper estuary channels and funnel is entirely of fluvial origin (Glangeaud, 1939). It is introduced into the estuary during river floods and gradually works its way through the upper estuary channel point bars to accumulate in the elongate tidal bars of the funnel zone. The downstream limit of the fluvial sand within the estuary is presently located about 25 km from the inlet (Cremer, 1977; Legigan and Castaing, 1981). A study of ancient maps indicates that the tidal bars (and the fluvial sand limit) have prograded seaward about 30 km since the middle ages (Allen, 1972).

A marked longitudinal sand grain size gradient is observed along the estuary, with grain sizes decreasing downstream

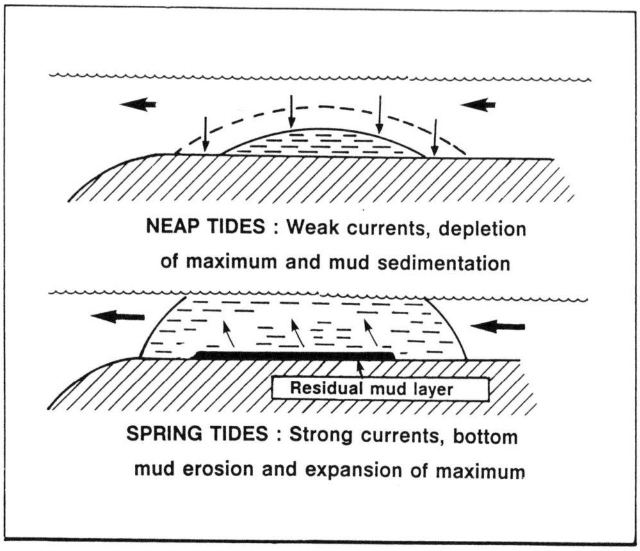

Figure 7. Schematic illustration of the fortnightly neap-spring-neap erosion and sedimentation cycle, which affects the turbidity maximum, and results in a residual mud layer accumulated at every neap tide.

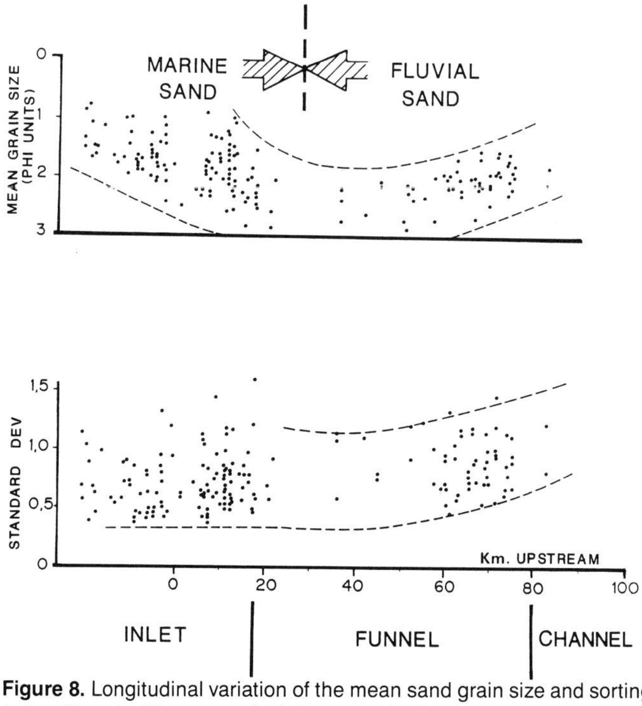

Figure 8. Longitudinal variation of the mean sand grain size and sorting in the Gironde. The zone of minimum grain size in the distal extremity of the estuarine funnel is located at the convergence of the fluvial and marine sand (see Fig. 2).

to the lower funnel zone, and increasing from the funnel to the inlet (Fig. 8). The zone of grain size minimum in the estuary funnel coincides with the downstream limit of fluvial sand. The seaward-decreasing grain size gradient in the upper estuary is a result of the tide current asymmetry (Fig. 4), with the stronger flood tide bottom currents in the upper estuary impeding the seaward transport of coarse sand and gravel.

Figure 2 illustrates the estuary-wide sediment patterns engendered by these processes. The upper estuary channels and funnel contain medium grained sand, which grades downstream to estuarine mud in the distal part of the funnel. The mud-dominant zone extends more than 20 km and passes downstream in the inlet to coarse sand with gravel and bioclastic debris.

Sedimentary Facies

Numerous cores and borings have been described in various parts of the estuary by several workers (Allen *et al.*, 1970; Feral, 1970; Assor, 1972; Tastet *et al.*, 1986). These studies have shown that sediment facies patterns parallel the tripartite geomorphological and energy zonation of the estuary.

Fluvial-Estuarine Transition

The fluvial deposits consist of gravel and coarse sand point bars, overlain by 3 to 5 m of silty clay containing erosional surfaces marking crevasse splay channels. The present fluvial section is about 10 to 15 m in thickness. Observations and borings along the Garonne River indicate that a pronounced facies transition occurs when the river becomes affected by tides (Allen and Truilhé, 1987). As illustrated in Figure 9, downstream from the maximum limit of tidal currents, the channel sediments grade from gravel to coarse sand, then to medium sand with abundant mud laminae. This transition is centered on the bayline zone, about 100 km landward from the inlet, and roughly coincides with the limit of tidal current reversal during high river discharge. Seaward of this transition, the upper estuary channels accumulate tidal estuarine point bars comprising medium grained, crossbedded to rippled sand, with abundant clay interbeds and flasers. The crossbedding contains clay drapes and occurs in 20 to 30 cm thick sets, separated by mud laminae up to several cm thick. These sands fine upward into rhythmic alternations of medium to fine grained, rippled sand with mud flasers. The upper 5 m of the section is within the intertidal zone and consists of massive clay with isolated ripples at the base, and silt stringers and roots at the top. Conspicuously absent are fluvial levees and crevasse splays, and facies are clearly tide-dominant and similar to those described in numerous tidal settings (de Boer *et al.*, 1988).

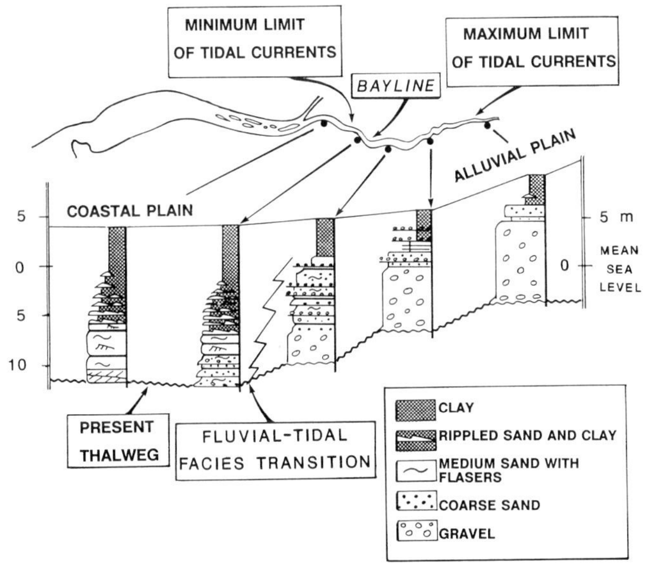

Figure 9. Longitudinal facies section at the transition between the fluvial and estuarine channels, based on cored borings. This facies transition occurs downstream from the maximum tide limit, and is centered on the bayline, that is, at the upstream limit of tidal currents during high river discharge.

This fluvial to tidal facies transition appears to be a common feature in meso- and macrotidal estuaries and delta channels, and similar features have been described in the Mahakam delta (Allen *et al.*, 1979), in Eocene fluvial and deltaic channels in northern Spain (Leo and Allen, 1984) and in modern estuaries in Ecuador and Washington State (Smith, 1988). Tidal estuarine point bars are distinguishable from fluvial point bars by their rhythmic sand-mud alternations, abundant mud drapes and interbeds, and lack of river flood overbank deposits. They form inclined heterolithic strata as described by Thomas *et al.* (1987).

These facies changes result from a combination of the three sedimentological processes described in the previous paragraphs: 1) the estuarine turbidity maximum, which furnishes a concentrated source of mud; 2) tide current asymmetry, which blocks the downstream transport of coarse fluvial bedload; and 3) attenuation of the fluvial floods in the estuary, which impedes the development of alluvial overbank deposits.

Estuary Funnel

The estuary funnel zone is characterized by the presence of tidal sand bars that prograde over estuarine mud. These bars form linear to lobate sand bodies 0.5 to 5 km wide, and 10 to 15 km long. In the upstream part of the estuary funnel, the bars coalesce to form large tidal sand bodies with complex internal architectures (Tastet *et al.*, 1986). Facies and bedding patterns are tidal and exhibit rhythmic alternations of medium grained sand and mud, reflecting tidal cycles on different time scales. Crossbed sets are 20 to 40 cm in thickness, and in any one section are unidirectional in the subtidal zone, and commonly bidirectional in the intertidal zone (H. Fenies, pers. comm.). The crossbeds exhibit a sigmoidal geometry (Mutti *et al.*, 1985), and tidal bundles (Boersma and Terwindt, 1981) separated by erosional reactivation surfaces. Clay drapes accumulate during tidal slacks; however, these are not frequently preserved, because of erosion by tidal currents. This erosion furnishes the abundant clay clasts interbedded within the foresets, or between crossbed sets (Tastet *et al.*, 1986). No trace of fluvial flood sedimentation is visible, since even the strongest river floods are damped out in this part of the estuary. Individual bars comprise juxtaposed ebb- and flood-dominant lobes and lineaments. Sediment transport patterns are complex, and are related to local ebb-flood transport asymmetry resulting from morphological shielding (Hayes, 1971).

The bars prograde seaward and laterally over a substratum of estuarine mud that accumulates in the distal part of the funnel as well as the inter-bar channels. They form a sandy, upward-thickening section 6 to 7 m thick. Maximum sand thickness and crossbedding scale occur at the top of the subtidal zone. No precise facies criteria distinguish the limit between the subtidal and intertidal zone, since clay drapes are not usually preserved. However, the sub-intertidal limit coincides with the limit between the upward-thickening bar facies, and the upward-fining and -thinning tidal flat deposits. In the intertidal zone, sand thickness decreases and the summit of the bars is capped by 2 to 3 m of massive mud flat and marsh clay. This marsh clay stabilizes the bar, which then becomes a permanent island in the estuary funnel. Because of the high water turbidity and strong tidal currents, burrowing fauna are absent in the subtidal estuary, and the muds of the funnel zone are not bioturbated.

The tidal flats and marshes that border the estuary funnel form a seaward-widening belt of intertidal, laminated mud. In the distal extremity of the funnel, waves introduce coastal sand onto the intertidal flats, which become sandier, bioturbated, and contain abundant pelecypod debris.

Estuary Inlet

The estuary inlet forms a mixed tide and wave environment depositing essentially sand and gravel, with minor amounts of mud accumulating in sheltered coastal embayments. The sources of this coarse sand and gravel, as mentioned previously, are the eroding beaches and shoreface adjacent to the estuary mouth. The inlet channel thalweg is deeply scoured by tidal currents and is paved with coarse gravel and shell debris over which large ebb- and flood-dominated tidal megaripples migrate (Castaing and Froidefond, 1978). Sand grain size directly reflects tidal current strength, with coarse sand and gravel in the channels, and medium sand on the shoals (Allen, 1971).

The erosional inlet channel abuts upstream and downstream into ebb and flood tidal delta-like shoals. The outer shoal is wave-dominated, and exhibits an arcuate morphology across the estuary mouth. The inner shoal constitutes a tide-dominated, flood tidal delta, and interfingers upstream with the estuarine mud of the lower funnel zone. A study of cores from the inlet (Braud, 1986) indicates that the sands are well-sorted, medium to coarse grained, and exhibit bidirectional cross-stratification, parallel stratification, and wave ripples. Erosion surfaces with pebble and bioclastic lags are common, and clay laminae and flasers are absent.

The maximum potential thickness of these inlet sands is in the order of a few tens of metres. They will be preserved as an erosional-based, upward-fining, sandy tidal inlet section similar to those described by Kumar and Sanders (1974). The erosional base will be paved with a conglomerate and shell lag, and overlain by coarse, tidal and wave stratified sand. The inlet sand will contrast with the tidal bar sand farther landward in the estuarine funnel since it is coarser grained and devoid of clay laminae.

Conclusions

The Gironde estuary sediments and facies are organized into a longitudinal pattern that reflects the seaward transition from fluvial, to tidal, to mixed tide and wave environments (Fig. 10). Many macrotidal estuaries exhibit a similar tripartite morphological zonation consisting of meandering upper estuary channels, a smooth, seaward-flaring estuary funnel, and an inlet zone. The characteristics of the inlet vary with the coastal wave regime and tide range. On high wave energy coastlines, the formation of spits and barriers constrict the estuary mouths of macrotidal estuaries such as the Gironde, causing a deepening of the inlet channel by tidal current scour.

The fluvial-tidal facies transition can be recognized by a marked decrease in channel sediment grain size, an appearance of numerous mud laminae and rhythmic sand-mud bedding, and a disappearance of alluvial flood and overbank deposits. This transition represents the landward extremity of estuarine facies, and occurs in the vicinity of the bayline, which roughly coincides with the tidal current limit during high river discharge. Each morphological zone of the estuary contains distinct facies types, which together comprise a characteristic longitudinal estuarine sequence (Fig. 10). The sinuous upper estuary channels are preserved as erosional-based, fining-upward estuarine tidal point bars; the funnel zone is characterized by seaward prograding, coarsening-upward tidal sand bar complexes in its upper end, and massive estuarine mud in its distal extremity. These deposits constitute a regressive sediment wedge sourced by fluvial influx, which is gradually filling the estuary funnel. Facies are typically tidal, with no evidence of fluvial floods, and the sediments are non-bioturbated. This particular feature is probably distinctive of high energy macrotidal estuaries. The inlet zone contains coarse sand and gravel, with mixed tidal and wave facies. Bioclastic debris is furnished by the oysters and other pelecypods that colonize the intertidal flats. These sands will form a tidal inlet section; that is, a section with an erosional, channelized base, fining up into beach or tidal flat and marsh deposits. All these facies will be sufficiently distinct to be readily identifiable in the field and in cores. The transition between the different facies assemblages will enable the identification of the major paleogeographic boundaries within the estuarine system (Fig. 10).

Estuaries result from the drowning of river valleys during periods of rapid sea level rise, and their stratigraphic organization reflects successive episodes of transgression and regression during periods of more or less rapid relative sea

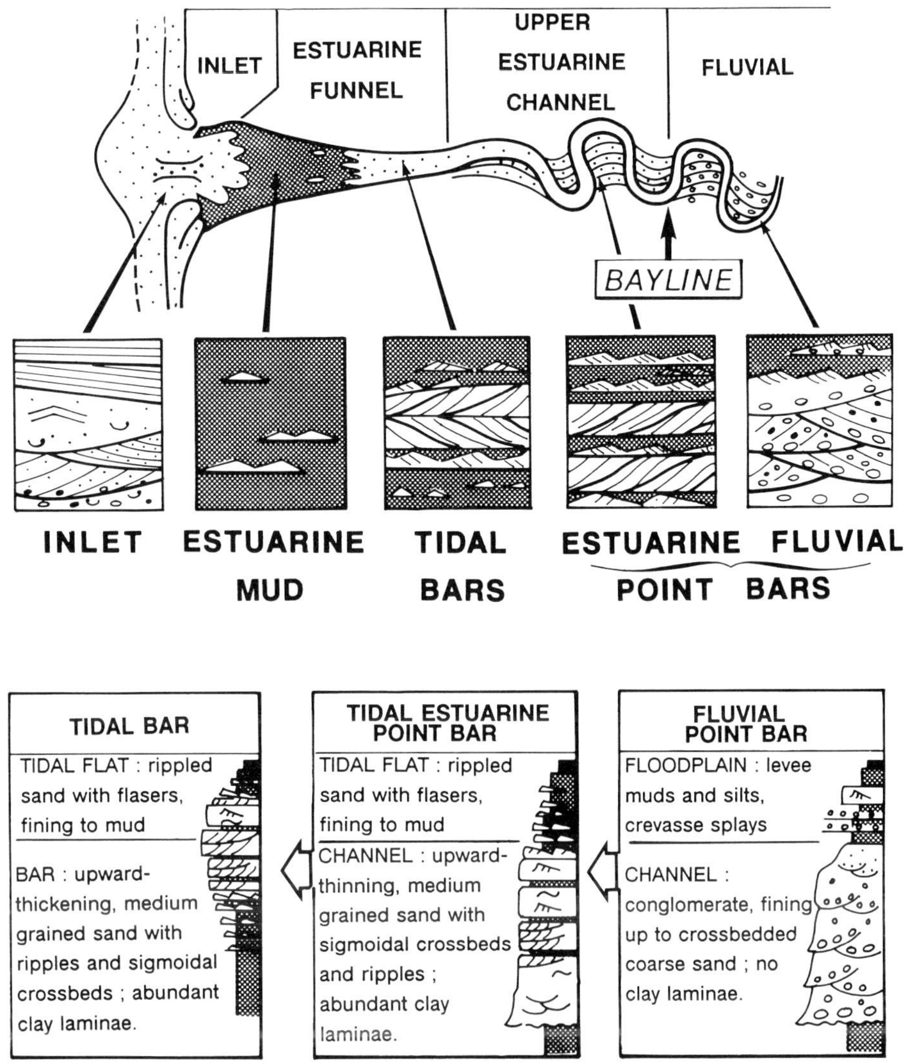

Figure 10. Schematic illustration of the longitudinal facies sequence within a Gironde-type macrotidal estuary. Each morphological zone is represented by a characteristic facies type, indicating a downstream evolution from fluvial to estuarine channels and bars.

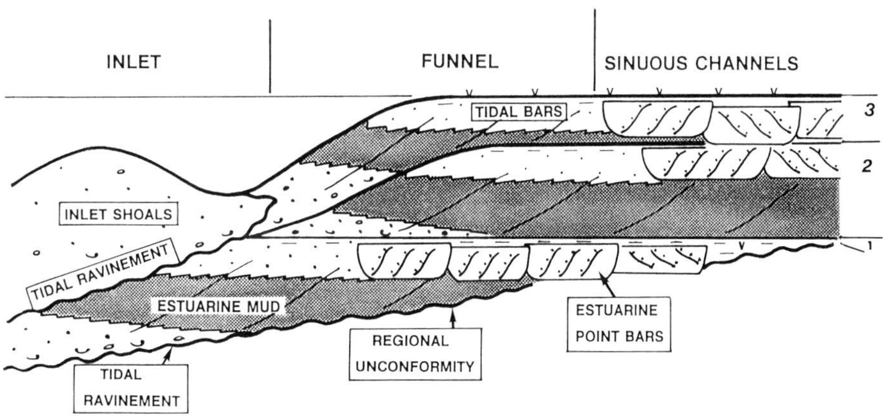

Figure 11. Idealized stratigraphic section through the estuary thalweg, illustrating a period of transgression (units 1, 2), followed by regression and estuary infill (unit 3). This reflects the stratigraphic pattern developed during and since the Holocene transgression. Note the marked tidal ravinement surface formed during transgression by the landward-migrating inlet tidal channel.

level rise. As illustrated in the Gironde, estuarine sediments are furnished by two sources: transgressive sediments composed of marine sand introduced into the inlet, and a regressive source comprising the sediments introduced by the river.

During transgression, the inlet deposits will form a sandy wedge overriding the estuarine muds and bars of the funnel zone. Tidal scour in the landward retreating inlet will result in a transgressive tidal ravinement surface incising the underlying estuarine funnel deposits (Fig. 11). During periods of regression, such as in the present Gironde, the funnel sediments will prograde seaward and downlap over the inlet sands. A schematic stratigraphic section is illustrated in Figure 11. The maximum thickness of the estuarine section is a function of the depth of scour of the inlet channel, which will vary directly with wave energy and tidal range. Macrotidal estuaries on high-wave coasts form a thick section, exceeding 20 to 30 m. As the estuary fills during a regressive phase, the funnel muds and tidal bars prograde over the marine sands of the inlet, and the upper estuary channels gradually extend downstream, incising the tidal bar deposits of the funnel. In time, these tidal bar deposits will be eroded and replaced by estuarine point bars. Since the maximum depth of sand in the tidal bars is in the order of 8 to 9 m below sea level, whereas the point bars attain depths of 10 to 12 m, the upper estuary point bars will eventually directly overlie the estuarine muds of the distal funnel zone.

References

Allen, G. P. 1971. Relationship between grain size parameter distribution and current patterns in the Gironde estuary (France). Journal of Sedimentary Petrology, v. 41, p. 74-78.

——— 1972. Etude des processus sédimentaires dans l'estuaire de la Gironde. Thèse Doct. es Sciences, Univ. Bordeaux 1, 314 p.

———, Castaing, P., Feral, A., Klingebiel, A. and Vigneaux, M. 1970. Contribution à l'étude des facies de comblement et interprétation péléogéographique de l'évolution des milieux sédimentaires récents et actuels de l'estuaire de la Gironde. Bulletin Institut Géologique du Bassin d'Aquitaine, Univ. Bordeaux, v. 8, p. 99-155.

———, Carbonel, P., Castaing, P., Gayet, J., Jouanneau, J. M., Klingebiel, A., Latouche, C., Legigan, Ph., Jujos, M. and Vernette, G. 1973. Environments and sedimentary processes of the North Aquitaine Coast. Guide Book, International Symposium on Interrelationships of Estuarine and Continental Shelf sedimentation, Institut de Géologie du Bassin d'Aquitaine, Bordeaux, July 1973.

———, Sauzay, G., Castaing, P. and Jouanneau, J. M. 1977. Transport and deposition of suspended sediment in the Gironde estuary, France. *In:* Estuarine Processes, M. Wiley (ed.). Vol. II, Academic Press, New York, p. 63-81.

———, Laurier, D. and Thouvenin, J. 1979. Etude sédimentologique du delta de la Mahakam. Notes et Mémoires v. 15, Compagnie Française des Pétroles, Paris, 156 p.

———, Salomon, J.C., Bassoulet, P., Dupenhoat, Y. and Degrandpre, C. 1980. Effects of tides on mixing and suspended sediment transport in macrotidal estuaries. Sedimentary Geology, v. 26, p. 62-90.

——— and Truilhe, G. 1987. Fluvial-tidal transition in alluvial and deltaic channels: diagnostic criteria for paleoenvironmental reconstruction (abs.) Society of Economic Paleontologists and Mineralogists, Annual Midyear Mtg., Aug. 1987, Austin, Texas.

——— and ——— 1988. Stratigraphic and facies model of a transgressing estuarine valley fill in the Gironde estuary (France) (abs.). *In:* Sequences, Stratigraphy, Sedimentology: Surface and Subsurface, D. P. James and D. A. Leckie (eds.). Proceedings of the Canadian Society of Petroleum Geologists Technical meeting. Memoir 15, Calgary, 1988, 575 p.

Assor, R. 1972. Interprétation paléogéographique des terrains sédimentaires de la presqu'île du Médoc. Thèse Doct. 3ème cycle, Univ. Bordeaux 1, 132 p.

Boersma, J. R. and Terwindt, J. H. J. 1981. Neap-spring tide sequences of intertidal shoal deposits in a mesotidal estuary. Sedimentology, v. 28, p. 151-170.

Bonnefille, R. 1971. Remarques sur les écoulements moyens à l'aval de la Gironde. Bulletin de l'Institut de Géologie du Bassin d'Aquitaine, Bordeaux, v. 11, p. 361-364.

Braud, F. 1986. La dynamique sédimentaire à l'embouchure d'un estuaire: Thèse Doct. 3ème cycle, n° 2117, Univ. Bordeaux 1.

Carbonel, P., Moyes, J. and Vigneaux, M. 1971. La répartition des thanatocoenoses d'Ostracodes dans l'estuaire de la Gironde et ses relations avec les courants. Compte Rendu de l'Académie des Sciences, Paris, t. 273.

Castaing, P. 1981. Le transfer à l'océan des suspensions estuariennes; cas de la Gironde. Thèse Doct. es Sciences, Univ. Bordeaux 1, 530 p.

——— and Allen, G. P. 1981. Mechanisms controlling seaward escape of suspended sediment from the Gironde; a macrotidal estuary in France. Marine Geology, v. 40, p. 101-118.

——— and Froidefond, J. M. 1978. Etude morphologique et dynamique des dunes hydrauliques à l'embouchure de la Gironde. Bulletin de l'Institut Géologie du Bassin d'Aquitaine, Bordeaux, v. 24, p. 131-145.

Coleman, J. M. and Wright, L. C. 1978. Sedimentation in an arid macrotidal alluvial river system: Ord river, Western Australia. Journal of Geology, v. 86, p. 621-642.

Cremer, M. 1977. Utilisation de la morphoscopie pour l'étude des transports de sables dans l'estuaire de la Gironde. Bulletin de l'Institut de Géologie du Bassin d'Aquitaine, Bordeaux, v. 22, p. 149-155.

De Boer, P. L., Van Gelder, A. and Nio, S. D. 1988 (Editors). Tide-influenced Sedimentary Environments and Facies: D. Reidel Publ. Co., Dordrecht, Holland, 530 p.

Dutrait, M. 1895. De mutationibus orae fluvialis et maritimae in Peninsula medulorum et garumnae fluminis ostrie. Thèse Docteur-es-Lettres, Université de Bordeaux, Imprimerie Cadouet, Bordeaux, 114 p.

Dugas, F. 1966. Etude du déplacement des sables sur la partie nord du littoral des Landes. Thèse Doct. 3ème cycle, Univ. Paris, 77 p.

——— and Glangeaud, L. 1969. Le mouvement des sables sur la partie nord du littoral des Landes de Gascogne. Revue de Géographie Physique et de Géologie Dynamique, Paris, v. 11, p. 385-391.

Feral, A. 1970. Interprétation sédimentologique et paléogéographique des formations alluviales de l'estuaire de la Gironde et de ses dépendances marines. Thèse Doct. 3ème cycle, Faculté des Sciences, Université de Bordeaux, n° 806, 158 p.

Gensous, B. 1971. Contribution à l'étude des sédiments meubles superficiels du plateau continental aquitain. Thèse Doct. 3ème cycle, Faculté des Sciences, Université de Bordeaux, 58 p.

Glangeaud, L. 1939. Le mouvement des sédiments et la formation des bancs, seuils et mouilles dans la Garonne et l'estuaire de la Gironde. Extrait du Congrès International de Géodésique et de Géophysique, Washington, 1939.

——— and Bonnichon, Y. 1939. Sur les causes dynamiques des variations de la turbidité et du débit solide dans la Gironde et la Garonne au cours des marées. Compte Rendu de l'Académie des Sciences, Paris. t. 208, 1072 p.

Hayes, M. O. 1971. Geomorphology and sedimentation of some New England estuaries. *In:* The Estuarine Environment; American Geological Institute lecture notes, v. 12, p. 1-71.

Harleman, M. and Ippen, A. T. 1969. Salinity intrusion effects in estuary shoaling. Proceedings of the American Society of Civil Engineers; Journal of the Hydraulics Division, v. 25, p. 9-27.

Howa, H. 1987. Le littoral du Nord Médoc (Gironde); évolution d'une côte sableuse en érosion. Thèse Doct. 3ème cycle, Univ. Bordeaux 1, n° 146, 258 p.

Kumar, N. and Sanders, J. E. 1974. Inlet sequence: a vertical succession of sedimentary structures and textures created by the lateral migration of tidal inlets. Sedimentology, v. 21, p. 491-532.

Leo, M. and Allen, G. P. 1984. An example of a succession of coarse and fine-grained point bars in the Montllobat Formation (North Spain): environmental implications and comparison with a modern analogue, the Garonne River (abs.). 5th European Regional Meeting of Sedimentology (abs.), International Association of Sedimentologists, Marseille, 255 p.

Legigan, P. and Castaing, P. 1981. La pénétration des sables marins dans les estuaires; cas de la Gironde. Compte Rendu de l'Académie des Sciences, Paris, t. 292, p. 207-212.

Lesueur, P., Tastet, J. P., Weber O. and Sinko, J. A. 1989. La vasière à l'ouest de la Gironde (France): corps sédimentaire pélitique d'âge historique. 2ème Congrès Français de Sédimentologie (abs.) Association des Sédimentologistes Français, Paris, 177 p.

Leckie, D. A. 1988. Sedimentology and sequences of the Paddy and Cadotte Members along the Peace River. Field trip guide to Sequences, Stratigraphy, Sedimentology: Surface and Subsurface; Technical Meeting Canadian Society of Petroleum Geologists, Calgary, 78 p.

Migniot, C. 1971. L'évolution de la Gironde au cours des temps. Bulletin de l'Institut de Géologie du Bassin d'Aquitaine, Bordeaux, France, v. 11, p. 221-281.

Mutti, E., Rosell, J., Allen, G. P., Fonnesu, F. and Sgavetti, M. 1985. The Eocene Baronia tide-dominated delta-shelf system in the Ager basin. *In:* 6th European Regional Meeting, International Association of Sedimentologists, Excursion Guidebook, M. D. Mila and J. Rosell (eds.). p. 579-600.

Nichols, M. M. and Biggs, R. B. 1985. Estuaries. *In:* Coastal Sedimentary Environments (2 ed.), R. A. Davis (ed.). New York, Springer-Verlag, p. 77-186.

Penin, F. 1980. Le prisme littoral aquitain: histoire et évolution récente des environnements morphosédimentaires. Thèse Doct. 3ème cycle, Univ. Bordeaux I, n° 1577, 129 p.

Posamentier, H. W. and Vail, P. R. 1988. Eustatic controls on clastic deposition II — Sequence and system tracts models. *In:* Sea Level Changes: an Integrated Approach, C. K. Wilgus *et al.* (ed.). Society of Economic Paleontologists and Mineralogists, Special Publication 42, p. 125-155.

Postma, H. 1967. Sediment transport and sedimentation in the marine environment. *In:* Estuaries, G. H. Lauff (ed.). American Association for the Advancement of Science, Publication 83, Washington, DC, p. 158-179.

Pritchard, D. W. 1955. Estuarine circulation patterns. Proceedings of the American Society of Civil Engineering, New York, v. 81, p. 1-11.

Pujos, M. 1971. Foraminifères et thécamoébiens de la Gironde: leur intérêt dans la mise en évidence des biotopes estuariens. Compte Rendu de l'Académie des Sciences, Paris, t. 273, p. 1095-1097.

Rahmani, R. A. 1988. Estuarine tidal channel and nearshore sedimentation of a Late Cretaceous epicontinental sea, Drumheller, Alberta, Canada. *In:* Tide Influenced Sedimentary Environments and Facies, P. L. de Boer *et al.* (eds.). p. 433-471, D. Reidel Publ. Co., Dordrecht.

Reinson, G. E., Clark, J. E. and Foscolos, A. E. 1988. Reservoir geology of Crystal Viking field, Lower Cretaceous estuarine tidal channel-bay complex, south-central Alberta. American Association of Petroleum Geologists, Bulletin, v. 72, p. 1270-1294.

Smith, D. G. 1988. Modern point bar deposits analogous to the Athabasca oil sands, Alberta, Canada. *In:* Tide-influenced Sedimentary Environments and Facies, P. L. De Boer, A. Van Gelder and S. D. Nio (eds.). Dordrecht, Holland, D. Reidel Publishing Company, p. 417-432.

Tastet, J. P., Fenies, H. and Allen, G. P. 1986. Facies, sequences et géométrie d'une barre tidale estuarienne; le banc de Trompeloup dans l'estuaire de la Gironde. Bulletin de l'Institut de Géologie du Bassin d'Aquitaine, Bordeaux, v. 39, p. 165-184.

Thomas, R. G., Smith, D. G., Wood, J. M., Visser, J., Calverley-Range, E. A. and Koster, E. H. 1987. Inclined heterolithic stratification-terminology, description, interpretation and significance. Sedimentary Geology, v. 53, p. 123-179.

Wood, J. M. and Hopkins, J. C. 1989. Reservoir sandstone bodies in estuarine valley fill: Lower Cretaceous Glauconitic member, Little Bow Field, Alberta, Canada. American Association of Petroleum Geologists, Bulletin, v. 37, p. 1361-1382.

Wright, L. D. 1977. Sediment transport and deposition at river mouths, a synthesis. Bulletin of the Geological Society of America, v. 88, p. 857-868.

KEY WORDS: GIRONDE ESTUARY, MACROTIDAL ESTUARINE SYSTEMS, LONGITUDINAL FACIES ZONATION, INCISED VALLEY

In: Clastic Tidal Sedimentology, D. G. Smith, G. E. Reinson, B. A. Zaitlin and R. A. Rahmani (eds.). Canadian Society of Petroleum Geologists. Memoir 16, p. 41-58.

Zonation and sedimentology of estuarine facies in an incised valley, wave-dominated, microtidal setting, New South Wales, Australia

SCOTT L. NICHOL

Coastal Studies Unit, Department of Geography
University of Sydney, Sydney, N.S.W., Australia, 2006

ABSTRACT

The southern sector of the New South Wales coast, Australia, delimited by the extent of the Lachlan Fold Belt and southern portion of the Sydney Basin structural provinces, provides an opportunity to examine estuarine depositional complexes in a number of bedrock-controlled valley settings. Sedimentation in these estuaries has progressed since the mid-Holocene under a process regime characterized by a high wave energy shoreface, a microtidal regime, relatively low fluvial inputs, and a relative sea-level stillstand.

A tripartite zonation of estuarine facies is shown to be a consistent feature of estuaries in the study area. The three facies zones are termed: barrier/inlet (Zone A); estuarine lagoon (Zone B); and tidal fluvial (Zone C). Each zone comprises an assemblage of morphostratigraphic units with a distinct sedimentological character interpreted as being intimately linked to the sediment source and the dominant processes operating within the zone.

Zone A sediments, sourced from marine and nearshore environments, are mature, quartz-dominant sands. Bedforms vary from three-dimensional dunes with negligible bioturbation near the estuary mouth, to two-dimensional, moderately bioturbated dunes in the medial portion of Zone A, to intensely bioturbated, incipient dunes at the distal part of Zone A. These variations are attributed to the landward decrease in tidal flow velocity. Zone B sediments are of terrestrial origin and are typically silty to sandy muds deposited in a low energy lagoon. The lagoon floor lacks dune bedforms, and sedimentary structures feature laminae, wavy beds and abundant burrow traces. Zone C sediments are also sourced from the estuary catchment, but are much coarser. Marked variations in textural properties are observed in both the surface and subsurface sediments; they incorporate muds, silty sands and sandy gravels commonly with a high amount of organic detritus. Grain size variations are interpreted as reflecting temporal and spatial fluctuations in the fluvial flow regime.

The sedimentological traits of each zone are found to be common to the two estuaries examined in detail, Wapengo Lagoon and Narrawallee Inlet, the former being partially filled and the latter completely filled. The Narrawallee sequence is interpreted as representing successive episodes, in the late Quaternary, of estuarine sedimentation during sea level highstand and fluvial scouring in the intervening glacial lowstand. Despite the complexity introduced by multiple filling and scouring in Narrawallee, facies zonation and character are maintained.

RÉSUMÉ

Le secteur sud de la côte de la Nouvelle-Galles du Sud, Australie, est délimité par deux provinces structurales, l'une étant la ceinture plissée de Lachlan, et l'autre, la partie sud du Bassin de Sydney. Ce secteur offre l'opportunité d'examiner les complexes de sédimentation estuarienne dans un nombre d'environnements de vallée contrôlés par le substratum. Depuis le mid-Holocène, la sédimentation de ces estuaires a progressé sous un régime caractérisé par une zone infratidale de haute vague-énergie, un régime microtidal, des apports fluviaux relativement faibles et un niveau marin relativement constant.

Une zonation tripartie des faciès estuariens s'avère être une caractéristique conséquente des estuaires de la région étudiée. Les trois zones de faciès sont appelées: barrière/petit bras de mer (Zone A); lagon estuarien (Zone B), et fluviatile de marée (Zone C). Chaque zone comprend un assemblage d'unités morphostratigraphiques avec une caractéristique sédimentaire qui serait intimement reliée à la provenance des sédiments et aux processus dominants dans la zone.

Les sédiments de la Zone A sont des sables matures dominés par le quartz originant des environnements marin et littoral. Les «morphologies de fond» («bedforms») de la Zone A varient de dunes à trois dimensions avec bioturbation négligeable à l'embouchure de l'estuaire, à des dunes à deux dimensions modérément bioturbées dans la partie médiane, à des amorces de dunes intensément bioturbées dans la partie distale. Ces variations sont attribuées à la diminution de la vitesse de l'écoulement de marée vers le continent. Les sédiments de la Zone B sont d'origine terrestre et sont typiques des boues silteuses à sableuses déposées dans un lagon de faible énergie. Le fond du lagon se caractérise par l'absence de dunes et par la présence de structures sédimentaires dont des laminations, des lits onduleux et des traces de fouisseurs en abondance. Les sédiments de la Zone C, nettement plus grossiers, proviennent également du système estuarien. Des variations importantes des propriétés texturales des sédiments sont observées tant en affleurements qu'en sous-surface. Ils contiennent des boues, sables silteux et graviers sableux, communément avec une grande quantité de débris organiques. Les variations de la taille des grains reflèteraient les fluctuations temporelles et spatiales de l'écoulement du régime fluvial.

Les caractères sédimentaires de chaque zone sont communs aux deux estuaires étudiés en détail: celui du lagon Wapengo et celui du petit bras de mer de Narrawalle. Le remplissage du premier est partiel tandis que celui du second est complet. La séquence de Narrawallee du Quaternaire tardif est interprétée comme représentant des épisodes successives de sédimentation estuarienne et d'érosion fluvial, dont la première correspondant au haut niveau marin, et la seconde, au bas niveau marin de la période glaciaire. Malgré la complexité produite par les multiples remplissages et érosions dans la séquence de Narrawallee, la zonation et le caractère des faciès sont maintenus.

Introduction

Estuaries mark the transition from terrestrial to marine environments, resulting in a high degree of spatial and temporal variance in their hydrological, sedimentological and biological regimes (Rochford, 1951, 1959; Bird, 1967a; Bowden, 1967; Caspers, 1967; Steers, 1967; Dyer, 1973). Interrelationships between these regimes are similarly variable and complex (Jennings and Bird, 1967; Wolfe and Kjerfve, 1986). From a geological viewpoint, however, estuaries may be regarded as well organized assemblages of sedimentary facies. Facies organization in estuaries has long been recognized in modern systems and is expressed in terms of a threefold, or tripartite, zonation (Bird, 1967a). Each zone occupies a specific area within an estuary and each has a distinct sedimentological character.

The spatial distribution and geometry of facies in coastal environments is often explained in terms of the association between coastal morphology and relative dominance of waves, tides or rivers (Davies, 1964, 1980; Galloway, 1975; Hayes, 1975, 1979; Heward, 1981; Davis and Hayes, 1984). The relationship has recently been detailed for estuarine systems specifically by Dalrymple and Zaitlin (1989) who highlighted the contrasting morphology of two end-members, wave-dominated and tide-dominated complexes, and stress that dominance of one process does not imply exclusion of the other. Both end-members feature a tripartite facies organization, a product essentially of sediment derived from strongly contrasting environments, fluvial and marine, being focussed toward the estuarine environment. However, the morphological and architectural characters of the facies zones differ between tide-dominated and wave-dominated systems, in response to their contrasting process regimes. The nature of contrasts between the two end members are detailed by Dalrymple and Zaitlin (1989) and Zaitlin and Shultz (1990).

The zonation appears to be applicable to a range of estuary types and has been specifically recognized in both modern and ancient deposits, including: Gippsland Lakes in Victoria (Bird, 1967a, 1978); Mallacoota Inlet, Victoria (Reinson, 1973, 1977); New South Wales estuaries (Roy, 1984a); the inner portion of the macrotidal Bay of Fundy (Zaitlin and Dalrymple, 1985); the James estuary in Virginia (Nichols *et al.*, 1989); the Gironde estuary in western France (Allen, 1989); the Lloydminster member of the lower Cretaceous Senlac heavy oil pool, Saskatchewan (Zaitlin and Shultz, 1984, 1990); the Upper Cretaceous Horseshoe Canyon Formation, Alberta, (Rahmani, 1988); and the Crystal field of the Lower Cretaceous Viking Formation, Alberta (Reinson *et al.*, 1988). That the tripartite zonation has been recognized in such a variety of deposits suggests that the pattern is independent of the local process conditions and geological settings.

Bird (1967a) was the first to recognize that the estuaries of the south coast of New South Wales (NSW) comprise three zones. They were defined according to tidal and salinity characteristics and include: a tideless, low salinity (0-10 ‰) zone located in the vicinity of river mouths; a central zone that has a small tidal range and low to moderate salinity (5-30 ‰); and a saline (30-35 ‰) tidal zone at the seaward end of the estuary (Bird, 1967a). In addition, Bird (1967a,b) noted that the zones differed in their ecological and geomorphic properties and that the area of each zone varied between estuaries as a function of the shape and dimensions of the estuary, river discharge, and volume of tidal exchange. The geomorphic character of the zones was described by Bird (1967a,b) in terms of the sources of sediment and general depositional processes acting to distribute the sediment within estuaries. Clastic sediment sources include the marine and terrestrial environments linked to estuaries via tidal entrances and rivers, respectively. Carbonaceous sediment is also present but is primarily produced in situ by estuarine organisms and is of relatively minor importance.

The intention of this paper is to apply the pre-existing tripartite zonation concept to two south coast estuaries, Wapengo Lagoon and Narrawallee Inlet, thereby demonstrating consistency of zonation among New South Wales estuaries and detailing sedimentological properties within and between zones.

The Study Area: Physical Setting and Previous Research

The research area encompasses a 250 km section of the New South Wales coast (Fig. 1), which is defined by the limits of the Lachlan Fold Belt and the southern tip of the Sydney Basin structural provinces. Lithologically, the Lachlan Fold Belt is dominated by Ordovician metasediments into which 55 coastal valleys have been incised. The 13 valleys of the southern Sydney Basin occur in Permian sandstones. The south coast was selected because of the relative uniformity of the dimensions and geology of estuary valleys and their catchments. Coastal embayments are relatively small and narrow, and are backed by catchments of high relief drained by small river systems. This is in contrast to estuaries of the central and northern New South Wales coast where the geological and topographic setting provides for considerable variation in embayment size and configuration, catchment relief and river size (Chapman *et al.*, 1982; Roy *et al.*, 1980). In the context of the estuary classification for the New South

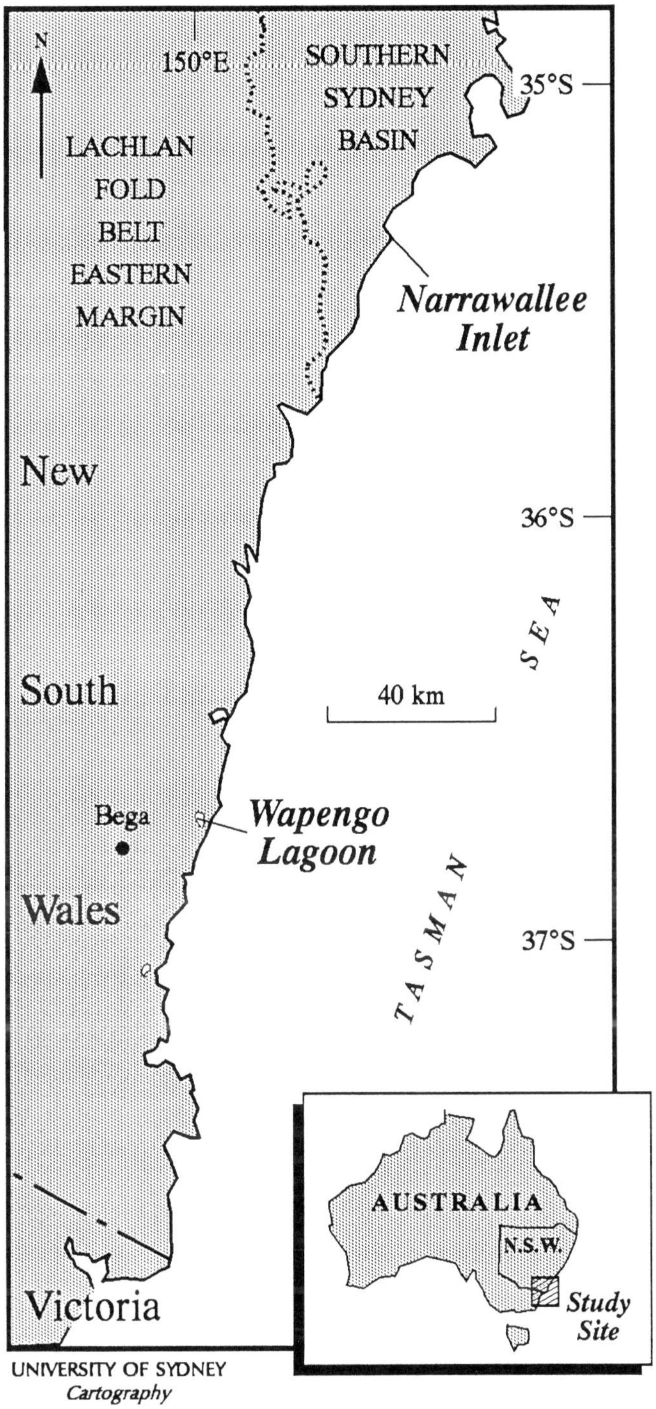

Figure 1. Location of the study area, south-east Australia.

Wales coast put forward by Roy (1984a), the south coast systems include bay-mouth barriers and saline coastal lakes. The distinction between the two types is in the condition of the entrance. The former have narrow, perennial entrance channels and the latter have closed or intermittent entrances.

Since the mid-Holocene, drowned valleys have been sites for the deposition of clastic sediments under estuarine conditions. These conditions have been controlled by: a wave dominated shoreface (modal wave height 1-2.5 m); a mean tidal range of 1.6 m, which is attenuated to less than 1 m within estuaries as a result of the presence of inlet shoals; a relatively low fluvial input, as evidenced by the paucity of fluvial sediment delivery to the coast (Bird, 1967b; Roy, 1977; Roy *et al.*, 1980); and sea-level stillstand since c. 6500 yrs. BP (Thom and Roy, 1983). In addition, the southeastern Australian coast occupies a passive continental margin and is, therefore, tectonically stable (Nakada and Lambeck, 1989). Collectively, these conditions provide a unique process setting for the study of estuarine sedimentation, particularly the relative stability of sea level, since previous investigations of incised valley, wave-dominated estuaries have concentrated on actively transgressing shorelines, such as the Atlantic coast of North America (*e.g.* Kraft, 1971; Armon, 1979; Kraft *et al.*, 1979, 1987; Boyd *et al.*, 1987; Carter *et al.*, 1989; Fletcher *et al.*, 1990).

The study area has received considerable attention from a number of workers, though previous geomorphic work has concentrated on documenting the general properties of surface sediments. Subsurface investigations have been relatively scarce. Bird (1967a,b) and Jennings and Bird (1967) undertook reconnaisance surveys in a number of estuaries, including Wagonga Inlet, Lake Illawarra, Lake Conjola, Burrill Lake, Narrawallee Inlet and Tuross Lake, noting the uniformity of geomorphic features such as river deltas, lagoons, barriers and entrance "thresholds".

Later, Roy and Peat (1975, 1976) sampled and mapped the bathymetry and bottom sediments of the larger south coast estuaries, including Tuross Lake, Coila Lake and Lake Illawarra. Kidd (1978) concentrated on the source and mineralogical properties of surface sediments from 14 south coast estuaries, emphasizing the differences between sediment of marine origin and that of terrestrial origin.

Other site-specific studies have been carried out by Reinson (1977) in Mallacoota Inlet, Eliot (1979) in Wallaga Lake, Bradshaw (1988) in St. Georges Basin, and Hunter (1989) in Lake Conjola and Narrawallee Inlet. All of these investigations have recognized the presence of three facies zones in the respective estuaries. Moreover, the sedimentological properties of each of the zones is common to all study sites. Summary depositional models for southeast Australian estuaries have been put forward by Reinson (1977) and Roy (1984a), the latter incorporating an evolutionary component. These models are derived primarily from surface sediment data. Of the available subsurface information, much of it pertains to the seaward (barrier/inlet) portion of estuaries (*e.g.*, Roy, 1984b).

Consequently, the models are restricted to generalized stratigraphic schemes with little detail regarding intrazone variability. The definition of assemblages of morphostratigraphic units within zones and the examination of respective sedimentological traits presented below address this problem.

Facies Terminology and Field Sites

Excellent examples of the tripartite arrangement of facies zones in wave-dominated estuaries are found along the southern coast of New South Wales. In order to document the organization of lithofacies in individual estuarine systems, it was necessary to define a set of geomorphic features common to all systems. The morphostratigraphic unit of Frye and Willman (1962) is thus adopted as the basic mapping unit. Morphostratigraphic units may be considered as distinct surface morphological and lithostratigraphic environments. A suite of 13 morphostratigraphic units were defined and mapped from aerial photographs of 68 New South Wales estuaries. Definitions of these units are presented in Table 1.

The tripartite character of New South Wales estuaries is apparent when the morphostratigraphic units are grouped according to general environments of deposition and lithofacies (Table 1). Thus, Zone A, extending landward from the estuary mouth, comprises those units associated with barrier and tidal inlet sedimentary processes and features a well sorted, sandy lithofacies. Morphostratigraphic units include barrier spit, back barrier flat, flood-tidal delta, tidal channel and beach ridge plain. Zone B, lying in the central portion of the estuary, consists of units related to low energy lagoonal depositional conditions with a mud to silty mud lithofacies. Morphostratigraphic units include barrier basin,

Table 1. Estuarine morphostratigraphic units and definition of terms

FACIES ZONE	MORPHOSTRATIGRAPHIC UNIT
A BARRIER/ INLET QUARTZOSE SANDS	*Barrier Spit: "A barrier island or barrier beach that is connected at one end to the mainland" and here represents the most seaward portion of paleovalley infill. *(Back) Barrier Flat: "A relatively flat area, often occupied by of water, separating the exposed or seaward edge of a barrier from the lagoon behind it" This unit may derive from either storm washover or be a relic flood-tidal delta. *Flood-Tidal Delta: "A delta formed at the mouth of a tidal inlet on the...lagoon side of a barrier island or baymouth bar by...tidal currents that sweep sand in the inlet" *Tidal Channel: "A major channel followed by the tidal currents, extending...landward...into a tidal marsh or tidal flat", or tidal delta. *Beach Ridge (Plain): "A low, essentially continuous mound of beach or beach-and-dune material...on the backshore of a beach beyond the present limit of storm waves or the reach of ordinary tides, and occurring singly or as one of a series of approximately parallel deposits...roughly parallel to the shoreline."
B ESTUARINE LAGOON MUDS AND SANDY MUDS	*Barrier Basin: "A basin produced by the formation of a...[bay-mouth] barrier." *Supratidal Shoreline: "Pertaining to the shore area marginal to the littoral zone, just above high tide level" Taken here to include deposits occurring adjacent to and above the barrier basin shore and resulting from wind-driven wave reworking of basin sediments. *Intertidal Shoreline: "Pertaining to the...depth zone between high water and low water", and here refers to deposits adjacent to and below the barrier basin shoreline. Also the product of wave reworking.
C TIDAL FLUVIAL SILTY SANDS AND GRAVELS	*Supratidal Fluvial Delta : "The low, nearly flat alluvial tract of land at or near the mouth of a river..." here taken to include the extension of land beyond the general trend of the basin shoreline. *Intertidal Fluvial Delta: The subaqueous extension of the supratidal portion of a fluvial delta, often present as a subaqueous levee. *Fluvial Channel: "The bed where a natural body of surface water flows...", and here includes medial channel bar deposits. *Point Bar: "low, arcuate ridge of sand and gravel developed on the inside of a growing meander" and attached to the convex side of a fluvial channel. Commonly unvegetated. *Floodplain: "The surface or strip of relatively smooth land adjacent to a river channel...", and here includes levee ridges and overbank depression deposits. The floodplain boundary is defined by the abrupt break in slope at the contact with bedrock. This boundary also defies the pre-transgression paleovalley.

SOURCE: Bates and Jackson

supratidal and intertidal shoreline. Zone C, at the landward end of the system, represents an assemblage of units resulting from tidally-influenced fluvial processes characterized by a moderately to poorly sorted silty muds, silty sand and gravel lithofacies. Morphostratigraphic units include supratidal and intertidal fluvial delta, fluvial channel, point bar and floodplain. Morphostratigraphic units from within the same zone are distinguished by their respective surface morphology, location and lithological characteristics, including sediment texture, sedimentary structures and gross mineralogy.

Two study sites have been selected for the documentation of the sedimentological characteristics of the estuaries of the southern New South Wales coast: Wapengo Lagoon and Narrawallee Inlet. Both are the bay-mouth barrier type of estuary, yet Narrawallee has achieved almost complete valley filling while Wapengo has not. Consequently, they provide an opportunity to compare and contrast the character of facies zones and component morphostratigraphic units and to test for consistency of tripartite facies zonation in valleys at different stages of infill.

Wapengo Lagoon is situated on the far south coast of New South Wales within a valley incised into metasedimentary rocks of the Lachlan Fold Belt (Fig. 1). The area available for estuarine sedimentation at present sea level is 7.86 km^2. Wapengo is appealing from the field sampling perspective because it features a diverse array of morphostratigraphic units readily observable at the surface (Fig. 2). The morphometric characteristics of Wapengo Lagoon's morphostratigraphic units and lithofacies zones are detailed in Table 2. These data show Wapengo Lagoon to be an estuary filled with predominantly terrestrially-sourced sediment, with Zone C constituting 53 per cent of the palaeovalley area and 48 per cent of the volume. Zone A, however, is not insignificant in contributing to valley fill, occupying 29 per cent of the estuary area and 39 per cent of the volume. The estuary is permanently open to the sea and is tidal to within 3 km of the head of the floodplain, where tidal range is less than 0.5 m. Only Zones A and C have well defined tidal channels, which terminate at the edge of the Zone B. The intermediate zone (B) is a broad, flat bottomed basin, but is able to convey tidal waters between Zones A and C.

Narrawallee Inlet occupies a valley cut into Permian sandstone and siltstone of the Sydney Basin (Fig. 1). The estuary has experienced a significant degree of valley filling, thereby providing an opportunity to investigate the nature of facies organization in a complete valley sequence. Eighty-seven per cent of the valley surface area (10.81 km^2) is occupied by fluvial floodplain deposits (Fig. 3). Volumetrically, Zone C deposits comprise 43 per cent of the valley fill (Table 2). The floodplain interfingers seaward with an assemblage of Zone A units, including flood-tidal delta, barrier flats and barrier spit. The Zone A assemblage occupies only 13 per cent of valley area but makes up 48 per cent of the fill volume.

Zone B facies are not surficially evident, however, subsurface investigations, to be discussed later in the paper, show Zone B sediments to be incorporated in the valley fill. The inlet is permanently open to the tidal exchange, with Croobyar and Currowar creeks tidal almost to the landward limit

Table 2. Morphometric data for Wapengo Lagoon and Narrawallee Inlet. Volume data were calculated by multiplying the surface area of each zone by the mean facies thickness taken from vibracore and drill information

MORPHOSTRATIGRAPHIC UNIT	WAPENGO		NARRAWALLEE	
	AREA (%)	ZONE (%)	AREA (%)	ZONE (%)
BARRIER SPIT	2.93		7.22	
BARRIER FLAT	6.49		3.88	
BEACH RIDGE PLAIN	00.00	A:29.01	00.00	A:12.59
FLOOD-TIDAL DELTA	19.57		1.20	
TIDAL CHANNEL	0.02		0.28	
BARRIER BASIN	14.89		00.00	
SUPRATIDAL SHORELINE	3.05	B:17.94	00.00	B:00.00
INTERTIDAL SHORELINE	00.00		00.00	
SUPRATIDAL FLUV. DELTA	5.60		00.00	
INTERTIDAL FLUV. DELTA	0.38		00.00	
FLUVIAL CHANNEL	2.54	C:53.05	2.49	C:87.41
FLUVIAL POINT BAR	00.00		00.00	
FLOODPLAIN	44.53		84.92	
ZONE A VOLUME (m^3x10^6)	13.68	A:39.55 %	64.69	A:48.72 %
ZONE B VOLUME (m^3x10^6)	4.23	B:12.23 %	10.89	B: 8.20 %
ZONE C VOLUME (m^3x10^6)	16.68	C:48.22 %	57.20	C:43.08 %

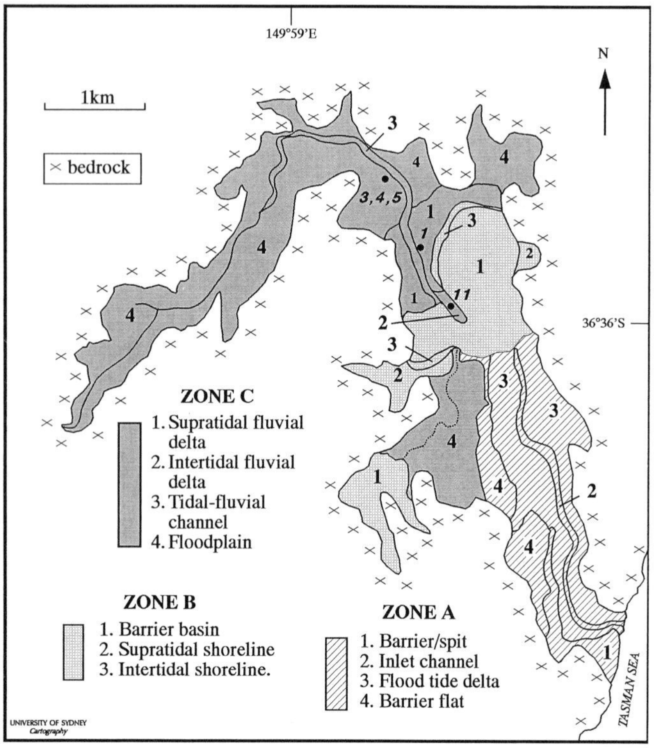

Figure 2. Map of Wapengo Lagoon showing facies zones and component morphostratigraphic units. Vibracore sites for cores discussed in text are also shown.

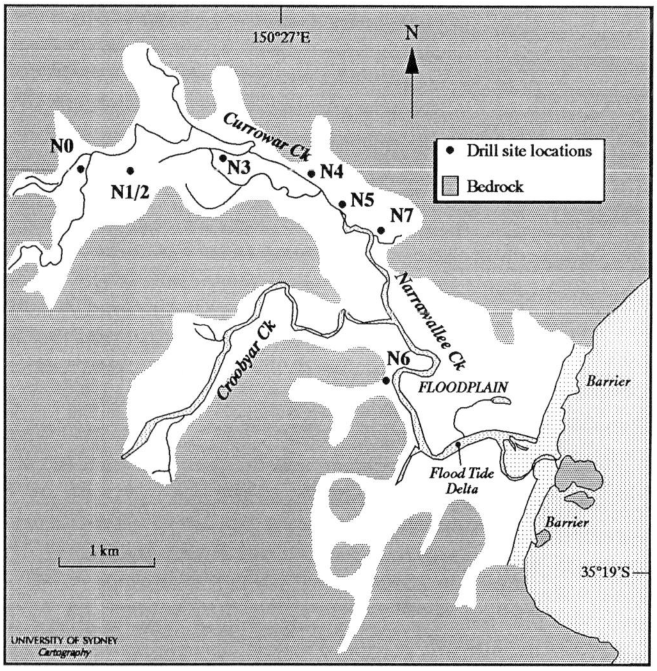

Figure 3. Map of Narrawallee Inlet indicating drill hole locations.

of the floodplain, where tidal range is attenuated to between 0.5 and 1m.

Field and Laboratory Methods

The contrasting infill of the two estuaries required the use of different sampling techniques. Thus in Wapengo, sampling was directed toward characterizing the system from the bedform specific scale to facies specific scale. Fifty-one hammer driven box-cores were taken from bedforms in the intertidal portion of Zone A (Fig. 4). Cores were taken in both flow-parallel and flow-normal orientations in order to ascertain the three-dimensional internal structure of bedforms. Nineteen hammer-driven cores (maximum penetration: 2 m) were taken from sites representative of the main morphological elements of Zone A (Fig. 4). Eleven vibracores were also collected (penetration range: 2 to 5 m) from Zones B and C along lateral and longitudinal transects, thereby providing a record of vertical facies sequences and in some cases evidence for the sequential burial of morphostratigraphic units.

Sampling methods in Narrawallee estuary consisted of sinking deep drill holes through the entire valley-fill sequence

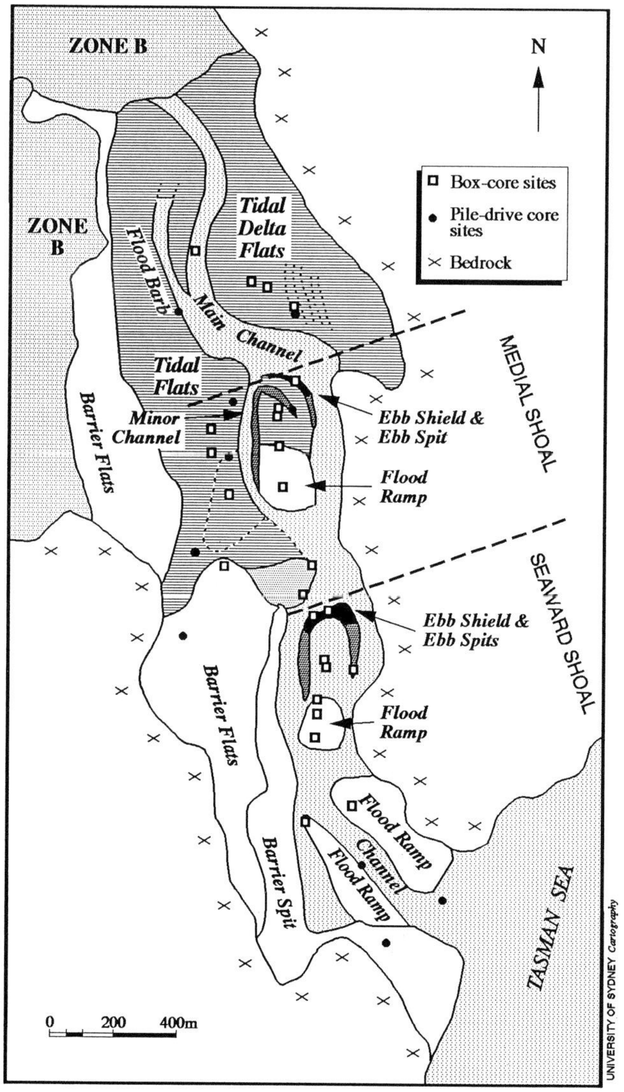

Figure 4. Flood-tidal delta morphology and core locations in Wapengo Lagoon.

at seven sites along a longitudinal transect (Fig. 3). Where possible, holes were taken to basement (ca. 30 m) though recovery was rarely complete, particularly in sandy sediment. Holes N1/2, N3, N5 and N6 were sampled using coring techniques. Holes N0, N4 and N7 were augered.

Araldite resin peels were taken from box-cores and pile-drive cores in order to highlight sedimentary structures. X-ray radiographs of peels were subsequently made to further enhance bedform and facies architecture. Split lengths of vibracores were photographed and X-rayed. All cores were subsampled for grain size and mineralogical analysis. Subsamples were washed through a 63 micron sieve to separate the sand and gravel fractions from the silt and clay fraction. Size analysis on the fines was undertaken using the pipette method described by Lewis (1984). An automated settling column was employed to generate grain size information for the sand fraction of samples. The maturity of mineralogical suites from each lithofacies was qualitatively assessed via microscope inspection of samples.

Monitoring of estuarine processes has not been undertaken for this study. Tidal range and velocity data have been collected from several other nearby south coast estuaries (Williams, 1983), which provide a general guide for the nature of variation in processes influencing the transport and deposition of sediment.

Facies Zonation in a Partially Filled Valley: Wapengo Lagoon

The description of the sedimentological character of the three facies zones present within Wapengo Lagoon draws upon observations related to the following parameters: bedform type, employing the revised terminology for subaqueous bedforms of Ashley (1990); physical and biogenic sedimentary structures; and maturity of mineralogical suites. Wapengo data provide the bulk of information pertinent to ongoing deposition since the site provides for specific sampling of active bedforms and facies.

The presentation of data is restricted to an analysis of those components of the facies zones considered most likely to be preserved in the geological record. Thus, emphasis is given to morphostratigraphic units lying within the modern intertidal to subtidal portions of the valley-fill sequence. In Zone A such units include the flood-tidal delta and associated channels, the main entrance channel and barrier flat units; in Zone B the barrier basin unit; and in Zone C the intertidal fluvial delta and fluvial channel units. These units are more likely to be preserved during a transgression than the supratidal and subaerial units, such as barrier/spit, beach ridge plain and shoreline deposits. Likewise, in a regressive situation, the deeper units of the sequence will have preferential preservation to the shallower units, though depending on the rate of sea level change, net preservation is likely to be less following a regression than after a transgression (Boothroyd, 1985; Davis and Clifton, 1987; Carter *et al.*, 1989). Moreover, the intertidal and subtidal deposits are volumetrically more significant than the subaerial components of each zone (Table 2).

Zone A

As previously noted, process monitoring has not been undertaken for this study. However, data collected in similarly configured inlets on the New South Wales south coast relate a consistent trend of a landward decrease in current velocities and tidal range (Williams, 1983). For example, tidal flows within Lake Illawarra inlet experience a landward velocity decrease of 48 per cent during flood stage and 69 per cent during the ebb (Williams, 1983). Figure 5 shows the majority of this decrease to occur over the outer (seaward) 800 m of the inlet and is attributed by Williams (1983) to the presence of large shoals in the inlet mouth. It is proposed that Illawarra inlet provides a reasonable analogy for the general nature of process variation in Wapengo since both are approximately 3 km in length, have large entrance shoals, and neither are artificially trained.

Surface sediments analyzed from the Wapengo flood-tidal delta range from well to very well sorted (moment standard deviation: 0.29-0.55 phi) medium (moment mean: 1.21-1.64 phi), subrounded to rounded, quartz dominant sands. The silt and clay fraction is negligible to absent, never exceeding 2 per cent of sample weight. The high degree of sorting and mature mineralogy of sediments analyzed from the Wapengo tidal delta is attributed to the high energy

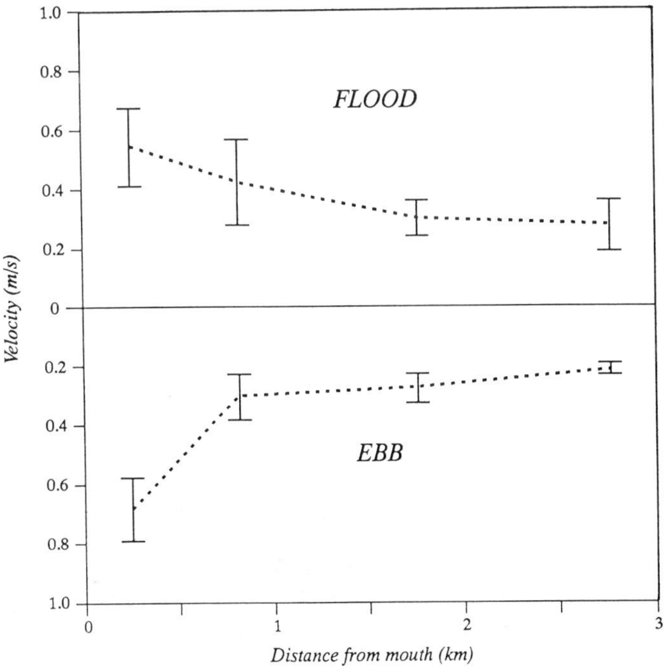

Figure 5. Longitudinal tidal current velocity profile in Lake Illawarra. Plot shows mean velocity with error bars representing one standard deviation for four flood-stage observations and six ebb-stage observations (source: Williams, 1983).

nature of the nearshore and surf zone environments from which they have been derived.

Field observations of intertidal bedforms reveal distinct changes in bedform type and associated internal structure along a longitudinal transect within the inlet. The change in bedform character along the Wapengo tidal delta entails a landward transition from a dominantly three-dimensional (3-D) morphology near the estuary mouth, to a predominantly two-dimensional (2-D) bedform shape on the medial portion of the tidal delta, to tidal flats devoid of dunes on the landward portion of the tidal delta and on back barrier flats. This transition occurs in association with morphological elements of the flood-tidal delta. Figure 4 shows the Wapengo flood-tidal delta to comprise a seaward shoal and an inner medial shoal, separated by flood and ebb channels. The innermost portion of the delta is occupied by topographically featureless intertidal flats. Both the seaward shoal and medial shoal have developed morphological elements of the model flood-tidal delta as defined by Hayes (1980). Thus, each shoal is subdivided into a flood ramp, ebb shield and ebb spit.

The seaward shoal of the flood-tidal delta features the greater variety of bedform character. Thus, bedforms range in size from small (spacing: <1 m; height: <0.11 m) to medium (spacing: 5-6 m; height: 0.4-0.5 m) two-and three-dimensional dunes. Three-dimensionality appears to be more prevalent in, though not restricted to, smaller scale dunes. The small- and medium-sized bedforms are predominantly ebb oriented (a function of field observations at low water) and are superimposed upon the stoss sides of large-scale dunes. Bedform crestlines of 3-D dunes show a greater tendency to be in phase for small dunes than for medium-sized dunes, though there is no direct correlation between dune size and crestline behaviour. The large dunes equate with aspects of the shoal morphology. Thus, the flood ramp is a composite of several large (spacing: ca. 10 m; height: ca. 0.70 m), flood oriented, two-dimensional dunes.

Sedimentary structures revealed in box-core peels taken from the stoss side of a large 3-D dune migrating transverse to the flood current are illustrated in Figure 6 A, B. The bedform was located on the flood ramp of the seaward shoal (Fig. 4). In flow parallel section the core features a variety of crossbedding structures, including: tabular to low-angle trough crossbeds that pinch out in a downcurrent direction (10-15 cm, Fig. 6A); small-scale foreset trough crossbedding (15-20 cm, Fig 6A) transitional up-core to convex and planar bedding; and well defined bounding surfaces that are irregular in dip pattern because of the reversing flow conditions. Similar structures are observed in a section normal to flow (Fig. 6B). The preservation of foreset crossbeds in flow normal section is interpreted as representing variation in flow direction through the tidal cycle, causing shifts in migration paths of bedforms. The concave form of some bedding planes, observed in both flow parallel and flow normal sections, is interpreted to have been derived from scour pits, and the convex bedding planes reflect the advancing sinuous bedform crest, characteristics inherent to 3-D dunes.

Figure 6 also shows the primary structures of a vertical sequence, as revealed in core peel, for the flood ramp of the seaward shoal of Wapengo tidal delta. Small-scale trough crossbedding and large-scale planar crossbeds are seen throughout the section. The dominant dip of foresets is in the flood (landward) direction, with only a minor indication of reversing flows. Planar foreset beds constitute the thicker (50-100 mm) beds of the sequence, whilst trough foresets occur occasionally as relatively thin beds (10-40 mm). This assemblage of structures suggests that larger scale dunes, represented by the thicker sets of crossbeds, possessed a 2-D morphology. Three-dimensionality appears to be confined to smaller scale dunes, which is in accord with overall bedform character discussed earlier. Furthermore, the limited occurrence of trough crossbedding is in line with the expected low preservation potential of small-scale bedforms.

Burrow traces are rare to absent in all cores taken from the seaward shoal. Soldier crabs (*Mictyris longicarpus*) were observed, however, to be active on the seaward shoal at low tide, suggesting that the seaward portion of the delta is too dynamic for evidence of biological activity to be incorporated into the sedimentary record.

The dominant bedforms across the surface of the medial shoal of the tidal delta, are medium (spacing: 5-10 m; height: <0.50 m), flood oriented, two-dimensional dunes. Again, these beds define the flood ramp element of the delta. Superimposed on the stoss side of the 2-D dunes are small ebb and flood oriented 2-D and 3-D dunes.

The internal structure of medial shoal 2-D dunes is illustrated from box-core peels in Figure 6 c, d. In flow-parallel section, physical sedimentary structures record the stoss-erosional migration behaviour of the bedform. Thus, large-scale, flood oriented crossbeds are in erosional contact with the underlying bedding plane. Crossbed shape changes up-sequence from shallow-concave to planar to shallow-convex, as testimony to the advancing crestline. The absence of bedding below the upper set is attributed to bioturbation, illustrating the relatively low preservation potential of physical sedimentary structures in this portion of the tidal delta.

Illustrations of bedform structure viewed facing the flood current again show the absence of physical structures below the upper set (Fig. 6 C,D). Within the upper set, structures are preserved as wavy, discontinuous beds. This trait is not considered to be due to partial bedform three-dimensionality, but to the presence of small-scale dunes on the stoss side of the main bedform, which produce minor irregularities in the along-crest topography.

Reference has been made in the preceding discussion to the potentially destructive influence of burrowing organisms upon primary physical structures. Within the medial shoal of the Wapengo tidal delta, bioturbation appears to be most complete below the troughs of active bedforms (Fig. 6 C,D). Inspection of peels of box cores and shallow cores taken on

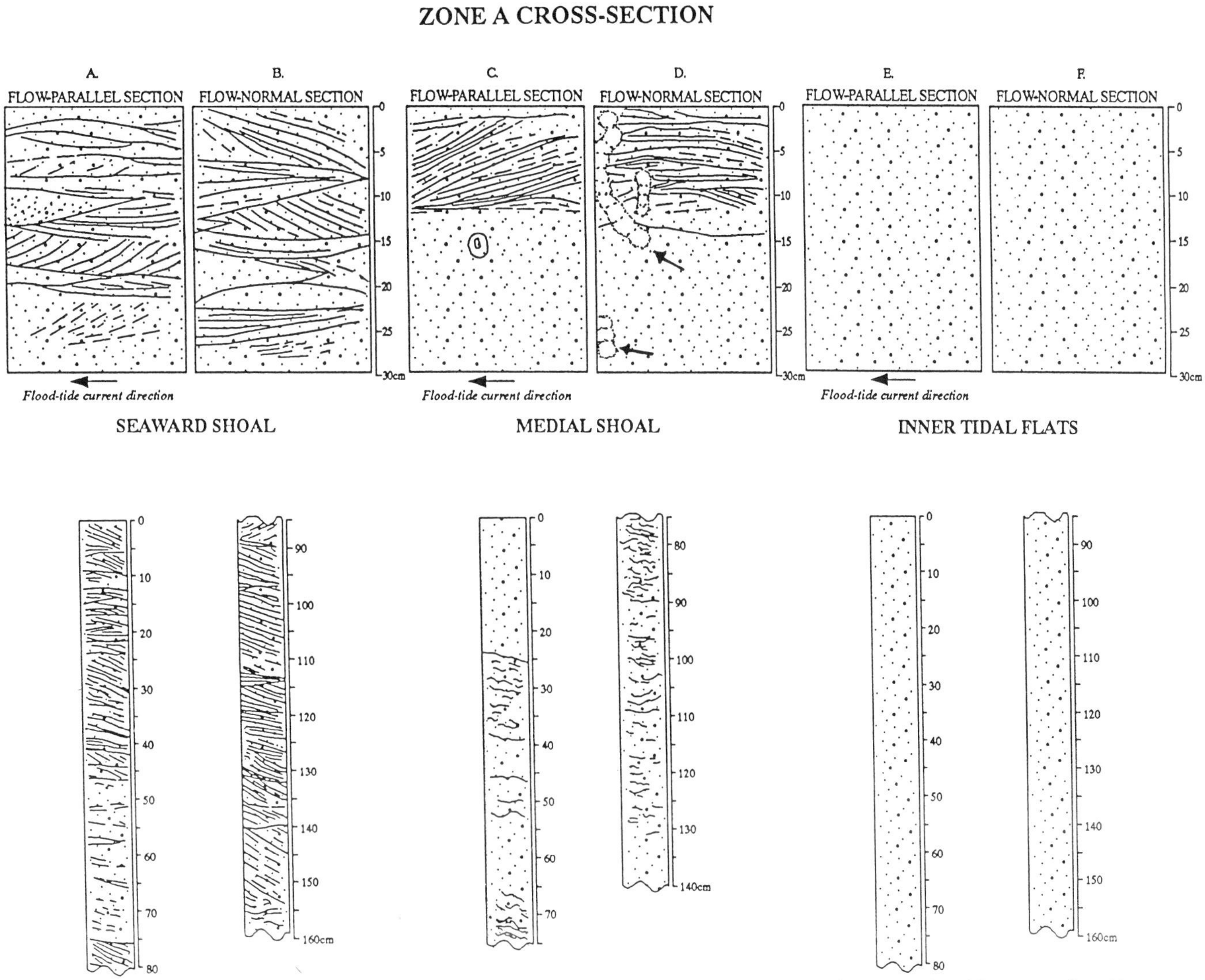

Figure 6. Line drawings of primary physical structures observed in resin peels from box core (upper) (flow parallel and flow normal) and hammer core (lower) from flood-tidal delta, Wapengo Lagoon: **A.-B.**, sections from a medium, three-dimensional dune migrating across seaward shoal flood ramp; **C.-D.**, sections from a medium, two-dimensional dune migrating across medial shoal flood ramp. Arrows indicate vertical escape burrows (*Skolithos*); **E.-F.**, sections from incipient, medium dunes located landward of the medial shoal.

the medial set reveal the co-preservation of physical and biogenic structures at depth (Fig. 6). The quality of both forms of sedimentary structure is generally poor. Physical structures appear discontinuous and wavy, while biogenic structures exist as ill-defined, sandy burrow fills, such as vertical escape burrows (*Skolithos*) formed by the soldier crab *Mictyris longicarpus*. The lack of dominance of one structural form over the other exemplifies the transitional nature of the medial shoal in the Wapengo flood-tidal delta.

Bedforms in the portion of the delta landward of the medial shoal are poorly developed, present only in incipient form. Surface topography consists of shallow (<5 cm) discontinuous, straight to wavy scour pits, separated by low energy 2-D dunes. Box core peels reveal subsurface sediments to have an overall massive structure, littered with fossil remains of resident organisms (Fig. 6 E,F). The barrier flat unit of Zone A is similarly void of well developed bedforms and is extensively bioturbated.

The sedimentological character of the intertidal and subtidal portion of Zone A as observed in Wapengo Lagoon may be summarized in the context of inferred tidal inlet processes. Inferring relative flow strength from bedform size, shape and internal structure in an environment influenced by reversing flows is difficult. Rubin (1987) observed that bedforms formed by reversing flows tend to be more 2-D in form than those resulting from unidirectional flows. Yet, Dalrymple (1987) contended that there are no fundamental differences between the geometry and internal structure of bedforms occurring in uni-and bidirectional flows. Nevertheless, the formation of 3-D bedforms has been well correlated with increased flow velocities at a fixed depth (Allen, 1968). Given that flow depth varies in tidal flows, it is postulated here that 3-D bedforms develop during the peak flow period and remain essentially unmodified throughout the remainder of the tidal cycle. Within Wapengo inlet the distribution of 2-D and 3-D bedforms is thus interpreted as a function of the inferred landward decrease in the relative strength and duration of tidal currents. In addition, lower current velocities not only retard bedform migration and maturation but also provide ideal conditions for intense bioturbation, hence the destruction of primary sedimentary structures. In this way, the process regime of the inlet imparts a clear signature upon the sedimentary record.

Zone B

The estuarine lagoon zone (B) of Wapengo Lagoon is an environment of predominantly low energy depositional conditions that is influenced by both tidal and fluvial processes. Fluctuations in water levels in the basin are on the order of 0.25 to 0.5 m, the tidal range having been attenuated by the flood-tidal delta. As a result of the absence of well defined channels in the basin, tidal currents are also reduced in magnitude. In Lake Illawarra, for example, mean tidal velocity at the seaward edge of the basin is 0.24 m/s, compared to 0.61 m/s near the mouth (Williams, 1983).

Zone B is also influenced by fluvial flow, which is subject to intermittent fluctuations in energy associated with periods of high and low river discharge. Fluctuations are both temporally irregular and of variable magnitude. Again, the lack of well defined channels in the basin results in dampening of the impact of extreme fluvial events. The accumulation of sediment in Zone B is therefore achieved under fluctuating energy conditions.

Sediments deposited in Zone B of Wapengo Lagoon are typically muds and silts. Figure 7 presents graphic logs and textural plots (sand-silt-clay) for three sample vibracores taken from Wapengo Lagoon. The basal unit in VC1 and the unit overlying weathered Pleistocene clays in VC11 are interpreted as representing the barrier basin of Zone B. Silt (4 phi-8 phi) is the dominant size fraction in both cores, ranging from 31 to 54 per cent of sample weight. Clay (<8 phi) constitutes between 21 and 47 per cent of sample weight and decreases upsequence, in line with shoaling of the basin. Conversely, sand content increases upsequence from 6 to 39

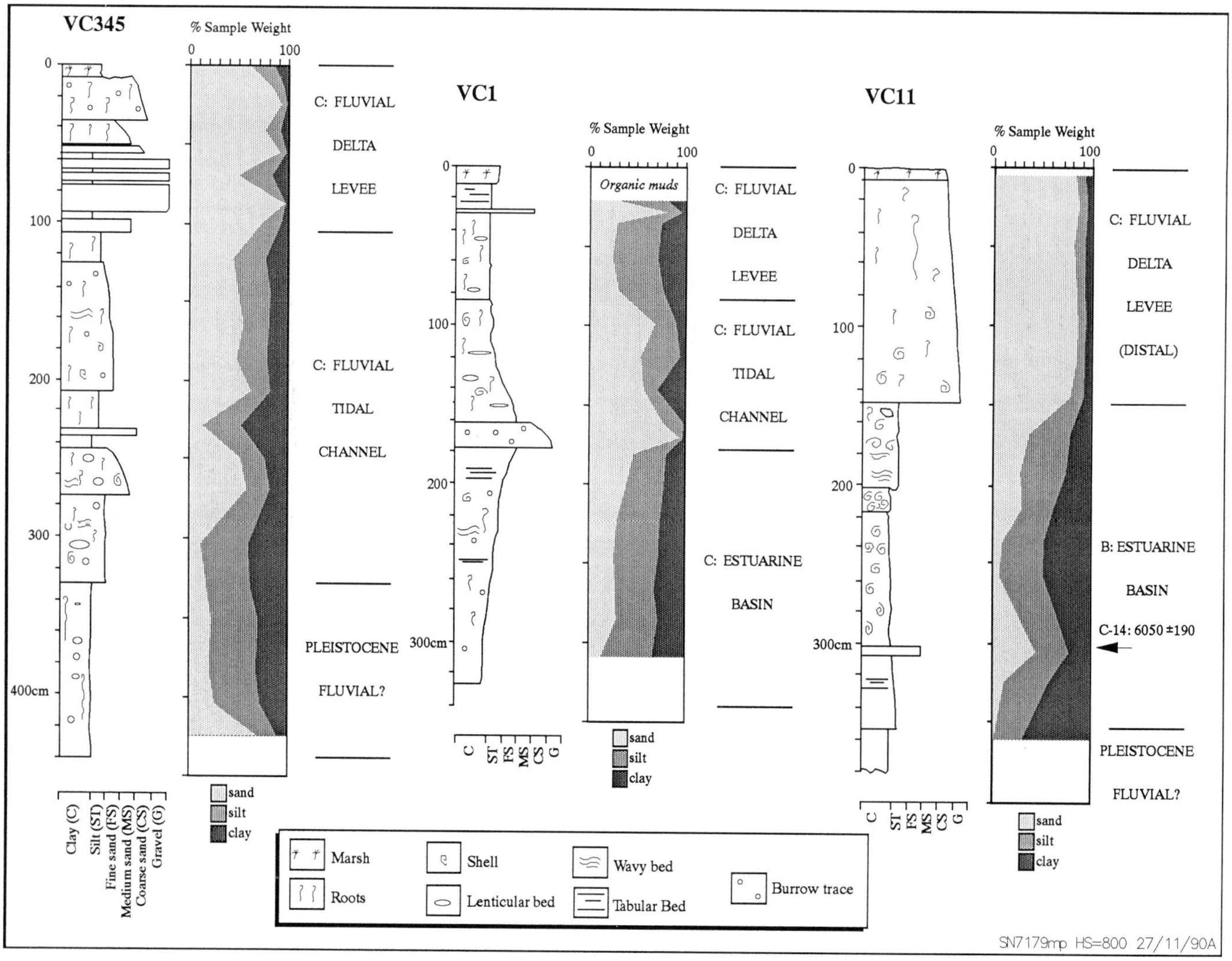

Figure 7. Graphic logs and sediment texture plots of three vibracores taken along a transect of Zone C, Wapengo Lagoon. Diagrams show overall coarsening-upward progradational sequence from barrier basin to fluvio-deltaic to floodplain facies. Mean sealevel corresponds to 0 cm in cores VC1 and VC11.

per cent (Fig. 7). Sands are moderately sorted (moment standard deviation: 0.58-0.96 phi) and fine to very fine grained (moment mean: 2.75-3.25 phi).

An apparent consequence of the low energy depositional conditions and fine texture of Zone B sediments is the absence of dune bedforms on the basin floor. Crossbedding is, therefore, absent from basin sediments. The dominant sedimentary structures are tabular and wavy laminations with abundant horizontal and vertical burrow traces (Fig. 7). The high degree of bioturbation of basin sediments is attributed to the relatively high concentration of organic material available for consumption by infauna.

Zone C

Depositional processes in Zone C of Wapengo Lagoon are characterized by a fluvial regime that is temporally and spatially variable and is overprinted by tidal processes. The effect of tidal processes is also spatially variable, decreasing in a landward direction. The core record from Zone C contains features that are in accord with fluctuations in the flow regime. Figure 7 presents graphic logs and plots of textural variations (sand/silt/clay) in sample cores from the supratidal portion of the Wapengo Lagoon fluvial delta. The cores exhibit an overall upward-coarsening trend, from the underlying muddy basin unit to the silty sands and sandy gravels of the fluvial channel and fluvial delta levee units. The contact between the lower facies to the overlying unit is typically erosional.

The sediments of the fluvial channel are sand dominant, ranging between 70 and 95 per cent sand in VC1 (Fig. 7). Silts and clays make up less than 40 and 29 per cent, respectively. Silt- and clay-rich beds are also present within the channel deposit (*e.g.*, 230 cm in VC345, Fig. 7). The sediments of the delta levee are also dominated by sands with a minor (<5 %) gravel component. For example, sand constitutes between 50 and 96 per cent of the levee unit of VC11 and VC345 (Fig. 7). Thin beds of sandy silt and clay are also preserved within the levee sands, but in lower concentrations than in the channel sediments (*e.g.*, 70 cm in VC345, Fig. 7).

The sand fraction of both the fluvial channel and delta levee is moderately sorted (moment standard deviation: 0.72-0.90 phi), subangular to angular and fine to medium in size (moment mean: 2.00-2.70 phi). The sands are composed of an immature mineralogical suite, reflecting their terrestrial origin. Quartz is the dominant mineral (80-90%), but feldspar (2-3%), lithic fragments (2-3%), micas (1-2%) and accessory minerals are in relative abundance (Kidd, 1978).

The predominant sedimentary structure within Zone C sediments are graded beds, ranging in thickness from one centimetre to tens of centimetres (Fig. 7). Grading is typically present as a fining-upward trend and appears confined to coarser sediments such as gravels and sands. Graded coarse beds are bound by massive beds of silty medium grained sands and the contact between graded and massive beds is commonly erosional. The upper muddy and silty marsh sediments are massive and marked by a high degree of rooting and bioturbation.

The presence of packages of sediment that contrast in bed thickness and grain size is interpreted as the product of fluctuations in fluvial flow energy. Thus, coarsest sediments are attributed to flows of flood magnitude and the graded nature of flood deposits is an expression of the waning flow following the flood peak. Massive, silty, medium grained sands are presumed to be deposits representative of "average" flow conditions. The absence of crossbedding in the core record reflects the paucity of dune bedforms in Zone C.

In summary, the overall sedimentary character of Zone C in Wapengo Lagoon consists of a vertical transition from barrier basin (Zone B) to fluvial channel and fluvial delta deposits that is interpreted as being a record of the development of a progradational (regressive) facies sequence.

Facies Zonation in a Completely Filled Valley: Narrawallee Inlet

Data collected from drill cores in the Narrawallee valley fill provide for the reconstruction of the history of the valley filling process, the testing for consistency of the tripartite zonation, and the characterization of buried, or relict, facies zones.

Zone A

Drill holes N5, N6 and N7 (Fig. 3) provide the only available record of preserved Zone A sediments for the Narrawallee valley fill. Figure 8 presents graphic core logs and plots of the gross textural properties of material recovered in core N5. It is acknowledged that core recovery is not complete, especially in sandy sediment. Recovery was particularly poor from hole N6, and data from that core and from auger hole N7 are used to assist in stratigraphic reconstructions only. Nevertheless, between -14 and -21 m, the N5 core yielded a relatively thick (7 m) unit of very well to well sorted (moment standard deviation: 0.31-0.48 phi), well rounded, quartz dominant, fine grained (moment mean: 2.12-2.67 phi) sand. Between -20 m and -21.2 m the sands are rich in fragmented shell and have a higher content of lithics (2-3%) and feldspars (1-2%) than the remainder of the unit. The sediment that was recovered from N6 and N7 also consisted of well sorted fine to medium quartzose shelly sands, with an increase in lithics and feldspar concentrations toward the base of the unit. The silt and clay fractions of these sediments are low to negligible, <2 per cent and <1 per cent, respectively.

Facies architecture is difficult to discern because of poor core recovery in sands. Small-to medium-scale planar crossbedding and flaser bedding is occasionally evident in resin peels (Fig. 8). The majority of core recovered from this facies, however, exhibits a massive structure. Burrow traces are rare and, where present, are poorly defined.

The mature mineralogy and overall textural properties observed from core N5, N6 and N7 subsamples are consistent

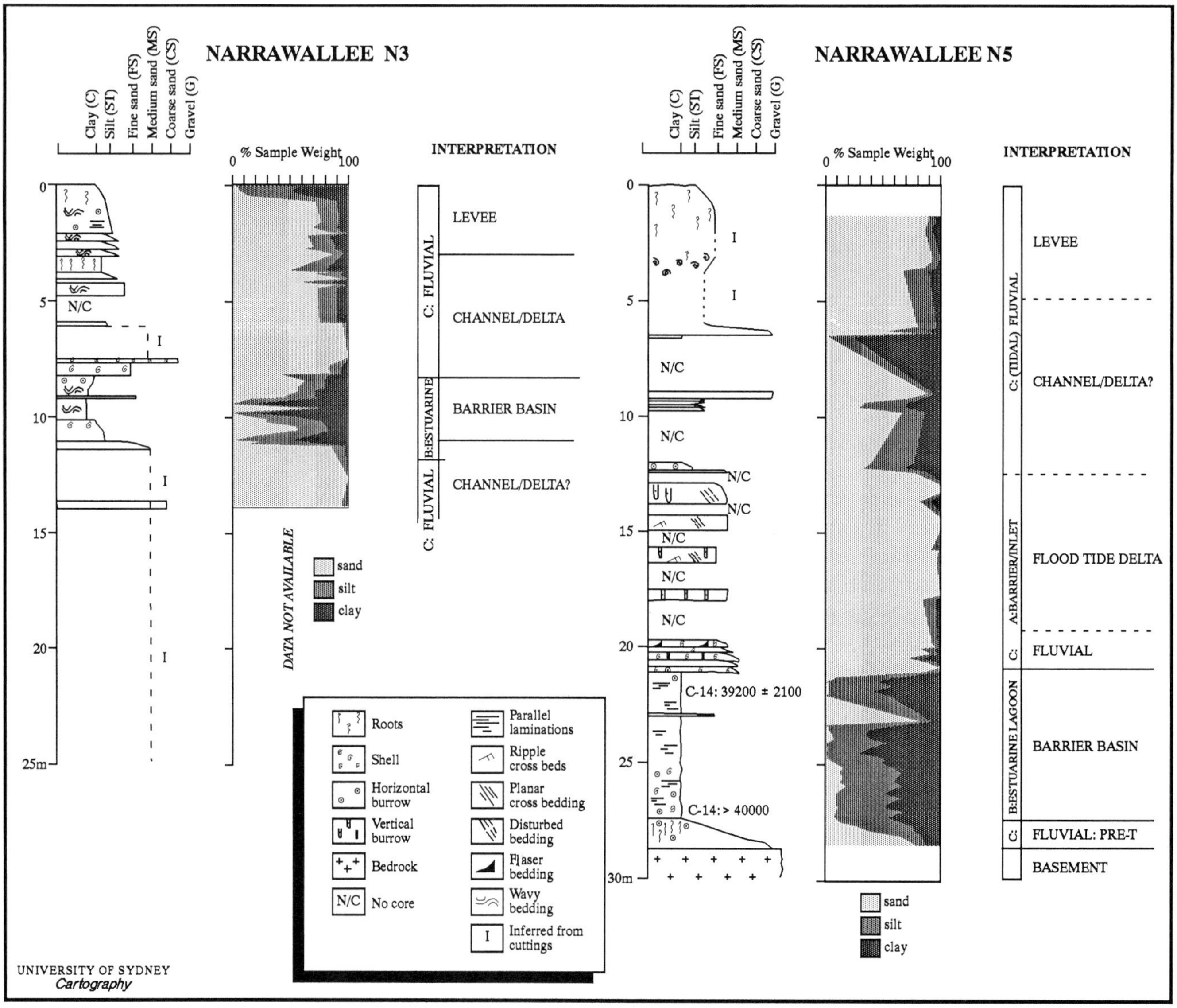

Figure 8. Graphic logs and sediment texture plots of drill cores N3 and N5 taken along a transect of Narrawallee Inlet.

with those documented for the intertidal environents of Zone A in Wapengo Lagoon. The coarser sands of this unit are interpreted as representing tidal channel deposits, while medium to fine sands are associated with flood-tidal delta shoals. Crossbedding in association with flaser bedding is also suggestive of tidally influenced deposition. The massive appearance of the majority of the unit is attributed to bioturbation. The presence of fragmented shell, lithics and feldspars in the lower portion of the Zone A unit from N5, N6 and N7 is likely to reflect reworking and mixing of fluvial with marine sediments during the Post-glacial Marine Transgression. Though no single attribute of the deposit provides conclusive evidence of tidal inlet depositional conditions, it is argued that collectively these traits are similar to those described for the medial to inner portion of the Wapengo tidal delta. Certainly, given the distal location of the core relative to the Narrawallee inlet mouth, deposits of the type described would be anticipated.

Zone B

Sediments interpreted as being indicative of facies Zone B were recovered in drill holes N0, N1/2, N3, N5 and N7 (Fig. 4). The most complete core recovery was obtained at sites N3 and N5 and detailed discussion is limited here to those cores. Core log diagrams and gross textural variations for cores N3 and N5 are shown in Figure 8.

Situated between -8.2 and -11.2 m in core N3 is a sequence of compacted, heavily oxidized, bioturbated silty clays and silty sands (Fig. 8). Sand content varies between 2 and 69 per cent, silt between 9 and 57 per cent, and clay between 18 and 85 per cent of sample weight. The sands are moderately sorted (moment standard deviation: 0.43-0.89), fine to very fine grained (moment mean: 2.72-3.00 phi) and occur as sharply defined beds within the silty clays. Overall, the unit appears as a series of three fining-upward sequences. These are in turn overlain by a unit of sandy silts and clays that coarsen upward to silty sands and gravels. A similar sequence was recovered from core N1/2 (Fig. 9).

Located 1.37 km seaward of N3 (Fig. 4), drill site N5 features 29 m of valley deposits. The graphic core log shows bedrock to be mantled by a thin unit of coarse muddy to silty sands (Fig. 8). These fine upward into a relatively thick unit of compact, silty clays, which occur between -21 and -27.5 m. The plot of sand-silt-clay ratios shows this unit to consist of repeated, small-scale fining upward cycles that collectively describe a progressive increase in clay content from 24 to 70 per cent (Fig. 8). This trend is interrupted by a thin bed of very fine grained, silty sand (moment mean: 3.77 phi) at -23 m. In contrast to the silty clay unit of N3, the fine grained unit in N5 appears unweathered (non-oxidized) with fine tabular and wavy laminae preserved throughout and burrow traces confined to the upper 3.5 m of the unit. Auger hole N7 yielded a sequence of similar thickness and character to that described from N5, with the exception that N7 features a second clay unit at the relatively shallow depth of 4.4 to 7.6 m (Fig. 9).

The fine grained texture, structural characteristics and stratigraphic position of the deposits recovered from N3 and N5 are used as a basis for associating them the with low energy depositional barrier basin of Zone B. The presence of thin sand beds within the clays is attributed to relatively higher energy depositional conditions bought about by fluvial floods. Further, the diagenetic properties of the shallower Zone B deposits suggest that they are of pre-Holocene age. This is supported by a series of C-14 calibrated amino acid racemization dates of fossil shell material recovered from the deposit. Full details of the amino acid dates are presented by Nichol and Murray-Wallace (*in press*); suffice to note here that amino acid D/L ratios suggest a last-interglacial (oxygen isotope substage 5e) age for the lower clay unit. The up per clay unit, recovered in N7, is most likely of Holocene age and a C-14 date from wood of 6320±100 yrs. BP (SUA-2907) supports this interpretation.

Zone C

Cores N1/2, N3, N5 and auger holes N0 and N7 all yielded deposits that may be associated with Zone C. Again, core recovery was most complete for N3 and N5 and facies description is restricted to these cores. Core N3 features a basal unit of well to moderately sorted (moment standard deviation: 0.39-0.53 phi) medium grained sands (moment mean: 1.11-1.60 phi) with a low silt (<10%) and clay (<5%) content (Fig. 8). The top of the unit occurs at -11 m and it is assumed to continue to bedrock at -25 m. The bedrock profile has been extrapolated from known depths measured at drill sites N0, N1/2 and N5 (Fig. 9). The mineral composition of the sands is predominantly quartzose (90%), with a minor component of lithic fragments, feldspars and micas. Core samples in the sands were disturbed, preventing analysis of sedimentary structures. This unit is overlain by the Zone B silty clays discussed above.

The remainder of the N3 core consists of a unit of gradually fining-upward sediments. Recovery between -4.5 and -7.5 m was poor in coarse to medium grained sands (moment mean: 0.21-1.98 phi). The sands grade progressively into organic-rich, fine to very fine grained sands (moment mean:

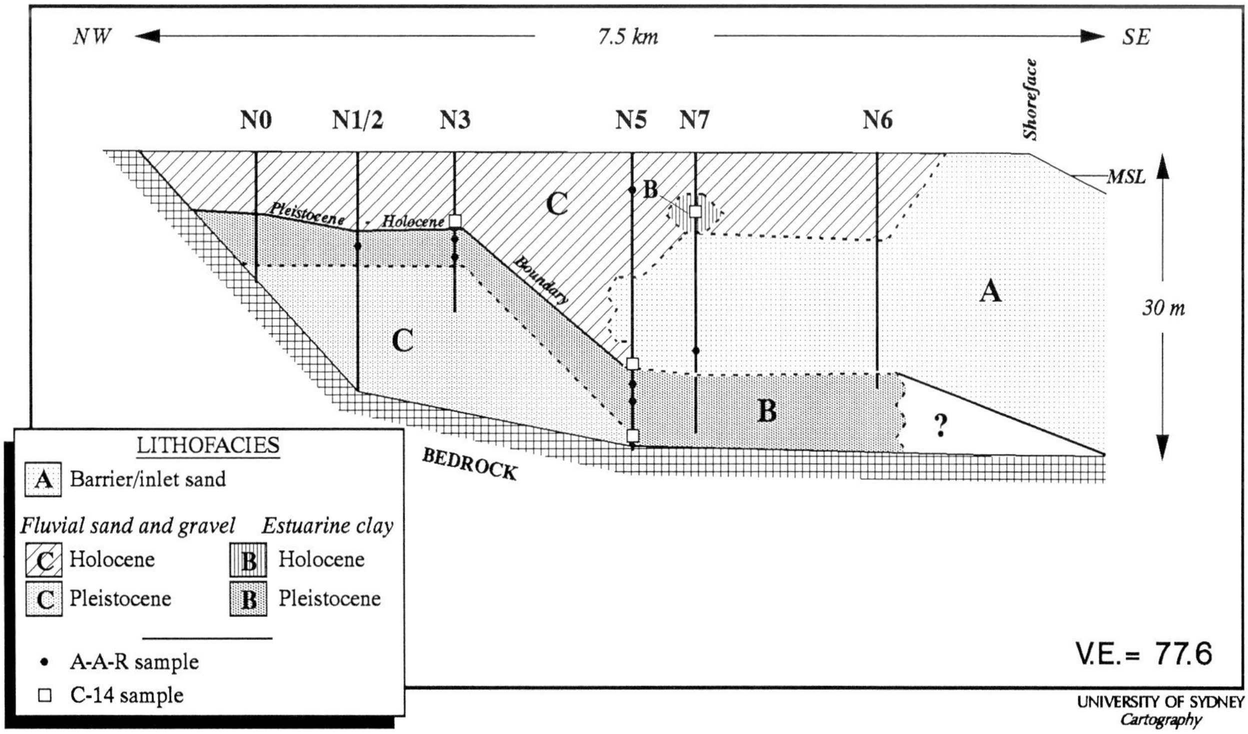

Figure 9. Longitudinal section of Narrawallee valley fill.

2.64-3.19 phi) mixed with increasing silt and clay fractions. Silt increases from 0 to 52 per cent and clay from 0 to 50 per cent (Fig. 8). Small-scale (decimetres in thickness) fining-upward sequences are repeated throughout the unit (Fig. 8). Sedimentary structure was not clear as a result of core disturbance.

The uppermost unit present in N5, between -0.0 and -12.4 m, consists of a fining-upward sequence of similar character to that observed in N3; that being, a series of fining-upward sequences consisting of coarse (moment mean: 0.36 phi) silty sands grading to rooted fine (moment mean: 2.08 phi) silty sands (Fig. 8). Sand is the dominant size fraction (35%-90%), with silt varying between 4 and 39 per cent and clay between 1 and 31 per cent.

In summary, the interpretation of the deposits recovered in N3 and N5 recognizes two units of fluvial facies (Zone C) assemblages and an estuarine lagoon facies (Zone B) (Fig. 9). Thus, the basal sands are associated with fluvial channel and/or fluvial delta depositional conditions. The overlying silts and clays relate to the barrier basin environment. The well defined beds of fine sands within basin deposits are attributed to discrete river flood events.

In N3, the barrier basin facies is capped by the second fluvial facies assemblage. The fining-upward trend within the unit is interpreted as representing the transition from channel/delta to overbank levee and floodplain depositional environments. The sequence described for N5 incorporates the same fluvial-estuarine-fluvial transition noted in N3 but is punctuated by a barrier/inlet facies (Zone A).

Narrawallee Valley Stratigraphy

Figure 9 illustrates the longitudinal distribution of major facies zones in the Narrawallee valley fill, incorporating data from all drill holes. The valley fill is complicated by the presence of multiple units of Zone B and C facies. These units are interpreted as the product of two sea-level highstand infill episodes. The episodes are preserved as a transgressive-to-highstand barrier basin facies bound by a regressive fluvial facies.

The basal regressive fluvial facies, preserved at the head of the valley, based on stratigraphic position, is inferred to be greater than 120,000 years old. It is overlain by a relatively thick unit of the barrier basin facies. The basin deposit is interpreted as representing a period of at least partial valley infilling during the last interglacial (ca. 120,000 BP). Support for this age approximation may be drawn from the weathering characteristics of the basin deposits. The up-valley deposits were presumably exhumed during the lowstand of the late Pleistocene, hence subjected to weathering processes. The lack of weathering of preserved N5 and N7 basin deposits is attributed to their greater burial depth than those in N3 and N1/2. Further credence for a last interglacial, rather than the last interstadial, age estimate is provided by C-14 calibrated amino acid racemization dates.

The second highstand episode, associated with present sea level, is recorded as a relatively thin (3 m), laterally discontinuous unit of the basin facies and a regressive fluvial deposit. Barrier basin sediments were recovered at only one drill site, N7 (Fig. 9). Lowstand scouring of the fill laid down during the last interglacial appears to have been minimal and concentrated in the seaward half of the valley, resulting in limited accommodation space for Holocene barrier basin sedimentation. The majority of Holocene infilling has, therefore, been achieved by the deposition of a regressive fluvial facies that occupies the upper 3 to 12 m of the sequence (Fig. 9).

The Narrawallee valley fill is interpreted as an example of a composite prograding sequence, described by Davis and Clifton (1987). Sequences of this type are deposited under conditions of relative sea level stability in a wave dominated setting, and comprise stratigraphic intervals equal to initial water depth (Davis and Clifton, 1987). The last interglacial deposits in Narrawallee are not the same thickness as paleowater depth which, based on the Quaternary sea level reconstructions of Chappell (1983), would have been some 30 to 40 m. Lowstand scouring during the last glacial period and reworking of the upper bounding surface during the subsequent marine transgression have attenuated the thickness of Pleistocene deposits to between 7.5 and 17 m. The composite form of the valley fill results from the preservation of the Pleistocene deposits along with the recent Holocene progradational sequence.

Summary and Conclusion

Modelling the organization of estuarine depositional complexes as a three-fold zonation of primary facies appears appropriate to the two, wave-dominated, microtidal estuaries examined in this paper. Both Wapengo Lagoon and Narrawallee Inlet exhibit a tripartite facies zonation, though regional differences in factors such as palaeovalley dimensions and preservation of Pleistocene deposits have resulted in contrasting stages of infill. Nevertheless, the sedimentological character of each zone is sufficiently unique to allow for their identification in the two estuaries.

The interaction of inlet hydrodynamic processes with the sedimentological and biological regimes of the barrier/inlet zone (A) of Wapengo Lagoon produce distinct spatial trends in the manner in which sediment is deposited. These trends are most evident in the intertidal environment, where process-sediment interaction is currently active and readily observable. Thus, the inferred gradient of decreasing tidal flow velocity in a landward direction has influenced the morphology and internal structure of intertidal bedforms by producing longitudinal variations in dominant bedform type. These variations are most evident when individual shoals of the delta are compared. Thus, a transition occurs from three-dimensional dunes on the seaward shoal, to two-dimensional dunes on the medial shoal. In addition, the waning of current velocities provides conditions suitable for the overprinting of primary physical traits by secondary biogenic markers. The degree of bioturbation also follows a transitional pattern, from nearly absent at the seaward end to pervasive at the landward limit of the delta.

In the Narrawallee sequence, it is argued that the gross textural and mineralogical character of sediment (and to a lesser extent the structural character) recovered from the seaward end of the valley fill, is consistent with that described from Zone A in Wapengo Lagoon. The dominance of sand-sized particles, particle roundness, and maturity of the mineralogical suite are taken as the prime indicators for barrier/inlet depositional conditions.

The estuarine/lagoon (Zone B) and tidal-fluvial (Zone C) zones of both Wapengo and Narrawallee are characterized by upward-coarsening barrier basin to fluvial channel to fluvial delta and floodplain sediments. These units are recognized on the basis of sediment texture, structural properties, mineralogy and stratigraphic position. Thus, the barrier basin is represented by fine grained laminated sediment (mud and silt) and occupies a basal position in a sequence. Older non-basin deposits may occur below the basin unit, as in Narrawallee. Fluvial channel, delta and floodplain sediments are made up of silty sands and gravels in massive (bioturbated) and graded beds, of immature mineralogy and typically in a vertical sequence from channel to delta to plain. Variations in sediment texture and structural features such as graded beds are interpreted as a record of fluctuations in relative fluvial flow energy. Collectively, the Zone B and C facies describe a progradational (regressive) sequence. In Wapengo, progradation is ongoing, while in Narrawallee it is complete insofar as there is no accommodation space for further deposition at present sea level.

Finally, the partial preservation of a highstand valley fill from the last interglacial stage at Narrawallee provides an indication of the preservation potential of each facies. Given that preservation is a function of the degree of scouring that occurs during sea-level lowstand and subsequent transgression, the highest preservation potential is afforded to facies located in the deepest and headward portion of the valley. In the Narrawallee case, those facies include both the estuarine lagoon and tidally influenced fluvial units. The preservation of multiple units of the barrier/inlet facies is not evident from the available data in the Narrawallee fill, suggesting that lowstand scouring of the Pleistocene deposit toward the seaward end of the valley was extensive. In contrast, the barriers of the Illawarra coast do have a Pleistocene component (Bradshaw, 1988; Roy, pers. comm.). Even so, Zone A is regarded as having a somewhat lower preservation potential than Zones B and C because of its seaward location. This is not to imply that barrier/inlet sediments will be absent in the geological record. The degree of preservation of all three zones is dependent upon the conditions prevailing during a change in relative sea level. Generally, the slower the change in sea level, the lower the preservation potential (Davis and Clifton, 1987).

The demonstration, herein, of consistent patterns of facies organization and sedimentological traits within study sites from the relatively unique estuarine setting offered by the New South Wales coast provides further evidence for the uniformity of the pre-existing concept of tripartite facies zonation, and support for it's incorporation into estuarine facies models.

Acknowledgements

This study represents part of the author's doctoral research and thanks are owed to my supervisors B. Thom and P. Roy who provided guidance for the project. E. Clifton, D. Smith, G. Reinson and P. Roy all provided constructive criticism of early drafts of the paper. Drilling was carried out with rigs provided by the New South Wales Department of Minerals and Energy, operated by P. Johnston, and the Department of Geology and Geophysics at The University of Sydney, operated by P. Manning. The vibracorer was provided courtesy of E. Frankel of the University of Technology. Amino acid racemization analysis on shell material was performed with the assistance of C. Murray-Wallace. Cartographic work was undertaken by P. Johnson.

References

Allen, G. P. 1989. Sedimentary facies and patterns in the Gironde estuary: a recent model of macrotidal estuarine sedimentation. *In:* Abstracts, Second International Research Symposium on Clastic Tidal Deposits, Calgary, Alberta, p. 2.

Allen, J. R. L. 1968. Current ripples, their relation to patterns of water and sediment motion. North-Holland Publishing Company, Amsterdam. 433 p.

Armon, J. W. 1979. Landward sediment transfers in a transgressive barrier island system, Canada. *In:* Barrier Islands: From the Gulf of St Lawrence to the Gulf of Mexico, S. P. Leatherman (ed.). Academic Press, New York. p. 65-80.

Ashley, G. M., Chairperson and others, 1990. Classification of large-scale subaqueous bedforms: a new look at an old problem. Journal of Sedimentary Petrology, v. 60, p. 160-172.

Bates, R. L. and Jackson, J. A., (eds.) 1980. Glossary of Geology. 2nd Edition. American Geological Institute.

Bird, E. C. F. 1967a. Coastal lagoons of southeastern Australia. *In:*Landform Studies from Australia and New Guinea, J. N. Jennings and J. A. Mabbutt (eds.). Cambridge University Press, p. 365-386.

______ 1967b. Depositional features in estuaries and lagoons on the south coast of New South Wales. Australian Geographical Studies, v. 5, p. 113-125.

______ 1978. The geomorphology of the Gippsland Lakes region. Ministry for Conservation, Victoria, Environmental Studies Series, Publication no. 186.

Boothroyd, J. C. 1985. Tidal inlets and tidal deltas. *In:* Coastal Sedimentary Environments, R. A. Davis (ed.). 2nd Ed. Springer-Verlag, New York. p. 445-533.

Bowden, K. F. 1967. Circulation and Diffusion. *In:* Estuaries, G. H. Lauff (ed.). American Association for the Advancement of Science, Publication no. 83. p. 15-37.

Boyd, R., Bowen, A. J. and Hall, R. K. 1987. An evolutionary model for transgressive sedimentation on the eastern shore of Nova Scotia. *In:* Glaciated Coasts, D. M. Fitzgerald and P. S. Rosen (eds.). Academic Press. p. 87-115.

Bradshaw, B. E. 1988. St. Georges Basin: Morphology and late Quaternary deposits. Unpublished B.Sc. Honours Thesis, Department of Geography, University of Sydney.

Carter, R. W. G., Forbes, D. L., Jennings, S. C., Orford, J. D., Shaw, J. and Taylor, R. B. 1989. Barrier and lagoon coast evolution under differing relative sea-level regimes: Examples from Ireland and Nova Scotia. Marine Geology, v. 88, p. 221-242.

Caspers, H. 1967. Estuaries: Analysis of definitions and biological considerations. *In:* Estuaries, G. H. Lauff (ed.). American Association for the Advancement of Science, Publication no. 83, p. 6-9.

Chapman, D. M., Geary, M., Roy, P. S. and Thom, B. G. 1982. Coastal Evolution and Coastal Erosion in New South Wales. Coastal Council of New South Wales.

Chappell, J. 1983. A revised sea-level record for the last 300,000 years from Papua New Guinea. Search, v. 14, p. 99-101.

Dalrymple, R. W. 1987. Classification of large-scale, transverse bedforms. *In:* Society of Economic Paleontologists and Mineralogists Research Group: Bedforms and Bedding Structures, Classification of Large-Scale Flow-Transverse Bedforms. Extended Abstract.

______ and Zaitlin, B. A. 1989. Tidal sedimentation in the macrotidal, Cobequid Bay-Salmon River estuary, Bay of Fundy. Canadian Society of Petroleum Geologists' field guide to: Second International Research Symposium on Clastic Tidal Deposits, Calgary, Alberta. 84 p.

Davies, J. L. 1964. A morphogenic approach to world shorelines. Zeitschrift fur Geomorphologie, v. 8, p. 27-42.

______ 1980. Geographical Variation in Coastal Development (2nd ed.). Longman, New York.

Davis, R. A. and Clifton, H. E. 1987. Sea-level change and the preservation potential of wave-dominated and tide-dominated coastal sequences. *In:* Sea-level Fluctuations and Coastal Evolution, D. Nummedal, O. H. Pilkey and J. D. Howard (eds.). Society of Economic Paleontologists and Mineralogists Special Publication 41. p. 167-178.

______ and Hayes, M.O. 1984. What is a wave-dominated coast? Marine Geology, v. 60, p. 313-329.

Dyer, K. R. 1973. Estuaries: A Physical Introduction. John Wiley and Sons, London.

Eliot, I. 1979. Lake sediments. *In:* University of Wollongong Study Group, Wallaga Lake: Report on it's Condition, Use and Development Potential; Unpublished Report, p. 4.2-4.13.

Fletcher, C. H., III, Knebel, H. J. and Kraft, J. C. 1990. Holocene evolution of an estuarine coast and tidal wetlands. Geological Society of America, Bulletin, v. 102, p. 283-297.

Frye, J. C. and Willman, H. B. 1962. Note 27- Morphostratigraphic units in Pleistocene stratigraphy. American Association of Petroleum Geologists, Bulletin, v. 46 p. 112-113.

Galloway, W. E. 1975. Process framework for describing the morphologic and stratigraphic evolution of deltaic depositional systems. *In:* Deltas, Models for Exploration, M. L. Brousard (ed.). Houston Geological Society. p. 87-98.

Hayes, M. O. 1975. Morphology of sand accumulation in estuaries: an introduction to the symposium. *In:* Estuarine Research II, L. E. Cronin (ed.). Academic Press, New York. p. 322.

______ 1979. Barrier island morphology as a function of tidal and wave regime. *In:* Barrier Islands: From the Gulf of St Lawrence to the Gulf of Mexico, S. P. Leatherman (ed.). Academic Press, New York. p. 1-28.

______ 1980. General morphology and sediment patterns in tidal inlets. Sedimentary Geology, v. 26. p. 139-156.

Heward, A. P. 1981. A review of wave-dominated clastic shoreline deposits. Earth Science Reviews, v. 17, p. 223-276.

Hunter, T. 1989. Late Quaternary evolution of two estuaries, southern New South Wales, Australia. Unpublished B.A. Honours Thesis, Department of Geography, University of Sydney.

Jennings, J. N. and Bird, E. C. F. 1967. Regional geomorphological characteristics of some Australian estuaries. *In:* Estuaries, G. H. Lauff (ed.). American Association for the Advancement of Science, Publication #83, p. 121-129.

Kidd, R. W. 1978. Estuarine sediment regimes, Far South Coast, New South Wales. Unpublished Ph.D. Thesis, Department of Geography, Macquarie University.

Kraft, J. C. 1971. Sedimentary facies patterns and geological history of a Holocene marine transgression. Geological Society of America, Bulletin, v. 82, p. 2131-2158.

______, Allen, E. A., Belknap, D. F., John, C. J. and Maurmeyer, E. M. 1979. Processes and morphologic evolution of an estuarine and coastal barrier system. *In:* Barrier Islands: From the Gulf of St Lawrence to the Gulf of Mexico, S. P. Leatherman (ed.). Academic Press, New York. p. 149-184.

______, Chrzastowski, M. J., Belknap, D. F., Toscano, M. A., Fletcher, C. N., III 1987. The transgressive barrier-lagoon coast of Delaware: Morphostratigraphy, sedimentary sequences and responses to relative rise in sea level. *In:* Sea-level Fluctuation and Coastal Evolution, D. Nummendal, O. Pilkey and J. Howard (eds.). Society of Economic Paleontologists and Mineralogists Special Publication 41. p. 129-143.

Lewis, D. W. 1984. Practical Sedimentology. Hutchinson Ross Publishing Company.

Nakada, M. and Lambeck, K. 1989. Late Pleistocene and Holocene sea-level change in the Australian region and mantle rheology. Geophysical Journal, v. 96, p. 497-517.

Nichol, S. L. and Murray-Wallace, C. V. (*in press*). A partially preserved last interglacial estuarine fill: Narrawallee Inlet, N.S.W. Australian Journal of Earth Sciences.

Nichols, M. M., Johnson, G. H. and Pebbles, P. C. 1989. Estuarine facies and sequences in a microtidal coastal plain estuary: I. Modern deposits from the James Bay, Virginia. *In:* Abstracts, Second International Research Symposium on Clastic Tidal Deposits, Calgary, Alberta, p. 68-69.

Rahmani, R. A. 1988. Estuarine tidal channel and nearshore sedimentation of a Late Cretaceous epicontinental sea, Drumheller, Alberta, Canada. *In:* Tide-Influenced Sedimentary Environments and Facies, P. L. de Boer, A. van Gelder and S. D. Nio (eds.). D. Reidel Publishing Co. p. 433-471.

Reinson, G. E. 1973. Aspects of weathering and sedimentation in the Genoa River basin and estuary, N.S.W.-Victoria. Unpublished Ph.D. Thesis, Department of Geology, Australian National University.

______ 1977. Hydrology and sediments of a temperate estuary-Mallacoota Inlet, Victoria. Department of National Resources, Bureau of Mineral Resources, Geology and Geophysics, Bulletin 178.

______, Clark, J. E. and Foscolos, A. E. 1988. Reservoir geology of Crystal Viking field, Lower Cretaceous estuarine tidal channel-bay complex, south-central Alberta, American Association of Petroleum Geologists, Bulletin, v. 72, p. 1270-1294.

Rochford, D. J. 1951. Studies in Australian estuarine hydrology. I: Introductory and comparative features. Australian Journal of Marine and Freshwater Research, v. 2, p. 1-117.

______ 1959. Classification of Australian estuarine systems. Archivo di oceanografie e limnologia, v. 11, p. 171-177.

Roy, P. S. 1977. Does the Hunter River supply sand to the New South Wales coast today? Proceedings of the Royal Society of New South Wales, v. 110, p. 17-24.

______ 1984a. New South Wales estuaries: their origin and evolution. *In:* Coastal Geomorphology In Australia, B. G. Thom (ed.). Academic Press, Sydney. p. 99-122.

______ 1984b. Holocene sedimentation histories of estuaries in south-eastern Australia. *In:* Estuarine environments of the Southern Hemisphere, E. P. Hodgkin (ed.). Department of Conservation and Environment, Western Australia, Bulletin 161, p. 23-61.

______ and Peat, C. 1975. Bathymetry and bottom sediments of Lake Illawarra. Records of the Geological Society of New South Wales, v. 17, p. 65-79.

______ and ______ 1976. Bathymetry and bottom sediments of Tuross estuary and Coila Lake. Records of the Geological Society of New South Wales, v. 18, p. 103-134.

______, Thom, B. G. and Wright, L. D. 1980. Holocene sequences on an embayed high-energy coast: an evolutionary model. Sedimentary Geology, v. 26, p. 1-19.

Rubin, D. M. 1987. Cross-bedding, Bedforms and Paleocurrents. Society of Economic Paleontologists and Mineralogists, Concepts in Sedimentology and Paleontology, v 1. 187 p.

Steers, J. A. 1967. Geomorphology and coastal processes. *In:* Estuaries, G. H. Lauff (ed.). American Association for the Advancement of Science, Publication no. 83. p. 100-108.

Thom, B. G. and Roy, P. S. 1983. Sea level change in New South Wales over the past 15,000 years. *In:* Australian Sea Levels in the last 15,000 years: A Review, D. Hopley (ed). Occasional Paper no. 3, James Cook University of North Queensland, Department of Geography, p. 64-85.

Williams, G. C. 1983. Hydraulic Characteristics of New South Wales Estuarine Lake Inlets. Unpublished Report for Master of Engineering Science, University of New South Wales.

Wolfe, D. A. and Kjerfve, B. 1986. Estuarine variability: an overview. *In:* Estuarine Variability, D. A. Wolfe (ed.). Academic Press, New York, p. 3-17.

Zaitlin, B. A. and Dalrymple, R. W. 1985. Depositional patterns and stratigraphic sequences from the inner part of a sand-dominated, macrotidal estuary, Cobequid bay-Salmon River estuary, Bay of Fundy. *In:* Abstracts to First Symposium on Modern and Ancient Clastic Tidal Deposits, August 1985, Utrecht, The Netherlands. p. 169-171.

______ and Shultz, B. C. 1984. An estuarine-embayment fill model from the Lower Cretaceous Mannville Group, west-central Saskatchewan. *In:* The Mesozoic of Middle North America, D. F. Stott and D. J. Glass (eds.). Canadian Society of Petroleum Geologists, Memoir 9, p. 455-469.

______ and ______ 1990. Wave-influenced estuarine sand body, Senlac heavy oil pool, Saskatchewan, Canada. *In:* Sandstone Petroleum Reservoirs-Casebooks in Earth Science, J. H. Barwis, J. G. McPherson and R. J. Studlick (eds.). Springer-Verlag, New York, p. 363-387.

KEY WORDS: ESTUARINE FACIES, MICROTIDAL SETTINGS, NEW SOUTH WALES, BEDROCK-CONTROLLED VALLEY

In: Clastic Tidal Sedimentology, D. G. Smith, G. E. Reinson, B. A. Zaitlin and R. A. Rahmani (eds.). Canadian Society of Petroleum Geologists. Memoir 16, p. 59-90.

The post-glacial evolution of Chignecto Bay, Bay of Fundy, and its modern environment of deposition

CARL L. AMOS

Geological Survey of Canada
Department of Energy, Mines and Resources
Bedford Institute of Oceanography
P.O. Box 1006, Dartmouth
Nova Scotia, Canada
B2Y 4A2

KIM T. TEE

Physical and Chemical Sciences
Department of Fisheries and Oceans
Bedford Institute of Oceanography
P.O. Box 1006, Dartmouth
Nova Scotia, Canada
B2Y 4A2

BRIAN A. ZAITLIN

Esso Resources Canada Ltd.
237, 4th Avenue S.W.
Calgary, Alberta, Canada
T2P 0H6

ABSTRACT

Chignecto Bay is a macrotidal estuary situated in the upper part of the upper Bay of Fundy. Its physiography is bedrock controlled though it is blanketed in the upper reaches by transgressive, silty, glacial and postglacial unconsolidated sediments. This sequence is largely preserved in the sublittoral portion of the bay. It is complicated by rapid changes in sea level (55 m above present sea level at 15,000 BP, 25 m below present sea level circa 7,000 BP) and large changes in tidal range (varying from less than 2 m to more than 12 m over the last 13,300 years). The fining-upward macrosequence and expansive sand flats, which typify wide, shallow, mesotidal and macrotidal estuaries such as Minas Basin, are largely absent due to the major changes in sea level and tidal range. The depositional character and trapping efficiency of the estuary has also changed dramatically. From 13,300 to 10,000 BP, the estuary was characterized by seabed erosion, by the dominance of macrotidal sand and gravel facies throughout the estuary, by a low trapping efficiency of fine grained sediment and by a net export of this material. Between 9,000 and 6,000 BP, the estuary was characterized by the rapid accumulation of sand/mud laminae in its central, eastern side, possibly within a mesotidal lagoon, and by a high trapping efficiency and a probable import of material suspended in the Bay of Fundy. Wave activity during this time was suppressed, possibly as a result of a barrier island across the mouth of the paleoestuary. The estuarine depocentre moved to the western side of the central and inner estuary between 6,000 and 3,000 BP. The dominant facies also changed to macrotidal lenticular and flaser bedding. From 3,000 BP to the present, the estuary returned to conditions interpreted as having prevailed during late Pleistocene. The depocentre for these Holocene sediments is along the eastern part of the central sublittoral estuary. Recent sublittoral sediments are dominated by sand and gravel facies. Recent littoral sediments form fringing mud flats and salt marshes in the inner estuary. The modern estuary is subject to a mean spring semidiurnal tidal range of 11.3 m and associated tidal currents in excess 1.5 m/s. These currents are eroding the previously-deposited muddy sediments and creating a transgressive sand unit dominated by large-scale bedforms (sand waves, 2-D and 3-D megaripples) and sand ribbons. The bay is presently exposed to a peak significant wave height of 2 m. Waves thus play a major role in reworking and resuspending seabed sediments and in releasing fine sand and silt (1.03×10^6 m^3/a) through cliffline erosion (up to 1.0 m/a). A total of 7.3×10^6 m^3 of sediment is mobilized each year; 99 per cent of this volume is in the silt-clay size range and is transported in suspension. Residual circulation near the estuary mouth is headward along the eastern margins (0.1 m/s) and seaward along the western margin. Suspended sediment is transported out of the system as a result of this residual circulation (6.3×10^6 m^3/a) and is subsequently deposited in the outer Bay of Fundy.

RÉSUMÉ

La baie de Chignecto, située dans la partie supérieure de la baie de Fundy, est un estuaire macrotidal dont la physiographie est contrôlée par le substratum malgré le revêtement aux points culminants d'une séquence transgressive. Celle-ci, composée de sédiments glaciaires et post-glaciaires, silteux et non consolidés, est principalement préservée dans la zone sublittorale de la baie. Des changements rapides du niveau de la mer (55 m au-dessus du niveau actuel, il y a 15,000 av. P., et 25 m au-dessous du niveau actuel, il y a 7,000 av. P.) et des changements importants des écarts de marée (variation de moins de 2 m à plus de 12 m durant les derniers 13,300 ans) compliquent la séquence transgressive. La macroséquence virtuelle (i.e. granulométrie décroissante de la base au sommet) et les vastes estrans sableux («sand flats»), caractéristiques des estuaires, mesotidal et macrotidal, larges et peu profonds, comme par exemple le bassin Minas, sont absents principalement à cause des changements majeurs du niveau marin et des écarts de la marée. L'aspect sédimentaire et l'efficacité de piégeage des sédiments dans l'estuaire ont également changé drastiquement. Par exemple, de 13,300 à 10,000 ans av. P., l'estuaire se caractérisait par une érosion des couches marines, une prédominance des faciès macrotidaux sableux et graveleux, une faible efficacité du piégeage du sédiment fin et un transport net de ce sédiment fin vers l'extérieur du système. Par contre entre 9,000 et 6,000 ans av. P., la partie centrale-est de l'estuaire se caractérisait par une accumulation

Geological Survey of Canada Contribution Number 26390

rapide de laminations sable/boue, caractérisant possiblement un lagon mesotidal, et par un grande efficacité du piégeage du matériel en suspension. Une partie du matériel en suspension provenait probablement de la baie de Fundy. Durant ce temps, la suppression de l'action des vagues résultait probablement de la présence d'une crête d'avant-plage émergée à l'embouchure de l'estuaire. Entre 3,000 et 9,000 ans av. P., les points de sédimentation maximum de l'estuaire se déplacèrent vers le côté ouest du centre et de la partie interne de l'estuaire. Aussi, le faciès dominant était remplacé par des litages macrotidaux lenticulaires et de type «flaser». Depuis 3,000 ans, l'estuaire serait retourné aux conditions qui auraient prévalu pendant le Pléistocène tardif. À l'Holocène, les points de sédimentation maximum se situent le long de la partie orientale du centre de l'estuaire sublittoral. Ces sédiments sublittoraux sont dominés par des faciès de sable et gravier. Dans la partie interne de l'estuaire, les sédiments littoraux holocènes forment les slikkes («mud flats») et marais salants périphériques. L'estuaire moderne est soumis à de grandes marées semi-diurnes d'une moyenne de 11.3 m, et à des courants de marée associés excèdant 1.5 m/s. Ces courants érodent les sédiments boueux antérieurement déposés, et sédimentent l'unité sableuse transgressive dominée par des «morphologies de fond» («bedforms») de grande échelle (rides de sable, et mégarides en 2-D et 3-D) et des sables rubannées. La baie est actuellement exposée à des vagues dont la hauteur atteint un maximum important de 2 m. Celles-ci jouent donc un rôle majeur dans le remaniement et la remise en suspension des sédiments marins, ainsi que le dégagement des sables fins et silts (1.03 x 10^6 m^3) de la falaise. Un total de 7.3 x 10^6 m^3 de sédiment est mobilisé chaque année; 99 pour cent de ce volume varient de la taille des silts à des argiles, et sont transportés par suspension. Près de l'embouchure de l'estuaire, la circulation résiduelle se déplace vers l'amont le long de la marge (0.2 m/s) orientale, et vers le large le long de la marge occidentale. Le sédiment en suspension est transporté hors du système par la circulation résiduelle (6.3 x 10^6 m^3), et est subséquemment décanté dans la partie externe de la baie de Fundy.

Introduction

The Chignecto Bay system is situated within a splay of the upper Bay of Fundy, which exhibits morphological characteristics of a glacially-excavated, elongated macrotidal estuary (Fig. 1). It hosts the largest expanse of salt marsh deposits in the Canadian Maritimes (147 km^2), most of which has been reclaimed for agricultural purposes. Despite this, the sedimentology of the Chignecto Bay system, in contrast to Minas Basin and Cobequid Bay, was virtually unknown prior to 1977 (Laming, 1967; Bay of Fundy Tidal Power Review Board, 1977; Amos *et al.*, 1980). Since that time, oceanographic and sedimentological studies have been carried out, the major part of which has been reviewed by Gordon and Dadswell (1984). These studies include a geophysical survey and sediment budget analysis (Amos and Asprey, 1979), an analysis of the surficial sediments (Amos *et al.*, 1980; Hildebrand *et al.*, 1980; Hussain, 1980; Bray *et al.*, 1982; Amos, 1984; Amos and Zaitlin, 1985) and a study of the suspended sediment flux in Cumberland Basin (Amos and Tee, 1989; Tee and Amos, *in press*). Preliminary results showed that the sedimentary character of Chignecto Bay differs from Minas Basin by hosting much higher quantities of fine grained sediment in suspension and within bed sediments. Indeed, Chignecto Bay is defined by Middleton (pers. comm., 1986) as a "muddy" estuary, whereas Minas Basin is defined as a "sandy" estuary (Dalrymple *et al.*, 1983). In this paper we examine the character and evolution of the unconsolidated sedimentary sequence in Chignecto Bay and the interpreted depositional environments. We then examine the present-day sediment budget, the sediment deposits and the sediment transport pathways and the processes that influence them. From this examination we hope to provide some insight into the reasons for the turbid character of this estuary within the Bay of Fundy system.

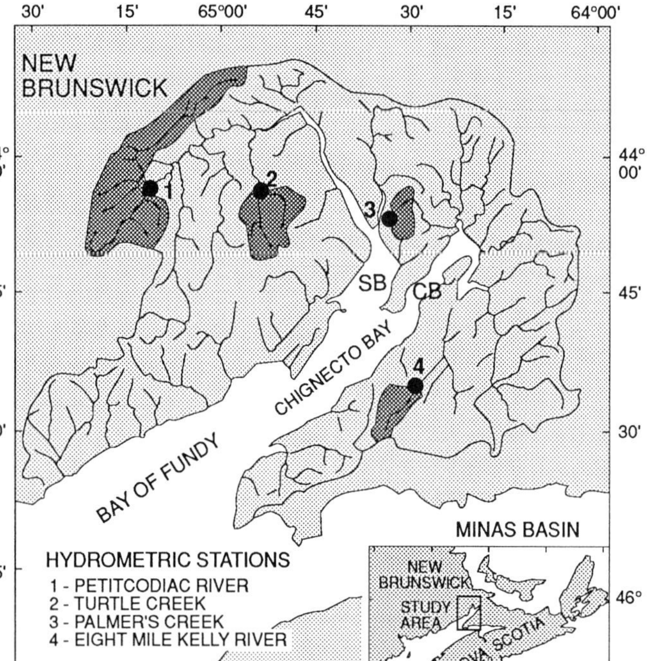

Figure 1. Locality map of the Chignecto Bay system illustrating the four river gauging stations employed in this study. SB, Shepody Bay; CB, Cumberland Basin.

Database and Methods

The database and methods used in the evaluation of the surficial geology, sediment budget and the sediment transport pathways of Chignecto Bay are given in Amos and Asprey (1979). A total of 587 line km of high resolution seismic surveying (shallow-towed sparker) and sidescan sonar (100 kHz) imagery was collected in order to define the nature and thickness of the unconsolidated sublittoral sediments (Amos and Asprey, 1979). A further 230 km of airgun seismics were collected in order to define the bedrock surface beneath the unconsolidated sediments. The granulometry of bottom sediments was determined from 225 samples collected with a 0.2 m^2 Van Veen grab sampler. The stratigraphy and Late Quaternary evolution of the surficial sequence were interpreted from nineteen vibracores between 0.82 and 5.7 m in length, and twelve gravity cores collected at sites along seismic sections. Lithofacies were defined in the cores using methods outlined by Anderton (1985) based on grain size, primary sedimentary structures, bioturbation, biogenic material and bedding characteristics.

The distribution and transport of suspended sediment, temperature, salinity and current velocity were determined at 28 anchor stations located throughout Chignecto Bay. Testing was carried out at 7 different times of the year (Fig. 2). Surface water samples were collected along each of the estuaries from helicopters in order to detect spatial variations in suspended sediment concentration. Long-term measurements of currents were made using Aanderaa RCM 4 current meters at twelve anchor stations for approximately 1 month. Wave climate was measured using Datawell 6000 waverider buoys at two sites (near Cape Enragé and Cape Maringouin) for approximately 6 months. Aanderaa tide gauges were deployed in Cumberland Basin and Shepody Bay for approximately 1 month to determine the magnitude and phase of the tidal constituents. The transport pathways of bedload were inferred from bedform orientation measured on sidescan records and from a heavy mineral analysis undertaken by Hussain (1980).

Fluvial input of sediment was measured at four gauging stations within the Chignecto Bay system drainage basin.

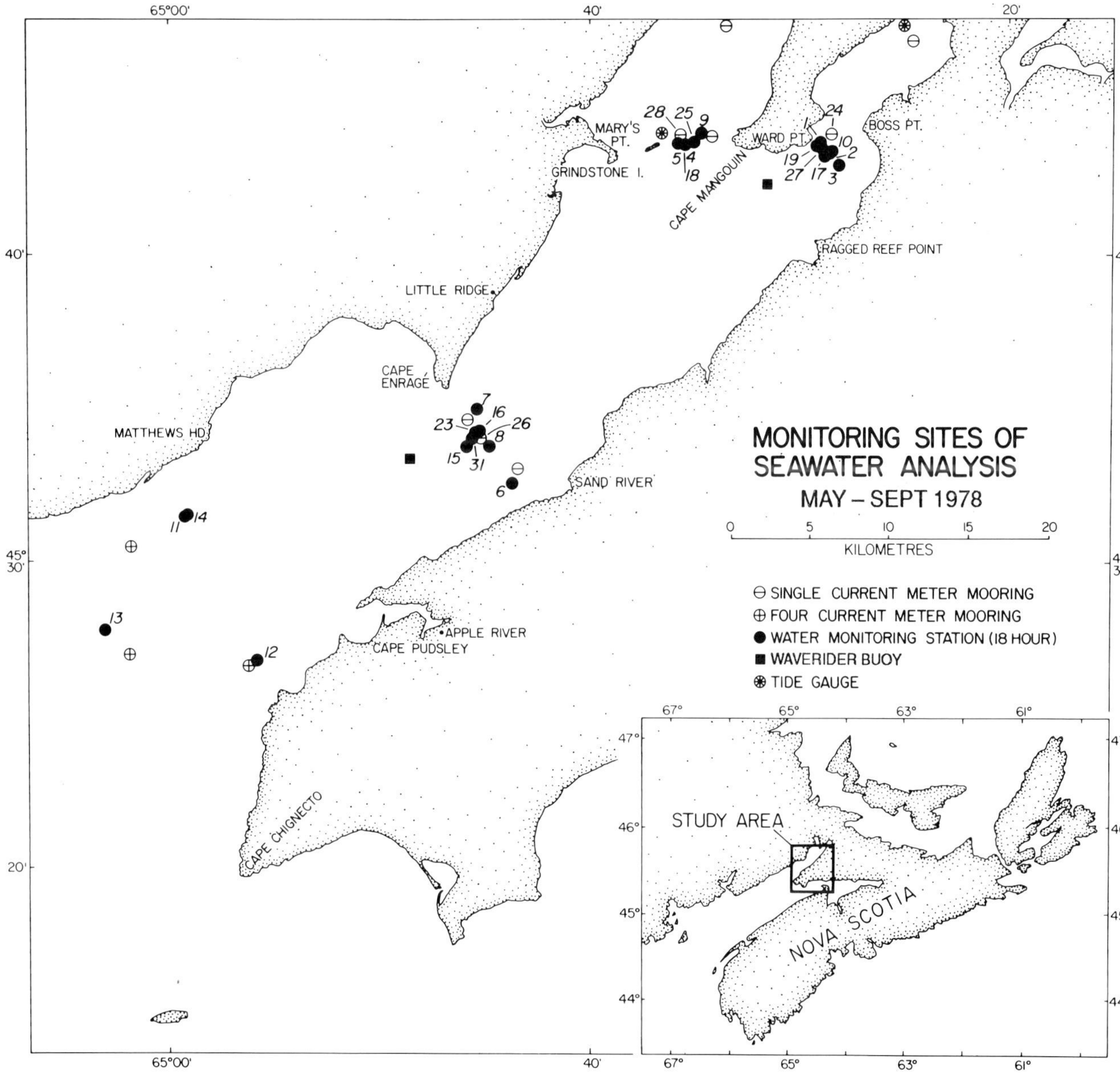

Figure 2. Monitoring sites for seawater analysis including stations 1 to 28, 12 current meter mooring sites, 2 tide gauge sites, and 2 waverider buoy sites.

Water samples were collected every two days for a period of approximately 9 months and analyzed by filtration for suspended sediment concentration and percentage of organic detritus. Coastal erosion was determined by comparison of orthophoto maps created from aerial photographs collected in 1947 and 1975, respectively. Volume input of sediment from this source was determined from cliff recession rates between the successive surveys and by scaling cliff height from large-format (105 mm), horizontal photographs of the cliffs. A review of the methods used, computational procedures, data collection and results is given in Hildebrand *et al.* (1980) and Amos and Asprey (1981).

Regional Setting and Geological Evolution

The Chignecto Bay system is subdivided into three physiographic provinces (Fig. 1): 1) Chignecto Bay, *sensu stricto*; 2) Shepody Bay; and 3) Cumberland Basin. A general summary of the dimensions of the Chignecto Bay system is given in Table 1. Chignecto Bay is a high-energy environment. It is dominated by semidiurnal tides that have a spring tidal range of 11.3 m and a neap tidal range of 7.2 m. The associated spring tidal prism is 5.56 km^2 causing peak currents between 1.0 and 2 m/s (Tee and Amos, *in press*). The bay is oriented parallel to the dominant northeast-southwest winds and is, therefore, subject to storm wave influences. This is apparent in its two coastal types: 1) a cliffed, high tide shoreline with a narrow intertidal zone in the central and outer parts of the estuary; and 2) a cliffed, salt marsh shoreline with an extensive intertidal zone in the inner parts of the estuary.

The gross morphology of Chignecto Bay is controlled by the Appalachian Orogenic Physiographic Province (Williams *et al.*, 1972), within which the bay is located. The system is largely controlled by development of the Chignecto-Chedabucto-Cobequid Fault Zone (C^3FZ; Poole *et al.*, 1976). Post-Acadian and Appalachian tectonic events (Carboniferous - Permian) along the C^3FZ resulted in the formation of a series of semi-connected depositional basins infilled with terrestrial, shallow marine and volcanic material (Klein, 1970). The bedrock geology of the region exhibits Triassic half-graben basinal development, which began 430 million years ago (Swift and Lyall, 1968; Stephens, 1977). The half-graben was formed in Precambrian volcanics and Paleozoic fluvial and deltaic sediments and volcanics and was subsequently filled with Triassic sandstones and volcanics (King and MacLean, 1976). Bedrock flexuring during earth movements upthrusted and folded much of the material that presently forms the eroded cliffs at Chignecto Bay and underlies the bay. This material was further dissected by Cretaceous and Tertiary fluvial development (Grant, 1970).

The bedrock is composed of friable siltstone and shale that disintegrates to provide large quantities of silt-sized sediment to the bay. The character of the unconsolidated surficial sediments of Chignecto Bay thus reflects the lithology of the local bedrock, although it is modified in part by glacial erosion and postglacial tidal deposition. Fine grained (clay-rich) glacial and postglacial sediments or muddy gravel constitute much of the seabed of the outer Bay of Fundy and Chignecto Bay (Pelletier and McMullen, 1972; Swift *et al.*, 1973; Fader *et al.*, 1977). These sediments are presently in a state of erosion and are a source of the fine grained suspended sediments in the bay (Amos and Topliss, 1985; Tee and Amos, *in press*).

Shepody Bay and Cumberland Basin are separated from Chignecto Bay by constricted mouths at Cape Maringouin. These estuaries are characterized by highly turbid waters (Gordon and Dadswell, 1984; Amos and Asprey, 1981) and differ from Chignecto Bay by possessing extensive fringing salt marshes and mud flats (coastal type 2). These inner estuaries have been extensively modified during the last 13,300 years as a result of a period of regression and subsequent transgression and by large changes in tidal range. Much of the record of these events is preserved in the sublittoral deposits of Chignecto Bay (Amos and Zaitlin, 1985). A synthesis of these changes is shown in Figure 3. It appears that Chignecto Bay was under the influence of glacial ice (the Chignecto Phase) prior to 13,300 BP (Scott and Medioli, 1980; Rampton *et al.*, 1984). Submergence of Chignecto Bay through ice loading was greatest (55 m) at approximately 15,000 BP (Quinlan and Beaumont, 1982). Sandy deltas were formed extensively along the front of the retreating Chignecto ice sheet through meltout (Wightman and Cooke, 1978) into the inundated Bay of Fundy (the De Geer Sea, after Lougee, 1953). Concomitant emergence, that took place at a rate of 106 cm/century, elevated the deltas above present sea level (Goldthwait, 1924). Foreset beds of these deltas presently exposed in cliff sections around Chignecto Bay, together with macrofaunal evidence, are diagnostic of microtidal conditions at the time (Grant, 1970; Wagner, 1980). Marine reworking and erosion of glacial deposits into gravel beds subsequent to deglaciation (13,300 to 10,000 BP) is suggestive of strong tidal currents. This is tentatively correlated with

Table 1. A summary of the general dimensions of the Chignecto Bay system

	Shepody Bay	Cumberland Basin	Chignecto Bay
Length of coastline (km)	94.0	68.6	106.7
Littoral area (km^2)	99	83	46
Sublittoral area (km^2)	76	38	469
Spring tidal prism (km^3)	1.43	0.90	5.56
Neap tidal prism (km^3)	0.91	0.57	3.54

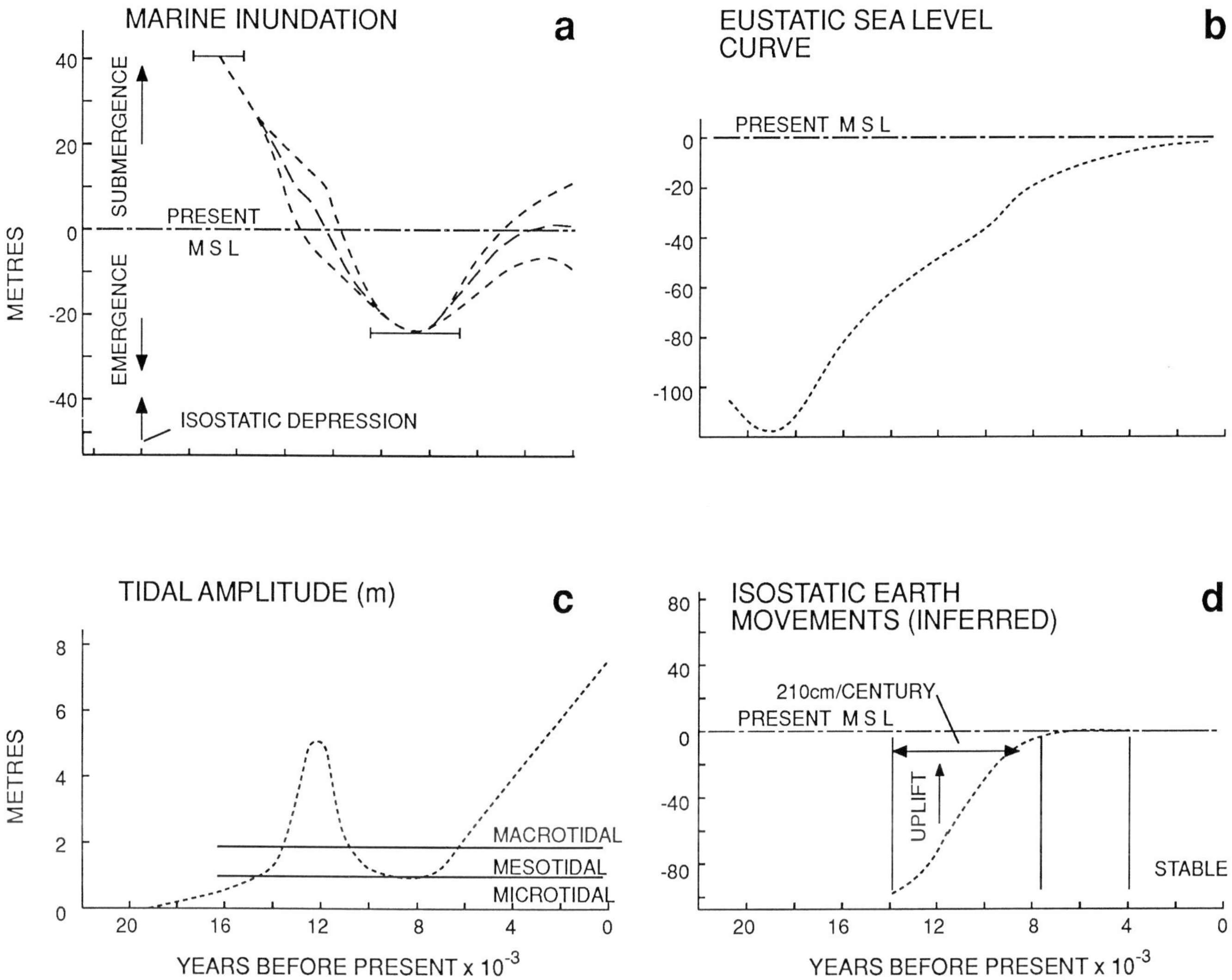

Figure 3. A schematic representation of marine inundation of Chignecto Bay over the last 20,000 years. The figure shows: **a.** the position of high water relative to present-day mean sea-level; **b.** a generalized eustatic sea level curve for eastern North America (after Curray, 1960); **c.** the amplitude of the Bay of Fundy tides (after Amos and Zaitlin, 1985); and **d.** the inferred isostatic movements derived by subtracting **a** and **b**.

a brief period of macrotidal conditions that peaked 12,000 BP as postglacial emergence caused sea level to approach that of today (Fig. 3A, C). The uplift rate, derived by summing the eustatic sea level curve of Milliman and Emery (1968; Fig. 3B) and the relative emergence curve (Fig. 3A), was approximately 2 m/century during this time. This rate is comparable with uplift rates during early deglaciation in the eastern Canadian Arctic (Stravers and Syvitski, *in press*).

Fresh water Holocene deposits found at a depth of 25 m below mean sea level at the mouth of Cumberland Basin demonstrate that glacial rebound resulted in the relative emergence of the Chignecto Bay system at least to this depth (note that the Holocene lowstand is corrected from 32 m below sea level reported by Amos and Zaitlin, 1985). This emergence is dated circa 7,000 BP (Quinlan and Beaumont, 1982; Grant, 1985; Amos and Zaitlin, 1985). During this time, Shepody Bay and Cumberland Basin were fluvial valleys and Chignecto Bay was the site of accumulation of estuarine, fine grained deposits (Amos and Zaitlin, 1985) Results of numerical simulations of the Chignecto Bay paleotides by Scott and Greenberg (1983) indicate that the tidal range 7,000 BP was between 1.6 and 2.0 m; that is, microtidal. Subsequently, mean sea level rose at a steady rate of 15 cm/century to its present level with an associated steady increase in tidal range of 30 cm/century (Harrison and Lyon, 1963; Grant, 1970; Scott and Greenberg, 1983). No evidence for isostatic earth movements are evident from 8,000 BP to present. Approximately 20 m of fine grained, "lagoonal" muds were deposited in the sub-estuaries during the Holocene transgression. These muds covered a basal forest horizon described by Chalmers (1894) and mapped by Noordijk and Pronk (1981) under much of the inner estuary.

Past Sedimentary Character

The majority of the unconsolidated sedimentary sequence of Chignecto Bay is preserved in the sublittoral, outer part of this estuary. Here, a sequence over 50 m thick has infilled a meandering, steep-sided paleochannel that runs along the median line of the system (Fig. 4). The sequence comprises five seismo-stratigraphic units (Gl to G4 and R) separated by four acoustic discontinuities (Fig. 5). These units are correlated with lithological and stratigraphic attributes from nineteen vibracores analyzed by Amos and Zaitlin (1985; Figs. 6, 7). The stratigraphic units consist of ten distinct lithofacies:

Facies M (clayey mud) — massive to indistinctly laminated

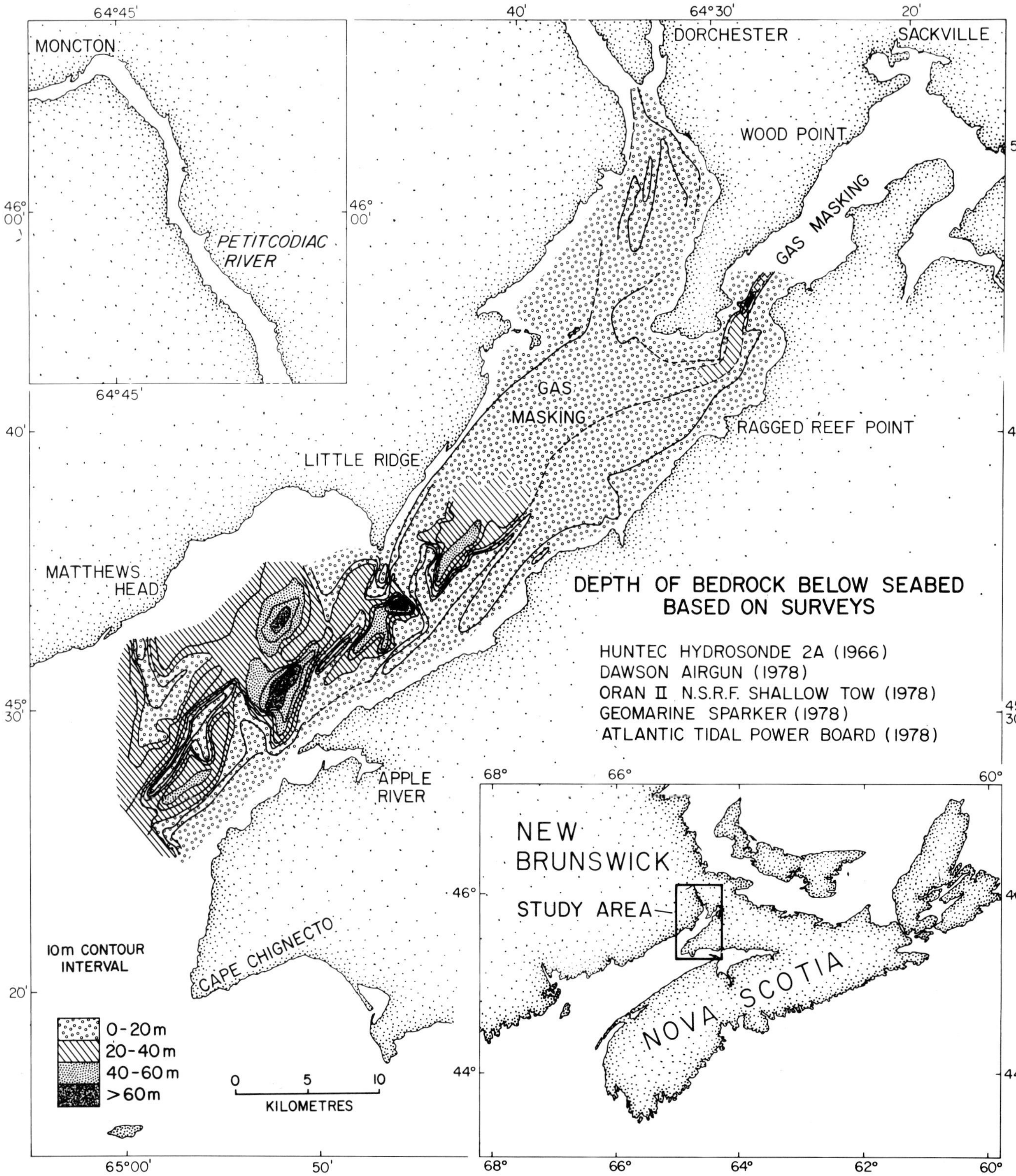

Figure 4. An isopach map of the surficial sediments of Chignecto Bay. The greatest thickness of sediments occurs in the buried channel system of the outer bay. Acoustic masking due to the presence of gas in surficial sediments prevents detailed mapping in the inner bay. Barriers across the estuary are evident off Matthews Head and Little Ridge. These barriers are composed of well-rounded sand and gravel from unit G3.

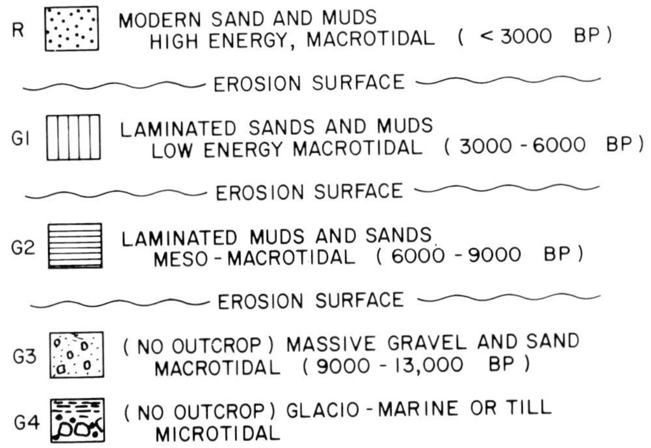

Figure 5. Generalized stratigraphy of the surficial sediments of Chignecto Bay, based on analyses of vibracore samples and seismic records. G4 is glacial in origin, G1 to G3 are postglacial estuarine deposits, and R is Recent (less than 3,000 years old).

red clay to silty/clayey mud (Fig. 8, core 3), ranging in thickness from 10 mm to 2 m. It is characterized by horizontal and vertical burrows and the presence of marine bivalve fragments.

Facies M_p (pebbly clayey mud) — massive to weakly laminated, red clayey mud with 10 to 30 per cent well-rounded pebbles or small cobbles in a poorly sorted matrix.

Facies M_b (bioturbated pebbly clayey mud) — similar to facies M_p but a high degree of bioturbation is evident (Fig. 9, core 13).

Facies L (lenticular-bedded sandy mud) — massive to well-laminated, bioturbated, red clayey mud (facies M) occupying 60 to 85 per cent of the unit, interbedded with rippled silts or fine sands (Fig. 8, core 10; Fig. 10, core 11). Sands have distinct erosional bases, well-defined tops and form pods or lenses 10 to 30 mm thick. Medium-to high-angle planar and trough crosslaminae are evident in the sands.

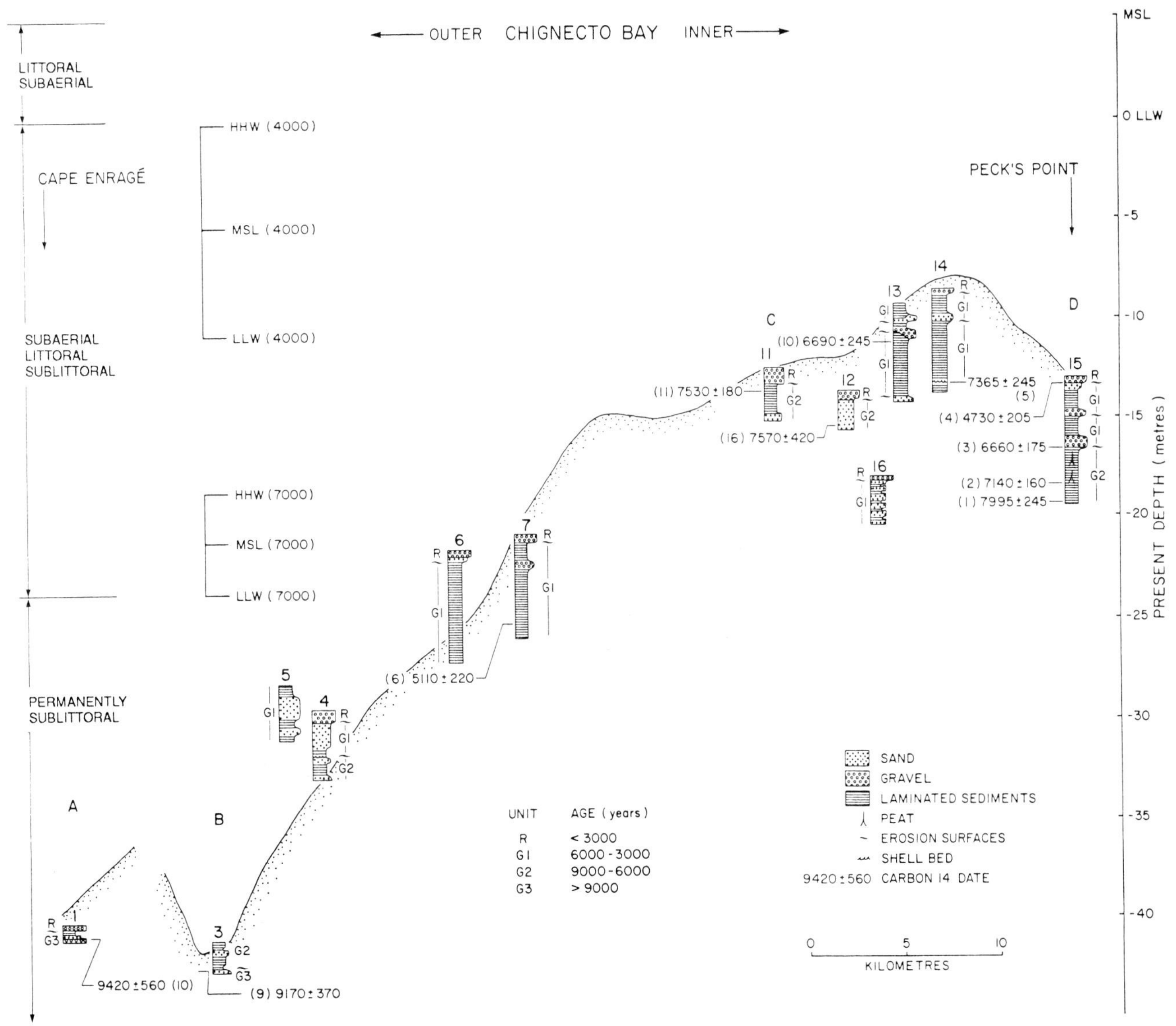

Figure 6. The generalized lithology and relative position of 12 vibracores collected along a median section of Chignecto Bay (after Amos and Zaitlin, 1985). The positions of highest high water (HHW), mean sea level (MSL) and lowest low water (LLW) 7,000 and 4,000 BP are shown for comparative purposes. A complex pattern of littoral and sublittoral settings has evolved, modified by drastic changes in tidal range.

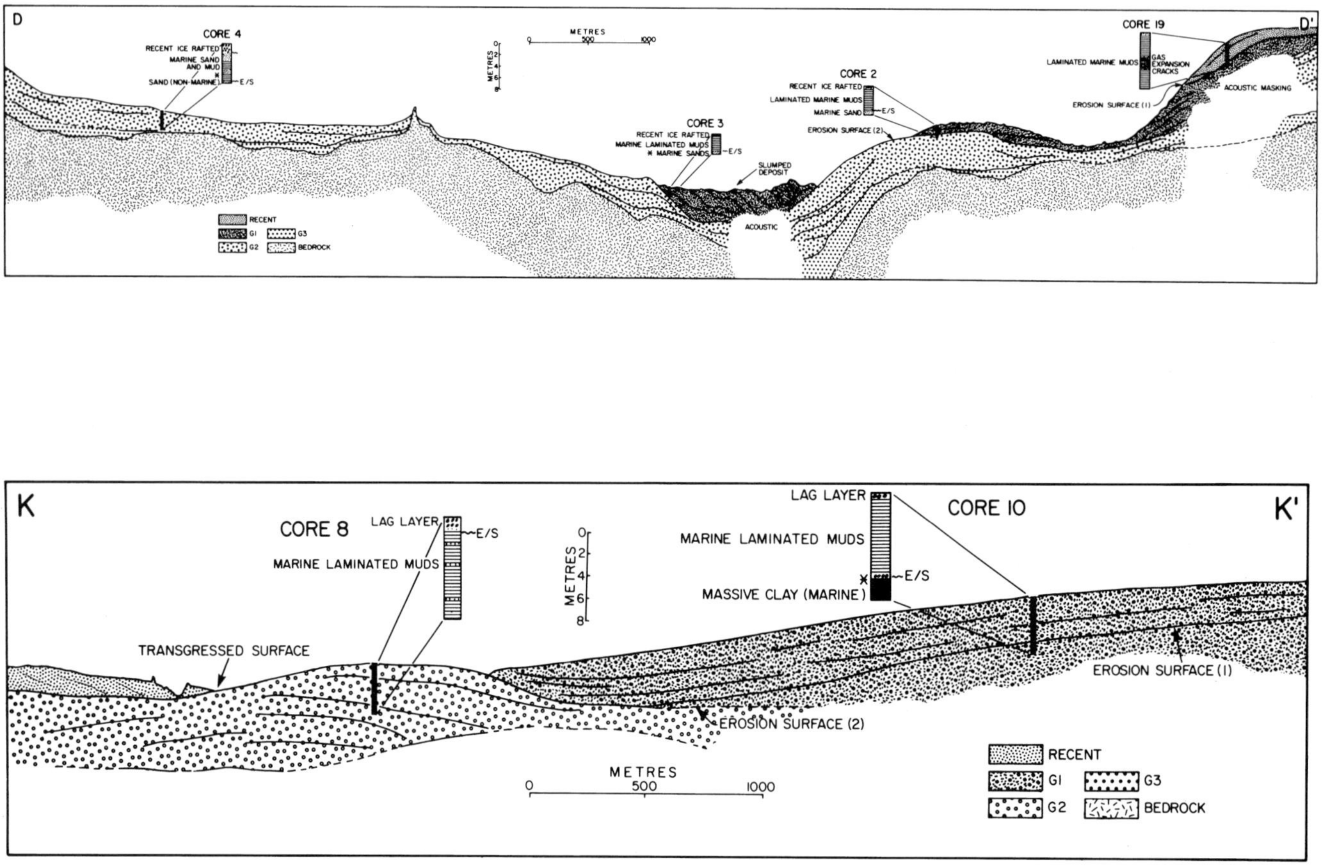

Figure 7. Two interpreted seismic sections through Chignecto Bay illustrating the seismo-stratigraphic units of Figure 5. The gross lithology measured from vibracores collected along the sections is also shown. D-D' is an east-west section near Cape Enragé (see Figure 13 for location) showing the central buried channel infilled with slumped deposits of unit G1 (3,000 to 6,000 BP). Erosion surfaces 1 and 2 are also visible in the section. Note that the present-day surface is covered by a thin veneer of gravelly sediment interpreted as a winnowed lag. Section K-K' is a northeast-southwest section off Cape Maringouin showing a good correlation between acoustic reflectors and sand beds within vibracores 8 and 10. The predominance of G1 and G2 is evident in the section as is the transgression surface that bounds them.

Facies W (wavy-bedded muddy sand) — wavy-bedded silt or medium sand, interbedded or interlaminated with facies M (Fig. 11, core 9). Sands have sharp, erosional bases and undulatory tops. Planar and trough crosslaminae occur in the sand, cut by low angle reactivation surfaces. Bioturbation is restricted to facies M.

Facies F (flaser-bedded muddy sand) — silty, very fine to medium rippled sand (60-85%) interlaminated with lenses of facies M (Fig. 11, core 5). The muds predominate in the troughs of the rippled sand and are bioturbated.

Facies S_c (coarse grained sand) — massive, medium to coarse grained sand forming laminae and beds 10 mm to 0.2 m in thickness (Fig. 11, cores 5, 9; Fig. 10, cores 7, 3). A distinct depositional cycle is evident in this unit, starting with an eroded base that is overlain by a fining-upward sequence of shelly sand.

Facies S_p (pebbly sand) — massive to poorly bedded granules or coarse sand beds up to 0.2 m thick, bearing pebble inclusions (15-25%; Fig. 8, core 3). The base of the sand is distinctly eroded and has 20 mm of relief. The unit fines upward into facies M, L or F.

Facies G (gravel) — massively bedded, well rounded pebbles of moderate sorting. A matrix of shelly sand forms 20 to 40 per cent of the facies (Fig. 10, core 3). It has an abrupt upper contact and is overlain by facies M or L.

Facies P (peat) — The base of the unit is characterized by rooted, grey-green, silty seatearth. It is overlain by laminated, grey-green to dark grey mud, containing roots in growth position (Fig. 12, core 15). This facies is associated with fresh water microfauna (A. Miller, pers. comm., 1986).

The seismo-stratigraphic sequence interpreted from acoustic data is shown in Figure 13. The base of the sequence is interpreted as being characterized by glaciomarine sediments or till deposited during the Chignecto Phase (acoustic unit G4). It is found only in the deeper parts of the bay infilling irregularities in the underlying bedrock surface (seen on original Huntec Hydronsonde 2A records of Swift *et al.*, 1973). This unit is up to 5 m thick. It is not known to outcrop in the bay.

The oldest marine deposits (acoustic unit G3) sampled were in cores 1 and 3 (Figs. 6, 8, 10). This unit is up to 30 m thick and generally rests on bedrock (Fig. 7, section D-

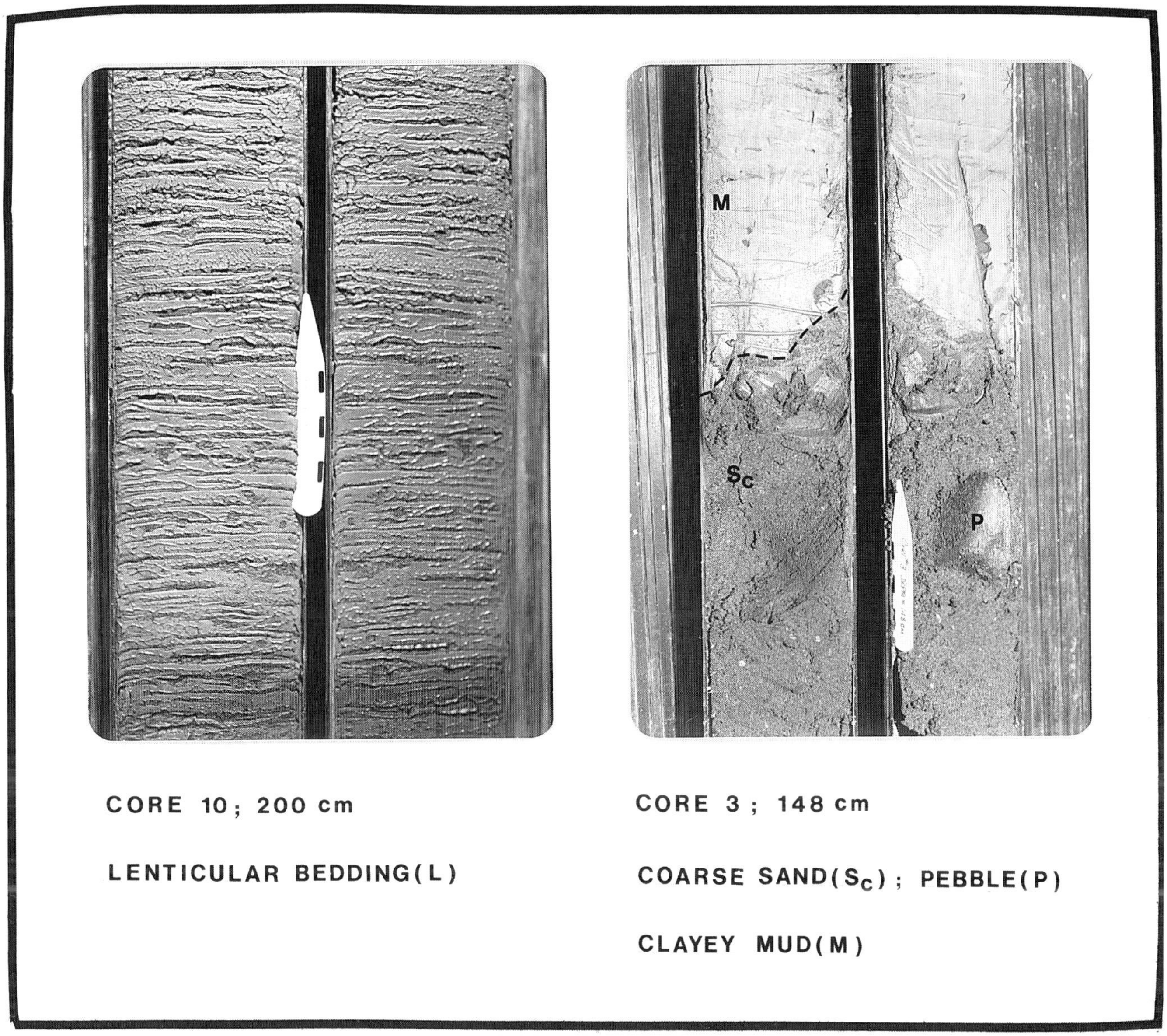

Figure 8. Core 10: lenticular bedding (L) at a depth of 2.00 m in unit G1 (see Fig. 7; section K-K'). Note the rhythmic bedding, the dominance of mud and the lack of any obvious cyclicity in laminae thickness. Core 3: the erosion surface separating units G1 (the clayey-mud facies, M) and G2 (the coarse sand facies, S_c; see Fig. 7). The irregular erosion surface is the result of G1 have slumped over G2 (section D-D'). The coarse facies is interpreted as representing a brief period of macrotidal activity approximately 9,000 BP. The scale in each plate is graduated in cm and the scale point indicates the top of the sequence.

D'). Sediments are composed of facies G, S_p and S_c, which were deposited approximately 9,000 to 13,300 BP within a sublittoral setting. A marine origin for the unit is evident by the presence of the macrofossils *Spisula* sp. and *Mytilus edulis* (Table 2). The system at that time was regressive. The lower part of the unit formed, therefore, at a depth no greater than 50 m below extant sea level. The dominance of well-rounded gravel in this unit is attributed to currents associated with a brief period of high-energy macrotidal conditions as sea level passed through sea level today (12,000 BP). The Chignecto Bay system during late Pleistocene was likely connected to the Gulf of St. Lawrence. It is proposed that the isthmus of Chignecto (that joins New Brunswick and Nova Scotia; see Fig. 2) emerged during this period of regression, thus isolating the Bay of Fundy from the Gulf of St. Lawrence and providing a suitable bay geometry for macrotidal conditions.

The upper part of unit G3 was deposited in a littoral setting in the form of a barrier at the mouth of the paleoestuary (Plate 3, core 1, C-14 dated at 9,420 BP *in* Amos and Zaitlin, 1985). The barrier developed at the paleoestuary mouth as sea level fell 20 m below that of today during late

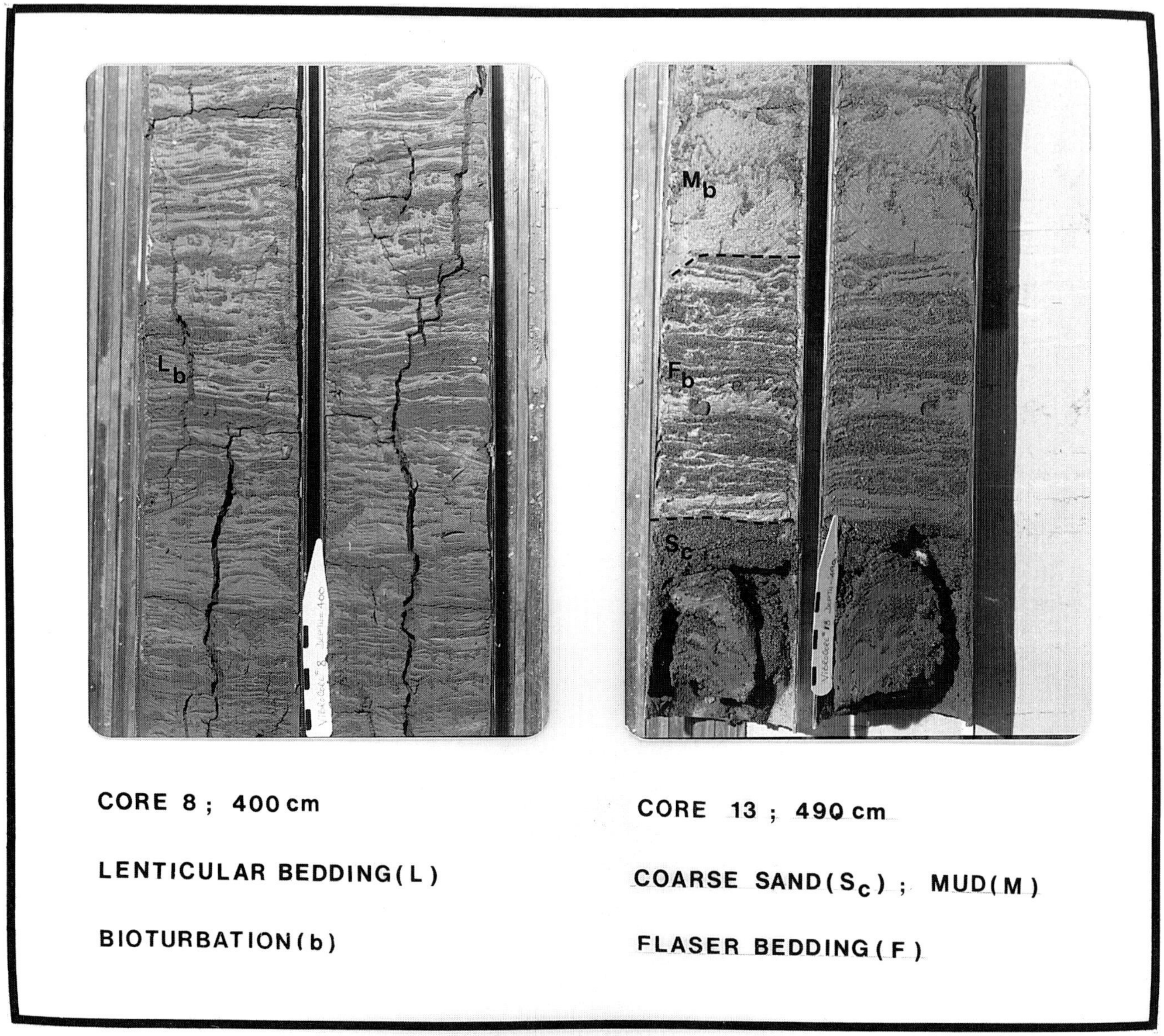

Figure 9. Core 8: highly bioturbated, lenticular bedding from the mesotidal deposits of unit G2 (Fig. 7, section K-K'). Core 13: highly bioturbated muds, flaser deposits and massive coarse sands from the macrotidal deposits of unit G1.

G3 time. We postulate that wave conditions at the time were able to create this barrier through reworking of underlying glacial debris in a fashion similar to cobble barriers across the mouth of estuaries on the Gulf of St. Lawrence (Syvitski *et al.*, 1987) and across fjords of Baffin Island (Taylor *et al.*, 1987).

Unit G2 overlies unit G3 and is separated from it by a widespread acoustic discontinuity that correlates with an eroded surface (Fig. 8, core 3). It was deposited between 9,000 and 6,000 BP. It is up to 10 m thick and underlies most of Chignecto Bay, outcropping in the eastern and extreme western parts of the bay (Fig. 13). It is thickest along the eastern margin of the central estuary. Off Cape Enragé it is highly variable in thickness, varying in a estuary-parallel pattern. It is dominated by a stiff, dry, muddy facies M (41%), L (39%), and P (7%) with only minor amounts of facies W, S_c, S_p and M_p. Massive muds predominate in the lower part of the sequence whereas fine (approximately 10 laminae per cm), lenticular laminae dominate the upper part. The change in facies is interpreted as a result of a decrease in water depth as G2 was deposited. There is an abundance of organic detritus and a distinct, well-preserved macrofauna uniquely characterized by the gastropods *Acteocina canaliculata* and *Crepidula fornicata* (Table 2). These fauna are diagnostic of shallow marine conditions subject to a fluctuating salinity. Facies L are intercalated with fresh water peats and salt marsh deposits (facies P) at the mouth of Cumberland Basin and are, therefore, littoral in origin (see

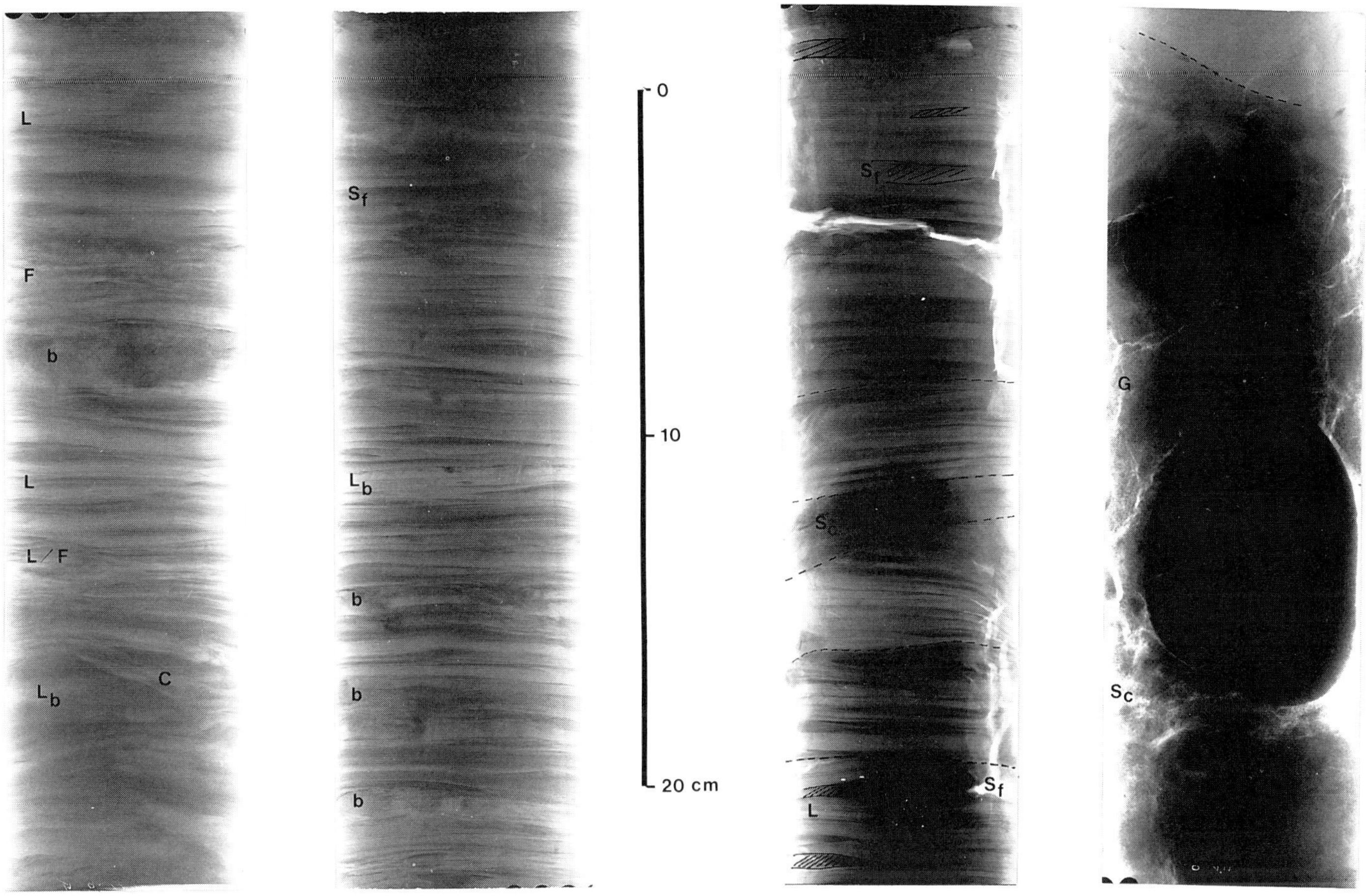

Figure 10. Four positive X-radiographs of vibracore samples. Core 11: lenticular bedding with evidence of bioturbation from unit G2. Core 7: laminated sand/fines from unit G1. Note the interbedded, tabular, ripple crosslaminated fine sand and the abundance of erosion surfaces. Core 3: well-sorted coarse sand interspersed with well-rounded pebbles and cobbles. The dashed line denotes the boundary between G3 (below) and G2 (above).

Figs. 6, 12, core 15). This interpretation is supported by the seismic character of the unit. The seismic section in Figure 7 (K - K') shows convex up, gently-dipping reflectors diagnostic of point bar lateral accretion beds associated with littoral creek migration (Nio and Yang, 1989). Dewatering structures (seen in core 11) are typical of present-day creeks in the Bay of Fundy that have developed in regions of high deposition.

The final stages of deposition of unit G2 coincided with a mid-Holocene lowstand in sea level (Fig. 3A). Microtidal to mesotidal conditions are interpreted to have prevailed based on the dominance of the fine grained facies. The salt marsh sequence is up to 2.5 m thick and yields a deposition rate of 0.1 mm/year (note that this rate does not account for the effects of sediment compaction). The overall lithofacies distribution of this unit is characterized by a transition from the dominance of facies M in the southwest to facies L in the central part of the bay, to a cyclic sequence of facies M - L - P in the northeast (Table 3). This sequence of facies is interpreted to represent the transition from a sublittoral zone at the mouth of Chignecto Bay, to a lower tidal flat in central Chignecto Bay, to an upper tidal flat in the region of Cape Maringouin. Shepody Bay and Cumberland Basin were subaerial at this time.

A well defined unit of mesotidal to low energy macrotidal muddy, lenticular, wavy and flaser bedded sands and muds (G1) overlies unit G2 and is separated from it by a widespread erosion surface (Fig. 8, core 10; Fig. 9, cores 8, 13). This unit is up to 12 m thick, and is thickest along the median line of the central estuary. It is characterized by facies L (72%) and F (16%) with minor anounts of facies M, W, S_c and G (Fig. 10, core 13). There is a significant

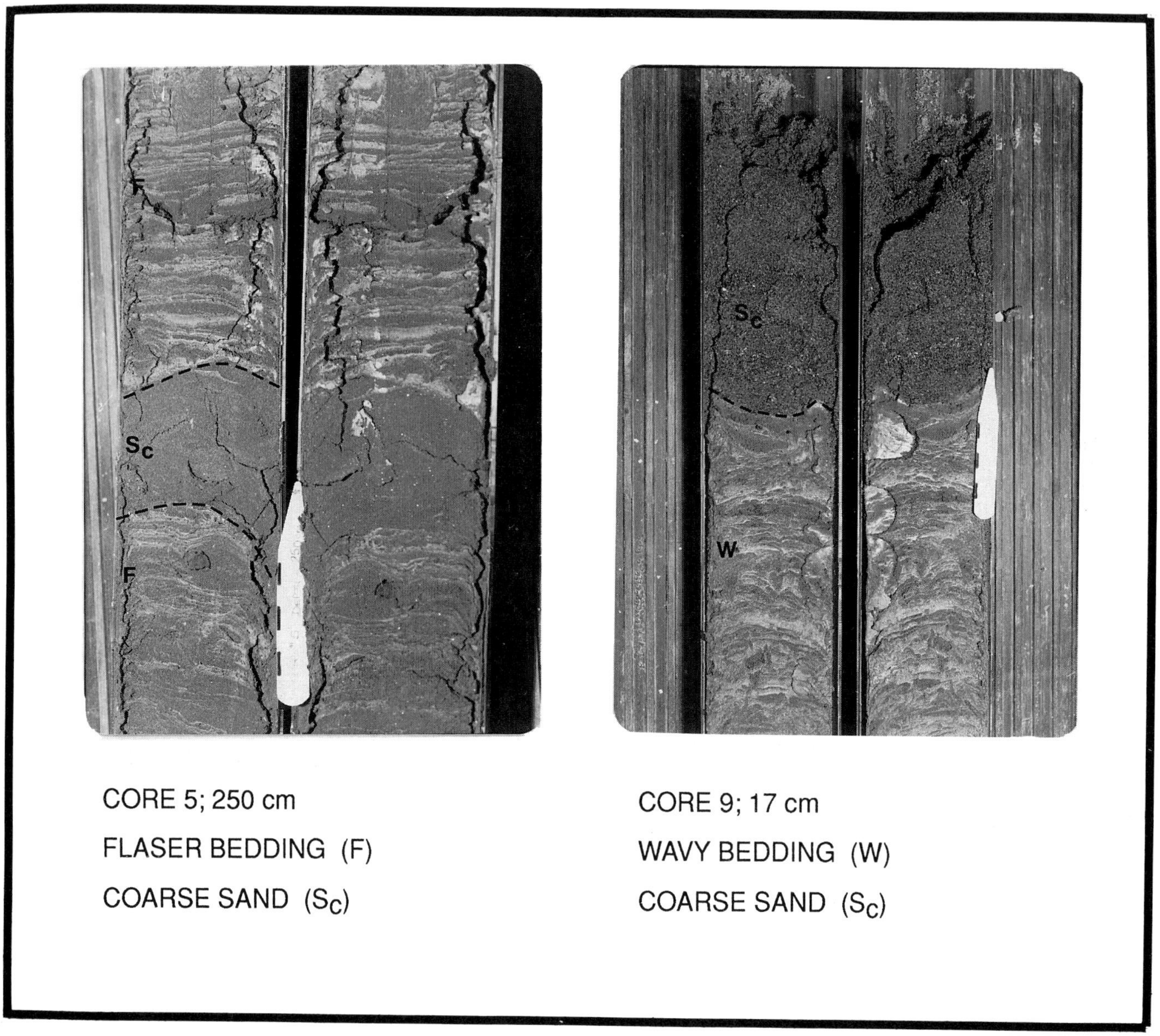

Figure 11. Core 5: flaser bedding and coarse sand from unit G1, outer Chignecto Bay. The sand shows evidence of planar ripple crosslamination; flaser facies are highly bioturbated. Core 9: coarse sand from unit R overlying abruptly wavy bedding from unit G1. The sand shows evidence of tabular crosslamination; the wavy bedding is highly bioturbated.

increase in the abundance of facies F and L over underlying units, an increase in the abundance of ripple crosslaminated sand and gravel and a decrease in bioturbation. The unit is characterized by a facies transition of L - F - M. Lamina thickness is greater (2-3 laminae per cm) than in G2, indicating higher transport and deposition rates (the mean deposition rate of the unit was 3.3 mm/a). The macrofaunal assemblage is unique though poorly developed. It is characterized by the pelycepods *Ensis directus* and *Thyasira flexuosa* and the gastropods *Oenopota harpularia, Lunatia pallida* and *Hydrobia totteni* (Table 2). This assemblage is diagnostic of a salt marsh environment with strong tidal currents. The doublets of sand and mud, common in this unit (Fig. 8, core 10; Fig. 10, core 7) are diagnostic of point bar lateral accretion under a tidally varying current (Nio and Yang, 1989). The convex-up attitude of the acoustic reflectors and the presence of low angle cut-and-fill structures indicate that the majority of this unit was reworked by creek migration. Thin beds of fine grained sand show tabular crosslamination diagnostic of straight-crested ripple migration. No herringbone crosslamination was seen; all cross-sets are oriented in the same direction.

The sediments of this unit are fully saturated and very soft, suggesting rapid deposition. The unit shows evidence of sediment deformation and massive slumping into the central channel (Fig. 7, D-D'). The slump illustrated in Figure 13 is 6 km long and up to 1 km wide and flowed down a slope of 1:100. The unit is dated to have been deposited between 3,000 and 6,000 BP and is, therefore, a transgressive deposit (see Fig. 3A). It occurs principally in inner Chignecto

Table 2. Macrofossil assemblage — Chignecto Bay.

MACROFOSSIL		0 - 3000	3000 - 6000	6000 - 9000	> 9000
Hiatella arctica	p	X			
Nucula proxima	p	X	X		
Yoldia limatula	p	X	X		
Yoldia myalis	p	X	X		
Ensis directus	p		X		
Thyasira flexuosa	p		X		
Oenopata harpularia	g		X		
Lunatia pallida	g		X		
Hydrobia totteni	g		X		
Mya arenaria	p		X	X	
Retusa obtusa	g		X	X	
Crepidula fornicata	g		X	X	
Nassarius trivittatus	g	X	X	X	
Mitrella lunata	p	X	X	X	
Cerastoderma pinnulatum	p	X		X	
Mulinia lateralis	p	X		X	
Macoma balthica	p	X		X	
Acteocina canaliculata	g			X	
Crepidula fornicata	g			X	
Spisula polynyna	p				X
Mytilus edulis	p	X	X	X	X

Bay, and along the axes of Shepody Bay and Cumberland Basin; that is, in the inner part of the paleostuary. The lower part of the deposit was formed in a littoral, low energy (4-6 m tidal range) macrotidal setting, whereas the upper part was deposited in a sublittoral, medium energy (6-8 m tidal range) setting in the central part of the paleoestuary. The unit outcrops in the eastern-central part of Chignecto Bay and along the central parts of Shepody Bay and Cumberland Basin (Fig. 13).

Modern, high energy (8-12 m tidal range), macrotidal sand and gravel deposits are being created under present-day tidal conditions (R). These sands are dominated by the

Table 3. The major facies cycles recognized in each of the four sampled seismo-stratigraphic units. G3 and R are dominated by sand and gravel facies (diagnostic of intense reworking) whereas units G1 and G2 are dominated by muddy facies (diagnostic of a dominantly depositional environment).

Major Cycles	
G3	G
G2	M – L – P
G1	L – F – M
R	Sp – Mpb – Sc

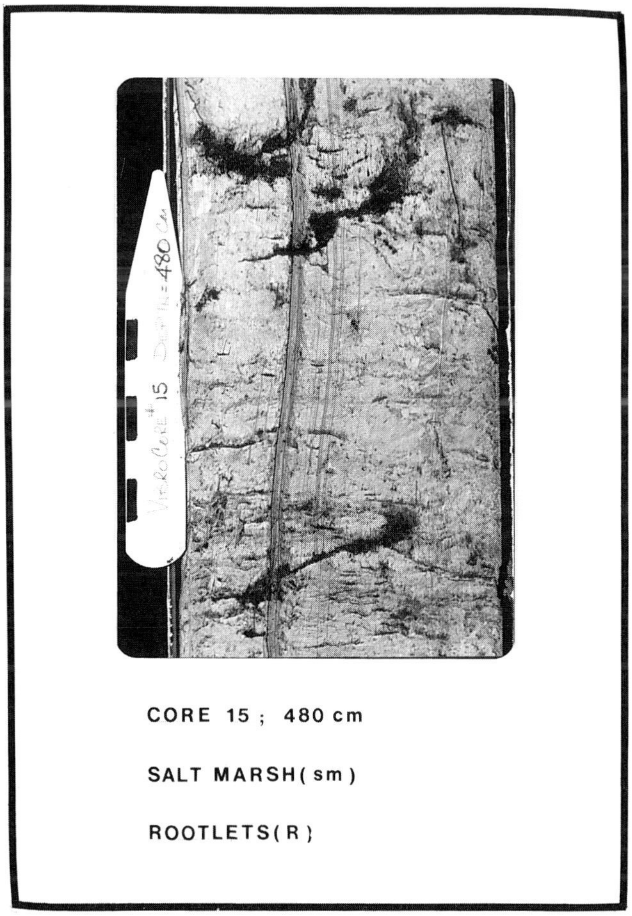

Figure 12. Core 15: salt marsh rootlets in growth position from a massive clay facies at a depth of 4.80 m below the seabed at the mouth of Cumberland Basin. The sample comes from unit G2 and formed approximately 7,000 BP.

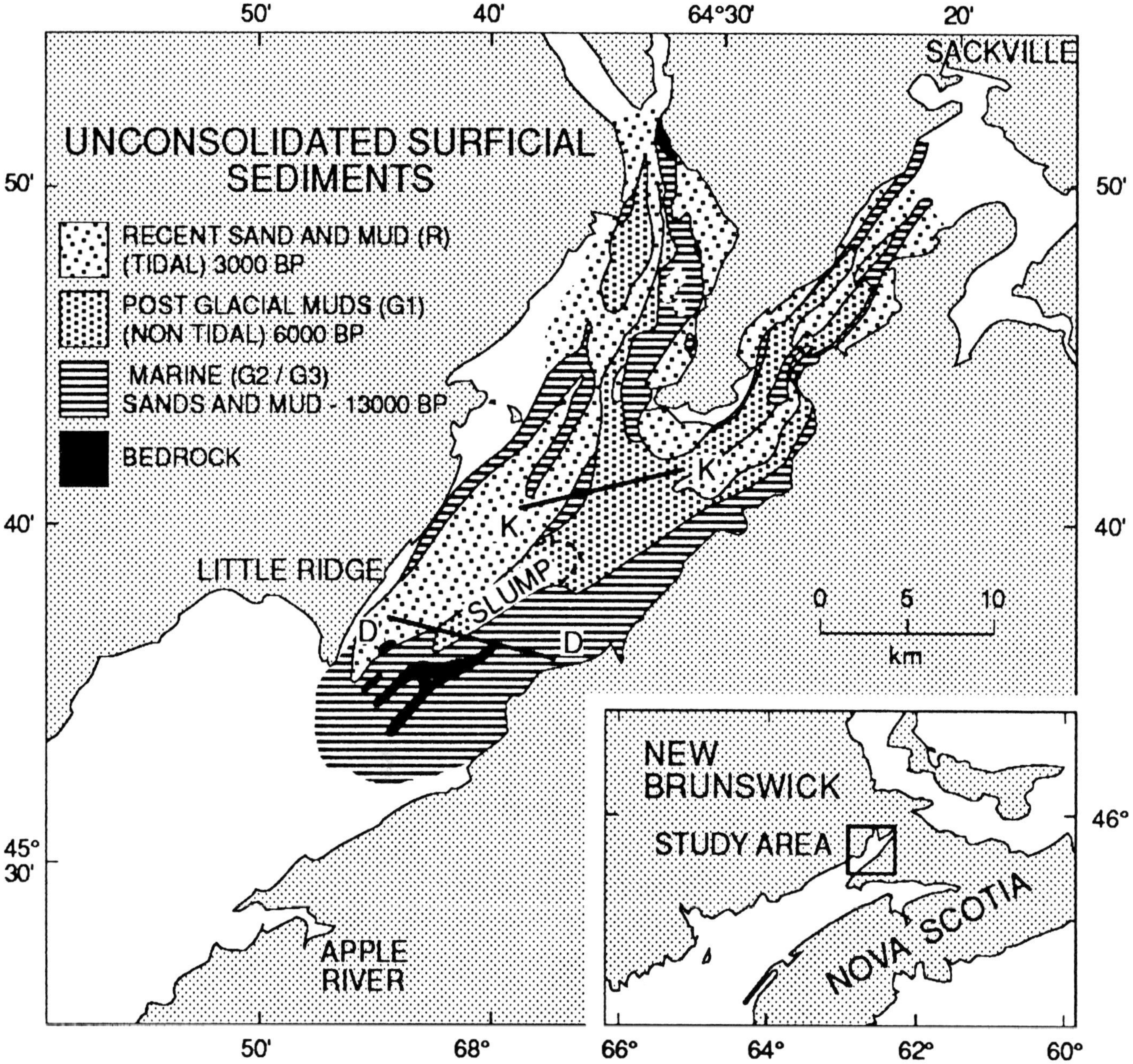

Figure 13. A seismo-stratigraphic map of the unconsolidated sedimentary sequence in Chignecto Bay illustrating the major acoustic units. Note the zone of slumped sediments, illustrated in section D-D' (Fig. 7). This slump is 6 km in width and approximately 1 km long. The origin of the slump is unknown.

facies S_c S_p and M_{pb} and are uniquely characterized by the pelycepod *Hiatella arctica.* Such conditions we consider to have prevailed for approximately 3,000 years. Much of this deposit is reworked from the underlying sequence and is separated from it by a well developed, transgressed, eroded surface (Fig. 7; K-K', core 8). Sublittoral sandy equivalents form an asymmetrical wedge that thins headward and to the west. It is thickest (2 m) along the extreme eastern portion of the central estuary. No sublittoral, fine grained deposits are known to be presently forming in the Chignecto Bay system. The mud facies of this unit were deposited in the present-day littoral parts of the system in a manner reviewed by Dalrymple *et al.*, (*this volume*).

Modern Hydrodynamic Processes

Tides

The tidal regime of Chignecto Bay is dominated by the lunar semidiurnal component because of its near-resonance with the Gulf of Maine/Bay of Fundy system (Garrett, 1972). The amplitudes of the diurnal K_1 and semidiurnal M_2 components are 0.15 and 4 m, respectively, illustrating the small diurnal contribution to tidal range. Similarly, the amplitude ratios of the N_2/M_2 and S_2/M_2 are about 0.22 and 0.16, respectively. The amplitude of the M_2 tide in Chignecto Bay increases from 4.2 m at the mouth to 4.4 m off Cape Enragé to 4.7 m at the mouths of Cumberland Basin and

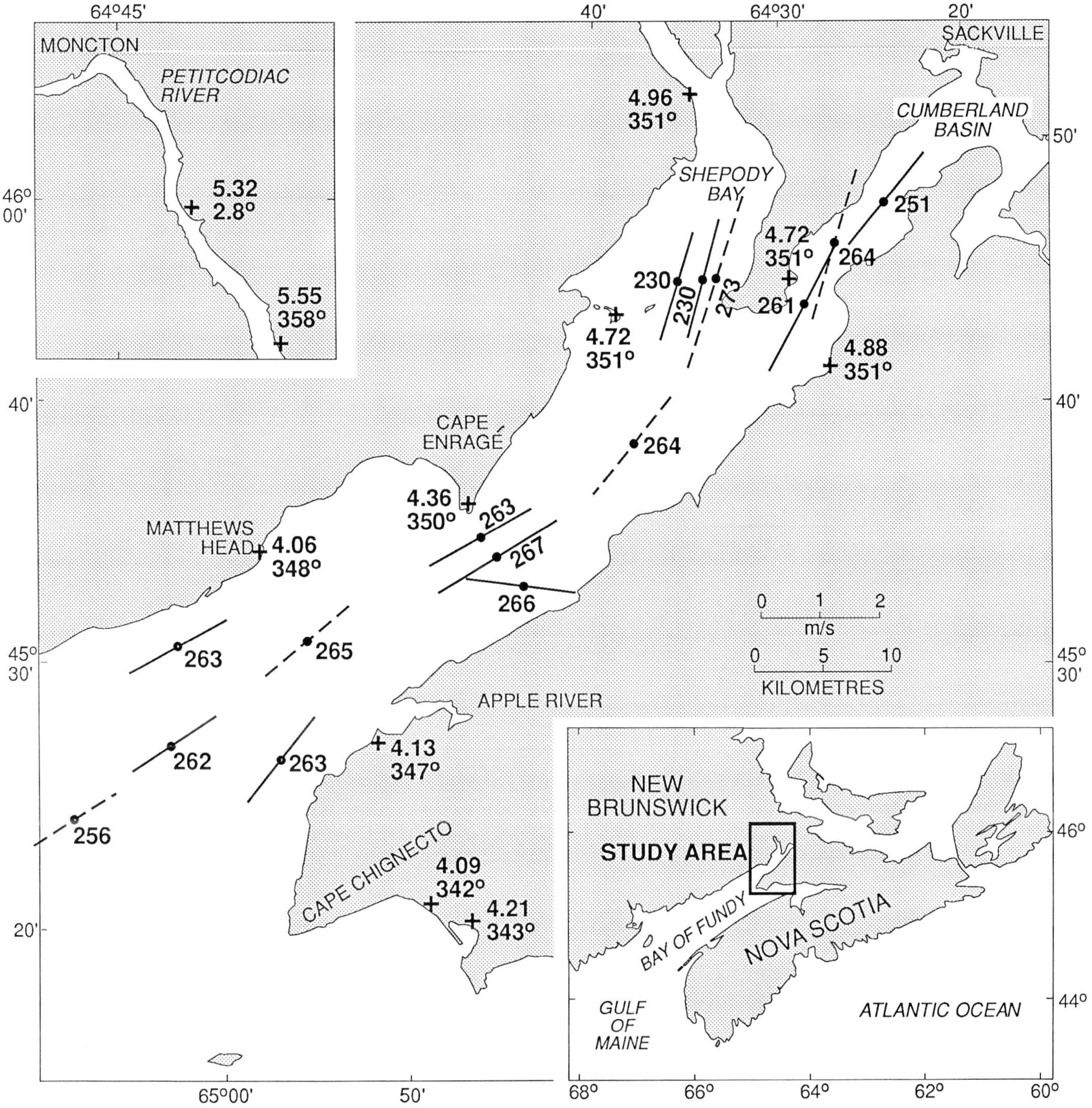

Figure 14. The major and minor axes of the observed tidal ellipses in Chignecto Bay. The data collected during this study are shown by solid lines. Data collected by Godin (1968) are illustrated as dashed lines. The numbers near the axes refer to station number. Peak currents vary between 1 and 2 m/s and generally parallel the estuary margins.

Shepody Bay (Godin, 1980; deMargerie and Dewolfe, 1987; Greenberg, pers. comm., 1990). The respective Greenwich phase lags systematically from 347° to 350° and 351°. This corresponds to a variation in the times of high and low water of 8.3 minutes. The neap-spring oscillation is 0.38 times the M_2 amplitude. The diurnal inequality (M_4/M_2) in tidal amplitude is about 0.02 at the mouth of Chignecto Bay, 0.03 at the mouths of Shepody Bay and Cumberland Basin, and 0.09 near the head of the system. This is very small compared to values reported for Dutch mesotidal estuaries (Nio and Yang, *this volume*).

Currents

Tidal Currents

The currents of the semidiurnal M_2 constituent dominate flow patterns in the Chignecto Bay system. The major and minor axes of the M_2 tidal ellipses are shown in Fig. 14 (solid lines). The figure also shows the results of the 1965 survey carried out by Godin (1968; stations 66-71; dashed lines). The depth-averaged semi-major axes of these ellipses increase from between 0.7 to 0.8 m/s near the mouth of the system to between 1.0 and 1.5 m/s at the mouths of Cumberland Basin and Shepody Bay and are greater than 1.5 m/s in the tidal channels at the estuary head. Currents are generally parallel to the channel margins in which they flow. The tidal ellipses are virtually rectilinear, with the semi-minor axes usually an order of magnitude smaller than the semi-major axes. Rotation of the current ellipses is mainly anticlockwise. The Greenwich phase lag of the semi-major axes varies between 260° and 270° and leads the tidal elevation by approximately 90° (Fig. 15C). Near-surface currents at the mouths of Shepody Bay and Cumberland Basin (stations 70-71) were considerably larger than near-bed ones (stations 291, 303, 304) and were consistent with short-term observations made at nearby anchor stations.

Examination of measurements of the current flow at the mouth of Chignecto Bay, made by Tee (1982; stations 283, 284, 306), showed that the semi-major axes decreased from surface to bottom. He found a typical ratio of near-bed to mid-depth currents to be between 0.65 and 0.75. Also, he found the phase lag to decrease toward the seabed between 5° and 10° with a change in rotation of the semi-minor axes from clockwise (negative) near the surface to anticlockwise (positive) near the seabed. By contrast, the orientation of the semi-major axes changed only slightly with depth.

Residual Currents

The residual current is the time-average mean of measures of speed and direction over a period long enough to eliminate tidal and storm-driven signals. Figure 16 shows the residual current so calculated from measurements in Chignecto Bay. Notice that the greater part of these residual currents, in contrast to the tidal currents, are oriented across the bay, not parallel to its margins. The maximum and minimum residual currents both exhibit this pattern. Minima approach zero at mid-depths near the mouth of the system (Fig. 14, stations 283, 284, 306). Maxima approach a velocity of 0.17 m/s and occur near Cape Enragé. Here, the residual current direction is to the southeast. This trend results from the presence of two, tidally-induced residual gyres; a clockwise gyre southwest of Cape Enragé and an anticlockwise gyre to the northeast (Greenberg, 1983; Tee and Amos, *in press*).

Residual current directions in Cumberland Basin (Fig. 9, stations 70, 290) are to the east and northeast, while nearer the mouth (station 291) they flow westward. The along-channel component of the residual current is upstream near the surface and downstream near the bed. This pattern does not conform with the classic Pritchard (1956) estuarine circulation model. Here, the residual currents are unlikely to be a result of water density variations. Rather they appear to result from effects of tidal propagation through complex channel geometry in shallow water.

Residual currents at the entrance to Shepody Bay are to the north and northwest. Observations are consistent with

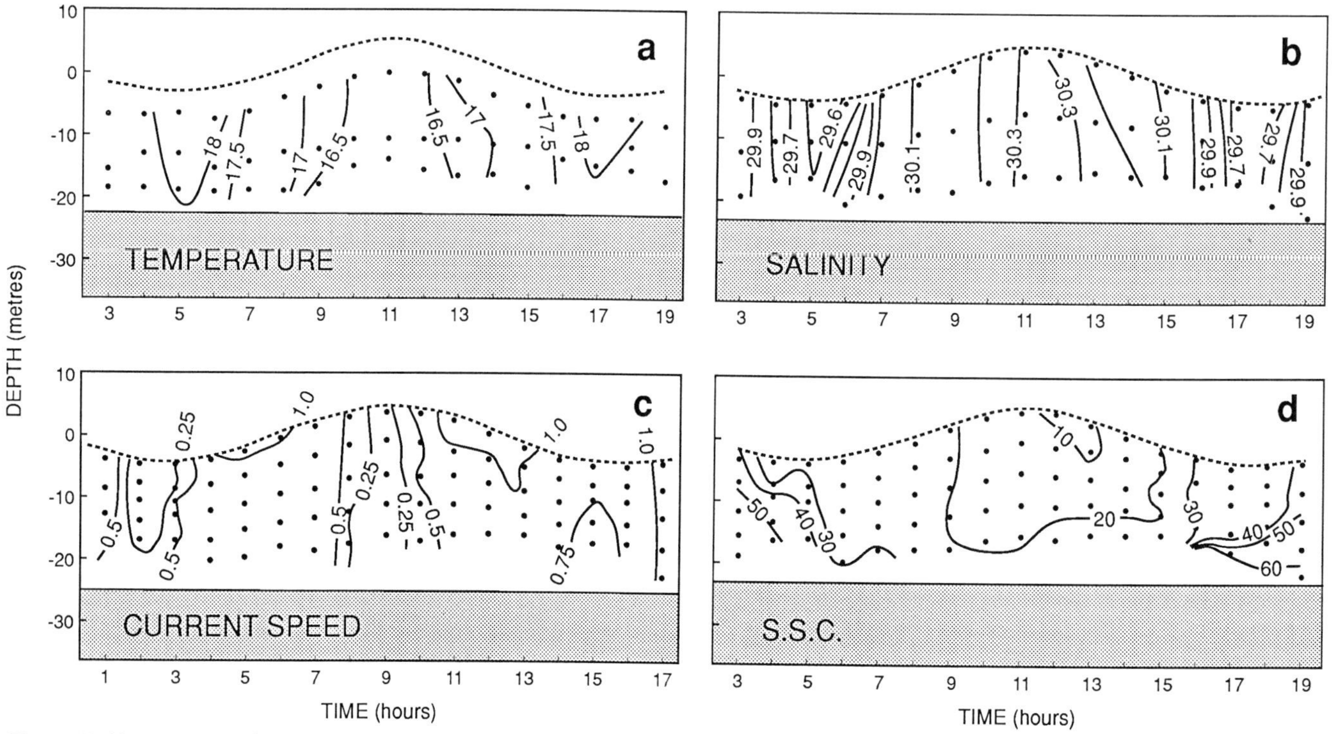

Figure 15. Measurements from station 9 (mouth of Shepody Bay, Fig. 2C) collected 8 August, 1978; **a.** temperature; **b.** salinity; **c.** current speed; and **d.** suspended sediment concentration. Observations throughout the water column were made hourly from 1 hour before low water to 1 hour after the successive low water (approximately 15 hours). Note in **d** the near-bed increase in SSC during slack tide, and the absence of stratification in **a** and **b**.

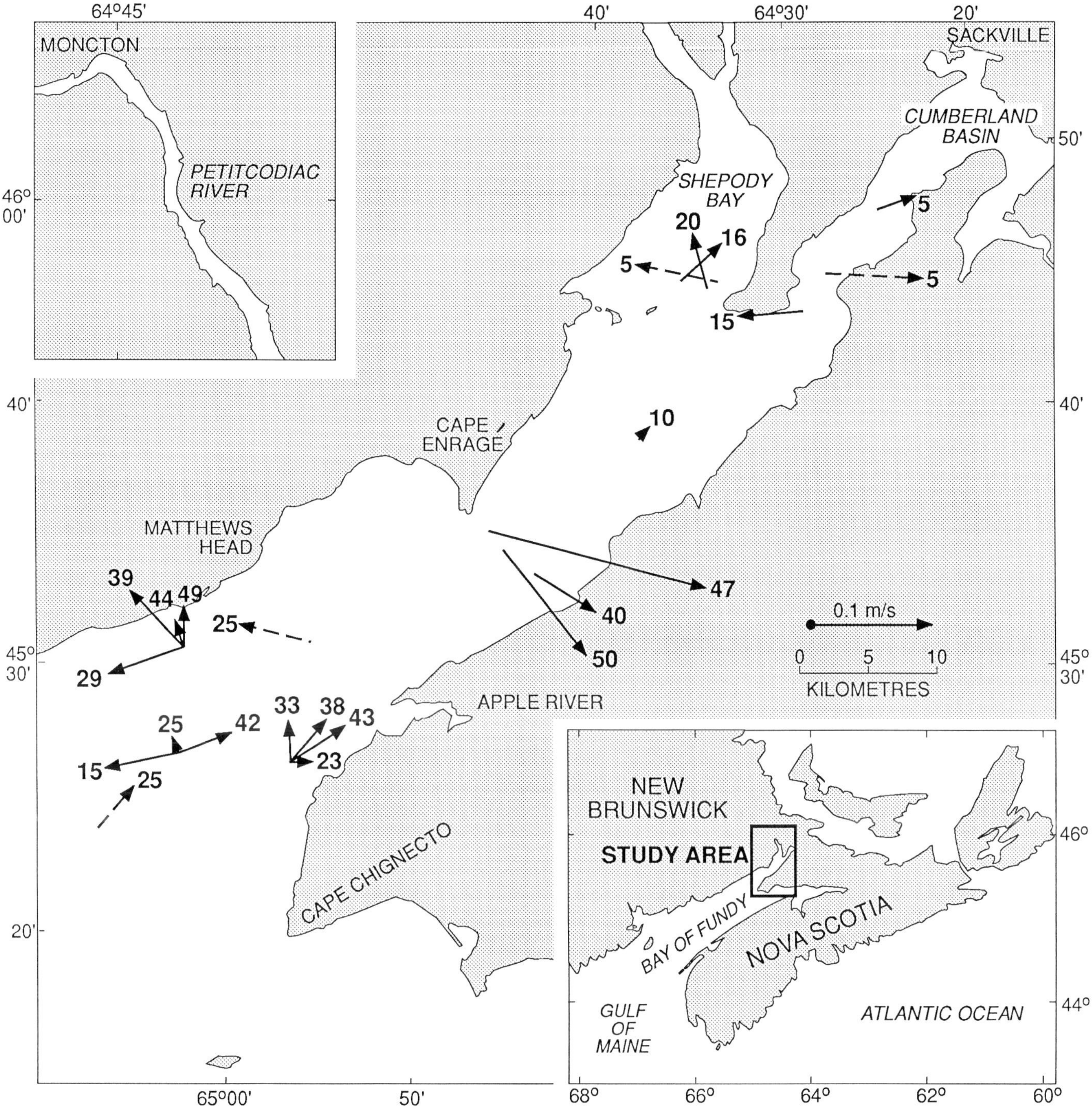

Figure 16. The measured residual currents in Chignecto Bay. Data collected during this study are shown as solid arrows; results of Godin (1968) are shown as dashed arrows. Numbers at the arrow heads denote the depth (in m) of the observation. The magnitude of residual currents rarely exceeds 0.15 m/s. Note the peculiar directions of residual flows; the result of gyres in the system.

the tidally-induced residual currents modelled by Greenberg (1983), who showed the existence of an anticlockwise gyre at this site. Measured residual currents at the mouth of Chignecto Bay are oriented in the along-channel direction. Surface measurements indicate a net outflow from the system whereas near-bed measurements indicate an inflow. Outflow is greatest in the western portion; inflow is greatest to the east. This pattern is likely a result of horizontal gradients in density.

Waves

The entire Chignecto Bay system is influenced by winds blowing up the Bay of Fundy from the southwest. These winds are responsible for the generation of locally-generated

Table 4: The measured wave climate recorded at Cape Maringouin and Cape Enragé between June, 1978 and February, 1979 (from Marine Environmental Data Service, 1979(a) and 1979(b)). Wind speed recorded at Cape Enragé is also shown (courtesy Atmospheric Environment Service, Bedford).

Month	Cape Maringouin Height(m)	Cape Maringouin Period(s)	Cape Enragé Height(m)	Cape Enragé Period(s)	Cape Enragé Wind Speed (km/h)
May, 1978	—	—	—	—	23.5
June	0.33	4.94	0.35	4.99	17.3
July	0.33	5.05	0.26	4.54	13.3
August	0.32	5.64	0.33	4.31	17.8
September	0.42	4.42	0.56	4.47	19.2
October	0.41	4.63	0.62	4.56	18.7
November	0.38	4.32	0.65	4.81	28.8
December	0.40	3.94	0.85	5.79	—
January, 1979	—	—	0.82	5.54	—
February	—	—	0.73	4.65	—

waves over a fetch of approximately 140 km. Waves propagate from the Bay of Fundy into Chignecto Bay. They are greatest in Chignecto Bay but are nevertheless significant at the heads of Shepody Bay and Cumberland Basin because of the alignment of the system with the dominant southwesterly winds. Table 4 synthesizes the wave measurements made in this study based on data presented by Marine Environmental Data Services (1979a, b). These data are shown graphically in Figure 17. Wave conditions are most severe at the mouth of the system and decrease landward. The peak significant wave height and period for the outer and inner stations are 3 m and 10 seconds and 2 m and 8 seconds, respectively. Waves 2 m in height have been observed at the heads of both Shepody Bay and Cumberland Basin breaking over mud flats and resulting in the production of salt marsh cliffs (Desplanque, pers. comm., 1989). Wave conditions are highly variable with time (Fig. 18). Most severe conditions occur during the passage of winter cyclones (Lewis and Moran, 1984) that occur principally during the months of November and January and last up to 2 days. In Figure 18, the measured significant wave height for the month of November, 1979 is illustrated together with the critical heights for sediment resuspension in the outer and inner bay. There were 9 resuspension events during that mouth in the outer bay and 10 in the inner bay. These events resulted in the release of fine grained sediments from the seabed.

Temperature and Salinity

Water temperature in Chignecto Bay varies by approximately 18°C throughout the year (Fig. 19). The greatest var-

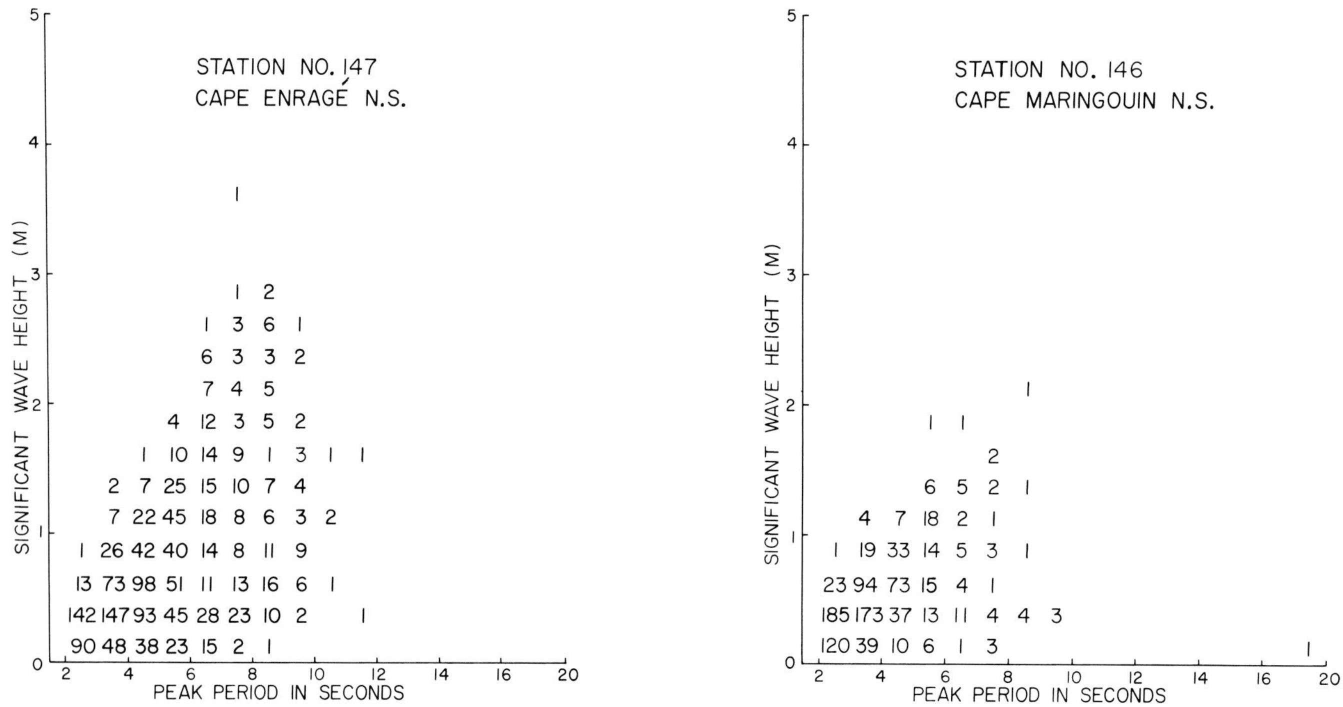

Figure 17. A scattergram of significant wave height and peak periods measured at the two waverider buoy sites (Fig. 2) during 1978. Note the significant attenuation in wave height from Cape Enragé to Cape Maringouin with no change in period.

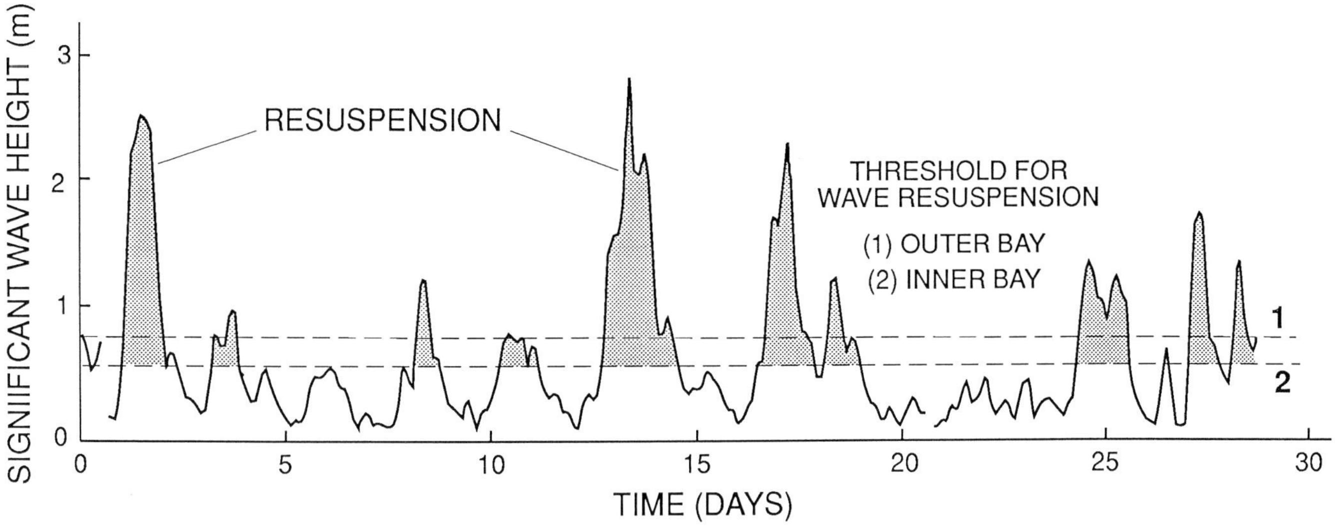

Figure 18. A time-series plot of the significant wave height for November, 1978 recorded at Cape Enragé with the observed threshold for sediment suspension superimposed. Note the irregular influences of waves and the approximate 1-day duration of most storms. This may explain the lack of cyclicity in the tidal rhythmites of sediments illustrated in Figure 8.

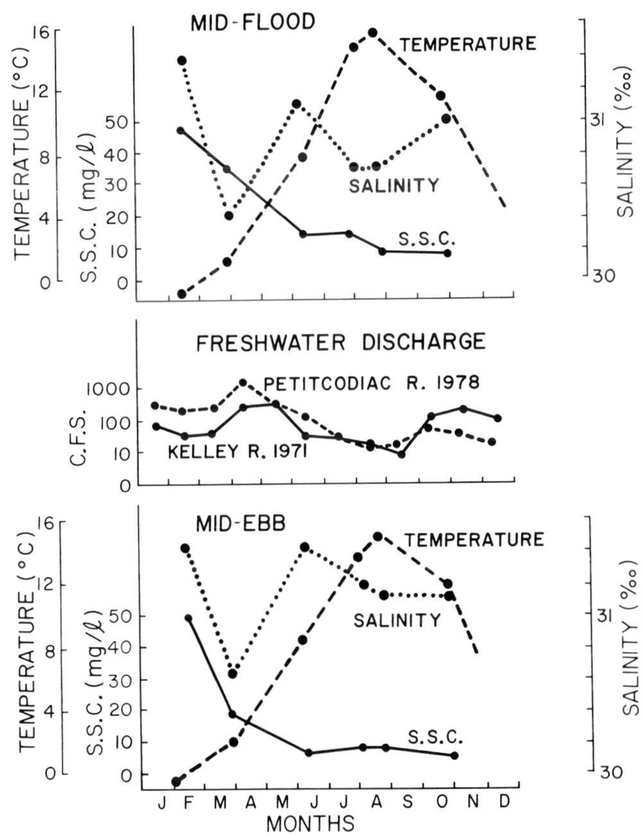

Figure 19. A synthesis of seasonal mean values of water temperature, salinity and suspended sediment concentration during mid-flood and mid-ebb conditions in Chignecto Bay. Seasonal inputs of fresh water from the two largest rivers entering Chignecto Bay (the Petitcodiac and Kelley rivers; see Fig. 1 for location) are also shown. Note that the highest values in SSC occur during late winter/early spring and decrease steadily throughout the year. Lowest salinities correspond to maximum river discharges.

iation occurs at the head of the system where temperature drops to −1.5°C during January/February and reaches 17°C during mid-August. This is because of heat transfer between the water column and expansive littoral zone. In outer Chignecto Bay, the temperature varies from −1.3°C to 14.5°C. Here, the variation over a tidal cycle is approximately 1°C; farther landward the variation is up to 2°C. During winter, there is a temperature gradient dipping landward that produces the highest water temperatures around the time of high water. During the summer, the converse is true; temperature decreases seaward and highest temperatures occur around low water due to solar heating of the tidal flats (Fig. 15A).

The vertical variation in temperature is small as a result of strong tidal mixing (Garrett *et al.*, 1978). In the inner bay, waters are well mixed except during stillstands of the tide, when surface solar heating can take place. In the outer bay, there is a systematic decrease in temperature from the surface to the seabed of approximately 1°C. This gradient is apparent throughout the entire tidal cycle.

Sea water density variations in Chignecto Bay are controlled largely by variations in salinity. Consequently, the effects of fresh water runoff have a strong impact on stratification, ice formation and ice melting (Gordon and Desplanque, 1983). The source of fresh water is in the inner bay. Consequently, the salinity gradient shows a steady increase down the estuary. This increase persists throughout the year. The lowest salinities occur during low water. Typical midsummer salinities are 31 ppt at the mouth of Chignecto Bay and 29 to 30 ppt at the mouths of Shepody Bay and Cumberland Basin (Fig. 15b; Holloway, 1981). Typical springtime salinities are 26 to 30 ppt. Salinity gradients are higher in the upper bay than in the lower bay. Salinity fluctuations over a tidal cycle during midsummer are 0.6 to 0.4

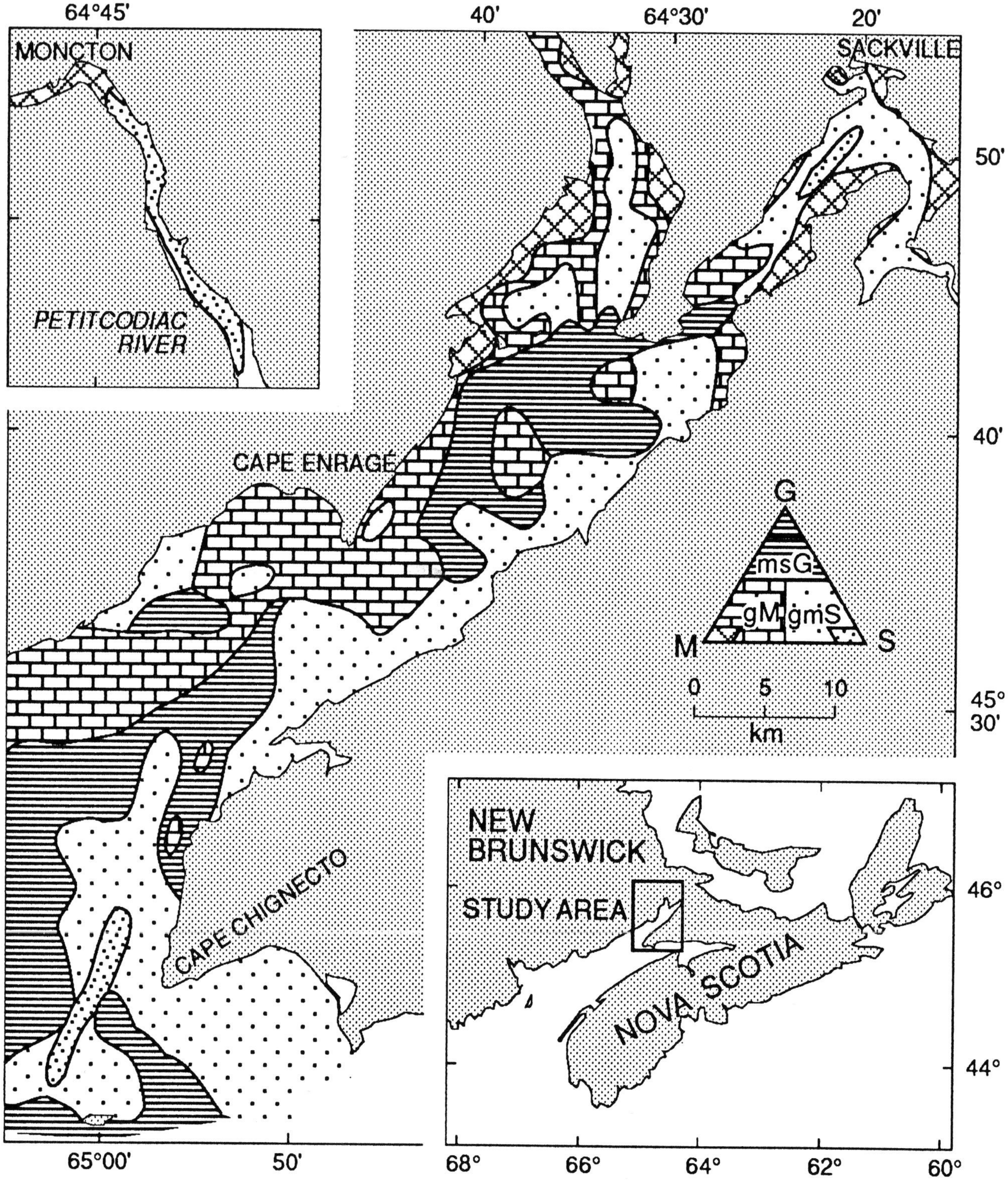

Figure 20. A textural classification of the bottom sediments in Chignecto Bay (M, mud; S, sand; G, gravel). The correspondence between this map and the surficial geological map (Fig. 13) is poor. This may be due to the dominating influences of ice-rafting of gravel over the hydrodynamic sorting processes.

ppt. A weak salinity stratification occurs throughout the system during the summer; surface salinity being 0.3 ppt less than that of bottom waters. At the mouth of Chignecto Bay, stratification is less near the coast than near its centre (Tee, 1982).

A large volume of fresh water flows into Chignecto Bay during early spring, significantly reducing salinity (Fig. 19). A curious mid-depth minima (by 2 ppt) also results (Fig. 22 *in* Amos and Asprey, 1981). This causes an instability of the water column of the inner bay (density variation approximately 1.5×10^{-9} g/m^3). The density variation due to changes in suspended sediment concentration with depth is small (2×10^{-10} g/m^3). The water column of the outer bay is stable at all times, with salinity increasing with depth.

Present Sedimentary Character

The Distribution of Sediments

Bottom Sediments

The distribution of bottom sediment in the Chignecto Bay system is shown in Figure 20. A complex distribution of varying sediment types is evident, which bears a weak relationship to the underlying surficial geology shown in Figure 13. Gravel and muddy, sandy gravel dominate the outer and axial parts of Chignecto Bay. Gravelly, muddy sands prevail along the eastern margin of Chignecto Bay and the axial parts of Shepody Bay and Cumberland Basin. Gravelly muds and sandy muds are found along the western margin of Chignecto Bay, throughout Shepody Bay and in isolated patches elsewhere. Muds predominate on the littoral margins of Shepody Bay and Cumberland Basin and are least abundant in the sublittoral regions. Clean sand occurs only in the Petitcodiac River region of Shepody Bay, along the eastern margin of Chignecto Bay and off Cape Chignecto. In general, the bottom sediments are poorly sorted and multimodal, and commonly contain a gravel-sized component. Figure 21A, B shows the percentages of sand and gravel, respectively, in bottom sediments of the Chignecto Bay system. Note the lateral asymmetry in sediment distribution: sand predominates in the eastern margin of the system and conversely, fines predominate in the western margin. Gravel is ubiquitous, though it maximizes along the central axis of the bay and at the mouth.

Heavy minerals in the sand-sized range constitute between 2 and 3 per cent by weight of bottom sediments and are made up of volcanic rock fragments and ferromagnesian minerals (principally pyroxenes, amphiboles and garnet; Hussain, 1980). No clear spatial segregation of heavy minerals is apparent within Chignecto Bay, and therefore the role of hydrodynamic processes in sorting sand-sized bed sediment is considered minor. Hussain (1980) has correlated many of the attributes of the heavy minerals to those in the rocks surrounding the bay. He showed that the heavy minerals were generally close to source. The abundance of the metastable minerals is diagnostic of an adjacent cliff source

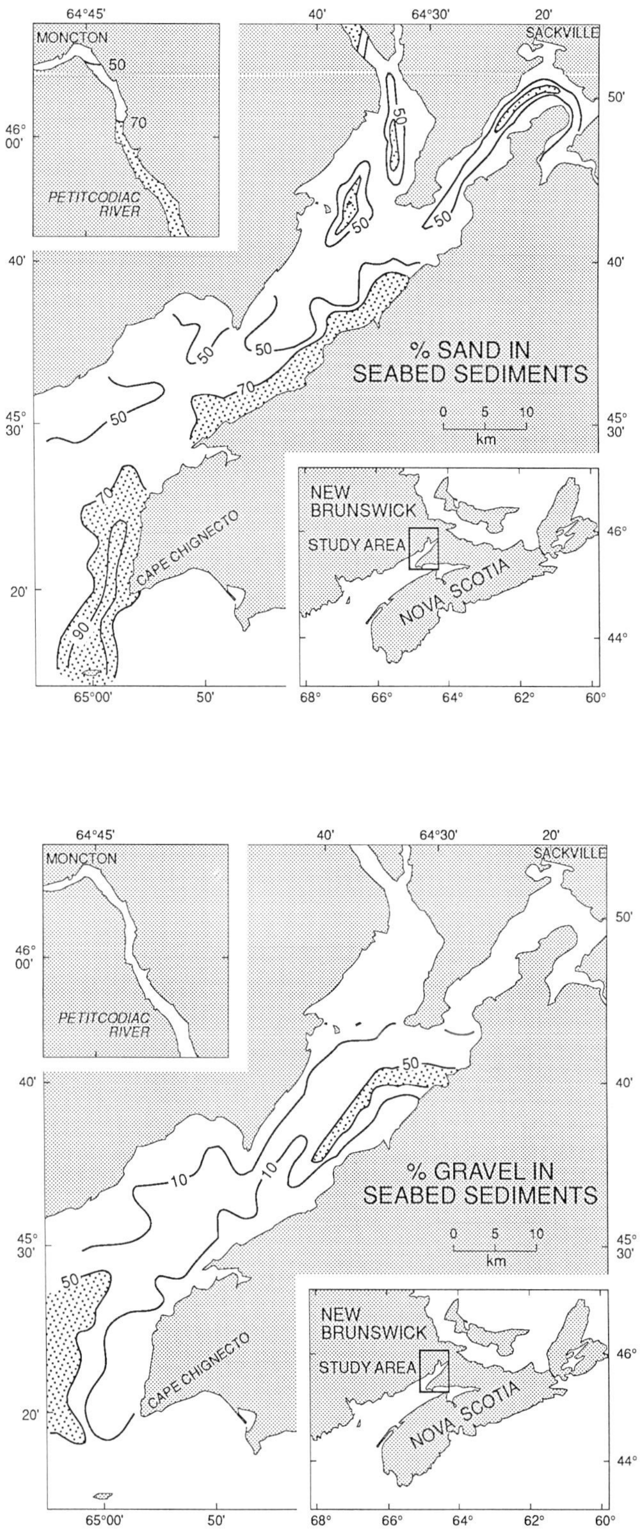

Figure 21. The percentages of sand, and gravel measured in seabed samples taken from Chignecto Bay. Sand predominates in the eastern portion of the bay and is derived from coastal erosion of glaciodeltaic deposits. Gravel is abundant along the median of the bay, where it forms a lag deposit derived by hydraulic winnowing of ice-rafted debris.

with little subsequent reworking. At the mouth of Chignecto Bay, heavy minerals comprise up to 10 per cent of the bottom sediments and are associated with well-sorted, wave-reworked sand or gravel. This illustrates that here, hydrodynamic sediment sorting and winnowing has taken place.

Suspended Sediments

The suspended sediments of Chignecto Bay form a complex pattern of turbid fronts, ribbons, plumes and boils diagnostic of resuspension, dispersion and advection. Under fair-weather conditions, the suspended sediment concentration (SSC) shows an exponential decrease seaward from a predicted maximum of around 68 g/L (Amos and Tee, 1989). The gradient in concentration has been observed by Keizer *et al.* (1984) and Dewolfe *et al.* (1990) in Cumberland Basin, by Topliss *et al.* (1990) in Shepody Bay and by Amos and Topliss (1985) in Chignecto Bay. It is also present in Minas Basin, and the Avon River Estuary (Amos and Tee, 1989). The exponential gradient persists in time, though the slope varies in a direct relationship with the concentration at the turbidity maximum and wave activity, which are both seasonal. The gradient appears to be a balance between sediment input at the head of the estuary (through river input and wave erosion) and the combined effects of dispersion seaward by longitudinal tidal mixing and deposition along the estuary (Amos and Tee, 1989). An example of the concentration gradient is shown in Figure 22. Here, the intrusion of the Cumberland Basin turbid plume into Chignecto Bay during the ebb tide is expressed by an offset in the sediment gradient at a turbid front (Amos, 1987). A similar turbid front occurs between the flooding water masses of

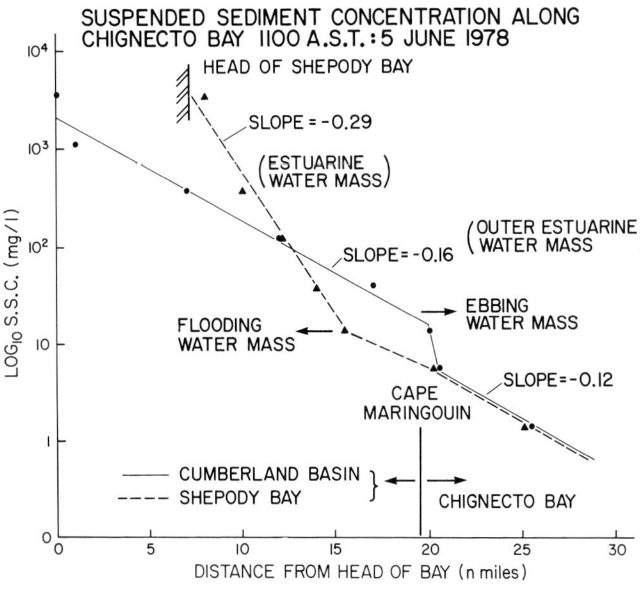

Figure 22. The width-averaged surface suspended sediment concentration measured along the axes of Shepody Bay, Cumberland Basin and Chignecto Bay at 1100 AST on 5 June, 1978. The exponential decrease in SSC is clear in the figure, together with the sharp inflection points at the boundary of water masses.

Shepody Bay and Chignecto Bay and results in an inflection point in the concentration gradient.

There is a large seasonal variation in SSC in all three bays. It is highest during ice breakup, which usually occurs in February or March, and least during the winter months of December and January (Fig. 19). SSC decreases steadily throughout the summer as a result of deposition on the littoral mud flats (Hildebrand *et al.*, 1980), but increases again during the autumn (October and November) as a result of wave resuspension. Observations made by Hildebrand *et al.*, (1980) in Cumberland Basin show a six-fold increase in SSC (from 500 to 3,000 mg/l) associated with an increase in wave height from 0.15 to 0.5 m. Waves may thus be very important to mud flat stability even in the inner parts of high-energy, macrotidal estuaries.

The spatial distribution of surface suspended sediments, discriminated using the Landsat MultiSpectral Scanner (Munday *et al.*, 1980), is illustrated in Figure 23. The figure shows examples of turbid fronts, turbid ribbons and turbid plumes. Turbid fronts and turbid ribbons are diagnostic of the boundaries of water masses commonly separated by shear planes. These planes were often detected during anchor stations. When they crossed the ship's location, significant and abrupt changes in the character of the water column occurred.

The vertical distribution in SSC varies considerably. In outer Chignecto Bay, stratification in SSC occurs throughout the entire tidal cycle and shows a decrease in concentration in proportion with the logarithm of height above the bed. Elsewhere in the system, stratification in SSC occurs immediately after the stillstand of tidal flow (high and low water), but the water column is mixed during peak flows (Fig. 15D). This settling can result in a 2 to 8 times enrichment in bottom water SSC over surface waters. The observed ratio of surface-to-depth-averaged suspended sediment concentration, averaged over all observations, was 0.57 ± 0.26 (Amos and Tee, 1989). This result is substantiated by work carried out throughut Cumberland Basin by Dewolfe *et al.* (1990).

Increases in SSC due to wave resuspension have been observed throughout the system. Wave heights in excess of 0.5 m resuspended material in the inner part of Chignecto Bay, whereas waves in excess of 0.7 m in height resuspended material in the outer estuary (Amos and Asprey, 1981). Still water settling experiments made on this material in a bottom-withdrawal tube showed that the concentration half-life was 6.4 ± 1.2 minutes and that particles coarser than 20 microns were first to deplete forming tidal couplets in regions of high turbidity (>200 mg/l).

Electron microscopic examination (using a Cambridge Stereoscan 180) of the material suspended in Shepody Bay and Cumberland Basin indicated that less than 5 per cent of particles are flocculated, and between 85 and 95 per cent are inorganic. The majority (70-90%) of this material comprises fine-silt-sized, angular, lithic fragments. Coulter Counter analysis of the size spectra showed the model size of

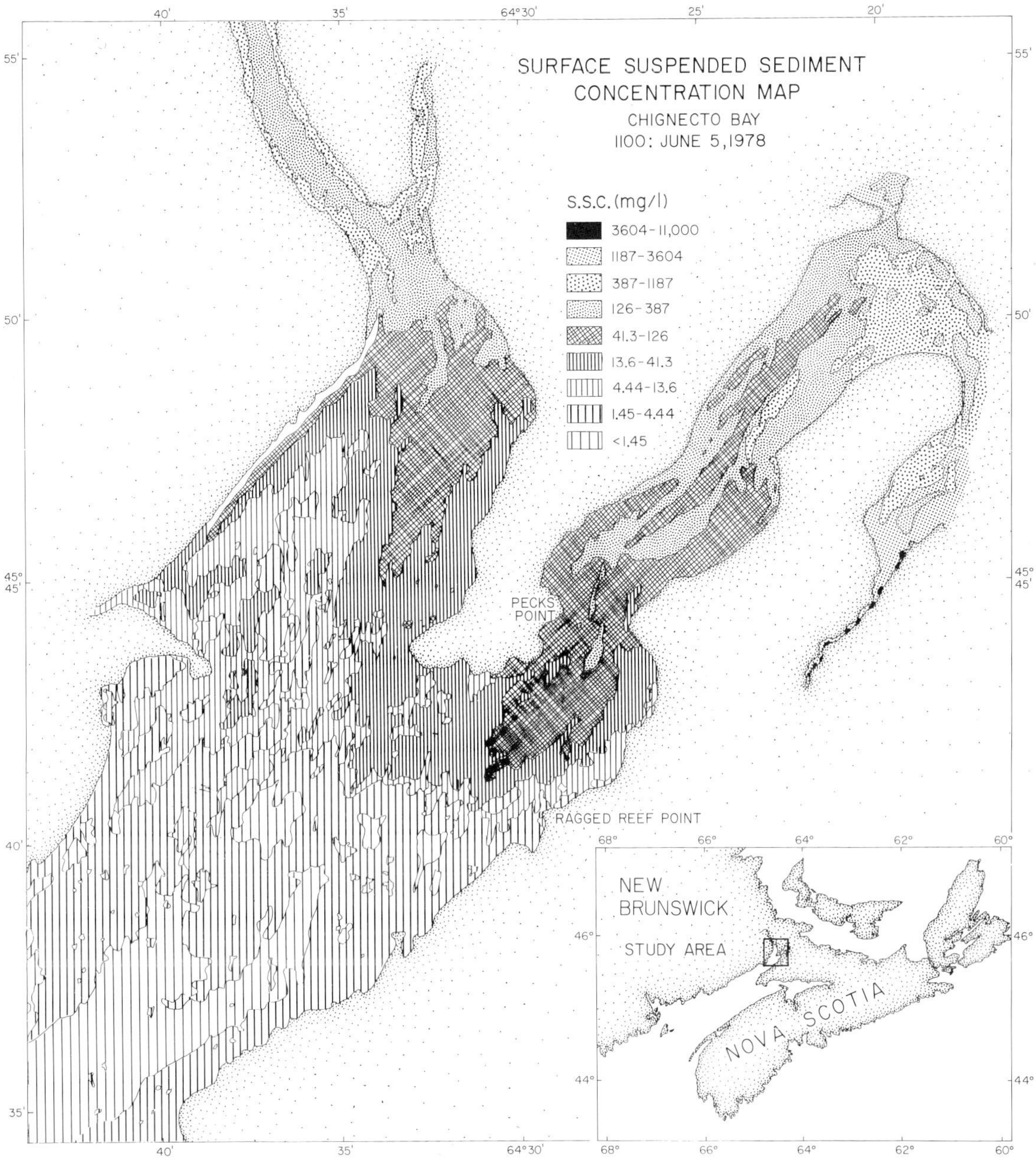

Figure 23. A thematic plot of surface suspended sediment concentration in the upper part of the Chignecto Bay system. The image was created from Landsat Multi-Spectral Scanner data recorded at 1100 AST on 5 June, 1978. The image coincides with the measurements shown in Figure 22. The isopleths of SSC show a seaward deflection in the zones of ebb dominance and a landward defection in zones of flood dominance. The irregular pattern of isopleths in Shepody Bay may result from localized resuspension of bed sediment.

suspended sediment to range from 10 to 30 microns over all stages of the tide (Figure 24). The clay content of suspended material, unlike other estuaries, is constant at 20 per cent along the length of Shepody Bay. Chignecto Bay, however, demonstrates a gradient in silt content of suspended material that decreases inward from a peak of 80 per cent at the mouth. Isolated measurements of suspended clay content made during October to November, 1978 in the inner, western part of the Chignecto Bay were up to 80 per cent. This anomaly is considered to be the result of wave resuspension (see Fig. 18) of early Holocene muds (G2) that here constitute the seabed. Analysis of particle character was made on near-surface and near-bed samples taken over a tidal cycle at station 19 (mouth of Cumberland Basin during calm conditions). There were no obvious correlations between grain size and stage of the tide (Fig. 25). However, an enrichment in silt content by 20 per cent was evident in the bottom waters. Note that no sand was suspended during the peak tidal flows (0.5 m/s) despite its presence at the bed.

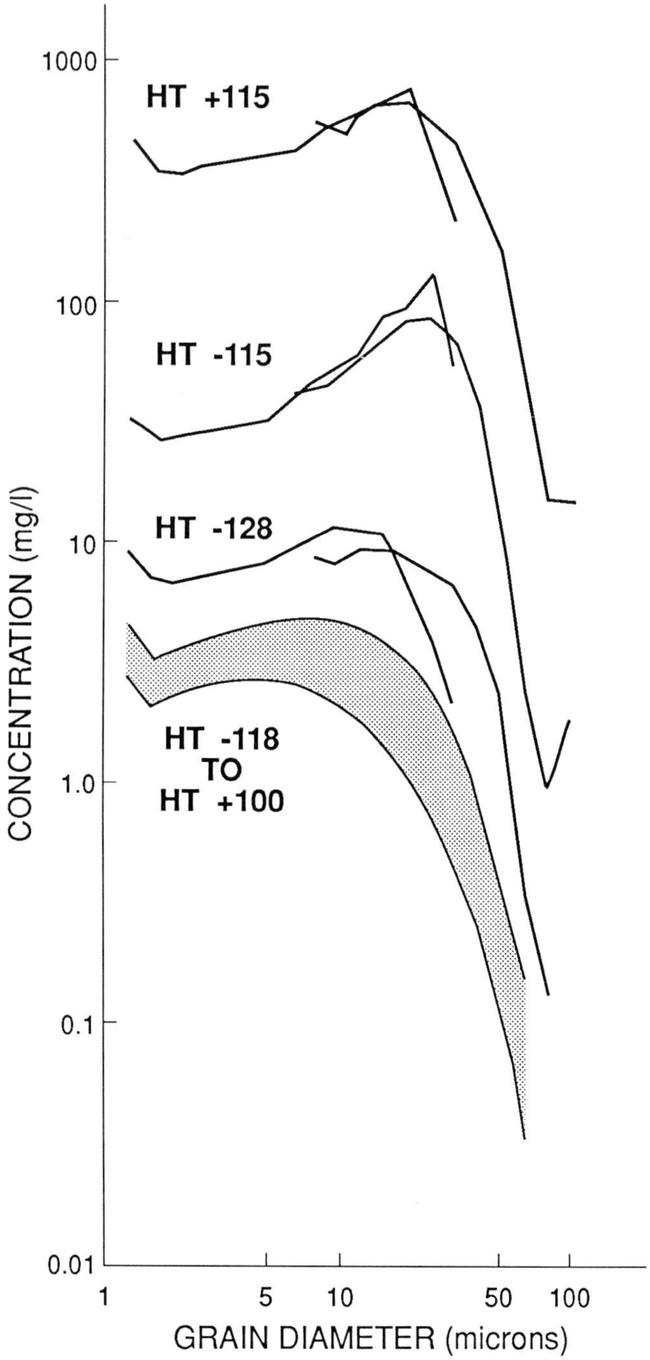

Figure 24. The grain size spectra of suspended sediments sampled over tidal flats near the mouth of Cumberland Basin on 7 November, 1979 at various times (in minutes) before and after two local high tides (HT). Note that the modes in the spectra occur at between 10 and 30 microns and a nearly even distribution in sizes finer than the mode. Almost no sand was detected in suspension.

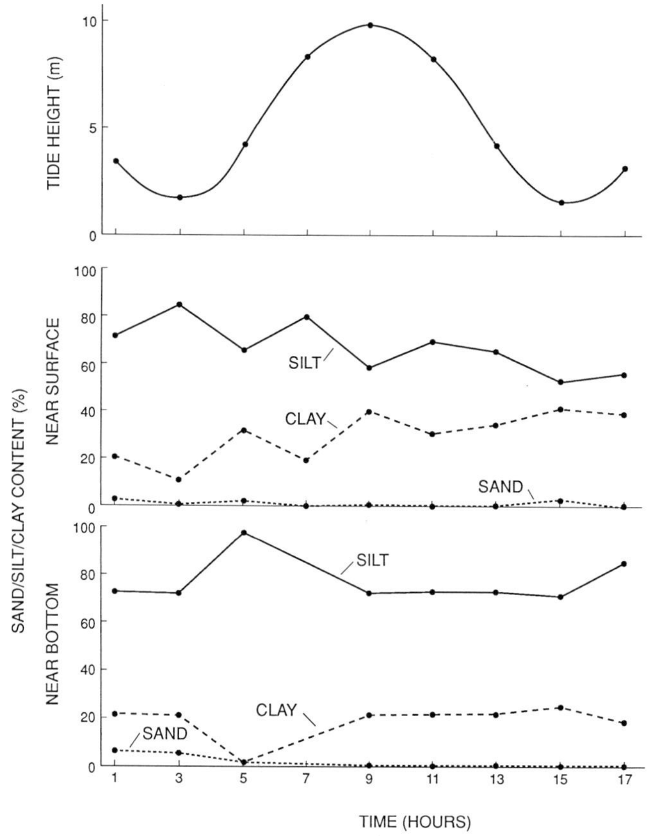

Figure 25. The percentages of suspended sand, silt and clay measured from scanning electron micrographs of surface and near-bed samples each taken two hours over a tidal cycle at the mouth of Cumberland Basin (Fig. 2, Station 19). Note the increase in silt content with depth, the lack of any obvious correlation between sediment texture and tidal stage, and the virtual absence of sand. No waves were present when these measurements were taken.

Sources of Sediment

Cliff Erosion

Wave erosion of the cliffs bordering Chignecto Bay provides an abundant source of sediments to the system. The mean recession rates of the cliffs bordering Shepody Bay, Cumberland Basin and Chignecto Bay, evaluated between 1947 and 1975, were 0.19, 0.20, and 0.37 m/a, respectively. The highest erosion rate (1.0 m/a) occurs in regions of greatest wave exposure, such as the western side of Chignecto Bay and the mouths of Cumberland Basin and Shepody Bay. The total volume of sediment released by coastal erosion is estimated to be 1.03×10^6 m^3/a (Fig. 26). The majority of this material is derived from the volcanic rocks of outer Chignecto Bay (20 $m^3/m/a$) where cliff heights (50 m) and wave exposure are greatest. The highest rate of cliff erosion occurs during periods of highest waves. This is typically during the months of December and January. Sediment so derived is predominantly in the silt size range (Chalmers, 1894).

Rivers

The total quantity of inorganic sediment entering the system from fluvial sources was determined from information gathered at the hydrometric stations illustrated in Figure 1. The figure also shows the catchment area of the system: 5014 km^2. The total input of fresh water is 3.14×10^5 m^3/a (Hildebrand *et al.*, 1980) and the maximum discharge of the largest river (the Petitcodiac River) is 37 m^3/s (Fig. 19).

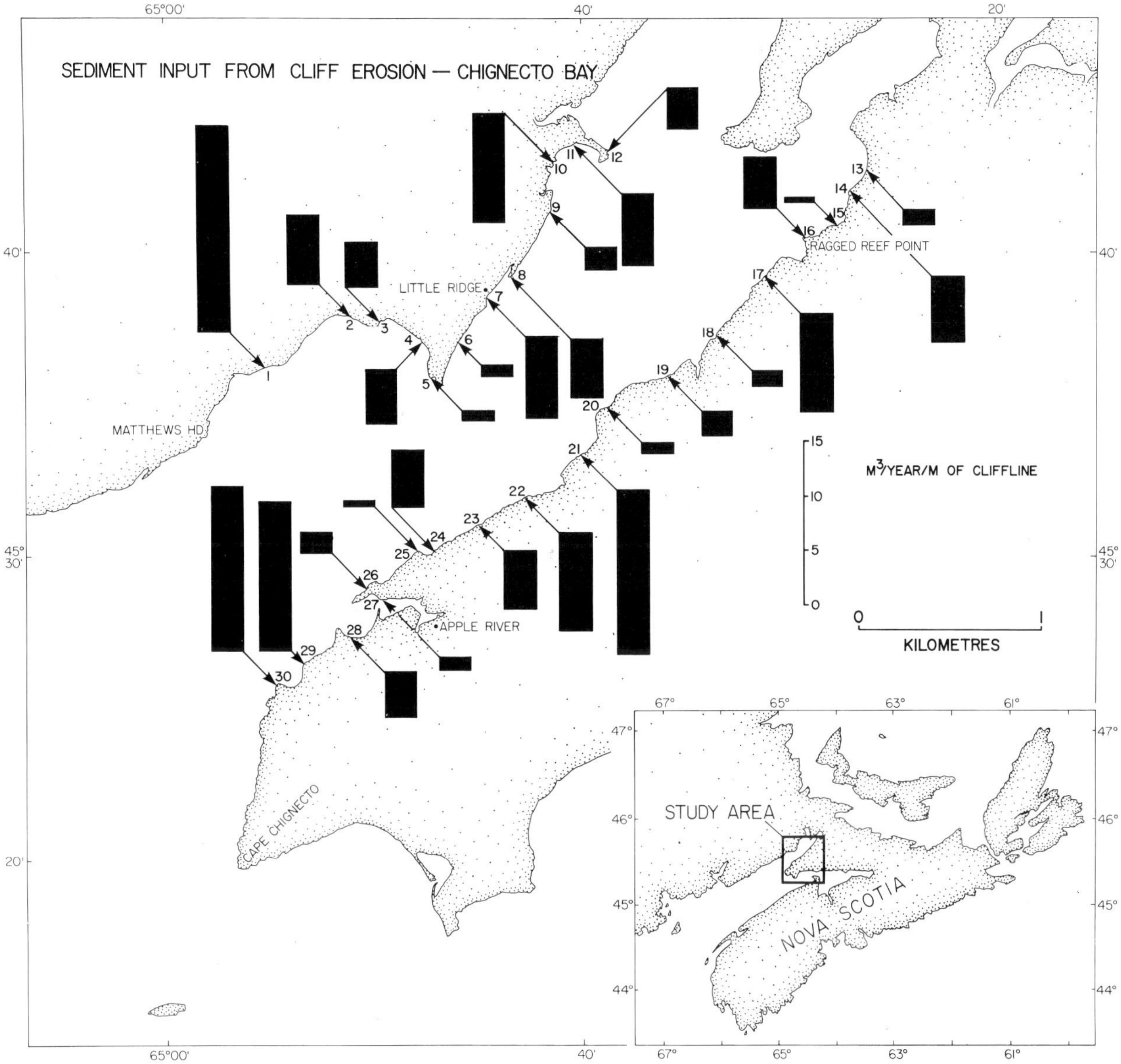

Figure 26. The sediment input to Chignecto Bay through cliffline erosion. Note that the majority of sediment is derived from the outer part of the bay where cliffs are highest and wave forces greatest.

Sediment concentrations in river water are typically less than 2 mg/l and rarely exceed 10 mg/l. The mean fluvial sediment input to the Chignecto Bay system is 3.5×10^5 m^3/a or 67 m^3/km^2. The majority of this is introduced to the system in suspension during spring floods in the months of March and April. Total fluvial input of sediment is 30 per cent of that derived from cliff erosion. The fluvial material is in the silt-clay size range.

The Bay of Fundy

An estimate of the exchange of sediment with the Bay of Fundy has been attempted by the establishment of 3 anchor stations at the mouth of Chignecto Bay (stations 12, 13, 14, Fig. 2). Residual fluxes of suspended sediment derived from these stations indicate a net export of sediment of 5.5×10^6 m^3/a during fair-weather conditions. According to Amos (1987) a further volume of 1.8×10^6 m^3/a is discharged during storms. A large part of this material is derived from seabed erosion off Cape Enragé. The bathymetry in this region shows a deepening by 20 m between the original Admiralty survey conducted in 1861 and the latest survey carried out during this project in 1979 (courtesy Canadian Hydrographic Service). Here, approximately 6.0×10^6 m^3 of fine material is released to the water column annually. A net export of sediment to the Bay of Fundy is verified qualitatively in Landsat and NIMBUS-7 images, which show turbid plumes extending from Chignecto Bay along the northwestern shore of the Bay of Fundy and over a region mapped by Fader *et al.* (1977), Swift *et al.* (1973) and Loring (1982) to be underlain by fine grained sediment.

Transport of Sediment

Bedload Transport and Deposition

The bedload transport of sediment was inferred from bedform type and orientation based on sidescan imagery of the seabed. A synthesis of the interpretation of this imagery is shown in Figure 27. Along the western margin of Chignecto Bay between Cape Enragé and Cape Maringouin, there is a suite of flood-oriented sand waves and 2-dimensional and 3-dimensional megaripples. This suite occurs in a region of peak tidal flow that varies between 1.0 to 1.5 m/s (Fig. 14). The central regions of the estuary are void of large-scale bedforms as a result of the paucity of sand. Flood-oriented comet marks are, however, abundant. This region is subject to peak tidal flows between 1.5 and 2 m/s. There is a trend of increasing bedload transport from the western margin of the bay (rippled plane bed) to its centre (sand waves and 2-dimensional megaripples) that parallels the increase in peak flow to 1.5 m/s. Above this flow, sand is presumed to be removed by intermittent suspension.

A band of irregularly spaced sand ribbons occurs parallel to the eastern margin of Chignecto Bay. Superimposed on the sand ribbons are poorly-developed, flood-oriented, 2-dimensional megaripples. The absence of well-defined slip faces and the high fines content of the constituent sediment suggests that the Chignecto Bay bedforms are at the feather-edge of unit R and are consequently sediment starved (the fines derived from the underlying unit). Large-scale bedforms in Shepody Bay and Cumberland Basin are restricted to the median line of the outer parts. Here, flood-oriented, 2-dimensional and 3-dimensional megaripples and comet marks attest to a headward transport of sand in regions where the peak flow is below 1.5 m/s. Above 1.5 m/s, sand flats occur as upper flow plane beds with flow-parallel lineations.

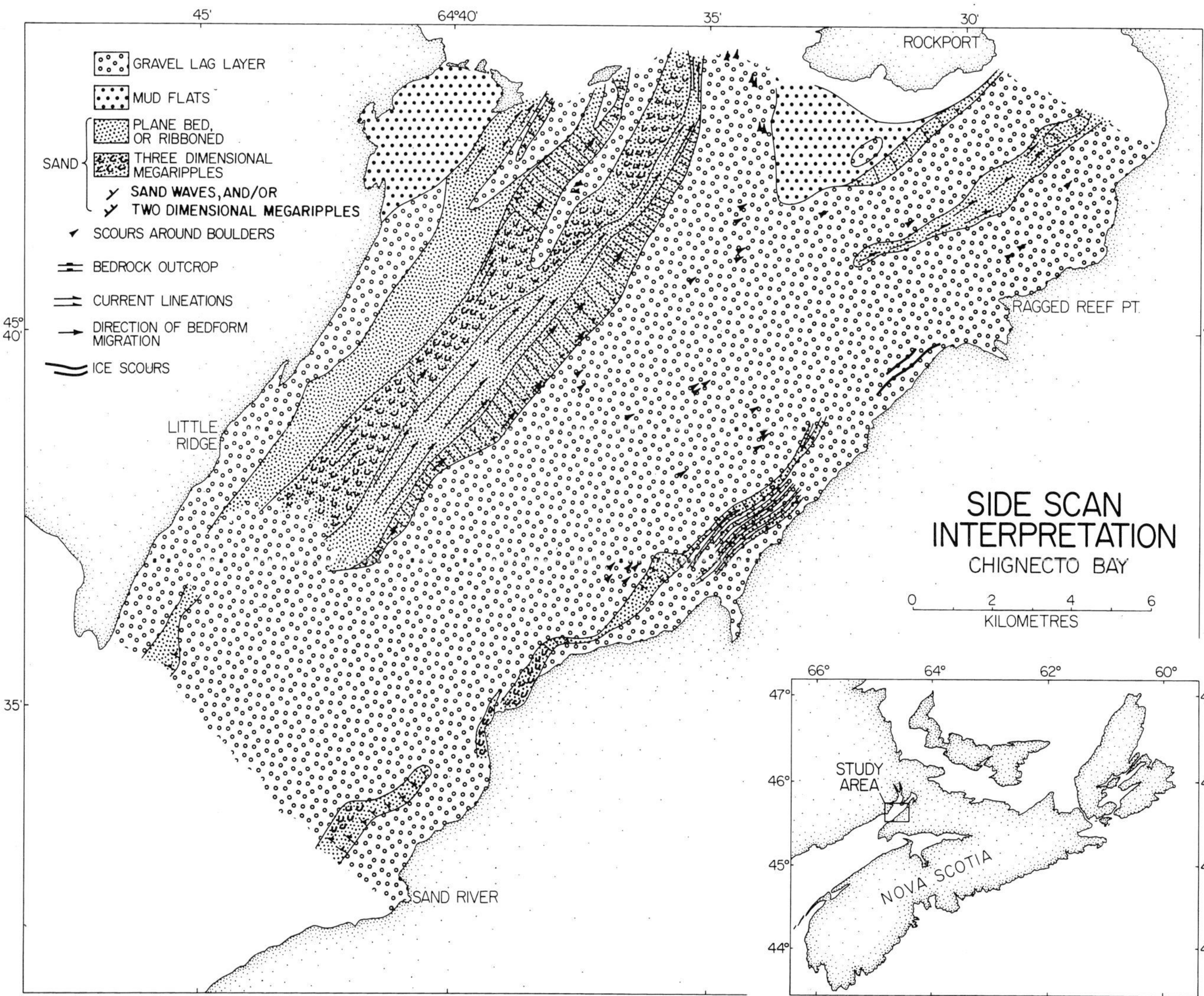

Figure 27. The interpretation of surficial bed features detected on sidescan sonograms of Chignecto Bay. The suite of flood-oriented bedforms along the western margin of the bay is evident. The remainder of the bay is devoid of bedforms with the exception of isolated sand ribbons along its eastern margin and at the mouth of Cumberland Basin.

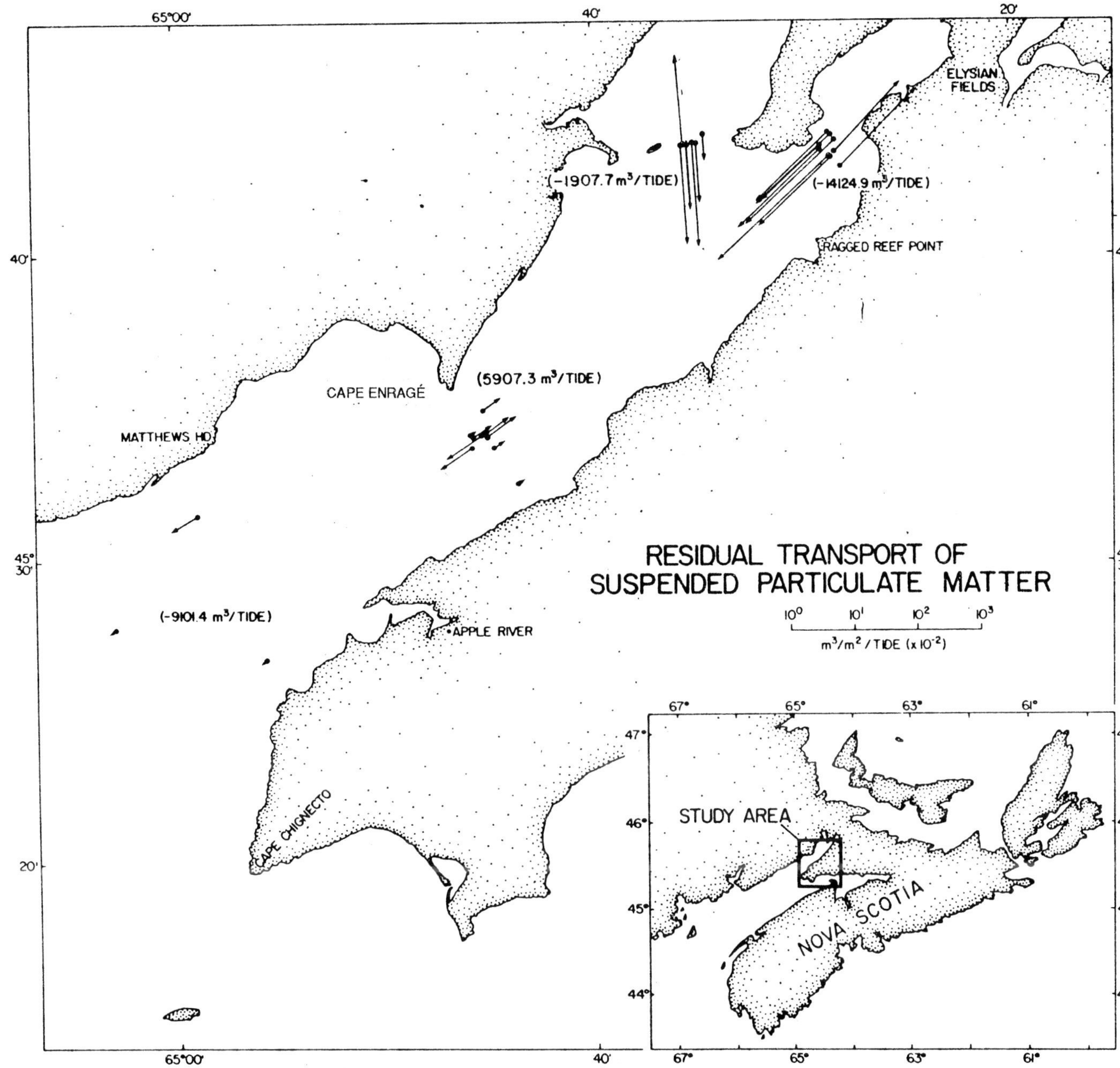

Figure 28. The residual transport of suspended particulate matter in the Chignecto Bay system measured over single tidal inundations at seven periods of the year (see Amos and Asprey, 1981 for details). Each value is the product of measured SSC and velocity integrated with depth and time for the duration of observations (normally 15 hours). The values in brackets denote first estimates of the mass balance of sediment across the sections on which they are found. Positive values denote import, negative values denote export. Note the strong export of sediment throughout the system except near Cape Enragé. Here, results are confused by the effects of wave resuspension. Almost 50 per cent of stations showed some wave effects on SSC.

Suspended Sediment Transport and Deposition

The flux of suspended sediment has been estimated for Chignecto Bay by Amos and Asprey (1981) for the stations illustrated in Figure 2. A brief review of these data is given by Amos (1987). A synthesis of the results is shown in Figure 28. Sediment budget assessments of Chignecto Bay indicate that a total of 7.3×10^6 m^3 of sediment is released annually, of which 1.3×10^6 m^3 is derived from external sources (rivers and eroding cliffs) and the remainder is derived from bed erosion. Based on the volume of the Recent deposits in Chignecto Bay, it is calculated that sand and gravel are being introduced into Chignecto Bay at a rate of 8×10^4 m^3/year. This is approximately 7 per cent of the annual influx of sediment to the system and only 1 per cent of the total volume of material in transport. Thus the major part of material in transport is in the silt or clay size range and has been demonstrated to be transported in suspension or intermittent suspension.

The trapping efficiency of the Chignecto Bay system (defined after Schubel and Carter, 1984) is very low as a result of the reclamation of fringing salt marshes (Amos and Tee, 1989) and high wave activity. That which is released into the system (by seabed erosion, fluvial discharge and cliff erosion) can be assumed to be ultimately exported to the Bay of Fundy.

In the inner estuary, the pattern of suspended sediment transport is unclear. Measurements at the mouths of Shepody Bay and Cumberland Basin show both import and export through time. This pattern is in response to changes in SSC and SSC gradients. Amos and Tee (1989) showed that the total suspended mass in Cumberland Basin remained remarkably constant throughout two years of observations (10^{11}g or approximately 3.8×10^4 m^3) despite changes in SSC patterns due to fluvial input, ice and wave reworking and mud flat sedimentation (Keizer *et al.*, 1984). We propose that exchanges at the mouth of Cumberland Basin occur that maintain the total mass in suspension and the equilibrium gradient constant and in proportion with the energetics of the system.

Summary and Conclusions

The Chignecto Bay system, Bay of Fundy, is dominated by fine grained sediments and suspended sediment transport because of the fine grained nature of the Paleozoic rocks on which the system sits and because of the present-day erosion of the fine grained, unconsolidated sediments that underlie much of the bay. It clearly differs from the Minas Basin, which is dominated by sand-sized sediment and bedload transport due to the erosion of Triassic sandstone cliffs surrounding that Basin (Amos and Long, 1980).

The unconsolidated sequence in Chignecto Bay is composed of up to 50 m of glacial and postglacial deposits that are predominantly fine grained in texture. These formed largely in a sublittoral setting under conditions of sea level, tidal range and wave exposure dramatically different from those of today. Two periods of microtidal conditions are recognized separating two periods of macrotidal conditions interpreted to have occurred during the last 13,300 years. A period of regression took place from 13,300 to 7000 BP. A brief period of macrotidal conditions 12,000 BP was responsible for the development of a sand and gravel unit (G3) that underlies much of the bay and is thickest between Matthews Head and Apple River (Fig. 4).

A lowstand of relative sea level is recognized to have occurred circa 7000 BP at 25 m below present sea level. At this time, an extensive low-gradient, mud-dominated littoral flat (G2) is interpreted to have developed north of Cape Enragé. This flat is dominated by point bar lateral accretion bedding associated with littoral creek migration (Smith, 1989). A lack of wave-formed structures in this unit suggests that the estuary was sheltered from the outer Bay of Fundy by a barrier at the estuary mouth. The deposition of fines would have taken place in the back-barrier lagoon situated in the central and inner paleoestuary. Here inclined heterolithic stratification predominated. A clay-rich salt marsh fringed the mud flats at the approximate positions of the mouths of Cumberland Basin and Shepody Bay. The facies of unit G2 bear only weak similarities to the tidal flats of the North Sea reported by Terwindt (1981) and Nio and Yang (1989). For example, we found no evidence of cyclicity in lamina thickness diagnostic of tidal bundles. Also, we found no evidence of large-scale crosslamination diagnostic of large-scale bedforms and the associated reactivation surfaces. This was likely due to the paucity of sand in the central estuary at the time. We presume the sand (if available) was deposited in the mouth of the paleoestuary (close to that of today) in the form of bars and ebb tidal deltas such as those reported from outer Cobequid Bay (Dalrymple, pers. comm., 1990).

A steady increase in tidal range prevailed throughout the late Holocene transgression of Chignecto Bay. Unit G1 was deposited during early stages of this transgression. The dominance of lenticular and flaser bedding is diagnostic of the increasing influence of tidal currents between 6,000 and 3,000 BP.

At 3,000 BP there was a significant change in depositional environment, possibly associated with overtopping of the outer estuary barrier and associated deposits. The depocentre shifted from the central estuary to the inner estuary, with littoral deposition of fringing mud flats and salt marshes. The central estuary became a region of erosion of units G1 and G2, and slow development of a sand and gravel facies (R).

The distribution of facies in the unconsolidated sequence in Chignecto Bay is summarized in Figure 29. Note the lateral asymmetry in facies distribution. The laminated facies associated with G2 point bar accretion deposits predominates in the eastern estuary and is virtually absent in the west. Lenticular and flaser bedding, generally associated with medium-energy, mesotidal deposits of unit G1 and diagnostic of tidal activity, prevail in the western part of the estuary and are virtually absent from the extreme east. The sand and gravel associated with modern-day processes does not correlate well with the acoustic unit R. The absence of this unit in eastern Chignecto Bay is a result of a lack of contrast with the underlying units likely due to the absence of large-scale bedforms. There is a reversal of the depocentre from that of unit G1 to the eastern part of the estuary. The reason for the asymmetry is unclear, but may be related to estuarine paleobathymetry and the resulting complex patterns of tidal flow.

Modern, sandy sediments account for only 1 per cent of the todal budget of sediment. Net input of Recent sediment is 1.3×10^6 m^3/a. Approximately 93 per cent of this volume is composed of fine grained sediment; the remainder is sand and gravel. The majority comes from cliffline erosion (1.03×10^6 m^3/a) while rivers supply 3.5×10^5 m^3/a. The fine grained sediment is quickly transported through the system in suspension by strong residual currents (Tee and Amos). Results of accretion/erosion on the mud flats at the

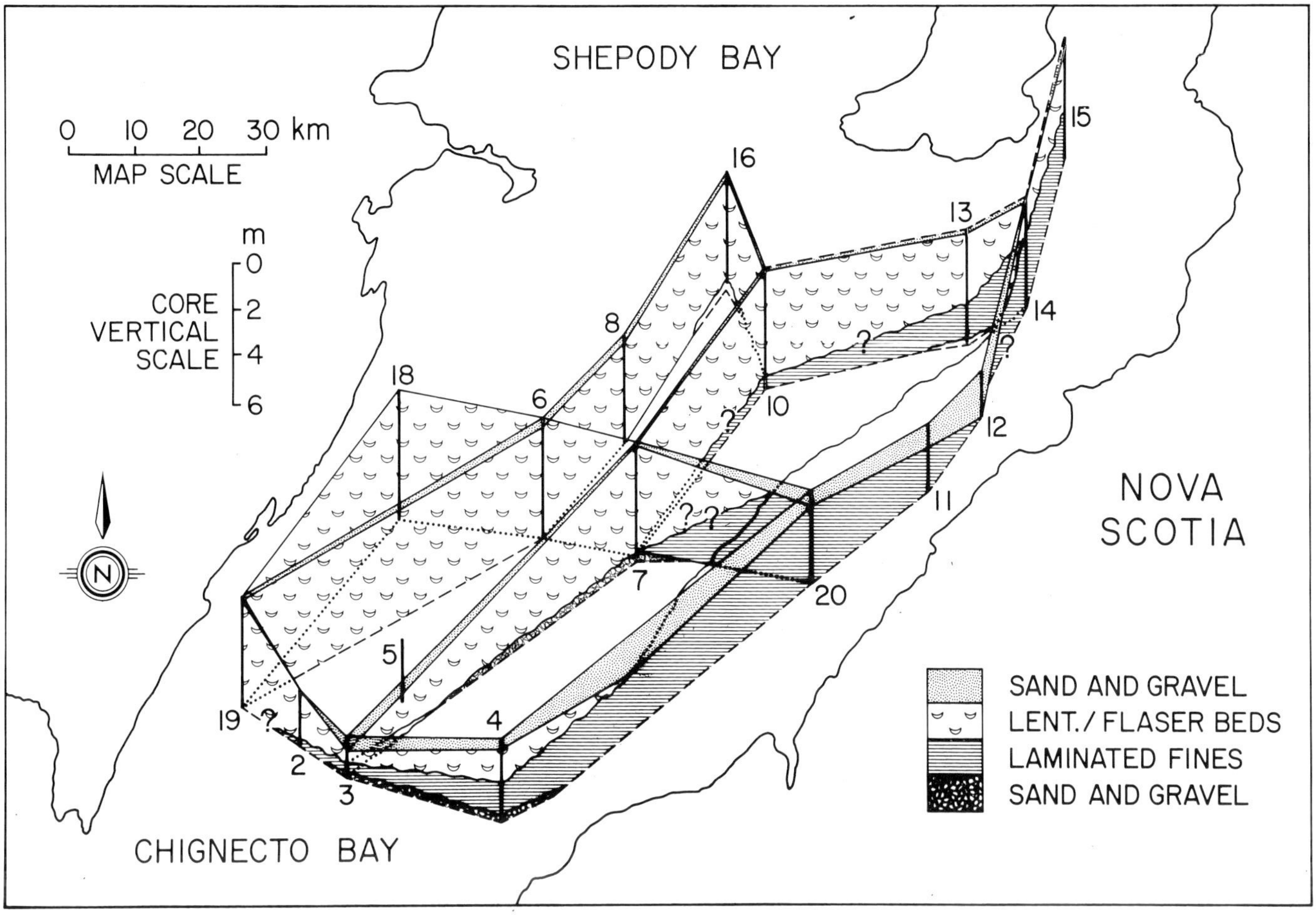

Figure 29. A 3-dimensional fence diagram of the facies distribution in unconsolidated sediments from inner Chignecto Bay, based on the examination of 16 vibracores. The facies show an asymmetrical distribution across the estuary. Laminated sand/fines prevail in the eastern side of the bay, whereas lenticular and flaser bedding prevail in the west. Modern sediments (sand and gravel) predominate in the eastern part of the bay due to a local supply (cliff erosion) of this material. Note that the sands on the eastern side of the bay are not resolved acoustically from the underlying units (G2).

head of Chignecto Bay by Dewolfe *et al.* (1990) show few areas of net accumulation. Seabed reworking introduces a further 6.0×10^6 m^3/a of fine grained sediment into suspension. Fines are, therefore, presumed to be exported to the Bay of Fundy at the rate of input from the rivers, cliffs and the seabed (7.3×10^6 m^3/a). This export was verified by direct measurements made by Amos and Asprey (1981) at the mouth of Chignecto Bay.

Bedload transport takes place at the margins of Chignecto Bay and along the median lines of Shepody Bay and Cumberland Basin. In all cases, the direction of transport is headward. Residual transport of suspended sediment in the inner bay is complex, and can reverse direction through time. Such reversals maintain an equilibrium suspended mass in the system and an equilibrium down-estuary concentration gradient.

Acknowledgments

These data were collected in cooperation with D. C. Gordon (D.F.O.) and G. R. Daborn (Acadia University) with support from K. W. Asprey, D. Beaver, R. B. Murphy and F. Jodrey of the Atlantic Geoscience Centre. Data reduction and plotting were carried out by L. P. Hildebrand, E. B. MacDormand, D. R. Nelson, C. G. Powell, N. A. Rodgers and J. A. Walker. Anchor station data were collected with the support of the officers and crew of CSS Dawson. Interpretation of the surficial sequence was made possible through discussions with R. W. Dalrymple. The updated bathymetry of Chignecto Bay was provided courtesy of the Canadian Hydrographic Service. The two waveriders were maintained and serviced, and the data processed by the Wave Climate Group (Marine Environmental Data Services, Ottawa). Stream gauging results and fluvial water samples were collected by Inland Waters, Dartmouth, courtesy of D. McBride. This work was carried out under Geological Survey of Canada Project number 780022.

References

Admiralty Survey 1861. Unpublished Field Sheets, British Admiralty, Admiralty House, Taunton, United Kingdom.

Amos, C. L. 1984. An overview of sedimentological research in the Bay of Fundy. *In:* Update on the marine environmental consequences of tidal power development in the upper reaches of the Bay of Fundy, D. C. Gordon and M. J. Dadswell (eds.). Canadian Technical Report of Fisheries and Aquatic Sciences, v. 1256, p. 31-43.

______ 1987. Fine-grained sediment transport in Chignecto Bay, Bay of Fundy, Canada. Continental Shelf Research, v. 7, p. 1295-1300.

______ and Asprey, K. W. 1979. Geophysical and sedimentary studies in the Chignecto Bay system, Bay of Fundy — a progress report. *In:* Current Research, Part B, Geological Survey of Canada, Paper 79-1B, p. 245-252.

______ and ______ 1981. An interpretation of oceanographic and sediment data from the upper Bay of Fundy. Bedford Institute of Oceanography Report Series BI-R-81-15, 143 p.

______ , Buckley, D. E., Daborn, G. R., Dalrymple, R. W., McCann, S. B. and Risk, M. J. 1980. Geomorphology and sedimentology of the Bay of Fundy. Geological Association of Canada Field Trip Guidebook, v. 23, 82 p.

______ and Long, B. F. N. 1980. The sedimentary character of the Minas Basin, Bay of Fundy. *In:* The Coastline of Canada, S. B. McCann (ed.). Geological Survey of Canada Paper 80-10, p. 123-152.

______ and King, E. L. 1984. Bedforms of the Canadian eastern seaboard: a comparison with global occurrences. Marine Geology, v. 57, 167-208.

______ and Tee, K. T. 1989. Suspended sediment transport processes in Cumberland Basin, Bay of Fundy. Journal of Geophysical Research, v. 94(C10), p. 14,407-14,417.

______ and Topliss, B. J. 1985. Discrimination of suspended particulate matter in the Bay of Fundy using the NIMBUS-7 Coastal Zone Color Scanner. Canadian Journal of Remote Sensing, v. 11(1) p. 85-92.

______ and Zaitlin, B. A. 1985. The effect of changes in tidal range on a sublittoral macrotidal sequence, Bay of Fundy, Canada. Geo-Marine Letters, v. 4, p. 161-169.

Anderton, R. 1985. Clastic facies models and facies analysis. *In:* Sedimentology, Recent Developments and Applied Aspects, P. J. Brenchley and B. P. J. Williams (eds.). Blackwell Scientific Publications, 31-47.

Bay of Fundy Tidal Power Review Board, 1977. Reassessment of Fundy tidal power: Reports of the Bay of Fundy Tidal Power Review Board and Management Committee. Published by Thorn Press Limited, 516 p.

Bray, D. I., Demarchant, D. P. and Sullivan, D. L. 1982. Some hydrotechnical problems related to the construction of a causeway in the estuary of the Petitcodiac River, New Brunswick. Canadian Journal of Civil Engineering, v. 9, p. 296-307.

Canadian Hydrographic Survey, 1979. Unpublished Field Sheets, Bedford Institute of Oceanography, Dartmouth, Nova Scotia, Canada.

Chalmers, R. 1894. Report on the surficial geology of eastern New Brunswick, north-western Nova Scotia and a portion of Prince Edward Island. Geological Survey of Canada Annual Report, v. 7, p. 124-131.

Curray, J. R. 1960. Sediments and history of Holocene transgression, continental shelf, northwest Gulf of Mexico. *In:* Recent Sediments, northwest Gulf of Mexico, F. B. Phleger and T. H. Van Andel (eds.). American Association of Petroleum Geologists, p. 221-226.

Dalrymple, R. W., Amos, C. L. and McCann, S. B. 1983. Beach and nearshore depositional environments of the Bay of Fundy and southern Gulf of St. Lawrence. International Association of Sedimentologists, Excursion 6A, 116 p.

______ , Makino, Y. and Zaitlin, B. A. (*this volume*). Temporal and spatial patterns of rhythmite deposition on mud flats in the macrotidal Cobequid Bay-Salmon River estuary, Bay of Fundy, Canada. *In:* Clastic Tidal Sedimentology. Canadian Society of Petroleum Geologists, Memoir 16.

deMargerie, S. and deWolfe, D. 1987. Development of a hydrodynamic model in Cumberland Basin for tidal power applications. Consultant Report prepared by ASA Consulting Ltd., Dartmouth, submitted to Department of Supply and Services, Dartmouth; 98 p.

deWolfe, D., MacNeil, M., deMargerie, S., Willis, D., Christian, H. and Amos, C. 1990. Sediment transport in Cumberland Basin — Field program data report. Consultant Report prepared by ASA Consulting Ltd., Dartmouth, submitted to Nova Scotia Tidal Power Corporation; 77 p.

Fader, G. B., King, L. H. and MacLean, B. 1977. Surficial geology of the eastern Gulf of Maine and Bay of Fundy. Geological Survey of Canada, Paper 76-17, 23 p.

Garrett, C. J. R. 1972. Tidal resonance in the Bay of Fundy and Gulf of Maine. Nature, v. 283, p. 441-443.

______ , Keeley, J. R. and Greenberg, D. A. 1978. Tidal mixing versus thermal stratification in the Bay of Fundy and Gulf of Maine. Atmosphere and Oceans, v. 16(4), p. 403-423.

Godin, G. 1968. The 1965 current survey of the Bay of Fundy — a new analysis of the data and the interpretation of the results. Marine Sciences Directorate Report, Ottawa, no. 8, 96 p.

______ 1980. Cotidal charts for Canada. Marine Sciences Directorate Report, Ottawa, no. 55, 93 p.

Goldthwait, J. W. 1924. Physiography of Nova Scotia. Geological Survey of Canada, Memoir 140, 179 p.

Gordon, D. C. and Desplanque, C. 1983. Dynamics and environmental effects of ice in the Cumberland Basin of the Bay of Fundy. Canadian Journal of Fisheries and Aquatic Sciences, v. 40(9), p. 1331-1342.

______ and Dadswell, M. J. 1984. Update on the marine environmental consequences of tidal power development in the upper reaches of the Bay of Fundy. Canadian Technical Report of Fisheries and Aquatic Sciences, v. 1256, 686 p.

Grant, D. R. 1970. Recent coastal submergence of the Maritime Provinces, Canada. Canadian Journal of Earth Sciences, v. 7, p. 676-689.

Grant, D. R. 1985. Glaciers, sediment and sea-level; northern Bay of Fundy, N.S. 14th Artic Workshop, Nova Scotia Field Trip B, 36 p.

Greenberg, D. A. 1983. Modeling the mean barotropic circulation in the Bay of Fundy and Gulf of Maine. Journal of Physical Oceanography, v. 13, p. 886-904.

Harrison, W. and Lyon, C. J. 1963. Sea-level and crustal movements along the New England-Acadian shore, 4,500-3,000 B.P. Journal of Geology, v. 71(1), p. 96-108.

Hildebrand, L. P., MacDormand, E. B., Nelson, D. R., Powell, C. G., Rodgers, N. A. and Walker, J. A. 1980. Activities of the Job Corps Program; Fundy Tidal Power Development. National Research Council of Canada, Halifax, Project 16-01-002N, 176 p.

Holloway, P. E. 1981. Longitudinal mixing in the upper reaches of the Bay of Fundy. Estuarine Coastal and Shelf Sciences, v. 13, p. 495-515.

Hussain, M. 1980. Textural and mineralogical analysis of Chignecto Bay sediments, Canada. Unpublished M. Sc. Thesis, Acadia University, 138 p.

Keizer, P. D., Gordon, D. C. and Hayes, E. R. 1984. A brief overview of recent research in the Bay of Fundy. *In:* Update on the marine environmental consequences of tidal power development in the upper reaches of the Bay of Fundy, D. C. Gordon and M. J. Dadswell (eds.). Canadian Technical Report of Fisheries and Aquatic Sciences, v. 1256, p. 45-63.

King, L. H. and MacLean, B. 1976. Geology of the Scotian Shelf and adjacent areas. Marine Sciences Paper no. 7, Geological Survey of Canada, Paper 74-31, 31 p.

Klein, G. deV, 1970. Depositional and dispersal dynamics of intertidal sand bars. Journal of Sedimentary Petrology 40, 1095-1127.

Laming, D. J. C. 1967. Sediment movement in Chignecto Bay. Unpublished report to Atlantic Tidal Power Programming Board, May, 1967, 26 p.

Lewis, P. J. and Moran, M. D. 1984. Severe storms off Canada's east coast: a catalogue summary for the period 1957 to 1983. Canadian Climate Centre Report, 84-13, C1.3 p.

Loring, D. H. 1982. Geochemical factors controlling the accumulation and dispersal of heavy metals in the Bay of Fundy sediments. Canadian Journal of Earth Sciences, v. 19(5), p. 930-944.

Lougee, R. J. 1953. A chronology of post-glacial time in eastern North America. Scientific Monthly, v. 76, p. 259-276.

Marine Environmental Data Services, 1979(a). Waves recorded off Shepody Bay, N.B., June 14, 1978 to December 13, 1978: Unpublished Report of Wave Climate Group, Ottawa, no. 146-1m.

______ 1979(b). Waves recorded off Cape Enragé, Chignecto Bay, June 13, 1978 to April 20, 1979. Unpublished Report of Wave Climate Group, Ottawa, no. 147-1m.

Milliman, J. D. and Enory, K. O. 1968. Sea levels during the past 35,000 years. Science, v. 162, p. 1121-1123.

Munday, J. C., Alfoldi, T. T. and Amos, C. L. 1980. Application of a system for automated multidate Landsat measurement of suspended sediment. Water Quality Bulletin, v. 5(1), p. 6-10.

Nio, S-D and Yang, C-S. 1989. Recognition of tidally-influenced facies and environments. Short course notes, series #1, International Geoservices BV, Leiderdorp, The Netherlands, 230 p.

______ (*this volume*). Diagnostic attributes of clastic tidal deposits: a review. *In:* Clastic Tidal Sedimentology, Canadian Society of Petroleum Geologists, Memoir 16.

Noordijk, A. and Pronk, T. 1981. De Holocene afzettingen in de dalen van de Missiguash, La Planche, en Nappan, Bay of Fundy, Canada: Unpublished Ph.D. Thesis, Free University, Amsterdam, 65 p.

Pelletier, B. R. and McMullen, R. M. 1972. Sedimentation patterns in the Bay of Fundy and Minas Basin. *In:* Tidal Power, T. J. Gray and O. K. Gashus (eds.). Plenum Publishing Corporation, New York, p. 153-187.

Poole, W. H., Sanford, B. V., Williams, H. and Kelley, D. G. 1976. Geology of southeastern Canada. *In:* Geology and Economic Minerals of Canada, R. J. W. Douglas (ed.). Geological Survey of Canada, p. 227-304.

Pritchard, D. W. 1956. The dynamic structure of a coastal plain estuary. Journal of Marine Research, v. 15, p. 33-42.

Quinlan, G. and Beaumont, C. 1982. The deglaciation of Atlantic Canada as reconstructed from the postglacial relative sea-level record. Canadian Journal of Earth Sciences, v. 19, p. 2232-2246.

Rampton, V. N., Gauthier, R. C., Thibauld, J. and Seaman, A. A. 1984. Quaternary geology of New Brunswick. Geological Survey of Canada, Memoir, v. 416, 77 p.

Schubel, J. R. and Carter, H. H. 1984. The estuary as a filter for fine-grained suspended sediment. *In:* The Estuary as a Filter, V. S. Kennedy (ed.). Academic Press, San Diego, p. 81-105.

Scott, D. B. and Greenberg, D. A. 1983. Relative sea-level rise and tidal development in the Fundy tidal system: Canadian Journal of Earth Sciences, v. 20(10), p. 1554-1564.

______ and Medioli, F. S. 1980. Post-glacial emergence curves in the Maritimes determined from marine sediments in raised basins. Proceedings of the Canadian Coastal Conference, Burlington, p. 428-446.

Smith, D. G. 1989. Comparative sedimentology of mesotidal (2 to 4 m) estuarine channel point bar deposits from modern examples and ancient Athabasca oil sands (Lower Cretaceous), McMurray Formation. *In:* Modern and Ancient Examples of Clastic Tidal Deposits — A Core and Peel Workshop, G. E. Reinson (ed.). Canadian Society of Petroleum Geologists, Special Publication S 31, p. 60-65.

Stephens, G. R. 1977. Geology and tectonic framework of the Bay of Fundy — Gulf of Maine region. *In:* Fundy Tidal Power and the Environment, G. R. Daborn (ed.). Acadia University Institute, Publication no. 28, p. 82-100.

Stravers, J. A. and Syvitski, J. P. M. (*in press*). Early Holocene land-sea correlations and deglacial evolution of the Cambridge Fiord Basin, northern Baffin Island. Quaternary Research.

Swift, D. J. P. and Lyall, A. K. 1968. Origin of the Bay of Fundy. Marine Geology, v. 6, p. 331-343.

______, Pelletier, B. R., Lyall, A. K. and Miller, J. A. 1973. Quaternary sedimentation in the Bay of Fundy. Geological Survey of Canada, Paper 71-23, p. 113-151.

Syvitski, J. P. M., Beattie, D. D., Praeg, D. B. and Schaffer, C. T. 1987. Marine Geology of Baie des Chaleurs. Geological Survey of Canada Open File 1375.

Taylor, R. B., Praeg, D. B. and Syvitski, J. P. M. 1987. Coastal morphology and sedimentation, eastern Baffin and Bylot Islands, NWT. *In:* Sedimentology of Arctic fjords experiment, Data report, volume 3, J. P. M. Syvitski and D. B. Praeg (eds.). Canadian Data Report of Hydrography and Ocean Sciences 54, p. 3-1 — 3-60.

Tee, K. T. 1982. The structure of the three-dimensional tide-generating currents: experimental verification of a theoretical model. Estuarine, Coastal and Shelf Science, v. 14, p. 27-48.

Tee, K. T. and Amos, C. L. (*in press*). Tidal and buoyancy driven currents in Chignecto Bay, Bay of Fundy. Journal of Geophysical Research.

Terwindt, J. H. J. 1981. Origin and sequences of sedimentary structures in inshore mesotidal deposits of the North Sea. International Association of Sedimentologists, Special Issue, 5, p. 4-26.

Topliss, B. J., Amos, C. L. and Hill, P. R. 1990. Algorithms for remote sensing of high concentration, inorganic suspended sediment. International Journal of Remote Sensing 11(6), p. 947-966.

Wagner, F. J. E. 1980. A faunal interpretation of tidal evolution in Minas Basin, Nova Scotia. Géographie, Physique et Quaternaire, v. 34, p. 253-257.

Williams, H., Kennedy, J. J. and Neale, E. R. W. 1972. The Appalachian structural province. *In:* Variations in tectonic style in Canada, L. A. Price and R. J. W. Douglas (eds.). v. II, p. 181-261.

Wightman, D. M. and Cooke, H. B. S. 1978. Postglacial emergence in Atlantic Canada. Geoscience Canada, v. 5, p. 61-65.

KEY WORDS: MACROTIDAL, MESOTIDAL, TURBID, ESTUARINE, SEDIMENT BUDGET, GLACIAL

In: Clastic Tidal Sedimentology, D. G. Smith, G. E. Reinson, B. A. Zaitlin and R. A. Rahmani (eds.). Canadian Society of Petroleum Geologists. Memoir 16, p. 91-106.

Lower Paleocene drowned valley and barred estuaries, Canadian Arctic Islands: aspects of their geomorphological and sedimentological evolution

BRIAN D. RICKETTS

Geological Survey of Canada, Cordilleran Division
100 West Pender St. Vancouver,
British Columbia, V6B 1R8
Canada

ABSTRACT

Barred, coastal plain estuaries and drowned valley estuaries comprise paralic, basal Paleocene strata in the eastern Arctic Islands of Canada. The estuarine deposits accumulated during initial transgression over a regional, sub-Paleocene unconformity, a surface corresponding to a third-order sequence boundary.

Fundamental differences exist between valley types containing the estuarine facies, depending on the nature of the bedrock. Shallow, broad valleys were eroded into nondeformed, moderately to weakly indurated rocks only slightly greater in age; these consequent drainage patterns are relatively young. Deep, narrower valleys are underlain by highly deformed, highly indurated, lower Paleozoic strata; drainage patterns here are very old. These geomorphic differences are reflected in the different estuarine facies and valley-fill stratigraphies. Shallow valleys were filled with barrier-inlet and washover sands, and finer grained lagoonal, marsh, tidal flat and estuarine channel deposits. The deeper, higher relief bedrock valleys are characterized by lower estuary spits and bars, middle estuary channels and tidal flats, and upper estuary-fluvial channels; coastline configuration was initially embayed, but became 'straighter' as bedrock relief was subdued. Both valley types existed at the same time in the basin.

It is maintained throughout the analysis that the geomorphic aspects of the antecedent valleys must be included in facies models that attempt to make the distinction between different types of estuary in the rock record, and even the more basic distinction between estuarine and other paralic deposits.

RÉSUMÉ

Dans la partie orientale des îles Arctiques canadiennes, les plaines côtières estuariennes et les vallées estuariennes inondées, toutes protégées par une barrière, comprennent des strates paraliques de la base du Paléocène. Durant la phase initiale d'une transgression, les dépôts estuariens se sont accumulés sur une discordance régionale sub-paléocène qui correspond à une limite de séquence de troisième ordre.

Les différences fondamentales entre les types de vallée contenant des faciès estuariens dépendent de la nature du substratum. Les vallées larges et peu profondes ont été érodées, et ont exposé des roches modérément à faiblement indurées, légèrement plus vieilles. Les patrons de drainage résultants étaient relativement jeunes. Au contraire, les vallées étroites et profondes surmontent des strates du Paléozoïques Inférieur, très déformées et fortement indurées. Dans ce cas, les patrons de drainage étaient très vieux. Les différences géomorphologiques sont reflétées dans les différents faciès estuariens et dans la stratigraphie des remplissages des vallées. Les vallées peu profondes ont été remplies par des sables remaniés provenant de la barrière («washover sands») et des sables du petit bras de mer recoupant la barrière, ainsi que par des sédiments plus fins de lagon, de marais, d'estran et de chenal estuarien. Les vallées plus profondes, dont le subtratum était à fort relief, sont caractérisées par des flèches et barres estuariennes basses, des chenaux estuariens et des estrans dans la partie médiane, ainsi que des chenaux fluviaux-estuariens dans la partie supérieure. Le littoral était initialement découpé mais le devenait moins avec l'affaiblissement du relief du substratum. Simultanément, les deux types de vallée existaient dans le bassin.

Il ressort de cette analyse que les aspects géomorphologiques des vallées antécédentes doivent être incluses dans les modèles de faciès tentant de séparer les différents types d'estuaire dans les archives géologiques, ou tout simplement pour distinguer les dépôts estuariens des autres dépôts paraliques.

Geological Survey of Canada Contribution Number 22890

Recognition of Ancient Estuaries

Estuarine deposits, being the products of transgression and sedimentation in antecedent valleys, are key elements in the unravelling of sedimentary basin architecture. They overlie subaerial unconformities, which constitute the boundaries of depositional sequences (*sensu* Vail *et al.*, 1977), or represent important surfaces within genetic sequences (Galloway, 1989). Regardless of the sequence analysis approach taken, sub-estuarine unconformities mark reversals in the polarities of base line (*e.g.*, shoreline) migrations. Furthermore, because estuaries fill pre-existing valleys, there is some built-in protection from subsequent erosional episodes such that their preservation potential is usually greater than is the case for other types of transgressive deposit.

Definitions of estuaries generally stress such factors as valley geomorphology and orientation, the physical and biological manifestation of the transition from fluvial to marine, and the relationship between fresh water input and sea water dilution (*e.g.*, Pritchard, 1967; Kjerfve and Magill, 1989); these definitions are based on modern analogues. In basin analysis, the recognition of ancient estuaries is based primarily on the identification of lithofacies and stratigraphic succession. However, a perusal of much of the literature that deals with ancient estuarine successions indicates that the putative facies models/hypotheses bear remarkable similar ities to successions described from shelf-coastal plain and shelf-barrier/lagoon settings. Perhaps we should not be too surprised at this. Many of the sedimentological and biological features that characterize estuaries overlap to a considerable degree those found in the other "paralic" environments (see Ward and Ashley, 1989, Fig. 2). For example, if we assume, in an overly simplistic fashion, that lagoons are essentially parallel to coastlines, and estuarine channels are normal to coastlines (*e.g.*, Kjerfve and Magill, 1989), then an assessment of the overall geometry of ancient analogues may be sufficient to distinguish between these depositional systems. But such distinctions become fuzzy in situations where both lagoons and intricate estuarine channel systems are intimately associated, such as at present along the coasts of North Carolina and Georgia (*e.g.*, Frey and Howard, 1986; Knebel *et al.*, 1988).

The essential problem is that there is a significant gap between our knowledge of modern estuaries, and the utility of the resulting facies models or hypotheses in unravelling the rock record. A glimmer of hope is afforded by attempts to classify estuarine lithofacies in terms of their geomorphic associations; again from modern analogues (*e.g.*, Nichols and Biggs, 1985; Allen, 1989; Nichol, 1989; Nichols *et al.*, 1989; Woodroffe *et al.*, 1989; Zaitlin and Dalrymple, 1989). Most proposed models describe three, longitudinally disposed regions: an upper estuary/fluvial realm, a middle estuary, and a lower estuary region that contains the transition to shelf, shoal or barrier island/lagoon settings. However, when applied to the rock record, with all its vagaries of preservation, exposure and subsurface control, one is still faced with the problem of distinguishing among a variety of possible coastal environments.

An estuary, if it is to exist at all, requires flooding or drowning of pre-existing river valleys. Thus, the least equivocal criterion for recognition of ancient estuaries is the antecedent valleys themselves; in other words, the relief on the sub-estuary surface. These valleys may be cut into bedrock as in the classical type of drowned valley along the modern southeast Australian coast (*e.g.*, Roy *et al.*, 1980), Severn Estuary (Harris, 1988), and northern New Zealand, or incised into low relief coastal plains as along much of the Atlantic North American coast (*e.g.*, Frey and Howard, 1986; Fletcher *et al.*, 1990). Ancient examples of the coastal plain type, wherein the paleovalley profile can be mapped out, are found in the Cretaceous of San Juan Basin (Tillman and Nummedal, 1989), and the Western Canada foreland basin (Leckie, 1988; Leckie and Staniland, 1988; Wood and Hopkins, 1989; where valley form can be established in subsurface structure and isopach maps and in outcrop).

In this paper, examples of both coastal plain — barred estuaries and drowned valley — bedrock estuaries from the Lower Paleocene of Ellesmere Island, Arctic Canada are illustrated (Fig. 1). Both types existed at about the same time but along different parts of the basin margin. From a procedural point of view, it should be stressed that the idea of these rocks representing ancient estuaries arose not from lithofacies analysis, but from the initial identification of relief on the sub-estuary unconformities. Subsequent facies analysis confirmed these early suspicions, but it should be noted that without the knowledge of valley geomorphology, the distinction between estuaries and some other paralic settings might have been less convincing.

Regional Setting

The Eureka Sound Group is an Upper Cretaceous-Paleogene siliciclastic succession that records the metamorphosis of sedimentary basins spanning the earliest to latest stages of Eurekan tectonism, a period of some 40 to 45 Ma (Balkwill, 1978; Miall, 1985; Ricketts, 1987, 1989). Current investigations (Ricketts, *op. cit.*, and work in progress; Stephenson and Ricketts, in press), provide the rationale for subdivision of the Eurekan succession into three, distinct, basinal phases:

1. A middle Campanian to Maastrichtian phase recording the end of the Sverdrup Basin regime of rifting and thermally dominated subsidence (*e.g.*, Stephenson *et al.*, 1987). Sedimentation at this stage was characterized by wave-dominated deltas and associated coastal and delta plain deposits.

2. A Paleocene to Middle Eocene phase, designated Fosheim Basin, partly superposed on Sverdrup Basin, but having greatly expanded depositional limits, overstepping deformed lower Paleozoic Franklinian bedrock. The Fosheim Basin is separated from the Sverdrup succession by a regional sub-Paleocene unconformity, below which some or all Maastrichtian rocks have been removed. Estuarine deposits that

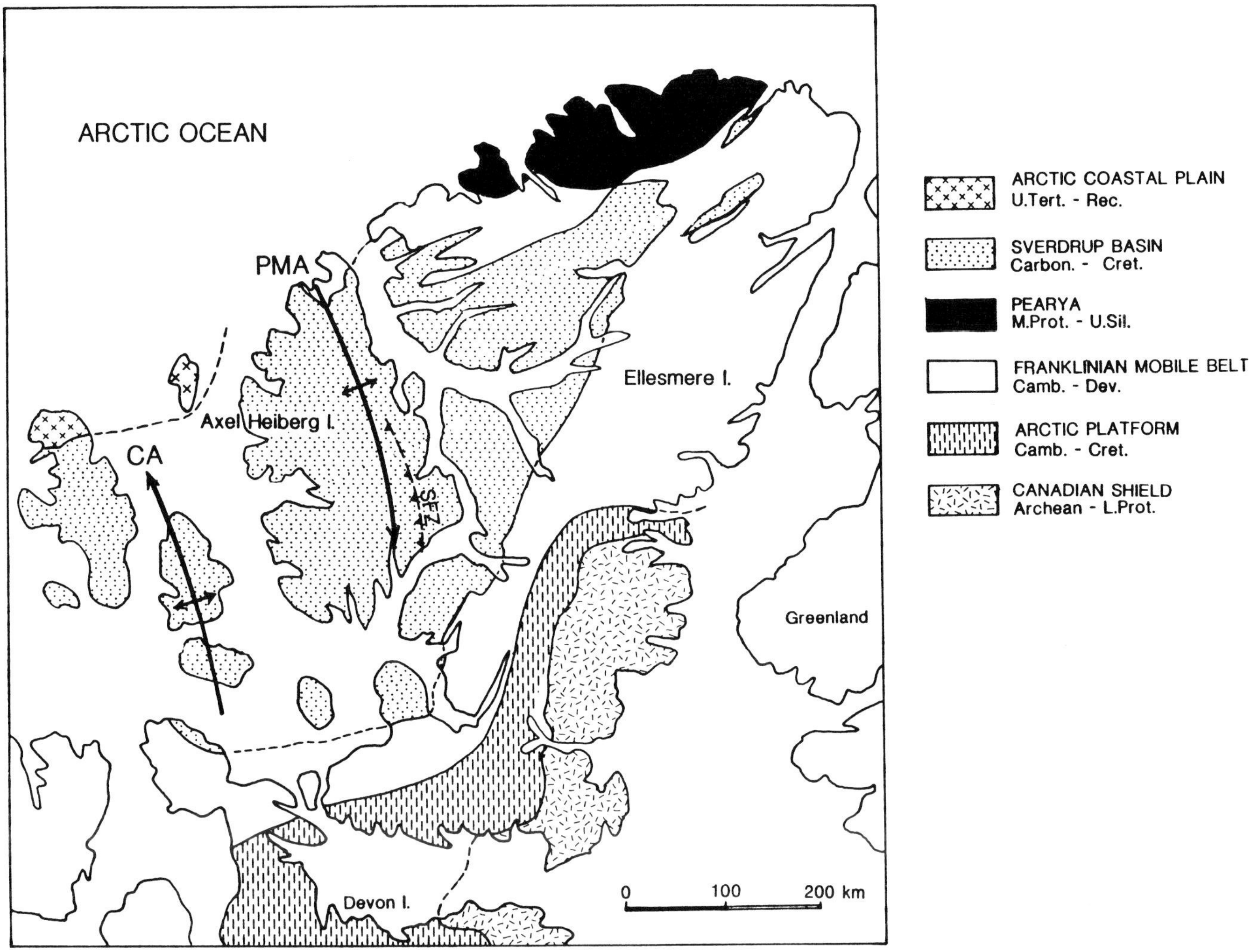

Figure 1. The simplified map of the eastern Arctic Archipelago shows the principal tectono-stratigraphic domains and major Eurekan structural elements: **CA**, Cornwall Arch; **PMA**, Princess Margaret Arch; **SFZ**, Stolz Fault (thrust) Zone. The Lower Paleocene-Eocene Fosheim Basin expanded well beyond the limits of the older Sverdrup Basin, encroaching upon the Franklinian Mobile Belt.

are the subject of this paper accumulated on this unconformity. The top of the Fosheim Basin succession also is defined by a profound discordance — the Eurekan unconformity. Igneous activity continued sporadically along the northern margin of the basin but instead of being dominated by tholeiitic magmas typical of the Sverdrup Basin phase, became bimodal and alkalic (Muecke *et al.*, 1990). Basin subsidence during this period, while having a thermal component, became increasingly influenced by compressional tectonism as the climactic phase of Eurekan deformation approached.

3. The principal phase of Eurekan deformation in the Middle Eocene (Ricketts, *in press* a), which fragmented the Sverdrup and Fosheim basins, producing several small, intermontane and related basins that commonly lie outboard of large, steeply dipping thrust faults.

Biostratigraphic Database

Good biostratigraphic control is now available for most of the Eurekan succession. For the critical interval in this study — Campanian to Early Paleocene — the database comprises primarily palynological studies (McIntyre and Ricketts, 1989; McIntyre *in press*), which provide temporal resolution of the sub-Paleocene unconformity over a wide area. Arenaceous foraminifera assemblages (Wall *et al.*, 1988; Wall, *in press*), and a sparse Inoceramid fauna (Ricketts, *in press* b) augment this data.

Depositional Sequences

The Eurekan succession has been divided into several third-order depositional sequences (1-10 Ma duration, *e.g.*, Vail *et al.*, 1977) that have basin-wide extent (Ricketts, 1989), and comprise a variety of delta, shelf, coastal and alluvial deposits (Fig. 2). These third-order sequences are in turn contained within much larger, second-order shoaling-upward cycles that are bound by major unconformities (Fig. 2); and are analogous to the T-R cycles recognized in Sverdrup Basin by Embry and Osadetz (1987).

A basin-wide sub-Paleocene unconformity forms the base

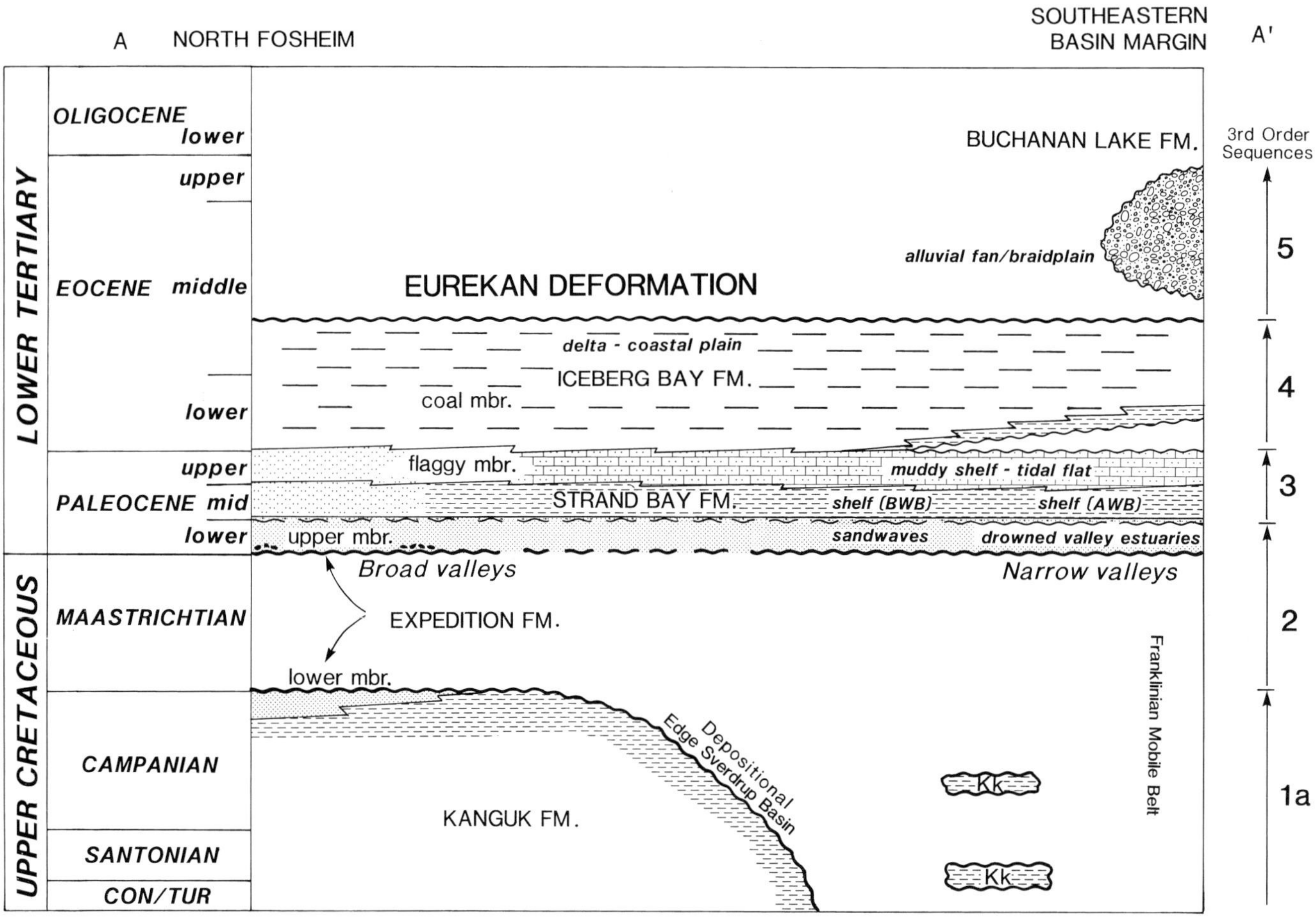

Figure 2. A schematic representation of the Eureka Sound Group time stratigraphy, sequence stratigraphy, and basin designation along transect AA', central Ellesmere Island (see Fig. 3). The depositional sequences are restored to their approximate positions before the main phase of Eurekan crustal shortening. The relative positions of the two types of estuarine valley fill are indicated with respect to the sub-Paleocene unconformity. Wavy lines indicate principal unconformities. Note that the syn- to late-orogenic conglomeratic facies of Sequence 5 does not actually occur along this transect, but along the west coast of Ellesmere Island — it is included here for completeness. BWB and AWB refer to shelf deposits below and above wave-base, respectively, in Sequence 3.

of Sequence 2; estuarine deposits accumulated on this surface (Fig. 2). In the eastern part of Fosheim Basin (Fig. 3) erosion removed part or all of the Maastrichtian component of the Expedition Formation (lower member). Here, estuarine deposits that initiated Early Paleocene sedimentation (basal part of the upper member, Expedition Formation), infill valleys eroded into relatively undeformed Sverdrup Basin strata, and herald an early Paleocene transgression. In contrast, approximately equivalent basal Paleocene estuarine deposits along the southern margin of Fosheim Basin directly overlie deformed Lower Paleozoic strata of the Franklinian Mobile Belt. The differences in bedrock lithology and structural style beneath the sub-Paleocene unconformity had a profound effect on the initial topographic relief, which in turn strongly influenced the paralic lithofacies composition and geometry. Some of these aspects are discussed below.

Paleogeomorphology

Fundamental differences in the lithology and structure of bedrock beneath the sub-estuary unconformity gave rise to different types of valley (Table 1).

Shallow Valleys

Broad, shallow valleys developed where the sub-Paleocene unconformity overlies relatively undeformed and weakly indurated Sverdrup Basin strata (geomorphic terms are intended to denote relative differences in relief only). Relief on the unconformity is identified where distinctive, white quartz sandstones making up the basal Paleocene succession are in contact with different stratigraphic levels of the subjacent units. For example, on north Fosheim Peninsula, the basal sandstones overlie (concordantly) different thicknesses of the Expedition Formation Lower member, and in places directly overlie Kanguk Formation shale (Figs. 4, 7). Where the lacuna is greatest, the basal Paleocene sandstone beds are

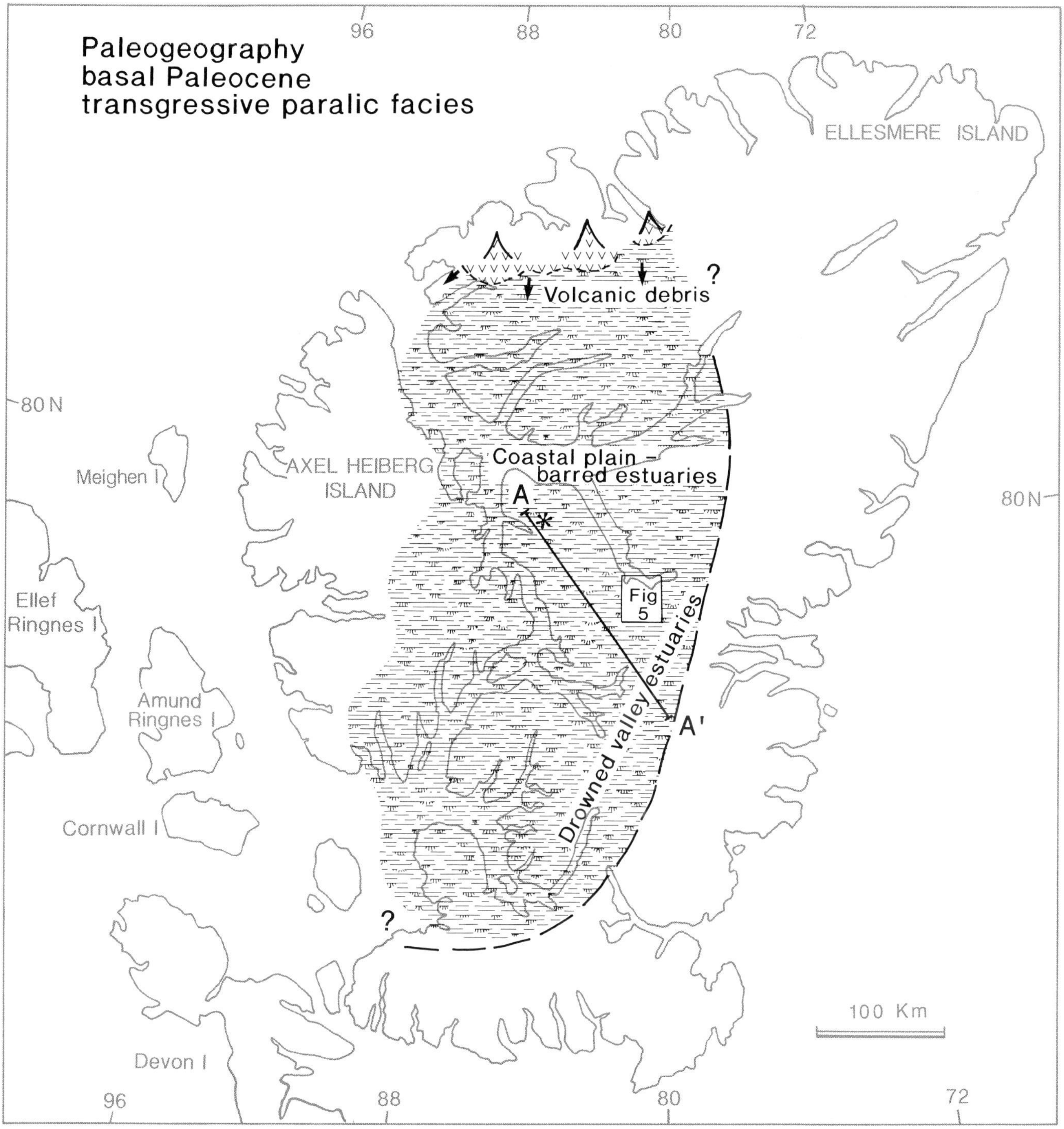

Figure 3. The paleogeography map of the basal, Lower Paleocene deposits of the Eureka Sound Group indicates schematically the distribution of the two principal valley types above the sub-Paleocene unconformity. Cross-section AA' is shown in Figure 2. Inset locates Figure 5. Asterisk locates Remus Creek.

thickest, and although along-strike exposures of this interval are discontinuous, the basal beds appear to thicken and thin in concert with the different levels of sub-Paleocene erosion. Thus, the valleys typically are kilometres to tens of kilometres wide and at least 100 m deep. They represent the product of relatively young, consequent drainage; drainage patterns were probably dictated by antecedent channels or distributaries in the older bedrock. In such cases, the sub-Paleocene unconformity is mostly an erosional vacuity.

Deep Valleys

Transgression and expansion of Fosheim Basin during the Early Paleocene (Ricketts, 1989) resulted in basal sediments extending well beyond the depositional limits of the previous

Table 1. Principal geomorphic characteristics of the two types of Lower Paleocene estuary valleys, Ellesmere Island, Arctic Canada

Broad, shallow valleys
Low relief
Kilometres wide and at least 100 m deep
Bedrock is undeformed Upper Cretaceous strata of the Fosheim Basin or Sverdrup Basin
Bedrock has low to moderate induration
No or very little microtopography (*i.e.*, local relief) on the sub-estuary surface
Tendency toward straightening of the coastline
Young, consequent drainage, probably controlled by antecedent fluvial or distributary channels

Narrow, deep valleys
(Relatively) high relief
100's of metres to kilometres wide and up to 200 m deep
Bedrock is lower Paleozoic, highly deformed, siliciclastic, carbonate and evaporitic rock
Bedrock generally is highly indurated
Pronounced local relief caused by structural discordances, or in the case of carbonate bedrock, local karst development
Embayed coastline
Very old consequent and subsequent drainage

Figure 4. The sub-Paleocene unconformity (arrow) is exposed at Remus Creek, north Fosheim Peninsula (Fig. 3), where all of the Maastrichtian section has been eroded. White sandstone and related units above the unconformity are part of the shallow valley-fill succession. The basal white sandstone unit (18 m thick — see Fig. 7) is underlain by a thin mudstone, which may represent the initial valley-drowning event.

Sverdrup Basin onto lower Paleozoic bedrock of the Franklinian Mobile Belt (Figs. 3, 5). The Paleozoic rocks were strongly deformed during the Devonian and early Carboniferous, and pre-Paleocene erosion patterns clearly were controlled by the antecedent structural patterns. The existence of considerable topographic relief on the sub-Paleocene surface can be observed in many localities, where basal Paleocene strata infill depressions in the bedrock and progressively overstep valley walls (Fig. 6). Distinctive sandstone units, that can be used as local stratigraphic markers, can be traced across paleovalleys to the point where they directly overlie bedrock.

Dissection of the lower Paleozoic bedrock into relatively narrow (hundreds of metres to kilometres wide), deep valleys took place over a protracted lacuna that involved several episodes of erosion and nondeposition. The drainage is therefore a mixture of much older consequent and subsequent patterns. Where basal Paleocene deposits overlie Paleozoic carbonate, limestone regoliths and shallow karst features (shallow dissolution joints) developed. Some of these weathering features were responsible for the micro-relief developed in the more deeply incised valleys (Fig. 6); this kind of relief is not observed in the broad, shallow valleys.

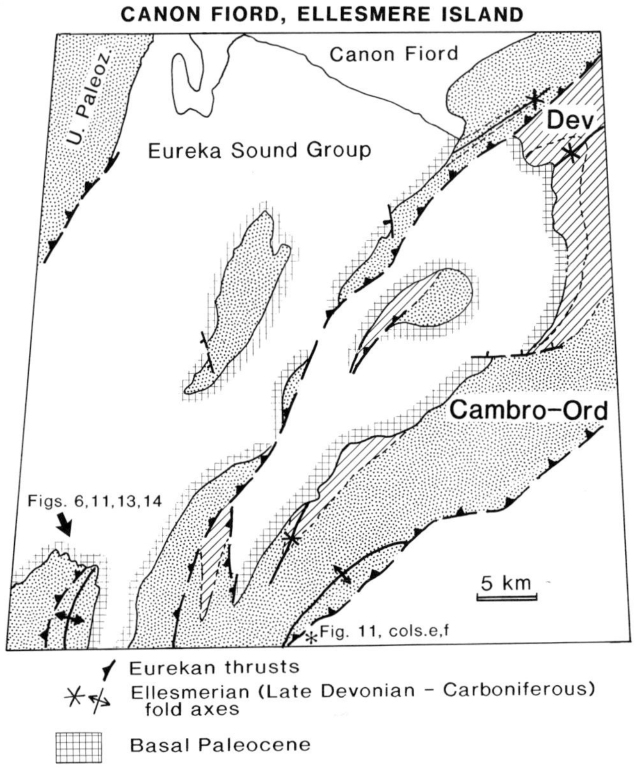

Figure 5. Basal Paleocene strata in the southeastern part of Fosheim Basin encroached upon a dissected landscape underlain by deformed Cambrian to Devonian bedrock. Blank regions in this sample of the Canon Fiord map area are mostly underlain by Eureka Sound Group. The relationship of the basal Paleocene contact to Ellesmerian deformation patterns is clearly illustrated here, with deep-valley estuarine fill located primarily in 'synclinal' valleys. Basal Paleocene units, especially the distinctive white sandstone beds, are observed to thin dramatically as they overstep and abut 'highs' in the bedrock, with angular discordances up to 25° (Fig. 6). Deformation of the Eurekan succession is relatively minor. Open folds (not included on this map) produced during Middle Eocene deformation, are to a large extent controlled by pre-existing Paleozoic structural trends. Dev, Devonian; Cambro-Ord, Cambrian to Ordovician bedrock. Bedrock geology simplified after Thorsteinsson (1972).

Shallow Valley-Fill Stratigraphy

The example illustrated in Figure 7 is situated at Remus Creek, near the north end of Fosheim Peninsula; examples of the various lithofacies are taken from other localities. Here, basal Paleocene strata, about 100 m thick, rest variably on upper Campanian coastal plain deposits (lower member, Expedition Formation) or directly overlie older Kanguk Formation shale. About 100 m of stratigraphic downcutting is estimated for this area over a lateral (parallel to strike) distance of 20 to 30 km, providing approximate limits for the depth of the valley. The basal contact is marked by distinctive quartz sandstone, locally containing unusual dacitic pebbles (Ricketts, 1987). Fluvial deposits gradationally overlie the paralic facies.

The valley-fill stratigraphy, lithofacies and inferred relative sea level changes are summarized in Figure 7.

White Crossbedded Sandstone Facies

Distinctive, clean, well sorted, white quartz sandstone at the base of the valley-fill succession, comprises a laterally extensive (although not necessarily continuous) unit, the thickness of which varies according to its relationship to levels of erosion below the sub-Paleocene surface (as noted above, thickest where the greatest amount of Maastrichtian strata was removed). The sandstones, up to 18 m thick, are characterized by abundant, large planar-tabular and trough crossbeds, up to a metre in set thickness. The planar-tabular varieties are usually arranged into stacked units, separated by pebble or mudchip lags, and having foreset azimuths toward the east and southeast; namely, toward the paleoshoreline. Reactivation surfaces and carbonaceous veneers are common, the latter separating couplets 1 to 2 cm thick. The trough crossbeds are usually contained within larger scour troughs that have coarse lags of pebbles and wood, and local scour relief up to 1.5 m (Fig. 8). Oversteepened crossbedding and ball-and-pillow structures are common. These larger scour or channel structures truncate the stacked planar-tabular crossbedded units. There are no discernible fining- or coarsening-upward trends.

Very large planar-tabular crossbed sets up to 6.5 m thick are encountered in a few places, within or overlying the white crossbedded sandstone facies (*e.g.*, 60 m level, Fig. 7). Most have a basal veneer of pebbles but show only minor scouring. Foresets usually have abrupt toesets and contain reactivation surfaces.

Carbonaceous mudstone and shaly coal overlie white crossbedded sandstone facies with abrupt but conformable contact. At the 30 m level in Figure 7, the crossbedded sandstones are overlain by slightly finer grained, more tabular layered sandstone containing low-angle planar crossbeds; this transition again is abrupt but conformable.

Coaly Mudstone Facies

Carbonaceous mudstone and thin, usually shaly coal are interbedded with thin, brown, fine grained sandstone, in units up to 5 m thick. Contact with the overlying or underlying white crossbedded sandstone facies generally is abrupt but with other facies is commonly gradational (*e.g.*, sandstone-mudstone facies). Most of the sandstones (only

Figure 6. Basal Paleocene estuarine deposits progressively overstep and fill (toward the left) a valley eroded into deformed Ordovician carbonate, south of Canon Fiord (see Fig. 5); note the microtopography on the unconformity (dots). A basal sandstone unit (inlet-spit facies) in the foreground and middleground (arrows), directly overlies bedrock on the left, but is separated from bedrock by increasingly thick sandstone and mudstone toward the right.

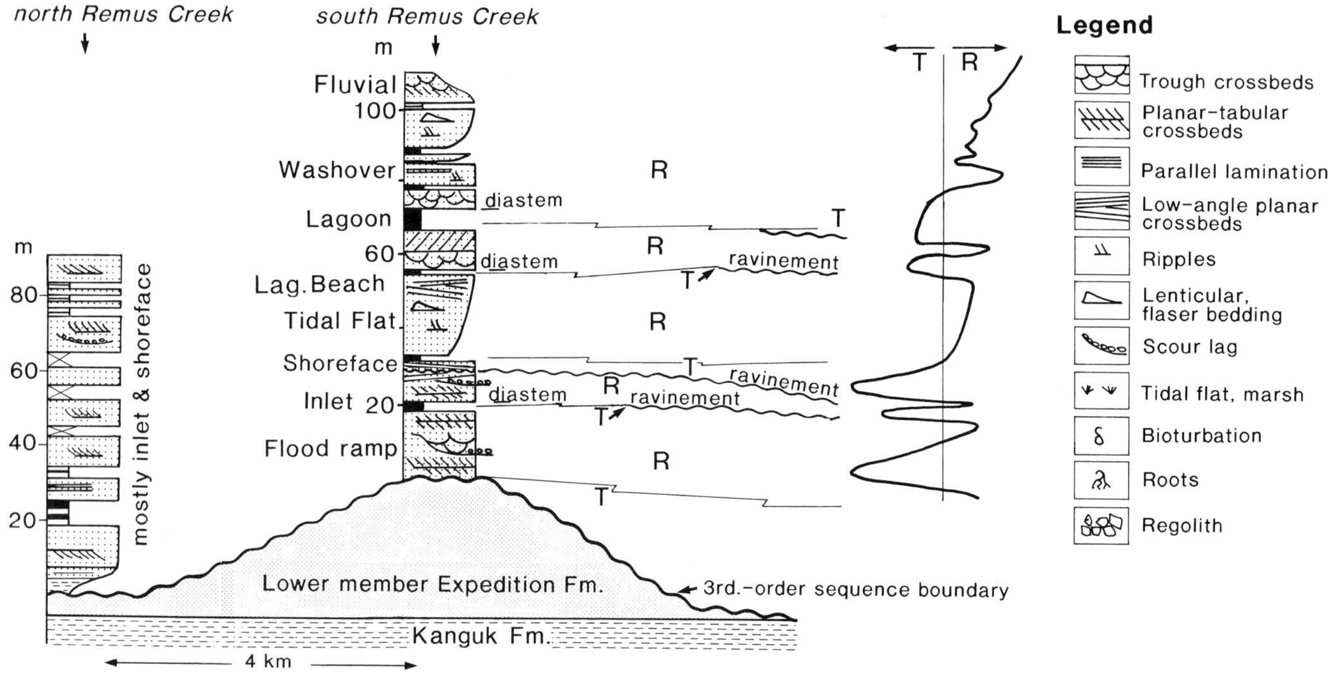

Figure 7. A schematic representation of the basal Paleocene stratigraphy in two sections at Remus Creek (Fig. 3), showing different levels of downcutting of sub-Paleocene strata (Lower member, Expedition Fm.), major facies, regressive and transgressive components, diastem and possible ravinement surfaces, and relative sea level changes. At the south Remus Creek section, 100 m of Lower member strata remain; at the north section about 10 m remain below the unconformity. Detailed facies assessment is possible only at the southern section. In the southern section there is a general stratigraphic trend toward shallower conditions. Except at the 30 m level, where shoreface sandstone abruptly overlies inlet facies (see Fig. 9a), the transition from transgression to regression occurs within the lagoonal facies because the limit of shoreface erosion lay seaward of the line of section. Lag. beach refers to the back-barrier beach facies at the 50 m level.

Figure 8. Large, stacked trough crossbeds, lined with pebble lags (arrows) from the barrier-inlet facies at Remus Creek (about the 10 m level in Figure 7), overlain by lagoonal marsh deposits.

a few centimetres thick) consist of carbonaceous, parallel laminae and any crossbedding is restricted to small ripples. Root structures are common; bioturbation is uncommon and form-genera are indeterminate. Some coaly layers contain well preserved *Metasequoia* leaves and cones.

SANDSTONE-MUDSTONE FACIES

Coarsening- and sandier-upward units of fine, brown, carbonaceous sandstone are in gradational contact with the subjacent coaly mudstone facies (Fig. 9a). Typically, the facies is thin bedded (less than 20 cm thick), the proportion of interbedded mudstone decreasing upward, in concert with an increase in the number and size of crossbeds. Flaser, lenticular crossbedded, and wavy bedded sandstone are common. Larger ripple structures (up to 20 cm thick) possess reactivation surfaces (Fig. 9b). A limited trace fossil assemblage occurs including *Skolithos, Thalassinoides, Chondrites* and *Planolites*. Some coarsening-upward units, for example at the 50 m level in Figure 7, are capped by relatively clean, low-angle planar crossbedded sandstone, similar to that which caps the white crossbedded sandstone facies at 30 m; the important difference in this facies is that the crossbedded sandstones are part of the coarsening-upward trend as shown in Figure 9c, rather than in abrupt contract (as in Fig. 9a).

Figure 9a. The transition from ravinement surface and overlying shoreface deposits (dotted line, lower left), to lagoonal and tidal flat facies in the back-barrier realm. Remus Creek, 30-40 m level, Figure 7. Hammer is 33 cm long.

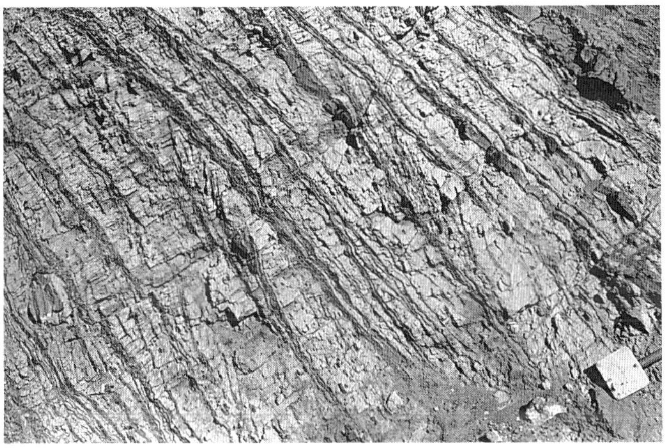

Figure 9b. Lenticular, flaser and wavy bedding typical of the tidal flat facies at Remus Creek, 40 m level, Figure 7. Larger crossbeds have reactivation surfaces and mud drapes. Scale (bottom right) is 15 cm wide.

Figure 9c. A coarsening, sandier-upward unit in the tidal flat back-barrier beach facies, near Vesle Fiord. Note the small sand-filled channel. This facies is also associated with barrier washover sandstone facies in Figure 10.

Sandstone Wedge Facies

This facies, although composed of clean, well sorted quartz sandstone, is arranged into beds generally less than 25 cm thick, separated by veneers or thin beds of black, carbonaceous mudstone. Overall, the facies has a distinct wedge geometry — individual sandstone beds converge and are replaced by mudstone like that of the coaly mudstone facies (Fig. 10). The general direction of pinchout is to the southeast. Internally, the sandstone beds are laminated and rippled, or contain small, stacked planar-tabular crossbeds; comminuted plant debris is common. Bioturbation is not common. Most examples of this facies overlie strata of the white crossbedded sandstone facies.

Figure 10. An example of the barrier washover facies exposed in the Vesle Fiord area. Note the progressive thinning to the right (eastward) and the interfingering lagoonal mudrocks. This example overlies barrier-inlet facies sandstone.

Fining-Upward Facies

At Remus Creek, fining-upward units of sandstone, mudstone and shaly coal occur only above the 100 m level (Fig. 7); at other localities the facies occurs lower in the succession and usually is associated with coaly mudstone or sandstone-mudstone facies. Trough crossbeds and ripples are common. Sand-mud couplets occur in some units, and where exposure is reasonable, can be seen to be arranged into large foresets, similar to inclined heterolithic stratification of Thomas *et al.* (1987).

Interpretation of Facies

Interpretation of lithofacies and their relationships to inferred excursions of the strand line are shown in Figure 7. The white crossbedded sandstone facies bears the hallmarks of tidal inlets (Kumar and Sanders, 1974; Hennessy and Zarillo, 1987). Direct evidence of tidal current asymmetry is found only in the reactivated surfaces of stacked planar-tabular crossbeds; paleocurrent directions (toward the east and southeast) indicate a predominance of flood tidal conditions. Most examples of inlets and associated tidal deltas occur within barrier-lagoonal or estuary-lagoonal settings.

An expected process in such settings involves the lateral migration of inlets resulting in laterally extensive sediment bodies (Reinson, 1984). Very large planar crossbed sets that cap some units may indicate local spit formation, perhaps migrating over the inlet deposits as the inlet itself migrated laterally.

Coaly mudstone facies indicate depositional regimes at the other extreme, being dominated by suspension and weak bedload transport conditions. The presence of root structures attests to the in situ accumulation of much of the plant debris in a lagoon-marsh setting.

Sedimentary structures in the sandstone-mudstone facies are typical of mixed sand-mud tidal flats (*e.g.*, Reineck and Wunderlich, 1968). Coarsening-upward trends and concomitant changes in crossbed types indicate the transition from lagoonal to paralic, or back-barrier shoreface and beach settings. Tidal flat deposits are usually separated from the flood ramp/inlet facies by a stage of lagoon deepening or by a minor stratigraphic discordance (*e.g.*, a channel diastem).

Large-scale wedges of white sandstone are interpreted as washover deposits; the overall geometry and interfingering relationship with lagoonal muds helps to distinguish these deposits from the tidal inlets and deltas (*e.g.*, Hennessey and Zarillo, 1987). Washover direction was lagoonward, and is consistent with the observed direction of flood-tide sediment transport. Most of the washover fan or lobe deposits overlie crossbedded inlet facies sandstones, and in such cases, may be all that remains of the subaerial component of a barrier island or spit.

The fining-upward facies compares with high sinuosity fluvial or tidal channel stratigraphic profiles. A similar facies is developed in the deep valley-fill succession, where it is also better exposed; more discussion of the facies is provided herein.

Excursions of the Strandline

Excursions of the strandline represented by the stratigraphic fill and in relation to various surfaces of erosion or hiatus, are indicated on Figure 7. Initial drowning of the drainage basin, and thus inception of estuary deposition, resulted in the development of a barrier-spit system, backed by river-fed lagoons, fluvial and tidal channels. Overall, the lower 100 m of the succession remains within the barrier-inlet-lagoon setting. Only in the latter stages of valley fill, when relief on the sub-Paleocene surface had been effectively subdued, did the fluvial component of estuarine channels prograde over the paralic deposits. Likewise, correlative offshore deposits on the adjacent shelf only occurred seaward of the preserved section (shelf deposits outcrop in other parts of the region). Relative stability of the strandline over the period represented by the basal Paleocene succession suggests that sedimentation rates kept pace with rising relative sea level. Nevertheless, the secular change in sea level (on the scale of third-order sequences) indicates overall progradation.

An important consequence of this stable shoreline condition concerns the possible paleogeographic positions of successive shoreface ravinement surfaces that were the product of repeated smaller-scale transgressions (Fig. 7). In the Remus Creek section, a possible ravinement surface is preserved at the 30 m level (Fig. 7), where shoreface deposits abruptly overlie the flood ramp/inlet facies. Tidal inlet deposits will be preserved during transgression if they are deposited below the level of shoreface retreat (Demarest and Kraft, 1987). In the above example the succeeding regressive phase produced lagoon-marsh, tidal flat-channel and back-barrier beach deposits. In all other cases where relative deepening is implied by abrupt stratigraphic changes in facies, the landward limit of shoreface ravinement may never have reached the line of section. However, any landward shift in the shoreline will produce changes in at least the local base levels, even if the ravinement surface did not encroach upon the line of section. If this situation existed in the Remus Creek sections, then the transition from transgression to regression probably occurs within the lagoonal mudstone unit as indicated in Figure 7.

Deep Valley-Fill Stratigraphy

Basal Paleocene strata along the eastern and southern margins of Fosheim Basin overlie bedrock ranging in age from Early Ordovician to Lake Devonian. The criteria for paleovalley recognition have been discussed above (Fig. 6). Lithofacies associations in a typical valley-ridge system are illustrated in Figures 11 and 12. A diagnostic feature of the facies is their abrupt lateral changes both within and between adjacent valleys. The valley-fill succession is overlain by laterally extensive coastal plain and delta plain deposits that make up the remainder of the Upper member, Expedition Formation.

Trough Crossbedded Facies

Trough crossbedded sandstone (and the commonly associated foreset-bedded facies described below) is similar in some respects to the white crossbedded sandstone facies described from the shallow valley estuarine fill successions: they are quartz rich, clean and well sorted. Nevertheless, there are significant differences in internal organization to warrant separate discussion. Like many of the facies in the deep valley successions, the sandstone beds (up to 10 m thick) overstep or abut the valley walls, eventually lying directly on bedrock.

Trough crossbedded sandstone forms isolated channel structures, ranging in width from a few metres to tens of metres, or comprises channels that are laterally associated with the foreset-bedded facies. Planar-tabular crossbeds also occur in stacked units. Isolated channel structures are commonly, but not always, spatially associated with local topographic highs and lows (microtopography) on the bedrock surface. Channels may be eroded into, or juxtaposed with strata of the coaly mudstone facies, or the (coarsening-upward) sandstone-mudstone facies (like those of the shallow valley deposits).

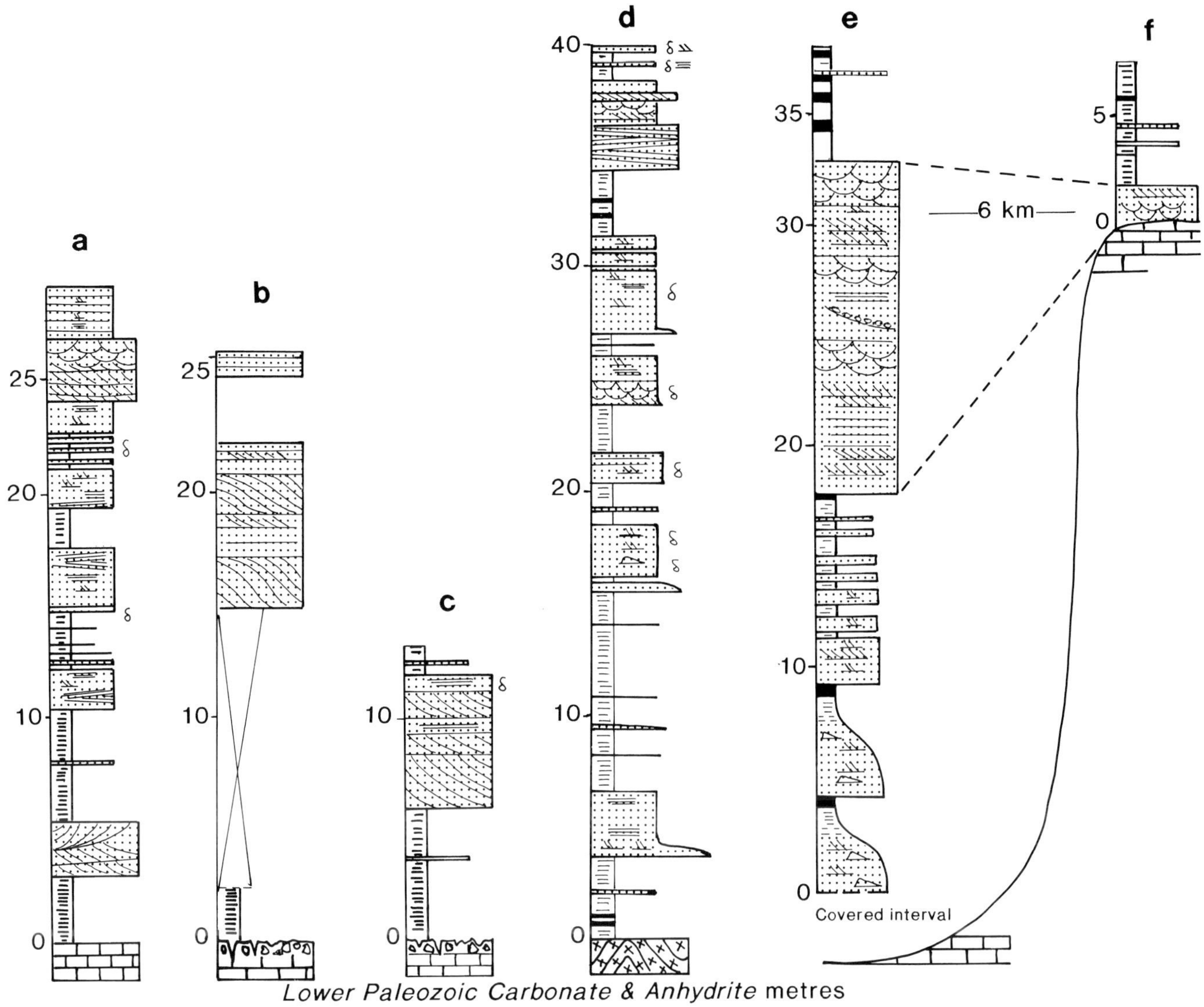

Figure 11. Simplified stratigraphic columns of the basal Paleocene, deep valley-fill from two areas south of Canon Fiord (Fig. 5), illustrating the abrupt lateral and vertical facies changes. Columns **a** to **d** are from the Mount Moore area (see Fig. 12) where key sandstone beds can be 'walked out' in relation to the bedrock paleotopography. Columns **e** and **f** are located in the footwall of Parrish Glacier Thrust — the distinctive white sandstone marker can also be traced along strike. A covered interval between the base of column **e** and the bedrock is a few metres thick. No specific correlation between the Mount Moore and Thrust localities is implied. The legend is the same as that for Figure 7.

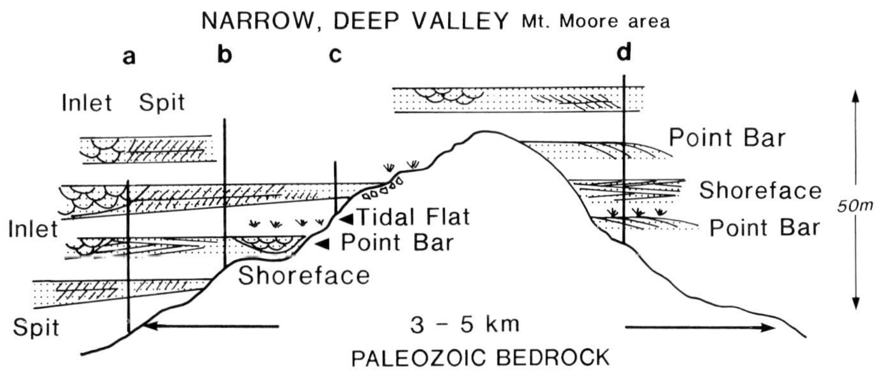

Figure 12. A schematic representation of valley-ridge topography and the positions of major sandstone bodies, as mapped south of Canon Fiord; columns **a** to **d** correspond to those in Figure 11 and are located on Figure 5. Total thickness here is about 50 m; the estuarine deposits are overlain by laterally extensive coastal plain strata.

Foreset-Bedded Facies

This, the most conspicuous sandstone facies in the deep valley-fill succession, consists of large planar-tabular crossbeds with individual sets up to 6 m thick, but more commonly 2 to 3 m thick (Fig. 13). Crossbed sets occur individually (*i.e.*, make up an entire bedding unit) or are stacked two to three sets deep. Individual crossbeds have an overall wedge geometry and can be traced laterally for 50 m and more. Contact with underlying coaly mudstone facies or the coarsening-upward sandstone-mudstone facies is usually abrupt and planar (non-scoured). A few of the larger foreset units are overlain by a topset sandstone that contains parallel laminae and small-scale trough crossbeds. Crossbed set contacts are veneered with mudchips and wood fragments. Reactivation surfaces are common; however, none of the foresets are draped with mud.

Figure 13. This example of the foreset bedded sandstone facies (Fig. 5) is 2.6 m thick and is overlain by laminated and rippled sandstone comprising a topset unit. The foreset unit thins abruptly to the right. Hammer (circled) is 33 cm long.

Trace fossils, including *Skolithos, Paleophycus, Planolites*, and indeterminate meniscate burrows, are only locally abundant and generally are restricted to the laminated sandstones. Pockets of *Skolithos* occur on the primary set contacts between large planar crossbeds.

Fining-Upward Facies

Fining-upward sandstone units up to 5 m thick contain a distinctive suite of sedimentary structures and overall geometry. The sandstone and mudstone interbeds are arranged into pairs and couplets, wherein the proportion of mudstone increases upward; many units are capped by shaly coal beds up to 50 cm thick (Fig. 14a, b). The sandstone bundles form wedges or inclined foresets, analogous to the inclined heterolithic stratification of Thomas *et al.* (1987). Parallel laminae, ripples and starved ripples are common; some bioturbation occurs but burrow form is usually indeterminate. Fining-upward units are locally truncated by the trough crossbedded sandstone facies, but are gradationally juxtaposed or superposed with the sandstone-mudstone and coaly mudstone facies.

Figure 14a. An example of the fining-upward facies (and inclined heterolithic stratification — arrow) a few metres above the bedrock contact. Photograph taken south of Canon Fiord — see Figure 5. The more massive sandstone immediately overlying the fining-upward unit here corresponds to the sandstone shown in Figure 6 (arrows) that directly overlies the bedrock.

Figure 14b. This closer view of the fining-upward facies shows a typical succession of sandstone bundles and some mudstone couplets with a veneer of sand (arrows). Field pack for scale, rests on scree, lower left.

Sandstone-Mudstone Facies and Coaly Mudstone Facies

Coarsening-upward units of interbedded sandstone and mudstone, similar to those encountered in the shallow valley-fill successions, are usually in gradational contact with the coaly mudstone facies. Several coarsening-upward units are capped by sandstone that contains low-angle planar cross-beds; *Skolithos* and *Muensteria* are locally abundant. Small channels, usually less than a metre thick, locally truncate these units. Contact with trough crossbedded sandstone is erosional; that with foreset bedded sandstone is abrupt but usually planar.

Interpretation of Facies

The internal organization of the trough crossbedded facies is similar to that of the white crossbedded sandstone facies, but the trough crossbedded facies is generally of more limited lateral extent, in places forming isolated channels. The only direct evidence of tidal current asymmetry is the presence of reactivation surfaces (reactivation surfaces have been recorded in fluvial deposits), although their association with what are clearly tidal flat deposits more convincingly points to a tidally influenced environment. Interpretation of the trough crossbedded facies is aided by its association with the foreset bedded facies. Comparison is made here with the Late Pleistocene spit systems of Denmark described by Nielsen *et al.* (1988) which consist of large planar foreset beds formed during progradation of a subaqueous spit platform, and a topset unit that represents the subaerial portion of the spit. Attachment of the basal Paleocene spits to bedrock headlands is clearly demonstrated here. Most foreset units appear to have prograded in the direction of valley deepening; in Figure 13 this is toward the northeast. Thus the spatial association of trough crossbedded and foreset bedded facies can be interpreted as a system of spits and tidal inlets. The dearth of trace fossils in the spit facies, including the more opportunistic species like *Skolithos*, may be a function of high sedimentation rates (unstable substrate) or reduced salinity. Stacked planar-tabular crossbeds in the inlet facies, like their counterparts in the shallow valley-fill successions, may represent tidal delta ramps; however, it is not known whether the ramps were flood- or ebb-dominated. Isolated channel structures are perhaps tidal channels and/or the lagoonward equivalents of the inlets.

The fining-upward facies, although superficially resembling high sinuosity fluvial deposits, contains two features that are diagnostic of estuarine point bars: accretionary foreset bedding of sandstone bundles and (carbonaceous) mud couplets (Fig. 14b), and bioturbation. Mud couplets and tidal bundles have been described from modern and ancient estuaries (Boersma and Terwindt, 1981; Smith, 1988), and represent the transition from a dominant flood or ebb flow to slack water (*i.e.*, tidal reversals). The point bar deposits are associated with tidal flat — back barrier beach (sandstone-mudstone) and lagoon-marsh (coaly mudstone) facies.

The overall setting is one of relatively small valleys, where the transition from fluvial and estuarine channels to the shallow shelf is marked by small, attached spits and bars and associated inlets. This facies configuration is typical of classical drowned bedrock valleys. Landward of the spits, the tidal flats, marshes and channels have a 'lagoonal' character, but are of more limited lateral extent than typical lagoons. Like the shallow valley-fill succession, successive estuaries developed in response to rising relative sea level. The fluvial-estuarine channel component is more commonly developed here, indicating greater seaward excursions of the shoreline during regressive stages. No shelf deposits have been found within these portions of the estuarine valley fill; arguments concerning the presence or absence of ravinement surfaces are basically the same as those presented for the shallow valley estuarine systems.

Discussion — A Comparison of Estuary Types

Following regional Maastrichtian uplift and emergence that created the sub-Paleocene unconformity, Early Paleocene transgression initiated deposition within Fosheim Basin by progressive eastward and southward onlap and infilling of coastal valleys. Despite the fact that individual small-scale cycles (a few metres thick) in the basal Paleocene deposits are regressive in nature, the stratigraphic package overall represents the transgressive component of a basin-wide, third-order stratigraphic sequence, the upper part of this sequence comprising regressive delta facies. In the portions of Fosheim Basin discussed here, there is no evidence for shelf sedimentation that succeeded the estuarine valley fill. This means that the putative submarine unconformity inferred in some Sverdrup Basin sequences (Embry and Podruski, 1988) — the product of mid- to outer-shelf erosion — is not present in the marginal zone of the Early Paleocene Fosheim Basin. The transition from a transgressive to regressive regime could in this case correspond to the highest ravinement surface in the estuarine succession, or landward of this zone, to a laterally equivalent conformity.

Deposition of basal Paleocene paralic sediments took place in two different geomorphic settings along the eastern and southern basin margins (Fig. 15):

1. A coastal region characterized by relatively young, broad, shallow valleys.
2. A second type that had significantly greater relief imposed by a system of deeper, narrower valleys — the products of a much older drainage pattern.

The principal attributes of lithofacies contained within each valley type are summarized in Table 2 and Figure 15.

The shallow estuarine valley systems can be regarded in a general way as part of a broad coastal plain, possessing a coastline that was effectively straightened by barrier island construction. In contrast, the higher relief bedrock valleys imparted an embayed character to the coastline, although even here coastline straightening would have eventually

Table 2. Principal lithofacies in the two types of basal Paleocene, estuarine valley fill

Broad, shallow valleys	Narrow, deep valleys
Laterally extensive facies	Laterally restricted facies
Barrier island – coastal plain	Spits (lower estuary), with or without tidal inlets, deltas
Tidal inlet and flood delta, shoreface-beach	Narrow tidal flats
Washover, lagoon, marsh, tidal flat, back-barrier beach	Point bars, mudstone couplets, tidal sandstone bundles
Lagoons transitional to tidal-influenced channels	

N.B. The similarity to shallow valley-fill facies and facies geometries may converge as bedrock relief is subdued and coastal segments become less embayed.

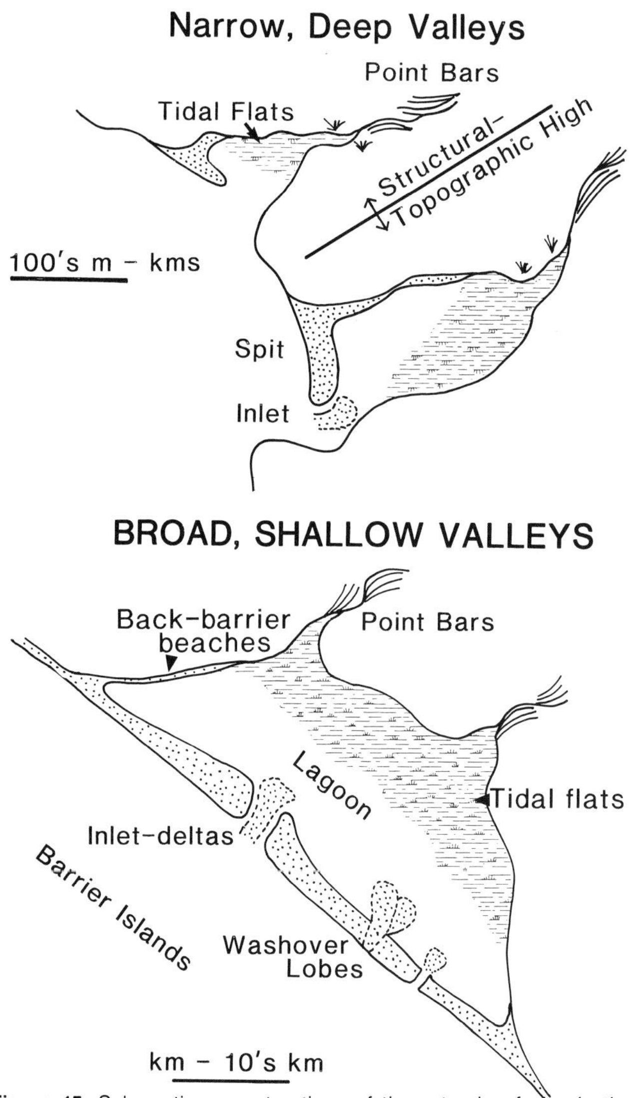

Figure 15. Schematic reconstructions of the estuarine facies in the shallow and deep valleys. The fundamental differences in geomorphic expression and scales between the valleys are emphasized (see also Tables 1 and 2).

occurred. The resulting sedimentary facies in each valley type, despite many similarities in their internal organization, display quite different geometries. Facies in the shallow valley-fill successions generally have greater lateral continuity than their counterparts in the higher-relief valleys. Clearly however, if knowledge of lateral facies extent and geometry of the sub-estuary surface are lacking, distinction between these two types of estuary would be difficult.

The tripartite subdivision of estuaries, alluded to in the introductory comments, can be envisaged for the two examples discussed here. In the deep bedrock valleys, the fluvial-estuarine component is commonly developed, and the tripartite subdivision of upper fluvial estuary — middle estuary (or funnel as referred to by Allen, 1989) — and lower estuary mouth or outer marine regions, seems to be applicable. For the shallow valley types, this three-fold subdivision seems somewhat forced in that the upper estuary-fluvial component is not always present (*i.e.*, the uppermost region of tidal influence in an estuary). When sedimentation rates within estuary-lagoon-barrier systems are capable of keeping pace with relative sea level rise, there may be no major seaward or landward shift in facies belts. This situation existed in the Remus Creek section; from a regional perspective this part of Fosheim Basin (underlain by thick Sverdrup Basin and Franklinian successions) was undergoing more rapid subsidence than the part of the basin that directly overlies the Franklinian rocks. Thus, where one of the components of the tripartite subdivision is missing, the distinction between different estuary types is difficult, and the more basic distinction between estuaries and other paralic settings even more problematic — unless there is independent knowledge of the nature of the sub-estuary surface.

Different types of estuary can exist at the same time within the same sedimentary basin, depending on the history of the antecedent drainage, the nature of the bedrock (differences in induration and structure), and different rates of subsidence (relative sea level change) and sediment supply. As drowned bedrock valleys, like those described here, are filled and the topographic relief subdued, subsequent estuaries will resemble those developed on shallow valley — coastal plains. Therefore, estuary types within a single valley system will change, in addition to the various estuary types that will form in response to different initial conditions along the basin margin.

Acknowledgments

This paper is an outgrowth of a general synthesis of what has been learned about Eurekan sedimentation over the last few years. Over this period I have benefited greatly from discussions with a large number of people, especially J. Dixon, A. F. Embry, and D. Leckie (ISPG, Calgary). D. J. McIntyre and J. H. Wall (ISPG) generously contributed biostratigraphic data. Reviewers A. F. Embry and J. R. Suter (Exxon, Houston) provided thoughtful critiques that resulted in an improved text.

References

Allen, G. P. 1989. Sedimentary facies and patterns in the Gironde Estuary: A recent model of macrotidal estuarine sedimentation. Second International Research Symposium on Clastic Tidal Deposits, August 22-25, 1989, Calgary, Program and Abstracts, p. 2.

Balkwill, H. R. 1978. Evolution of Sverdrup Basin, Arctic Canada. American Association of Petroleum Geologists, Bulletin, v. 62, p. 1004-1028.

Boersma, J. R. and Terwindt, J. H. J. 1981. Neap-spring tide sequences of intertidal shoal deposits in a mesotidal estuary. Sedimentology, v. 28, p. 151-170.

Demarest, J. M. II and Kraft, J. C. 1987. Stratigraphic record of Quaternary sea levels: Implications for more ancient strata: *In:* Sea-Level Fluctuation and Coastal Evolution, D. Nummedal, O. H. Pilkey and J. D. Howard (eds.). Society of Economic Paleontologists and Mineralogists, Special Publication 41, p. 223-239.

Embry, A. F. and Osadetz, K. G. 1987. Stratigraphy and tectonic significance of Cretaceous volcanism in the Queen Elizabeth Islands, Canadian Arctic Archipelago. Canadian Journal of Earth Sciences, v. 25, p. 1209-1219.

Embry, A. F. and Podruski, J. A. 1988. Third-order depositional sequences of the Mesozoic succession of Sverdrup Basin. *In:* Sequences, Stratigraphy, Sedimentology: Surface and Subsurface, D. P. James and D. A. Leckie (eds.). Canadian Society of Petroleum Geologists, Memoir 15, p. 73-84.

Fletcher, C. H. III, Knebel, H. J. and Kraft, J. C. 1990. Holocene evolution of an estuarine coast and tidal wetlands. Geological Society of America, Bulletin, v. 102, p. 283-297.

Frey, R. W. and Howard, J. D. 1986. Mesotidal estuarine sequences: A perspective from the Georgia Bight Sedimentology, v. 56, p. 911-924.

Galloway, W. E. 1989. Genetic stratigraphic sequences in basin analysis I: architecture and genesis of flooding-surface bounded depositional units. American Association of Petroleum Geologists, Bulletin, v. 73, p. 125-142.

Harris, P. T. 1988. Large-scale bedforms as indicators of mutually evasive sand transport and the sequential infilling of wide-mouthed estuaries. Sedimentary Geology, v. 57, p. 273-298.

Hennessy, J. T. and Zarillo, G. A. 1987. The interrelation and distinction between flood-tidal delta and washover deposits in a transgressive barrier island. Marine Geology, v. 78, p. 35-56.

Kjerfve, B. and Magill, K. E. 1989. Geographic and hydrodynamic characteristics of shallow coastal lagoons: *In:* Physical processes and sedimentology of siliciclastic-dominated lagoonal systems, L. G. Ward and G. M. Ashley (eds.). Special Issue, Marine Geology, v. 88, p. 187-199.

Knebel, H. J., Fletcher, C. H. III and Kraft, J. C. 1988. Late Wisconsinian-Holocene paleogeography of Delaware Bay; A large coastal plain estuary. Marine Geology, v. 83, p. 115-133.

Kumar, N. and Sanders, J. E. 1974. Inlet sequence: a vertical succession of sedimentary structures and textures created by the lateral migration of tidal inlets. Sedimentology, v. 21, p. 491-532.

Leckie, D. 1988. Sedimentology and sequences of the Paddy and Cadotte members along Peace River: Field Trip Guide to: Sequences, Stratigraphy, Sedimentology: Surface and Subsurface. Canadian Society of Petroleum Geologists, 78 p.

______ and Staniland, M. 1988. Regional outcrop and subsurface geology. *In:* Leckie, D., Sedimentology and sequences of the Paddy and Cadotte members along Peace River. Field Trip Guide to: Sequences, Stratigraphy, Sedimentology: Surface and Subsurface. Canadian Society of Petroleum Geologists, p. 59-70.

McIntyre, D. J. (*in press*). Appendix 3. Palynology: *In:* Delta evolution in the Eureka Sound Group, western Axel Heiberg Island: The transition from wave-dominated to fluvial-dominated deltas, B. D. Ricketts (ed.). Geological Survey of Canada, Bulletin 402.

McIntyre, D. J. and Ricketts, B. D. 1989. New palynological data concerning Cornwall Arch from Cornwall and Amund Ringnes islands, District of Franklin. *In:* Current Research, Part G. Geological Survey of Canada Paper 89-1G, p. 199-202.

Miall, A. D. 1985. Stratigraphic and structural predictions from a plate tectonic model of an oblique-slip orogen: the Eureka Sound Formation (Campanian-Oligocene), northeast Canadian Arctic Islands. *In:* Strike-Slip Deformation, Basin Formation, and Sedimentation, K. T. Biddle and N. Christie-Blick (eds.). Society of Economic Paleontologists and Mineralogists, Special Publication 37, p. 361-374.

Muecke, G. K., Reynolds, P. H. and Avison, H. A. 1990. $^{40}Ar/^{39}Ar$ geochronology of episodic magmatism during the late phase of Sverdrup Basin development, Canadian Arctic Islands. GAC/MAC Annual Meeting, Vancouver, May 16-18, 1990, Program with Abstracts, p. A93.

Nichol, S. 1989. Morphometric and sedimentological characteristics of estuarine lithofacies along the micro-tidal, wave-dominant coast of New South Wales (N.S.W.), Australia. Second International Research Symposium on Clastic Tidal Deposits, August 22-25, Calgary, Program and Abstracts, p. 66-67.

Nichols, M. M. and Biggs, R. B. 1985. Estuaries, *In:* Coastal Sedimentary Environments, R. A. Davis Jr. (ed.). Second Edition, Springer-Verlag, New York, p. 77-186.

______ , Johnson, G. H. and Pebbles, P. C. 1989. Estuarine facies and sequences in a microtidal coastal plain estuary: 1. Modern deposits from the James Bay, Virginia. Second International Research Symposium on Clastic Tidal Deposits, August 22-25, Calgary, Program and Abstracts, p. 68-69.

Nielsen, L. H., Johannessen, P. N. and Surlyk, F. 1988. A Late Pleistocene coarse-grained spit-platform sequence in northern Jylland, Denmark. Sedimentology, v. 35, p. 915-937.

Pritchard, D. W. 1967. What is an estuary: Physical standpoint? *In:* Estuaries, G. H. Lauff (ed.). American Association for Advancement of Science, Publication 83, p. 3-5.

Reineck, H.-E. and Wunderlich, F. 1968. Classification and origin of flaser and lenticular bedding. Sedimentology, v. 11, p. 99-104.

Reinson, G. E. 1984. Barrier-island and associated strand-plain systems: *In:* Facies Models, R. G. Walker (ed.). Geoscience Canada Reprint Series 1, p. 119-140.

Ricketts, B. D. 1986. New formations in the Eureka Sound Group, Canadian Arctic Islands. *In*: Current Research, Part B, Geological Survey of Canada Paper 86-1B, p. 363-374.

______ 1987. Princess Margaret Arch: re-evaluation of an element of the Eurekan Orogen, Axel Heiberg Island, Arctic Archipelago. Canadian Journal of Earth Sciences, v. 24, p. 2499-2505.

______ 1989. Prediction of Late Cretaceous-Tertiary depositional sequences below the polar continental shelf: Constraints from onshore synorogenic basins, eastern Arctic Islands. Forum '89, Geological Survey of Canada, Program and Abstracts, p. 24.

______ (*in press a*). Sedimentation, Eurekan tectonism and the fossil forest succession on eastern Axel Heiberg Island, Arctic Archipelago. Geological Survey of Canada Bulletin.

______ (*in press b*). Delta evolution in the Eureka Sound Group, western Axel Heiberg Island: The transition from wave-dominated to fluvial-dominated deltas. Geological Survey of Canada, Bulletin 402.

Roy, P. S., Thom, B. G. and Wright, L. D. 1980. Holocene sequences on an embayed high-energy coast: An evolutionary model. Sedimentary Geology, v. 26, p. 1-19.

Smith, D. G. 1988. Tidal bundles and mud couplets in the McMurray Formation, northeastern Alberta, Canada. Bulletin of Canadian Petroleum Geology, v. 36, p. 216-219.

Stephenson, R. A. and Ricketts, B. D. 1990. Bouguer gravity anomalies and speculations on the regional crustal structure of the Eurekan Orogen, Arctic Canada. Marine Geology, v. 93, p. 401-420.

Stephenson, R. A., Ricketts, B. D., Cloetingh, S. A. and Beekman, F. 1990. Lithosphere folds in the Eurekan Orogen, Arctic Canada? Geology, v. 18, p. 603-606.

Stephenson, R. A., Embry, A. F., Nakiboglu, S. M. and Hastaoglu, M. A. 1987. Rift-initiated Permian to Early Cretaceous subsidence of the Sverdrup Basin. *In:* Sedimentary Basins and Basin-Forming Mechanisms, C. Beaumont and A. J. Tankard (eds.). Canadian Society of Petroleum Geologists, Memoir 12, p. 213-231.

Thomas, R. G., Smith, D. G., Wood, J. M., Visser, J., Calverley-Range, E. A. and Koster, E. H. 1987. Inclined heterolithic stratification — terminology, description, interpretation and significance. Sedimentary Geology, v. 53, p. 123-179.

Thorsteinsson, R. 1972. Geology, Canon Fiord, District of Franklin. Geological Survey of Canada, Map 1308A.

Tillman, R. W. and Nummedal, D. 1989. Sequence stratigraphy and facies of Cretaceous estuarine sandstones, New Mexico. Second International Research Symposium on Clastic Tidal Deposits, August 22-25, Calgary, Program and Abstracts, p. 91.

Vail, P. R., Mitchum, R. M., Todd, R. G., Widmier, J. M., Thompson, J. S., Sangree, J. B., Bubb, J. N. and Hatlelid, W. G. 1977. Seismic stratigraphy and global changes in sea level. *In:* Seismic Stratigraphy — Applications to Hydrocarbon Exploration, C. E. Payton (ed.). American Association of Petroleum Geologists, Memoir 26, p. 49-212.

Wall, J. H. (*in press*). Appendix 2. Micropaleontology. *In:* Delta evolution in the Eureka Sound Group, western Axel Heiberg Island: The transition from wave-dominated to fluvial-dominated deltas, B. D. Ricketts (ed.). Geological Survey of Canada, Bulletin 402.

______, McNeil, D. H., Ricketts, B. D. and McIntyre, D. J. 1988. Paleocene foraminifera from the Eureka Sound Group, eastern Sverdrup Basin, Canadian Arctic Archipelago. Geological Society of America, Abstracts with Program, 1988, p. A222.

Ward, L. G. and Ashley, G. M. 1989. Introduction: Coastal lagoonal systems. *In:* Physical processes and sedimentology of siliciclastic-dominated lagoonal systems, L. G. Ward and G. M. Ashley (eds.). Special Issue, Marine Geology, v. 88, p. 181-185.

Wood, J. M. and Hopkins, J. C. 1989. Reservoir sandstone bodies in estuarine valley fill: Lower Cretaceous Glauconitic Member, Little Bow Field, Alberta, Canada: American Association of Petroleum Geologists, Bulletin, v. 73, p. 1361-1382.

Woodroffe, C. D., Chappell, J., Thom, B. G. and Wallensky, E. 1989. Depositional model of a macrotidal estuary and floodplain, South Alligator River, Northern Australia. Sedimentology, v. 36, p. 737-756.

Zaitlin, B. A. and Dalrymple, R. W. 1989. A conceptual framework for estuarine complexes. Second International Research Symposium on Clastic Tidal Deposits, August 22-25, Calgary, Program and Abstacts, p. 98.

KEY WORDS: BARRED ESTUARY, DROWNED VALLEY ESTUARY, COASTAL PLAIN, BEDROCK GEOMORPHOLOGY, CONSEQUENT DRAINAGE, PALEOCENE, LITHOFACIES, UNCONFORMITY, RAVINEMENT, TIDAL FLAT, WASHOVER, LAGOON, SPIT, POINT BAR, TRANSGRESSION, SEQUENCE BOUNDARY

In: Clastic Tidal Sedimentology, D. G. Smith, G. E. Reinson, B. A. Zaitlin and R. A. Rahmani (eds.). Canadian Society of Petroleum Geologists. Memoir 16, p. 107-122.

Tidal estuary and marine facies of the Glauconitic Member, Drayton Valley, central Alberta

STEVE BROWNRIDGE

Total Petroleum Canada Ltd.
Calgary, Alberta
T2P 0M9

THOMAS F. MOSLOW

Department of Geology
University of Alberta
Edmonton, Alberta
T6G 2E3

ABSTRACT

The Lower Cretaceous Glauconitic Member near Drayton Valley, Alberta, consists of regionally extensive marine strata incised by laterally confined, estuarine channel fills. Marine facies consist of burrowed and laminated mudstone, muddy sandstone, sandy mudstone, and clean sandstone. These facies comprise three, widespread, coarsening-upward lithosomes, which represent three transgressive-regressive shelf-to-shoreface depositional Cycles (Cycles 1, 2 and 3). Estuarine channel-fill sequences are fining-upward successions of crossbedded sandstone, interbedded sandstone and mudstone, and massive mudstone deposited in tidally-influenced point-bar, channel margin, and abandoned channel environments. Evidence of tidal estuarine processes in the channel deposits includes: 1) rhythmically-bedded inclined heterolithic stratification (IHS), and wavy, lenticular, and flaser bedding, which infer a prominence of tidal currents; and 2) restricted suites of ichnofossils and microfossils, which indicates brackish water conditions.

Petrographic and sedimentological evidence suggest contemporaneous deposition of the tidal estuarine channel fills and marine facies within the two youngest depositional Cycles (Cycles 1 and 2). Evidence for this genetic relationship between tidal estuarine and marine facies includes: 1) the oldest channel fill and the middle marine succession (Cycle 2) are quartzose-lithic sandstones, whereas the youngest channel fill and the uppermost marine succession (Cycle 3) are both feldspathic-lithic sandstones; and 2) cored sequences document the presence of an outer estuary facies that is transitional with middle and upper shoreface facies of depositional Cycle 2. Outer estuary facies comprise burrowed, muddy sandstones that occur in channel-fill sequences and crossbedded shoreface sandstones that contain wood and shale intraclasts.

Cycle 3 represents the last stage of tidal estuarine deposition in the Glauconitic Member of the study area, and is characterized by feldspathic-lithic tidal point-bar deposits. Significant reworking of the sedimentary record by channel scour and migration occurred at the close of Cycle 3 deposition. It is likely that a number of tide-influenced rivers representing the mid to upper portions of an estuary extended across the entire study area.

RÉSUMÉ

Près de Drayton Valley en Alberta, le Membre Glauconitique du Crétacé Inférieur consiste en strates marines régionalement étendues et ravinées en marges latérales par des chenaux d'estuaire comblés. Les faciès marins se composent de mudstones laminés et creusés par des terriers, de grès boueux, de mudstones sableux et de grès pures. Ces faciès comprennent trois lithosomes très étendus dont la granulométrie s'accroît de la base au sommet. Ils correspondent à trois cycles sédimentaires transgressifs-régressifs (Cycles 1, 2, et 3) représentés par des milieux variant de la plate-forme à la zone infratidale. Les séquences de remplissage des chenaux estuariens sont des successions de grès à stratifications obliques à granulométrie décroissante vers le haut; elles s'interstratifient avec des grès et mudstones, et des mudstones massifs déposés dans des environnements: de barres d'accrétion latérale influencées par les marées; de chenaux marginaux; et de chenaux abandonnés. Dans l'estuaire, les évidences de processus de marée dans les dépôts sont: 1) des séquences rythmiques de lits à stratifications hétérolithiques inclinées (SHI), et des lits onduleux, lenticulaires et de type «flaser» qui indiquent une prédominance des courants de marée; 2) des suites restreintes d'ichnofossiles et de microfossiles indiquant des conditions d'eau saumâtre.

Dans les Cycles 1 et 2, les deux cycles les plus jeunes, les évidences pétrographiques et sédimentaires suggèrent une sédimentation contemporaine des remplissages du chenal de marée de l'estuaire et des faciès marins. Les évidences pour cette relation génétique entre les faciès estuariens de marée et marins sont: 1) le plus vieux remplissage du chenal et la succession marine-médiane (Cycle 2) sont des grès quartzeux-lithique, tandis que le plus jeune remplissage et la succession marine supérieure sont tous les deux des grès feldspathiques-lithiques; 2) des séquences carottées documentent la présence d'un faciès estuarien externe qui est transitionnel avec les faciès moyen et supérieur de la zone infratidale du Cycle sédimentaire 2. Le faciès estuarien externe comprend des grès boueux avec des terriers, rencontrés dans les séquences de remplissage de chenal, et des grès à stratifications obliques de la zone infratidale, contenant des intraclastes de «shale» et de bois d'origine terrestre.

Dans la région étudiée, le Cycle 3 représente le dernier stade de sédimentation estuarienne de marée dans le Membre Glauconitique. Il est caractérisé par des faciès feldspathiques-lithiques de remplissage de chenal. La provenance de ces dépôts diffère de celle du Cycle 2, ce qui suggère une épisode distincte de sédimentation estuarienne influencée par les marées, ainsi qu'un remaniement significatif des archives sédimentaires par l'érosion et la migration du chenal vers la fin du Cycle sédimentaire 3. Durant la sédimentation du Cycle 3, il est probable qu'un certain nombre de chenaux de marée existait dans les parties médiane et supérieure d'un estuaire s'étendant au delà de la région étudiée.

Introduction

This paper examines deposits of the Lower Cretaceous Glauconitic Member in the subsurface of central Alberta, interpreted as marine shoreface and tidal estuarine in origin. The objective is to examine sedimentary facies in cores that characterize these environments, and to develop a stratigraphic model based on subsurface facies relationships. These strata tend to exhibit complex sequences of sedimentary facies, and specific indicators of tidal processes, such as tidal re-activation surfaces, and cyclicly bundled beds, are not readily detectable in core. Thus, the environmental interpretations presented here are based on documented observations of sedimentary facies and their spatial relationships in modern settings. Important aspects of the interpretations are sedimentary structures indicating a prominence of tidal currents, and ichnofossils and microfossils demonstrating marine and brackish water environments existing in close spatial relationship.

Study Area and Database

This paper is based on data derived from a subsurface study of the Glauconitic Member within a 1,100 km^2 area in central Alberta (Fig. 1). The study area is located approximately 40 km southwest of Edmonton and includes Townships 50 to 52 and Range 9 through Range 6 west of the 5th meridian. The database includes petrophysical well log responses in 318 boreholes and 72 subsurface cores. Cores were described in detail for the purpose of identifying sedimentary facies and their lateral variability. The comparison of cored sequences to accompanying well-log responses facilitated correlations of sedimentary units between boreholes lacking cores. Thin sections of 36 sandstone beds, 3 siltstone, and 1 micrite bed were studied petrographically. Thirty-three of the sandstone samples were point-counted to determine modal grain compositions. Fifteen mudstone samples were collected and examined for foraminifera and dinoflagellate species.

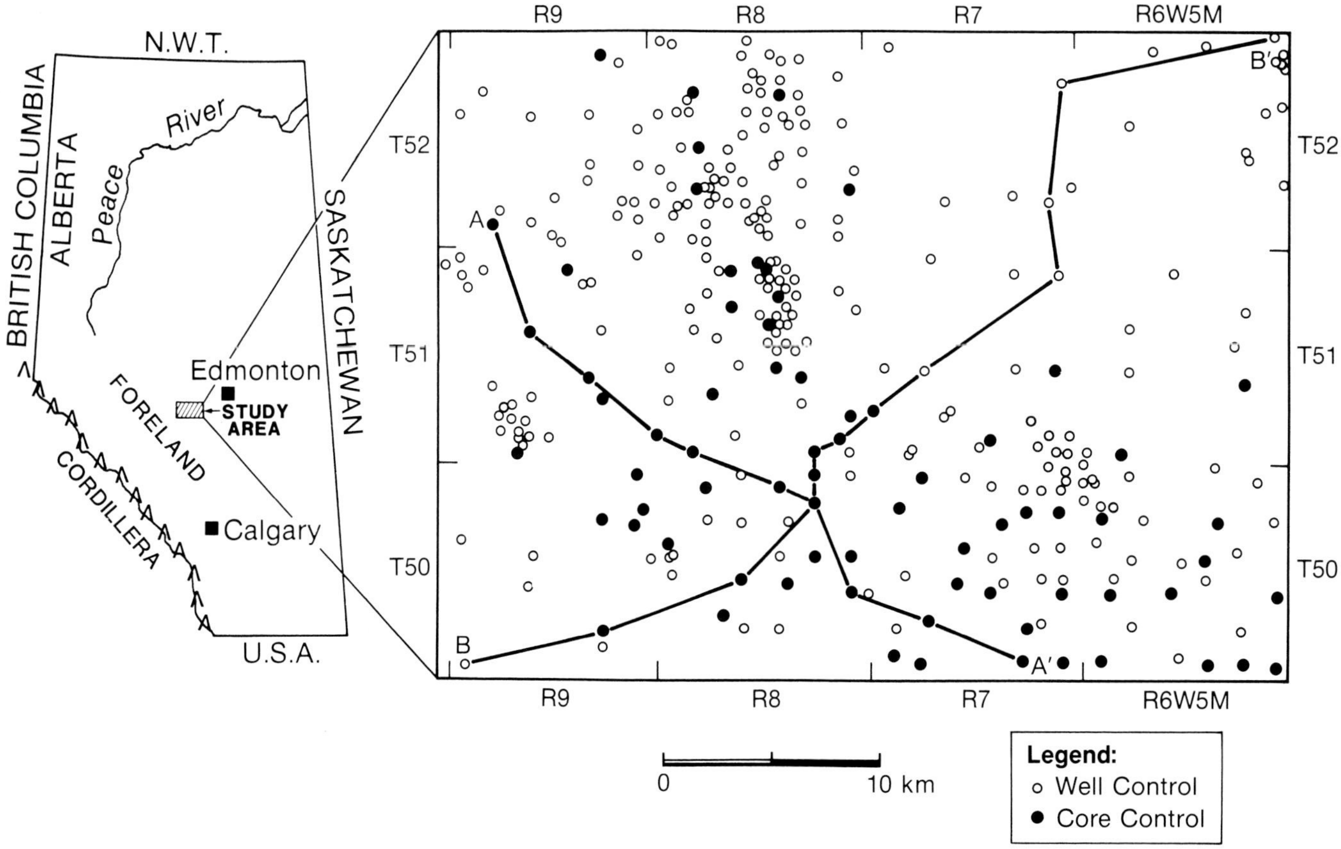

Figure 1. Study area showing location of cored and non-cored wells used in this investigation and orientation of regional cross-sections A-A′ and B-B′.

Regional Stratigraphy and Paleogeography

The Glauconitic Member is part of the Lower Cretaceous Mannville Group, a succession of siliciclastics occurring directly above a regional Lower Cretaceous unconformity in the subsurface of Alberta (Fig. 2). Glaister (1959) first applied the term "Mannville" to these strata in southern and central Alberta after recognizing their lithostratigraphic similarity to the previously defined Mannville Group of eastern Alberta. Glaister (1959) separated the Mannville into upper and lower formations, based on a quartz dominance in sandstones in the lower part of the group, and a strong feldspathic component in sandstones in the upper part. The Glauconitic Member is a prominent sandstone unit at the base of the Upper Mannville Formation. Subsequent work (James, 1985; Hopkins and Farshori, 1987; and Rosenthal, 1988) has confirmed these observations, and also revealed the stratigraphic complexities of the Lower and Upper Mannville contact.

The Mannville Group represents synorogenic deposition in a foreland basin that developed inboard of the Western Canadian Cordillera beginning in the Late Jurassic. Paleogeographic analyses of Lower Cretaceous strata in the Cordilleran Foothills, and in boreholes in the adjoining plains (McLean and Wall, 1981), indicate the southerly transgression of a boreal sea during Lower Mannville deposition. Maximum transgression occurred during either Ostracode Zone (Jackson 1984), or Glauconitic Member deposition (Rosenthal 1988). A series of northeast-trending Glauconitic Member shorelines occur in central Alberta, which Rosenthal (1988) interpreted as the result of episodic northerly retreat of the boreal sea. The overlying Upper Mannville sediments in the region are a product of continental deposition following withdrawal of this seaway (Jackson, 1984).

Glauconitic Member in the Drayton Valley Area

Three regional successions, termed the Glauconite A, B, and C, were identified by Rosenthal (1988) within the Glauconitic Member of central Alberta. Each unit represents a transgressive-regressive depositional cycle consisting of inner-shelf-to-shoreface strata which host laterally confined, channel-fill deposits. This study focuses on the detailed stratigraphy of those strata in the vicinity of Drayton Valley, Alberta, with particular emphasis on the depositional processes and timing of the channel-fill sequences.

Four distinct lithosomes, representing discrete lithostratigraphic components of the Glauconitic Member, were correlated throughout the study area (Fig. 3A, B). These units are recognized in core as facies associations and are correlated with the aid of laterally extensive stratal horizons that are distinctive in cores and on well-logs. Three key horizons, termed the lower, middle, and upper shale markers, represent the basal units of three coarsening upward lithosomes informally termed Cycles 1, 2, and 3 (Fig. 3A, B). By contrast, the fourth lithosome consists of a laterally confined channel-fill facies association, which truncates Cycles 1, 2, and 3.

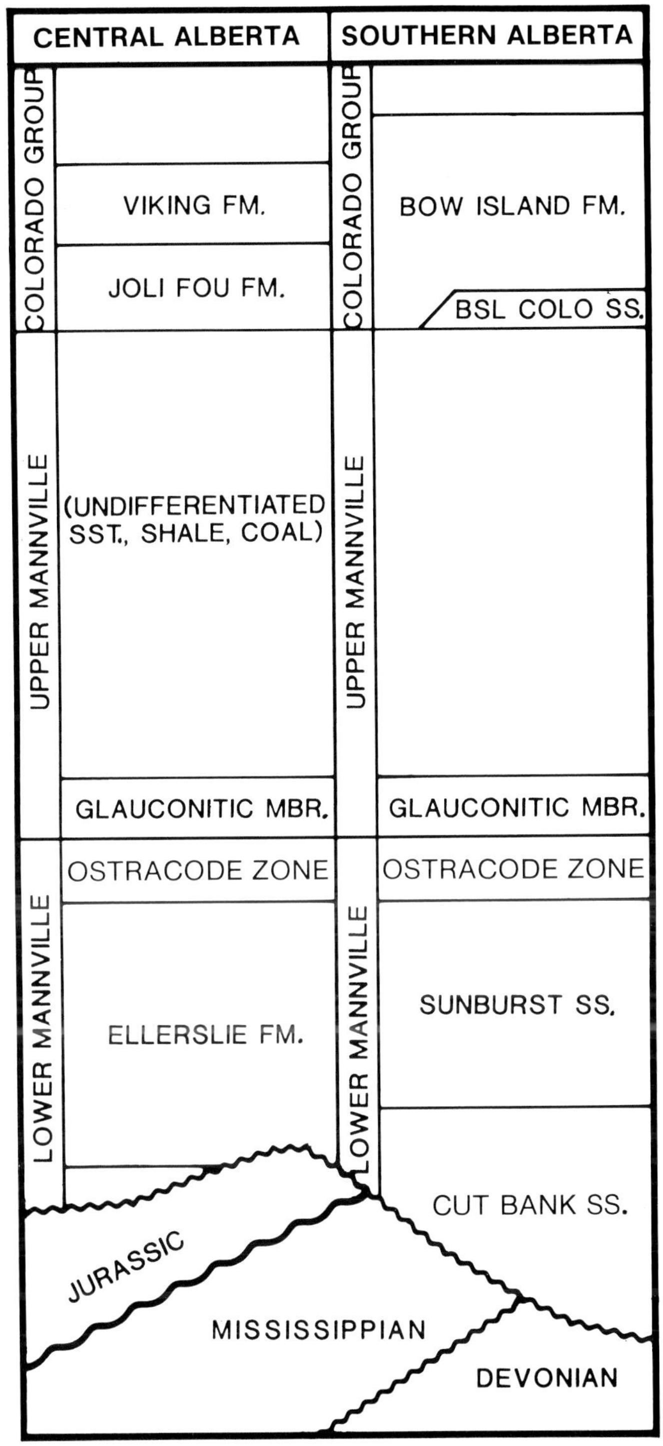

Figure 2. Stratigraphic column of the Lower Cretaceous Mannville and Colorado groups in southern and central Alberta (after Rudkin, 1964). The Glauconitic Member is the basal lithostratigraphic unit of the Upper Mannville.

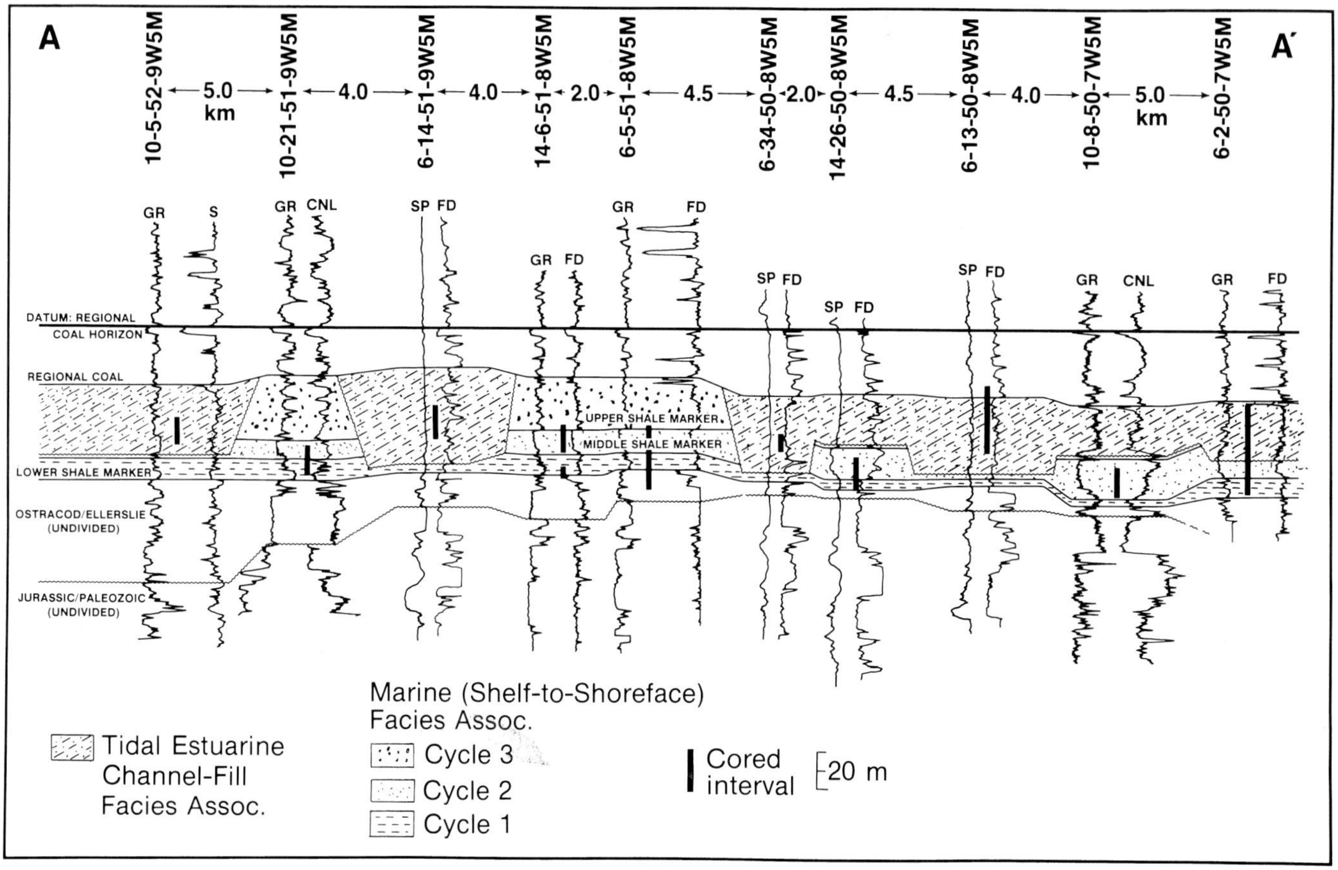

Figure 3A. Regional cross-section A-A′ of the four major lithostratigraphic units of the Glauconitic Member in the study area. Cross-section is oriented subparallel to depositional dip as shown in Figure 1. Note the predominance of tidal estuarine channel fill crosscutting the three progradational cycles of marine facies associations. Key to lithologies is shown in Table 1. Well logs are identified by the following: GR, Gamma Ray; SP, Spontaneous Potential; S, Sonic; CNL, Compensated Neutron; FD, Formation Density.

Cycle 1

Cycle 1 represents the earliest phase of Glauconitic Member deposition in the study area. The basal portion of Cycle 1 consists of 4 to 8 m of fissile shale containing pelecypod grainstone and siltstone beds and is referred to as the lower shale marker (Fig. 3A, B). Overlying this are laterally discontinuous deposits of laminated to bioturbated mudstone, siltstone, and very fine grained sandstone. Cycle 1 is capped by a thin (0.5-2.0 m), but relatively continuous deposit of glauconitic, medium grained sandstone.

Cycle 1 is equivalent to Rosenthal's (1988) Glauconite-C succession, which is interpreted as the initial transgressive-regressive phase of Glauconitic deposition. In the Drayton Valley area the succession is relatively thin and fine grained, and appears to terminate vertically at a minor disconformity produced by a period of shoreface retreat. This disconformity is expressed by the lateral discontinuity of facies horizons near the top of the cycle, and the relatively persistent, medium grained sandstone deposit, which is interpreted as a transgressive lag. The high glauconite content of the lag deposit suggests this interval is a condensed horizon (Loutit *et al.*, 1988).

Cycle 2

Cycle 2 is a 10 to 25 m coarsening-upward succession that culminates in a prominent sandstone informally termed the "Bigoray Sandstone" (Jackson, 1984). The succession is correlatable on logs as a smooth cleaning-upward response (Fig. 3A, B). Bedding contacts between sedimentary facies observed in cores are typically gradational.

At the base of Cycle 2, the middle shale marker consists of several metres of fissile shales grading upward to sandy mudstones (Fig. 4). The shales contain grainstone beds comprising disarticulated pelecypod valves and normally graded silt laminae. The sandy mudstones are bioturbated to such a degree that physical structures are obliterated, and individual burrow traces are difficult to discern. However, *Teichichnus, Planolites, Palaeophycus, Terebelina, Asterosoma*, and *Chondrites* were observed.

The Cycle 2 sequence coarsens upward to several metres of burrowed to laminated, muddy sandstones. Planar to curvilinear laminated sandstone beds 2 to 30 cm thick (Fig. 5A) make up 5 to 60 per cent of this facies. Basal contacts of these beds are erosive and load deformed, and upper contacts are diffuse and burrowed. The intervening burrowed

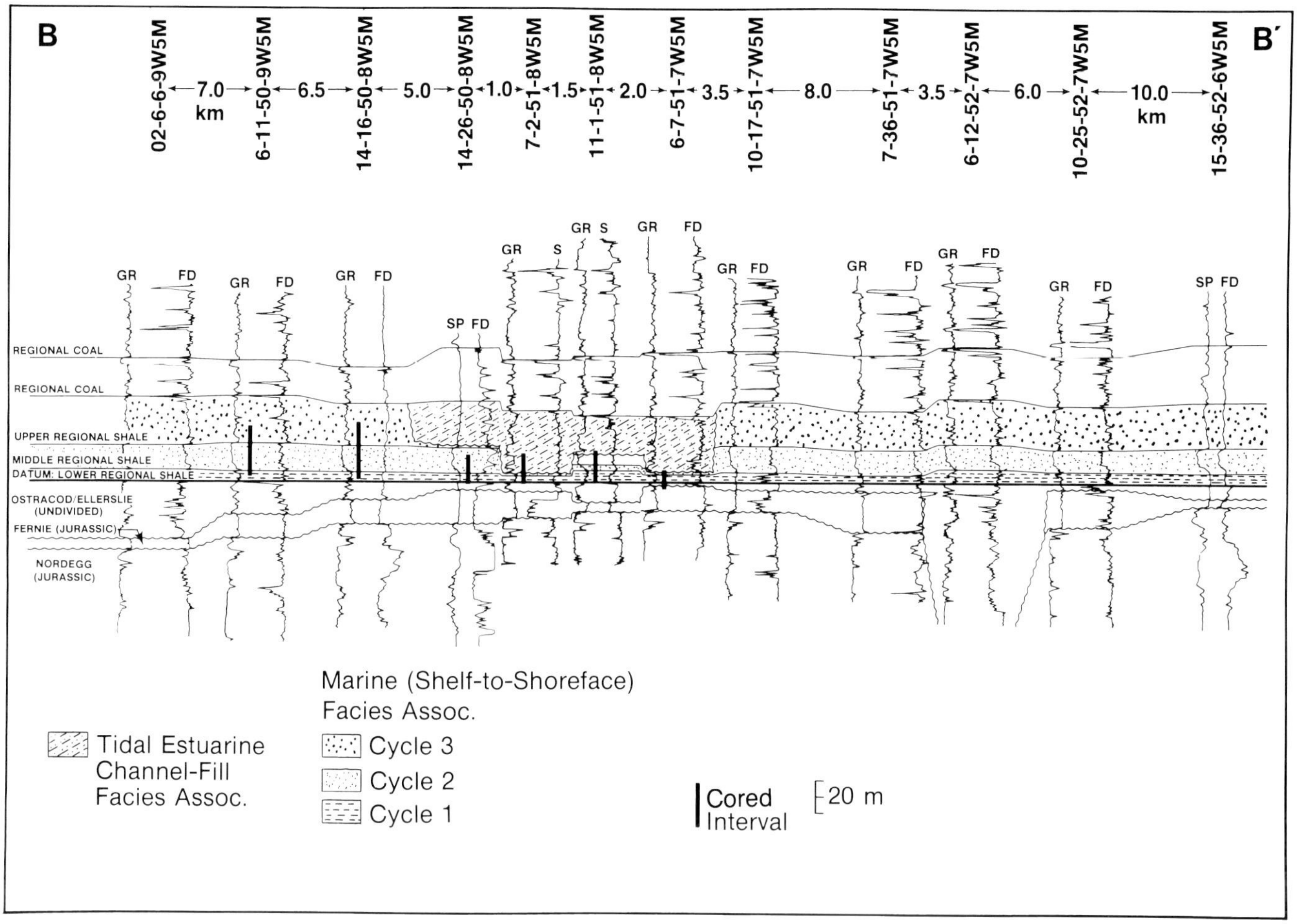

Figure 3B. Regional lithostratigraphic cross-section B-B′ of the Glauconitic Member in the study area. Cross-section is oriented subparallel to depositional strike as shown in Figure 1. Note lateral continuity of marine Cycles 1, 2, and 3 and the lower, middle and upper shale marker beds. Key to lithologies is shown in Table 1. Well log abbreviations are the same as in Figure 3A.

intervals are muddy and contain abundant *Asterosoma* and *Palaeophycus* traces. *Terebelina* is densely concentrated at the top of the facies, and *Teichichnus, Planolites*, and *Chondrites* also occur. Carbonaceous laminae and large wood fragments are common.

Occurring at the top of Cycle 2 is the Bigoray Sandstone, a sparsely burrowed, massive to crossbedded sandstone that is 5 to 15 m thick (Fig. 5B). The lowermost sandstone beds are slightly argillaceous, fine grained, and planar- to curvilinar-laminated. *Terebelina, Palaeophycus*, and *Asterosoma* are common trace fossils. The deposit grades upward into fine to medium grained, crossbedded sandstones with rare, argillaceous lenses. Recognizable trace fossils in this interval are abundant *Macaronichnus*, common *Palaeophycus*,

Table 1. Key to lithological and sedimentological symbols and ichnofauna abbreviations used in cross-sections and core sequences

Lithology	Bedding, Structures, Contracts	Ichnofauna
Sandstone	Crossbedded	T - *Terebellina*
Mudstone	Ripple lamination	Tc - *Teichichnus*
Interbedded Sandstone/Mudstone, mm-dm Scale, (inclined)	Wavy Bedding	P - *Palaeophycus*
Muddy Sandstone	Planar Bedding	Pl - *Planolites*
Limestone	Flaser Bedding	A - *Asterosoma*
Coal	MD Mud Drapes	Sk - *Skolithos*
	Burrows	Sk(L) - *S. Linearis*
	Bioturbation	Ch - *Chondrites*
	Micro Faults	Mac - *Macaronichnus*
	Erosive Bedding Contact	**Acc. Components**
	Internal Erosional Contact	Mudclasts
		Wood Fragments
		Bivalve Fragments

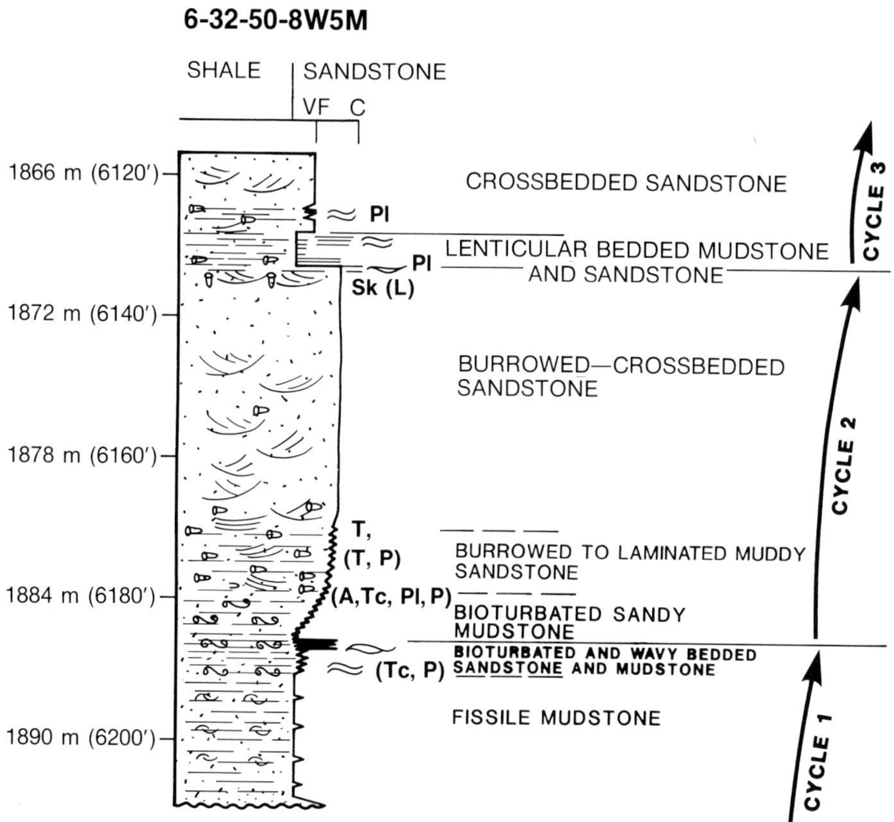

Figure 4. Sedimentological description for a Glauconitic Member core from Well 6-32-50-8WM showing sedimentary facies and vertical succession common to progradational Cycles 1, 2, and 3. Well location is shown in Figure 14. Key to lithological and sedimentological symbols and abbreviations for ichnofauna is shown in Table 1.

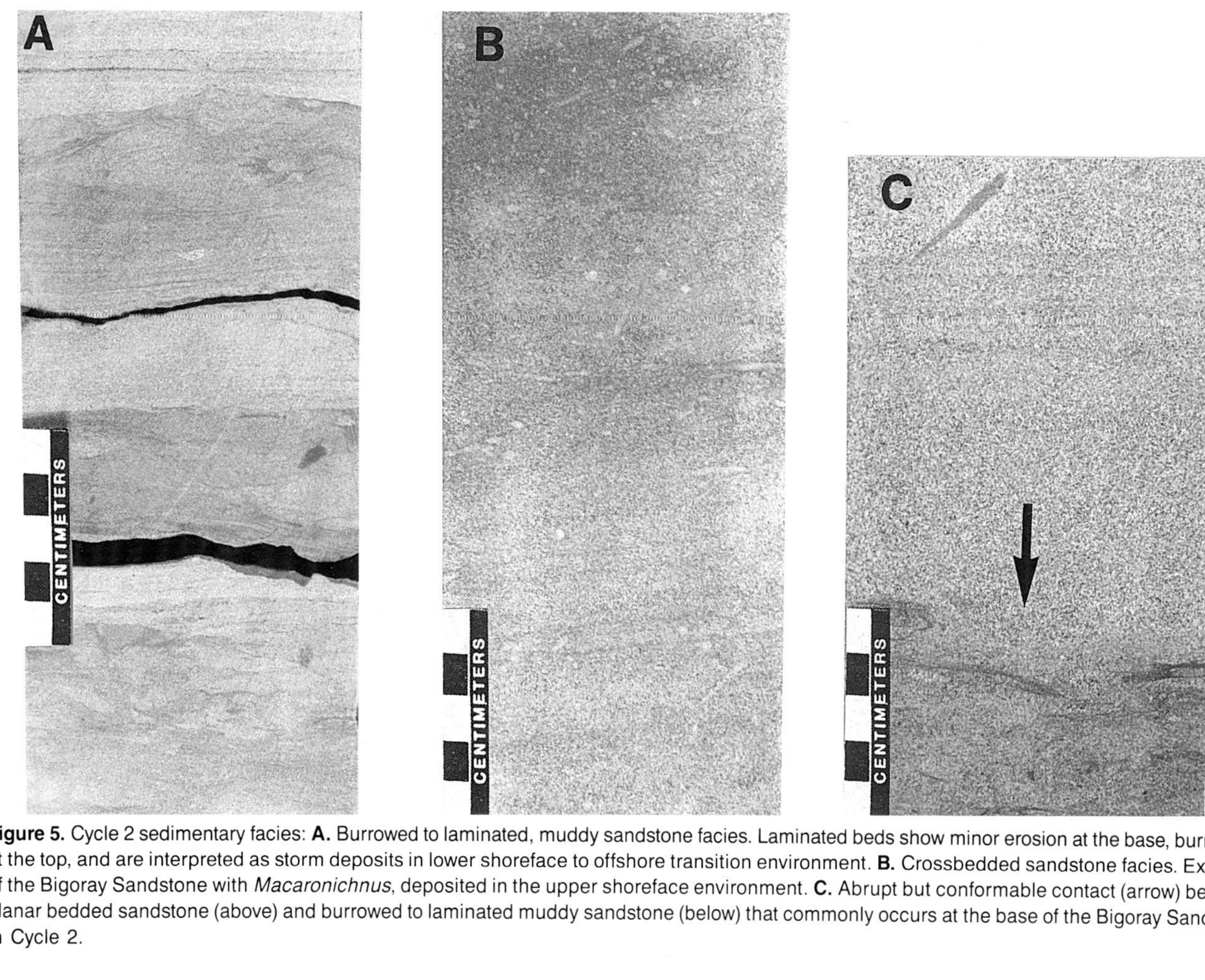

Figure 5. Cycle 2 sedimentary facies: **A.** Burrowed to laminated, muddy sandstone facies. Laminated beds show minor erosion at the base, burrowing at the top, and are interpreted as storm deposits in lower shoreface to offshore transition environment. **B.** Crossbedded sandstone facies. Example of the Bigoray Sandstone with *Macaronichnus*, deposited in the upper shoreface environment. **C.** Abrupt but conformable contact (arrow) between planar bedded sandstone (above) and burrowed to laminated muddy sandstone (below) that commonly occurs at the base of the Bigoray Sandstone in Cycle 2.

and rare *Rhizocorallium*. The uppermost sandstones are fine to medium grained, and massive to crossbedded with rare, low-angle planar lamination. *Skolithos linearis* burrows greater than 20 cm in length are common in this interval. The top few centimetres of the succession typically contain lithoclasts, coarse grained sandstone laminae, and centimetre-scale shale lenses.

In several cores, the lower contact of the Bigoray Sandstone is abrupt, and the entire deposit consists of fine to medium grained, planar to crossbedded sandstone (Fig. 5C). Mudstone clasts up to 1 cm in diameter and wood fragments are common. Burrow traces are very rare. In core, all of these deposits are truncated by channel fills (Fig. 3B).

Cycle 2 is interpreted as a second transgressive-regressive phase of sedimentation in the Glauconitic Member. Laminated mudstones of the middle shale marker are a product of suspension sedimentation, likely in a deep water, low energy environment. This is inferred to reflect offshore-to-shelf deposition following transgression of the Cycle 1 shoreline. Silt laminae and pelecypod beds, which are the only indications of substrate reworking in this horizon, represent possible storm deposits. In the sandy mudstones the degree of bioturbation and diverse burrow assemblages suggest low energy ecologically hospitable marine substrates (Frey and Pemberton, 1985).

The burrowed to laminated muddy sandstones of Cycle 2 were deposited in a storm-influenced lower shoreface environment. Ichnofossils in these sediments are diverse, indicating a hospitable marine environment, but their reduced density suggests more frequent substrate agitation (Ekdale *et al.*, 1984). The sharp-based laminated beds indicate substrate agitation, probably as a result of storms. In modern lower shoreface environments, storm deposited sands occur in thin (centimetre to decimetre), sharp based, planar to curvilinear laminated beds (Morton, 1981). During intervals between storms, benthic fauna rework upper parts of these beds along with overlying, fine grained, suspension deposited sediment. The resultant interbedded laminated and burrowed muddy sandstones represent a balance of storm and biogenic processes distinctive of storm influenced environments below "normal" wave base (Howard and Nelson, 1982).

Sandstones at the top of Cycle 2 are interpreted as middle to upper shoreface deposits. Ichnofossils indicate frequently agitated, unconsolidated marine substrates (Ekdale *et al.*, 1984). Planar and curvilinear laminated beds near the base of the sandstones represent possible storm-generated hummocky bedforms in a middle shoreface zone. Crossbedding in middle and upper intervals can be attributed to migrating megaripples and surf-zone bars generated by shoaling waves in an upper shoreface (Clifton *et al.*, 1971;

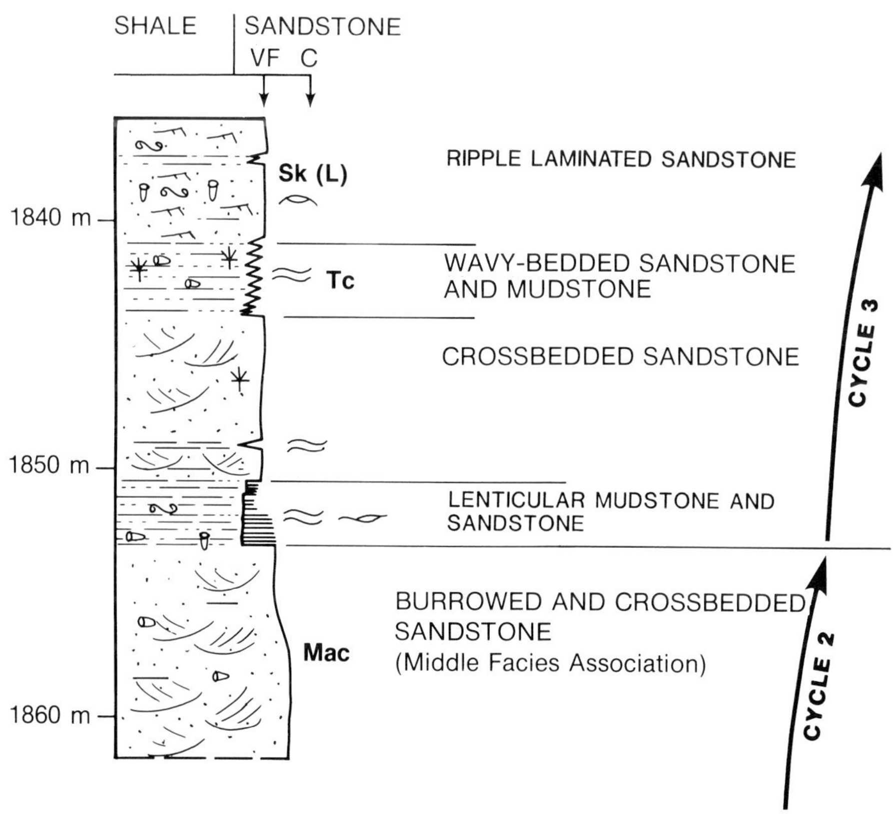

Figure 6. Sedimentological description of core containing sedimentary facies in the upper part of Cycle 2 and base of Cycle 3 from Well 14-16-50-8W5M. Well location is shown in Figure 14. Refer to Table 1 for key to lithological and sedimentological symbols.

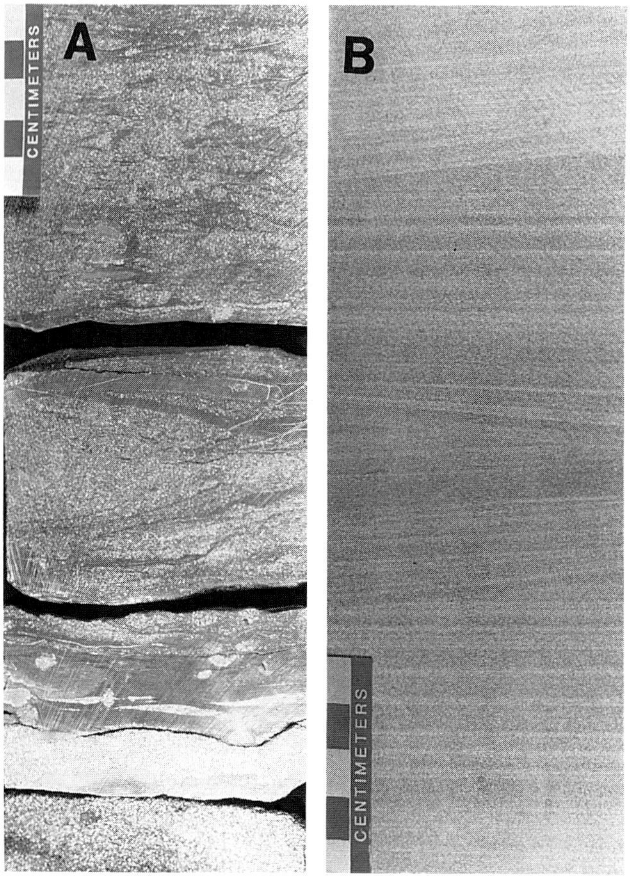

Figure 7. Close up core photographs of Cycle 3 sedimentary units and facies. **A.** Lenticular bedding in glauconitic sediments at the basal contact of the upper shale marker. *Planolites* burrows occur in the mudstone beds (dark tones). **B.** Crossbedded sandstone facies (upper shoreface depositional environment) directly overlying the upper shale marker.

Davidson-Arnott and Greenwood, 1976). Low angle laminae attributable to wave swash in a foreshore zone are rare, and are confined to the very top of the sequence.

Cycle 3

Cycle 3 is the uppermost regional succession in the Glauconitic Member of the Drayton Valley area. It is overlain by widespread coal horizons, which define the base of the Upper Mannville Formation in central Alberta (Rosenthal, 1988). Inferences about the variations of sedimentary facies and their lateral continuity are dependent upon correlation of well logs, as Cycle 3 is rarely cored in the study area.

At the base of Cycle 3, the upper shale marker immediately overlies the Bigoray Sandstone (Fig. 6). The basal contact is abrupt and in one core, truncation of underlying laminae is evident. The lowermost beds of the marker are characterized by lenses of medium grained glauconitic sandstone (Fig. 7A), and *Planolites* and *Skolithos* burrow traces. The upper shale marker grades upward into millimetre to centimetre beds of massive mudstone and normally graded siltstone.

Abruptly overlying the upper shale marker is a crossbedded sandstone facies with abundant, fine grained carbonaceous debris, large wood fragments, and mudstone clasts (Fig. 7B). The consistent occurrence of this facies in core, and a pervasive, low radiation, gamma-ray deflection directly above the upper shale marker (Fig. 3), suggests the sandstone facies is a laterally extensive deposit 2 to 5 m thick. Well-log deflections above this strata suggest a laterally extensive interval of argillaceous deposits (Fig. 3A, B). Within this interval, several metres of wavy-bedded sandstone and mudstone were observed in three cores (Fig. 6). An impoverished ichnofacies consisting of *Skolithos linearis, Paleophycus,* and *Planolites* was observed in these sediments. Logs indicate a laterally extensive sandstone at the top of Cycle 3. In northern and western parts of the area, this sandstone facies is greater than 15 m thick (Fig. 3A, B), however, it has not been cored. Coal horizons marking the top of the Glauconitic Member immediately overlie this sandstone.

The upper shale marker is inferred to represent a third transgression in the study area. The rarity of swash laminae at the top of the underlying Bigoray Sandstone suggests that foreshore deposits were reworked into extensive shoals as the Cycle 2 shoreline was transgressed. The "transgressive lag" thus consists of crossbedded sandstones in the top of the Bigoray Sandstone, and lenticular sandstones at the base of the upper shale marker. The upper shale marker beds are interpreted as having been deposited in a low energy shelf environment.

The overlying strata are interpreted as a product of a third progradational event. They are capped by laterally extensive coal beds, indicating widespread establishment of terrestrial environments at the close of Cycle 3 deposition. These strata are equivalent to Rosenthal's (1988) Glauconite-A sandstone, which is interpreted as representing a regionally progradational event that precedes Upper Mannville deposition in central Alberta. Deposition of these beds in a marginal marine setting is interpreted from the impoverished ichnofossil assemblage, suggesting stressful marine conditions (Ekdale *et al.*, 1984), and an abundance of wood fragments, which suggests proximity to terrestrial environments.

Tidal Estuarine Facies Association

The tidal estuarine strata are characterized by complex facies relationships, and extreme lateral and vertical facies variability. More than 90 per cent of the facies association is characterized by fining-upward successions interpreted as tide-influenced channel fills. Channel fills contain both single fining-upward successions (up to 35 m thick), and multiple successions separated by erosive contacts (Fig. 8). The tidal estuarine facies association contains five sedimentary facies, described below.

Crossbedded Sandstone Facies

The base of all fining-upward successions is a crossbedded, fine to medium grained sandstone with an erosive basal contact marked by mud clasts and wood fragments (Fig. 9A). Tangential foreset and bottomset laminae display superimposed current ripples and evenly spaced mud drapes. Inter-

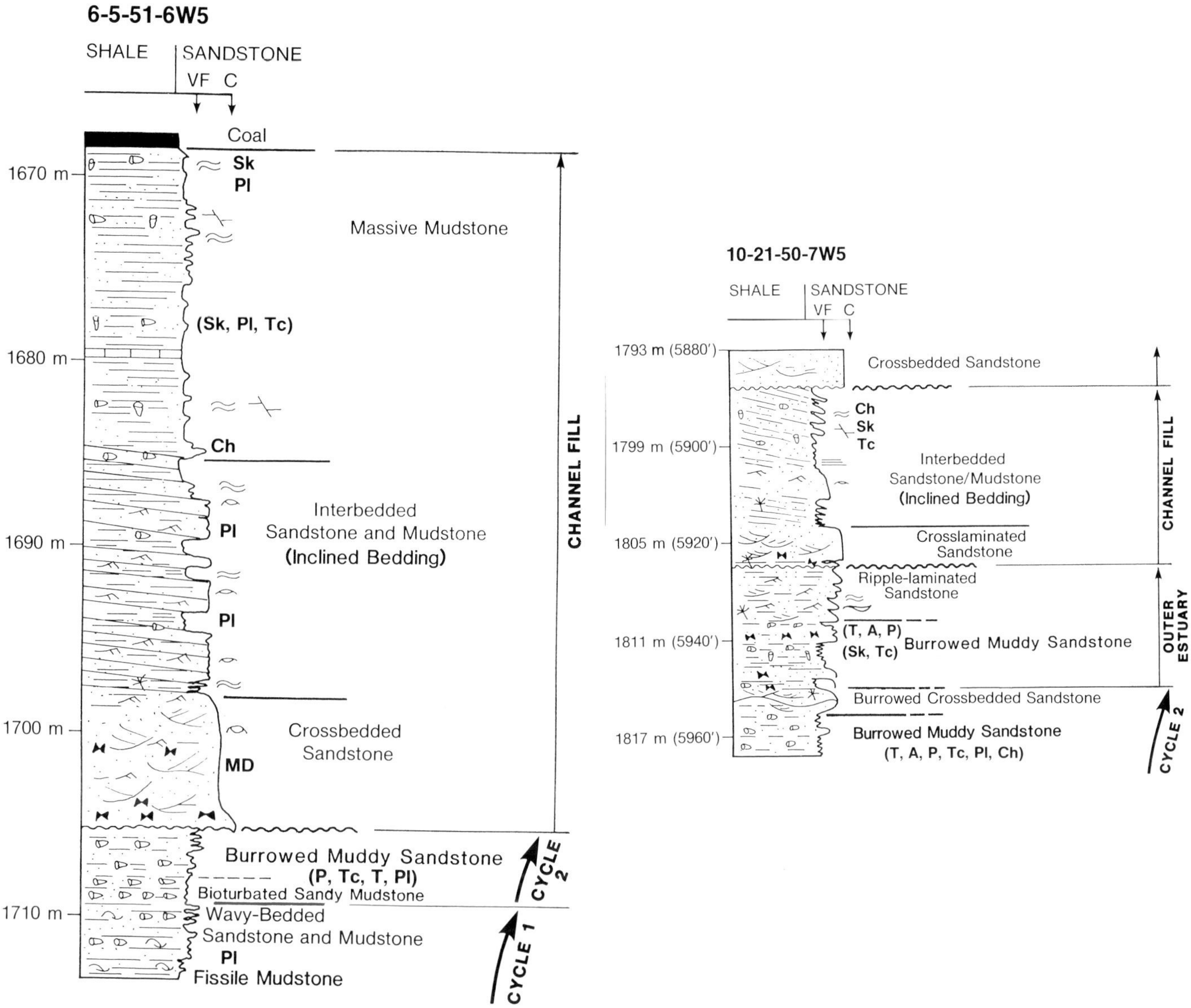

Figure 8. Sedimentological descriptions of cored sedimentary facies from tidal estuarine deposits: **(Left)** Single 35 m thick channel-fill fining-upward sequence erosionally overlying burrowed to laminated sediments of Cycle 2, and consisting of three distinct facies: 1) crossbedded sandstone; 2) interbedded sandstone and mudstone: and 3) massive mudstone. The channel fill is conformably overlain by a coal horizon at the top of the Glauconitic Member. Refer to Table 1 for key to symbols and abbreviations. Cored intervals are contained in cross-section C-C′ (Fig.13). See Figure 15 for well location. **(Right)** A truncated fining-upward tidal channel fill, which erosionally overlies a non-fining-upward sequence of estuarine facies conformable with the middle-upper shoreface (crossbedded sandstone) facies of Cycle 2. Refer to Table 1 for key to symbols and abbreviations. Cored intervals are contained in cross-section C-C′ (Fig. 13). See Figure 15 for well location.

vals of climbing ripples, locally with flasers, also occur. Thickness of individual sandstone beds ranges from 0.5 to 1.8 m. Centimetre-scale muddy beds or erosional contacts commonly separate amalgamated sandstones. Wood fragments and angular mudstone clasts of various sizes (less than 1 cm to greater than 20 cm) are abundant throughout. Fine grained carbonaceous debris is a dominant component of argillaceous laminae. Biogenic structures include rare *Palaeophycus*, and *Planolites* burrows, which occur in muddy interbeds.

Mud-draped foresets and muddy interbeds suggest slack water interludes, possibly caused by tidal processes (deRaaf and Boersma, 1971). Variation of the flow regime and/or suspended load is also indicated by intervals of climbing ripples (Reineck and Singh, 1980) and tangential foreset laminae (Jopling, 1965). While these characteristics could reflect seasonal or tidal effects in numerous channel environments, the monospecific *Planolites* assemblages in the muddy interbeds suggest marine to brackish water influence (Wightman *et al.*, 1987).

Interbedded Sandstone/Mudstone Facies

This facies consists of regularly alternating sandy and muddy beds, and bedsets gradationally to abruptly overlying the basal sandstone facies. Overall sand content in these beds varies from 90 to 10 per cent, and sand size is very fine

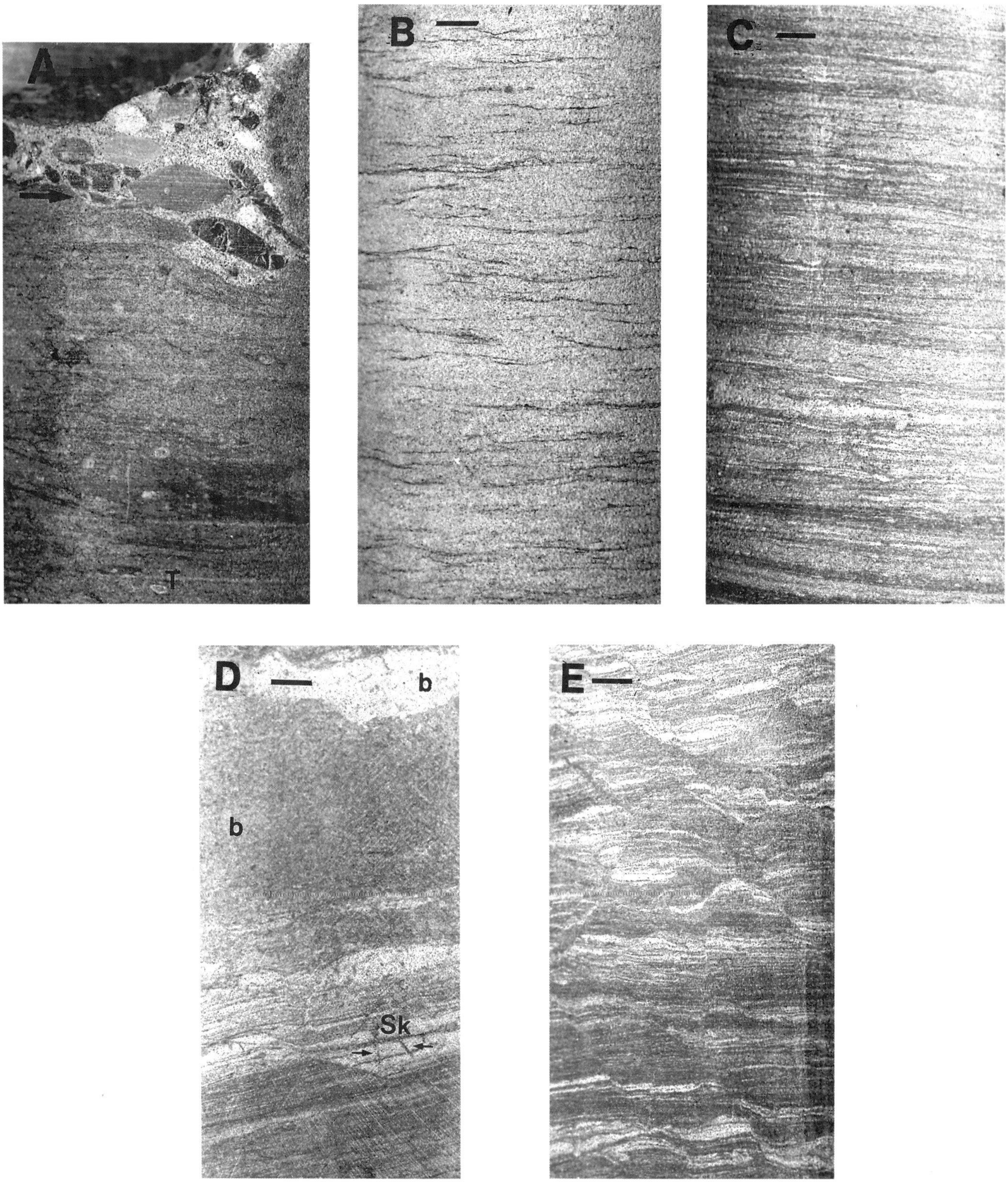

Figure 9. Core photographs of sedimentary facies from tidal estuarine channel fills: **A.** Crossbedded sandstone facies above a basal erosional contact overlying the burrowed to laminated (lower shoreface) facies of Cycle 2 (*T=Terebellina*). Location: 6-5-51-6W5; 1705 m on Figure 8A. **B.** Interbedded sandstone and mudstone facies showing sandy beds containing ripple lamination, mud drapes and flaser bedding. (Location: 6-5-51-6W5; 1695.4 m on Figure 8A). **C.** Interbedded sandstone and mudstone facies showing muddy bedset consisting of gently inclined wavy-bedded sandstone and mudstone (Location: 6-5-51-6W5; 1697.6 m on Figure 8A). **D.** Interbedded sandstone and mudstone facies showing muddy bedset consisting of inclined wavy- to parallel-bedded sandstone and mudstone with thoroughly bioturbated intervals (b), and individual *Skolithos* burrows (Sk). **E.** Massive mudstone facies showing inclined and microfaulted sandstone beds. All core slabs are 7.5 cm in width.

to upper fine grained. Bedsets are centimetres to metres thick and have abrupt to erosional basal contacts. Sandy bedsets are crossbedded, ripple laminated, or massive, and contain mud drapes, flasers, and thin wavy mudstone beds (Fig. 9B). Muddy bedsets consist of massive mudstone and wavy to planar bedded mudstone and sandstone. Finer grained examples of this facies are progressively thinner bedded, down to millimetre-scale, wavy to planar bedded mudstone and sandstone (Fig. 9C). Bedding is characteristically inclined from several degrees to greater than 20 degrees, with the steepest beds occurring in the finest grained units. Evenly spaced layering between beds is characteristic. Biogenic structures comprise less than 10 per cent of the facies overall, but some beds are intensely burrowed (greater than 90 per cent). Ichnofossils occur in low diversity, and burrow forms are small (Fig. 9D). *Chondrites, Skolithos, Planolites, Paleophycus* and *Teichichnus* were observed.

The interbedded sandstone and mudstone facies is interpreted as representing epsilon cross-stratification (ECS) (Allen, 1963), or inclined heterolithic stratification (IHS), (Thomas *et al.*, 1987). Criteria supporting this inference are inclined muddy and sandy interbeds, lower order stratification within inclined beds, and erosional to discordant bedset contacts. The IHS occurs within fining-upward successions and contains abundant traction deposited structures (wavy bedding, ripple laminae, and crosslaminae). These features are associated with IHS deposition on laterally migrating point-bars (Thomas *et al.*, 1987). Flasers, mud drapes, and wavy bedding suggest influences of tidal processes (deRaaf and Boersma, 1971). In addition, evenly-bedded IHS is typical of modern tide-influenced point bar deposits (Smith, 1988), and the fine grained, thinly bedded intervals closely resemble lateral accretion bedding in modern intertidal creeks (Reineck and Singh, 1980; deMowbray, 1983). Low diversity assemblages of both vertical and horizontal ichnofossils within the facies are indicative of brackish water conditions found in modern estuaries (Wightman *et al.*, 1987). Burrowed horizons in modern estuarine point bars contain simple vertical and horizontal tubes (Howard *et al.* 1975), similar to burrowed intervals in the IHS.

Massive Mudstone Facies

Massive mudstone deposits abruptly to gradationally overlie either the interbedded sandstone and mudstones or crosbedded sandstones. This facies is typically 1 to 5 m thick and contains intervals with thin (millimetre to centimetre), inclined and microfaulted sandstone beds and laminae (Fig. 9E). Wood fragments are abundant, and *Planolites* and *Skolithos* burrows are common. Low diversity assemblages of dinoflagellates and agglutinated foraminifera were collected from samples of this facies.

The massive mudstone facies contains little evidence of bedload deposition and is interpreted as abandoned channel and channel margin deposits. Inclined and microfaulted sandstone beds in the facies may represent deposition on abandoned point bar surfaces by overflow from adjacent channels during spring tides or seasonal floods. In modern estuaries, thick muddy deposits also occur along the channel margins (Dorjes and Howard, 1975). The impoverished ichnofossil and dinoflagellate assemblages suggest low salinity waters typical of estuaries or lagoons (Wightman *et al.*, 1987). An assemblage of agglutinated foraminifera also suggests estuarine conditions (Haynes, 1981).

Ripple Laminated Sandstone Facies

Ripple laminated sandstones occur in thick bedsets (greater than 4 m) in three cores. The facies is dominated by climbing asymmetrical ripple laminae with flasers and thin muddy laminae. Subordinate structures are: planar lamination, crosslamination, symmetrical ripples, and wavy to lenticular beds. The facies consists mostly of thin (decimetre-scale) beds displaying these sedimentary structures in vertical successions indicative of waning of flows upward. Angular mudstone clasts and large wood fragments are abundant, and biogenic structures are absent.

The absence of biogenic structures and the abundance of climbing ripple laminae imply rapid deposition with a large component of suspension sedimentation (Reineck and Singh, 1980). The waning flow characteristic of individual bedding sequences and abundant intraclasts suggests episodic deposition. These characteristics suggest a fluvially influenced setting in which unidirectional currents and seasonal flooding are important processes and bioturbation is rare. Ripple laminated sediments of this type are observed in upper point bar and natural levee deposits, where they are related to seasonal flooding (Coleman and Wright, 1965). Such deposits could be expected to be associated with tidally-influenced rivers.

Burrowed Muddy Sandstone Facies

Burrowed muddy sandstones occur in 1 to 2 m thick beds that contain 20 to 50 per cent mud, and are greater than 80 per cent bioturbated. Several of these beds have diverse ichnofossil assemblages typical of the *Cruziana* ichnofacies (Seilacher, 1967), and are in all respects identical to burrowed muddy sandstone beds of Cycle 2. One bed contains *Skolithos linearis* in addition to the *Cruziana* assemblage (Fig. 10), and several beds contain only a monospecific assemblage of *Chondrites*. The facies commonly occurs interbedded with the crossbedded sandstone facies in the lowermost intervals of channel fills. In one core (10-21-50-7W5, Fig. 8), beds of significantly different biogenic character (*i.e.*, diverse versus monospecific assemblages of ichnotaxa) make up the basal 4 m of the tidal estuarine succession. This deposit also contains abundant mudstone clasts, and several unburrowed decimetre- to centimetre-scale beds that are ripple laminated, planar laminated, and reverse and normally graded.

Burrowed muddy sandstones containing diverse *Cruziana* ichnotaxa assemblages are indicative of shallow marine environments and unconsolidated substrates (Ekdale *et al.*, 1984). Interbedding of these sediments with basal channel

Figure 10. Burrowed muddy sandstone facies deposited in an outer estuarine environment of deposition. Bioturbated intervals contain the ichnofossils *Skolithos* (S), *Teichichnus* (T), and *Asterosoma* (A). (Location: 10-21-50-7W5; 1857 m on Figure 8B). All core slabs are 7.5 cm in width.

sandstones indicates that marine conditions prevailed during periods of low channel flow, or following channel abandonment. This suggests a marine or lower coastal plain setting for these channels. Interbedded muddy sandstones with highly variable ichnofossil characteristics that occur in one core (10-21-50-7W5) indicate that differing ecological conditions that existed in adjacent environments influenced the populations of benthic organisms. Some modern outer estuarine environments can support diverse populations of burrowing organisms similar to the adjacent marine shoreface (Dorjes and Howard, 1975). However, adjacent estuarine environments, which are more terrestrially influenced, are brackish and support only restricted benthic populations. Similarly, substrates constantly exposed to current activity (tidal, fluvial, or wave generated) are dominated by physical sedimentary structures (Frey and Howard, 1986). The burrowed muddy sandstone facies is interpreted to represent several outer estuary environments including seaward reaches of estuarine channels, and low energy areas of estuary inlets and ebb-tidal deltas.

Fining-upward Tidal Channel Fills

The tidal estuarine facies association conforms closely with modern mesotidal estuary deposits. Particularly significant attributes are: 1) abundant tidally influenced point-bar deposits, 2) diverse sediment textures and facies, 3) common facies transitions, and 4) ichnofossils and microfossils indicative of brackish waters (Frey and Howard, 1986). The fining-upward successions are primarily composed of the crossbedded sandstone facies (base), interbedded sandstone and mudstone facies (middle), and massive mudstone facies (top). These deposits are inferred to represent tidal creeks and tide-influenced rivers in upper to middle estuary settings. Individual successions range from <10 per cent to > 90 per cent mudstone, and cores typically contain amalgamated successions with basal truncations. Basal crossbedded sandstones directly overlain by the massive mudstone facies are also common. These stratal characteristics reflect processes of erosion and lateral accretion due to channel migration, and vertical accretion following channel abandonment. The fine grained, steeply inclined IHS deposits represent lateral accretion in intertidal to subtidal creeks, which are maintained primarily by tidal currents. The thicker and sandier IHS suggest point-bar deposition in tide-influenced fluvial rivers. The insertion of the ripple laminated facies in some of these successions is also evidence of a fluvial component in some channels. Thick, crossbedded sandstone deposits (> 10 m) may represent semipermanent estuarine sandbars similar to those of the Gioronde and Ogibeeche estuaries (*i.e.,* Jouanneau and Latouche 1981; Greer 1975), rather than point-bar associated sandstones.

Sandstone Petrography

Modal grain compositions of clean, fine to medium grained sandstones of Cycles 2 and 3, and tidal channel fills were determined through point counting. Cycle 1 strata were not included in this analysis because comparable sandstones do not occur. All of the samples examined contain substantial fractions of lithic fragments and could be classified as sub-litharenites. However, distinction can be made between the sandstones based on the feldspar content.

Cycle 2 sandstones sampled for petrographic analysis were taken from the uppermost Bigoray Sandstone and are upper fine to medium grained. They are characterized by very low feldspar contents (less than 5 per cent), and accordingly the modal grain plots occupy a narrow field along the quartz-lithic fragment axis on ternary diagrams (Fig. 11). Tidal channel-fill sandstones, specifically upper fine to medium grained basal channel deposits, make up two compositional populations. One group contains less than 5 per cent feldspar and is compositionally indistinguishable from Cycle 2 sandstones, while the second group has a distinctly higher feldspar content (Fig. 11). Cycle 3 samples were collected from crossbedded sandstones overlying the upper shale marker. Modal grain plots of these Cycle 3 sandstones fall within the compositional field encompassing the feldspathic group of tidal channel-fill sandstones.

Sandstones of the Glauconitic Member in the Drayton Valley area comprise two petrographic populations. Stratigraphically lower sandstones (Cycle 2) fall into a non-feldspathic group, while stratigraphically higher sandstones

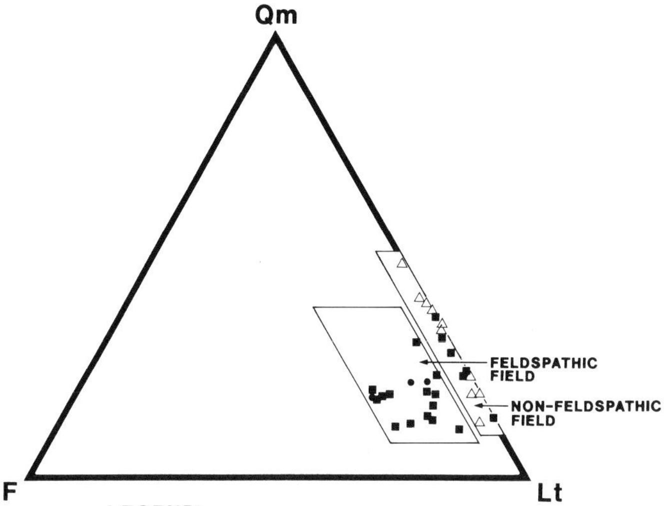

Figure 11. QmFLt diagram for upper fine to medium grained sandstones of the Glauconitic Member in the Drayton Valley area (Qm, monocrystalline quartz; F, K-feldspar + plagioclase; Lt, lithic fragments + chert + polycrystalline quartz).

(Cycle 3) fall into a feldspathic group. Both populations also include channel-fill sandstones. The variation of sandstone composition with stratigraphic position in the Glauconitic Member is common in the subsurface of south and central Alberta (Rosenthal, 1988; James, 1985). It is interpreted to result from changes in provenance related to the Lower Cretaceous structural evolution of the Western Canadian Cordillera and adjacent foreland basin. In central Alberta, the oldest sandstones of the Glauconitic Member are quartzose, middle sandstones are quartzose-lithic, and the youngest sandstones are feldspathic-lithic (Rosenthal, 1988). In this study, Cycle 2 and 3 sandstones represent the middle and youngest provenance categories; respectively, the Glauconite B and A successions of Rosenthal (1988). Tidal channel fills include sandstones in both categories.

Discussion

Stratigraphic Relationships and Timing of Estuary and Shoreface Facies Deposition

Facies relationships and sandstone petrography suggest the tidal channel-fill deposits represent two episodes of estuarine sedimentation that are stratigraphically equivalent to the marine facies of Cycles 2 and 3. Evidence for the contemporaneous deposition of tidal estuarine and marine facies includes: 1) the occurrence of two identical petrographic classes of marine and estuarine sandstones; and 2) the presence of outer estuary deposits adjacent to marine shoreface strata.

The similar compositions of non-feldspathic tidal

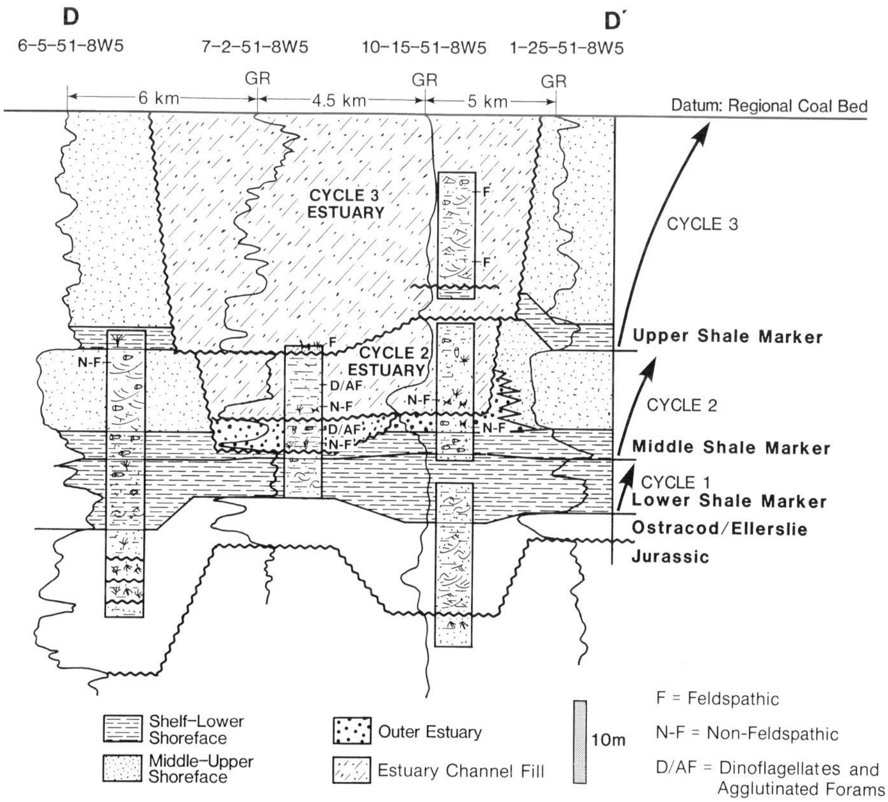

Figure 12. Stratigraphic cross-section D-D′ showing vertical and lateral relationships of incised tidal estuary channel fill, shoreface and outer estuary facies. Tidal channel fill that contains burrowed muddy sandstones is included as outer estuary facies in 7-2-51-8W5. Cross-section located on Figure 14. Refer to Table 1 for key to lithologies.

channel fills and Cycle 2 sandstones suggest a depositional link between the estuary system and marine shoreface. Coeval deposition within these two systems is also indicated by the facies relationships. Non-feldspathic tidal channel fills were observed to be incised into Cycle 1 and Cycle 2 strata, but do not occur adjacent to Cycle 3 strata (Fig. 12). This indicates the channels are age equivalent to slightly younger than Cycle 2 strata. Wood fragments and mudstone clasts observed in several cores of the Bigoray Sandstone may represent estuary-derived clasts that were rapidly buried in the shoreface environment. The sharp but apparently conformable contact with lower shoreface facies at these locations also suggests relatively rapid progradation. These sandstones are possible outer estuary ebb-tidal shoals that prograded onto the marine shoreface. Outer estuary burrowed muddy sandstones occurring within non-feldspathic tidal channel fills also suggest the proximity of marine environments at the time of deposition (Fig. 12). These stratigraphic relationships indicate the existence of linked estuary, estuary throat, and marine shoreface environments during Cycle 2 deposition. During submergence of the Cycle 2 shoreline however, the non-feldspathic estuarine system was backfilled with the massive mudstone facies, and abandoned.

The sandstone petrography and facies architecture suggest a connection between feldspathic tidal channel fills and Cycle 3 strata. Feldspathic tidal channel fills are incised into strata of all three marine cycles, but are conformably overlain by coals, which also cap Cycle 3 (Fig. 13). This indicates the channels were synchronous with, or are slightly younger than Cycle 3 strata. Abundant mudstone clasts and wood fragments within Cycle 3 deposits suggest the coexistence of estuarine and marine environments. The impoverished ichnofacies of Cycle 3 is also an indication of brackish conditions, which would be expected adjacent to an active estuary. Feldspathic channel fills dominate the tidal estuarine facies association indicating widespread establishment of Cycle 3 associated estuary channels (Figs 14, 15). The dominance of lateral accretion facies in this strata reflects extensive reworking of pre-existing deposits by middle to upper estuary channels that eventually transected the entire study area. Deeply incised point-bar (fining-upward) successions that extend up to coal horizons at the top of the Glauconite Member (Figs. 8a, 15) represent the final stage of Cycle 3 estuarine deposition. At this time, tide-influenced rivers representing upper estuaries extended across the study area. Coal beds directly overlying both the channel fill and marginal marine strata reflect flanking coastal plain backswamps.

Summary and Conclusions

The Glauconitic Member near Drayton Valley, Alberta, consists of regionally extensive marine strata incised locally by laterally confined tidal estuarine channel fills. The marine facies associations comprise three coarsening-upward lithosomes that represent three transgressive-regressive shelf to shoreface depositional cycles. Channel fills are fining-upward successions inferred to represent tide-influenced point-bar, channel margin, and channel abandonment deposits within estuaries. Evidence of estuarine conditions in the channel facies includes: 1) rhythmically bedded

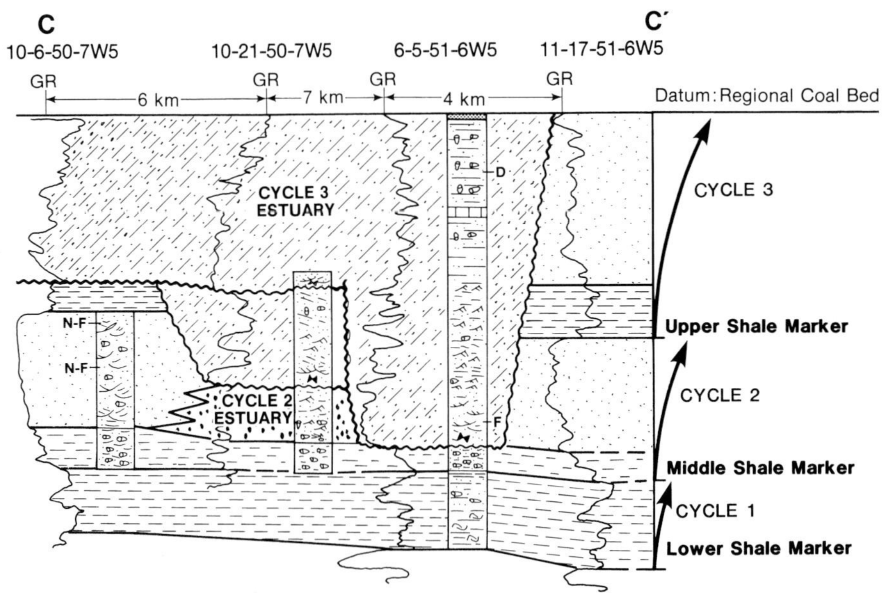

Figure 13. Stratigraphic cross-section C-C′ showing incisement of Cycle 3 tidal estuary channels into Cycle 2 and 3 shoreface and outer estuary facies. Cross-section location shown on Figure 15. Refer to Table 1 for key to lithologies.

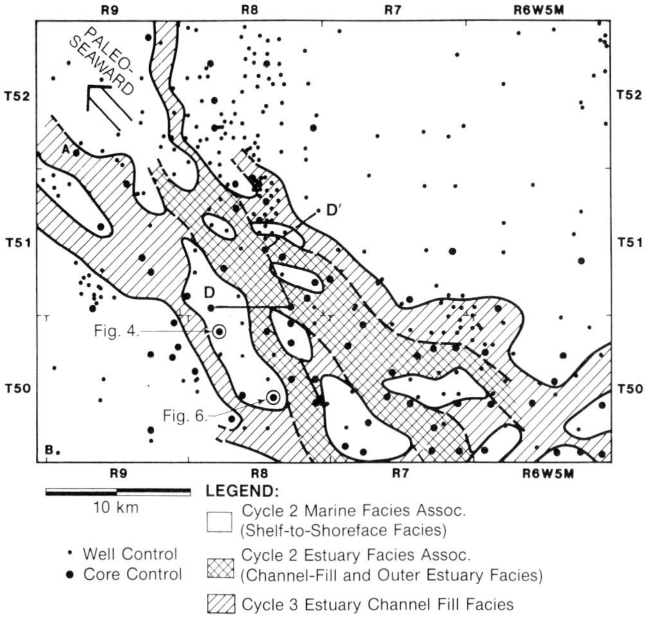

Figure 14. Map of Cycle 2 and 3 tidal estuary facies associations incised into the Cycle 2 marine facies association. This map confines the spatial limits of tidal estuarine channels contemporaneous with Cycle 2 deposition.

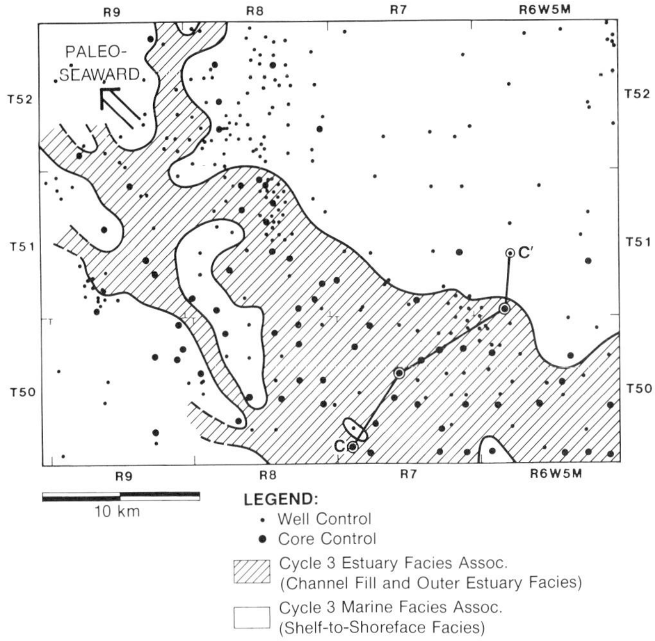

Figure 15. Map of Cycle 3 estuary facies association that is incised into Cycle 3 marine facies association, showing an increase in areal extent of estuarine sedimentation (compare to Figure 14).

inclined heterolithic stratification, and wavy, lenticular, and flaser bedding, which infer tidal current processes of deposition; and 2) impoverished suites of ichnofossils, dinoflagellates, and agglutinated foraminifera, which imply brackish water depositional environments.

A genetic relationship between the tidal channel fills and strata of the two youngest marine cycles is indicated by the occurrence of quartzose-lithic sandstones in the oldest channel fills and in Cycle 2 marine facies, and feldspathic-lithic sandstones in the youngest channel fills and the Cycle 3 marine succession. A further link between the oldest channel fills and adjacent marine strata is indicated by the presence of outer estuary deposits that are transitional with Cycle 2 middle and upper shoreface deposits. Most estuary sediments however, are feldspathic-lithic point-bar deposits of middle to upper estuary channels. This suggests late stage reworking by middle and upper estuary channels, which extended across the entire study area at the close of Cycle 3 deposition. The end of estuarine deposition is marked by extensive coal deposits that cap the Glauconitic Member, and represent coastal backswamps.

Acknowledgments

Financial and logistical support for this study was provided by Total Petroleum Canada Ltd. and the Natural Sciences and Engineering Research Council of Canada (NSERC). The authors especially thank J.-C. Beauvilain and P. Ward of Total Petroleum for their interest and support. A. Jenkins, of Associated Biostratigraphic Consultants, provided palynological analyses. M. Gallagher identified the foraminifera. The manuscript benefitted greatly from the critical review of J. Hopkins, B. Hayes, P. Putnam and D. Smith. The advice and encouragement of G. Reinson is also gratefully acknowledged.

References

Allen, J. R. L. 1963. The classification of cross-stratified units, with notes on their origin. Sedimentology, v. 2, p. 93-114.

Clifton, H. E., Hunter, R. E. and Phillips, R. L. 1971. Depositional processes and structures in the non-barred high-energy nearshore. Journal of Sedimentary Petrology, v. 41, p. 651-670.

Coleman, J. M. and Wright, L. D. 1965. Sedimentary structures: Mississippi River deltaic plain. *In:* Primary sedimentary structures and their hydrodynamic interpretation, G. V. Middleton (ed.). Society of Economic Paleontologists and Mineralogists, Special Publication no. 12, p. 133-148.

Davidson-Arnott, R. G. D. and Greenwood, B. 1976. Facies relationships on a barred coast, Kouchibouguac Bay, New Brunswick, Canada. *In:* Beach and Nearshore Sedimentation. Society of Economic Paleontologists and Mineralogists special publication no. 24, p. 149-168.

deMowbray, T. 1983. The genesis of lateral accretion deposits in recent intertidal mudflat channels, Solway Firth, Scotland. Sedimentology, v. 30, p. 425-435.

deRaaf, J. F. M. and Boersma, J. R. 1971. Tidal deposits and their sedimentary structures. Geologie en Minjnbouw, v. 50, p. 479-504.

Dorjes, J. and Howard, J. D. 1975. Estuaries of the Georgia coast, U.S.A.: sedimentology and biology. IV. Fluvial-marine transition indicators in an estuarine environment, Ogeechee River-Ossabaw Sound. Senckenbergiana Maritima, v. 7, p. 137-179.

Ekdale, A. A., Bromley, R. G. and Pemberton, S. G. 1984. Ichnology: trace fossils in sedimentology and stratigraphy. Society of Economic Paleontologists and Mineralogists, Short Course no. 15, 317 p.

Frey, R. W. and Howard, J. D. 1986. Mesotidal estuarine sequences: a perspective from the Georgia Bight. Journal of Sedimentary Petrology, v. 56, 911 p.

______ and Pemberton, S. G. 1985. Biogenic structures in outcrops and cores I, Approaches to ichnology. Bulletin of Canadian Petroleum Geology, v. 33, p. 72-115.

Glaister, R. P. 1959. Lower Cretaceous of southern Alberta and adjoining areas. American Association of Petroleum Geologists, Bulletin, v. 43, p. 590-640.

Greer, S. A. 1975. Estuaries of the Georgia coast, U.S.A.: sedimentology and biology. III Sandbody geometry and sedimentary facies at the estuary-marine transition zone, Ossabaw Sound, Georgia: a stratigraphic model. Senckenbergiana Maritima, v. 7, p. 181-203.

Haynes, J. R. 1981. Foraminifera. Macmillan Ltd., Basingstoke, England. 433 p.

Hopkins, J. C. and Farshori, Z. 1987. Fluvial and lacustrine sandstone bodies of the Middle Mannville: exploration significance of the Ostracode beds and Sunburst (Basal Quartz) beds, Alberta and Montana. Canadian Society of Petroleum Geologists, Reservoir, v. 14, p. 1-3.

Howard, J. D., Elders, C. A. and Heinbokel, J. F. 1975. Estuaries of the Georgia coast, U.S.A.: sedimentology and biology. V. Animal-sediment relationships in estuarine point-bar deposits, Ogeechee River-Ossabaw Sound, Georgia. Senckenbergiana Maritima, v. 7, p. 181-203.

______ and Nelson, C. H. 1982. Sedimentary structures on a delta-influenced shallow shelf, Norton Sound, Alaska. *In:* The Northeastern Bering Shelf: new perspectives of epicontinental shelf processes and depositional products, C. H. Nelson and S. D. Nio (eds.). Geologie en Mijnbouw, v. 61, p. 29-36.

Jackson, P. C. 1984. Paleogeography of the Lower Cretaceous Mannville Group of Western Canada. *In:* Elmworth-case study of a deep basin gas field, J. A. Masters (ed.). American Association of Petroleum Geologists, Memoir 38, p. 49-77.

James, D. P. 1985. Stratigraphy, sedimentology and diagenesis of Upper Jurassic and Lower Cretaceous (Mannville strata), south-western Alberta. PhD thesis, University of Oxford, Oxford, England, 225 p.

Jouanneau, J. M. and Latouche, C. 1981. The Gironde estuary. Contributions to sedimentology 10; Stuttgart, E Schweizerbart'che Verlagsbuchhandlung, 115 p.

Jopling, A. V. 1965. Hydraulic factors controlling the shape of laminae in laboratory deltas. Journal of Sedimentary Petrology, v. 35, p. 777-791.

Loutit, T. S., Hardenbol, J., Vail, P. R. and Baum, G. R. 1988. Condensed sections: the key to age dating and correlation of continental margin sequences. *In:* Sea-level Changes, an Integrated Approach, C. K. Wilgus, B. S. Hastings, C. G. St.C. Kendall, H. W. Posamentier, C. A. Ross and J. C. Van Wagoner (eds.). Society of Economic Paleontologists and Mineralogists, Special Publication no. 42, p. 183-216.

McLean, J. R. and Wall, J. H. 1981. The Early Cretaceous Moosebar Sea in Alberta. Bulletin of Canadian Petroleum Geology, v. 29, p. 334-377.

Morton, R. A. 1981. Formation of storm deposits by wind forced currents in the Gulf of Mexico and the North Sea. *In:* Holocene Marine Sedimentation in the North Sea Basin, S. D. Nio (ed.). International Association of Sedimentologists, Special Publication no. 5, p. 385-396.

Reineck, H. E. and Singh, B. 1980. Depositional Sedimentary Environments. Springer-Verlag, New York, 549 p.

Rosenthal, L. 1988. Wave dominated shorelines and incised channel trends: Lower Cretaceous Glauconite Formation, west-central Alberta. *In:* Sequences, Stratigraphy, Sedimentology: Surface and Subsurface, D. P. James and D. A. Leckie (eds.). Canadian Society of Petroleum Geologists, Memoir 15, p. 207-220.

Rudkin, R. A. 1964. Lower Cretaceous. *In:* Geological History of Western Canada, R. G. McCrossan and R. P. Glaister (eds.). Alberta Society of Petroleum Geologists, p. 156-169.

Seilacher, A. 1967. Bathymetry of trace fossils. Marine Geology, v. 5, p. 413-428.

Smith, D. G. 1988. Modern point bar deposits analogous to the Athabasca oil sands, Alberta, Canada. *In:* Tide Influenced Sedimentary Environments and Facies, P. L. deBoer, A. van Gelder and S. D. Nio (eds.). Reidel, Dordrecht, p. 417-432.

Thomas, R. G., Smith, D. G., Wood, J. M., Visser, J., Calverly-Range, E. A. and Koster, E. H. 1987. Inclined Heterolithic Stratification: Terminology, description, interpretation and significance. Sedimentary Geology, v. 53, p. 123-179.

Wightman, D. M., Pemberton, S. G. and Singh, C. 1987. Depositional modelling of the upper Mannville (Lower Cretaceous), East Central Alberta: implications for the recognition of brackish water deposits. *In:* Reservoir Sedimentology, R. W. Tillman and K. Weber (eds.). Society of Economic Paleontologists and Mineralogists, Special Publication no. 40, p. 190-220.

KEY WORDS: ESTUARY AND MARINE FACIES, GLAUCONITIC MEMBER, DRAYTON VALLEY, CENTRAL ALBERTA

In: Clastic Tidal Sedimentology, D. G. Smith, G. E. Reinson, B. A. Zaitlin and R. A. Rahmani (eds.). Canadian Society of Petroleum Geologists. Memoir 16, p. 123-134.

An exhumed, Lower Cambrian tidal flat: the Antietam Formation, central Virginia, U.S.A.

EDWARD L. SIMPSON

Department of Physical Sciences
Kutztown University
Kutztown, PA 19530
U.S.A.

ABSTRACT

At Natural Bridge Quarry, Virginia, tidal-flat facies are preserved within the Antietam Formation, of the Upper Precambrian-Lower Cambrian Chilhowee Group. The Antietam Formation is composed of two storm- and wave-dominated progradational parasequences; the uppermost of these is capped by a sequence boundary overlain by tidal-flat facies. Shoreface deposits are in erosional contact above and below the tidal-flat sediments.

Preserved sedimentary structures within the tidal-flat assemblage permit construction of a model for the ebb-through-flood phase. During the flood phase, flood channels and topographic highs were migration sites of small- to medium-scale, subaqueous dunes with superimposed current ripples. Current, oscillatory, and combined-flow ripples developed marginal to the flood channels, followed by suspension settling of mud during the flood stillstands. Peak ebb flow was marked by development of dunes mainly in ebb channels. Both flood- and ebb-oriented dunes and current ripples were modified during ebb runoff and emergence. Runoff along current ripple and dune troughs downcut troughs and resulted in deposition of runoff microdeltas. Near the ebb slack water, the tidal flat became emergent; lee sides and troughs of dunes were either covered with current ripples, erosively steepened, infilled with sand or crosscut by rill features.

Tidal processes developed during transgression across exposed shoreface deposits of the initial sequence of the passive margin. The tidal-flat facies was deposited probably in an estuarine setting. An upper microtidal to lower mesotidal setting is indicated by preserved paleorelief, scale of bedforms, trace fossil distribution and thickness of tidal-flat facies. High frequency sea level changes or autocyclic processes within the near-shore system are possible mechanisms that caused preservation of the bedforms. In other outcrops of the Chilhowee Group, preservation of tidal-flat deposits is rare because shoreface retreat eliminated more proximal facies, resulting in amalgamation of progradational and transgressive shoreface deposits.

RÉSUMÉ

À la carrière Natural Bridge en Virginie, des faciès d'estran sont préservés dans la Formation Antietam du Groupe Chilhowee (Précambrien Supérieur-Cambrien Inférieur). La Formation Antietam se compose de deux paraséquences progradationnelles, l'une dominée par les tempêtes, et l'autre par l'action des vagues. Une limite de séquence sépare la paraséquence supérieure des faciès de slikke sus-jacents. Les sédiments intertidaux sont à la base et au sommet en contact d'érosion avec les sédiments de la zone infratidale.

La préservation des structures sédimentaires dans les dépôts d'estran a permi d'établir un modèle pour la phase complète flux-reflux. Durant la marée ascendante, la migration des dunes subaquatiques de dimensions petites à moyennes, superposées de rides de courant, prenaient place dans les chenaux inondés et les hauts points topographiques. Des rides de courant, d'oscillation et d'écoulement combiné s'édifiaient en marge des chenaux inondés, suivies d'une sédimentation des particules fines en suspension durant la période plus calme de l'invasion marine. Principalement dans les chenaux de marée, le point culminant de la marée descendante se soulignait par le développement de dunes formées durant le reflux. Les dunes orientées durant le flux et le reflux ainsi que les rides de courant ont été modifiées durant l'émergence et le retrait des eaux. L'écoulement le long des rides de courant et des fosses des dunes recoupait les fosses et construisait des microdeltas d'écoulement. Près de l'étale de la marée descendante, l'estran émergeait; les fosses et les côtés abrités des dunes étaient, soit recouverts de rides de courant, soit érodés abruptement, soit remplis de sable, ou encore recoupés par des structures en rigole.

Durant la transgression, les processus de marée se développèrent sur l'ensemble des dépôts exposés de la zone infratidale, appartenant à la séquence initiale de la marge passive. Les faciès d'estran ont été déposés probablement dans un environnement estuarien. La préservation du paléorelief, l'envergure des «morphologies de fond» («bedforms»), la distribution des traces fossiles et l'épaisseur des faciès de la zone intertidale indiquent un environnement microtidal supérieur à mésotidal inférieur. La grande fréquence des changements du niveau marin ou encore les processus autocycliques dans un système littoral sont probablement des mécanismes responsables de la préservation des «morphologies de fond» («bedforms»). La rare préservation des dépôts d'estrans à d'autres affleurements du Groupe Chilhowee s'explique par le retrait de la zone infratidale, éliminant ainsi les faciès plus proximaux d'où l'amalgame de dépôts infratidaux de transgression et de progradation.

Introduction

Tidal currents have been interpreted as playing an important role in the production of Cambrian-age quartz arenites throughout the central and southern Appalachian Mountain belt (Schwab, 1970, 1971, 1972; Mack, 1980). Since these interpretations were made, detailed investigations of Holocene and ancient tidal deposits have identified new sedimentological criteria that permit more detailed identification and discrimination of tidal processes. Applying these new standards to the Chilhowee Group in Tennessee and southern and central Virginia, Cudzil and Driese (1987), Simpson (1987), Walker *et al.* (1988) and Simpson and Eriksson (1990) demonstrated that tidally-influenced sedimentation was less widespread than previously envisioned.

This study applies modern diagnostic criteria to identify intertidal deposits within the Antietam Formation of the Chilhowee Group exposed in the Natural Bridge quarry in Virginia (Fig. 1). Superb bedding plane exposures of exhumed, tidally generated bedforms and sedimentary structures provide a unique opportunity to reconstruct the processes active on an Early Cambrian tidal flat and permit evaluation of paleotidal range. Understanding processes active on this and many other Cambrian tidal flats is necessary to evaluate the effects of land vegetation on types and distribution of facies.

Stratigraphic Setting

The Weaverton, Harpers and Antietam formations (Fig. 1) compose the Chilhowee Group. Equivalent stratigraphic units in southern Virginia are the Unicoi, Hampton and Erwin formations. Central Virginia nomenclature is used in this paper. In the study area, the Weaverton Formation unconformably overlies Grenville-age basement consisting of diorite, granodiorite, and granite (Brown and Spencer, 1981);

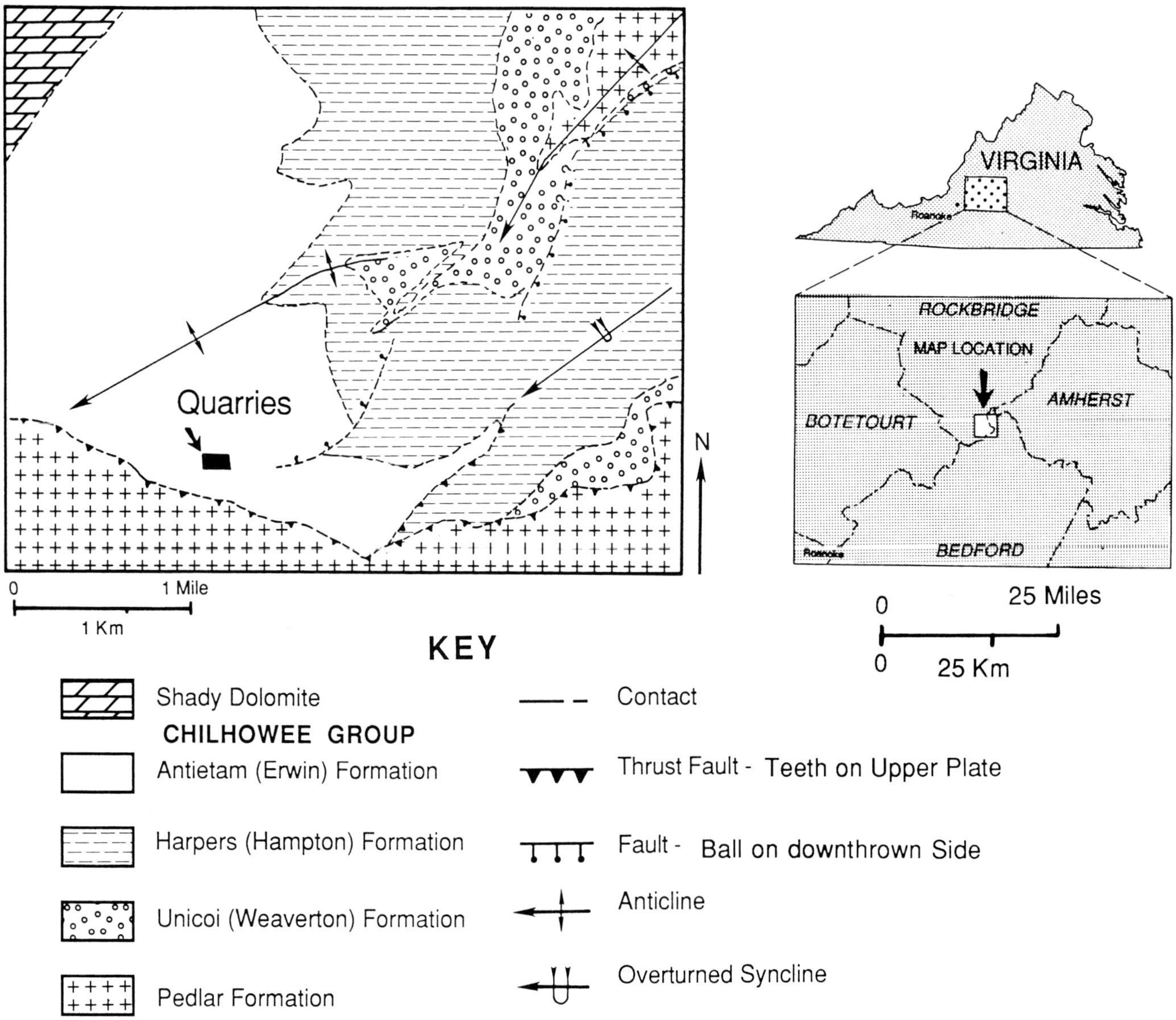

Figure 1. Geological map of the Natural Bridge Quarry Area. Map modified from Brown and Spencer (1981).

the Antietam Formation is overlain by the Shady Dolomite. The age of the Chilhowee Group is poorly constrained.

Discovery of the trilobite trace fossil *Rusophycus* from lower Weaverton Formation equivalents in southern Virginia indicates that most of the Chilhowee Group is of Early Cambrian age (Simpson and Sundberg, 1987); the Early-Middle Cambrian boundary is located in the upper Shady Dolomite (see Barnaby and Read, 1990 and references therein).

Strata of the Chilhowee Group reflect rift-related and passive-margin sedimentation associated with the opening of the Iapetus Ocean (Simpson, 1987). Feldspathic and lithic arenites, conglomerates, and subordinate basalts of the lower Weaverton Formation record rifting (Simpson and Eriksson, 1989). Quartz and minor feldspathic arenites of the upper Weaverton Formation preserve the incipient passive margin development (Simpson and Eriksson, 1989). During the initial transgression of the rift sediments, quartz arenites in the upper Weaverton Formation developed under a strong tidal influence; as the transgression continued, wave dominance supplanted the strong tidal influence (Simpson and Eriksson, 1989). Mudstones and subordinate wackes characterize the Harpers Formation, while quartz arenites and lesser amounts of mudstone compose the Antietam Formation (Schwab, 1970, 1971, 1972; Simpson and Eriksson, 1990); these formations record progradational outbuilding of the passive margin under storm- and wave-dominated conditions with only minor tidal overprint (Cudzil and Driese, 1987; Simpson, 1987; Walker *et al.*, 1988; Simpson and Eriksson, 1990).

The lowermost sequence of the passive margin starts in the upper Weaverton Formation and terminates in a type 2 sequence boundary in the upper Antietam Formation (Simpson and Eriksson, 1990). Superimposed on this sequence are five parasequences. Facies in the Natural Bridge quarry record the transition from the lower to an overlying sequence that continues into the Shady Dolomite (Simpson and Eriksson, 1990; Barnaby and Read, 1990).

At the Natural Bridge Quarry, Virginia, three facies are recognized: proximal inner shelf, shoreface, and tidal-flat facies (Figs. 1, 2). Only general observations of the shoreface and proximal-inner shelf facies are given below; additional details are discussed in Simpson (1987) and Simpson and Eriksson (1990). Bedding planes of the tidal-flat facies were marked with 7.62 m-square grids; structures were identified and mapped using the grid base (Figs. 3, 4).

Proximal Inner Shelf Facies

Description

Proximal inner shelf facies at the Natural Bridge Quarry consists of medium bedded, medium grained quartz arenites (Fig. 2). Trough cross-stratification and tabular(?)-

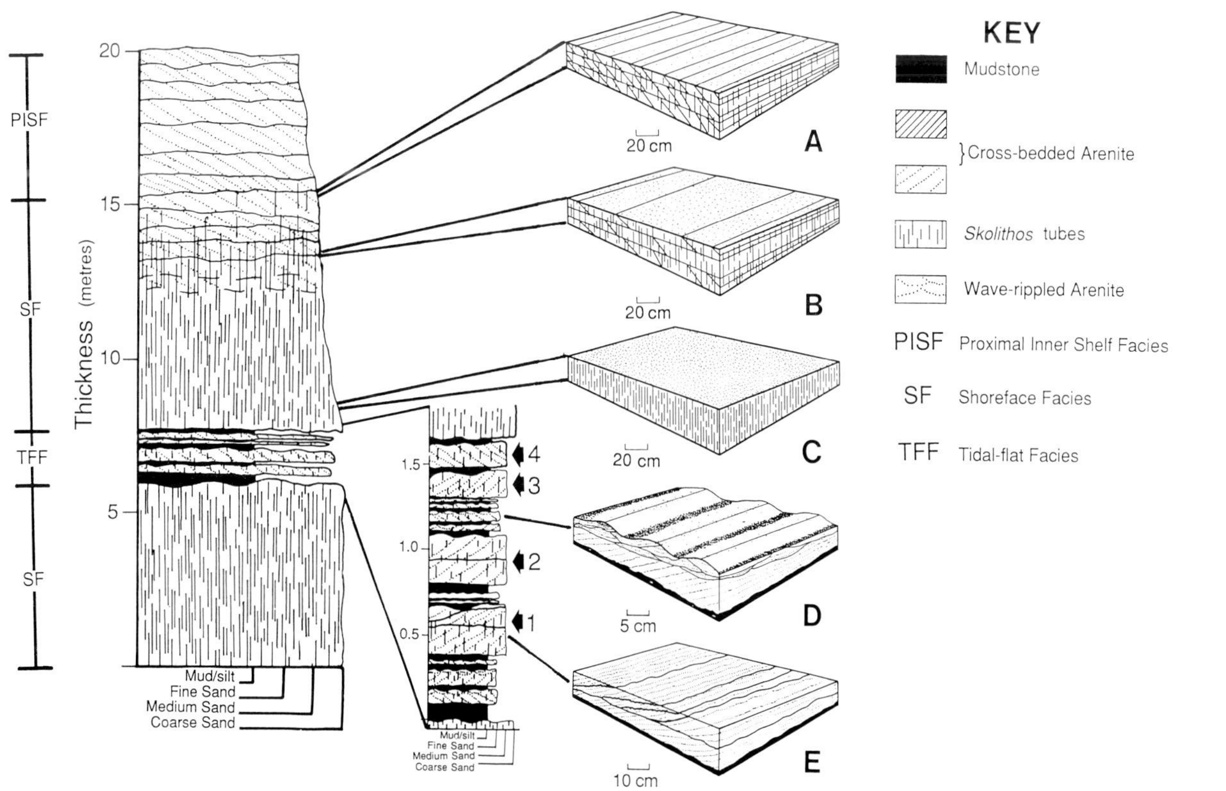

Figure 2. Stratigraphic section at the Natural Bridge Quarry with enlarged stratigraphic section of tidal-flat facies and block diagrams. **A.** Tabular-planar foresets with low concentrations of *Skolithos* tubes. **B.** Tabular, cross-stratified sets with diffuse foresets and moderate concentrations of *Skolithos* tubes. **C.** Cross-stratified sets with dense concentrations of *Skolithos* tubes. **D.** Unidirectional, cross-stratified base overlain by bidirectional bundled lenses of cross-stratification capped by form-discordant, combined-flow ripples. **E.** Cross-stratified set with mud-draped pause planes in medium bedded arenite. Medium bedded arenites are labeled 1 through 4.

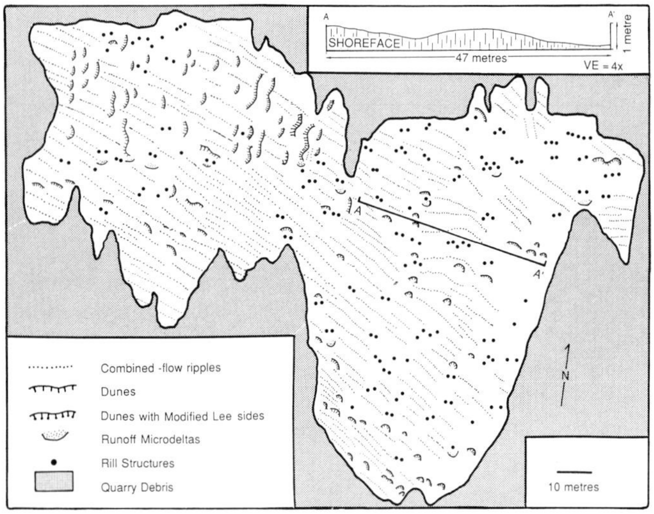

Figure 3. Distribution of dunes, ripples and other sedimentary structures on the main quarry surface. Cross-section illustrates preserved paleotopography. Note vertical exaggeration.

tangential cross-stratification are the dominant sedimentary structures. Trough axes yield a north-to-northwest paleocurrent mode (Simpson and Eriksson, 1990).

Interpretation

This facies is assigned to the proximal inner-shelf setting on a storm- and wave-dominated coast (Simpson and Eriksson, 1990). Trough and tabular-tangential cross-stratification were generated by relatively high current velocities during onshore-directed flow produced during initial coastal setup or after cessation of storm conditions by the fair-weather wave spectrum (Clifton *et al.*, 1971).

Shoreface Facies

Description

The shoreface facies at the Natural Bridge Quarry consists of very thick bedded, coarse grained quartz arenites. Intense bioturbation by *Skolithos linearis* tubes characterize the shoreface facies (Fig. 2A, B, C). Primary structures are absent when *Skolithos linearis* density is high (5 of Droser and Bottijer, 1989) and grain size is coarse (Fig. 2C). Tabular set boundaries and diffuse planar and tangential foresets appear with a reduction in grain size and *Skolithos* tube density (4 of Droser and Bottijer, 1989; Fig. 2B). Set thickness varies from a few centimetres to 25 cm. Low *Skolithos* concentrations (2 to 3 of Droser and Bottijer, 1989) permit complete preservation of tabular-tangential, tabular planar, and trough cross-stratification (Fig. 2A). At other localities, Simpson and Eriksson (1990) report northeast and southwest paleocurrent modes with a subordinate mode to the southeast.

Interpretation

This facies is interpreted as being deposited in a high energy shoreface setting influenced by shore-parallel currents. Intense post-depositional bioturbation within the majority of beds precludes detailed process interpretation. A longshore current system is favored over a tidally-influenced shoreface (Simpson and Eriksson, 1990) because of the vertical association with proximal inner-shelf deposits and the absence of definitive tidally produced structures, such as mud drapes and bundled foresets (e.g., Visser, 1980; Allen, 1981; Boersma and Terwindt, 1981; de Mowbray and Visser, 1984; Kriesa and Moiola, 1986) in beds with low concentrations of *Skolithos* tubes. *Skolithos linearis* is diagnostic of the

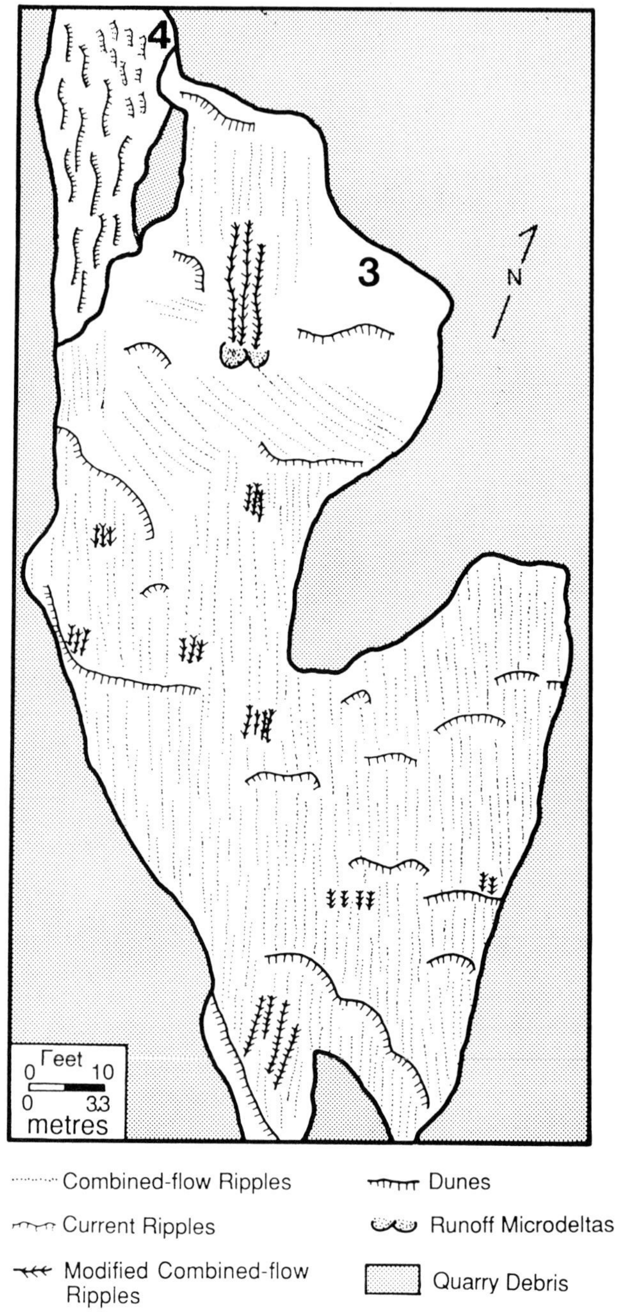

Figure 4. Distribution of dunes, ripples and other sedimentary features on medium bedded arenite beds 3 and 4.

Skolithos ichnofacies, which is best developed in shallow-marine, subtidal to intertidal settings on a mobile substrate (Seilacher, 1969; Frey and Pemberton, 1984).

Tidal-Flat Facies

Description

Thin to medium bedded, fine to medium grained quartz arenites (99% quartz plus 1% heavy minerals) interbedded with very thin bedded mudstones characterize this facies (Fig. 2). Paraconformities above and below the tidal-flat facies mark the transition to and from the shoreface facies. Medium bedded arenites occur in four beds (1-4; Fig. 2) that vary from 23 to 35 cm in thickness. Thin and medium beds of arenites exhibit a tabular geometry in the quarry.

Exposures of the tidal-flat facies are restricted to the Natural Bridge Quarry. The overall geometry of the tidal-flat lithosome is imprecisely known. Outcrops approximately 6 km northward along strike, within the same thrust sheet, display a vertical change in ichnofauna from *Skolithos linearis* to *Monocraterion* sp. back to *Skolithos linearis* at the sequence boundary; thin to medium bedded arenites separated by mudstones are absent (Simpson, 1987). This reflects the northernmost extent of the tidal-flat facies. Southward along strike, the Antietam Formation is truncated by a thrust fault (Fig. 1).

Quarrying operations have exhumed three types of bedding planes. These include: 1) thin bedded arenites covered by ripples 2) medium bedded arenites capped by small- to medium-scale dunes, ripples, and runoff microdeltas; and 3) the lower paraconformity surface topped with small- to medium-scale dunes, ripples, rills, runoff microdeltas and preserved paleotopography (Fig. 3). The main quarry surface exposes sediments above the lower paraconformity (Fig. 3), whereas higher, smaller quarry surfaces reveal bedding planes of thin to medium bedded arenites (Fig. 2 inset; Fig. 4).

Small- to Medium-scale, Subaqueous Dunes

Small to medium, two-dimensional dunes and small, three-dimensional dunes (*cf.* Ashley, 1990) with well developed scour pits cover the main quarry surface (Figs. 3, 5, and 6A) and bed number 3 (Fig. 4) The bedforms have: 1) wavelengths that vary from 1.50 to 10.00 m; 2) amplitudes that range from 0.10 to 0.37 m; 3) ripple indices (wavelength/amplitude) that vary from 8.8 to 18.8; and 4) ripple asymmetry indices (crest to trough length/wavelength) that vary from 0.04 to 0.26.

Dunes on the main quarry surface yield a bimodal distribution of paleocurrents consisting of northwest and southwest modes (Fig. 7A). Northwest-oriented dunes are restricted to the northwestern quadrant of the quarry surface (Fig. 3). Most dunes have lee-side inclinations of approximately 20°; lee sides of some northwest-directed dunes are steeper, with a maximum angle of 40° (Fig. 5). Small- to medium-scale, lunate shaped dunes with scour pits give

Figure 5. Photographs of sinuous-crested dune with steepened lee face. Arrows point to current ripples in trough and outlined runoff microdelta. Backpack for scale.

a southwest paleocurrent mode (Fig. 7A) and are found across the entire surface, but with only a limited number in the northwest quadrant (Fig. 3). Several lee faces of southwest-directed dunes pass laterally into mudstone followed by a 5 to 10 cm-thick, sigmoidal shaped sand fill.

Small- to medium-scale, sinuous-crested and lunate shaped dunes cap the medium bedded arenites of beds 1 to 3, whereas small, lunate shaped dunes cap bed 4 (Fig. 4). Dunes yield a bimodal paleocurrent distribution. Dunes on the tops of beds 1 through 3 are south- to west-directed, whereas dunes on bed 4 are east-directed (Figs. 4, 8A). The amplitudes of the lee faces in bed 4 are less than those in beds 1 to 3 on the main quarry surface. Cross-sections through beds reveal that medium bedded arenites internally possess trough, tabular-planar and tangential cross-stratified sets with pause-plane surfaces; some pause-plane surfaces are draped by mudstone (Fig. 2E).

Ripples

Ripples cover all sandstone surfaces. Slightly asymmetrical ripples with wavelengths of approximately 15 cm are the

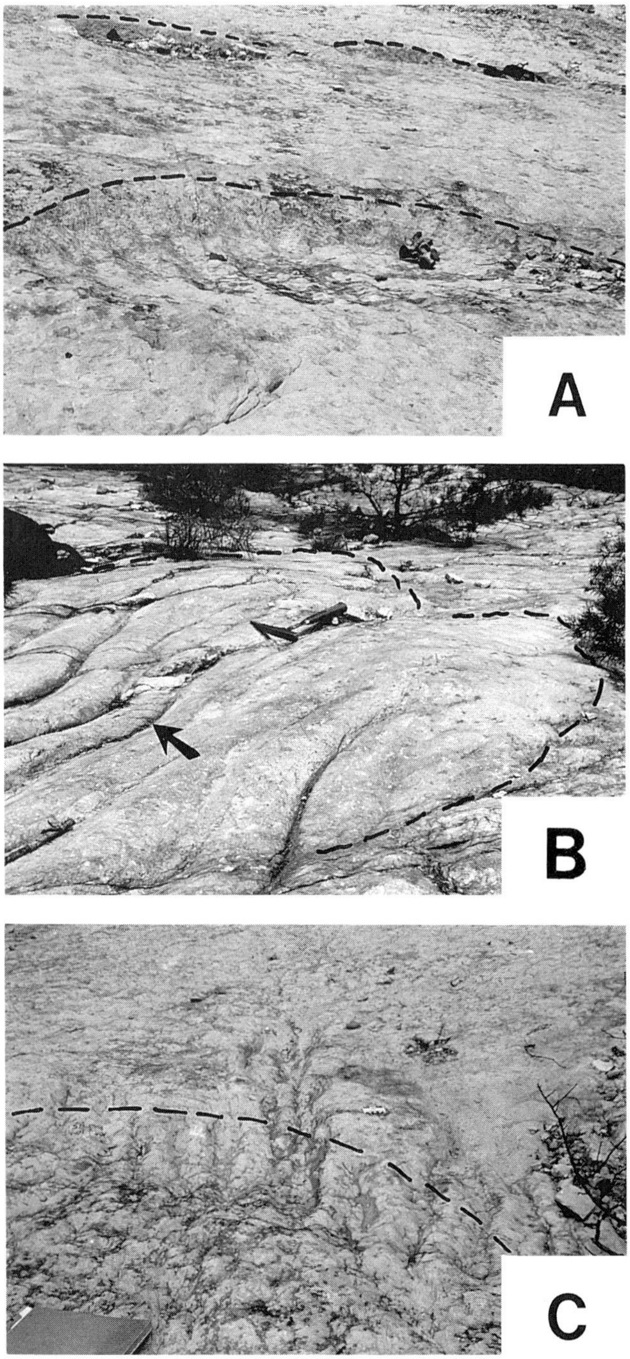

Figure 6. Photographs of dunes and sedimentary structures on bedding planes. **A.** Dunes with scour pits. Dashed line marks crest line. Gloves for scale. **B.** Runoff microdelta with partially filled troughs of combined flow ripples (arrows). **C.** Braided rill marks crosscutting dune crest line outlined by dashed line.

most common ripple type and cover the main quarry surface and tops of the medium bedded arenites (1-3; Fig. 8B). Crestlines trend northwest-southeast, but display a dominantly asymmetrical direction to the southwest with a lesser number of lee sides to the northeast (Fig. 6A). Strongly asymmetrical ripples oriented subparallel or parallel to the dune crestlines partially cover the stoss sides of northwest-directed dunes (Fig. 7B). In troughs and on the steepened

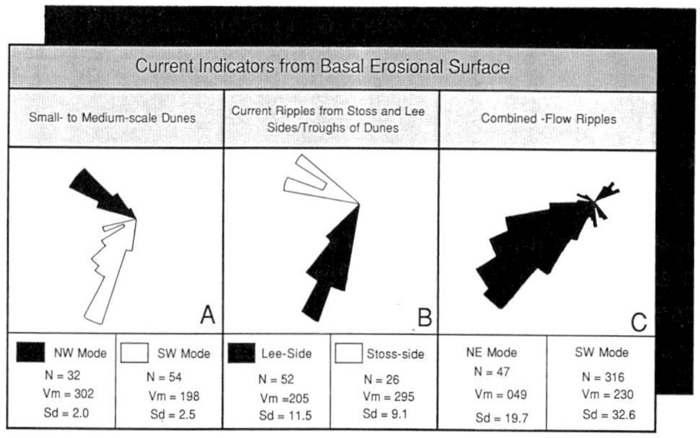

Figure 7. Paleocurrents from structures on the basal erosional surface. N, number of measurements; Vm, vector mean; Sd, standard deviation.

lee sides of northwest-directed dunes are southwest-directed, strongly asymmetrical ripples (Fig. 7B). West-directed, slightly asymmetrical ripples dominate tops of medium bedded arenites 1 to 3 (Fig. 8B). On the top of bed 3, west-directed ripples are steepened and have cuspate profiles in trough-to-trough cross-sections (Fig. 6B). On bed 4, west-directed, strongly asymmetrical ripples oppose the eastward mode of the dunes (Figs. 4, 7A, B).

Slightly asymmetrical ripples or symmetrical ripples with an equant, hexagonal pattern and foresets with a random orientation cap thin arenite beds. Internally, thin bedded arenites display the following vertical transitions: 1) unidirectional, ripple crosslaminations; 2) chevon-bundled, ripple crosslaminations; 3) unidirectional, crosslaminated base overlain by chevron-bundled, crosslaminated tops; and 4) bidirectional, bundled crosslaminations capped by form-discordant, slightly asymmetrical ripples (Fig. 2D).

Additional Sedimentary Structures

Bedding plane tops of medium bedded arenites and the main quarry surface display sedimentary structures in

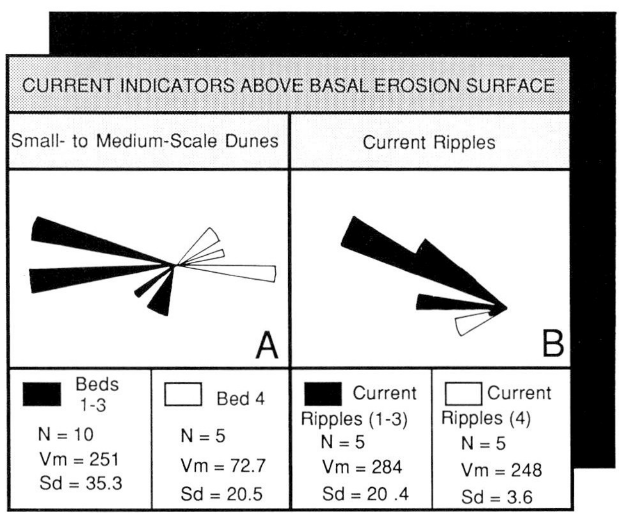

Figure 8. Paleocurrents from medium bedded arenites 1-4. N, number of measurements; Vm, vector mean; Sd, standard deviation.

addition to dunes and ripples. When troughs of ripples with cuspate profiles (bed 3) and lee sides of dunes with steepened lee faces are traced southward, they terminate in structureless, lobate shaped arenite bodies (Fig. 6B). Approaching the structureless, arenite mass, ripple troughs become shallower and broader and eventually merge with the arenite mass. Braided rill marks characterized by a decrease in the number of branches in a southerly direction are found across the main quarry surface except in the northwest-directed dune cluster (Figs. 3, 6B). Rills are perpendicular to and crosscut stoss and lee faces of southwest-directed dunes (Fig. 6C). Some rills pass into structureless, lobate sandstone mounds. Discontinuous dikelets with a crude, orthogonal geometry are commonly present on the soles of medium- and thin-bedded arenites.

Ichnofauna in the tidal-flat facies are of low diversity. Funnel-shaped *Monocraterion sp.* burrows are restricted to bed 4, and horizontal burrows (*Planolites sp.*) are found within mudstones. *Skolithos linearis* is distributed across the main quarry surface and the medium bedded arenites. *Skolithos* density is significantly less in the medium bedded arenites compared to densities in the shoreface deposits. Broken pieces of *Skolithos* tubes are incorporated into sediment, forming dunes.

Paleotopography

Paleotopography is discernible on the eastern half of the main quarry surface, where two depressions trending approximately north-south are separated by a ridge (Fig. 3). The depressions merge in the southern part of the quarry, and the intervening ridge disappears. Paleorelief in the quarry is approximately 1 m. Northwest-directed bedforms appear to be restricted to topographic highs. Additional paleorelief s problematic west of the termination of cross-section A-' (Fig. 3) because the quarry surface is warped slightly by olding (Fig. 1). South-directed bedforms are confined mainly to topographic lows in the cross-section and are only rarely developed on topographic highs. Orientation of asymmetrical ripples changes over the topographic highs.

Mudstones

Mudstone beds range from 0.5 to 3.0 cm in thickness. The mudstones are massive, and crosscut by sinuous, horizontal burrows and dikelets from overlying arenite beds.

Interpretation

Construction of a model for the ebb and flood tidal phases is possible by analysis of superposed sedimentary structures and analogy with modern Holocene tidal flats (*cf.* Klein, 1963, 1977; Evans, 1965, 1975; de Raaf and Boersma, 1971; Reineck and Singh, 1980; Dalrymple, 1984a, 1984b; Terwindt, 1988). Processes responsible for the genesis of a particular sedimentary structure can be assigned to one of four broad subdivisions (phases) of the tidal cycle: flood, flood slack-water, ebb, and ebb slack-water. Structures preserved within the quarry developed during more than a single tidal cycle. The following model is a generalized tidal cycle for all preserved structures.

Symmetrical, slightly asymmetrical, and strongly asymmetrical ripples are interpreted to be the product of oscillatory, combined, and unidirectional flow, respectively (Harms *et al.*, 1982).

Flood Phase

The northwest-directed dunes on the main quarry surface (Figs. 3, 9a) are assigned to the flood stage for the following reasons: 1) dunes are onshore-directed relative to the northeast-southwest paleoshoreline established by Brown (1970), Schwab (1970, 1971, 1972), Wehr and Glover (1985), and Simpson and Eriksson (1990); 2) northwest-directed dunes occupy topographic highs; and 3) paleocurrents from dunes oppose the southward-tilting, local paleoslope determined from emergence structures (discussed later). Dunes record maximum flood velocity (Dalrymple *et al.*, 1977; Harms *et al.*, 1982; Dalrymple, 1984a). Current ripples on the stoss sides of the dunes (Fig. 9b) developed coeval with or after dune migration as speeds decreased toward the flood, slack-water phase. Parallelism of dune and ripple crestlines indicates flow was toward the northwest even during waning stages of the flood phase (Fig. 7A, B).

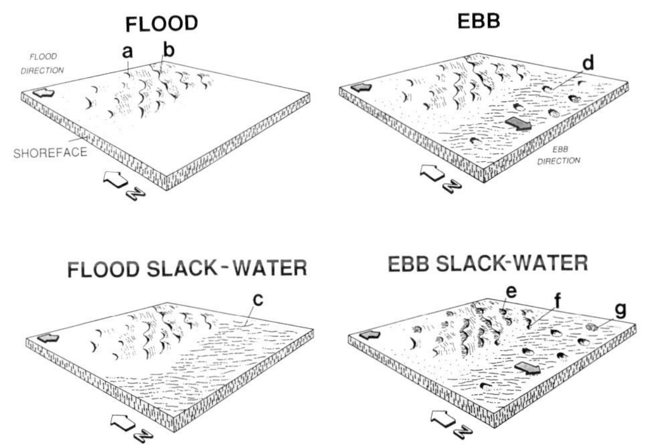

Figure 9. The model for ebb through flood tidal cycle for the main quarry surface. **a)** Dune field consisting of northwest-directed, 2- and 3-dimensional dunes; **b)** ripples partially covering stross sides of dunes; **c)** combined-flow ripples; **d)** south-directed dunes; **e)** south-directed current ripples on lee sides and in troughs of dunes; **f)** runoff microdelta; **g)** rill structures.

Flood Slack-water Phase

Maximum flooding of the tidal flat with reduction or elimination of tidally generated currents characterized the flood slack-water stage. Muds settling from suspension draped pause planes of bedforms in areas sheltered from wave action. In exposed areas, waves molded sediment into oscillatory and combined-flow ripples. On the main quarry surface, wave-induced current velocities lacked sufficient strength and duration to modify dunes, but generated combined-flow ripples across most of the quarry surface. Flood-produced ripples, with the exception of ripples near the crests of dunes, were reworked. The asymmetrical direction of the combined-flow ripples paralleling the ebb drain direction indicates that the unidirectional component in the combined-flow current began during the initiation of the ebb phase. Shoaling waves probably produced the observed minor northeast asymmetry component of the combined-flow ripples and the change in orientation of ripple crestlines across topographic highs.

Structures on tops of the thin arenite beds developed peripheral to the main channels. Basal portions of thin arenite beds characterized by unidirectional crosslaminae (Fig. 2D) were produced during either the previous flood or ebb phase. During flood slack-water, most tops were reworked by oscillatory or combined-flow currents into chevron-bundled ripple crosslaminae and bidirectional, bundled crosslaminae capped by form discordant, wave- or combined flow-rippled tops (*cf.* Raaf *et al.*, 1977). Complex interaction of oscillatory or oscillatory and unidirectional currents resulted in ripples with an equant, hexagonal pattern (Allen, 1983). Thin-bedded arenites resemble distal storm beds associated with shelf deposits, but can be distinguished by ripple type on tops of beds.

Ebb Phase

As water levels began to recede, ebb channels became active. Some oscillatory reworking may have continued into the ebb phase. Ebb channels probably were separate from flood channels because of the scarcity of concurrent, opposed, sedimentary structures. On bed 4 (Fig. 4), opposed flood-produced dunes and ebb-generated current ripples record limited ebb drainage in the flood channels. Discrete channels with flood-tide dominating ebb-tide or vice versa are common in modern tidal inlet settings (Hayes, 1980). Ebb-related dunes are confined to the ebb channels, with a few dunes developing on topographic highs (Fig. 9d). Flood slack-water mud drapes were covered by sand transported by ebb currents, producing sigmoidal-shaped tidal bundles. Tidal bundles of 5 to 10 cm thickness are consistent with observed bundle thickness ranges from Holocene and ancient deposits (*cf.* McCave, 1971; Allen and Homewood, 1984; Kreisa and Moiola, 1986). Neap-spring tidal sequences could not be positively identified because of the limited lateral exposure.

Ebb Slack-water Phase

Shallow-water runoff features and modification of existing bedforms record emergence of the tidal flat. Shallow ebb runoff funneled preferentially into troughs of dunes and ripples. Late-stage ebb runoff attained sufficient velocity to develop ripples in flood-produced dune troughs (Fig. 9e) and steepened lee faces. Shallow-water runoff modified ripple troughs and crests into a cuspate profile. Sand was transported southward along troughs of dunes and deposited as sandlobes (Fig. 6B, 9f); sandlobes are analogous to runoff microdeltas of Dalrymple (1984b). Late-stage runoff generated rills that modified bedforms (Fig. 6C, 9g). Rills crosscut dune crestlines as the water level fell below dune height; continued reduction in water level caused rills to incise deeper into the crestline. Water in some rills transported sufficient sand to generate runoff microdeltas. Southerly drain directions from rills, the position of runoff microdeltas at the southern ends of dunes and combined-flow ripple troughs, and the merging of topographic depressions southward all imply that the local paleoslope tilted toward the south.

Sandstone dikelets on bases of beds may have been produced by either synaeresis or subaerial desiccation of the mudstone beds. If subaerial desiccation is responsible, then the incomplete nature of the cracking may be due to only brief exposure of the muds during the tidal cycle (*cf.* Allen, 1983).

Paleotidal Range

Klein (1971) and Terwindt (1988) enumerated criteria that can be used to reconstruct the tidal ranges of ancient deposits. Paleorelief, bedform amplitude, ichnofossil distribution and thickness of tidal-flat facies are evaluated as tools to determine tidal range within the Natural Bridge quarry.

At minimum a microtidal setting is required to develop the 1 m of paleorelief exposed within the quarry (Fig. 3). Low-tide line must be below the topographically lowest point on the main quarry surface because emergence structures cover the entire surface. Flood-produced bedforms present on topographic highs indicate that the high-tide line is at a greater elevation than the highest topographic point.

Terwindt (1988) reported that small amplitude lee faces are associated with lesser water level fluctuations, while larger lee-face amplitudes are produced with greater water level variations within a tidal setting. Maximum bedform amplitudes of 0.37 m develop in water depths of possibly 3 to 4 m during bedform migration, (*cf.* Allen, 1968, Fig. 6-4; Allen, 1983 Fig. 8.20) supporting a mesotidal interpretation. Bedform amplitude is problematic; amplitude may be controlled by velocity and water depth, not tidal range (*cf.* Harms *et al.*, 1982).

Ichnofauna distribution is consistent with an intertidal interpretation but is of limited utility in constraining high- and low-tide lines. The *Skolithos*-producing organism was possibly a filter-feeding polychaete (Hallam and Swett, 1966; Goodwin and Anderson, 1974), a phoronid (Sundberg, 1983), or some unknown taxon (Sundberg, 1983). The polychaetes *Diopatra cuprea* and *Lanice conchilega* are modern analogues for *Skolithos* (Barwis, 1985; Carey, 1987).

Studies of the distribution of modern tube-building polychaetes *Diopatra cuprea* and *Lanice conchilega* are ambiguous (*cf.* Barwis, 1985; Carey, 1987). At the entrance of the Tay estuary in England, Carey (1987) reported that *Lanice conchilega* clusters are restricted to the low-water, neap-tide level and below. On Kiawah Island of South Carolina, *Diopatra cuprea* is limited to the intertidal zone, with *Monocraterion*-like funnels formed near the mean, high-water mark (Barwis, 1985).

Working in Pennsylvania with stratigraphic equivalents

of the Weaverton Formation, Goodwin and Anderson (1974) assigned *Skolithos* to high energy, low flats and *Monocraterion* to low energy, high flats. Analogy with *Lanice* (Carey, 1987) requires that the main quarry surface with the highest *Skolithos* burrow densities accumulated below the low water, neap-tidal level; comparison with the distribution of *Diopatra* indicates an intertidal setting. *Monocraterion* found in bed 4 records the high-tide level according to Goodwin and Anderson (1974) and Barwis (1985).

The thickness of the tidal-flat facies from the main quarry surface (approximate low-water mark) to the top of bed 4 is 1.6 m (Fig. 2). Preserved thickness (1.6 m) approximates the tidal range and denotes a microtidal setting (*cf.* Klein, 1971; Terwindt, 1988). This value is problematic because it assumes: 1) constant sea level; 2) lack of subsidence by thermal processes, and sediment loading and sediment compaction; and 3) a conformable succession.

Transgressions or regressions cause a decrease or increase, respectively, in the accommodation space for sediment, resulting in a preserved stratigraphic thickness that is different from the actual tidal range. In the Antietam, deviation from the true thickness is dependent on the rate of change in relative sea level rise associated with the transgression. Greater rates of sea level change result in an overestimation of the tidal range. The tidal-flat facies developed during transgression. Thus the actual tidal range may have been somewhat less than 1.6 m. Over- or underestimation of the tidal range is reduced with a shorter duration deposit.

Mudstones compose approximately 0.5 m of the tidal-flat facies (Fig. 2). Decompaction of mudstones and sandstones expands the section to approximately 2.9 m. Thermal subsidence and sediment loading increase preserved thickness, causing an overestimation of the tidal range. Thermal subsidence rates of 10 cm/1000 yrs. for the early phase of the passive margin have been reported from the overlying Shady Dolomite (Barnaby and Read, 1990). This effect can not be evaluated because the accumulation rate of the tidal-flat facies is unknown.

Structures and bedforms from the tidal-flat facies are similar to lower sand flats from modern tidal flats (Evans, 1965, 1975; Dalrymple, 1984a). Many modern tidal flats consist of lower sand flats that pass landward into mud flats (Reineck, 1963; Evans, 1965, 1975; Semeniuk, 1981; Yeo and Risk, 1981). Mid to upper mud-flat deposits are missing in the tidal-flat facies; these deposits have lower preservation potential (Nio *et al.*, 1980; Terwindt, 1988). If developed in the Antietam Formation, upper tidal flats were eliminated by shoreface retreat during the transgression (*cf.* Nummedal and Swift, 1987). The erosional bedding plane separating tidal-flat facies from the overlying shoreface deposit records the shoreface retreat unconformity. Therefore, the preserved thickness of the tidal-flat facies records a minimum value for the tidal range.

The best approximation for the tidal range from the discussion above is, at minimum, an upper microtidal to lower mesotidal setting.

Discussion

Estuaries are produced during transgression by the flooding of river mouths. An estuarine setting is indicated for the tidal-flat facies by limited lateral distribution of facies (approximately 6 km), similarity of sedimentary structures with modern and ancient estuarine sediments (Evans, 1965, 1975; Dalrymple, 1984a, b), and its association within the transgressive portion of the facies succession (Fig. 2).

Preservation of dunes, ripples and ephemeral features with low preservation potential, such as microdeltas and rills (Terwindt, 1988), necessitates an abrupt cessation of sediment supply and current velocity. Autocyclic and allocyclic (*cf.* Beerbower, 1964) processes are capable of causing preservation of bedforms.

An allocyclic process, such as small-scale, high frequency, sea level fluctuations, could initiate a change within the depositional systems and cause preservation of bedforms on tops of beds. The thicknesses of medium bedded arenite and mudstone couplets in the tidal-flat facies (approximately 40 cm) are less than those reported for beds deposited as a result of high frequency, sea level changes in siliciclastics (*e.g.*, Eriksson and Simpson, 1990; Simpson and Eriksson, submitted this volume) but are within the range recorded in carbonates (Read *et al.*, 1986; Goldhammer *et al.*, 1990).

Autocyclic processes including bar migration and growth, channel abandonment, and possibly a change in position of the turbidity maximum zone in the estuary can cause preservation of bedforms. Alternatively, seasonal floods introducing large volumes of fresh water into the estuary could have expanded the turbidity maximum zone onto the tidal flat, also pushing it seaward, cutting off sand supply from the shoreface and depositing the mud drape across the sandstone beds. Lack of land vegetation during the Cambrian Period enhanced flooding. (Schumm, 1968). An autocyclic process, such as bar migration into the area, could change the local tidal circulation pattern, reduce the tidal current strength, or pond water. A sheltered area would allow sedimentation of muds on abandoned bedforms and sedimentary structures. As the bar advanced, sand supply and current velocity was restored and another medium bedded arenite body formed. None of the above mechanisms can be identified positively as generating the exhumed bedforms.

Conclusions

1. Exhumed bedforms and sedimentary features permit reconstruction of flood through ebb phases of sedimentation on a Lower Cambrian tidal flat within the Antietam Formation.

2. Migration of small- to medium-scale, subaqueous dunes and ripples characterized the flood phase.

3. During flood slack-water, ripples were generated by oscillatory and combined-flow currents.

4. Migration of dunes in ebb channels typified the ebb phase.

5. Falling water levels near the ebb slack-water modified

some flood and ebb bedforms by erosion and transport of material parallel to the crestlines, resulting in the generation of runoff microdeltas. Rills crosscut ebb-produced bedforms.

6. Paleorelief, scale of bedforms, trace fossil distribution, and preserved thickness of tidal-flat facies suggest, at minimum, an upper micro- to lower mesotidal setting.

Acknowledgments

Initial stages of this project were conducted with grants from the Appalachian Basin Industrial Associates, the Geological Society of America, Mobil Foundation and Sigma Xi. The final stage was undertaken using equipment purchased through a grant from the Kutztown University Research Committee. Thanks are extended to K. A. Eriksson, C. W. Harris, J. G. Patterson, J. DeLuca, S. Dreise, and R. Scully for visiting the quarry and refining the various ideas presented in this paper. F. Sundberg and R. K. Bambach made critical observations and aided the ichnofaunal interpretation. This paper has benefitted from reviews by C. W. Harris, J. G. Patterson, W. Simpson, and C. S. P. G. reviewers, R. Dalrymple and D. Leckie. L. House, C. W. Harris, and W. Simpson assisted in mapping.

References

Allen, J. R. L. 1968. Current Ripples. Amsterdam. North Holland, 433 p.

______ , 1981. Lower Cretaceous tides revealed by cross-bedding with mud drapes. Nature, v. 289, p. 579-581.

______ 1983. Sedimentary structures: their character and physical basis. Developments in Sedimentology no. 30, Elsvier, New York, New York, 663 p.

Allen, P. A. and Homewood, P. 1984. Evolution and mechanics of a Miocene tidal sandwave. Sedimentology, v. 31, p. 63-81.

Ashley, G. M. 1990. Classification of large-scale subaqueous bedforms: a new look at an old problem. Journal of Sedimentary Petrology, v. 60, p. 160-172.

Barnaby, R. J. and Read, J. F. 1990. Evolution of a Lower to Middle Cambrian carbonate platform: Shady Dolomite, Virginia Appalachians. Geological Society of America, Bulletin, v. 102, p. 391-404.

Barwis, J. H. 1985. Tubes of the modern *Diopatra cuprea* as current velocity indicators and as analogs for *Skolithos - Moncraterion. In:* Biogenic structures: Their use in interpreting depositional environments, H. A. Curran, (ed.). Society of Economic Paleontologists and Mineralogists, Special Publication Number 35, p. 225-235.

Beerbower, J. R. 1964. Cyclothems and cyclic depositional mechanisms in alluvial plain sediments. *In:* Symposium on cyclic sedimentation, D. F. Merriam (ed.). Kansas State Geological Survey, Bulletin, v. 169, p. 31-42.

Boersma, J. R., and Terwindt, J. H. J. 1981. Neap-spring tidal sequences of intertidal shoal deposits in a mesotidal estuary. Sedimentology, v. 28, p. 151-170.

Brown, C. E., and Spencer, E. W. 1981. Geologic map of the James River Face Wilderness, Bedford and Rockbridge counties, Virginia. United States Geological Survey, Miscellaneous Field Studies Map MF-1337-A.

Brown, W. R. 1970. Investigations of the sedimentary record in the Piedmont and Blue Ridge of Virginia. *In:* Studies of Appalachian Geology: central and southern, G. W. Fischer, F. J. Pettijohn, J. C. Reed and K. N. Weaver (eds.). Interscience, New York, p. 335-349.

Carey, D. A. 1987. Sedimentological effects and paleoecological implications of the tube-building polychaete *Lanice conchilega* Pallas. Sedimentology, v. 34, p. 49-66.

Clifton, H. E., Hunter, R. E. and Phillips, R. L. 1971. Depositional structures and processes in non-barred high-energy nearshore. Journal of Sedimentary Petrology, v. 41, p. 651-670.

Cudzil, M. R. and Driese, S. G. 1987. Fluvial, tidal and storm sedimentation in the Chilhowee Group (Lower Cambrian), northeastern Tennessee, U.S.A. Sedimentology, v. 34, p. 861-883.

Dalrymple, R. W. 1984a. Morphology and internal structure of sandwaves in the Bay of Fundy. Sedimentology, v. 31, p. 365-382.

______ 1984b. Runoff microdeltas: a potential emergence indicator in cross-bedded sandstones. Journal of Sedimentary Petrology, v. 54, p. 825-830.

______ , Knight, R. J. and Lambiase, J. J. 1977. Bedforms and their hydraulic stability relationships in a tidal environment, Bay of Fundy, Canada. Nature, v. 275, p. 100-104.

Droser, M. L. and Bottijer, D. J. 1989. Ichnofabric of sandstones deposited in high-energy nearshore environments: Measurement and utilization. Palaios, v. 4, p. 598-604.

Eriksson, K. A. and Simpson, E. L. 1990. Recognition of high-frequency sea-level fluctuations in Proterozoic siliciclastic tidal deposits, Mount Isa, Australia. Geology, v. 18, p. 474-477.

Evans, G. 1965. Intertidal flat sediments and their environments of deposition in the Wash. Quarterly Journal of the Geological Society of London, v. 121, p. 209-245.

______ 1975. Intertidal flat deposits of the Wash, western margin of the North Sea. *In:* Tidal Deposits: a Casebook of Recent Examples and Fossil Counterparts, R. N. Ginsberg (ed.). Springer Verlag, Berlin, p. 13-20.

Frey, R. W. and Pemberton, S. G. 1984. Trace fossils facies models. *In:* Facies Models, R. G. Walker (ed.). Geosciences Reprint Series 1, p. 189-207.

Goldhammer, R. K., Dunn, P. A. and Hardie, L. A. 1990. Depositional cycles, composite sea-level changes, cyclic staking patterns, and hierarchy of stratigraphic forcing: examples from the Alpine Triassic platform carbonates. Geological Society of America, Bulletin, v. 102, p. 535-562.

Goodwin, P. W. and Anderson, E. J. 1974. Associated physical and biogenic structures in environmental subdivision of a Cambrian tidal sand body. Journal of Geology, v. 82, p. 779-794.

Hallam, A. and Swett, K. 1966. Trace fossils from the Lower Cambrian pipe rock of the north-west Highlands. Scottish Journal of Geology, v. 2, p. 101-106.

Harms, J. C., Southard, J. B. and Walker, R. G. 1982. Structures and sequences in clastic rocks. Society of Economic Paleontologists and Mineralogists, short course 9, Calgary. 251 p.

Hayes, M. O. 1980. General morphology and sediment patterns in tidal inlets. Sedimentary Geology, v. 26, p. 139-156.

Klein, G., de V. 1963. Bay of Fundy intertidal zone sediments. Journal of Sedimentary Petrology, v. 33, p. 84-854.

______ 1971. A sedimentary model for determining paleotidal range. Geological Society of America, Bulletin, v. 89, p. 2585-2592.

______ 1977. Clastic tidal facies. Continuing Education Publishing Company, Champaign, Illinois, 149 p.

Kreisa, R. D. and Moiola, R. J. 1986. Sigmoidal tidal bundles and other tide-generated sedimentary structures of the Curtis Formation, Utah. Geological Society of America, Bulletin, v. 97, p. 381-387.

Mack, G. H. 1980. Stratigraphy and depositional environments of the Chilhowee Group (Cambrian) in Georgia and Alabama. American Journal of Science, v. 280, p. 497-517.

McCave, I. N. 1971. Sandwaves in the North Sea off the coast of Holland. Marine Geology, v. 10, p. 199-225.

Mowbray, T. de and Visser, M. J. 1984. Reactivation surfaces in subtidal channel deposits, Oosterschelde, southwest Netherlands. Journal of Sedimentary Petrology, v. 54, p. 811-824.

Nio, S-D, Berg, J. H. Van Den, Goesten, M. and Smulders, F. 1980. Dynamics and sequential analysis of a mesotidal shoal and intershoal channel complex in the eastern Scheldt (southwestern Netherlands). Sedimentary Geology, v. 26, p. 263-279.

Nummedal, D. and Swift, D. J. P. 1987. Transgressive stratigraphy at sequence-bounding unconformities: Some principles derived from Holocene and Cretaceous examples. *In:* Sea-level fluctuations and coastal evolution, D. Nummendal, O. H. Pilkey and J. D. Howard (eds.). Society of Economic Paleontologists and Mineralogists, Special Publication No. 41, p. 249-260.

Pittman, W. C., III 1978. Relationship between eustacy and stratigraphic sequences of passive margins. Geological Society of America, Bulletin, v. 89, p. 1389-1403.

Raaf, J. M. F. de and Boersma, J. R. 1971. Tidal flat deposits and their sedimentary structures. Geologie en Mijnbouw, v. 50, p. 470-504.

______ , ______ and Van Gelder, A. 1977. Wave-generated structures and sequences from a shallow marine succession, Lower Carboniferous, county Cork, Ireland. Sedimentology, v. 24, p. 451-483.

Read, J. F., Grotzinger, J. P., Bova, J. A. and Koerschner, W. F. 1986. Models for generation of carbonates cycles. Geology, v. 14, p. 107-110.

Reineck, H. E. 1963. Sedimentgefuge im Bereich der Sudliche Nordsee. Abhandl. Senckenberg. Naturforsch Gesellschaft 505, 138 p.

______ and Singh, I. B. 1980. Depositional sedimentary environments. Springer Verlag, New York, 549 p.

Schumm, 1968. Speculations concerning paleohydrological controls of terrestrial sedimentation. Geologial Society of America, Bulletin, v. 79, p. 1573-1588.

Schwab, F. L. 1970. Origin of the Antietam Formation (Late Precambrian?-Lower Cambrian), central Virginia. Journal of Sedimentary Petrology, v. 40, p. 354-366.

______ 1971. Harpers Formation, central Virginia: A sedimentary model. Journal of Sedimentary Petrology, v. 41, p. 139-149.

______ 1972. The Chilhowee Group and late Precambrian, early Paleozoic sedimentary framework in the central and southern Appalachians. *In:* Appalachian Structures: Origin, Evolution and Possible Potential for Exploration Frontiers, a Seminar, P. Lessing, R. I. Hayhurst, J. A. Barlow and L. D. Woodford (eds.). West Virginia University and West Virginia Geological and Economic Survey, p. 59-101.

Seilacher, A. 1969. Bathymetry of trace fossils. Marine Geology, v. 5, p. 413-428.

Semeniuk, V. 1981. Sedimentology and stratigraphic sequence of a tropical tidal flat, NW Australia. Sedimentary Geology, v. 29, p. 195-221.

Simpson, E. L. 1987. Sedimentology and tectonic implications of the late Proterozoic and Early Cambrian Chilhowee Group in southern and central Virginia (unpublished Ph.D. Thesis). Virginia Polytechnic Institute and State University, Blacksburg, VA, 298 p.

______ and Eriksson, K. A. 1989. Rift to incipient passive-margin sedimentation within the Late Proterozoic to Early Cambrian Unicoi Formation of southern and central Virginia. Geological Society of America, Bulletin, v. 101, p. 42-56.

______ and ______ 1990. Early Cambrian progradational and transgressive sedimentation patterns in Virginia: An example of the early history of a passive margin. Journal of Sedimentary Petrology, v. 60, p. 84-100.

______ and ______ (this volume) Depositional facies and parasequences in silciclastic tidal deposits from the Lower Proterozoic upper Mount Guide Quartzite, Mount Isa Inlier, Australia: Evidence for high frequency sea level oscillations. *In:* Clastic Tidal Sedimentology, Canadian Society of Petroleum Geologists, Memoir 16.

______ and Sundberg, F. A. 1987. Early Cambrian Age from synrift deposits of the Chilhowee Group of southwestern Virginia. Geology, v. 12, p. 123-126.

Sundberg, F. A. 1983. *Skolithos linearis* Haldeman from the Carrara Formation (Cambrian) of California. Journal of Paleontology, v. 57, p. 145-149.

Terwindt, J. H. J. 1988. Paleotidal reconstructions in inshore tidal depositional environments. *In:* Tide-influenced Sedimentary Environments and Facies, P. L. de Boar, A. van Gelder and S.-D. Nio (eds.). D. Reidel Publishing Company, Boston, Mass., p. 233-263.

Visser, M. J. 1980. Neap/Spring cycles reflected in Holocene subtidal large-scale bedforms deposits, a preliminary note. Geology, v. 8, p. 543-546.

Walker, D., Skelly, R., Cudzil, M. R. and Dreise, S. G. 1988. The Chilhowee Group of east Tennessee: Sedimentology of the lower Cambrian fluvial-to-marine transition. *In:* Depositional History of Paleozoic sequences, southern Appalachians, S. G. Driese and D. Walker (eds.). University of Tennessee, Studies in Geology Number 14, p. 30-61.

Wehr, F. and Glover, L., III 1985. Stratigraphy and tectonics of the Virginia-North Carolina Blue Ridge: Evolution of a late Precambrian-early Paleozoic hinge zone. Geological Society of America, Bulletin, v. 96, p. 285-295.

Yeo, R. K. and Risk, J. R. 1981. The sedimentology, stratigraphy, and preservation of intertidal deposits in the Minas Basin System, Bay of Fundy. Journal of Sedimentary Petrology, v. 51, p. 245-260.

KEY WORDS: TIDAL FLAT, LOWER CAMBRIAN, ANTIETAM FORMATION, CHILHOWEE GROUP, PROGRADATIONAL PARASEQUENCES

Tidal Rhythmites

In: Clastic Tidal Sedimentology, D. G. Smith, G. E. Reinson, B. A. Zaitlin and R. A. Rahmani (eds.). Canadian Society of Petroleum Geologists. Memoir 16, p. 137-160.

Temporal and spatial patterns of rhythmite deposition on mud flats in the macrotidal Cobequid Bay-Salmon River estuary, Bay of Fundy, Canada

Robert W. Dalrymple
Department of Geological Sciences
Queen's University
Kingston, Ontario, Canada K7L 3N6

Yasuhiko Makino
Faculty of Education
Ibaraki University
Mito 310, Japan

Brian A. Zaitlin
Esso Resources Canada Ltd.
237, 4th Avenue S.W.
Calgary, Alberta, Canada T2P 0H6

Abstract

In the inner part of the macrotidal, Cobequid Bay-Salmon River estuary, the fundamental sedimentation unit in the mixed flats and mud flats is a sand-mud couplet that is deposited during the flood or ebb half of a tidal cycle. The thickness of the mud layer is relatively constant between couplets, whereas the thickness of, and structure in, the sandy portion varies significantly because of differences in the current speed and tidal range between tides. Thus the deposits of successive tidal cycles are alternately thicker and thinner as a result of the diurnal inequality, and show cyclic variations in sand-layer thicknesses and structures because of the 14-day, neap-spring cycle. These structures are the mud flat equivalent of tidal bundles, and are a modern, process analogue for the tidal rhythmites described by various authors. In areas of slower sedimentation, the deposits of individual tides are not recognizable; instead, annual deposits are developed that consist of alternating, bioturbated (summer-fall) and laminated (springtime) sandy muds. Ice-produced deformation and ice-rafted pebbles mark the winter period.

Because the width of the muddy flats decreases into the estuary, the flats become progressively sandier headward, and short-term sedimentation rates increase from near zero at the seaward end, to 5 to 30 cm/year in the middle estuary, and to several metres per year in the inner, tidal-fluvial transition. Therefore, individual tidal couplets, the diurnal inequality, and neap-spring cycles are clearly visible only in the inner estuary where bioturbation is almost non-existent because of the low salinity and high sedimentation rates. The annual deposits are best developed in the middle estuary, whereas intensely bioturbated muds characterize the outer estuary. These longitudinal changes may provide a means of identifying ancient estuarine deposits, and of locating oneself within them.

The architecture of the fringing muddy sediments along the entire length of the tidal-fluvial transition consists of elongate, wedge-shaped bodies of essentially flat-lying tidal sediments separated from each other by steep erosional surfaces. Vertical aggradation predominates over lateral accretion, even on laterally migrat[illegible]nks, and inclined heterolithic stratification (IHS) is rarely developed.

Rés[illegible]

Dans la partie interne de la zone macrotidale de l'estuaire d[illegible] Salmon, baie Cobequid, l'unité sédimentaire fondamentale des estrans mixtes et des slikkes est le couple sable-boue déposé durant le flux ou le reflux d'un cycle de marée. L'épaisseur de la couche de boue est relativement constante entre les couples, tandis que l'épaisseur et la structure de la couche gréseuse varient d'une façon significative. Ces variations résultent des différences d'amplitude de la marée (ou marnage) et des vitesses de courant entre les marées. Par conséquent, les dépôts cycliques de marée alternent en épaisseur à cause de l'inégalité diurne. Les variations cycliques de l'épaisseur et de la structure de la couche de sable sont produites par la durée de 14 jours du cycle de marée de vive-eau et morte-eau. Ces structures de slikkes sont les équivalents des «faisceaux» («bundles») de marée, et résultent d'un processus moderne analogue à celui des séquences rythmiques de marée, décrites par plusieurs auteurs. Dans les régions de faible sédimentation, les dépôts associés à chaque marée ne sont pas reconnaissables. Par contre, les dépôts annuels sont constitués d'une alternance de sables boueux bioturbés (automne-été) et laminés (printemps). La période hivernale est marquée par les effets produits par la glace: déformation et transport de galets.

La largeur des slikkes («mud flats») décroît de l'embouchure de la rivière vers l'estuaire, et les estrans deviennent progressivement plus sableux vers le continent dans cette direction. La vitesse de sédimentation à court terme s'accroît de près de zéro du côté de la mer, à 5 à 30 cm par année dans la partie médiane de l'estuaire, à plusieurs mètres par année dans la partie interne de la transition fleuve-marée. Donc, les couples individuels de marée, les inégalités diurnes et les cycles de marées de vives-eaux et de mortes-eaux sont clairement visibles seulement dans la partie interne de l'estuaire, où la bioturbation est presque inexistante à cause de la faible salinité et du haut taux de sédimentation. Les dépôts annuels sont mieux développés dans la partie médiane de l'estuaire alors que les boues intensivement bioturbés caractérisent la partie externe. Ces changements longitudinaux peuvent permettre de reconnaître les anciens dépôts d'estuaire et de se situer parmi ceux-ci.

La disposition des sédiments boueux périphériques tout le long de la transition marée-fleuve forme des corps prismatiques allongés composés essentiellement de sédiments d'estran disposés en couches planes. Ces couches sont séparées entre elles par des surfaces d'érosion escarpées. L'alluvionnement vertical prédomine sur l'alluvionnement latéral, même là où les rivages migrent latérallement. La stratification hétérolithique inclinée (SHI) ne s'est pas développée.

Introduction

Muddy intertidal flats are a common component of many low energy coastal environments, but their physical sedimentology has been neglected by geologists in recent years. An examination of recent textbooks (*e.g.*, Weimer *et al.*, 1982; Klein, 1985; Elliott, 1986) shows, in fact, that the most frequently cited papers on the physical structures of mudflats (flaser, tidal, and lenticular bedding, etc.) remain those written in the 1950's and 1960's (*e.g.*, van Straaten, 1954, 1961; Reineck, 1967; Reineck and Wunderlich, 1968). It is especially notable that the concept of tidal bundles and neap-spring cycles (*e.g.*, Visser, 1980; Allen, 1982; de Boer *et al.*, 1989), which has revolutionized the recognition and interpretation of tidal crossbedding, has not been widely applied to the muddy tidal-flat environment (Tessier and Larsonneur, *in* Larsonneur, 1988; Dalrymple and Makino, 1989; Tessier and Gigot, 1989). Research has instead focused on the biological reworking that occurs (Howard and Frey, 1975; Dorjes, 1978; Risk and Yeo, 1980; Frey *et al.*, 1989), and on the physical and biological processes that control sediment erosion and deposition (*e.g.*, Carling, 1982; Anderson, 1983; Amos and Mosher, 1985; Amos *et al.*, 1988). As a result, it is possible to get the impression that muddy tidal flats experience slow or negligible vertical aggradation (*cf.* van Straaten, 1954, 1961), and that the physical structures are largely destroyed by bioturbation. This is obviously not true in all cases, however, and particularly not on the banks of rapidly migrating channels (*e.g.*, Reineck, 1960, 1967; LeFournier and Friedman, 1974).

In this paper we describe the internal structures of the mixed flats and mud flats (collectively termed mud flats) from the inner part of the Cobequid Bay-Salmon River (CB-SR) macrotidal estuary, Bay of Fundy (Fig. 1), based on the detailed study of three representative sections and reconnaissance examination of numerous other localities. These observations reveal that the sediments that accumulated by vertical aggradation are not extensively bioturbated, and that sedimentation units representing several tidal and seasonal time intervals can be recognized. The variations in the thickness of these temporal sedimentation units, both vertically in any section and along the length of the estuary, provide important information on the fundamental patterns and processes of mud flat deposition. The objectives of this paper are, therefore, to describe the various temporal sedimentation units, and then to use them to analyze the spatial patterns of deposition throughout the estuary. The observations are directly relevant to the upper muddy portion of the facies model that has been constructed for the study area (Dalrymple *et al.*, 1990).

Recent work in the CB-SR estuary (Zaitlin, 1987; Dalrymple and Zaitlin, 1989; Dalrymple *et al.*, 1990) has resulted in the developing of a facies zonation of the sandy sediments occupying a belt along the axis of the estuary (Fig. 1): zone 1, the elongate, tidal sand bars at the seaward end; zone 2, the braided, upper-flow regime sand flats of the middle estuary; and zone 3, the single channel, tidal fluvial transition of the inner estuary. While this zonation is not directly applicable to the fringing muddy sediments, it does provide a convenient reference for longitudinal positions within the estuary and will be used here to facilitate cross-referencing

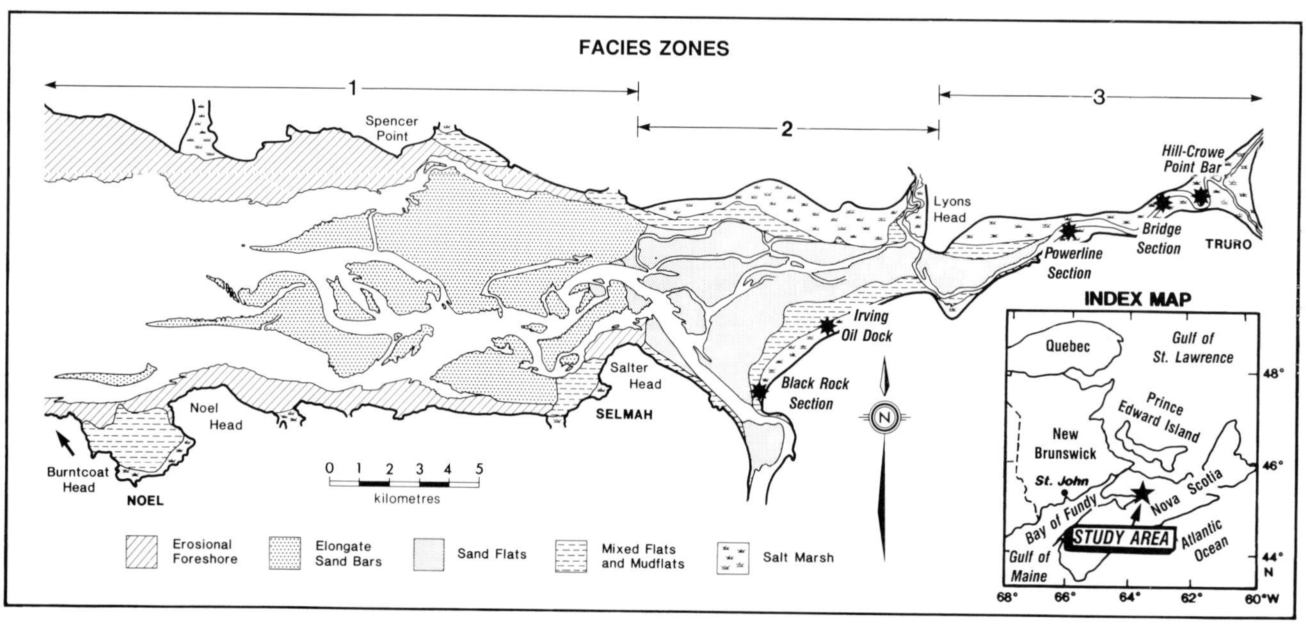

Figure 1. Facies map of the study area, showing the location of the measured sections and other sites mentioned in the text. The facies zones discussed in the text are indicated by arrows at the top; see Dalrymple and Zaitlin (1989) and Dalrymple *et al.* (1990) for additional descriptions. The outline of the intertidal areas in zone 1 (elongate tidal sand bars) is based on 1973 airphotos, and on 1982 airphotos in zones 2 and 3. The tidal limit (the inland extent of tidal rise and fall of water level) is at the right-hand edge of the map. After Dalrymple *et al.* (1990, Fig. 2). The inset map shows the general location of the study area.

with previous papers. In this study, the emphasis is placed on the mud flats located adjacent to zones 2 and 3, as these have not been studied in detail before.

Setting of the Study Area

Mud Flat Distribution and Morphology

In the CB-SR estuary, the bank-to-bank width at the mean high tide elevation increases exponentially through the first 20 km seaward of the tidal limit (Fig. 2A; Zaitlin, 1987), giving this area the classic "funnel-shaped" estuarine geometry (Pillsbury, 1939; Langbein, *in* Myrick and Leopold, 1963; Wright *et al.*, 1973). In this region (*i.e.*, zones 2 and 3), the valley width is greater than the bank-to-bank width, and salt marshes and depositional mud flats are extensively developed along the margins of the estuary (Figs. 1, 2). By contrast, the valley width limits the extent of the system seaward of the 20 km point (Fig. 2A) because it increases at a slower rate than the bank-to-bank width. As a result of this confinement, the shorelines of zone 1 are either nondepositional or erosional (Amos and Long, 1980; Dalrymple *et al.*, 1990), and fine grained sediments are of limited extent, occurring mainly as a veneer (<0.5 m thick) on the upper foreshore (Knight and Dalrymple, 1975). These muddy veneers have been studied extensively by Risk and others (*e.g.*, Risk *et al.*, 1977; Risk and Yeo, 1980; Yeo and Risk, 1979, 1981), but the more volumetrically important, muddy deposits of the inner estuary, which are the subject of this paper, have received only limited attention (Dalrymple *et al.*, 1982, 1990; Dewing, 1986; Dalrymple and Makino, 1989).

The mud flats adjacent to zones 2 and 3 experience alternate periods of erosion and deposition as a result of the movement of channels in the axial sandy facies. The effects on mud flat morphology vary in degree, although not in kind, along the length of the estuary because of differences in the widths of the mud flats and the rate of channel switching. In zone 2, the main channel of the Salmon River switches its path infrequently; in the 1950's, this channel lay along the south shore, but in the late 1950's or early 1960's it adopted its present path along the north shore (Fig. 1). As a result of this switch, the mud flats and salt marsh along the northern margin have undergone severe erosion, due primarily to undercutting by tidal currents but with a minor contribution by wave action near the high-water elevation, such that a 4 m high cliff has been formed at the outer margin of the salt marsh. At the same time, an aggradational

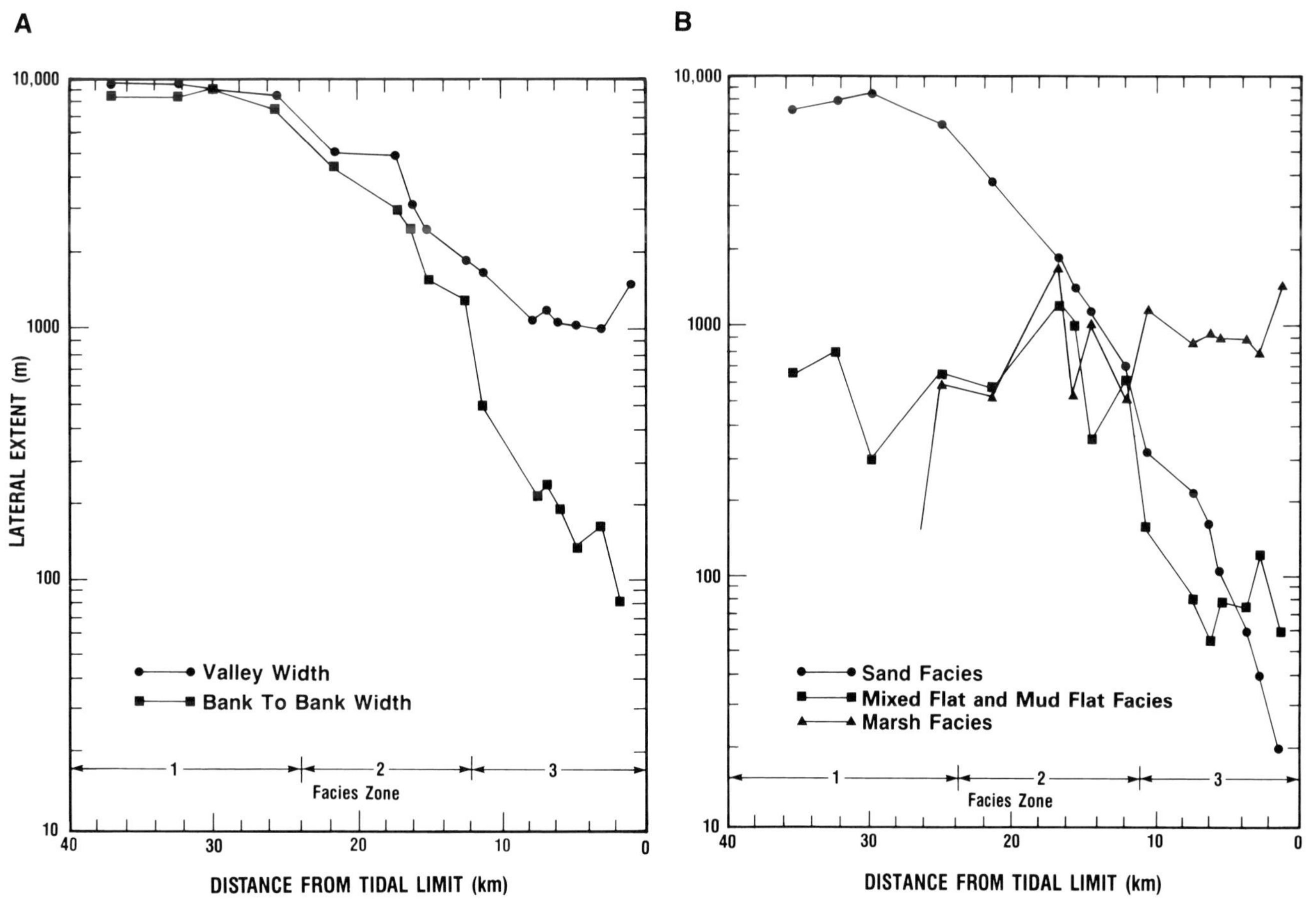

Figure 2. A. "Bank-to-bank width" at mean high tide, and valley width (lateral extent of all Holocene tidal sediments) as a function of distance from the tidal limit. **B.** Lateral extent of the sand, mixed flat plus mud flat, and salt marsh facies as a function of distance from the tidal limit. The facies zones at the bottom are as shown on Figure 1. After Dalrymple *et al.* (1990, Fig. 3).

wedge of sand flat, mud flat and salt marsh sediment up to 6 m thick and 2 km wide has accumulated along the south shore (Fig. 1). The facies in this area are broad (Fig. 2B) and their boundaries are generally gradational (Fig. 3A). The impingement of small, sand-flat braid channels against the outer edge of the muddy flats causes local erosion and the formation of small cliffs. Later abandonment or the northward migration of the channels cause deposition to resume. Each cliff remains visible for a considerable time as an abrupt step between adjacent facies (Fig. 3B), but such breaks are relatively uncommon and widely spaced in this part of the estuary.

Small changes in the path of the Salmon River thalweg also have a dramatic effect on the narrow (<100 m wide; Fig. 2B) muddy flats that border the channel in zone 3. As a result of frequent channel movement, the muddy banks show a stepped profile that consists of 2 to 5 terraces separated by steep rises that represent former erosional cliffs (Fig. 3C, D). The terraces reach 50 m in width, and vary from a few tens to several hundreds of metres in length. Their vertical separation ranges from a few decimetres to as much as 3 m, and is commonly sufficient to permit different facies to co-exist on adjacent terraces (Fig. 3C, D). Observations suggest that the terraces form because the channel switches location rapidly, allowing the site of the former channel to become depositional. This area then accretes vertically with an essentially level surface.

Because of the alternation of erosion and deposition (predominantly vertical aggradation), the architecture of the mud flat deposits consists of valley-parallel, elongate wedges of flat-bedded sediment, separated by steep erosional surfaces that are mantled by mud-pebble conglomerates (the "bluff breccias" of Clifton, 1983; Fig. 3B) which formed during bank erosion. The spatial scale and duration of the

Figure 3. A. Broad, gradational transition from mud flats to salt marsh along the southern margin of Zone 2 (Irving Oil Dock site; Fig. 1). **B.** Erosional scarp separating salt marsh from mud flats at the Black Rock section (Fig. 1). The cliff was initially formed by channel erosion, but continues to retreat as a result of wave action, even though an aggradational mud flat has been re-established in front of it. **C.** Two mud flat terraces at the Powerline section (Fig. 1). Note how the lower terrace slopes gently toward the channel whereas the upper one (left-hand edge of photo) is nearly level. The erosional surface represented by the left-hand cliff extends beneath the outer terrace. The location of tile 6 (Table 2) is indicated. Another tile was located to the right of the stake at the edge of the cliff (arrow), but has been removed by bank erosion even though deposition continues on the terrace surface. **D.** Mud flat terraces at the Bridge section (Fig. 1); the measured section is located at the outer edge of the higher terrace (arrow). The sediments of the lower terrace onlap the erosional surface represented by the cliff at the right. Tiles 15 and 16 are located on this terrace, the outer edge of which is slightly raised relative to the inner margin.

depositional wedges decreases headward as the number of erosional surfaces and their frequency of formation increases. Lateral accretion surfaces (inclined heterolithic strata (IHS); Smith, 1987, 1988; Thomas *et al.*, 1987) with set thicknesses approaching the channel depth are rarely developed; the only strata that dip appreciably (>2-3°) toward the channel are localized zones where the tidal bedding drapes the adjacent erosion surface (*e.g.*, Fig. 3C, D). The larger geometry does, however, resemble the "imbricate IHS" of Thomas *et al.* (1987). Similar erosional scarps are common in back-barrier, mesotidal settings such as the Wadden Sea (Davies, 1980; Weimer *et al.*, 1982) and Willapa Bay (Clifton, 1983; see additional references in Allen, 1987). The terraced morphology described here is identical to that reported from the macrotidal, Severn River estuary, England (Allen, 1987; Allen and Rae, 1988), although the causes of the erosional events appear to differ (wave erosion in the Severn *versus* channel erosion here). Consequently, this architecture may not be unique to the study area.

Physical and Biological Processes

The Salmon River, although one of the largest entering the Minas Basin and Cobequid Bay, is relatively small, and the fluvial discharge is less than 10 m^3/s (*i.e.*, $<2 \times 10^5$ m^3 per half tidal cycle) for much of the year (Fig. 4). Low-tide water depths in the thalweg are generally less than 1 to 1.5 m at times of low river flow. By comparison, water depths at high tide vary from 7 to 8 m in zone 2, to approximately 4 m in the middle part of zone 3. Therefore, the combined tidal prism of zones 2 and 3 (*i.e.*, the inner 20 km of the estuary) is on the order of 4×10^7 m^3 per half tidal cycle, of which approximately 6×10^6 m^3 enters zone 3 (the innermost 12 km; Fig. 1). As a result, sedimentation to within a few hundred metres of the tidal limit is tidally dominated when the river discharge is low. Current measurements taken in the channel at such times show that flood-current speeds generally exceed those of the ebb (Fig. 5A); consequently, there is a net headward transport of sediment and a gradual aggradation of the channel in the inner estuary (Dalrymple and Zaitlin, 1989). Indeed, the main sediment source for the entire estuary is the eroding shoreline in areas seaward of zone 1 (Amos and Long, 1980). Despite the high speeds of the flood-tidal currents (>2 m/s; Fig. 5A), the mean grain size of the channel sands throughout zones 2 and 3 is fine to very fine (Fig. 5B; Maher, 1986; Zaitlin, 1987) because a hydraulic sorting process caused by the headward termination of the flood channels in zone 1 prevents coarser sand from entering the inner estuary (Dalrymple *et al.*, 1990). At times of peak river discharge, a small amount of coarse sand to cobble-size material is introduced to the head of the estuary by the river, but this material is not transported more than 3 to 4 km below the tidal limit (Zaitlin, 1987; Dalrymple and Zaitlin, 1989). At the same time, some or all of the tidal sediment that had accumulated in the channel during low river flow is transported seaward.

The amount of suspended sediment in the water column

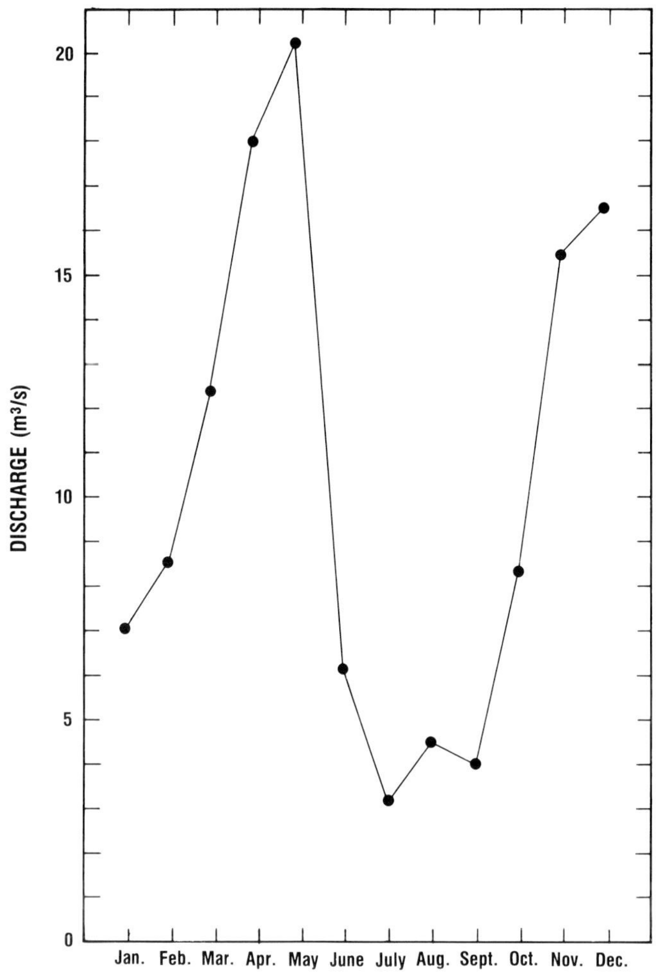

Figure 4. Monthly mean discharge of the Salmon River, based on data for the period from 1964 to 1972; data from Inland Waters Directorate (1980).

is high at all times (Amos and Alfoldi, 1979; Greenberg and Amos, 1983; Zaitlin, 1987), with suspended sediment concentrations (SSC) exceeding 200 mg/l throughout the study area. Measurements made during a period of average river discharge suggest that the turbidity maximum (SSC nearly 2000 mg/l) lies in zone 3 at high tide (Zaitlin, 1987). Simultaneous salinity measurements show a rapid headward decrease from approximately 22‰ at the seaward end of zone 2, to less than 1‰ within 2 to 3 km of the tidal limit (Fig. 5C; Zaitlin, 1987). Although data have not been obtained during the period of maximum fluvial discharge because of the presence of ice, the turbidity maximum and salinity gradient are presumably displaced seaward.

Because of the longitudinal gradient of salinity (Fig. 5C) (and perhaps also the variation in sedimentation rates as discussed below), the diversity of benthic organisms decreases headward (Fig. 5D), as does the number of individuals of each species (Dewing, 1986; Dewing *et al.*, 1986). In this regard, this area is similar to other estuaries (*e.g.*, Howard and Frey, 1975; Barnes, 1984; Wightman *et al.*, 1987). As a result, there is a headward decrease in the degree of

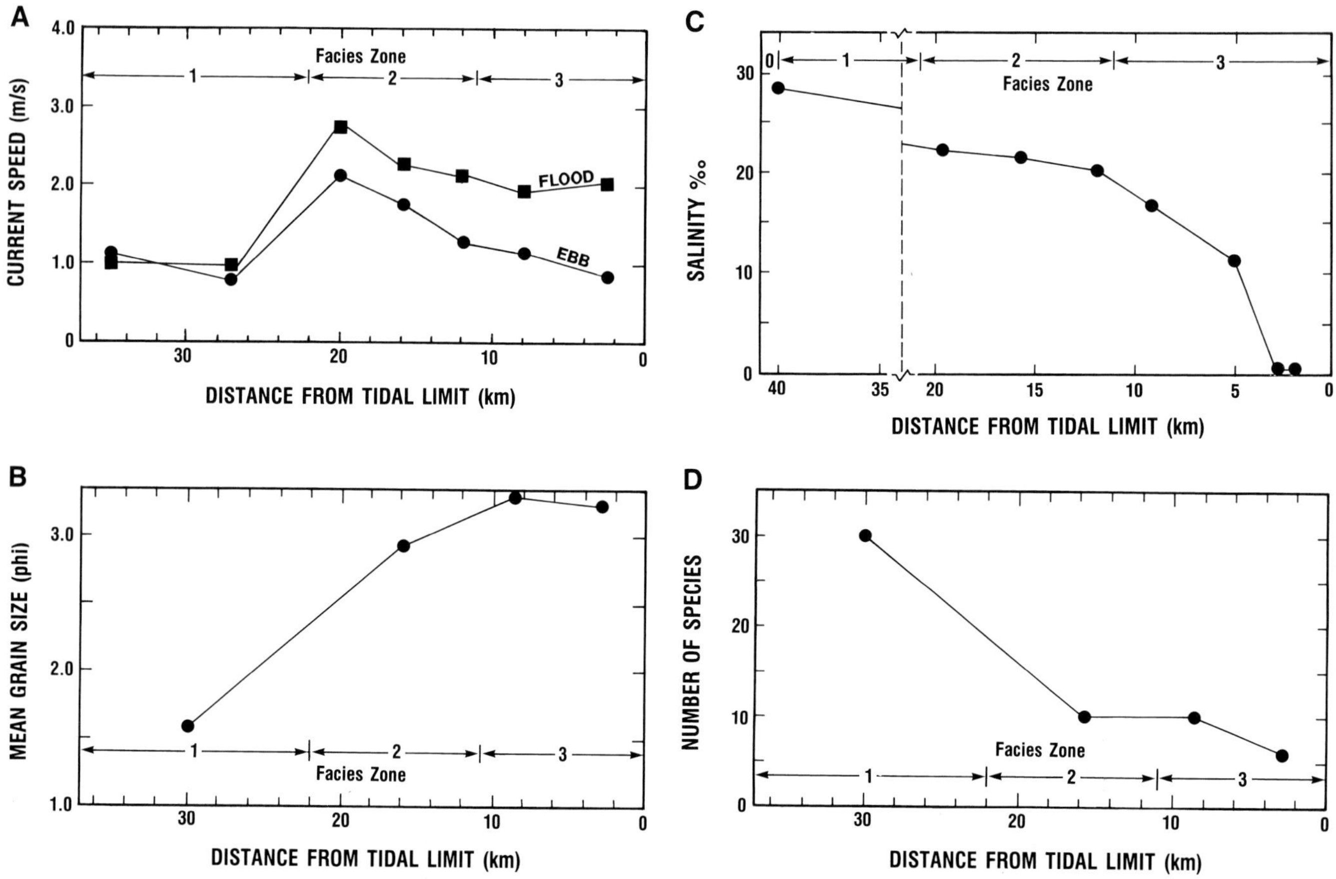

Figure 5. A. Longitudinal variation of average maximum current speeds (average of the three highest values in each half tidal cycle) over the sand facies, as measured 100 cm above the bed. Data from Dalrymple (1977) and Knight (1977) for zone 1, and from Zaitlin (1987) for zones 2 and 3. Values are averaged over all current meter stations in each area to remove the local effects of ebb or flood dominance. **B.** Longitudinal variation of mean grain size in the sand facies; data from Maher (1986). The values are the average of all analyzed samples in each area. **C.** Longitudinal distribution of salinity at high tide, as measured at a time of average river discharge (data from Zaitlin, 1987). **D.** Number of Benthic species (all facies) as a function of distance from the tidal limit. Data from Craig (1977) and Dewing (1986).

bioturbation, such that physical sedimentary structures in the mud flats adjacent to zone 2 are completely obliterated at some times of the year, whereas those farther headward are generally unaffected by biological action (see descriptions below). In addition, the large variations in temperature (Fig. 6) produce marked seasonal changes in the activity of the organisms (Craig, 1977; Yeo, 1978). During the winter, many of the intertidal organisms die or are in a state of suspended animation, and bioturbation is minimal. As temperatures warm through the spring and summer (Fig. 6), the number of organisms present and the rate of bioturbation increase dramatically.

The occurrence of below freezing temperatures from late December to March (Fig. 6) also causes large amounts of ice to form. In zone 3, the narrow width of the channel (Fig. 2) and the low current speeds (Fig. 5A) and salinity (Fig. 5C) allow an unbroken ice sheet to form across the channel (*cf.* Gordon and Desplanque, 1983; Zaitlin, 1987). In zone 2, the greater width and higher current speeds cause the ice to form as separate blocks and cakes ranging in diameter from a few centimetres to more than 2 m (Fig. 7A; Knight and Dalrymple, 1976; Zaitlin, 1987). The presence of an ice crust on the intertidal surface (Knight and Dalrymple, 1976; Zaitlin, 1987) limits the sedimentological impact of the drift ice throughout most of the winter, but significant amounts of ice gouging may occur in the muddy sediments of zone 2 (Fig. 7B; *cf.* Dionne, 1985) when warm temperatures cause the ice crust to melt.

Astronomically-produced variations in the tidal range also cause diurnal and neap-spring changes in the depositional conditions on the mud flats. Because the low-water level in zones 2 and 3 is controlled by the river discharge, the difference in the tidal range and current speed between tides is determined primarily by variations in the high-water elevation (Dalrymple and Makino, 1989). The predicted high-tide elevations for Burntcoat Head (Fig. 1; Canadian Hydrographic Service, pers. comm., 1988) show that the high-water level varies by up to 3.9 m (30% of the mean range) between neap and spring tides, while the difference between successive high tides (a result of the diurnal inequality) reaches 0.8 m (Fig. 8). Thus, all of the salt marsh and large portions of the mud flats remain exposed during neap high tides. These tidal height differences produce corresponding variations in the maximum tidal current speeds as a result of

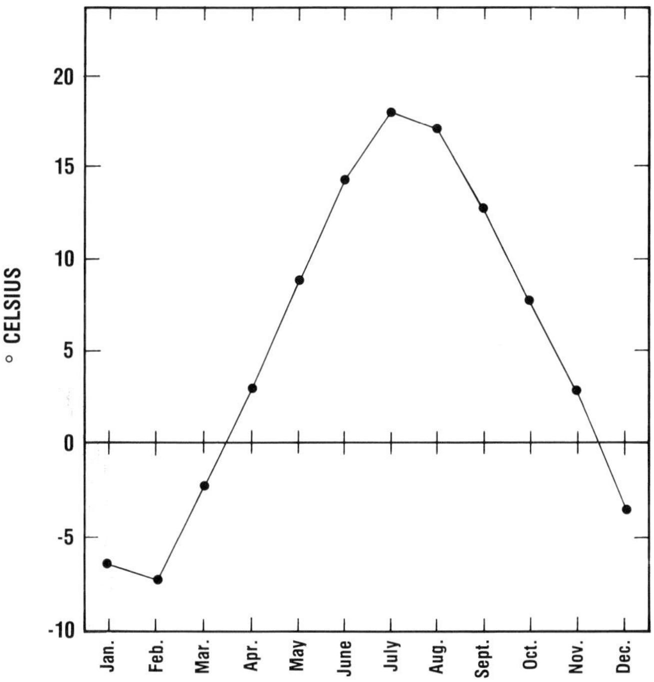

Figure 6. Annual variation of mean monthly temperature at Truro, Nova Scotia. Data from Atmospheric Environment Service (1982).

Figure 7. A. Blocks and cakes of sediment-laden drift ice (background and left foreground), stranded on the mixed flats at low tide. **B.** Mud flat surface with striated grooves created by drift ice. The presence of a frozen crust beneath the surficial mud layer limits the amount of deformation. Both photos taken near Black Rock (Fig. 1) in February, 1983.

changes in the tidal prism; consequently, maximum current speeds vary on both the neap-spring and diurnal time scales (*cf.* de Boer *et al.*, 1989). In fact, current meter data in the channel of zone 3 indicate that differences of more than 0.5 m/s may occur between successive flood tides (Fig. 9); even greater differences exist between neap and spring tides (Zaitlin, 1987).

The presence of Salter Head at the seaward end of zone 2 (Fig. 1), combined with the rapid narrowing of the estuary, limits the effect of wave action throughout zones 2 and 3, and waves have rarely been observed to exceed 0.25 m in height. The low slope of the depositional mud flats, together with the rapid rate of water level rise and fall, further reduces their effect. As a result, the influence of waves and storms is less than it is farther seaward (*cf.* Yeo and Risk, 1979). Waves have their greatest effect near the high-tide level because the fetch is greatest, and may contribute to the modification of cliffs that were initially formed by channel erosion (Fig. 3B). However, no beach facies is developed.

Mud Flat Structures

Section Locations and Methodology

Reconnaissance observations throughout zones 2 and 3 show that consistent changes occur within the mud flat structures along the length of the estuary, despite considerable local variability. In order to document the regional pattern, three representative sections were selected (Fig. 1): Black Rock, 19 km seaward of the tidal limit (outer end of zone 2; Fig. 3B); Powerline, 8.5 km from the tidal limit (outer part of zone 3; Fig. 3C); and Bridge, 5 km from the tidal limit (inner part of zone 3; Fig. 3D). These three sections were all situated in the middle to upper part of the elevation range of the mud flats (Table 1), and 1 to 3 m above the channel sand facies, so that the differences in the structures are not a result of differences in relative elevation.

The three sections, which ranged in thickness from 1.1 to 1.8 m (Table 1), were logged on a millimetre scale (Figs. 10, 12, 18). The thickness and internal structure of each unit more than 1 mm thick was recorded, as were any other sedimentary structures (desiccation cracks, burrows, etc.) and the nature of all bed contacts. The lateral continuity of laminae and beds was assessed by tracing them along the length of the exposure. In order to obtain an independent measure of the rate of deposition, 30 cm × 30 cm floor tiles were placed at various elevations near the measured sections and at other locations in the inner estuary. The position of each tile was marked by a stake, and its elevation relative to highest high water was determined (Table 2). The thickness of sediment that accumulated on each tile was measured with a small probe at intervals of several tidal cycles throughout the study period, which spanned 8 to 17 tidal cycles during a time of low river discharge (July and August, 1987).

From the detailed sedimentological observations and accretion rate data it is possible to recognize several temporal

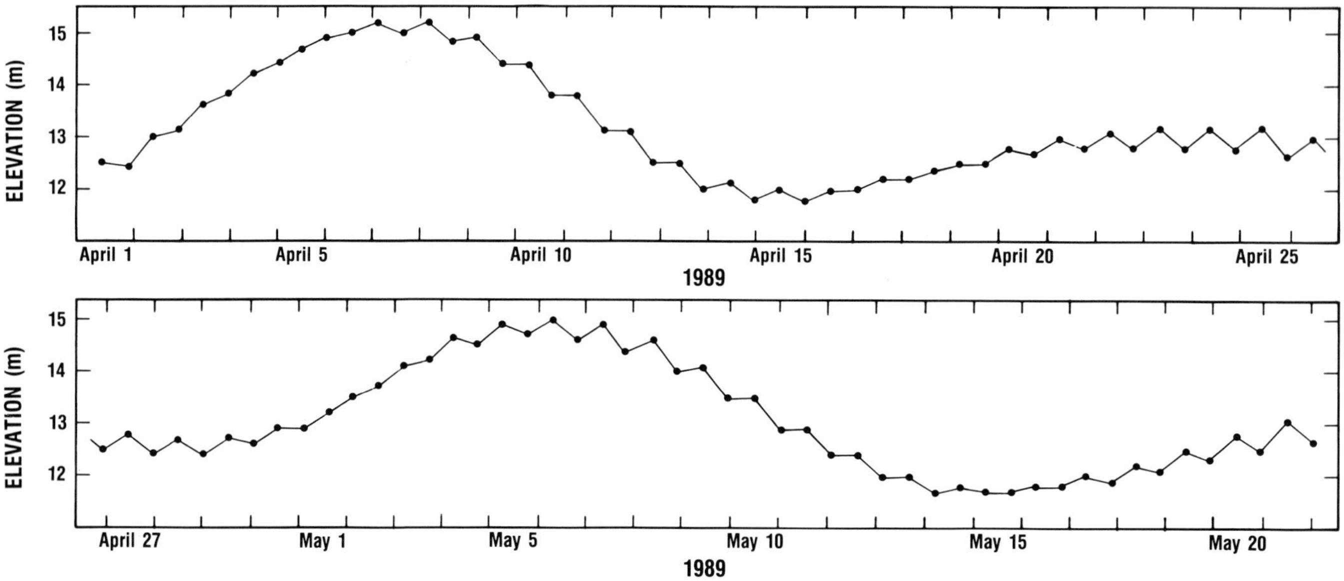

Figure 8. Variation of high-water elevation at Burntcoat Head, the nearest tidal reference station, as predicted by the Canadian Hydrographic Service. Note the marked difference between successive tides (diurnal inequality) and over the 14-day, neap-spring cycle. Elevations relative to chart datum, which is approximately equal to mean low water.

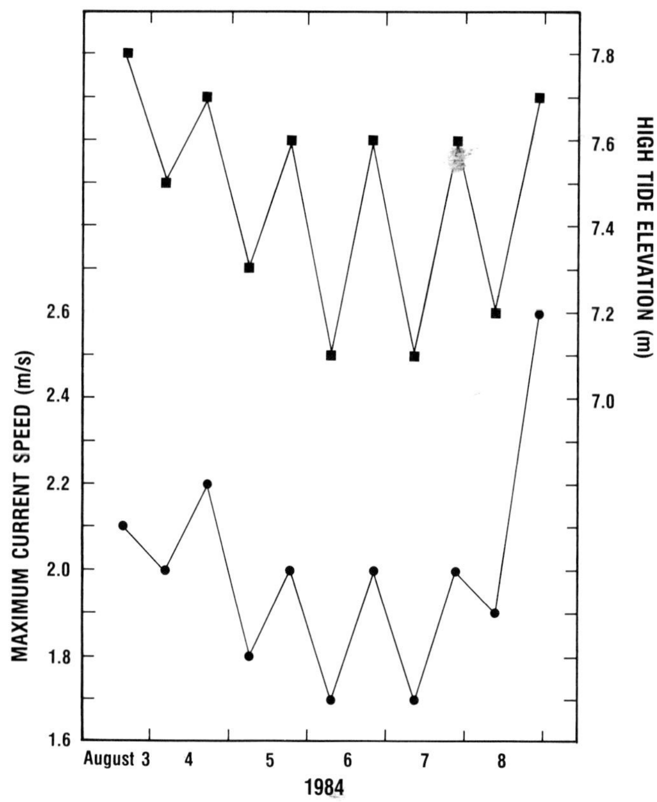

Figure 9. Maximum-recorded current speed in successive dominant (flood) tides (lower plot) compared with predicted high-tide elevations for the same flood tides (upper plot). Tidal-current data obtained at a height of 0.8 m above the bed in the Salmon River channel, several hundred metres from the Powerline section during a neap-tide period. Data from Zaitlin (1987). Tidal heights are values predicted for Saint John, New Brunswick, at the mouth of the Bay of Fundy (Fig. 1; Canadian Hydrographic Service, 1984).

periodicities of mud flat sedimentation: individual ebb and flood tides; complete tidal cycles; the diurnal inequality; neap-spring cycles; and annual deposits. In the following sections, each of these depositional increments will be described and its origin discussed. The vertical (within section) and horizontal (between section) variations will be considered later in the paper.

Sand-Mud (Tidal-Cycle) Couplets

Description

As documented in greater detail by Dalrymple and Makino (1989), the basic sedimentation unit in the measured sections is a couplet consisting of a sand layer overlain by mud (Figs. 10-13, 14, 19A, B). The sand portion of these couplets ranges from <1 to 38 mm in thickness, whereas the mud laminae vary between <1 and 18 mm thick; complete couplets range from 1 to 45 mm thick. Individual laminae and couplets are laterally extensive (Fig. 11A, B), and can

Table 1. Location of measured sections (see also Figure 1)

	Bridge Section	Powerline Section	Black Rock Section
Measured thickness	1.8 m	1.1 m	1.8 m
Elevation of top below marsh	1.5 m	1.4 m	0 m
Elevation of top below H.H.W.	2.9 m	2.7 m	4 m
Elevation of top above sand facies	2 m	3 m	4 m
Distance from tidal limit	5 km	8.5 km	19 km

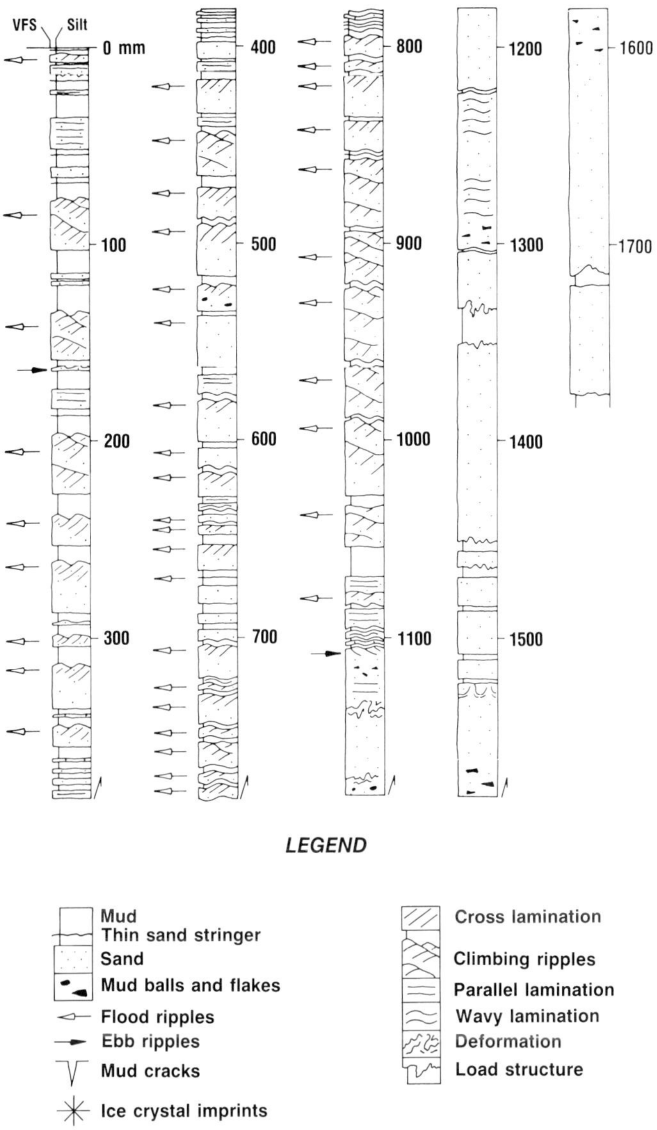

Figure 10. Graphic log of the Bridge section (in millimetres). VFS = very fine sand.

be traced along the entire length of a mud flat terrace (>100 m in some cases), with only slight changes in their character. Because of this continuity, they form wavy or tidal bedding (Reineck and Wunderlich, 1968). Lenticular bedding is present in some thin intervals where the sand layers consist of isolated ripple formsets (Fig. 13D). Flaser bedding is rare because of the high elevation and muddy nature of the sections.

The base of each sand layer is sharp but not obviously erosional (Fig. 11), and may show micro-load structures (relief 1-4 mm; spacing <20 mm; Fig. 13E) that in plan view resemble the wrinkle marks described by Allen (1984). The coarse layers generally consist of very fine grained sand with a variable content (15-35%) of silt and clay. The grain size becomes finer, either gradually or abruptly, at the top of the sand layer (Fig. 11) because of the addition of more silt. This silty sand-sandy silt transition ranges in thickness from <1 mm to nearly 1 cm, and is overlain by the mud layer, which is composed of clayey silt or silty clay (8-52% clay). In total, clay constitutes less than 20 per cent of the mud flat sediments (Zaitlin, 1987).

In general, the mud layers show little variation in their physical character within or between the measured sections, except for the local presence of one or more thin (a few grain-diameters thick), sand stringers (Figs. 11A, B, 13E). By contrast, the sand layers contain a variety of sedimentary structures that allow four couplet types to be distinguished (Dalrymple and Makino, 1989): 1) those containing flood-oriented current ripples or crosslamination; 2) those with ebb-oriented current ripples or crosslamination; 3) couplets containing parallel lamination; and 4) those with a thin (commonly only a few grain-diameters thick), structureless sand.

Flood-ripple (type 1) couplets (Figs. 11, 13D) are abundant in the Bridge and Powerline sections (Figs. 10, 12) because of the flood-current dominance of the inner estuary (Fig. 5A). The sand layer in these couplets is generally thick, reaching up to 38 mm, and makes up 50 to 90 per cent of the total couplet thickness. In cases where the sand layer is less than 3 to 5 mm thick, disconnected ripple formsets (starved ripples) occur, producing lenticular bedding (Fig. 13D). In the thicker sand layers, climbing ripples are common, with both stoss-erosional and stoss-depositional varieties occurring (Fig. 11C). The ripples in these layers are commonly overlain by silty sand or sandy silt containing undulatory laminae that drape the ripple topography, in a succession that is reminiscent of the c, d and e divisions of the Bouma sequence.

Ebb-ripple (type 2) couplets (Fig. 13D) are less abundant, and occur most commonly in the Powerline section (Figs. 10, 12). The sand layers are thinner on average than those containing flood ripples, reaching a maximum thickness of only 8 mm. As a result, type 2 couplets commonly contain disconnected, ripple formsets (*e.g.*, Fig. 13D). Ebb-ripple couplets have never been observed to have the silty sand transition between the sand and mud layers, and the capping mud layer is thinner than that in the flood-ripple couplets. These features suggest that type 2 couplets are formed by weaker currents than the type 1 couplets (Dalrymple and Makino, 1989).

In both of these couplet types, the silty sand drape and/or mud cap preferentially fill the troughs of the underlying ripples. If relief remains, it acts as the nucleus for ripples in the overlying sand (Figs. 11A, B, D, 13D). In packages of stacked flood-ripple couplets, this produces sequences of ripples that climb continuously through as many as 12 to 15 couplets (Fig. 11D), whereas in the case where ebb- and flood-ripple couplets alternate, the overlying ripple lies on the stoss side of the preceding ripple in an overlapping relationship (Fig. 13D).

Parallel-laminated (type 3) couplets are rare, being restricted to a few horizons in the Bridge section (Figs. 10, 11C). Such sand resemble the silty-sand to sandy-silt intervals

Table 2. Summary of measured accretion rates (see Figure 1 for location of measurement sites)

Location	Distance from tidal limit (km)	Tile number	Elevation below H.H.W. (m)	Accretion rate (mm/tidal cycle)
Black Rock Section	19	1	4.3	0.0
		2	7.1	0.0
Irving Oil Dock	16	9	0.5*	0.0
		24	1.4*	0.0
		25	3.2	1.0
Powerline Section	8.5	11	2.3*	0.0
		10	2.4*	0.0
		8	3.2	2.6
		7	3.3	4.6
		6	4.1	6.4
		4	4.7#	22.2
Bridge Section	5	14	2.4*	0.0
		13	2.6*	0.0
		12	2.8*	0.0
		15	3.9	5.8
		16	3.9	11.3
		18	4.2	9.7
		17	4.7	9.5
Hill-Crowe Point Bar	3	23	1.7*	0.0
		22	2.2*	0.0
		21	2.2*	0.0
		20	3.8	31.1
		19	4.0	24.9

* Because the measurements were made during neap tides, these tiles were never submerged, or were inundated fewer than 3 times during the study period.
# This tile was lost as a result of bank erosion (Fig. 3C) prior to completion of the levelling; consequently, its elevation is only approximate. Note that rapid deposition continued right up the time of removal.

that drape some flood-ripple couplets, but are distinguished from them by the lack of silt, the greater thickness of the sand layer (>10 mm) and the absence of undulating lamination. These parallel-laminated sand layers are interpreted as having been formed under upper-flow-regime conditions, in contrast to the draping laminae, which were deposited from suspension at speeds below the threshold of ripple generation (Figs. 14A, B; Dalrymple and Makino, 1989).

The structureless sand (or sand-stringer) couplet type (type 4) is the most abundant variety in the muddy flats (Figs. 10-13), but becomes relatively more common in the seaward direction; it is the only type recognized at Black Rock. It is the thinnest of the couplet types, ranging from approximately 1 to 5 mm thick. It is also the muddiest of the couplet types, with the sand layer generally thinner than the mud layer (Figs. 11A, B, 13A, C, E). The sand layers are usually laterally continuous and uniform in thickness, and lack current-generated structures.

Examination of the sections indicates that the four couplet types occur in a small number of preferred combinations (Fig. 14; Dalrymple and Makino, 1989). The most common ones consist of either a flood-ripple (type 1) or parallel-laminated (type 3) couplet overlain by a sand-stringer (type 4) couplet, producing 1-4 and 3-4 combinations. In many of these, the sand layer in the upper, type 4 couplet is so thin that the combination initially appears as a single type 1 or 3 couplet; it is only upon close inspection that the sand stringer can be seen, dividing the mud layer into two parts (Figs. 10, 11A, B). In other instances, the flood-ripple or parallel-laminated couplet may be succeeded by an ebb-ripple (type 2) couplet, producing 1-2 (Figs. 12, 13D) and 3-2 combinations (Fig. 10). In other cases, 4-4 combinations are observed, commonly in repeated sequences (4-4-4- . . .; Figs. 10, 11A, B, 12, 13A, C). This latter arrangement comprises all of the nonbioturbated portions of the Black Rock section (Fig. 19A, B).

Origin

The structures within the thicker sand-mud couplets indicate that each couplet was deposited by a waning current

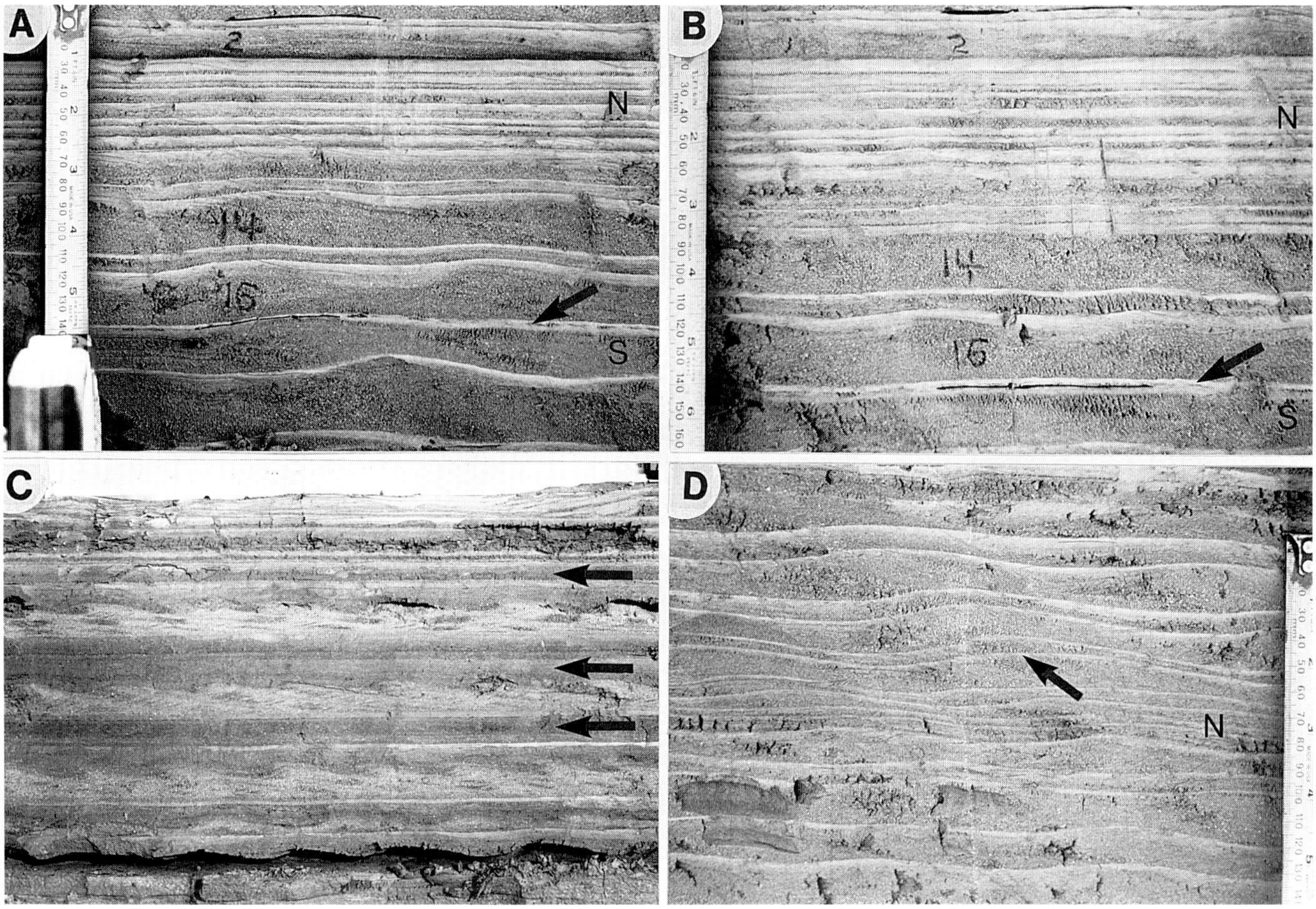

Figure 11. Sedimentary structures in the Bridge section. The field of view in **A**, **B** and **D** is approximately 30 cm wide; in **C** it is approximately 75 cm wide. Flood flow is to the left in all photos. **A** and **B**. Detailed views of sand-mud couplets taken 7 m apart, showing the lateral continuity of individual units; the numbers inscribed on the section indicate correlative sand layers. The thick sands near the base, which contain climbing, flood current ripples (type 1 couplets), were deposited during spring tides (S), whereas the thin, structureless sands at the top (type 4 couplets) were deposited during neap tides (N). The mud layer below couplet 16 contains a sand stringer (arrows) deposited during the ebb tide. The diurnal inequality is clearly shown by the alternation of thick and thin couplets in the spring-tide interval. **C.** Climbing, flood-oriented current ripples (type 1 couplets) alternating with parallel-laminated (type 3) couplets (arrows) in a spring-tide interval. **D.** Type 1 (flood ripple) couplets showing progressive climbing of the ripples (arrow) through 12 couplets. Note the alternation between thick and thin couplets as a result of the diurnal inequality. A neap-tide interval (N) is present just below centre.

with significant amounts of deposition from suspension. The lack of oppositely-directed current ripples within a single couplet, and the succession of structures in the various couplet combinations led Dalrymple and Makino (1989) to conclude that each couplet is the deposit of a single ebb or flood phase of the tidal cycle, and that the combinations represent the deposits of a complete tidal cycle, with the different combinations reflecting differences in the relative and absolute speeds of the flood and ebb currents (Fig. 14; Dalrymple and Makino, 1989, Fig. 22). In this interpretation, the (thicker) type 1 or 3 couplet at the base of a combination is deposited during the waning stage of the (dominant) flood tide as sand, carried in suspension, settles onto the mud flat (Fig. 14A-D). (The preservation of the underlying mud layer indicates that little if any erosion occurs during the accelerating portion of the flood tide, presumably because the mud has become at least partially consolidated during the lengthy exposure (Amos and Mosher, 1985).) If the peak speed of the flood currents is in the ripple stability field, climbing ripples followed by draping silty laminae are produced (Fig. 14B), whereas parallel lamination (a type 3 couplet) is produced if the maximum current speeds are in the upper flow regime (Fig. 14A). This parallel lamination passes directly into draping silty laminae rather than current ripples, as might be expected, presumably because the currents decelerate too quickly for ripple formation. At high slack water, the overlying mud lamina is formed. During the succeeding subordinate tide, the ebb ripples of the overlying type 2 couplet are formed if the current speed is sufficiently fast (Fig. 14A, B), again with little apparent erosion of the underlying mud drape. In the many cases where the ebb currents are below the threshold of ripple formation, only a thin layer of sand (a type 4 couplet) is deposited from suspension (Fig. 14C). This sand, which

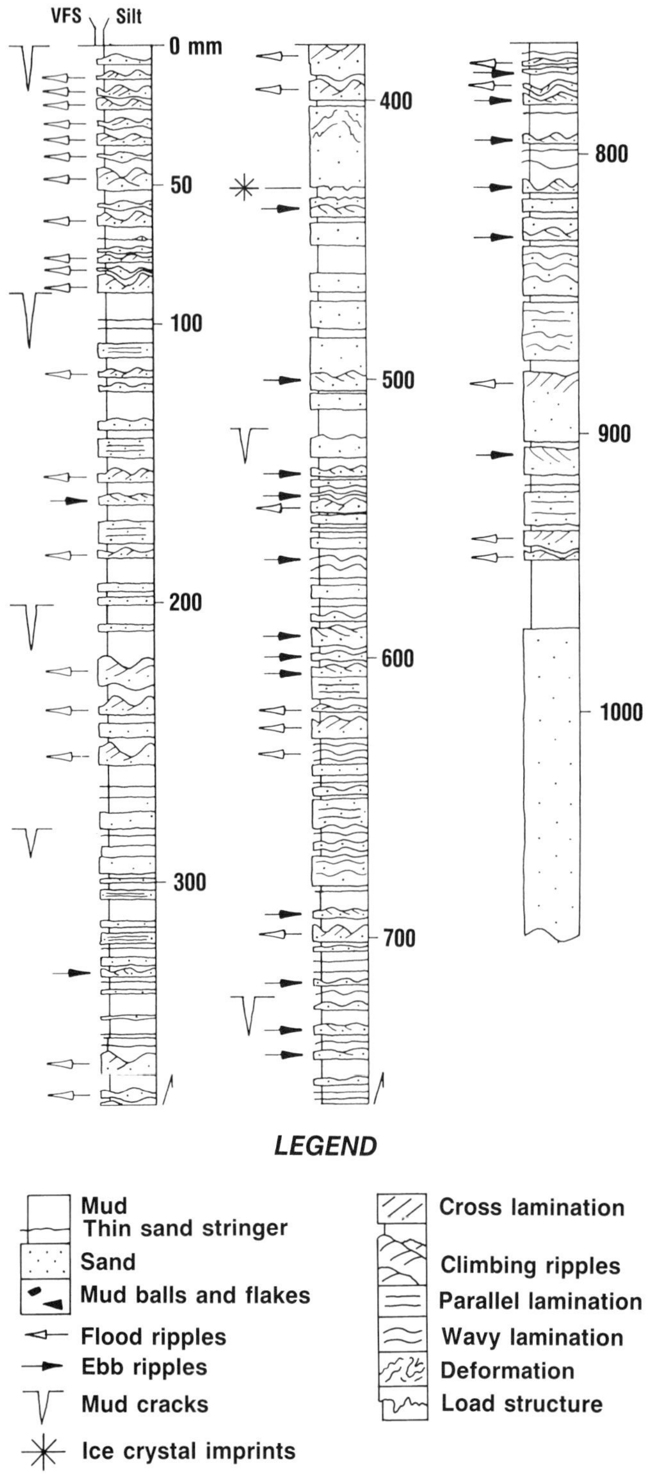

Figure 12. Graphic log of the Powerline section (in millimetres).

was placed in suspension by the stronger currents in the adjacent channel, appears to be carried onto the mud flat, even though the local currents are below the threshold of sand movement, because of the sand's "settling lag" (Postma, 1961, 1967). If the subordinate currents are too weak to retain sand in suspension, or the site is too far from the sandy channel, no sand is brought onto the mud flats, and any mud deposited during emergence is amalgamated with the underlying high-tide mud drape (Fig. 14D, F). This origin for the couplets and combinations is an elaboration of that proposed for tidal bedding by Reineck (1960) and Reineck and Wunderlich (1968).

Direct evidence of the tidal cycle origin of the couplets was obtained by excavating the sediment deposited on some of the tiles and counting the number of couplets deposited. In one such example (Fig. 15), the sediment that accumulated during 13 tidal cycles contained one 1-4 combination and 12 type 1 or 4 couplets. Thus, each dominant tide was represented by a separate couplet, whereas only one of the ebb tides left a discrete sand layer (*cf.* Figs. 14D, F). In addition, the average accretion rate at each of the three measured sections (Table 2) corresponds closely with the average thickness of the couplet combinations in that section: Bridge section — average accumulation rate 9.0 mm/tidal cycle versus 15 mm average combination thickness; Powerline section (excluding the anomalous rates on tile 4, which result from its extreme proximity to an eroding channel bank; see further discussion below) — 4.5 mm/tidal cycle versus 6 to 8 mm thick; and Black Rock — <1 mm/tidal cycle versus 1 to 2 mm thick for recognizable (*i.e.*, thicker than average) couplets.

The suggestion that sand layers are not deposited by every tide (Fig. 14G) raises the possibility that some mud layers may have been formed over several tidal cycles as suggested by McCave (1970, 1971). Such cases can be identified, however, by utilizing the analytical procedure of McCave (1970, equation 4) to calculate the thickest mud layer that is likely to have accumulated in a single slack water period. Using known values of suspended sediment concentration (0.3-1.9 g/l; Zaitlin, 1987) and *in situ* particle settling speed (2.1×10^{-3} m/s; Amos and Mosher, 1985), and assuming a bulk density of 1.5 g/cm^3 for the recently deposited mud (McCave, 1970), mud-layers 0.5 to 3.2 mm thick (uncompacted) can be deposited during a slack water period of 20 minutes. These values closely match the typical (semi-compacted) thickness of individual mud layers, and indicate that layers that are in this range are likely to be the product of a single slack water period, whereas mud layers thicker than 3 mm probably required two or more slack water periods to form. Indeed, some mud layers that are thicker than 3 mm can be traced laterally into 4-4-. . . sequences, in which the sand layers are only one or two grain diameters thick. This clearly shows that they are amalgamated. Such composite mud layers become more abundant toward the top of the Bridge and Powerline sections (Figs. 16B, 17B), particularly at elevations at or above the neap high-water level, and are very common at Black Rock.

Diurnal Inequality and Neap-Spring Cycles

Plots of sand layer thicknesses within the Bridge and Powerline sections (Figs. 16A, 17A) show that the thickness varies on two time scales in both data sets, although this is most clearly displayed at the Bridge section. The first of these

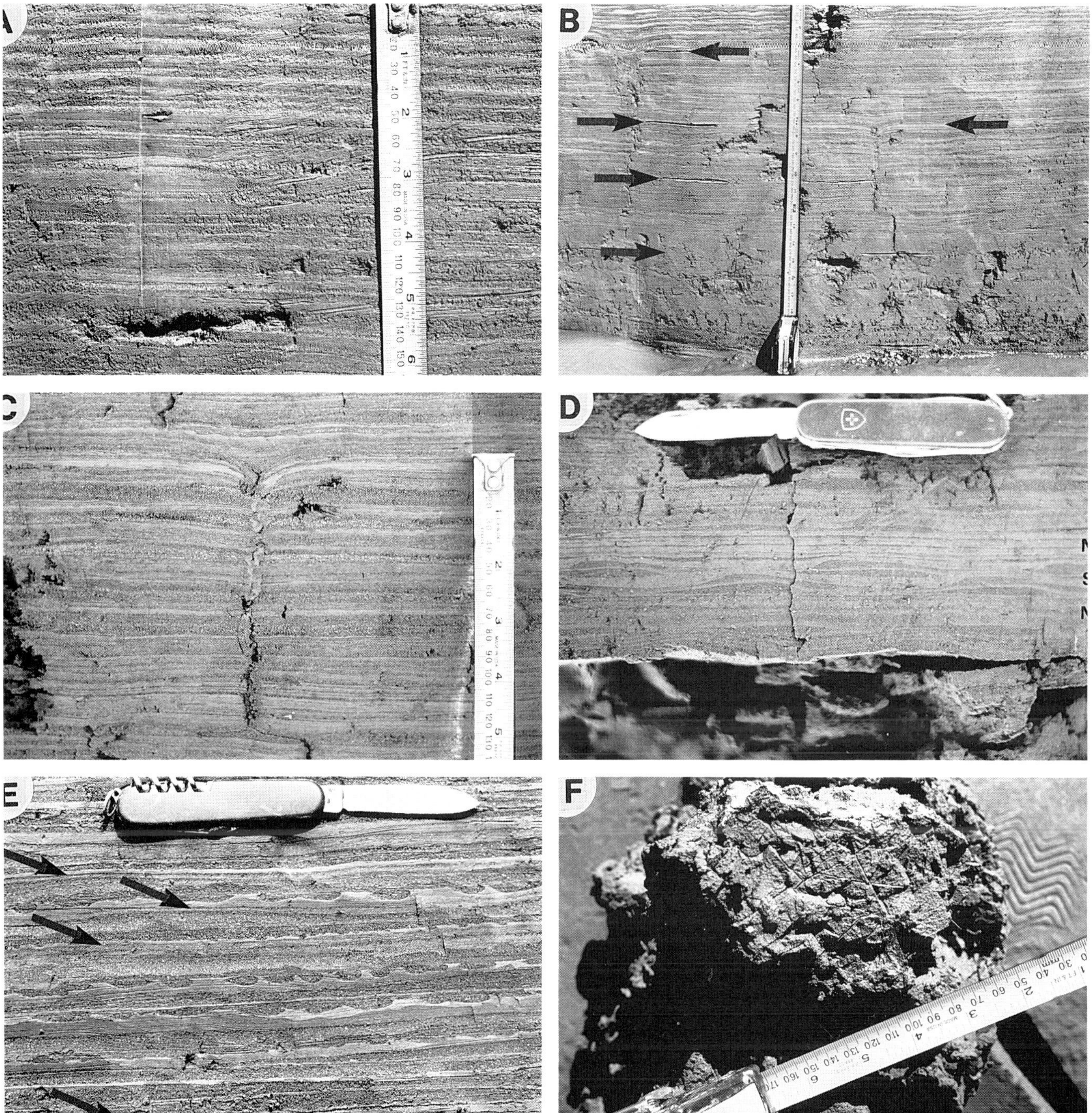

Figure 13. Sedimentary structures in the Powerline section. **A.** General view of the upper 170 mm of the measured section (perpendicular to the flow). Many of the couplets are type 1, but some are type 4 (see Fig. 12). Scale in millimetres and inches. **B.** General view of the upper 0.5 m of **A** showing four horizons (arrows) of fossil desiccation cracks. Each of these levels corresponds with a slightly muddier, neap-tide interval. **C.** Close-up of fossil desiccation crack from the second-highest level in **B** (to the right of the tape measure blade). Scale in millimetres and inches. **D.** Two sets of paired flood (type 1) and ebb (type 2) couplets from the 750 to 800 mm level of the section (Fig. 12). Flood flow to the left. These thicker couplets are a spring-tide deposit (S), and are over- and underlain by type 4 couplets deposited during neap tides (N). The knife handle is 6 cm long. **E.** Micro load structures. Note that several of the mud layers (arrowed) are doublets due to the presence of an internal sand stringer. **F.** Ice-crystal casts from the 420 mm level of the measured section (Fig. 12; lowermost blocky interval of Fig. 13B).

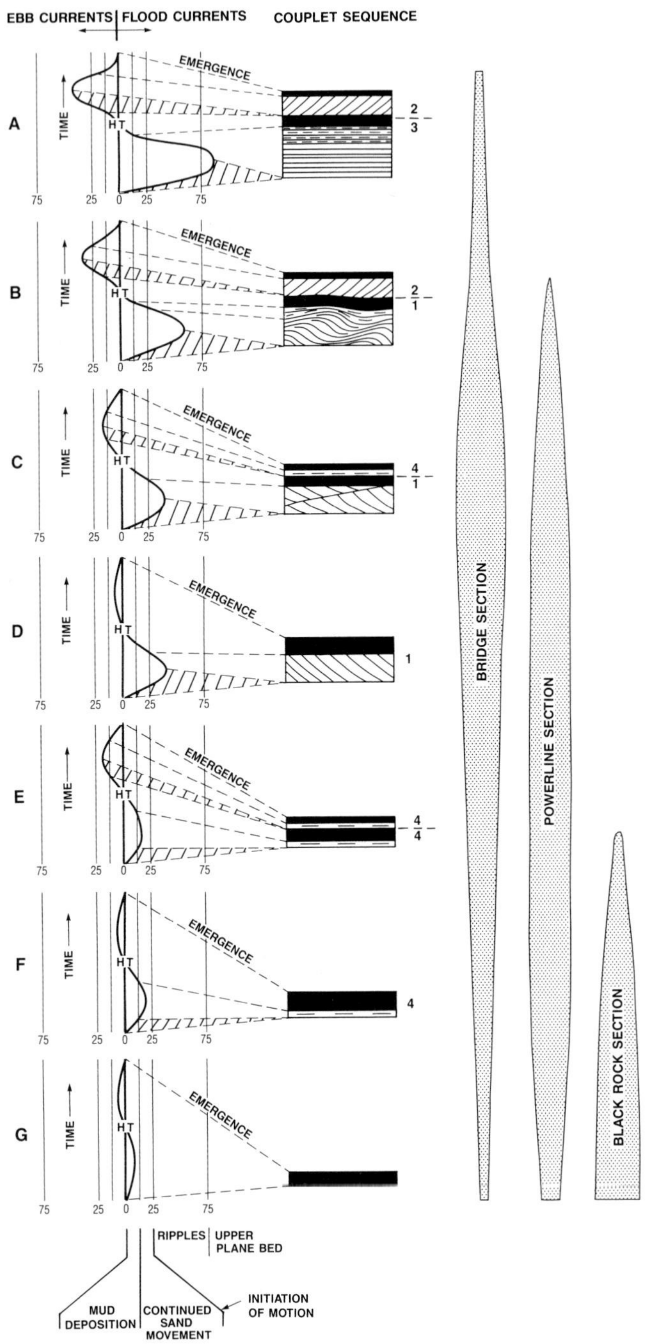

Figure 14. Schematic model of the tidal-current conditions (left side) under which the component parts of several common couplet sequences (middle) originate. The crosshatched regions represent times of nondeposition or erosion. In drawing this, it is assumed that deposition occurs only during the decelerating portion of each half tidal cycle; this need not be the case if spatial deceleration dominates over temporal acceleration (Rubin and Hunter, 1982). Further work is needed to refine the correlations shown. The current speeds are in cm/s, and the speed divisions are for fine and very fine sand in water 1 m deep (Harms *et al.*, 1982). "Continued sand movement" refers to current speeds between the threshold of movement ("Initiation of Motion") and the critical speed for the onset of mud deposition (*cf.* Amos and Mosher, 1985); in this range of speeds, sand will be deposited from suspension without the formation of traction structures. The widths of the bar graphs at the right indicate the relative abundance of the various couplet sequences in the measured sections.

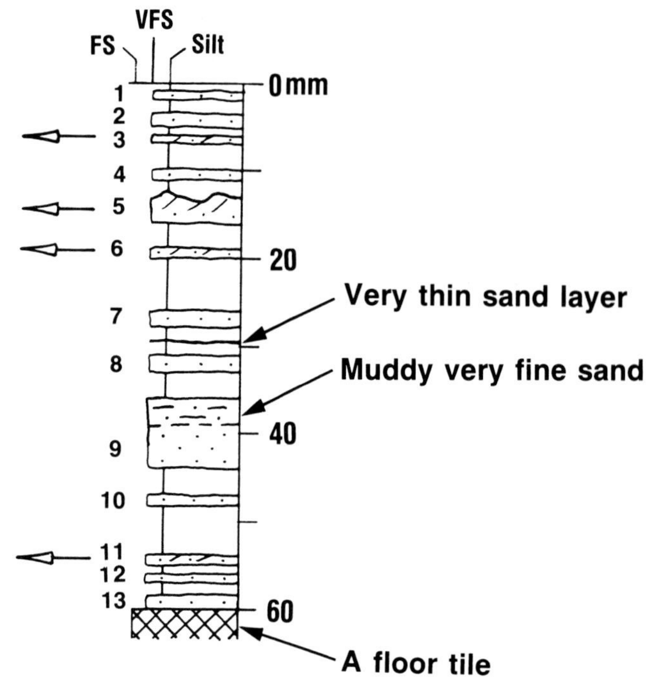

Figure 15. Sequence of tidal couplets (numbered at left) deposited on tile 15 (see Fig. 3D and Table 2 for location) during a 13 tidal-cycle period. FS = fine sand; VFS = very fine sand. See Figure 10 for legend.

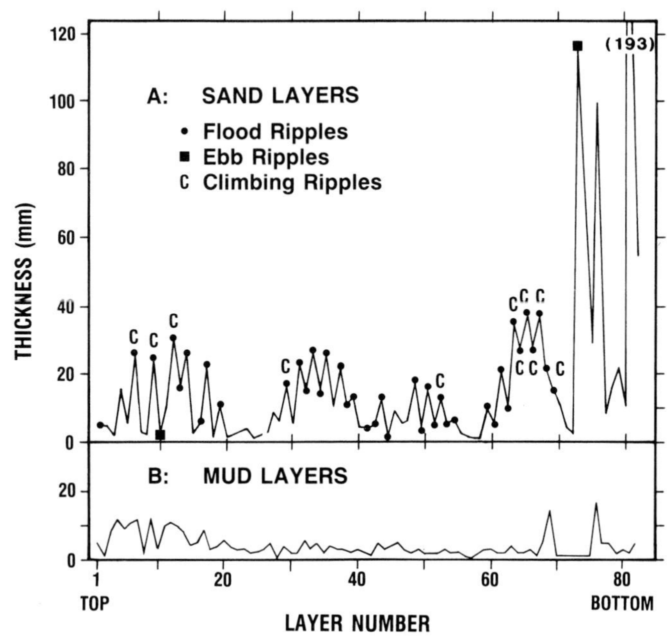

Figure 16. Variation in **A.** sand-layer thickness and **B.** mud-layer thickness in the Bridge section (see Fig. 10). Only individual sand layers more than 1 mm thick have been recorded. Note the pronounced alternation between thick and thin sand layers caused by the diurnal inequality, and the longer period (15-23 layers) cyclicity due to the neap-spring variation in current strength (Fig. 9). Neither of these features is strongly developed in the plot of mud-layer thickness. The very thick sands at the base of the section are sand-flat deposits and are probably amalgamated as a result of the erosion of mud drapes.

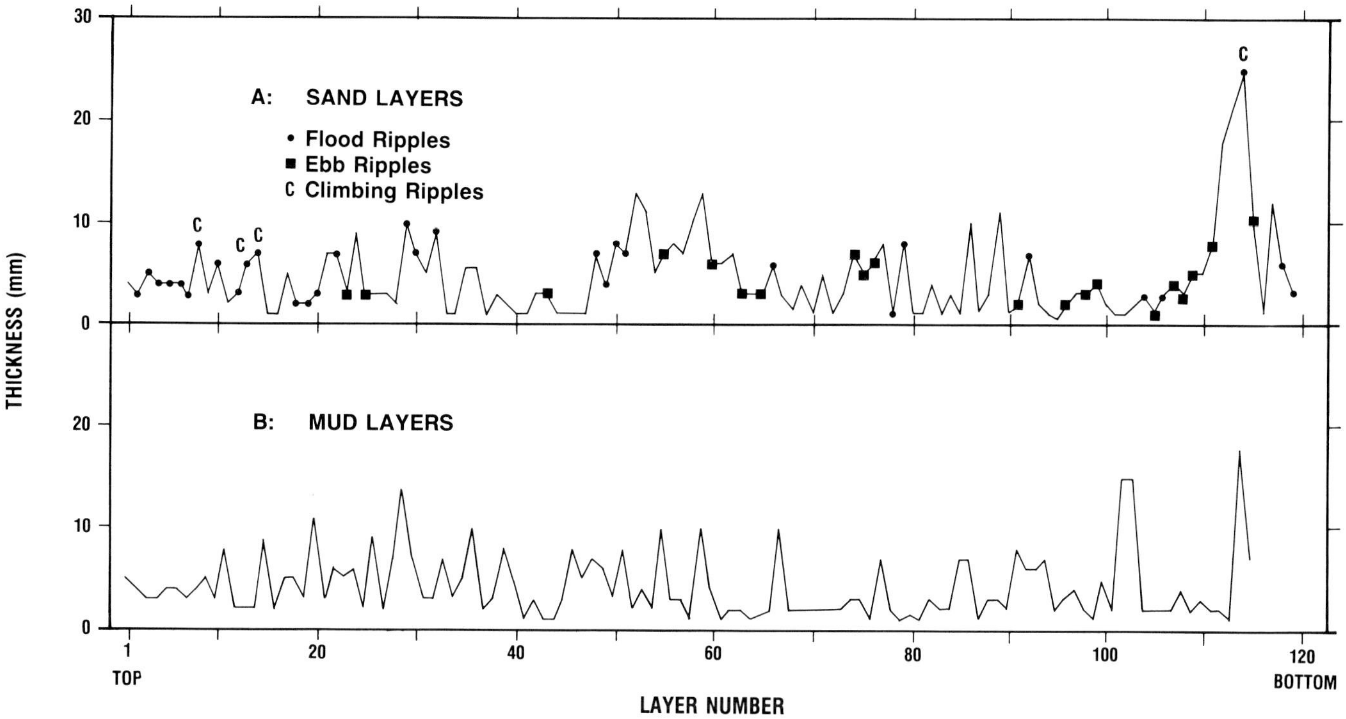

Figure 17. Variation of **A.** sand-layer thickness and **B.** mud-layer thickness in the Powerline section (see Fig. 12). The plot of sand-layer thickness shows a subtle cyclicity with a period of 15 to 25 layers representing the neap-spring variation in current speed. The diurnal inequality is evident in the plot of mud-layer thicknesses, but the neap-spring cycle is not well developed.

is the alternation between thicker and thinner, flood-rippled sand layers (Figs. 11, 16A), a pattern that gives the thickness versus depth plots a prominent saw-tooth appearance. This pattern mimics the variation of maximum flood-current speeds in the adjacent channel (Fig. 9), which vary as a result of the diurnal inequality of the high-water elevation. Presumably the maximum current speeds control the thickness of the sand layers by determining the amount of sand placed in suspension and transported onto the elevated surface of the mud flat.

The second temporal scale of sand thickness variation consists of a gradual increase, followed by a gradual decrease, in the sand layer thickness through 15 to 25 flood-ripple couplets (Figs. 16A, 17A). This cyclicity causes both sections to alternate between sandier and muddier intervals over a vertical distance of 5 to 40 cm (*e.g.*, Fig. 11A, B). These thickness variations are also associated with cyclic changes in the couplet combinations present. In the Bridge section, the sandy portions of the cycles contain 1-2 and even 3-2 combinations (Figs. 10, 11). Above or below this, where the sand layers are thinner, the ebb ripple couplet at the top of each combination is replaced by a type 4 (sand stringer) couplet, and the parallel-laminated, type 3 couplet is replaced by a flood-ripple couplet, yielding 1-4 combinations, presumably because of a decrease in the speed of both the flood and ebb tidal currents (Fig. 14). In the muddiest portions of the cycles even the flood ripples may disappear, so that only 4-4 combinations remain, indicating that the tidal current speeds are always below the threshold of ripple formation (Fig. 14E, F). In part of the Powerline section, 1-2 combinations occur in the sandy portion of the cycle, and change first to 4-2 combinations with the loss of the flood-current ripples, and then to 4-4 combinations (Fig. 13D). Desiccation cracks are also present at several levels in the Powerline sections (Figs. 12, 13B, C), but occur only at the horizons with the thinnest and muddiest couplets. They indicate that the mud flat was exposed continuously for several days at the times when the muddy intervals were deposited.

When the number of measurable sand layers (15-25) in these cycles is coupled with the observation that the muddier intervals contain desiccation cracks and evidence of weaker currents, it is clear that these larger cycles are a result of the neap-spring tidal cycle (Fig. 8). These variations in sand layer thickness are directly analogous to the neap-spring cycles described from tidal crossbedding (Visser, 1980; Allen, 1982; et Boer *et al.*, 1989), and indicate that the flood (dominant current) sand layers are the mud flat equivalent of tidal bundles (Boersma, 1969; Visser, 1980).

Remarkably similar tidal rhythmites have been described from the estuarine zone of the Baie du Mont Saint-Michel, France (Tessier and Larsonneur *in* Larsonneur, 1988), and ancient analogues that may also be of estuarine origin are known from the Miocene of the French Alps (Tessier and Gigot, 1989) and the Pennsylvanian of Indiana (Kvale *et al.*, 1989). In all of these examples, the sand layers commonly contain ripples, at least in the spring-tide layers, so that type 1 and 2 couplets are abundant; note that the terminology of couplet types used here differs from that employed by Tessier and Gigot (1989). The CB-SR rhythmites are also dynamically equivalent to the rhythmites described by Baird *et al.*

(1986), Kreisa *et al.* (1989), Leithold *et al.* (1989) and Williams (1989), although these examples are interpreted to have formed in a subtidal deltaic environment. The low energy conditions that characterize this setting result in a limited range of couplet types, with type 4 couplets overwhelmingly predominant. Figure 14 (this paper) and Figure 22 of Dalrymple and Makino (1989) provide a generalized framework for the interpretation of tidal rhythmites in any environment where wave action is not significant.

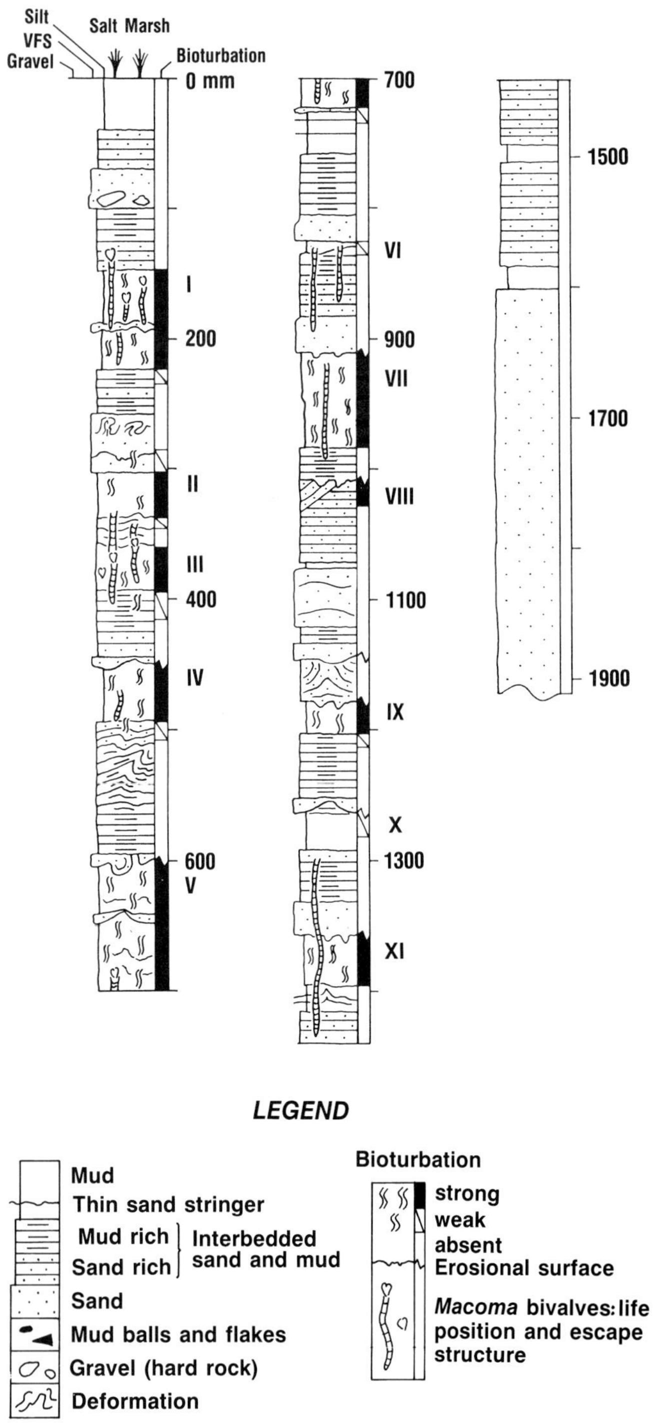

Figure 18. Graphic log of the Black Rock section (in millimetres). The Roman numerals to the right of the log indicate the top of successive annual cycles.

Annual Deposits

The Black Rock section differs significantly from either of the other sections because of the thinness of the sand-mud couplets (generally less than 1-2 mm), and the absence of structures that clearly reflect the tidal periodicities discussed above. Instead, the Black Rock section is characterized by an alternation of laminated and burrowed sandy muds that produce cycles 5 to 30 cm thick (Figs. 18, 19A, B). The laminated intervals range in thickness from 2 to 20 cm (average 6.5 cm), and consist of thinly laminated, very fine sand/coarse silt and silty mud; sand laminae rarely exceed 1 mm in thickness, whereas the mud laminae may be more than 5 mm thick. As already discussed, the thinnest sand-mud alternations are type 4, tidal-cycle couplets, while the thicker mud layers are amalgamated. In some of the laminated intervals, 0.5 to 2 cm thick cyclic variations in the prominence of sand laminae may represent neap-spring cycles (Fig. 19A, B), but the number of couplets present cannot be counted to confirm this. Biogenic structures are rare in the laminated intervals, and where present consist of escape traces made by the clam *Macoma balthica* (Figs. 18, 19A, B).

The burrowed units, which are 2 to 15 cm thick, gradationally overlie the laminated intervals. All degrees of bioturbation, from a few discrete burrows in laminated sediments to thoroughly homogenized sandy mud, are present, but some faint stratification generally remains (Fig. 19A-C). The most abundant traces are the U-shaped burrows of the amphipod *Corophium volutator* and the polychaete *Nereis virens*, as well as dwelling burrows made by *Macoma balthica* (Fig. 19B; Dewing, 1986). Large-scale convolutions are common in the upper portion of burrowed intervals, and may extend down into the underlying laminated material (Fig. 19C). In cross-section, the convolutions typically consist of sharp-crested anticlines separated by broad synclines. The vertical relief is 5 to 10 cm, whereas the lateral spacing between anticlines ranges from several decimetres to a few metres. The convolutions are everywhere overlain by laminated sediment. The contact is commonly erosional, and may be marked by out-sized gravel clasts up to 10 cm in diameter (Fig. 19F). In other cases, the relief of the convolutions dies out upward into the overlying laminated sediment.

Bedding plane exposures of the tops of burrowed intervals indicate that the deformation structures were produced by blocks of drift ice that were blown onto the mud flats and stranded during the falling tide (Fig. 17A). In plan, the anticlines partially or completely surround equant depressions or pits (Fig. 19D) that mark the location where ice blocks came to rest on the mud flat (*cf.* Dionne, 1985, Fig. 4). If the rim does not completely surround the depression, the non-rimmed side faces offshore and is commonly attached to a straight to gently curved, striated groove (Figs. 7B, 19D; *cf.* Dionne, 1985, Fig. 3). Some of these drag marks

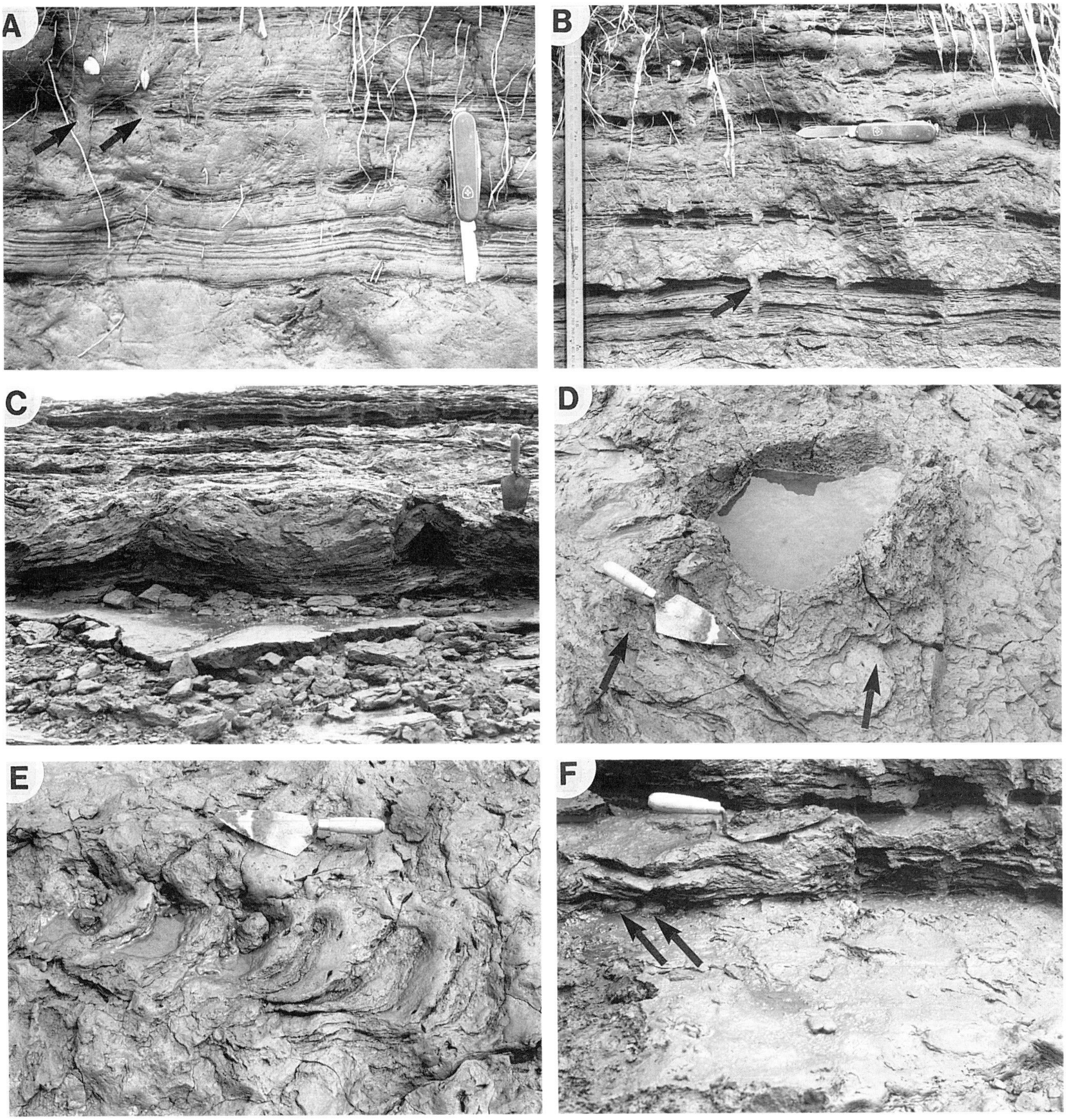

Figure 19. Sedimentary structures in the Black Rock section. **A** and **B**. General views of annual deposits consisting of alternating laminated (spring) and burrowed (summer-fall) sediment. The knife handle is 6 cm long. The stratified intervals contain a cyclic variation in sand content over a thickness of 0.5 to 2 cm (best seen behind the knife blade in **A** that may represent neap-spring cycles. Ice-produced deformation is present in the lower two bioturbated intervals in **A.** Note the *Macoma* shells in life position in **A**, overlying vertical escape traces (arrows); additional escape traces are visible in **B** (*e.g.*, at arrow). **C.** Large-scale, ice-produced deformation structures in the upper part of an annual deposit. Note that the overlying laminated sediment erosionally overlies the deformed, bioturbated material. **D.** Exhumed upper surface of a bioturbated interval showing a pit formed by an ice block, surrounded on the left, rear and right sides by a rim of sediment displaced from the pit. The side in the foreground is attached to a shallow, elongate groove (borders marked by arrows) formed as the ice block moved onto the mud flat in the direction indicated by the arrows. The trowel is 25 cm long. **E.** Exhumed upper surface of a bioturbated interval with a "beaded" groove produced by the periodic, wave-induced impact of an ice block with the bed. Trowel points in the direction of movement. The 1 cm long, oval slits are *Macoma balthica* burrows. **F.** Ice-rafted pebbles (arrows) at the contact between bioturbated and laminated sediments.

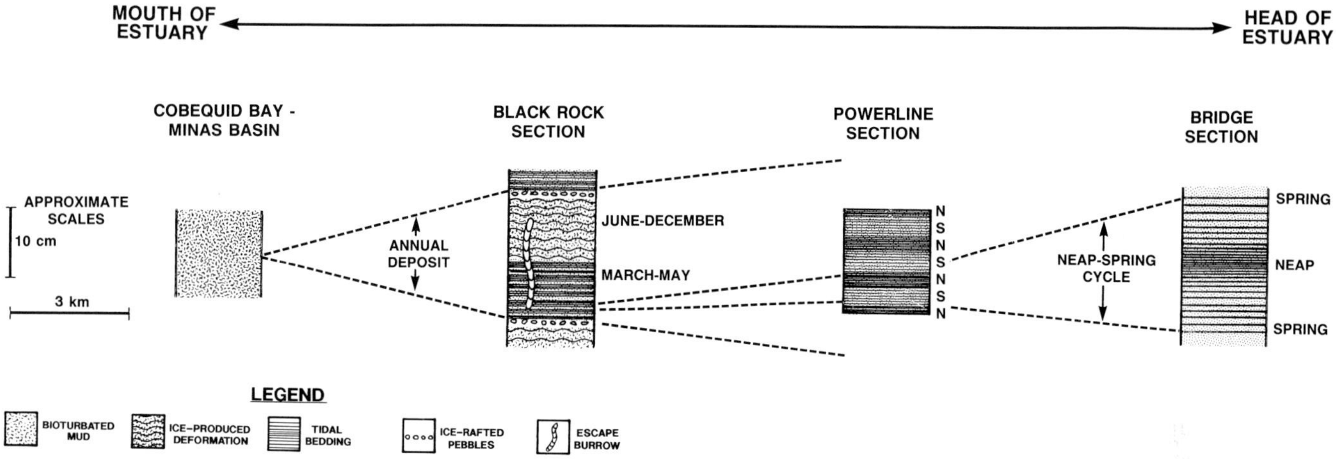

Figure 20. Schematic representation of the longitudinal changes in mud flat structures and sedimentation rates within the Cobequid Bay-Salmon River estuary.

also contain regularly-spaced pits, presumably caused by wave-induced motion of the ice block (Fig. 19E; *cf.* Dionne, 1985, Fig. 6). Features that would be expected of downslope slumping or flowage of sediment are not present.

The above observations indicate that the laminated to burrowed cycles (Figs. 18, 19) are annual deposits. The deformation structures can only have formed during the winter or early spring when drift ice is abundant in the bay (Fig. 7A; Knight and Dalrymple, 1976; Zaitlin, 1987). The outsized gravel clasts were also ice-rafted onto the mud flats at this time; such clasts do not occur at any other level in the laminated to burrowed units. The overlying laminated sediment accumulated in the spring, during and following the production of the deformation features, but before the organism numbers increased sufficiently to cause significant bioturbation of the sediments. The presence of ice-crystal casts throughout the entire thickness of the laminated interval in some cycles suggests that these laminated units were deposited in March, April and early May when air temperatures are frequently below freezing at night. The bioturbated sediment capping the cycles represents the material deposited during the summer and fall when the higher temperatures (Fig. 6) promoted a rapid increase in the number of organisms inhabiting the mud flats. These laminated to burrowed cycles are, therefore, similar in origin to the alternating, laminated and bioturbated intervals described from the Oosterscheldt estuary (van den Berg, 1981). Annual deposits also occur in the Baie du Mont Saint-Michel (Tessier and Larsonneur *in* Larsonneur, 1988) but in this case they are caused by seasonal variations in wave energy.

If the laminated sediments were deposited in a two- to three-month period in the spring, and the bioturbated intervals accumulated in approximately 5 months during the summer and fall, then the average accumulation rates during these periods are slightly more than 2 cm/month and less than 1 cm/month respectively; thus sedimentation rates are higher in the spring, an observation that has also been reported from the outer part of Cobequid Bay (Yeo, 1978). It is likely that this seasonal difference is a result of the higher suspended sediment concentrations that occur in the spring (Amos and Alfoldi, 1979; Greenberg and Amos, 1983; Amos and Tee, 1989). This in turn is probably a result of: 1) the large quantity of fine grained material that is released from the drift ice as it melts (Knight and Dalrymple, 1976; Amos and Tee, 1989); and 2) the sediment that is flushed from the inner part of the estuary by the springtime flood of the Salmon River (Fig. 4) (Dalrymple and Zaitlin, 1989).

The scarcity of bioturbation at all times of the year and the high rates of deposition (Table 2) that characterize the Bridge and Powerline sections do not allow readily-apparent annual cycles to form in the inner estuary; nevertheless, some indications of seasonality are evident at the Powerline location. Firstly, ice-crystal imprints (Fig. 13F) were observed near the base of the section (Fig. 12). On the basis of the number of neap-spring cycles in the overlying sediments, it is estimated that the ice-crystal imprints formed in April or May when nighttime temperatures were below freezing. In addition, the desiccation cracks at the neap-tide horizons become progressively deeper and wider in each higher neap-tide interval (Fig. 13B), presumably because the temperature, and thus the amount of desiccation, increased from spring into summer.

Temporal and Spatial Patterns of Sedimentation

Observations

The vertical (within section) and horizontal (between section) differences in the thickness of the various depositional units described above, together with the accretion rate data (Table 2), provide an opportunity to assess the temporal and spatial patterns of mud flat sedimentation.

Within each section, there is a general, upward decrease in the sedimentation rate, as shown by both the accretion rate data (Table 2) and the upward decrease in the thickness of the temporal sedimentation units (Figs. 12, 18). As the sedimentation units become thinner, they also become more difficult to distinguish. At Black Rock, the depth of bioturbation remains more or less constant while the annual deposits become thinner; as a result, the laminated intervals at the top of the exposure become completely reworked, causing biological amalgamation. In the Powerline section, the neap-spring cycles lose their identity because the upward decrease in the sand content of the spring-tide couplets causes them to resemble neap-tide deposits. Amalgamation of mud layers also increases in frequency upward, and the number of recognizable couplets in each neap-spring cycle decreases, especially at elevations above the neap high-tide level.

Relatively few data exist on the variation of sedimentation rate with distance from (*i.e.*, perpendicular to) the channel that are independent of elevation, as elevation generally increases away from the channel. However, a comparison of accretion rates from tiles at approximately the same elevation on inner and outer portions of terraces (Table 2), shows that the rates may either increase (tiles 15-16 and 19-20), stay constant (tiles 17-18) or decrease (tiles 7-8) as the distance from the channel increases. Observations on the change of surface slope as terraces aggrade suggest, however, that it is more common for the rate of deposition to decrease away from the channel: terraces at low elevations exhibit a marked slope toward the channel (Fig. 3C), whereas those at higher elevations are either nearly horizontal (Fig. 3D) or have a levee-like ridge along their outer edges. Sections cut perpendicular to the channel also show that the tidal-cycle couplets become thinner and muddier as distance from the channel increases.

Finally, there is a pronounced seaward decrease in both the measured sedimentation rates (Table 2) and those deduced from the measured sections: approximately 7 m/year at the Bridge section; about 3 m/year at the Powerline section; and only 10 to 20 cm/year in the Black Rock area. Measurements from areas seaward of Black Rock indicate that the accumulation rate continues to decrease, and approaches zero on the foreshores adjacent to zone 1 (Amos and Joice, 1977; Yeo, 1978). As a result, the mud flat deposits there are extensively bioturbated (Yeo and Risk, 1981). These longitudinal differences in the sedimentation rate are, in fact, responsible for the changes in the sedimentary character of the muddy flats along the length of the estuary, as shown schematically in Figure 20.

Depositional Model

The vertical and horizontal variations in sedimentation rate occur primarily because of differences in the amount of sand deposited, there being a strong positive correlation between the sedimentation rate and sand content: the sand content increases downward in each measured section (Figs. 10, 12, 18); each mud flat terrace becomes sandier toward the channel (Fig. 3C, D); and the sand content of the mud flats increases headward (Figs. 10-13, 18). Even on the level of individual tidal-cycle couplets, the difference in thickness between couplets is determined almost entirely by the amount of sand present, the thickness of the mud laminae being relatively constant within and between sections (Figs. 10-14, 16, 17).

The contrasting behaviour of the sand and mud fractions is a direct consequence of their different settling rates. Because the sand settles out of (intermittent) suspension much more rapidly than the silt and clay, its distribution on the mud flats is more strongly dependent on local current speeds and proximity to the source (the channel); thus, the sands are restricted to areas close to the channel, both vertically and laterally, whereas the muds are more widely and uniformly dispersed. These ideas provide a conceptual framework for summarizing the temporal and spatial patterns of mud flat sedimentation in the estuary (Fig. 21).

As the elevation of the mud flat surface increases through time (Fig. 21A), the amount of sand carried onto the flats decreases for two reasons: first, the maximum current speed usually decreases because the depth of inundation decreases and inundation occurs closer to high water when current speeds are lower; and second, the concentration of sand in suspension in the channel decreases upward through the flow so that the portion of the flow that spills out onto the mud flat will carry less sand as the terrace elevation increases. By contrast, the net accumulation rate of mud probably reaches a maximum at an intermediate elevation (Fig. 21A), because of the partial removal of mud drapes by stronger currents at lower elevations, and less frequent inundation once the mud flat rises above the level of high water during neap tides. The combined result of these various trends (Fig. 21A) is the production of a fining-upward deposit.

The plot of sedimentation rate versus elevation or time (Fig. 21A) indicates that the elevation of the mud flat rises most rapidly when deposition begins, but that upbuilding slows as the surface approaches the level of high water (Fig. 22). Once the mud flat has reached this elevation, it will remain there indefinitely, unless it is eroded or a rise in the high-water level permits continued slow deposition. (This latter condition applies in the study area, as the high-tide level is rising at a rate of approximately 2 mm/year (Scott and Greenberg, 1983). This behaviour suggests that the history of a mud flat can be divided into two phases: a short-lived phase of rapid aggradation, when sediments containing one or more of the temporal sedimentation units described above accumulate, followed by a quasi-equilibrium state of long duration in which deposition is slow or negligible and the deposits are more intensely bioturbated. This type of behaviour has been documented on the ''Windsor Causeway mudflat'' in the Avon River estuary (Bay of Fundy) (Amos, 1977; Amos *et al.*, 1980).

The decrease in sedimentation rate and sand content away from the channel (Fig. 21B) is clearly brought about by the dumping of the sands on the outer edges of the mud flat, presumably because of the rapid drop of current speed and turbulence as the flow passes onto the shallowly-submerged flats. Only the silt and clay remain in suspension long enough to be carried to the more distant parts of the flats, causing the deposits there to be muddier and to accumulate more

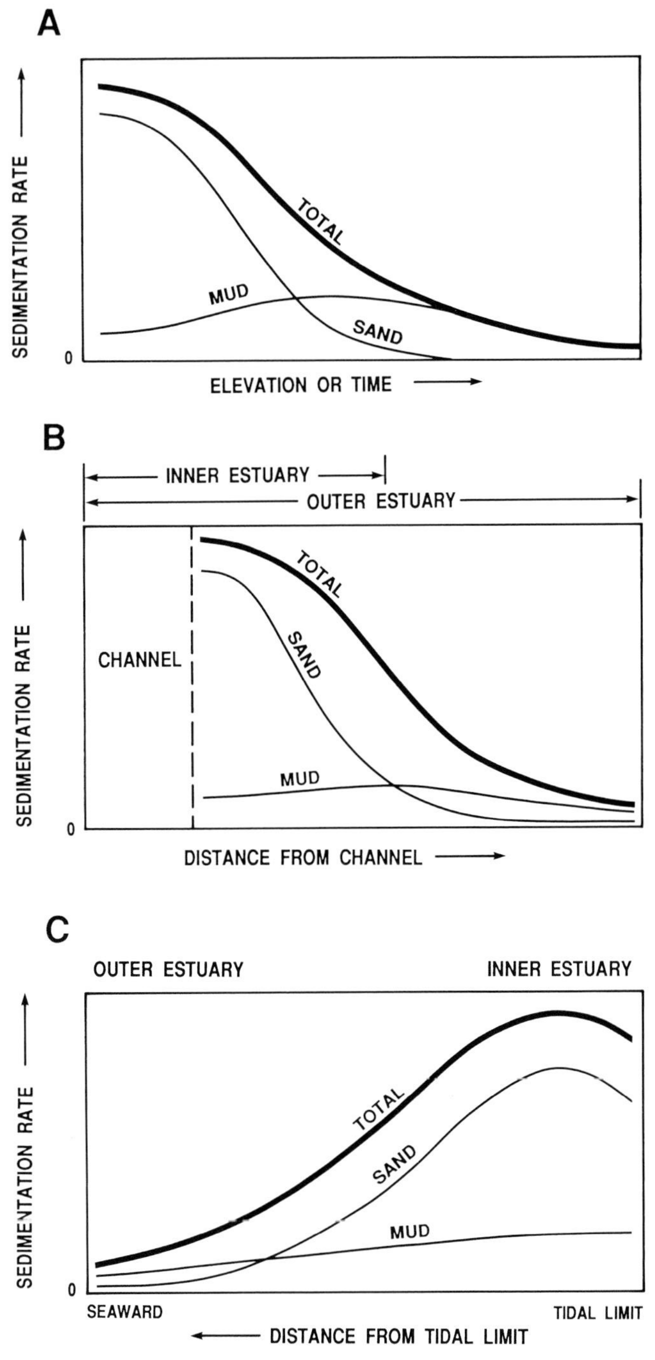

Figure 21. Conceptual models of sand, mud and total sediment accumulation rates on the unvegetated, muddy flats in the study area as a function of: **A.** mud flat elevation or time since the onset of deposition; **B.** lateral distance from the channel; and **C.** distance from the tidal limit along the length of the estuary. None of the diagrams is to scale. The arrows at the top of **B** indicate the relative lateral extent of the muddy flats in the inner (zone 3) and outer (outer zone 2 and zone 1) parts of the estuary. See text for discussion.

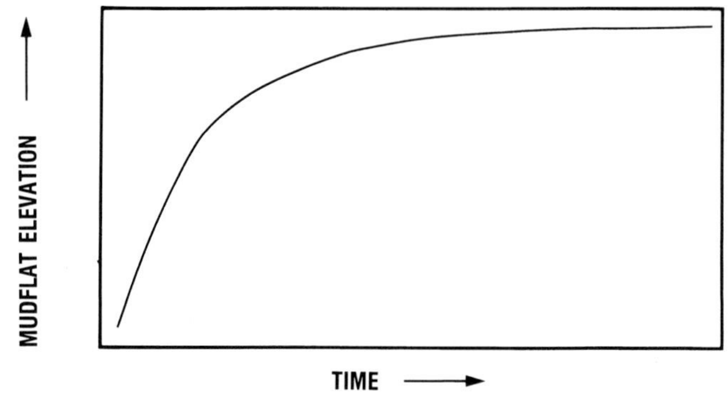

Figure 22. Plot of mud flat elevation against time. Sedimentation is initially rapid, but decreases through time to near zero when the mud flat surface reaches a quasi-equilibrium elevation. Most of the deposits accumulate in a short period of time, which is followed by a (potentially) longer period of relative stability.

slowly. Movement of the channel and the adjacent area of rapid sand deposition (Fig. 21B) toward an aggrading mud flat can reverse the anticipated fining-upward trend. In the Bridge section for example, the sand layers generally become thinner upward from layer 85 to approximately layer 40 (Fig. 16A), but above this, there is an upward coarsening as the sand-layer thicknesses increase. At the time of the measurements, the outer margin of the terrace was being eroded (Fig. 3D) and it is likely that the upward increase in sand content and sedimentation rate is a result of the channel moving closer to the measured section. Similarly, the anomalously high sedimentation rates measured at tile 4 (Table 2) were measured immediately before erosion of this tile (Fig. 3C); indeed, the rate of sediment deposition of the mud flat increased as the eroding bank approached the measurement site. Conversely, the fining-upward trend would be accentuated if the channel moved away from a depositional mud flat. Tessier and Larsonneur (*in* Larsonneur, 1988) note similar effects on estuarine mud flats in the Baie du Mont Saint-Michel.

The seaward decrease in sedimentation rate and sand content (Fig. 20) can also be explained in terms of the differential rates of sand and mud deposition (Fig. 21C). The amount of mud that accumulates on each tide probably increases headward (Fig. 21C) because the turbidity maximum is located near the Bridge section at high tide (Zaitlin, 1987). However, it is believed that sand deposition increases headward at a faster rate (Fig. 21C) because the width of the mud flats decreases (Fig. 2B) and they are situated closer to the channel on average (compare Fig. 3A with Figs. 3C, D). In the inner part of the estuary, the entire width of the mud flats (Fig. 3B) lies within the zone of rapid sand deposition so that the deposition rate and sand content are high (Fig. 21B, top). By contrast, the mud flats farther seaward are much wider (Fig. 3B) and extend beyond the zone of maximum sand deposition (Fig. 21B). On average, therefore, these flats will be muddier and will accrete more slowly than those farther headward.

Discussion

Previous studies of mud flat sedimentation (*e.g.*, van Straaten, 1954, 1961; Carling, 1982; Anderson, 1983; Amos *et al.*, 1988; Frey *et al.*, 1989) have generally found that vertical aggradation rates are low, with sedimentation proceeding only at the rate allowed by the long-term rise in the high-tide elevation. In the shorter term, deposition and erosion alternate in time and space because of the complex interaction of a large number of physical and biological variables (*e.g.*, Anderson, 1983). As a result of their relative stability, biological reworking is commonly intense (Howard and Frey, 1975; Dorjes, 1978; Reineck, 1978; Risk and Yeo, 1980; Frey *et al.*, 1989) and physical structures are rarely preserved. By comparison, the mud flats described here accrete rapidly, and their deposits are characterized by physically-generated structures in which bioturbation is limited in time (summers) and space (seaward part of the study area). Furthermore, the main controls on sedimentation are the migration of, and current speeds in, the adjacent channel, and the seasonal variations in temperature, ice action and river discharge; the other variables considered to be important by workers such as Anderson (1983) and Amos *et al.* (1988) appear to have relatively little effect because of the high rate of deposition.

A comparison of the mud flats described here with macrotidal flats from elsewhere in the Bay of Fundy (Yeo and Risk, 1981; Amos *et al.*, 1988) and along the Korean coast (Frey *et al.*, 1989) shows that the above differences are not due to the macrotidal nature of the study area. Instead, the major difference is that the present study area lies in the confined, inner portion of an estuary, whereas most of the areas studied previously are situated in back-barrier lagoons, sheltered coastal settings and the seaward parts of estuaries where the tidal flats are broad and slow sedimentation is to be expected (Figs. 20, 21C). Indeed, scattered observations suggest that high sedimentation rates may be a general feature of inner-estuary tidal flats. Spectacular examples of this are given by mud flats that have accumulated on the seaward sides of impermeable causeways constructed across estuaries bordering the Bay of Fundy; for example, the "Windsor Causeway mudflat" in the Avon River accreted at a rate of up to 0.15 m/month (average 50 mm/month) before stabilizing at an elevation approximately 3.5 m below highest high water (Amos, 1977; Amos *et al.*, 1980). In the Baie du Mont Saint-Michel, the tidal rhythmites described by Tessier and Larsonneur (*in* Larsonneur, 1988) indicate that deposition rates may exceed 5 mm/tidal cycle in the inner "estuarine" zone, whereas slow accumulation and thorough bioturbation dominate in seaward areas. High rates of sedimentation also occur on the macrotidal flats at the head of Cook Inlet in Alaska (Bartsch-Winkler *et al.*, 1983; Bartsch-Winkler, 1988); 1.4 m of sediment accumulated in a 9-year period (approximately 15 cm/year) following subsidence caused by a large earthquake, and even higher rates of deposition are associated with changes in channel location (Bartsch-Winkler, 1988). Data from mesotidal and microtidal areas are less abundant, but Smith (1987, 1988) and Thomas *et al.* (1987) report that point bars in mesotidally influenced rivers contain millimetre-thick laminae that may represent the deposits of individual tidal cycles, and decimetre-thick sand-mud alternations that may be annual deposits. The deposition rates inferred from such observations clearly exceed those that typify the more seaward parts of microtidal and mesotidal systems (van Straaten, 1954, 1961; Carling, 1982; Anderson, 1983).

If the headward increase in mud flat sedimentation rates is a general feature of estuaries, then it might have a significant influence on the biology of the inner estuary. It is well known that the diversity of organisms decreases markedly into estuaries (Fig. 5D; Howard and Frey, 1975; Barnes, 1984; Wightman *et al.*, 1987). This is usually ascribed to the headward decrease in salinity (Fig. 5C), but it is also known that few benthic organisms can withstand high rates of deposition (Yeo, 1978; Risk and Yeo, 1980; Rhodes *et al.*, 1985). Therefore, it is possible that the low species diversity of estuarine environments is due to the combined influence of the low salinity and high rates of deposition.

Finally, it must be noted that the high sedimentation rates recorded here represent short-term values and do not necessarily reflect the pattern of net deposition on the time span of decades or centuries, because the rapid rate of accumulation, which characterizes the inner estuary, is almost matched by the rate of erosion due to channel movement. However, the CB-SR system as a whole is progradational at present (Dalrymple *et al.*, 1990) and it is anticipated that the entire, funnel-shaped geometry of the estuary will translate seaward, with the highest rates of long-term deposition occurring where the estuary broadens most rapidly (*i.e.*, adjacent to zone 2). In the process, channel-margin mud flat deposits with the characteristics described above have a very high chance of being preserved, even though the short-term preservation potential of any individual mud flat is low.

Summary

Observations in the Cobequid Bay-Salmon River estuary, Bay of Fundy, show that the sedimentary facies consist of an axial belt of sandy sediments, bordered along either shore in the middle and inner parts of the system by mixed flats, mud flats and salt marsh (Fig. 1). The width of the mixed flats and mud flats decreases headward through this area (Fig. 2B) because of the decrease in the tidal prism. Episodic changes in the position of the channel cause the muddy flats to experience alternate periods of erosion and deposition, producing a distinctive terraced morphology to the margins of the estuary (Fig. 3). As a result, the architecture of the mud flat deposits consists of elongate bodies of flat-lying tidal sediments, each of which corresponds to a level terrace, that are separated from each other by steep erosional scarps produced by channel erosion. The size of the lenses becomes smaller toward the head as the lateral extent of the muddy flats decreases (Fig. 3).

Accretion rate measurements (Table 2) and examination of the sedimentary sturctures within the mixed flats and mud flats (Figs. 10-13, 19) show that, on average, short-term sedimentation rates increase headward (Fig. 20, 21C), from 5 to 30 cm/year in the middle estuary, to 7 m/year in the inner part. This is caused by an increase in the sand content of the deposits, which is, in turn, a result of the headward decrease in the width of the flats and their greater average proximity to the channel. Similar patterns of sedimentation rate may also characterize other estuaries.

Because of these high sedimentation rates, clearly-defined temporal sedimentation units with a variety of periods are well developed. In the inner part of the estuary, sand-mud couplets (Figs. 11, 13) up to 3.5 cm thick represent the deposits of individual flood or ebb tides; those produced by the dominant tide are the mud flat equivalent of tidal bundles. Differences in the thicknesses of the sand layers and the sedimentary structures in them (suspension drapes, current ripples or upper-flow-regime parallel lamination) between couplets record the variation in current speed between tides (Fig. 14) that result from differences in the tidal range. As a result, neap-spring cycles are represented by cyclic changes up to 40 cm thick between muddier (neap tides) and sandier (spring tides) couplets (Figs. 10-12, 13A, B, D, 16, 17). The deposits of successive tides are also alternately thicker and thinner as a result of the diurnal inequality. Bioturbation is negligible in these deposits because of the rapid sedimentation rates and low salinities that characterize the inner part of the estuary (Fig. 5C). These couplets provide a process analogue for the rhythmites described by Baird *et al.* (1986), Kvale *et al.* (1989), Williams (1989) and Tessier and Gigot (1989).

Farther seaward, where sedimentation rates are lower, tidal couplets and neap-spring cycles cannot be distinguished readily; instead, annual deposits are well developed. These consist of alternating laminated (spring) and bioturbated (summer and fall) sandy muds (Figs. 18, 19A-C). Ice-rafted pebbles and deformation structures produced by grounded ice blocks commonly occur at the contact between bioturbated sediments and the overlying laminated material (Fig. 19D-F). The longitudinal variation in the nature of the mud flat structures, if found to be a common feature of estuarine environments, will provide a means of identifying and determining one's position within ancient estuarine deposits.

Acknowledgments

Research in the Bay of Fundy by Dalrymple and Zaitlin has been supported by grants from the Natural Sciences and Engineering Research Council of Canada (A7553), the Advisory Research Committee of Queen's University, Imperial Oil Ltd., Gulf Canada Ltd., Mobil Oil Ltd., Shell Oil Ltd., Canterra Energy Ltd. and the Geological Society of America. Makino's participation was made possible by funding from the Japanese Ministry of Education, Science and Culture, and the hospitality of A. Miall, Department of Geology, University of Toronto. The Bedford Institute of Oceanography provided much appreciated logistical support for Zaitlin, and Esso Resources Canada Ltd. prepared the figures. Reviews of the manuscript by C. Amos, G. Ashley and G. Middleton lead to its improvement. The paper is dedicated to C. Amos for his logistical support of the research, and for prompting a closer examination of the mud flat deposits.

References

Allen, J. R. L. 1982. Mud drapes in sand-wave deposits: physical model with application to the Folkestone Beds (Early Cretaceous, southeast England). Philosophical Transactions of the Royal Society, v. A306, p. 291-345.

______ 1984. Wrinkle marks: an intertidal sedimentary structure due to aseismic soft-sediment loading. Sedimentary Geology, v. 41, p. 75-95.

______ 1987. Reworking of muddy intertidal sediments in the Severn Estuary, southwestern U.K. — a preliminary survey. Sedimentary Geology, v. 50, p. 1-23.

______ and Rae, J. E. 1988. Vertical salt-marsh accretion since the Roman period in the Severn estuary, southwest Britain. Marine Geology, v. 83, p. 225-235.

Amos, C. L. 1977. Effects of tidal power structures on sediment transport and loading in the Bay of Fundy — Gulf of Maine system. *In:* Fundy Tidal Power and the Environment, G. R. Daborn (ed.). Acadia University Institute Publication 28, Wolfville, p. 233-253.

______ and Alfoldi, T. T. 1979. The determination of suspended sediment concentration in a macrotidal system using LANDSAT data. Journal of Sedimentary Petrology, v. 49, p. 159-174.

______ and Joice, G. H. E. 1977. The sediment budget of the Minas Basin, Bay of Fundy, N.S. Bedford Institute of Oceanography, Data Series BI-D-77-3, 411 p.

______ and Long, B. F. N. 1980. The sedimentary character of the Minas Basin, Bay of Fundy. *In:* The Coastline of Canada, S. B McCann (ed.). Geological Survey of Canada, Paper 80-10, p. 123-152.

______ and Mosher, D. C. 1985. Erosion and deposition of fine-grained sediments from the Bay of Fundy. Sedimentology, v. 32, p. 815-832.

______ and Tee, K. T. 1989. Suspended sediment transport processes in Cumberland Basin, Bay of Fundy. Journal of Geophysical Research, v. 94, p. 14407-14417.

______ , Buckley, D. E., Daborn, G. R., Dalrymple, R. W., McCann, S. B. and Risk, M. J. 1980. Geomorphology and sedimentology of the Bay of Fundy. Geological Association of Canada Guidebook, Trip 23, Annual Meeting, Halifax, N.S., 82 p.

______ , van Wagoner, N. A. and Daborn, G. R. 1988. The influence of subaerial exposure on the bulk properties of fine-grained intertidal sediment from Minas Basin, Bay of Fundy. Estuarine, Coastal and Shelf Science, v. 27, p. 1-13.

Anderson, F. E. 1983. The nothern muddy intertidal: seasonal factors controlling erosion and deposition- a review. Canadian Journal of Fisheries and Aquatic Sciences, v. 40, Supplement 1, p. 143-159.

Atmospheric Environment Service, 1982. Monthly Record — Meteorological Observations in Eastern Canada. Canada Department of the Environment, v. 67, nos. 1-12.

Baird, G. C., Sroka, S. D., Shabica, C. W. and Kuecher, G. V. 1986. Taphonomy of middle Pennsylvanian Mazon Creek area fossil localities, northeast Illinois: significance of exceptional fossil preservation in syngenetic concretions. Palaios, v. 1, p. 271-285.

Barnes, R. S. K. 1984. Estuarine Biology (second edition). London, Edward Arnold Limited, 76 p.

Bartsch-Winkler, S. 1988. Cycle of earthquake-induced aggradation and related tidal channel shifting, upper Turnagain Arm, Alaska, USA. Sedimentology, v. 35, p. 621-628.

______ , Ovenshine, A. T. and Kachadoorian, R. 1983. Holocene history of the estuarine area surrounding Portage, Alaska as recorded in a 93 m core. Canadian Journal of Earth Science, v. 20, p. 802-820.

Berg, J. H., van den, 1981. Rhythmic seasonal layering in a mesotidal channel fill sequence, Oosterschelde Mouth, the Netherlands. *In:* Holocene Marine Sedimentation in the North Sea Basin, S.-D. Nio, R. T. E. Shuttenhelm and Tj. C. E. van Weering (eds.). International Association of Sedimentologists, Special Publication 5, p. 147-159.

Boer, P. L, de, Oost, A. P. and Visser, M. J. 1989. The diurnal inequality of the tide as a parameter for recognizing tidal influences. Journal of Sedimentary Petrology, v. 59, p. 912-921.

Boersma, J. R. 1969. Internal structures of some tidal megaripples on a shoal in the Westerschelde estuary, the Netherlands. Geologie en Mijnbouw, v. 48, p. 409-414.

Canadian Hydrographic Service, 1984. Canadian Tide and Current Tables, Volume 1, Atlantic Coast and Bay of Fundy. Ottawa, Department of the Environment, 53 p.

Carling, P. A. 1982. Temporal and spatial variation in intertidal sedimentation rates. Sedimentology, v. 29, 17-23.

Clifton, H. E. 1983. Discrimination between subtidal and intertidal facies in Pleistocene deposits, Willapa Bay, Washington. Journal of Sedimentary Petrology, v. 53, p. 353-369.

Craig, H. D. 1977. Intertidal animal and trace zonations in a macrotidal environment, Minas Basin, Bay of Fundy. Unpublished M.Sc. thesis, McMaster University, Hamilton, Ontario, 232 p.

Dalrymple, R. W. 1977. Sediment dynamics of macrotidal sand bars, Bay of Fundy. Unpublished Ph.D. thesis, McMaster University, Hamilton, Ontario, 635 p.

______, Amos, C. L. and McCann, S. B. 1982. Beach and nearshore depositional environments of the Bay of Fundy and southern Gulf of St. Lawrence. International Association of Sedimentologists, Guidebook, Excursion 6A, 11th Congress, Hamilton, Ontario, 116 p.

______, Knight, R. J., Zaitlin, B. A. and Middleton, G. V. 1990. Dynamics and facies model of a macrotidal sand bar complex, Cobequid Bay-Salmon River estuary (Bay of Fundy). Sedimentology, v. 37, p. 577-612.

______ and Makino, Y. 1989. Description and genesis of tidal bedding in the Cobequid Bay-Salmon River estuary, Bay of Fundy, Canada. *In:* Sedimentary Facies in the Active Plate Margin, A. Taira, and F. Masuda (eds.). Tokyo, Terra Scientific Publishing Company, p. 151-177.

______ and Zaitlin, B. A. 1989. Tidal deposits in the macrotidal, Cobequid Bay-Salmon River estuary, Bay of Fundy. Canadian Society of Petroleum Geologists, Field Guide, Second International Research Symposium on Clastic Tidal Deposits, August 22-25, Calgary, Alberta, 84 p.

Davies, J. L. 1980. Geographical Variation in Coastal Development (second edition). New York, Longman, 212 p.

Dewing, K. E. 1986. Organism zonation in a macrotidal estuary, Cobequid Bay-Salmon River, Bay of Fundy, Nova Scotia, Canada. Unpublished B.Sc. thesis, Queen's University, Kingston, Ontario, 88 p.

______, Zaitlin, B. A., Narbonne, G. M. and Dalrymple, R. W. 1986. Zonation of trace-producing organisms from the Cobequid Bay-Salmon River estuary, Bay of Fundy. Geological Association of Canada, Program with Abstracts, v. 11, p. 63.

Dionne, J. C. 1985. Forms, figures et facies sedimentaires glaciels dans des estrans vaseux des régions froides. Palaeogeography, Palaeoclimatology, Palaeoecology, v. 51, p. 415-451.

Dorjes, J. 1978. Das watt als lebenstraum. *In:* Das Watt, H.-E. Reineck, (ed.). Frankfurt am Main, Kromer, p. 107-143.

Elliott, T. 1986. Siliciclastic shorelines. *In:* Sedimentary Environments and Facies (second edition), H. G. Reading (ed.). Boston, Blackwell Scientific Publications, p. 155-188.

Frey, R. W., Howard, J. D., Han, S.-J. and Park, B.-K. 1989. Sediment and sedimentary sequences on a modern macrotidal flat, Inchon, Korea. Journal of Sedimentary Petrology, v. 59, p. 28-44.

Gordon, D. C., Jr. and Desplanque, C. 1983. Dynamics and environmental effects of ice in the Cumberland Basin of the Bay of Fundy. Canadian Journal of Fisheries and Aquatic Sciences, v. 40, p. 1331-1342.

Greenberg, D. A. and Amos, C. L. 1983. Suspended sediment transport and deposition modeling in the Bay of Fundy, Nova Scotia — a region of potential tidal power development. Canadian Journal of Fisheries and Aquatic Sciences, v. 40, p. 20-34.

Harms, J. C., Southard, J. B. and Walker, R. G. 1982. Structures and Sequences in Clastic Rocks. Society of Economic Paleontologists and Mineralogists, Short Course 9.

Howard, J. D. and Frey, R. W. 1975. Estuaries of the Georgia coast, U.S.A.: Sedimentology and biology, II Regional animal-sediment characteristics of Georgia estuaries. Senckenbergiana Maritima, v. 7, p. 33-103.

Inland Waters Directorate, 1980. Historical stream flow summary- Atlantic Provinces to 1979. Water Resources Branch, Water Survey of Canada, 188 p.

Klein, G. deV. 1985. Intertidal flats and intertidal sand bodies. *In:* Coastal Sedimentary Environments (second edition), R. A. Davis, Jr. (ed.). New York, Springer-Verlag, p. 187-224.

Knight, R. J. 1977. Sediments, bedforms and hydraulics in a macrotidal environment, Cobequid Bay (Bay of Fundy). Unpublished Ph.D. thesis, McMaster University, Hamilton, Ontario, 693 p.

______ and Dalrymple, R. W. 1975. Intertidal sediments from the south shore of Cobequid Bay, Bay of Fundy, Nova Scotia, Canada. *In:* Tidal Deposits — A Casebook of Recent Examples and Fossil Counterparts, R. N. Ginsburg (ed.). New York, Springer-Verlag, p. 47-55.

______ and Dalrymple, R. W. 1976. Winter conditions in a macrotidal environment, Cobequid Bay, Nova Scotia. Revue de Géographie de Montréal, v. XXX, p. 65-85.

Kreisa, R. D., Leithold, E. L. and Bourgeois, J. 1989. Tidal signature in mixed-energy deltaic facies. Canadian Society of Petroleum Geologists, Abstracts, Second International Research Symposium on Clastic Tidal Deposits, August 22-25, Calgary, Alberta, p. 49.

Kvale, E. P., Archer, A. W. and Johnson, H. R. 1989. Daily, monthly, and yearly tidal cycles within laminated siltstones of the Mansfield Formation (Pennsylvanian) of Indiana. Geology, v. 17, p. 365-368.

Larsonneur, C. 1988. La Baie du Mont Saint-Michel. Caen, Université de Caen, 85 p.

LeFournier, J. and Friedman, G. M. 1974. Rate of lateral migration of adjoining sea-marginal sedimentary environments shown by historical records, Authie Bay, France. Geology, v. 2, p. 497-498.

Leithold, E. L., Bourgeois, J. and Kreisa, R. D. 1989. Genesis of tidal rhythmites of an Eocene delta slope, S.W. Oregon, U.S.A. Canadian Society of Petroleum Geologists, Abstracts, Second International Research Symposium on Clastic Tidal Deposits, August 22-25, Calgary, Alberta, p. 55.

Maher, A. J. 1986. A statistical analysis of grain size distribution, Cobequid Bay-Salmon River estuary (Bay of Fundy), Nova Scotia. Unpublished B.Sc. thesis, Queen's University, Kingston, Ontario, 120 p.

McCave, I. N. 1970. Deposition of fine-grained suspended sediment from tidal streams. Journal of Geophysical Research, v. 75, p. 4151-4159.

______ 1971. Wave effectiveness at the sea bed and its relationship to bedforms and deposition of mud. Journal of Sedimentary Petrology, v. 41, p. 89-96.

Myrick, R. M. and Leopold, L. B. 1963. Hydraulic geometry of a small tidal estuary. United States Geological Survey, Professional Paper 422-B, 18 p.

Pillsbury, G. B. 1939. Tidal Hydraulics. United States Army Corps of Engineers, Professional Paper 34, 281 p.

Postma, H. 1961. Transport and accumulation of suspended matter in the Dutch Wadden Sea. Netherlands Journal of Sea Research, v. 1, p. 148-190.

______ 1967. Sediment transport and sedimentation in the marine environment. *In:* Estuaries, G. H. Lauff (ed.). American Association for the Advancement of Science, Publication 83, p. 158-179.

Reineck, H.-E. 1960. Uber die Enststehung von Linsen- und Flaserschichten. Abhandlichen der Deutschen Akademie der Wissenschaften zu Berlin, v. 3, p. 370-374.

______ 1967. Layered sediments of tidal flats, beaches and shelf bottoms of the North Sea. *In:* Estuaries, G. H. Lauff (ed.). American Association for the Advancement of Science, Publication 83, p. 191-206.

______ 1978. Tidal-flat geology. *In:* The Encyclopedia of Sedimentology, R. W. Fairbridge and J. Bourgeois (eds.). Stroudsburg, Dowden, Hutchinson and Ross, p. 799-801.

______ and Wunderlich, F. 1968. Classification and origin of flaser and lenticular bedding. Sedimentology, v. 11, p. 99-104.

Rhodes, D. C., Boesch, D. F., Tang, Z., Xu, F., Huang, L. and Nilsen, K. J. 1985. Macrobenthos and sedimentary facies on the Changjiang delta platform and adjacent continental shelf, East China Sea. Continental Shelf Research, v. 4, p. 189-213.

Risk, M. J. and Yeo, R. K. 1980. Animal-sediment relationships in the Minas Basin, Bay of Fundy. *In:* The Coastline of Canada, S. B. McCann (ed.). Geological Survey of Canada, Paper 80-10, p. 189-194.

______, Yeo, R. K. and Craig, H. D. 1977. Aspects of the marine ecology of the Minas Basin relevant to tidal power development. *In:* Fundy Tidal Power and the Environment, G. R. Daborn (ed.). Acadia University Institute, Publication 28, p. 164-179.

Rubin, D. M. and Hunter, R. E. 1982. Bedform climbing in theory and nature. Sedimentology, v. 29, p. 121-138.

Scott, D. B. and Greenberg, D. A. 1983. Relative sea-level rise and tidal development in the Fundy tidal system. Canadian Journal of Earth Science, v. 20, p. 1554-1564.

Smith, D. G. 1987. Meandering river point bar lithofacies models: modern and ancient examples compared. *In:* Recent Developments in Fluvial Sedimentology, F. G. Ethridge, R. M. Flores and M. D. Harvey (eds.). Society of Economic Paleontologists and Mineralogists, Special Publication 39, p. 83-91.

______ 1988. Modern point bar deposits analogous to the Athabasca Oil Sands, Alberta, Canada. *In:* Tide-Influenced Sedimentary Environments and Facies. P. L. de Boer, A. van Gelder and S. D. Nio (eds.). Boston, D. Reidel, p. 417-432.

Straaten, L. M. J. U., van, 1954. Composition and structure of recent marine sediments in the Netherlands. Leidse Geologische Mededelingen, v. XIX, p. 1-110.

______ 1961. Sedimentation in tidal flat areas. Journal of the Alberta Society of Petroleum Geologists, v. 9, p. 203-226.

Tessier, B. and Gigot, P. 1989. A vertical record of different tidal cyclicities: an example from the Miocene Marine Molasse of Digne (Haute Provence, France). Sedimentology, v. 36, p. 767-776.

Thomas, R. G., Smith, D. G., Wood, J. M., Visser, J., Calverley-Range, E. A. and Koster, E. H. 1987. Inclined heterolithic stratification — terminology, description, interpretation and significance. Sedimentary Geology, v. 53, p. 123-179.

Visser, M. J. 1980. Neap-spring cycles reflected in Holocene subtidal large-scale bedform deposits: a preliminary note. Geology, v. 8, p. 543-546.

Weimer, R. J., Howard, J. D. and Lindsay, D. R. 1982. Tidal flats and associated tidal channels. *In:* Sandstone Depositional Environments, P. A. Scholle and D. Spearing (eds.). American Association of Petroleum Geologists, Memoir 31, p. 191-245.

Wightman, D. M., Pemberton, S. G. and Singh, C. 1987. Depositional modelling of the Upper Mannville (Lower Cretaceous), east-central Alberta: implications for the recognition of brackish water deposits. *In:* Reservoir Sedimentology, R. W. Tillman and K. J. Weber (eds.). Society of Economic Paleontologists and Mineralogists, Special Publication 40, p. 189-220.

Williams, G. E. 1989. Late Precambrian tidal rhythmites in South Australia and the history of the Earth's rotation. Journal of the Geological Society, London, v. 146, p. 97-111.

Wright, L. D., Coleman, J. M. and Thom, B. G. 1973. Processes of channel development in a high-tide-range environment: Cambridge Gulf-Ord River delta, Western Australia. Journal of Geology, v. 81, p. 15-41.

Yeo, R. K. 1978. Animal-sediment relationships and the ecology of the macrotidal mudflat environment, Minas Basin, Bay of Fundy, Nova Scotia. Unpublished M.Sc. thesis, McMaster University, Hamilton, Ontario, 396 p.

______ and Risk, M. J. 1979. Intertidal catastrophes: effect of storms and hurricanes on intertidal benthos of the Minas Basin, Bay of Fundy. Canadian Journal of Fisheries and Aquatic Sciences, v. 36, p. 667-669.

______ and ______ 1981. The sedimentology, stratigraphy, and preservation of intertidal deposits in the Minas Basin System, Bay of Fundy. Journal of Sedimentary Petrology, v. 51, p. 245-260.

Zaitlin, B. A. 1987. Sedimentology of the Cobequid Bay-Salmon River estuary, Bay of Fundy, Canada. Unpublished Ph.D. thesis, Queen's University, Kingston, Ontario, 391 p.

KEY WORDS: RHYTHMITES, MACROTIDAL ESTUARY, BAY OF FUNDY, MUD FLATS, SAND-MUD COUPLETS

In: Clastic Tidal Sedimentology, D. G. Smith, G. E. Reinson, B. A. Zaitlin and R. A. Rahmani (eds.). Canadian Society of Petroleum Geologists. Memoir 16, p. 161-178.

Upper Proterozoic tidal rhythmites, South Australia: sedimentary features, deposition, and implications for the earth's paleorotation

George E. Williams

Department of Geology and Geophysics, University of Adelaide G.P.O. Box 498, Adelaide, South Australia 5001, Australia, **and** *Geosedimentary Services, P.O. Box 394, Marden South Australia 5070, Australia*

Abstract

Tidal rhythmites of laminated siltstone and very fine grained sandstone from Upper Proterozoic (~650-800 Ma) formations in South Australia (Reynella Siltstone, Elatina Formation, Chambers Bluff Tillite) record a wide range of paleotidal cycles that provide unique information on the earth's paleorotation and the past lunar orbit. The rhythmites evidently were built by vertical accretion in littoral settings ranging from estuarine to upper and lower delta slope and distal ebb-tidal delta. Periodic changes in the thickness of semidiurnal and diurnal laminae, and of conspicuous fortnightly groups of such laminae (lamina-cycles), are taken to reflect variations in the amount of fine clastic material entrained and deposited mainly from suspension by tidal currents in response to periodic changes in the velocity and range of paleotides.

Stratigraphic series of lamina and lamina-cycle thickness measurements obtained from drill core of the approximately 10-m thick rhythmite member of the Elatina Formation contain strong periodicities that are evident visually or revealed by Fourier spectral analysis. These data represent an unsurpassed paleotidal record of about 60 years' continuous deposition; supplemented by observations from the Reynella Siltstone, they allow determination of paleotidal and paleorotational periods with an accuracy previously unobtainable for any geological interval. The Late Proterozoic (~650 Ma) year contained 13.1±0.1 lunar months and 400±7 solar days, and the Late Proterozoic lunar month 30.5±0.5 solar days. The periods of the lunar apsides (perigee) and lunar nodal cycles were then 9.7±0.1 years and 19.5±0.5 years, respectively. The data clearly establish the synodic character of the paleotidal regimes. The implied low mean rate of lunar recession for the past ~650 m.y. militates against a close approach of the moon at any time during the Proterozoic.

Overall, the Upper Proterozoic rhythmites in South Australia provide valuable data that support and extend information on the earth's paleorotation and the past lunar orbit obtained from the growth increments of fossils. Study of sedimentary tidal rhythmites may allow the evolving dynamics of the earth-moon system to be traced over most of geological time.

Résumé

Les séquences rythmiques de marée, à siltites laminés et à grès à grain très fin, des formations du Protérozoïque supérieur (~650-800~M.a.) de l'Australie-Méridionale (Siltstone Reynella, Formation Elatina, Tillite Chambers Bluff) ont enregistré une grande variation des cycles de paléomarées; elle fournit des données importantes sur la paléorotation de la Terre et l'ancienne orbite de la Lune. Évidemment, ces séquences rythmiques ont résulté de l'accrétion vertical d'alluvions dans les environnements littoraux variant de l'estuaire au talus deltaïque supérieur et inférieur, au delta distal de marée descendante. Les changements périodiques de l'épaisseur des laminations semi-diurnes et diurnes ainsi que des groupes bimensuels de ces laminations (lamina-cycles), très remarquables, sont sensés refléter les variations du matériel clastique fin, principalement en supension, entraîné et déposé par les courants de marée répondant aux changements périodiques de la vélocité et du marnage des paléomarées.

Des mesures d'épaisseur de séries stratigraphiques de laminations et de cycles de laminations ont été obtenues à partir d'une carotte de 10 m d'épaisseur, forée dans une séquence rythmique du membre de la Formation Elatina. Ces séries contiennent de fortes périodicités qui sont évidentes visuellement ou qui ont été révélées par l'analyse spectrale de Fourier.

Ces données représentent environ 60 ans de sédimentation continue, soit un record de paléomarée non surpassé. Supplémentées par des observations faites sur le Siltstone Reynella, ces données permettent la détermination exacte des périodes de paléomarée et de paléorotation jusqu'alors impossible à obtenir pour n'importe lequel des intervalles géologiques. Au Protérozoïque supérieur (~650 M.a.), une année se subdivisait en 13.1±0.1 mois lunaires et 400±7 jours solaires, et le mois lunaire comptait 30.5±0.5 jours solaires. Les périodes des apsides lunaires (périgées) et les cycles lunaires nodales étaient alors respectivement de 9.7±0.1 années et 19.5±0.5 années. Les données établissent clairement le caractère synodique des anciens régimes de marée. La supposition d'une moyenne faible du taux de récession lunaire pour les dernières ~650 M.a. milite contre un rapprochement étroit de la Lune pour tout le Protérozoïque.

Les séquences rythmiques du Protérozoïque supérieur de l'Australie-Méridionale fournissent des données importantes sur la paléorotation de la Terre et l'ancienne orbite lunaire qui supportent et ajoutent à l'information obtenue à partir de l'augmentation de la croissance des fossiles.

Introduction

An important development in tidal sedimentology during the past few years has been the recognition of tidal rhythmites — packages of vertically accreted, laminated to thin bedded, medium to very fine grained sandstone, siltstone and mudstone that display cyclic variations in lamina and bed thickness and grain size that are attributable to a strong tidal influence on sedimentation. Such clastic tidal rhythmites are now known from the Upper Proterozoic (Williams, 1987, 1988, 1989a-c; von der Borch *et al.,* 1989), the Carboniferous (Baird *et al.,* 1985; Tankard, 1986; Broadhurst, 1988; Kvale *et al.,* 1989), the Cretaceous (Kreisa *et al.,* 1989), the Tertiary (Kreisa *et al.,* 1989; Leithold *et al.,* 1989; Tessier and Gigot, 1989; Tessier *et al.,* 1989), and modern deposits (Dalrymple and Makino, 1989; Tessier *et al.,* 1989). Inferred paleoenvironments range from estuarine to upper and lower delta slope and distal ebb-tidal delta.

Tidal rhythmites are of importance in paleoenvironmental interpretations, as they commonly represent the more distal complement of the well-known crossbedded tidal "bundle" sequences and afford evidence of tidal rhythms and marine influence in fine grained facies otherwise difficult to interpret. The recognition of tidal rhythmites thus helps complete a "tidal facies model" for littoral deposits. Furthermore, well preserved tidal rhythmites can provide valuable information on paleotidal periods, the earth's paleorotation, and the ancient lunar orbit. Such quantitative information supports and extends previous findings on the earth's rotational history obtained from the study of growth increments in marine invertebrate fossils.

This paper reviews tidal rhythmites of the Upper Proterozoic Reynella Siltstone, Elatina Formation and Chambers Bluff Tillite in the Adelaide Geosyncline and eastern Officer Basin, South Australia, and discusses their environment of deposition, encoded paleotidal periods, and importance to paleogeophysics.

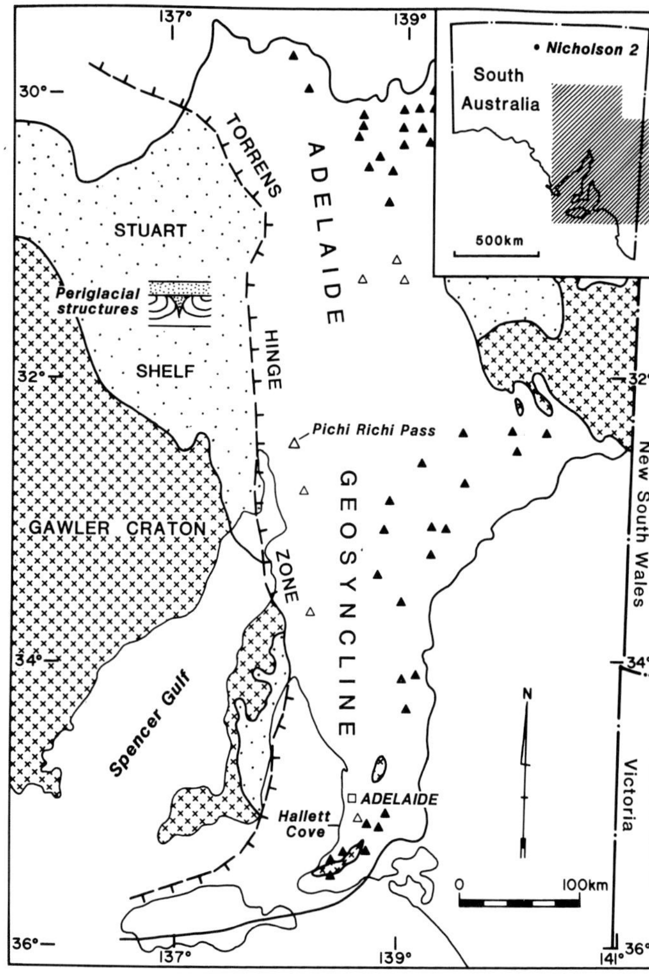

Figure 1. Location map of the Adelaide Geosyncline region in South Australia, showing the distribution of glacial and periglacial facies of the Late Proterozoic Marinoan Glaciation. Solid triangles represent tillitic facies; open triangles, dropstone facies; stipple, stable shelves; crosses, cratonic regions. The location of Comalco Nicholson No. 2 drill hole is shown in the inset. Reproduced with permission of the Geological Society, London.

Upper Proterozoic Tidal Rhythmites in South Australia

Upper Proterozoic cyclic rhythmites of proposed tidal origin in South Australia (Williams, 1987, 1988, 1989a-c) are best seen in the Reynella Siltstone along coastal sections at Hallett Cove near Adelaide, and in the correlative Elatina Formation at Pichi Richi Pass in the Flinders Ranges some 300 km to the north (Fig. 1). Both these formations were deposited in the Adelaide Geosyncline during the Marinoan Glaciation 650 Ma (Preiss, 1987). Postulated tidal rhythmites also are displayed in siltstones and fine grained sandstones immediately overlying the Chambers Bluff Tillite (between 800 and 650 Ma) in Comalco Nicholson No. 2 drillhole (Comalco, 1983) in the eastern Officer Basin in northern South Australia (inset, Fig. 1).

The rhythmites typically comprise graded laminae of very fine grained sandstone through siltstone to mudstone, grouped in "lamina-cycles" that range from less than 1 mm to more than 6 cm in thickness. Lamina-cycles contain from 8 to 26 or more laminae. Lamina thickness is maximal near the stratigraphic center of lamina-cycles, and minimal near cycle boundaries, where thinner, more clayey laminae may crowd together to form conspicuously darker, clayey bands.

As previously discussed (Williams, 1987, 1988, 1989a-c), the laminae are interpreted either as semidiurnal or diurnal increments, and the lamina-cycles as groupings of such increments that represent the lunar fortnightly cycle. Longer tidal cycles are recorded by systematic vertical change in lamina-cycle thickness. Thicker laminae within lamina-cycles are ascribed to the spring phase of the tidal cycle, and the clayey bands bounding lamina-cycles are interpreted as mud drapes deposited at neap tides. These rhythmites provide a rich paleotidal record that has contributed substantially to our understanding of the Earth's rotational history.

Reynella Siltstone

The Reynella Siltstone in the Hallett Cove area consists of massive and laminated red-brown siltstones and fine to

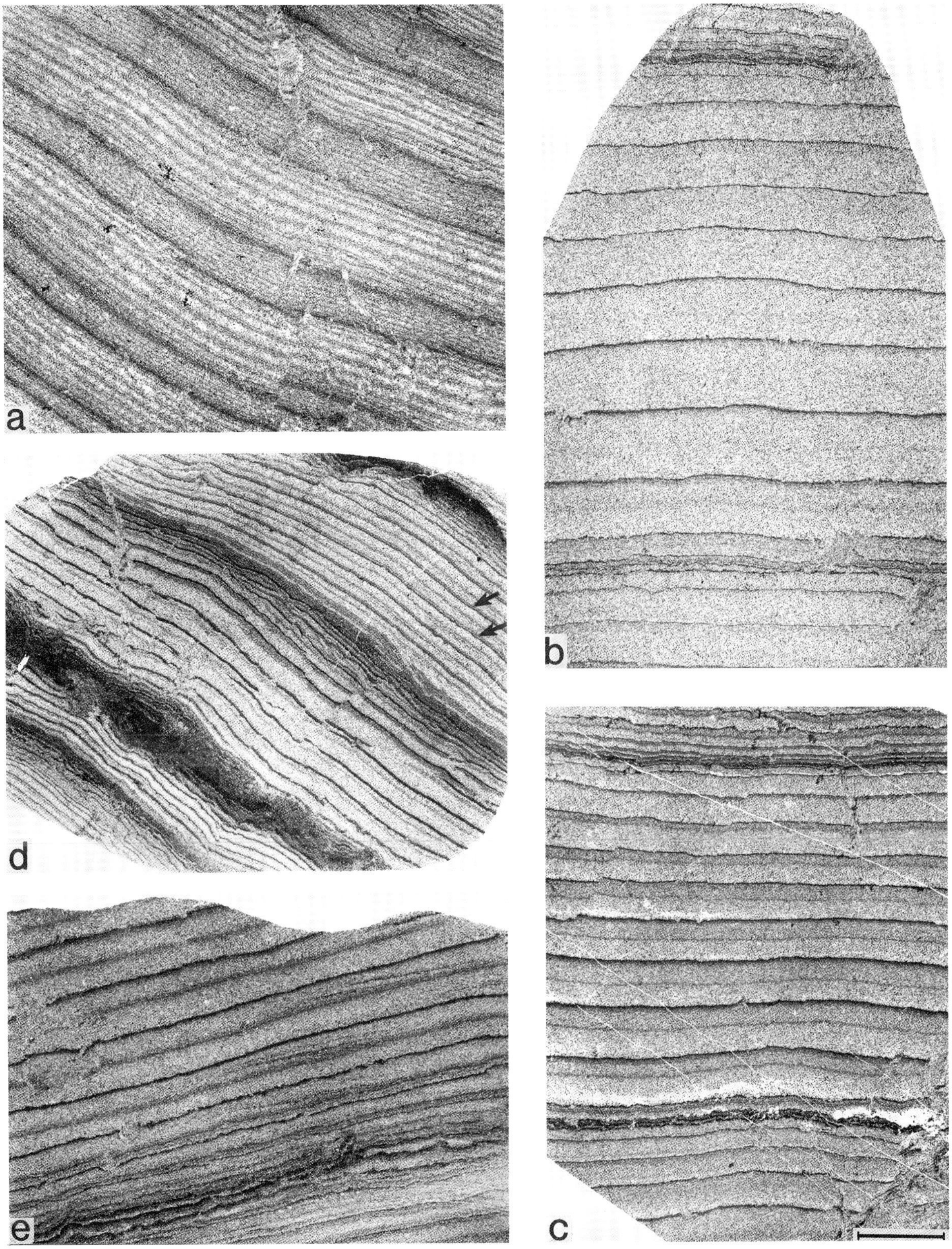

Figure 2. Thin sections of rhythmites from the Reynella Siltstone, Hallett Cove, viewed with transmitted light; opaque clayey material appears darker than translucent sandy layers. The fortnightly lamina-cycles shown comprise weakly graded diurnal laminae and semidiurnal sublaminae of very fine grained sandstone with clayey tops. Scale-bar 1 cm for all photographs. **a.** Nine alternately thick and thin lamina-cycles, indicating the monthly inequality. **b.** One complete lamina-cycle containing diurnal laminae up to 8 mm thick (see Fig. 3a). **c.** One complete lamina-cycle containing diurnal laminae up to 6 mm thick (see Fig. 3b); semidiurnal sublaminae also are evident. **d.** The uppermost lamina-cycle comprises diurnal laminae up to 3 mm thick marked by alternate darker bands, and semidiurnal sublaminae of about equal thickness; the arrows mark double mud drapes. In the cycle immediately below, the lower semidiurnal sublamina in each pair is the thicker, indicating the diurnal inequality. **e.** Portions of two lamina-cycles; in the upper part of the photograph, diurnal laminae up to 6 mm thick marked by alternate darker bands contain semidiurnal sublaminae of about equal thickness. Reproduced with permission of the Geological Society, London.

very fine grained sandstones, channel sandstones, pebbly mudstones, possible pedogenic carbonates, and replaced stromatolites and evaporites (Dyson and von der Borch, 1986; Preiss, 1987). An overall prograding sequence of distal fan delta and lacustrine to estuarine and inner marine shelf environments is envisaged. The tidal rhythmites occur within laminated siltstones and fine grained sandstones totalling several tens of metres in thickness in the lower half of the formation.

Lamina-cycles from the Reynella Siltstone (Fig. 2) show a wide range in thickness. Short sequences commonly display an alternation of relatively thick and thin lamina-cycles (Fig. 2a); this pattern records the alternation of successive high- and low-amplitude spring tidal cycles that reflects the monthly inequality arising from an eccentric lunar orbit (Williams, 1989a-c). Other, thicker lamina-cycles also are present (Fig. 2b,c). The stratigraphic relationship among the several types of lamina-cycle is not clear, however, as much of the coastal cliff section is inaccessible. A full understanding of the rhythmite stratigraphy of the Reynella Siltstone at Hallett Cove must await the obtaining of a complete drill core through the section.

A distinctive feature of the rhythmites in the Reynella Siltstone is the presence of two graded layers or coarse-fine couplets in many laminae (Fig. 2c-e). The laminae and constituent sublaminae are interpreted as diurnal and semidiurnal increments, respectively, of the lunar day (Williams, 1988, 1989a-c). The common alternation of thick and thin sublaminae represents the diurnal inequality of a mixed tidal pattern. Rarely, double mud drapes are associated with the sublaminae (Fig. 2d), indicating that ebb and flood tides as well as the semidiurnal constituents effected the deposition of clastic laminae.

Thick lamina-cycles that appear to be unabbreviated at neaps (Fig. 2b, c) consistently give counts of 14 or 15 diurnal laminae per fortnightly lamina-cycle. The plot of lamina thickness shows that the thick lamina-cycles usually are positively skewed (Fig. 3). Such asymmetrical sedimentary cycles might form in a tidal setting when coarser sediment was coated by algae and mud during neaps, thus raising the threshold current speed for movement of sediment in the neap to the spring part of the tidal cycle (see de Boer, 1981; Allen, 1982a).

The Reynella Siltstone at Hallett Cove also displays a variety of other flat-bedded facies, including sandstones and siltstones with rare beds of coarse grained to granule grade sand (Fig. 4a), and graded laminae and beds of fine grained sandstone and mudstone showing delicate flame structures (Fig. 4b), crosslaminae, irregular internal laminae, and cut-and-fill structures (Fig. 4c, d). The detailed stratigraphic relationships of these facies to the cyclic rhythmites is not clear and understanding must await drilling of the sequence. In addition, herringbone crossbedding, wavy bedding and flaser bedding (Fig. 4e, f) occur in sandstones stratigraphically above the rhythmites.

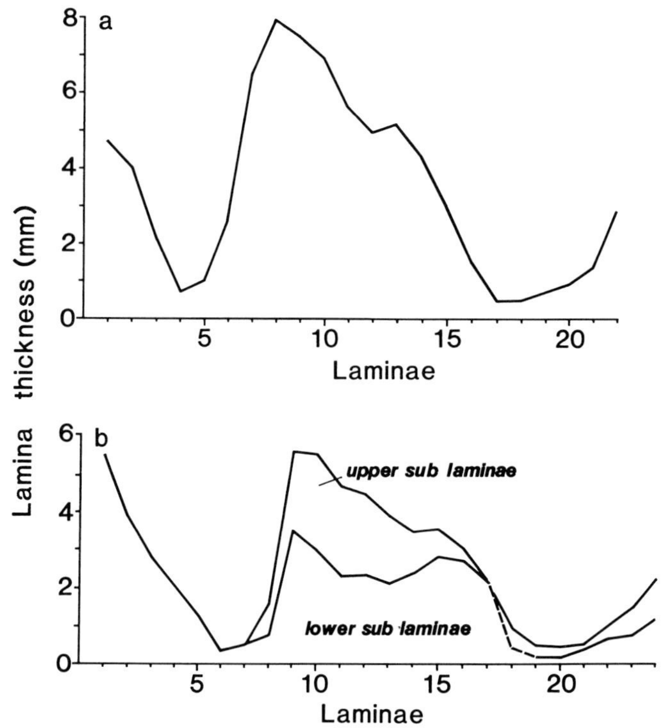

Figure 3. Thickness of diurnal laminae and semidiurnal sublaminae for lamina-cycles from the Reynella Siltstone (lamina number increases upsequence). **a.** Lamina-cycle shown in Figure 2b; **b.** lamina-cycle shown in Figure 2c. Reproduced with permission of the Geological Society, London.

Elatina Formation

The tidal rhythmites in the Elatina Formation at Pichi Richi Pass (Fig. 5) form an approximately 10 m thick member, representing more than 60 years of continuous deposition, within red-brown, crosslaminated, fine to medium grained sandstones. The contacts of the rhythmite member with the sandstones are gradational. Very thin laminae at the base of the member grade downward into unlaminated clayey, fine grained sandstone. The rhythmites are more sandy at the very top of the member and grade into the overlying sandstones over 20 cm.

Because the outcrop is limited, drill core of the rhythmite member was obtained from three vertical holes drilled up to 200 m apart. Throughout its full thickness the rhythmite member displays conspicuous lamina-cycles, interpreted as fortnightly groupings of laminae, up to 2 cm thick; the alternation of thick and thin lamina-cycles (Fig. 5b, c), reflecting the monthly inequality in the amplitude of spring tidal cycles, is characteristic. Very thin (0.5 to 3mm) lamina-cycles (Fig. 5d) are restricted to the base of the member.

The clastic laminae typically comprise a lower, paler, sandy layer and an upper, silty layer. The contact between such layers usually is gradational, whereas contacts between adjacent laminae are relatively sharp. Microscopically, the clastic laminae consist mainly of grains of quartz (dominant) and feldspar in a clay rich matrix. Grading results from an

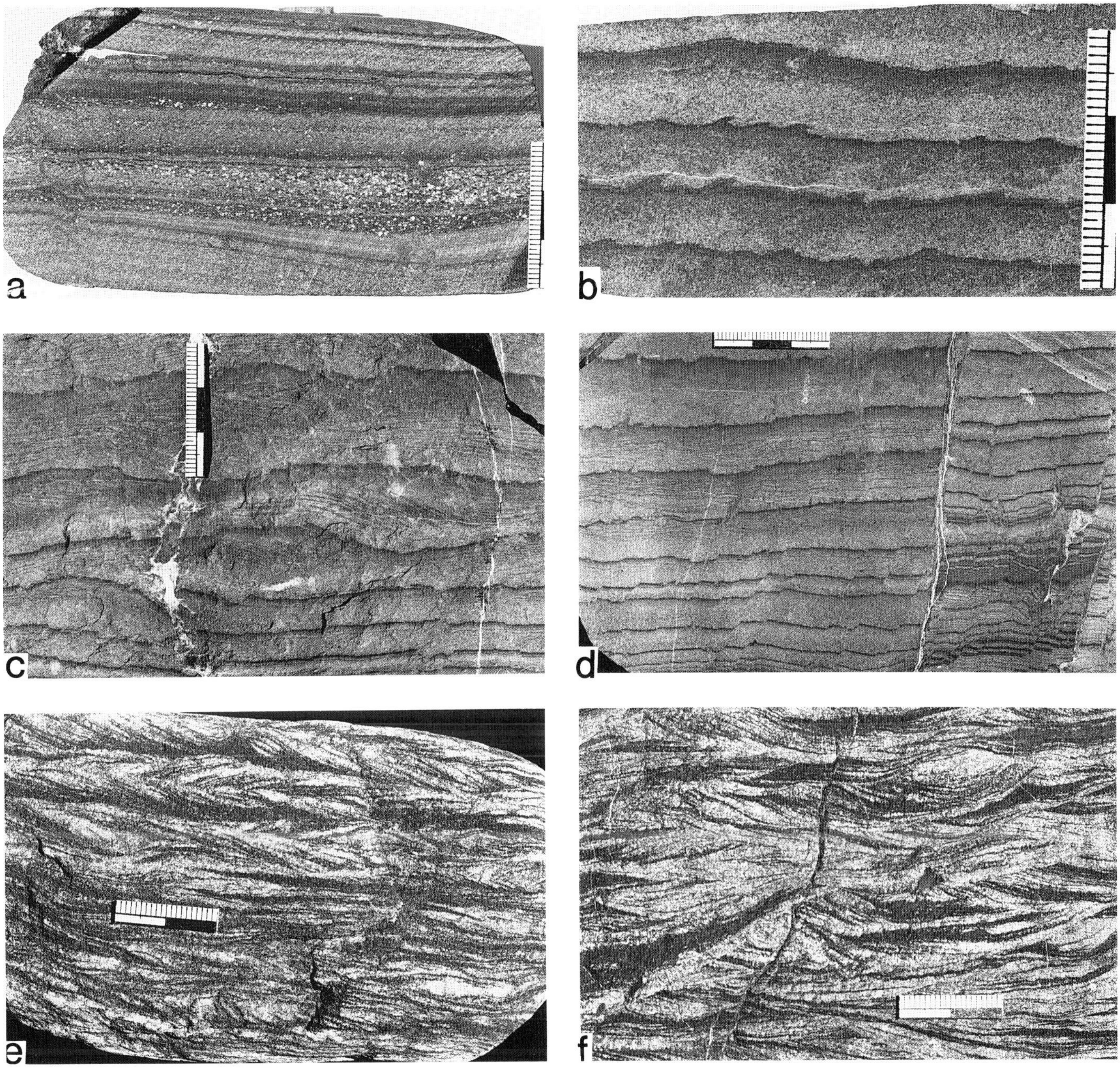

Figure 4. Flat bedded and crossbedded facies from the Reynella Siltstone. Scales graduated in millimetres and centimetres. **a.** Flat bedded siltstone and fine to coarse grained sandstone. **b.** Graded laminae of fine grained sandstone and mudstone showing delicate flame structures. **c.** and **d.** Graded laminae and thin beds of fine grained sandstone and mudstone showing internal laminae and crossbedding, and local cut-and-fill structure; **d** overlies **c. e.** Herringbone crossbedding and wavy bedding in fine grained sandstone and sandy mudstone. **f.** Flaser bedding (Reineck and Wunderlich, 1968), comprising crosslaminated, fine grained sandstone with flasers of mudstone in troughs.

upward reduction in grain size from 20 to 100 μm (silt to a very fine sand grade) at the base to 20 to 50 μm (silt) at the top of each lamina and an accompanying upward increase in the proportion of matrix.

The constituent laminae are regarded as diurnal increments except in rare instances where the presence of sublaminae of unequal thickness (Figs. 5e, 6) indicates semidiurnal increments and a mixed tidal pattern. Apparently the sedimentary system usually filtered out the lesser semidiurnal constituent of the diurnal inequality (Williams, 1989c). It would seem that the tidal ranges and current velocities associated with the lesser semidiurnal constituent usually were too small for sediment to be carried to the site of deposition at Pichi Richi Pass. Such filtering out of high frequency tidal signals in other tidal deposits is discussed by de Boer *et al.* (1989).

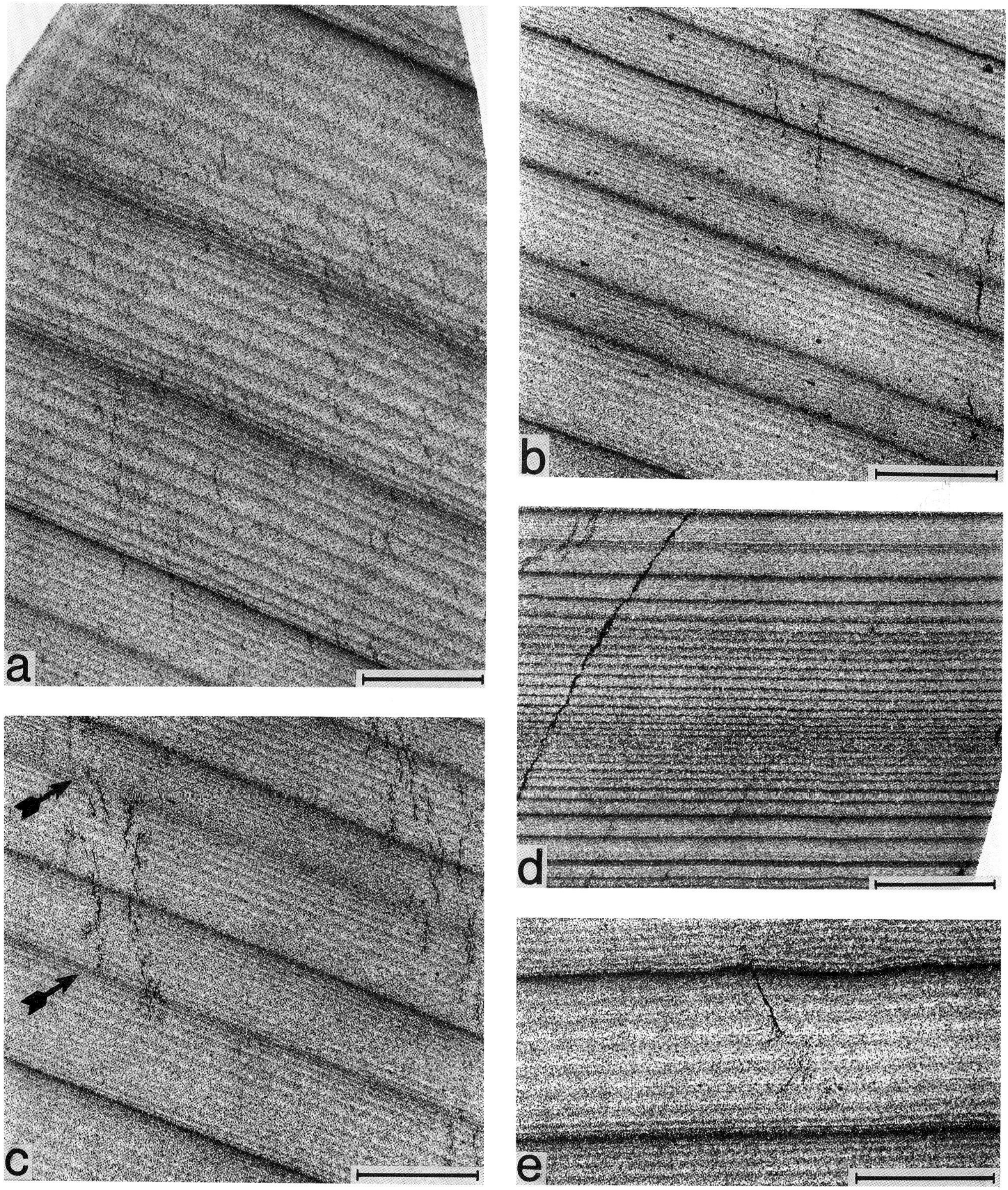

Figure 5. Thin sections of rhythmites from the Elatina Formation, Pichi Richi Pass, viewed with transmitted light; opaque clayey material appears darker than translucent sandy layers. The specimens are from drill core. Scale-bars 1 cm. **a.** Graded (upward-fining) diurnal laminae up to 3 mm thick comprising a lower, paler layer of very fine grained sandstone and an upper layer of clayey siltstone; four complete fortnightly lamina-cycles of about 10 to 14 laminae are bounded by conspicuous, dark, mud drapes. **b.** Thinner laminae and lamina-cycles, with alternate thick and thin lamina-cycles indicating the monthly inequality. **c.** Alternate boundaries between lamina-cycles represented by very thin silty laminae (arrows) rather than mud drapes; the lamina-cycles appear unabbreviated at these horizons. **d.** Very thin (0.5 to 3 mm) fortnightly lamina-cycles near the base of the rhythmite member; internal lamination between mud drapes is only discernible in a few places. **e.** Rare lamina-cycle that comprises laminae as well as sublaminae (best seen looking along the bedding) that are interpreted as semidiurnal increments.

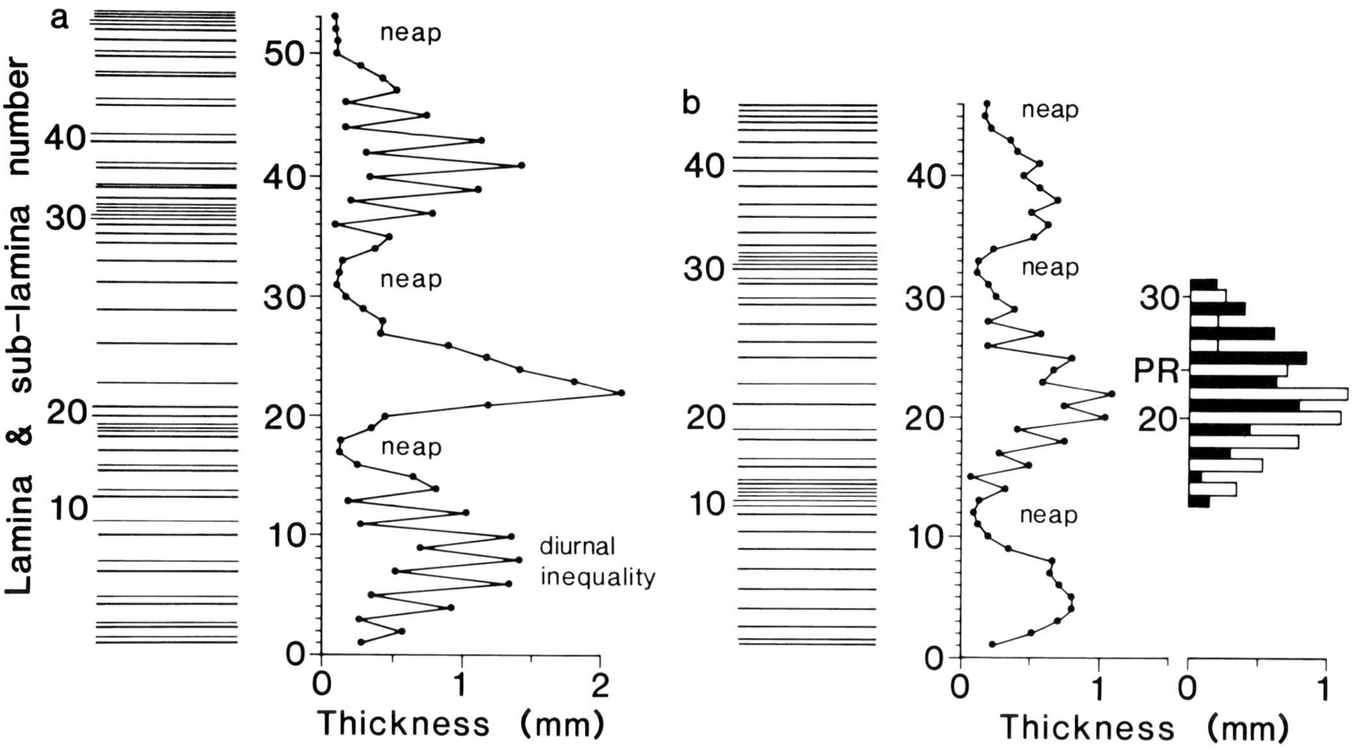

Figure 6. Two short sequences from the rhythmite member of the Elatina Formation that locally show a sawtooth pattern of alternate thick and thin semidiurnal sublaminae reflecting the diurnal inequality. A 180° phase-reversal (PR) in the sawtooth pattern, characteristic of the diurnal inequality, is shown in **b**. Lamina number increases upsequence.

The lamina-cycles contain from 8 to 16 laminae (average = 12) that are interpreted as diurnal increments, indicating that many lamina-cycles are abbreviated though nondeposition of clastic laminae at neaps, when muddrapes were laid down. At regular vertical intervals, however, alternate mud drapes between lamina-cycles are not well formed and are represented by very thin, silty laminae (Fig. 5c); the lamina-cycles appear largely unabbreviated at these horizons. The periodicity and origin of this feature is discussed in the second last section of this paper.

Laminae and lamina-cycles typically are planar and can be traced laterally over hundreds of metres within the limits of outcrop and between drill holes. This planarity is locally disturbed by small-scale ripple structures and soft-sediment deformation features. Isolated small-scale ripples that preserve the mud drapes between lamina-cycles (Fig. 7a) invariably indicate paleocurrents directed eastward, away from the presumed paleoshoreline along the nearby western margin of the Adelaide Geosyncline. Rare, very small-scale cross-laminae that occur in sets no more than about 1.5 mm thick (Fig. 7b) record weak, variable bottom currents. Elsewhere, rare undulations in laminae are truncated by succeeding laminae (Fig. 7c). In contrast to the rhythmites, small-scale crossbedding is common in the contiguous sandstone facies (Fig. 7d).

Local gravity slides are identifiable in the drill core. Because precise correlations of lamina-cycles and laminae were obtained between cores from the three drill holes, it was possible to identify where several isolated lamina-cycles were missing in the sequence from one western drill hole. In the core from another drill hole about 50 m due east, an unlaminated graded bed a few centimetres thick occurred as an additional increment at the exact level where a lamina-cycle was missing in the sequence from the western hole. Evidently, sporadic slumping of individual lamina-cycles, presumably lubricated by the immediately underlying mud drape, took place down an east-facing paleoslope soon after their deposition.

The rhythmites at Pichi Richi Pass also display cuspate structures that involve as may as 28 or more superimposed lamina-cycles (Fig. 8) and in plan have the appearance of near-parallel ripples spaced decimetres apart. The structures are unrelated to local tectonic fold trends. The general cuspate form of these structures bears some similarity to wave-oscillation ripples that occur in tidal rhythmites of the Eocene Elkton Formation in Oregon, U.S.A. (Leithold *et al.,* 1989). Unlike those of the Elkton Formation, however, the structures do not drape wave-rippled sandstone. The cuspate structures in the Elatina Formation are best attributed to soft-sediment deformation, perhaps gravity slumping on a sloping sea floor.

Chambers Bluff Tillite

The rhythmites overlying the Chambers Bluff Tillite (Fig. 9) also show an overall regular change in thickness of laminae and a pattern of alternate thick and thin lamina-cycles.

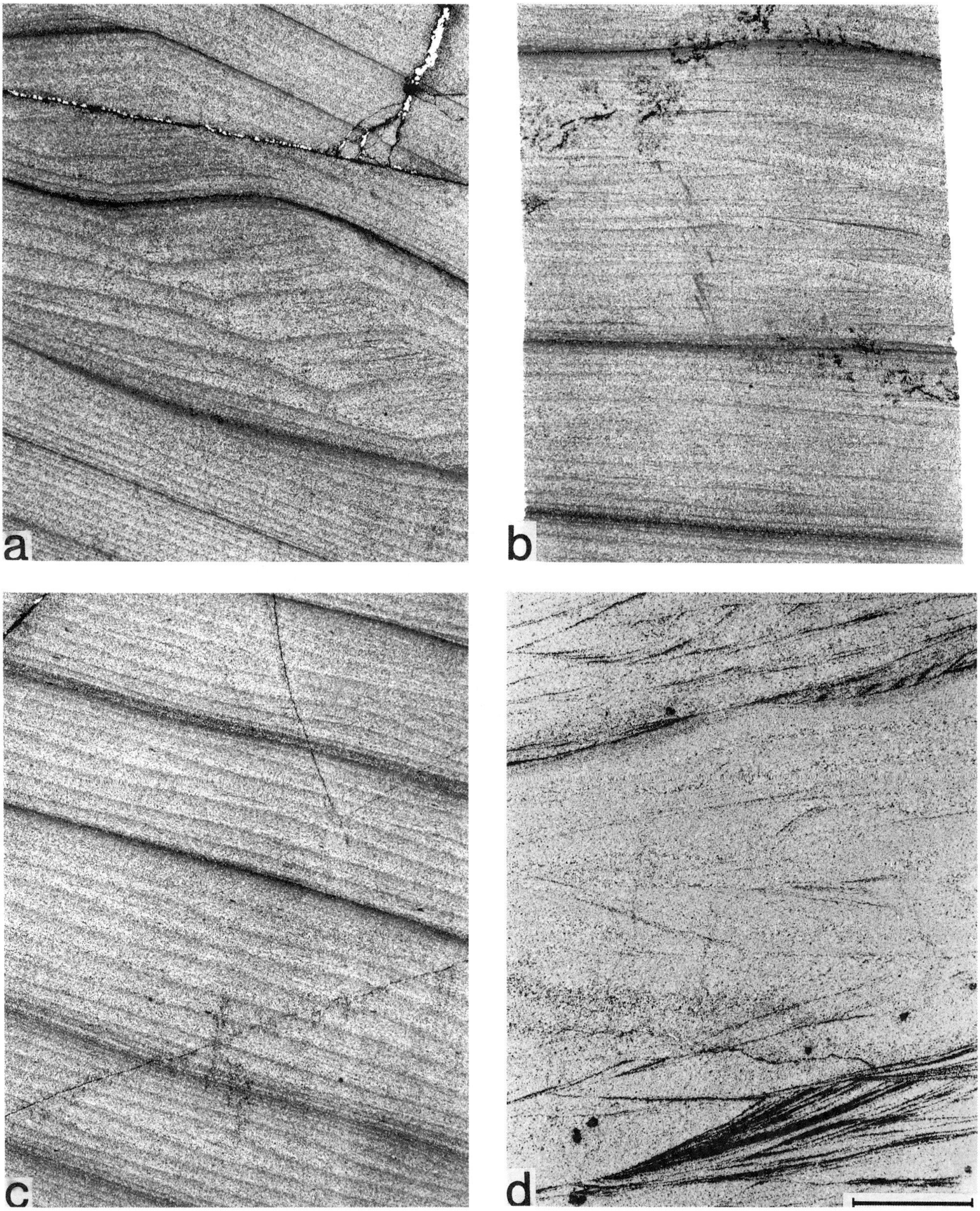

Figure 7. Thin sections of the Elatina Formation, viewed with transmitted light. Scale bar 1 cm for all photographs. **a.** Fortnightly lamina-cycles of diurnal laminae that pass laterally into isolated, small-scale ripples with mud drapes. **b.** Rare, very small-scale crosslaminae in opposed sets no more than about 1.5 mm thick within lamina-cycles. **c.** Rare undulations in diurnal laminae, some truncated by succeeding laminae. **d.** Crosslaminated, fine grained sandstone that overlies the rhythmite member; the bedding structures are marked by opaque heavy minerals, mainly hematite.

Figure 8. Rhythmites of the Elatina Formation, showing a cuspate structure of probable soft-sediment origin that involves at least 28 successive fortnightly lamina-cycles. Scale 3 cm long.

The tendency for relatively thick and thin laminae to alter nate (Fig. 10) suggests the influence of a weak diurnal inequality on deposition, and the pattern of lamina-cycle thickness the influence of the monthly inequality in spring-tidal amplitudes.

Synsedimentary scouring occurs at the tops of the lamina-cycles, and the thinner laminae in such places are commonly truncated. In a measured sequence of nine lamina-cycles the number of laminae per cycle ranges from 15 to 25, averaging 20.0. Because of the scouring, this mean must be viewed as a minimum estimate of cycle period. In view of the number of laminae per cycle, and the apparent signature of the diurnal inequality, the laminae are interpreted as semidiurnal increments. A dominantly semidiurnal tidal pattern prevailed during rhythmite deposition.

Depositional Environment of the Rhythmites

Any tidal facies model proposed for deposition of the rhythmites in the Elatina Formation must accommodate: 1) the lateral persistence of graded laminae and mud drapes; 2) the movement of sediment by unidirectional paleocurrents directed eastward away from the presumed paleoshoreline, down an east-facing paleoslope; and 3) the lack of reworking of the laminae by wave action. As discussed by Williams (1988, 1989a-c), these observations and the cyclicity displayed by the rhythmites can be explained by deposition from turbid ebb-tidal currents in a distal, ebb-tidal setting (Fig. 11). Thus, fine grained sediment is entrained by ebb-tidal currents in a tidal inlet and is transported mainly in suspension by strong, turbulent ebb-tidal currents or "jets" to deeper water offshore (see Özsoy, 1986). There, suspended sediment would settle to form graded laminae, as well as local, aggrading small-scale ripples directed offshore, in a distal ebb-tidal delta (delta front) setting below effective wave base.

Variation in thickness of semidiurnal and diurnal laminae and of fortnightly lamina-cycles in taken to reflect

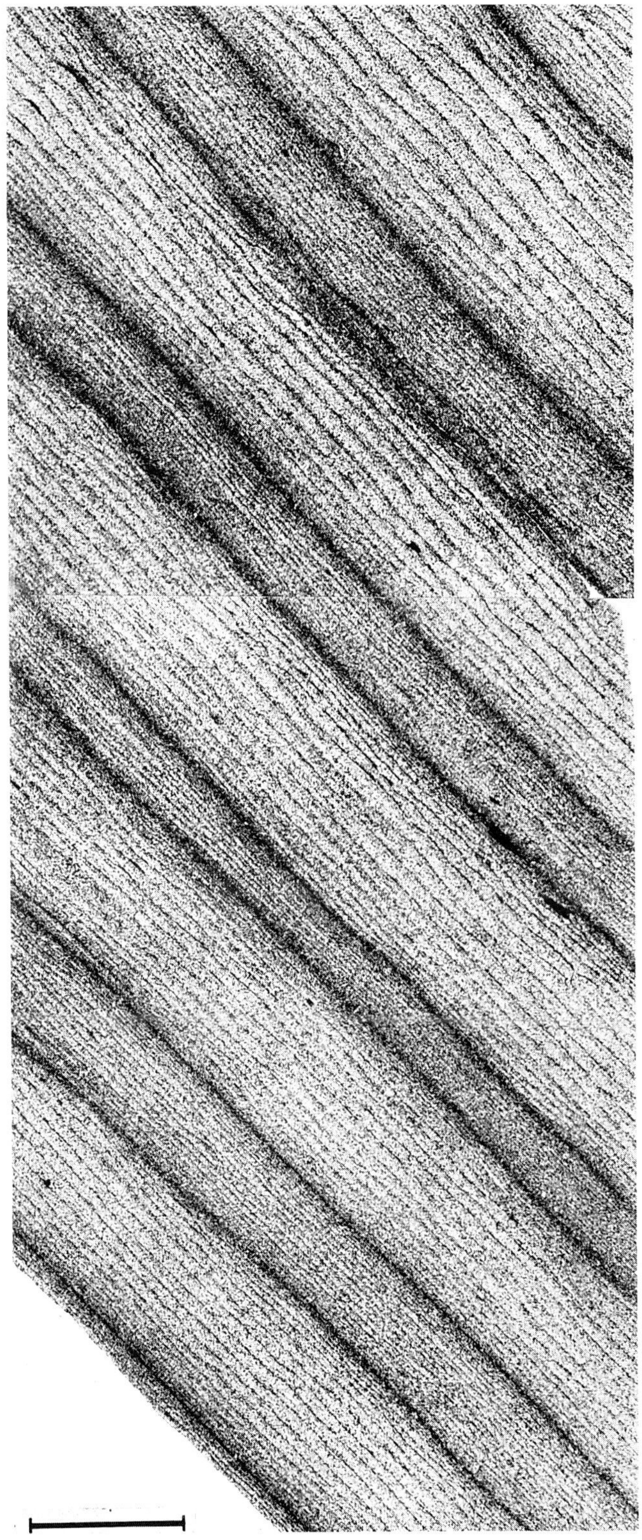

Figure 9. Thin section of rhythmite overlying the Chambers Bluff Tillite (Nicholson No. 2 drill hole), viewed with transmitted light; opaque, clayey material appears darker than translucent sandy layers. The specimen shows true dip; scale-bar 1 cm. Graded laminae of very fine grained sandstone and clayey siltstone are grouped in nine alternately thick and thin fortnightly lamina-cycles containing from 15 to 25 semidiurnal laminae. Laminae are eroded in places below the dark mud drapes bounding the cycles. Reproduced with permission of the Geological Society, London.

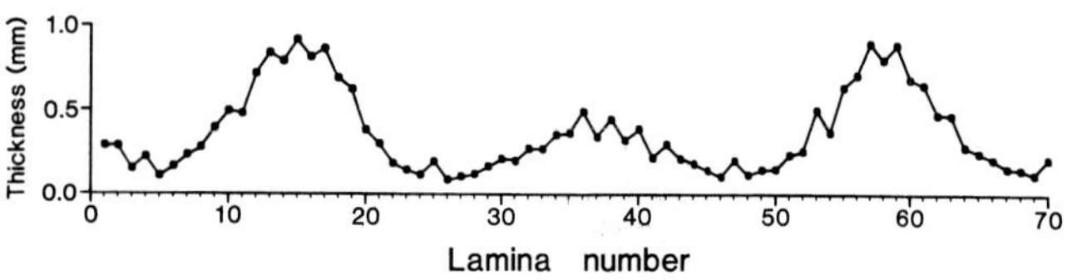

Figure 10. Thickness of laminae for the three lowermost lamina-cycles from the rhythmites overlying the Chambers Bluff Tillite shown in Figure 9 (lamina number increases upsequence). The laminae are interpreted as semidiurnal increments; the local sawtooth pattern reflects a weak diurnal inequality. Reproduced with permission of the Geological Society, London.

variation in the velocity of the ebb-tide and in tidal range. The patterns for thickness variations of diurnal laminae and of fortnightly lamina-cycles from the Elatina rhythmites (Figs. 12, 13) are indeed remarkably similar to maximum daily and maximum fortnightly tidal heights, respectively, for the mixed, synodic tides at Townsville, Queensland. (Maximum tidal heights are plotted in Figs. 12 and 13, as they are a relative measure of tidal range.) The similarities imply a direct relationship between paleotidal parameters (range, current velocity) and lamina and lamina-cycle thicknesses. Both these tidal parameters (and by implication, tidal height) are directly related to potential sediment movement on modern ebb-tidal deltas (FitzGerald and Nummedal, 1983).

Close inspection of the aggrading ripples (Fig. 7a) throws light on the mechanism of deposition of the rhythmites. The relatively thick laminae of very fine grained sandstone, which show thicker foresets containing sublaminae, distinct breaks of slope at the top and bottom of foresets, and shallow troughs at the bases of foresets suggestive of small lee-side eddies, were deposited from turbid, vertically-aggrading currents that reached the sea floor. Rees (1966) demonstrated experimentally that a ripple system can develop from bottom currents carrying an excess of very fine grained material in suspension. By contrast, the thinnest sandy laminae and the mud layers at lamina-cycle boundaries uniformly drape the thicker foresets, indicating that the turbid currents then did not reach the sea floor and that deposition occurred solely by vertical settling from suspension. Thus, at the site of Pichi Richi Pass, the turbid ebb-tidal currents expanded distally and reached the sea floor during spring ebb tides, and contracted to plumes above the sea floor around neap ebb tides.

As the settling velocity of very fine grained sand is only about 1 cm/sec (Allen, 1982b, vol. II, p. 312), turbulent ebb-tidal currents would have readily transported in suspension the very fine grained sand, silt and clay constituting the rhythmites of the Elatina Formation. Several factors would have enhanced the capacity of ebb-tidal currents to carry

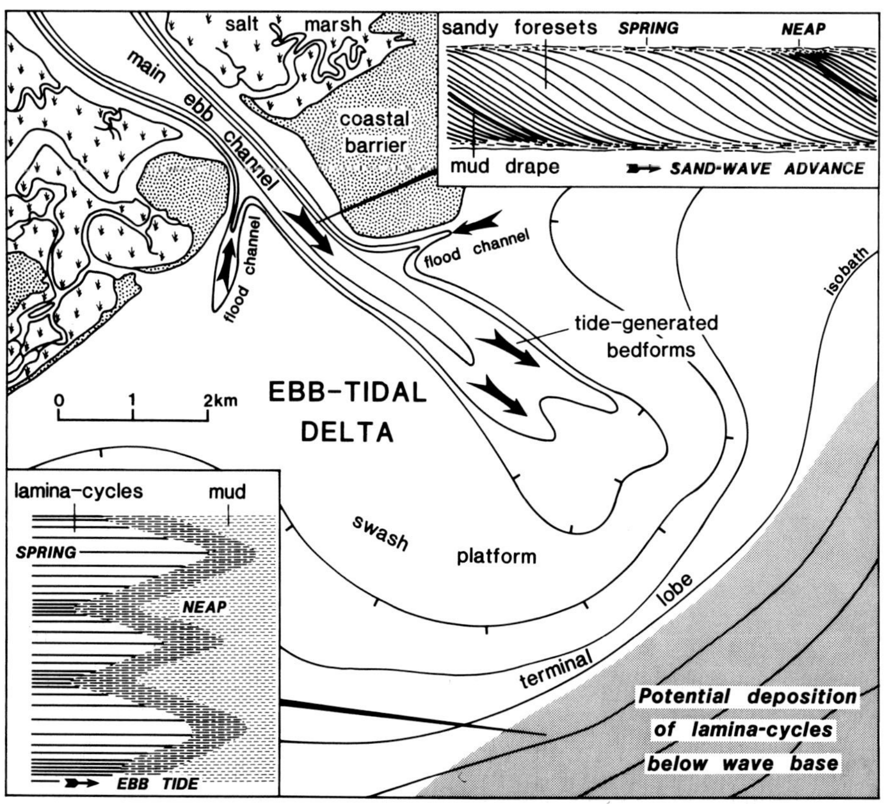

Figure 11. Postulated environment of deposition for lamina-cycles of the Elatina Formation, employing a hypothetical ebb-tidal delta modified from Imperato *et al.* (1988). Tidal "bundle" sequences of crossbedded sand (top inset) related to bars and sand waves occur in proximal tidal channels, whereas lamina-cycles of very fine grained sand and silt (shown schematically in bottom inset) are deposited from suspension in a distal, offshore setting below wave base. Distally, lamina-cycles become progressively more abbreviated at neaps and eventually pass into shelf muds. Reproduced with permission of the editor of Episodes.

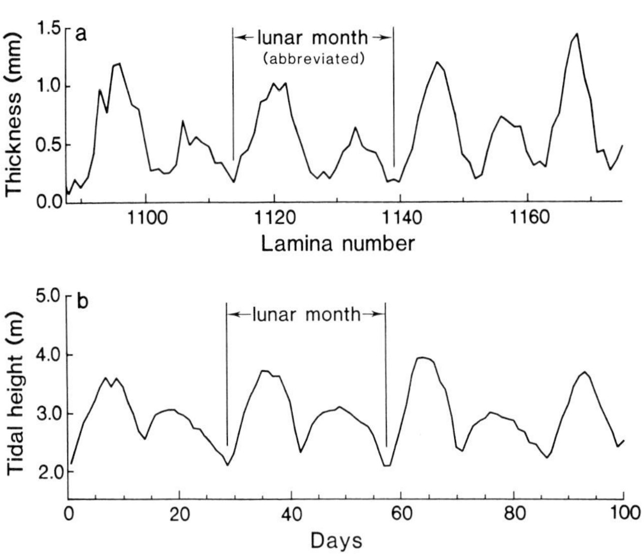

Figure 12. a. Thickness of diurnal laminae from the Elatina rhythmites, showing lamina-cycles of alternate high and low amplitude. Lamina number increases upsequence. **b.** Maximum daily tidal heights from January 1 to April 10, 1966, for Townsville, Queensland, showing fortnightly tidal cycles of alternate high and low amplitude. Pairs of successive cycles in each curve represent the lunar month, although the lamina-cyles are abbreviated at thickness minima. (Tidal data for Figures 12b, 13d, e, f and 14b supplied by the Tidal Laboratory of the Flinders Institute for Atmospheric and Marine Sciences, Flinders University of South Australia, copyright reserved.) Reproduced with permission of the editor of Episodes.

material in suspension to significant distances (kilometres) offshore. Given the prevailing glacial conditions, the sea water was likely very cold and its resulting high viscosity (see Allen, 1982b, vol. I, p. 2) would have assisted suspension transport. The arid periglacial setting in the hinterland (Williams, 1986; Williams and Tonkin, 1985) would have produced an abundance of fine grained clastic material that could have been transported in suspension. In addition, tidal ranges and the velocity of tidal currents may have been amplified by tidal resonances within the gulf-like Adelaide Geosyncline.

Remarkably, the rhythmites of the Elatina Formation, representing 60 to 70 years of continuous deposition (see following section), show no evidence of disturbance by wave activity. The site of Pichi Richi Pass may then have been an area of low wave energy, lying within the gulf-like Adelaide Geosyncline and probably sheltered from long-period waves (although not so restricted a setting that the semidiurnal tidal constituents were entirely filtered out). In addition, the water may have deepened rapidly beyond the terminal lobe of the ebb-tidal delta (Fig. 11) to the requisite depth for no wave disturbance; relatively rapid deepening is plausible, as the site lay near the fault-bounded western margin of the Adelaide Geosyncline.

The virtually noise-free character of the Elatina rhythmite data (see following section, and Williams, 1988, 1989a-c) indicates that fluvial of other random influences

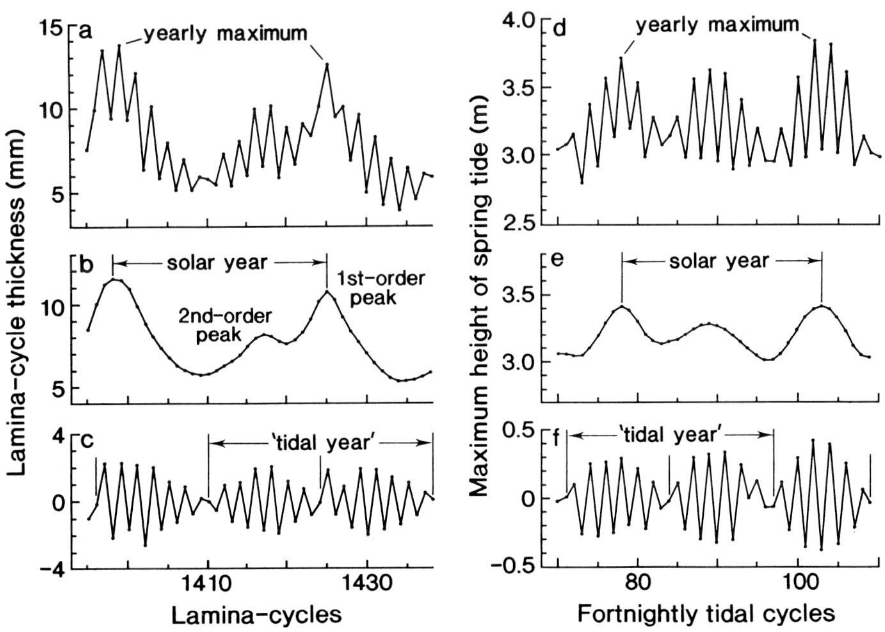

Figure 13. a-c. Thickness of fortnightly lamina-cycles from the Elatina series (lamina-cycle number increases upsequence). **a.** Unsmoothed curve; first-order peaks define yearly maxima in thickness. **b.** Smoothed curve, showing the solar year. **c.** Residual curve (**a** minus **b**) showing the "tidal year"; the vertical lines mark 180° phase-reversals in the sawtooth pattern. **d-f.** Tidal patterns for Townsville. **d.** Maximum height of the fortnightly tidal cycle from October 19, 1968, to June 3, 1970. **e.** Smoothed curve, showing the solar year. **f.** Residual curve (**d** minus **e**) showing the "tidal year"; the vertical lines mark 180° phase-reversals in the sawtooth pattern.

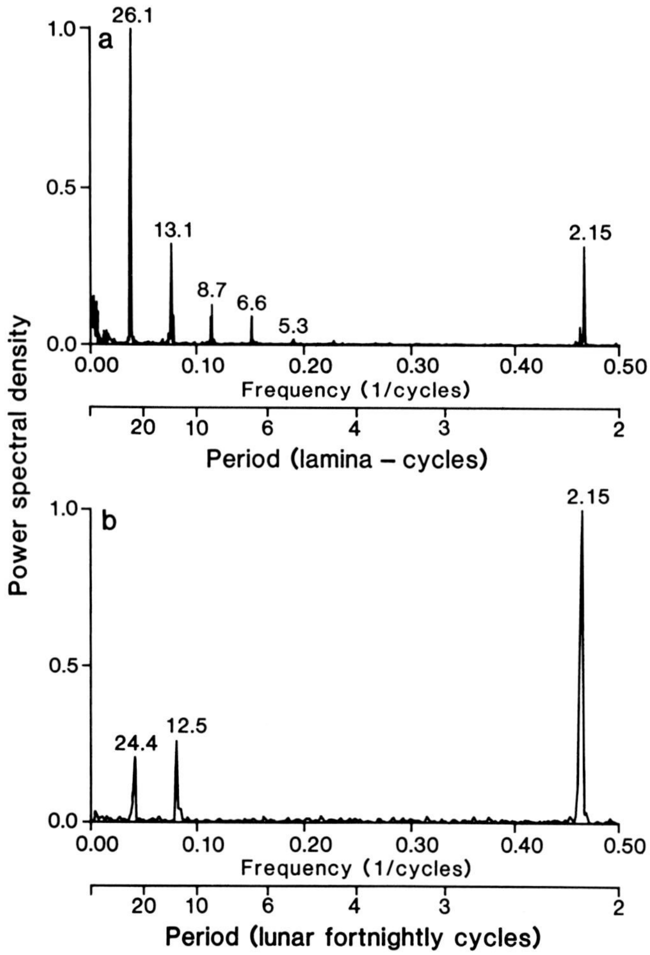

Figure 14. Fast Fourier transform smoothed spectra, normalized to unity for the strongest peak in each spectrum and with linear frequency scales. **a.** Spectrum for the Elatina sequence of 1580 fortnightly lamina-cycle thickness measurements. The strong period of 26.1 lamina-cycles represents a yearly signal, with additional harmonics at 13.1 (half-year), 8.7, 6.6 and 5.3 lamina-cycles. The peak near 2 lamina-cycles reflects the monthly inequality of alternate thick and thin fortnightly lamina-cycles. **b.** Spectrum for the maximum heights of 495 spring tides between January 1, 1966, and December 31, 1985, for Townsville, Queensland. The periods of 24.4 and 12.5 fortnightly cycles represent yearly and half-yearly signals. The peak near 2 fortnightly cycles reflects the monthly inequality of alternate high and low spring tides.

on rhythmite deposition were negligible; tidal influences overwhelmingly prevailed. This conclusion is supported by the nature of the adjacent hinterland during deposition of the Elatina Formation — an arid, periglacial land surface blanketed by wind-blown sands and with no perennial rivers (Williams, 1986; Williams and Tonkin, 1985). The dominance of tidal processes also may imply that deposition of the rhythmites was related to an episode of relative sea level rise, when tide-dominated facies form and are preferentially preserved (Nio and Yang, 1989).

It is noteworthy that the duration of rhythmite deposition at Pichi Richi Pass is almost exactly the period of a geomorphological cycle of about 70 years that regulates the deposition of ebb-tidal delta shoals seaward of Texel Inlet, The Netherlands (Sha, 1989). The tendency for an ebb-tidal delta to migrate laterally with its tidal inlet (Hayes, 1980) may restrict the continuous deposition of fine grained clastic material in a distal ebb-tidal setting to a relatively thin package at a particular site; the approximately 10 m thick rhythmite member of the Elatina Formation at Pichi Richi Pass may exemplify such a package. By contrast, thicker delta sequences could be expected to form in largely prograding, Fluvially influenced tidal deltas.

Overall, the envisaged paleoenvironment for the rhythmites at Pichi Richi Pass involves deposition from suspension on the seaward slope of an ebb-tidal delta on which mass movement of sediment occurred sporadically. A distal tidal setting seems appropriate also for the delicate rhythmites overlying the Chambers Bluff Tillite.

The Reynella Siltstone at Hallet Cove, containing some relatively thick, unabbreviated lamina-cycles (Fig. 2b, c), thinner, apparently more distal lamina-cycles (Fig. 2a), and a variety of continental to tidal flat and marginal-marine facies, indicates a more variable and at times a more proximal setting than that envisaged for the Elatina rhythmites. The Reynella rhythmites may have accumulated on upper and lower delta slopes and in estuarine settings. The sandstone-mudstone couplets in flat bedded facies from the Reynella Siltstone (Fig. 4b-d) are, with their load casts, flame structures, and lenticular, wavy and rippled bedding, similar to modern tidal bedding in the Cobequid Bay-Salmon River estuary, Bay of Fundy (Dalrymple and Makino, 1989). The variability of tidal facies at Hallett Cove may reflect an influence of fluvial processes on sedimentation. The envisaged environment has some similarity to that proposed for deposition of tidal rhythmites in the Eocene Elkton Formation, southwest Oregon (Leithold *et al.,* 1989).

Late Proterozoic Paleotidal and Paleorotational Periods

To determine the full complement of periods recorded by the Elatina rhythmites, two stratigraphic series ("time series") of measurements of lamina and lamina-cycle thicknesses were obtained from the drill core from Pichi Richi Pass.

A piece of core 93 cm long that contained well preserved laminae was selected for the lamina-thickness series. Overlapping thin sections were cut and a 4.4× photographic enlargement made of each. The boundaries of laminae were then carefully ruled on the photographs with a very fine pen; such marking proved necessary because the granularity of the rocks made the boundaries appear diffuse when viewed through the eyepiece of the measurer. Boundaries appeared sharper and easier to mark by sighting along the laminae on the photographs. Some very thin laminae, possible random layers, and rare soft-sediment deformation made counting uncertain in places, and the count of 1337 measurements has an estimated accuracy of ±5 per cent.

To obtain the stratigraphic series of lamina-cycle thickness measurements, strips of clear adhesive tape were first placed on the drill cores and the true thickness of lamina-cycles marked on the tape at the positions of the conspicuous mud drapes. The strips of tape were then transferred to transparent sheets to build up a sequence of lamina-cycle thicknesses. Overlaying the completed logs from the three drill holes allowed correlation between holes and the bridging of gaps caused by core breakage. The end result was a stratigraphically continuous log, 9.39 m long and comprising 1580 lamina-cycles, that could be measured precisely. Measurements for both stratigraphic series were made with a precision of 0.01 mm using the increment measurer at the Laboratory of Tree-ring Research, University of Arizona, Tucson. Extracts from the two stratigraphic series are shown in Figures 12 and 13, together with comparable plots for modern synodic tides.

As previously discussed (Williams, 1988, 1989a-c), time-series analysis of the Elatina stratigraphic series, particularly the series of lamina-cycle thickness, and study of the rhythmites from the Reynella Siltstone, have provided a full spectrum of paleotidal and paleorotational periods for ~650 Ma (Table 1). The derivation of these periods and the recognition of tidal type for the Elatina and Reynella paleotides are discussed briefly below.

The estimated number of lunar days per lunar month was based largely on observations of the Reynella Siltstone. Diurnal laminae and semidiurnal sublaminae are readily distinguishable in the thick lamina-cycles from this formation (Figs. 2, 3). Accurate counts of identifiable diurnal laminae in apparently unabbreviated lamina-cycles were employed, and the values of 14 to 15 diurnal laminae per lunar fortnightly cycle agreed well with the maximum value of 29 identifiable diurnal laminae in two successive lamina-cycles, representing the lunar monthly cycle (Fig. 12a), in the Elatina series. Maximum counts, taking due care to distinguish diurnal laminae from semidiurnal sublaminae, must of course be employed for the Elatina Formation because of the common abbreviation of lamina-cycles at neaps. These figures suggest 29 to 30 lunar days per lunar month and in turn imply around 30 to 31 solar days per lunar month (employing the present ratio of lunar-day to solar-day durations of 24.8/24 hours = 1.03).

A yearly, non-tidal signal, probably reflecting seasonal changes in sea level and weather, is evident in the Elatina rhythmites through periodic variation in the thickness of lamina-cycles (Fig. 13a). The repetition of this annual cycle indicates that the measured rhythmite sequence spans about 60 years of continuous deposition, and the full rhythmite member between 60 and 70 years. The markedly periodic, virtually noise-free nature of these data is demonstrated by the strong, narrow spectral lines obtained for the yearly, half-yearly and monthly signals (Fig. 14a). The spectrum for the maximum heights of spring tides at Townsville (Fig. 14b) likewise shows clear yearly, half-yearly and monthly peaks. The Late Proterozoic year contained 26.1 or 26.2 (depending on the method of spectral analysis; Williams, 1989a, b) ±0.2 lunar fortnightly cycles and 13.1±0.1 lunar (synodic) monthly cycles. Hence, at ~650 Ma there were 400±7 solar days per year and 21.9±0.4 hours per day.

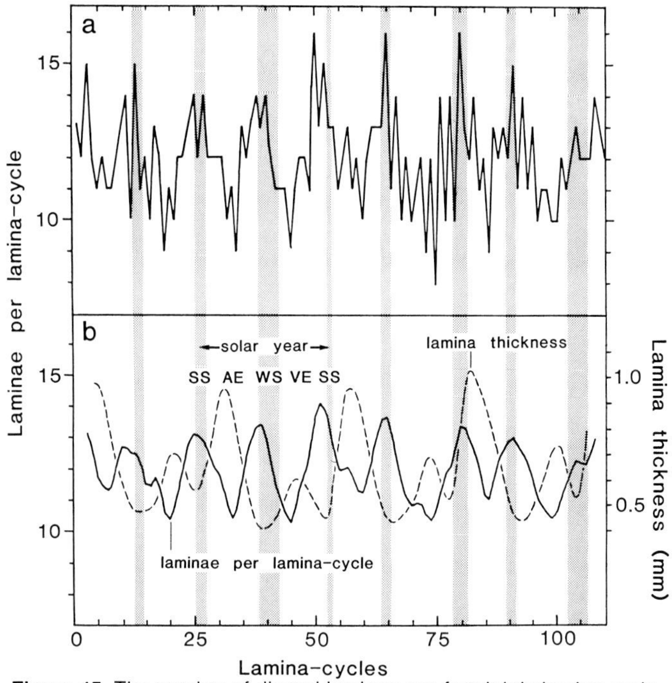

Figure 15. The number of diurnal laminae per fortnightly lamina-cycle counted between lamina thickness minima, for 110 lamina-cycles from the Elatina rhythmites (lamina-cycle number increases upsequence). **a.** Unsmoothed curve. **b.** Smoothed curve (5-point filter weighted, 1,4,6,4,1), revealing a strong modulation of lamina-cycle counts by a half-yearly period (about 13 lamina-cycles); the dashed curve shows lamina thickness for the same interval (smoothed by 111-point filter). The shaded bands mark the stratigraphic positions of lamina-cycles bounded by very thin silty laminae rather than mud drapes (Fig. 5c). SS, summer solstice; AE, autumnal equinox; WS, winter solstice; VE, vernal equinox. See text for full explanation.

The second harmonic of the yearly signal (Fig. 14a) has a period of 13.1 fortnightly cycles, and represents the half-yearly tidal cycle. This oscillation, through its modulation of tidal range, influenced neap-tidal pauses in the deposition of diurnal clastic laminae in the Elatina Formation, thereby modulating the number of laminae deposited per lamina-cycle (Fig. 15). The solid curve in Figure 15b therefore may be regarded as an approximation to the half-yearly tidal cycle. Significantly, the stratigraphic positions of lamina-cycles that are bounded by very thin silty laminae rather than mud drapes, where lamina-cycles appear to be unabbreviated (Fig. 5c), coincide quite closely with peaks in the number of diurnal laminae per lamina-cycle (Fig. 15b). Evidently the deposition of clastic laminae at the Pichi Richi Pass locality was not interrupted during the half-yearly maxima in tidal range at neaps, thereby causing lamina-cycles with the greatest number of diurnal laminae to be deposited. By comparison with tidal data for Townsville and annual sea level variations (Pariwono *et al.,* 1986; Komar and Enfield, 1987), the half-yearly peaks in lamina number (solid curve in Fig. 15b) would record maximum neap-tidal ranges at solstices, and the high peaks in lamina thickness (dashed curve) high annual sea level in the autumn. The positions of summer and winter solstices and autumnal and vernal equinoxes therefore can be identified in the Elatina data (Fig. 15).

A tendency for fortnightly lamina-cycles comprising relatively thick diurnal laminae to contain fewer laminae than average, and vice versa (Williams, 1988, 1989a, b), can be explained by the moon's more rapid orbital motion at perigee when the tide-generating force is greatest. At such times the spring-tidal cycle displays large tidal ranges but is of relatively short duration.

The synodic character of the Elatina fortnightly pattern and of the Townsville tidal pattern with which it is compared is demonstrated in Figure 16. Modern synodic tides display a characteristic variation in the amplitude of the lunar fortnightly cycle (Figs. 13f, 16b):

1. As discussed earlier, alternation of high- and low-amplitude fortnightly cycles to give a "sawtooth" pattern, or monthly inequality, resulting from the eccentricity of the lunar orbit.

2. Systematic modulation of the amplitude of the sawtooth pattern, with 180° changes of phase (that is, a reversal in the sequence of high- and low-amplitude fortnightly cycles) occurring where the amplitude of the pattern is minimal.

As shown in Figure 16a, b, maximum amplitude of the sawtooth pattern (that is, maximum monthly inequality) occurs when the earth, moon and sun all lie along the major axis of the elliptical lunar orbit, and 180° changes of phase of the sawtooth pattern at minimal monthly inequality when all bodies are aligned perpendicular to the major axis. The mean period for a 360° change of phase of the sawtooth pattern, which may be termed the "tidal year" (Williams 1989b, c) is longer than the solar year (13.95 and 12.37 synodic months, respectively) because the moon's perigee (the minimum distance between the earth and the moon) rotates in space in the same sense as the earth's orbital motion. The same pattern (but with respective periods of 14.6 and 13.1 synodic months; Williams, 1988, 1989a-c) is displayed by the Elatina series, as expressed by variation in the thickness of fortnightly lamina-cycles (Figs. 13c, 16c); the pattern also is shown by variation in the amplitudes of such cycles.

Hence both the semidiurnal (Fig. 6) and fortnightly (Fig. 16) signatures in the Elatina series demonstrate unequivocally the synodic character of the Elatina paleotides. The rhythmites of the Reynella Siltstone and Chambers Bluff Tillite also were deposited under synodic paleotidal regimes.

The lunar apsides cycle, or period of rotation of the lunar perigee (P), is given by the equation $P = Yt/(Yt - Ys)$, where Yt and Ys are the durations of the tidal year and solar years, respectively. The mean periods for the solar and tidal year at ~650 Ma indicate a lunar apsides cycle of 9.7±0.1 years (Williams, 1989a-c).

The period of the lunar nodal cycle — arising from precession of the lunar orbital plane relative to the ecliptic — also is recorded for ~650 Ma by the Elatina series.

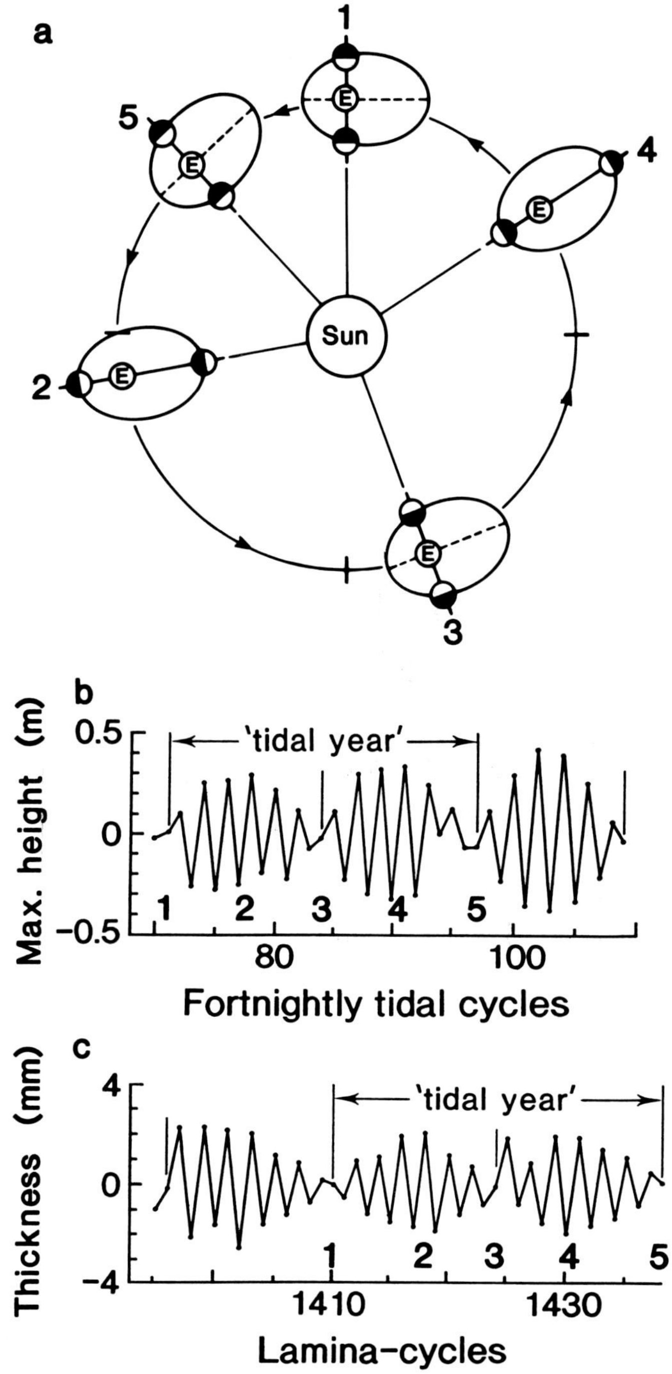

Figure 16. Synodic fortnightly tidal patterns for one "tidal year". **a.** Schematic luni-solar conjunctions: 1, 3 and 5, minimum monthly inequality of spring-tidal amplitudes, and phase reversal in the sawtooth pattern of such amplitudes; 2 and 4, maximum monthly inequality. The "tidal year" (1-5) is longer than the solar year because of prograde rotation of the lunar perigee. **b.** Relative amplitude (maximum height) of spring tides at Townsville over one "tidal year" (see Fig. 13f); nos. 1-5 refer to luni-solar conjunctions shown in **a. c.** Relative thickness of fortnightly lamina-cycles in the Elatina series over one "tidal year" (see Fig. 13c); nos. 1-5 refer to luni-solar conjunctions shown in **a**. The vertical lines in **(b)** and **(c)** mark positions of 180° phase-reversals in the sawtooth patterns.

Amplitude-modulation of the half-yearly paleotidal cycle, as recorded by the second-order peak of the yearly signal (Fig. 13b), indicates a long-term period of 19.5±0.5 years (Williams, 1989a-c). As shown by Williams (1989c), the curve of this modulation is very similar to that of the annual mean tidal range for Boston, which is dominated by the 18.6-year lunar nodal cycle (Kaye and Stuckey, 1973). The period near 19.5 years in the Elatina series is interpreted as that of the paleo-lunar nodal cycle.

Implications for the History of the Earth's Rotation and the Lunar Orbit

The implications of the Upper Proterozoic tidal-rhythmite data for the dynamic evolution of the earth-moon system are discussed by Williams (1989a-c). The paleotidal and paleorotational values provided for Late Proterozoic time (Table 1) are consistent with the hypothesis of lunar tidal friction, and indicate an earth-moon distance of about 58 earth radii (some 96% of the present distance) at ~650 Ma. The implied low mean rate of lunar recession for the past ~650 Ma militates against a close approach of the moon at any time during the Proterozoic.

The indicated value of 13.1±0.1 lunar months per year at ~650 Ma, which is well constrained by the Elatina data, accords with the small decrease in the number of presumed lunar months per year during the course of the Phanerozoic determined from fossil data (Fig. 17). The value of 400±7 solar days per year at ~650 Ma indicated by the rhythmites is consistent with a concomitant secular decrease in the number of days per year, in agreement with the hypothesis of lunar tidal friction.

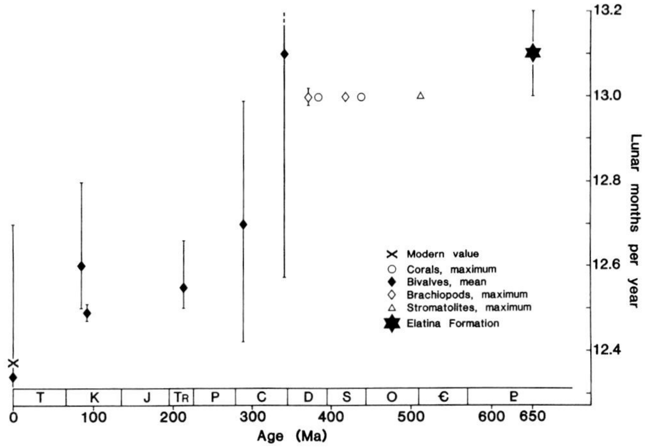

Figure 17. Plot of presumed lunar (synodic) months per year as compiled from paleontological data (Scrutton, 1978; also Berry and Barker, 1968; Mazzullo, 1971; Pannella, 1972), together with the value (with error estimate, see Table 1) indicated by the Upper Proterozoic Elatina Formation.

Table 1. Late Proterozoic (~650 Ma) and modern tidal and rotational parameters

Parameter	Late Proterozoic[1]	Modern
Solar days in lunar month	30.5±0.5	29.53
Lunar months in year	13.1±0.1	12.37
Lunar apsides cycle (years)	9.7±0.1	8.85
Lunar nodal cycle (years)	19.5±0.5	18.61
Solar days in year	400±7	365.2
Length of day (hours)	21.9±0.4	24.0
Earth-moon distance (earth radii)	~58.3	60.27

[1] Parameters indicated by the rhythmites of the Elatina Formation and Reynella Siltstone (Williams, 1987, 1988, 1989a-c); error estimates for all parameters except days/month and nodal cycle are standard deviations based on fast Fourier transform spectra.

The rhythmite value for days per year at ~650 Ma seems at odds, however, with some paleontological data for the lower Paleozoic (Fig. 18). As the number of solar days per year determined from the rhythmite data likely falls within the error limits shown, the early Paleozoic values for days per year must be viewed with caution. Indeed, the available ranges of values for the fossil data are large, two other values are only approximate, and five of the highest values are for maximum rather than mean counts. The remaining plotted bivalve data of Pannella (1972) also must be viewed with reservation, as Pannella (1975) concluded subsequently that a complete restudy of all the paleontological material is needed. Quite possibly, therefore, at least some of the paleontological data plotted in Figure 18 overestimate the number of days per year for the early Paleozoic. Sedimentological studies of Paleozoic tidal rhythmites are required to provide independent values for the number of days and months per year during the Paleozoic.

The rhythmites contain additional information of importance to paleogeophysics. The full spectrum of paleotidal periods recorded indicates the presence of open oceans in Late Proterozoic time. This observation, and the paleorotational data presented in Table 1 and Figures 17 and 18, appears difficult to accommodate in any hypothesis of significant earth expansion during the Phanerozoic. On the other hand, as the diurnal inequality is caused ultimately by the obliquity of the ecliptic, and the half-yearly tidal cycle by variation in solar declination also arising from the obliquity of the ecliptic, the clear signatures of these two tidal rhythms in the Upper Proterozoic rhythmites indicate a significant obliquity at ~650 Ma. The strong yearly, non-tidal signal in the rhythmites (Fig. 14a) accords with a substantial obliquity at that time. Williams (1975) postulated an increased obliquity of the ecliptic during the Late Proterozoic

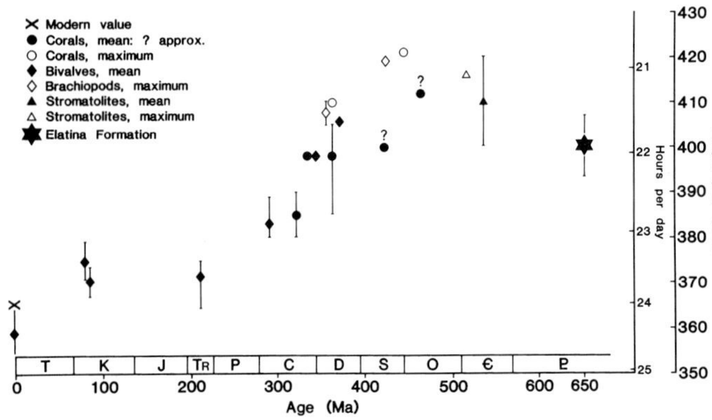

Figure 18. Plot of presumed solar days per year based on paleontological data (Scrutton, 1978; also Berry and Barker, 1968; McGugan, 1968; Wells, 1970; Mazzullo, 1971; Pannella, 1972), together with the value (with error estimate, see Table 1) indicated by the Upper Proterozoic Elatina Formation and Reynella Siltstone.

to explain the indicated strongly seasonal, in situ glacial and periglacial paleoclimate near sea level in low paleolatitudes; the paleogeophysical data provided by the tidal rhythmites of the Elatina Formation and Reynella Siltstone therefore cannot be viewed as inconsistent with that hypothesis.

Conclusions

The recognition of tidal rhythmites, as exemplified by Upper Proterozoic estuarine, upper and lower delta slope, and distal ebb-tidal delta deposits in South Australia, helps complete a "tidal facies model" for littoral deposits. The similarities between the structure and relative periods of cycles in the Upper Proterozoic rhythmites and in modern synodic tidal cycles argue strongly that the rhythmites indeed record a unique set of paleorotational and paleotidal periods for ~650 Ma. As such, the data have rejuvenated geochronometric study of the earth-moon system.

As paleotidal deposits are widely distributed in the stratigraphic record, it may be expected that many additional examples of vertically-accreted rhythmites will be discovered. Geologists accordingly are encouraged to seek comparable tidal rhythmites throughout the stratigraphic record to allow determination of paleotidal and paleorotational data for other geological times. Such studies promise to advance both tidal sedimentology and geochronometry.

Addendum

Since this paper was prepared, F.-L. Deubner, of the Institute of Astronomy and Astrophysics, University of Würzburg, Germany, has applied Kepler's third law and the principles of conservation of angular momentum of the earth-moon system to derive the earth-moon distance (relative to the present distance) at ~650 Ma, using the Late Proterozoic values for lunar months per year and solar days per year given in Table 1 (F.-L. Deubner, Journal of the Geological Society, London, v. 147, p. 1083-1084, 1990). As Deubner points out, his figures of 0.967 ± 0.005 and 0.968 ± 0.007, respectively, agree closely with the value of 0.969 ± 0.017 derived from the observed paleo-lunar nodal period (Williams, 1989a). Deubner states his study "confirms that the relatively low spin-down rate of the Earth inferred from the lamina counts, with the assumption of a tidal origin, is in accordance with a Precambrian semi-diameter of the lunar orbit that is only a few percent less than the present value." The excellent agreement among estimates of the earth-moon distance derived from three such widely separated and independent rhythmite periods lends strong support to the tidal interpretation given here and the accuracy of the Late Proterozoic tidal and rotational parameters listed in Table 1.

Acknowledgments

I thank K. Lambeck (Research School of Earth Sciences, Australian National University), G. W. Lennon and W. Mitchell (Flinders Institute for Atmospheric and Marine Sciences, Flinders University of South Australia), W. V. Preiss (South Australian Department of Mines and Energy), and C. T. Scrutton (Department of Geology, University of Durham) for assistance and helpful discussions, and two referees for constructive comments.

References

Allen, J. R. L. 1982a. Mud drapes in sand-wave deposits: a physical model with application to the Folkestone Beds (Early Cretaceous, southeast England). Philosophical Transactions of the Royal Society, London, A, v. 306, p. 291-345.

______ 1982b. Sedimentary Structures: Their Character and Physical Basis, Vols. I and II. Elsevier, Amsterdam, 593 p. and 663 p.

Baird, G. C., Shabica, C. W., Anderson, J. L. and Richardson, E. S., Jr. 1985. Biota of a Pennsylvanian muddy coast: habitats within the Mazonian delta complex, northeast Illinois. Journal of Paleontology, v. 59, p. 253-281.

Berry, W. B. and Barker, R. M. 1968. Fossil bivalve shells indicate longer month and year in Cretaceous than present. Nature, v. 217, p. 938-939.

Borch, C. C. von der, Grady A. E., Eickhoff, K. H., Dibona, P. and Christie-Blick, N. 1989. Late Proterozoic Patsy Springs Canyon, Adelaide Geosyncline: submarine or subaerial origin? Sedimentology, v. 36, p. 777-792.

Broadhurst, F. M. 1988. Seasons and tides in the Westphalian. *In:* Sedimentation in a Synorogenic Basin Complex, B. M. Besly and G. Kelling (eds.). Blackie, Glasgow, p. 264-272.

Comalco Aluminium Ltd. 1983. Well-completion report, Nicholson No. 2. South Australian Department of Mines and Energy, Adelaide, Open-file Report no. 3938.

Dalrymple, R. W. and Makino, Y. 1989. Description and genesis of tidal bedding in the Cobequid Bay-Salmon River estuary, Bay of Fundy, Canada. *In:* Sedimentary Facies in the Active Plate Margin, A. Taira and F. Masuda (eds.). Terra, Tokyo, p. 151-177.

de Boer, P. L. 1981. Mechanical effects of micro-organisms on intertidal bedform migration. Sedimentology, v. 28, p. 129-132.

Oost, A. P. and Visser, M. J. 1989. The diurnal inequality of the tide as a parameter for recognizing tidal influences. Journal of Sedimentary Petrology, v. 59, p. 912-921.

Dyson, I. A. and von der Borch, C. C. 1986. A field guide to the geology of the late Precambrian Wilpena Group, Hallett Cove, South Australia. *In:* One Day Geological Excursions of the Adelaide Region, A. J. Parker (comp.). Geological Society of Australia, South Australian Division, Adelaide, p. 17-40.

FitzGerald, D. M. and Nummedal, D. 1983. Response characteristics of an ebb-dominated tidal inlet channel. Journal of Sedimentary Petrology, v. 53, p. 833-845.

Hayes, M. O. 1980. General morphology and sediment patterns in tidal inlets. Sedimentary Geology, v. 26, p. 139-156.

Imperato, D. P. Sexton, W. J. and Hayes, M. O. 1988. Stratigraphy and sediment characteristics of a mesotidal ebb-tidal delta, North Edisto Inlet, South Carolina. Journal of Sedimentary Petrology, v. 58, p. 950-958.

Kaye, C. A. and Stuckey, G. W. 1973. Nodal tidal cycle of 18.6 yr. Geology, v. 1, p. 141-144.

Komar, P. D. and Enfield, D. B. 1987. Short-term sea-level changes and coastal erosion. *In:* Sea-level Fluctuation and Coastal Evolution, D. Nummedal, O. H. Pilkey and J. D. Howard (eds.). Society of Economic Paleontologists and Mineralogists, Special Publication 41, p. 17-27.

Kreisa, R. D., Leithold, E. L. and Bourgeois, J. 1989. Tidal signature in mixed-energy deltaic facies. Second International Research Symposium on Clastic Tidal Deposits, University of Calgary, August 1989, Abstracts p. 49.

Kvale, E. P., Archer, A. W. and Johnson, H. R. 1989. Daily, monthly and yearly tidal cycles within laminated siltstones of the Mansfield Formation (Pennsylvanian) of Indiana. Geology, v. 17, p. 365-368.

Leithold, E. L., Bourgeois, J. and Kreisa, R. D. 1989. Genesis of tidal rhythmites of an Eocene delta slope, S. W. Oregon, U.S.A. Second International Research Symposium on Clastic Tidal Deposits, University of Calgary, August 1989, Abstracts p. 55.

Mazzullo, S. J. 1971. Length of the year during the Silurian and Devonian Periods. Geological Society of America, Bulletin, v. 82, p. 1085-1086.

McGugan, A. 1968. Possible use of algal stromatolite rhythms in geochronology. Geological Society of America, Special Paper 115, p. 145.

Nio, S. D. and Yang, C. S. 1989. Recognition of tidally-influenced facies and environments. International Geoservices, Leiderdorp, The Netherlands, 230 p.

Özsoy, E. 1986. Ebb-tidal jets: a model of suspended sediment and mass transport at tidal inlets. Estuarine, Coastal and Shelf Science, v. 22, p. 45-62.

Pannella, G. 1972. Paleontological evidence on the Earth's rotational history since early Precambrian. Astrophysics and Space Science, v. 16, p. 212-237.

______ 1975. Palaeontological clocks and the history of the Earth's rotation. *In:* Growth Rhythms and the History of the Earth's Rotation, G. D. Rosenberg and S. K. Runcorn (eds.). Wiley, London, p. 253-284.

Pariwono, J. I., Bye, J. A. and Lennon, G. W. 1986. Long-period variations of sea-level in Australasia. Geophysical Journal of the Royal Astronomical Society, v. 87, p. 43-54.

Preiss, W. V. (comp.) 1987. The Adelaide Geosyncline. Bulletin 53, South Australian Department of Mines and Energy, Adelaide, 438 p.

Rees, A. I. 1966. Some flume experiments with a fine silt. Sedimentology, v. 6, p. 209-240.

Reineck, H. E. and Wunderlich, F. 1968. Classification and origin of flaser and lenticular bedding. Sedimentology, v. 11, p. 99-104.

Scrutton, C. T. 1978. Periodic growth features in fossil organisms and the length of the day and month. *In:* Tidal Friction and the Earth's Rotation, P. Brosche and J. Sündermann (eds.). Springer-Verlag, Berlin, p. 154-196.

Sha, L. P. 1989. Cyclic morphologic changes of the ebb-tidal delta, Texel Inlet, The Netherlands. Geologie en Mijnbouw, v. 68, p. 35-48.

Tankard, A. J. 1986. Depositional response to foreland deformation in the Carboniferous of eastern Kentucky. The American Association of Petroleum Geologists, Bulletin, v. 70, p. 853-868.

Tessier, B. and Gigot, P. 1989. A vertical record of different tidal cyclicities: an example from the Miocene Marine Molasse of Digne (Haute Provence, France). Sedimentology, v. 36, p. 767-776.

______ , Monfort, Y., Gigot, P. and Larsonneur, C. 1989. Enregistrement des cycles tidaux en accrétion verticale, adaptation d'un outil de traitement mathématique. Exemples en baie du Mont-Saint-Michel et dans la molasse marine miocène du bassin de Digne. Bulletin des Sociétés Géologique de France, t.v. (8), no. 5, p. 1029-1041.

Wells, J. W. 1970. Problems of annual and daily growth-rings in corals. *In:* Palaeogeophysics, S. K. Runcorn (ed.). Academic Press, London, p. 3-9.

Williams, G. E. 1975. Late Precambrian glacial climate and the Earth's obliquity. Geological Magazine, v. 112, p. 441-465.

______ 1986. Precambrian permafrost horizons as indicators of palaeoclimate. Precambrian Research, v. 32, p. 233-242.

______ 1987. Cosmic signals laid down in stone. New Scientist, v. 114, no. 1566, p. 63-66.

______ 1988. Cyclicity in the late Precambrian Elatina Formation, South Australia: solar or tidal signature? Climatic Change, v. 13, p. 117-128.

______ 1989a. Late Precambrian tidal rhythmites in South Australia and the history of the Earth's rotation. Journal of the Geological Society, London, v. 146, p. 97-111.

______ 1989b. Precambrian tidal sedimentary cycles and Earth's paleorotation. Eos, v. 70, p. 33, 40-41.

______ 1989c. Tidal rhythmites: geochronometers for the ancient Earth-Moon system. Episodes, v. 12, p. 162-171.

______ and Tonkin, D. G. 1985. Periglacial structures and palaeoclimatic significance of a late Precambrian block field in the Cattle Grid copper mine, Mount Gunson, South Australia. Australian Journal of Earth Sciences, v. 32, p. 287-300.

KEY WORDS: TIDAL RHYTHMITES, LATE PROTEROZOIC, REYNELLA SILTSTONE, ELATINA FORMATION, CHAMBERS BLUFF TILLITE, ADELAIDE GEOSYNCLINE, OFFICER BASIN, EBB-TIDAL DELTA, PALEOTIDAL PERIODS, PALEOROTATION OF EARTH

In: Clastic Tidal Sedimentology, D. G. Smith, G. E. Reinson, B. A. Zaitlin and R. A. Rahmani (eds.). Canadian Society of Petroleum Geologists. Memoir 16, p. 179-188.

Characteristics of two, Pennsylvanian-age, semidiurnal tidal deposits in the Illinois Basin, U.S.A.

ERIK P. KVALE

Indiana Geological Survey
Bloomington, Indiana 47405
USA

ALLEN W. ARCHER

Department of Geology
Kansas State University
Manhattan, Kansas 66506
USA

ABSTRACT

Pennsylvanian-age tidal bedforms have been generally undocumented in the Illinois Basin of the eastern U.S.A. It has been argued by some researchers that tidal currents were not strong enough to transport and deposit sediments in such a cratonic basin. Two Pennsylvanian-age tidal deposits in the Illinois Basin are described herein, both formed in strongly asymmetrical tidal regimes. One is characterized by large-scale tidal bundles (1.5 m thick); the other is characterized by planar-laminated siltstones and small-scale ripple-sized tidal bundles. Bedforms associated with these deposits were little modified by subordinate tidal currents. Analyses of the tidal bundle sequences indicate that both units were deposited in semidiurnal tidal systems, one from ebb-dominant tides, the other from flood-dominant tides. These deposits are an indication of the size range of tidal bedforms in the Illinois Basin.

RÉSUMÉ

Dans le bassin d'Illinois de l'est des États-Unis, les «morphologies de fond» («bedforms») associées aux marées pennsylvaniennes ont été peu documentées. Selon certains chercheurs, les courants de marée dans ce bassin cratonique n'étaient pas assez importants pour transporter et sédimenter les particules détritiques. Deux dépôts de marée d'âge pennsylvanien, tous les deux formés sous un régime d'intense marée asymétrique, sont ici décrits. L'un se caractérise par des «faisceaux» («bundles») de grandes dimensions (1.5 m d'épaisseur), et l'autre, par des siltites à laminations planaires et des «faisceaux» («bundles») de marée de petites dimensions avec rides. Les formes de ces dépôts ont été légèrement modifiées par des courants de marée subordonnés. L'analyse des séquences de «bundles» de marée indique que les deux unités ont été déposées dans un système de marée semi-diurne. Cette étude permet d'établir un ordre de grandeur des «morphologies de fond» des dépôts de marée du bassin d'Illinois.

INTRODUCTION

Until recently, tidal deposits have not been widely recognized in the Pennsylvanian rocks of the Illinois Basin (Fig. 1). This is in part because of the influential work of Shaw (1964). He argued that in inland cratonic basins, bottom friction would so dampen tidal currents they would become too weak to transport sediments. In addition, the occurrence of coals (especially those that contain less than 1 per cent sulfur), presence of rooted horizons, abundance of plant fossils, and general absence of marine megafossils within much of the middle to lower Pennsylvanian siliciclastic sequence were taken to imply that depositional settings were dominantly fluvial to upper delta plain. Lower delta plain and interdistributary bay deposits within these intervals were recognized mainly by the presence of marine- to brackish-water fossils or limestones.

The intent of this paper is to demonstrate how tidal processes can be interpreted from two rock sequences of Pennsylvanian age in the Illinois Basin. Both are roughly age equivalent (Morrowan), and are found in an interval once considered to be fluvially dominated. These deposits lack obvious bipolar current indicators, such as herringbone cross-stratification or the classic, sigmoidally shaped, large-scale tidal bundles described from Holocene and ancient deposits (*e.g.*, Visser, 1980; Allen, 1981; Homewood and Allen, 1981; Kreisa and Moiola, 1986). In addition, they lack the characteristic marine traces and macrofossils commonly associated with tidal environments. Even though tidal deposits made up of unidirectional cross-stratification are common (Terwindt, 1988), an emphasis is often placed on the recognition of bipolar paleocurrent indicators in the rock record in order to justify a tidal interpretation. Despite the absence of these structures in our Pennsylvanian deposits, a tidal influence can be clearly demonstrated.

The first deposits to be discussed are the Hindostan Whetstone beds of southern Indiana. These deposits were first described as having a subtidal to intertidal origin by Kvale and others (1989). They consist of 10 m of laminated siltstone with intercalated claystone and are sandwiched by coals. Well-preserved plant fossils are abundant in the Whetstone beds. Archer and Kvale (1989) suggested that the lower 6 m of the Whetstone beds may record as much as 6 years of nearly continuous tidal sedimentation.

The second unit is a 10 m thick sandstone (informally referred to here as the Abbott sandstone) in the Abbott Formation in southern Illinois. The sandstone overlies interbedded sandstones and mudstones interpreted as being of interdistributary bay origin (Gopinath *et al.*, 1973). The middle 3.5 m of the sandstone are characterized by three stacked sets of large-scale planar foresets bounded by foreset-parallel reactivation surfaces. These reactivation surfaces are

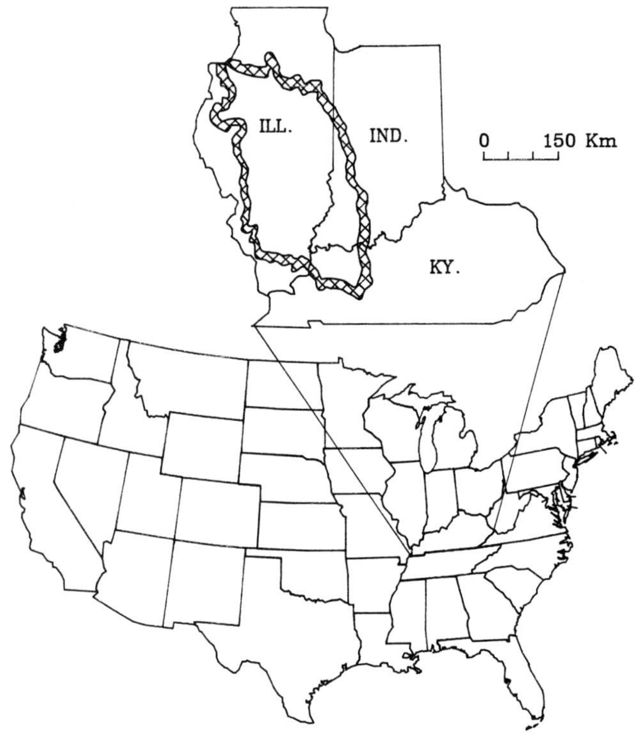

Figure 1. Distribution of Pennsylvanian rocks in the Illinois Basin.

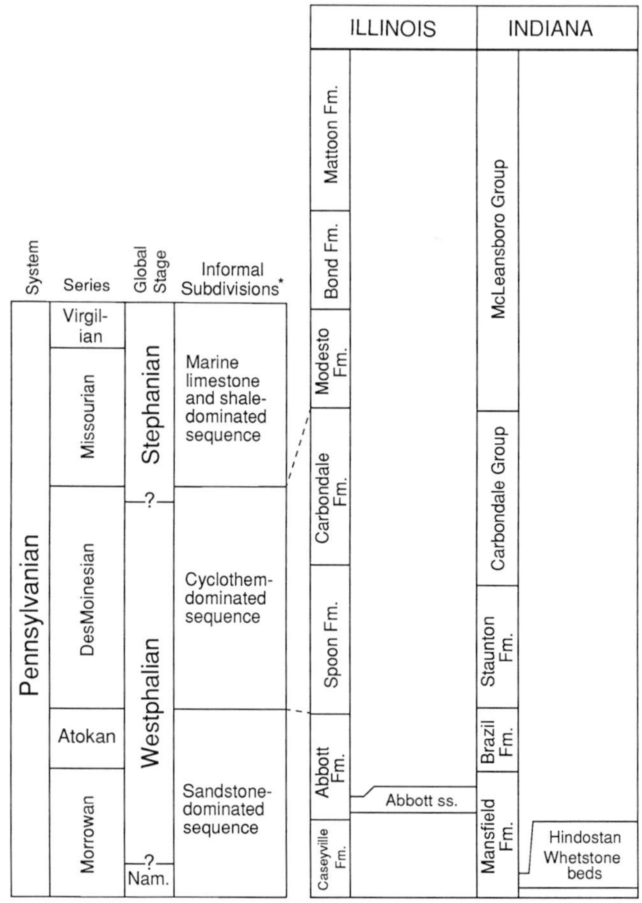

* from Damberger, 1989

Figure 2. Generalized diagram relating the Pennsylvanian mid-continent series and northwest Europe subsystems.

interpreted as being of tidal origin (Kvale and Uhlir, 1988; Archer *et al.*, 1989).

In the Whetstone beds and Abbott sandstone there exists a pronounced systematic thickness variation of vertically stacked laminae (Whetstone beds) and downcurrent accreting foreset packages (Abbott sandstone). We interpret these variations in thickness as neap-to-spring tidal events in a semidiurnal tidal system. What is unusual about both deposits is the very strong asymmetry of the tidal currents responsible for their deposition. Each was deposited in a semidiurnal system by either flood- or ebb-dominant tides with little to no sediment input from subordinate tidal currents. This is important because examples of this sort of system have not been generally recognized in the rock record of shallow cratonic basins and illustrate the potential range of scale of tidal bedforms within such settings.

General Geology

The Illinois Basin (Fig. 1) is situated in central and southern Illinois, southwestern Indiana, and western Kentucky. The basin is spoon-shaped, extending northwest-southeast for 600 km, and with a maximum width exceeding 320 km. During the Paleozoic, the basin was situated about 300 km north of the craton margin (Collinson *et al.*, 1988). Morrowan-age rocks are the oldest Pennsylvanian deposits in the basin (Fig. 2) and outcrop in the southern part. Morrowan- and Atokan-age rocks, until recently considered to be fluvially dominated, are now recognized as having significant marine influence (Damberger, 1989).

Hindostan Whetstone Beds

The Hindostan Whetstone beds are an informal unit in the Mansfield Formation (Shaver *et al.*, 1985), which outcrops in Orange County, Indiana (Fig. 3). Exposures are limited to abandoned quarries, most of which were closed at the turn of the century. Briefly, the Whetstone sequence consists of about 10 m of siltstone intercalated with claystone (Fig. 4). Planar and laterally continuous siltstone laminae (average lamina thickness of 2.7 mm) occur throughout the lower 6.9 m of the Whetstone sequence; the upper parts of the Whetstone sequence display mainly small-scale trough cross-stratification and discontinuous planar laminae. Sulfur and ash values of the coal that underlies the Whetstone beds in the Dishman Quarry are 2.08 per cent and 9.8 percent (moisture free), respectively. Sulfur and ash analyses on the upper coal are not available.

The Whetstone beds were studied in detail at three quarries in Orange County (Dishman, Old Braxton, and Hopper quarries; Fig. 3). To study laminae sequences, two cores 5 cm in diameter were drilled 6 m apart at the Dishman Quarry, and an additional core was provided by Amax Coal Company Indianapolis, Indiana. A flat surface was created on

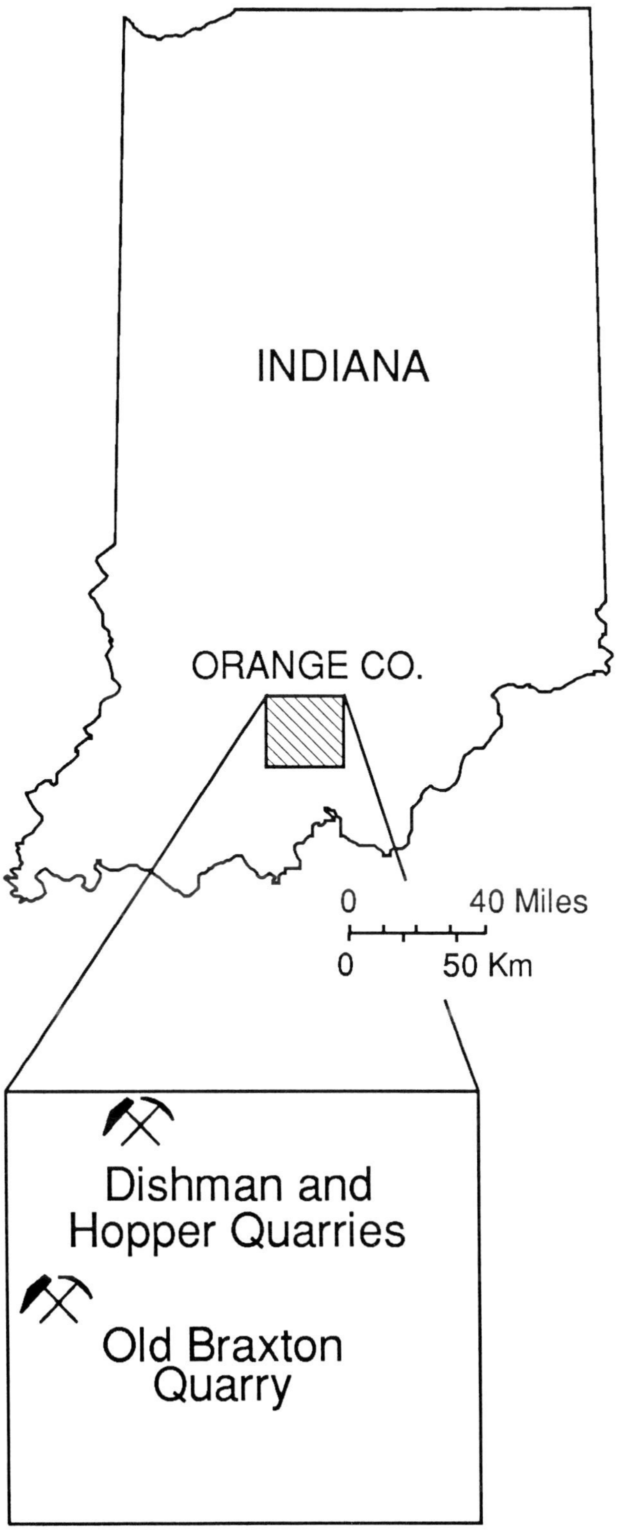

Figure 3. Location of Whetstone quarries investigated.

each Whetstone core with a belt sander; the flat surfaces were than stained red with food coloring to enhance the laminae.

Paleontology

Bedding planes of the laminated siltstones reveal a variety of trace fossils, the most common forms being *Plangtichnus eraticus, Treptichnus bifurcus*, and *Haplotichnus idianensis* (Archer and Maples, 1984). Although about 20 ichnogenera occur, and are common in places, there is virtually no bioturbation of the laminae (Archer and Maples, 1984; Maples and Archer, 1987; Archer, unpublished data). However, locally, dewatering structures and soft sediment deformation may disrupt laminae.

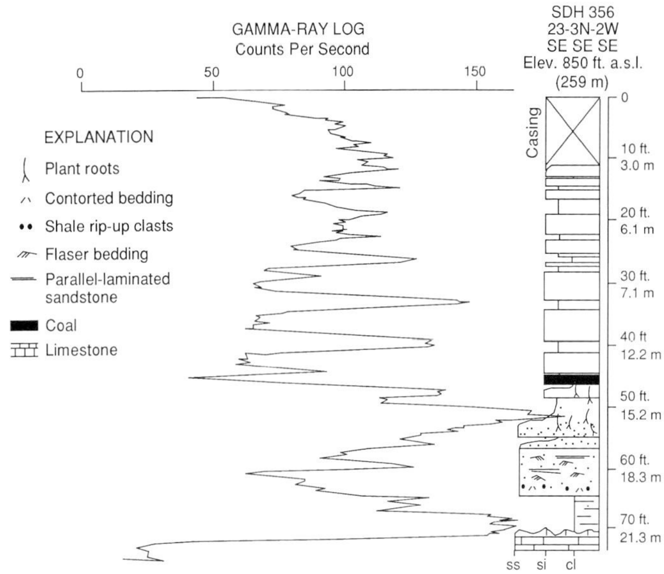

Figure 4. Gamma-ray log and columnar section from Indiana Geological Survey drill hole SDH 356.

Plant fossils found in the Whetstones are principally lycopods, including well-preserved trunks (*Lepidodendron*), cones (*Lepidostrobus*), and leaves (*Lepidopholius*). Lycopod trunks as much as 50 cm in diameter and 2.5 to 5.5 m in height have been preserved in an upright position. They appear to be rooted in the coal underlying the Whetstones. Other significant fossils include two well-preserved insect wings discovered in a Whetstone quarry in the late 1800's (Smith, 1871; Scudder, 1885) and an amphibian trackway (N. G. Lane, Indiana University, pers. comm., 1988). In addition, Carl Rexroad (Indiana Geological Survey, pers. comm., 1988) has recovered the conodonts *Streptognathodus* and *Cavusgnathus* found in intercalated claystones collected from Whetstones in the Hopper Quarry.

Sedimentology

Planar-laminated siltstone. The laminae thicknesses range from 0.5 mm to slightly more than 1 cm. The laminae are commonly conspicuously paired into couplets consisting of a thicker and a thinner lamina (Fig. 5). In addition, the laminae are separated by very thin (thickness scale of μ to mm) but laterally continuous clay drapes. Thin-section analyses reveal that an individual lamina grades from coarse silt at the base to finer silt at the top. Well-developed laminae within

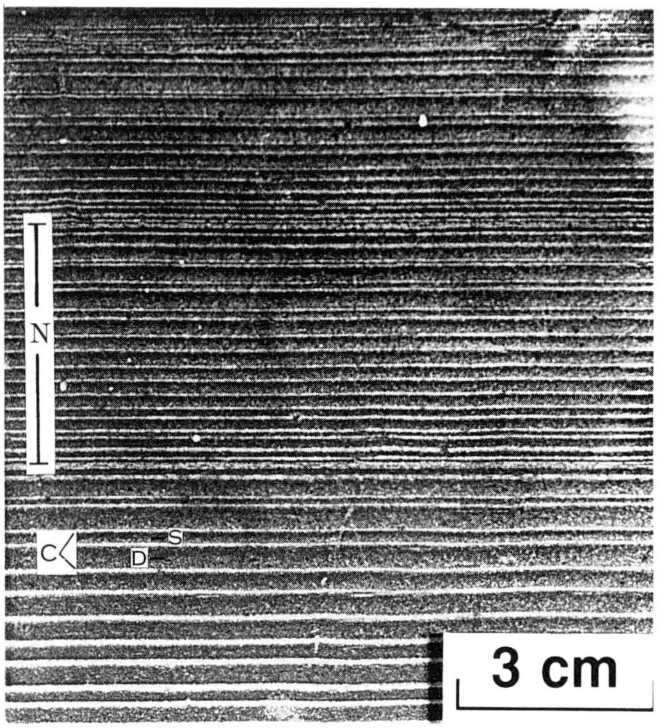

Figure 5. Sequence of laminae within a core slab of the planar-laminated siltstone facies, illustrating couplet *(C)* relationship of thinner lamina *(S)* and thicker lamina *(D)*. *N* delineates a tidal cycle of a half lunar month. Block collected at Hopper Quarry.

Figure 6. Cross-sectional view of a horizontally oriented lycopod pith cast. Note the preservation of the thick-thin couplets. Top of the photo is stratigraphic up.

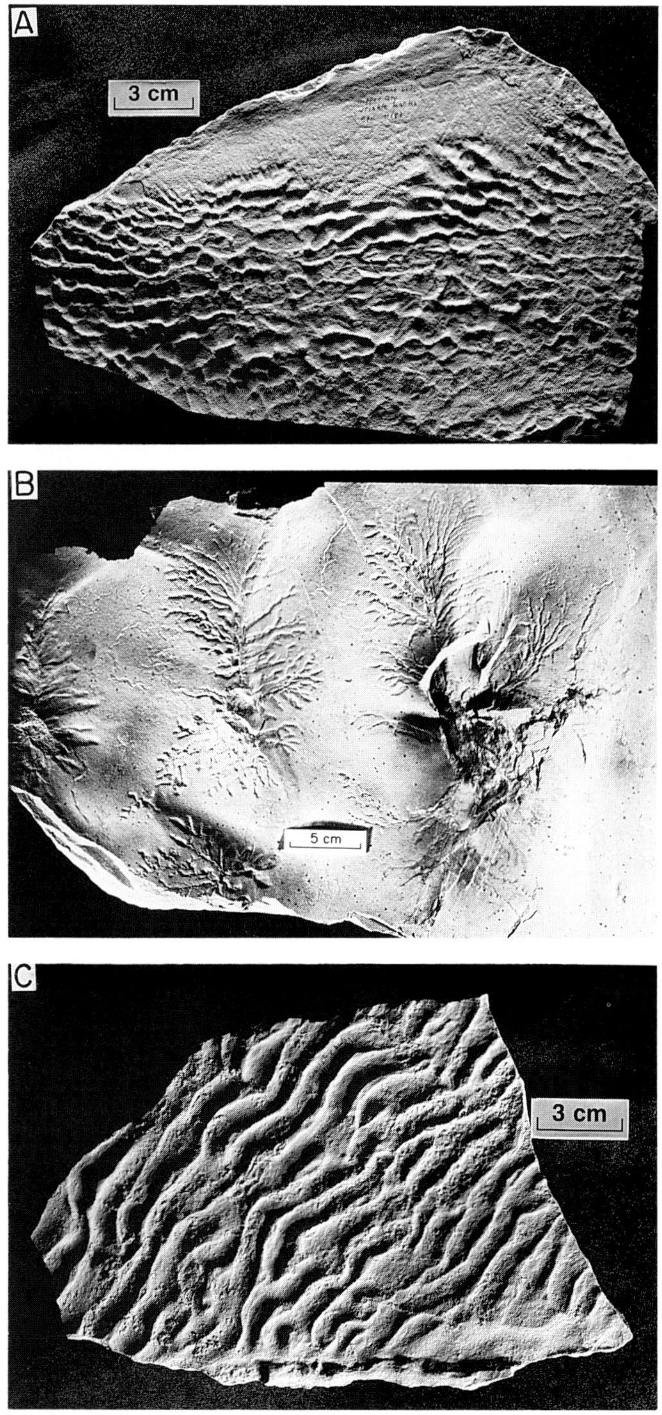

Figure 7. Bedding-planes of oriented blocks of Whetstones illustrating examples of: **A.** runzel marks, **B.** bifurcating micro rills with flow away from the center of silt volcanoes formed by dewatering or degassing of the sediment, and **C.** mm-sized ripples with flattened crests. All of these structures are useful indicators of the periodic subaerial exposure of the Whetstone beds.

pith casts of horizontally oriented lycopod branches indicate that silts were deposited by bottom-hugging density flows capable of filling hollow branches lying on the Whetstone surface (Fig. 6). Erosional contacts between successive laminae are rarely present in the lower 6.9 m of the laminated sequence, but become more common upward in conjunction with the occurrence of small-scale trough cross-stratification. In addition, a number of sedimentological structures indicate periodic emergence during deposition of the Whetstone sediments; these include runzel marks, drain features, and mm-sized ripples with flattened crests (Fig. 7).

Measurements of laminae in the Whetstones indicate cyclical thickening and thinning (Fig. 8). When sequential laminae are numbered and their thicknesses plotted on a

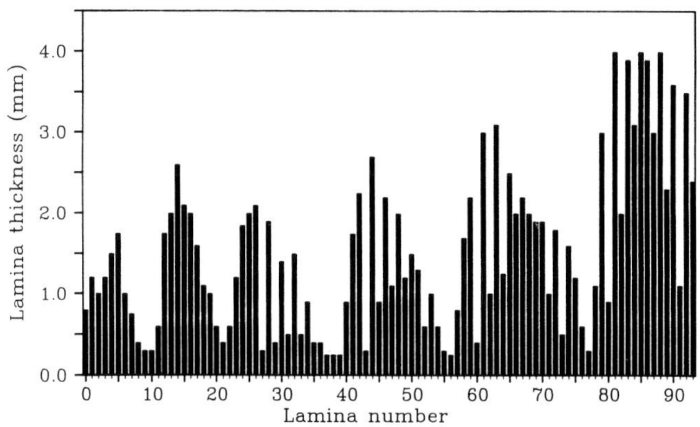

Figure 8. Systematic thickness variations of Whetstones laminae. Modified from Kvale *et al.* (1989).

histogram, a series of sinusoidal curves exhibiting a periodicity of 30 or fewer siltstone laminae is generated. Each periodicity is interpreted as representing a semidiurnal to mixed tidal cycle of a half lunar month. Similar curves have been generated from classic, large-scale tidal bundles of Holocene, Cretaceous, and Jurassic deposits from the United States and Europe (Visser, 1980; Allen, 1981; Homewood and Allen, 1981; Uhlir *et al.*, 1988).

Complex rippled siltstone. In places, laminae in several horizons can be traced laterally into climbing oscillatory ripples. These structures are variable, three-dimensional bedforms, are slightly coarser grained (coarse siltstone to very fine grained sandstone), and are nearly symmetrical, with ripple indices of 7 to 7.5. Harms *et al.* (1982) have shown that for silt sizes, an increase in flow velocity will involve essentially four phase fields: no movement, small ripples, upper plane bed, and antidunes. The absence of large-scale cross-stratification within the Whetstone interval is consistent with the absence of what Harms and others term a large, ripple-phase field within this size interval. Thus, the laminated siltstones may reflect deposition in the upper plane bed and the ripple forms a localized decrease in flow velocities.

The internal structures of these ripples are rather complex, yet the couplet relationship of thicker and thinner laminae is preserved within the ripple form (Fig. 9a). A detailed study of a set of climbing ripples shows that at the ripple crests, thicker laminae are separated from thinner laminae by very thin, discontinuous wedges of reverse-dipping foresets (Fig. 9b). It is unlikely that these smaller scale foresets are the result of backflow in the sense of Boersma *et al.* (1968), because they occur on the crests of the larger scale ripples rather than within the troughs where reverse-flow eddies most commonly occur; thus, these smaller scale foresets must represent the reversal events of the subordinate flow. Therefore, the stacked complex ripples, as well as the planar-laminated forms, are examples of a very small type of tidal-bundle sequence generated by a dominant flow ("D" lamina), followed by a weak subordinate-current flowing in the opposite direction (reverse-dipping foresets); in turn succeeded by a second dominant flow ("S" lamina), weaker than the initial flow, yet moving in the same direction as the first dominant flow. Since clay drapes are not preserved within the ripple sequence, either significant amounts of suspended materials were not deposited at slack water, or the drape was removed by the next flow event.

Similar ripple bedforms have been described elsewhere in Indiana (Fishbaugh *et al.*, 1989; Kvale and Archer, 1990); from the Tyee Formation in the state of Oregon (Leithold and Bourgeois, 1984); and from recent sediments in the Bay of Fundy, Canada (Dalrymple and Makino, 1989; Dalrymple and Zaitlin, 1989). In the example from Leithold and Bourgeois, the ripples, which also record a neap-spring thickening and thinning, are interpreted as having been formed by tidal restriction of ebb-dominant flow at the mouth of an estuary. However, the Canadian examples were interpreted to have been deposited by flood-dominant tidal currents on mud flats bordering a river channel in the fluvial-tidal transition zone at the head of the Bay of Fundy.

Tidal Regime

Two flow events (a strong dominant event followed by a weaker dominant event) in the same direction indicate diurnal inequality of tidal currents in a semidiurnal tidal regime. Because a "couplet" of thick and thinner laminae does not always occur (Fig. 8), this suggests that at times the tides may have been diurnal. Conversely, the diurnal inequality may have become so extreme (such as during times of maximum lunar declination) that the secondary high tide did not result in a distinguishable lamina. Nonetheless, given the very regular pattern of couplet development (Fig. 8), the Whetstone beds as a whole may best be considered as a mixed, predominantly semidiurnal tidal regime. In addition, the evidence of subaerial exposure from the penecontemporaneous structures indicates that the Whetstones were at least periodically intertidal.

The Whetstone laminae are interpreted as flood dominant, in part because an inland depositional setting is indicated by the upright trees within the Whetstone beds and by the coals immediately above and below the beds. Inland tidal settings (*e.g.*, inland tidal creeks and flats, and the uppermost reaches of estuaries) tend to be more flood-dominated because tides rise quickly and fall slowly in such settings (Dyer, 1986). In addition, paleoflow directions, as determined from the morphology of the ripple forms, were in a generally easterly direction. Inferred paleogeographic reconstructions indicate that this was landward.

The laterally continuous horizontal nature of the laminae indicates that the Whetstones were deposited on a surface with little relief, such as a tidal flat. However, vertical accretion on this tidal flat was as much as a m/yr. (Kvale *et al.*, 1989). Without invoking large tidal ranges (which the nature of the Whetstone deposits argues against), it is difficult to explain the association of intertidal features and the thick package of sediment. The presence of low-ash and moderate-sulfur coal directly beneath the Whetstone beds may be significant in terms of how so many vertically

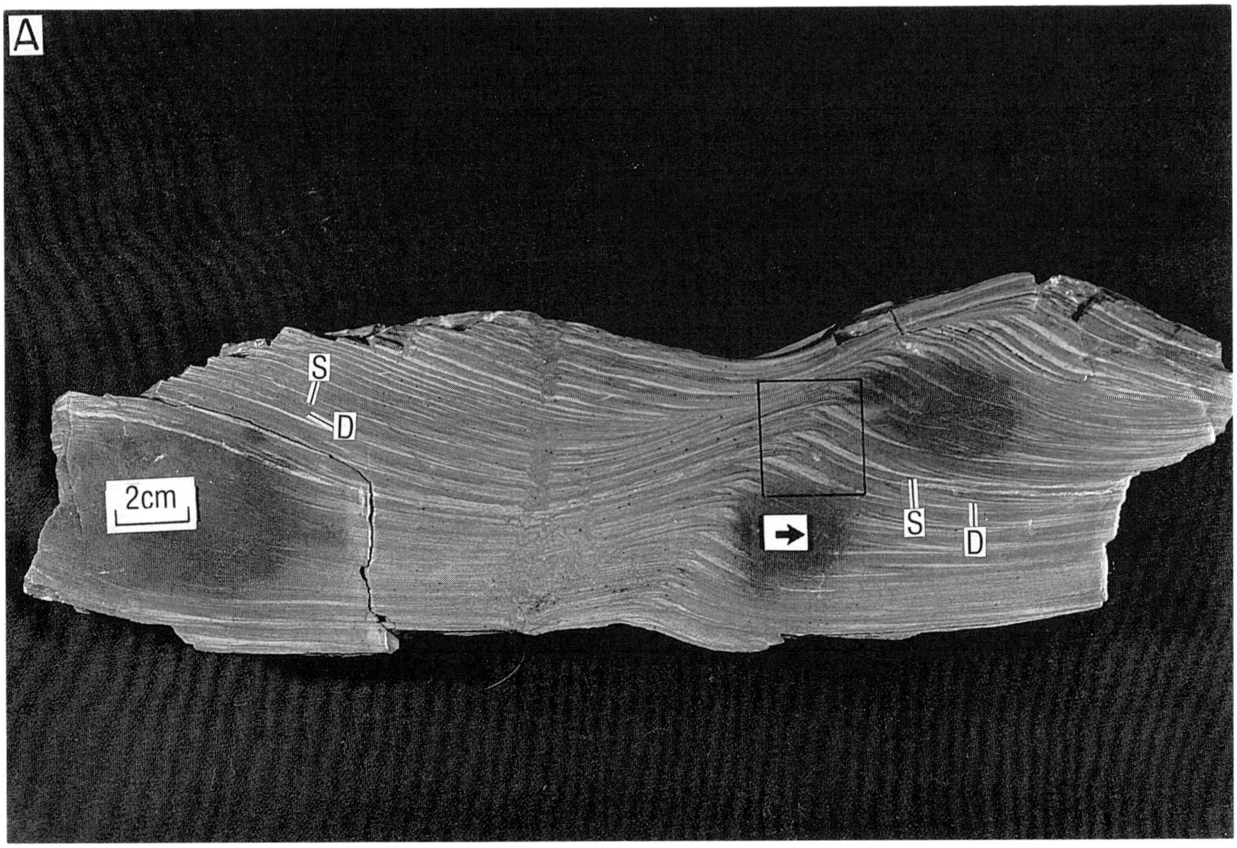

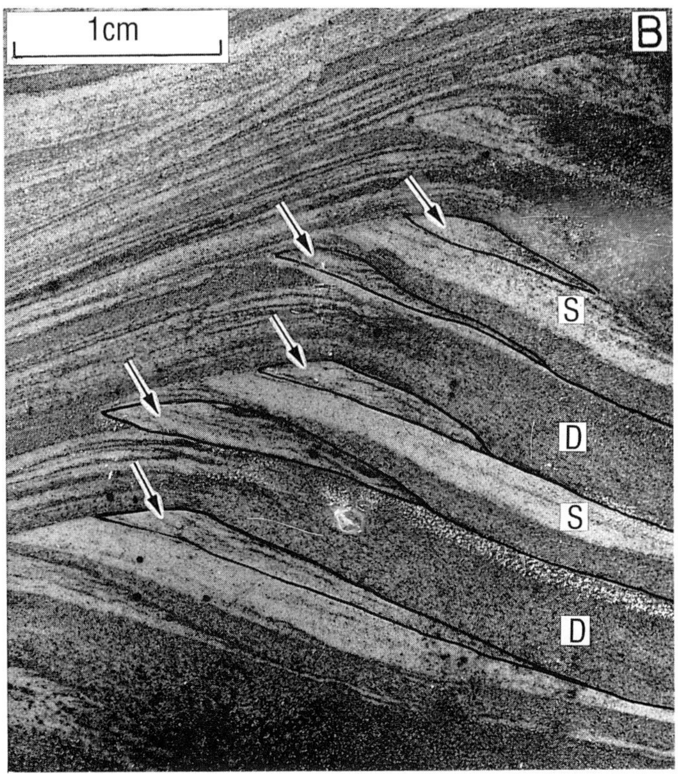

Figure 9. A. Polished slab of climbing oscillatory ripples. The dominant flow direction is indicated by the large arrow. Note that the couplet relationship of thick-thin laminae is preserved in the bedform. The laminae are disrupted by vertical dewatering or degassing structures. **B.** An enlargement of the rectangular area outlined in **A** shows discontinuous wedges of reverse-dipping foresets (small arrows) that separate the thicker laminae from the thinner laminae. The wedges were formed by subordinate tidal currents flowing in the opposite direction.

accreted Whetstone laminae were preserved. Lower Pennsylvanian low-ash (less than 10%) and low-sulfur coals (less than 1%) in the central Appalachian basin of the eastern U.S.A., have recently been attributed to ombrogenous (domed) peat deposits formed in an ever-wet climate (Cecil *et al.*, 1985). Similar climatic conditions have been evoked for the Lower Pennsylvanian rocks of the Illinois Basin. Recent investigations of modern domed peats present along the coasts of Borneo and Sumatra have revealed that the bases of some of these peats are subtidal, yet remain relatively uncompacted (J. Cobb, 1989, Kentucky Geological Survey, pers. comm.). Hypothetically, peat compaction driven by sediment loading of a transgressed coastal domed peat could provide the necessary accomodation space to preserve the essentially uninterrupted sedimentation as indicated by the lower 6.9 m of Whetstone laminae. If the rate of compaction were equal to the rate of sedimentation, the depositional environment could remain in an intertidal setting within a system of limited tidal range. However, many questions remain regarding the exact nature of this deposit.

Similarly laminated siltstones, also interpreted as being of tidal origin, have been described in Illinois from the Frances Creek Shale (Desmoinesian), which directly overlies the Colchester Coal (Baird *et al.*, 1986), and from the Precambrian Elatina Formation of Australia (Sonett *et al.*,

1988; Williams, 1989). The latter was interpreted by Williams to be subtidal in origin and preserved in a distal ebb-tidal delta. Other examples exist in the Glen Eyrie "member" of the Fountain Formation (Pennsylvanian) of Colorado and the Caseyville Formation (Pennsylvanian) of Iowa, both of which are underlain by coals (Archer and Kvale, unpublished data). Smith *et al.* (1990) describe laminated, fine grained sands and silts formed in subtidal settings at the base of two glacier-fed deltas in Glacier Bay, Alaska. The laminations form when intertidal flats are drained during ebb-tide events. The "tidal drawdown" (terminology of Smith *et al.*, 1990) initiates weak sediment gravity flows that transport sediments as much as 1 km seaward from the delta lip. It is not clear if a similar process could have generated the Whetstone laminae; however, a "tidal drawdown" mechanism does seem reasonable for those deposits that are clearly subtidal.

Abbott Sandstone

The Abbott Formation outcrops in southern Illinois in a roadcut along Interstate 57 between the Goreville and Lick Creek exits in Johnson County (Fig. 10). The lowest part of the Abbott Formation in this area consists of interbedded sandstone and mudstone interpreted by Gopinath and Ethridge (1976) as being interdistributary bay in origin (Fig. 11). This unit is dominated by wavy-to flaser-bedded oscillatory current- and (?)wave-generated sandstones. Ripple indices range from 7 to 18. Interbedded within these deposits are scour-based, lenticular and elongate, massive, sandstone-filled channel bodies with widths of 1 m or more and thicknesses generally less than 40 cm. Gutter cast orientations at the bases of these sandstones indicate a west-southwesterly paleoflow. Similar paleoflow directions were obtained by measuring dip directions of the ripple foreset laminae in the flaser- to wavy-bedded sandstones. The basal contact of the Abbott is gradational with the underlying Pounds Sandstone Member of the Caseyville Formation, a sandstone interpreted to have been deposited in a distributary channel complex (Gopinath and Ethridge, 1976).

Figure 10. Location of Abbott Formation outcrop studied.

A 10 m thick coarse to medium grained sandstone (Abbott sandstone), that is in part characterized by stacked, large-scale, planar foresets lies in erosional contact on the interbedded sandstone and mudstone unit. This sandstone is thought to laterally interfinger with the interbedded sandstone and mudstone unit (Devera, 1989). It is this sandstone that will be discussed in greater detail.

Paleontology

Finely disseminated plant fragments were the only megafossils found in the Abbott sandstone. A high abundance of moderate to low diversity ichnofossils occur within the underlying interbedded sandstone and mudstone unit (Archer *et al.*, 1989; Devera, 1989). Unnamed, shrimp-like arthropod traces are quite common and are mostly oriented perpendicular to the sandstone ripple crests (Devera, 1989), and parallel to the dominant southerly dip direction of the Abbott sandstone foresets. Other traces include the almond-shaped *Lockeia* sp., the repichnia *Uchirites* sp. and *Kouphichnium* sp. These types of traces are common in brackish water environments (Devera, 1989).

Sedimentology

Gopinath *et al.* (1973) divided the Abbott sandstone into three units (Fig. 11). The lowest unit is dominated by large-scale trough cross-stratification and rests erosively on the underlying interbedded sandstone and mudstone. The uppermost sandstone is likewise dominated by large-scale trough cross-stratification. This sand appears to be gradational with an overlying interbedded sandstone and mudstone. The medial unit is 3.5 m thick and is characterized by three stacked sets of large-scale planar foresets. The thickest set exceeds 1.5 m with the planar foresets being only slightly tangential (Fig. 12). The average dip of the foresets is 23° to the southeast. Differential weathering along goethite-cemented boundaries imparts a flaggy appearance to the outcrop. Close examination of the goethite-cemented boundaries reveals that they delineate reactivation surfaces. The reactivation surfaces actually parallel the foresets, which

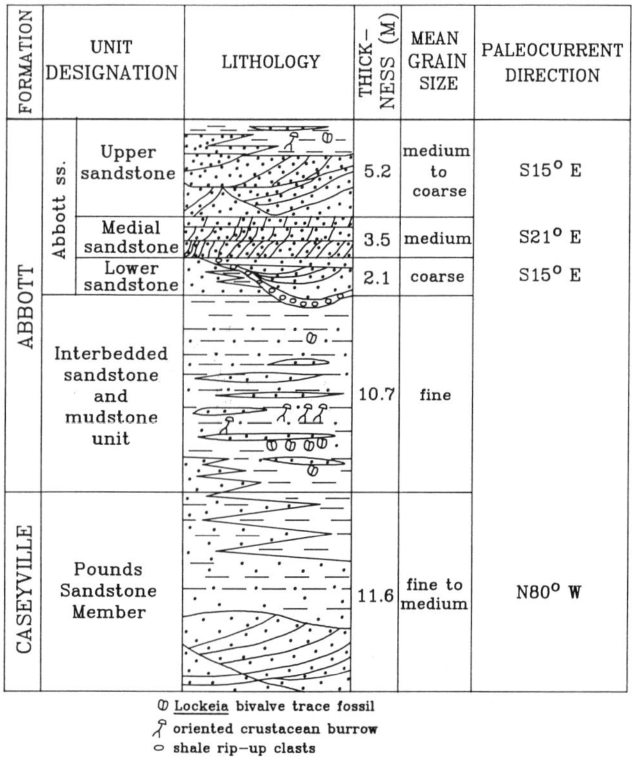

Figure 11. Columnar section of the Abbott sandstone outcrop studied. Modified from Gopinath *et al.* (1973) and Archer *et al.* (1989).

occur between the goethite-cemented boundaries, with very little apparent truncation. The distance between two reactivation surfaces (hereafter referred to as a tidal bundle) is on the average 8 cm (measured at right angles to the reactivation surface). The goethite appears to have replaced mud drapes deposited along these boundaries. Of particular interest is that thick tidal bundles alternate with thinner tidal bundles and form couplets not unlike the couplets described above from the Whetstone beds (Fig. 12). Superimposed on

Figure 12. Photograph of the lower set of the medial Abbott sandstone. Note the flaggy appearance of the thick-thin couplets of foreset packages. Differential weathering and parting is commonly most pronounced along goethite-cemented reactiviation surfaces. However, some weathering does occur along foresets between reactivation surfaces.

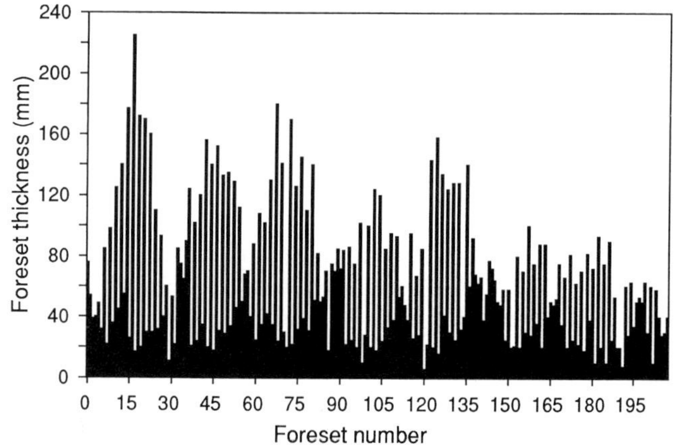

Figure 13. Systematic thickness variations of the medial Abbott sandstone tidal bundles. Note the semidiurnal pattern of thick bundle (strong ebb-dominant event) alternating with thin bundle (weaker ebb-dominated event). Subordinate (flood) deposits are not preserved. Bundles are numbered from upstream (left) to downstream (right).

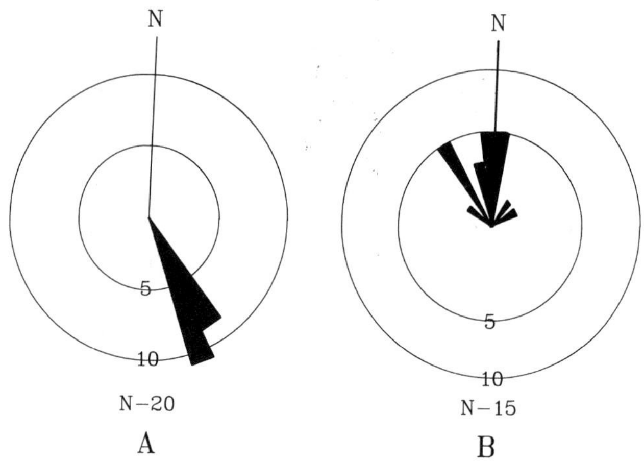

Figure 14. Rose plots illustrating the dip directions of the medial Abbott sandstone reactivation-foreset surfaces (**A**) and the paleomigration directions of the bar-top subordinate-flow linguoid ripples (**B**).

the cyclic thick-thin alternations is a larger-scale thickening and thinning in a downcurrent direction (Fig. 13). Cyclic thickening and thinning is continuous and best developed in the lowermost set where the thickest individual tidal bundle occurs every 29th depositional event. Such cyclicity is best explained by neap-spring tidal current fluctuations during a lunar month. Individual neap-spring cycles in the lowest medial sandstone set on the west side of the interstate highway right-of-way can be correlated with neap-spring cycles on the east side of the right-of-way, a distance of 50 m. The lateral continuity of this bedform width and the absence of scour pits at the bases of the foresets indicate that the medial sandstone was deposited in a tidal sandwave (used in the sense of Harms *et al.*, 1975) with a straight to slightly sinuous crestline. Additional evidence of tidal influence includes northward-dipping (opposite to the planar foreset migration direction) linguoid ripples formed as the sandwave top was reworked following its formation (Fig. 14). Overall there is a decrease in tidal bundle thicknesses downcurrent (Fig. 13).

Tidal Regime

The medial Abbott sandstone formed in a predominantly unidirectional tidal environment. Such an interpretation is consistent with the observation of Boothroyd and Hubbard (1975) that tidal sandwaves occur in areas with a dominant tidal current. The Abbott sandstone couplet, consisting of a thick tidal bundle followed by a thin tidal bundle, similar to the Whetstone beds, records deposition in a predominantly semidiurnal system. As in the Whetstone beds, two flow events (a strong dominant event followed by a weaker dominant event) in the same direction are recorded. The evidence for this conclusion is twofold. First, de Mowbray and Visser (1984) documented that the geometries of reactivation surfaces formed in bidirectional flows are largely a function of the strength of the subordinate current. Very weak subordinate currents cause very little erosion or modification of the dominant-current bedform. In the Abbott sandstone, the dips of the reactivation surfaces are essentially the same as those of the foresets within the tidal bundles, indicating little to no modification of the barform front by a subordinate flow. Secondly, both the thick and the thin tidal bundles contain foresets that dip in the same direction and both extend up the entire height of the avalanche face with no significant thinning at either the toe or top of the face. Such foresets could only have been generated by flow in the same direction. Thus, diurnal inequality and the neap-spring cycles are preserved in the Abbott sandstone without significant modification by a subordinate flow.

In terms of which flow (flood or ebb) was the dominant current in the Abbott sandstone, all that can be said is that a southerly paleoflow in this part of the Illinois Basin is considered to be seaward (Potter, 1962). Thus, it appears that the dominant current was generated by the ebb tide.

The exact origin of the Abbott sandstone is somewhat problematic, mainly because the only exposure of the sandstone is the roadcut and, in part, because the sedimentology is still incompletely documented. Devera (1989) interpreted the interbedded sandstone and mudstone unit to have been formed in a subtidal environment that during low tide may have been cut off from normal marine conditions. Subsurface information that would allow for a more detailed interpretation of the 3-dimensional aspects of the sandstone body is also lacking. Based on the Interstate 57 exposures, Gopinath *et al.* (1973) originally interpreted the medial Abbott sandstone as a Gilbert-type delta formed in a previously abandoned distributary channel (lower sandstone body). The medial sandstone is in turn overlain by what Gopinath *et al.* (1973) interpreted to be distributary channel deposits (upper sandstone). The evidence of tidal influence and lack of a well developed topset sequence precludes explaining these sandstones by using the classic Gilbert-type model. One of the interesting aspects of the Abbott sandstone is the thinness of the tidal bundles, especially when one considers the thickness of the barform (up to 1.5 m). Tidal bundles, similar in scale to the medial Abbott sandstone, have been reported elsewhere (Smith, 1984; Rahmani, 1988; Banerjee, 1989; and Reinson, 1989). In all cases, these examples have been associated with estuarine channel sytems. Smith and Rahmani suggested that their bundles formed in at least mesotidal conditions.

An estuarine setting opening into a brackish embayment seems most likely for the Abbott sandstone. Given the lack of obvious storm- or wave-influenced sands, it appears that the Abbott was deposited in an ebb-dominated, middle to upper estuarine setting rather than a lower estuarine setting.

Conclusions

The Abbott sandstone and the Whetstone beds illustrate that thick-thin tidal bundles are not necessarily generated by dominant-subordinate (flood-ebb) tidal events in megaripples or sand waves as interpreted for Abbott-like tidal bundles in Canada (Smith, 1988; Rahmani, 1988). Such sequences may simply reflect the daily inequality of semidiurnal tides. Recent work by deBoer *et al.* (1989) suggested that such thick-thin variations, when developed to a significant extent in a tidal bundle sequence, could only have been formed in semidiurnal systems.

Acknowledgments

We wish to thank J. Devera, J. Nelson, and D. Uhlir for discussions on the Illinoian portion of this project. We also acknowledge the encouragement and helpful comments of J. Harms, N. Hester, and R. Hunter, and thank L. Krystinik for his review of the manuscript. Funding was provided, in part, by two Indiana University Dean of Faculties research grants and by the Indiana and Illinois State Geological surveys. Drafting and photographic services were provided by the Indiana Geological Survey. Photography and drafting were done by B. Hill and R. Hill, respectively.

References

Allen, J. R. L. 1981. Lower Cretaceous tides revealed by cross-bedding with mud drapes. Nature, v. 289, p. 579-581.

Archer, A. W. and Kvale, E. P. 1989. Seasonal and yearly cycles within tidally laminated sediments: an example from the Pennsylvanian of Indiana, U.S.A. *In:* Geology of the Lower Pennsylvanian in Kentucky, Indiana, and Illinois, J. Cobb (ed.). Illinois Basin Consortium, no. 1, Indiana Geological Survey, Bloomington, p. 45-56.

______ and Maples, C. G. 1984. Trace-fossil distribution across a marine-to-nonmarine gradient in the Pennsylvanian of southwestern Indiana. Journal of Paleontology, v. 58, p. 448-456.

______, Devera, J. Kvale E. and Nelson J. 1989. Stop 9 — Upper Caseyville/Lower Tradewater Roadcut on Interstate 57, southern Illinois: tidally influenced deposits of Early Pennsylvanian age. *In:* Carboniferous Geology of the Eastern United States, C. B. Cecil and C. Eble (eds.). American Geophysical Union, Field Trip Guidebook T143, Washington, D.C., p. 30-32.

Baird, G. C., Sroka, S. D., Shabica, C. W. and Kuecher, C. J. 1986. Taphonomy of middle Pennsylvanian Mazon Creek area fossil localities, Northeast Illinois: significance of exceptional fossil preservation in syngenetic concretions. Palaios, v. 1, p. 271-285.

Banerjee, I. 1989. Tidal structures in the Glauconitic Sandstone, Countess Field, Southern Alberta, Canada. *In:* Modern and Ancient Examples of Clastic Tidal Deposits — a Core and Peel Workshop, G. E. Reinson (ed.). Canadian Society of Petroleum Geologists, Second International Research Symposium on Clastic Tidal Deposits, Calgary, Alberta, p. 89-97.

Boersma, J. R., Van de Meene, E. A. and Tjalsma, R. C. 1968. Intricated cross-stratification due to interaction of a megaripple with its lee-side system of back-flow ripples (upper-point-bar deposits, Lower Rhine). Sedimentology, v. 11, p. 147-162.

Boothroyd, J. C. and Hubbard, D. K. 1975. Genesis of bedforms in mesotidal estuaries. *In:* Estuarine Research II, L. E. Cronin (ed.). London, p. 217-234.

Collinson, C., Sargent, M. and Jennings, J. R. 1988. Illinois Basin region. *In:* Sedimentary Cover — North America Craton, L. L. Sloss (ed.). Geological Society of America, The Geology of North America, v. D-2, p. 383-426.

Cecil, C. B., Stanton, R. W., Neuzil, S. G., Dulong, F. T., Ruppert, L. F. and Pierce, B. S. 1985. Paleoclimate controls on Late Paleozoic sedimentation and peat formation in the central Appalachian basin (U.S.A.). *In:* Paleoclimatic controls on coal resources of the Pennsylvanian System of North America, T. L. Phillips and C. B. Cecil (eds.). International Journal of Coal Geology, v. 5, p. 195-230.

Dalrymple, R. W. and Makino, Y. 1989. Description and genesis of tidal bedding in the Cobequid Bay — Salmon River estuary, Bay of Fundy, Canada. *In:* Sedimentary Facies in the Active Plate Margins, A. Taira and F. Masuda (eds.). Tokyo, Terra Scientific Publishing Company, Tokyo, p. 151-177.

Dalrymple, R. W. and Zaitlin, B. A. 1989. Field trip guidebook to tidal sedimentation in the macrotidal, Cobequid Bay — Salmon River estuary, Bay of Fundy. Second International Research Symposium on Clastic Tidal Deposits, Calgary, Alberta, Canada, 84p.

Damberger, H. H. 1989. Overview of the Pennsylvanian in the Illinois Basin. *In:* Carboniferous Geology of the Eastern United States, C. B. Cecil and C. Eble (eds.). American Geophysical Union, Field Trip Guidebook, T143, Washington, D.C., p. 20-26.

De Boer, P. L., Oost, A. P. and Visser, M. J. 1989. The diurnal inequality of the tide as a parameter for recognizing tidal influences. Journal of Sedimentary Petrology, v. 59, p. 912-921.

Devera, J. A. 1989. Ichnofossil assemblages and associated lithofacies of the Lower Pennsylvanian (Caseyville and Tradewater formations), Southern Illinois. *In:* Geology of the Lower Pennsylvanian in Kentucky, Indiana, and Illinois, J. Cobb (ed.). Illinois Basin Consortium, no. 1, Indiana Geological Survey, Bloomington, p. 57-83.

Dyer, K. R. 1986. Coastal and Estuaring Sediment Dynamics. New York, John Wiley and Sons, 342 p.

Fishbaugh, D. A., Kvale, E. P. and Archer, A. W. 1989. Association of tidal and fluvial sediments within Lower Pennsylvanian rocks, Turkey Run State Park, Parke County, Indiana — guidebook for 1989 American Association of Petroleum Geologists Eastern Section Meeting and Fieldtrip. Indiana Geological Survey Guidebook, 46p.

Gopinath, T. R. R. and Ethridge, F. G. 1976. Upper Caseyville Formation (Pennsylvanian) roadcut on Interstate Highway 57 north of Lick Creek. *In:* Variability within a high constructive lobate delta: the Pounds Sandstone of southeastern Illinois, F. G. Ethridge and G. Fraunfelter (eds.). Great Lakes Section, Society of Economic Paleontologists and Mineralogists, Sixth Annual Field Conference, p. 56-61.

______ , ______ and Kolodziej, A. 1973. Stop 5, Upper Caseyville Formation (Pennsylvanian) roadcut on Interstate Highway 57 north of Lick Creek. *In:* Depositional environments of selected Lower Pennsylvanian and Upper Mississippian sequences of southern Illinois, F. G. Ethridge, G. Fraunfelter and J. Utgaard (eds.). 37th Annual Tri-State Field Conference, Carbondale, Southern Illinois University, p. 74-78.

Harms, J. C., Southard, J. B., Spearing, D. R. and Walker, R. G. 1975. Depositional environments as interpreted from primary sedimentary structures and stratification sequences. Society of Economic Paleontologists and Mineralogists, Short Course 2, Dallas, Texas, 161 p.

______ , ______ and Walker, R. G. 1982. Bed configurations. *In:* Structures and sequences in clastic rocks. Society of Economic Paleontologists and Mineralogists, Short Course No. 9, p. (2-1)-(2-55).

Homewood, P. and Allen, P. 1981. Wave-, tide-, and current-controlled sand bodies of Miocene Molasse, western Switzerland. American Association of Petroleum Geologists, Bulletin, v. 65, p. 2534-2545.

Kreisa, R. D. and Moiola, R. J. 1986. Sigmoidal tidal bundles and other tide-generated sedimentary structures of the Curtis Formation, Utah. Geological Society of America, Bulletin, v. 97, p. 381-387.

Kvale, E. P. and Archer, A. W. 1990. Tidal deposits associated with low-sulfur coals, Brazil Fm. (Lower Pennsylvanian), Indiana. Journal of Sedimentary Petrology, v. 60, no. 4, p. 563-574.

______ , ______ and Johnson, H. R. 1989. Daily, monthly, and yearly tidal cycles within laminated siltstones (Mansfield Formation: Pennsylvanian) of Indiana. Geology, v. 17, p. 365-368.

Kvale, E. P. and Uhlir, D. M. 1988. Tidal influence in an Upper Caseyville (Lower Pennsylvanian) Gilbert-type deltaic sequence, southern Illinois. Geological Society of America, Abstract with Programs, v. 20, no. 5.

Leithold, E. L. and Bourgeois, J. 1984. Characteristics of coarse-grained sequences deposited in nearshore, wave-dominated environments — examples from the Miocene of south-west Oregon. Sedimentology, v. 31, p. 749-775.

Maples, C. G. and Archer, A. W. 1987. Redescription of Early Pennsylvanian trace-fossil holotypes from the nonmarine Hindostan Whetstone beds of Indiana. Journal of Paleontology, v. 61, p. 890-897.

Mowbray, T. de and Visser, M. J. 1984. Reactivation surfaces in subtidal channel deposits, Oosterschelde, SW Netherlands. Journal of Sedimentary Petrology, v. 54, p. 811-824.

Potter, P. E. 1962. Regional distribution patterns of Pennsylvanian sandstones in Illinois Basin. American Association of Petroleum Geologists, Bulletin, v. 46, p. 1890-1911.

Rahmani, R. A. 1988. Estuarine tidal channel and nearshore sedimentation of a Late Cretaceous epicontinental sea, Drumheller, Alberta, Canada. *In:* Tide-influenced sedimentary environments and facies, P. L. de Boer, A. van Gelder and S. D. Nio (eds.). Boston, D. Reidel Publishing Co., p. 433-471.

Reinson, G. E. 1989. Tide-influenced channel deposits in the Lower Cretaceous Glauconitic Member, Southern Alberta. *In:* Modern and Ancient Examples of Clastic Tidal Deposits — a Core and Peel Workshop, G. E. Reinson (ed.). Canadian Society of Petroleum Geologists, Second International Research Symposium on Clastic Tidal Deposits, Calgary, Alberta, p. 98-104.

Scudder, S. H. 1985. *Dictyoneura* and the allied insects of the Carboniferous Epoch. Proceedings, American Academy of Science (new series), v. 20, p. 167-173.

Shaver, R. H., Ault, C. H., Burger, A. M., Carr, D. D., Droste, J. B., Eggert, D. L., Gray, H. H., Harper, D., Hasenmueller, N. R., Hasenmuller, W. A., Horowitz, A. S., Hutchinson, H. C., Keith, B. D., Keller, S. J., Patton, C. B., Rexroad, C. B. and Wier, C. E. 1986. Compendium of Paleozoic Rock-Unit Stratigraphy in Indiana — a revision. Indiana Department of Natural Resources, Geological Survey Bulletin 59, 203 p.

Shaw, A. B. 1964. Time in Stratigraphy. New York, McGraw-Hill, 365 p.

Smith, D. G. 1988. Tidal bundles and mud couplets in the McMurray Formation, Northeastern Alberta, Canada. Bulletin of Canadian Petroleum Geology, v. 36, p. 216-219.

Smith, N. D., Phillips, A. C. and Powell, R. D. 1990. Tidal drawdown: a mechanism for producing cyclic sediment laminations in glaciomarine deltas. Geology, v. 18, p. 10-13.

Smith, S. I. 1871. Notice of a fossil insect from the Carboniferous Formation of Indiana. American Journal of Science (ser. 3), v. 1, p. 44-46.

Sonett, C. P., Finney, S. A. and Williams, C. R. 1988. The lunar orbit in the late Precambrian and the Elatina sandstone laminae. Nature, v. 335, p. 806-808.

Terwindt, J. H. J. 1988. Paleo-tidal reconstructions of inshore tidal depositional environments. *In:* Tide-influenced sedimentary environments and facies, P. L. de Boer, A. van Gelder and S. D. Nio (eds.). Boston, D. Reidel Publishing Co., p. 233-263.

Uhlir, D. M., Akers, A. and Vondra, C. F. 1988. Tidal inlet sequence Sundance Formation (Upper Jurassic) north-central Wyoming. Sedimentology, v. 35, p. 739-752.

Visser, R. 1980. Neap-spring cycles reflected in Holocene subtidal large-scale bedform deposits: a preliminary note. Geology, v. 8, p. 543-546.

Williams, G. E. 1989. Late Precambrian tidal rhythmites in South Australia and the history of the Earth's rotation. Journal of the Geological Society London, v. 146, p. 97-111.

KEY WORDS: SEMIDIURNAL TIDAL DEPOSITS, TIDAL BUNDLES

In: Clastic Tidal Sedimentology, D. G. Smith, G. E. Reinson, B. A. Zaitlin and R. A. Rahmani (eds.). Canadian Society of Petroleum Geologists. Memoir 16, p. 189-196.

Analysis of modern equatorial tidal periodicities as a test of information encoded in ancient tidal rhythmites

ALLEN W. ARCHER
Department of Geology
Thompson Hall
Kansas State University
Manhattan, KS 66506-3201

ERIK P. KVALE
Indiana Geological Survey
611 North Walnut Grove
Bloomington, IN 47405

HOLLIS R. JOHNSON[1]
Department of Astronomy
Indiana University
Bloomington, IN 47405

ABSTRACT

Modelling of tidal rhythmites based on modern tidal data, including a spectrum of diurnal to semidiurnal settings, allows a greater understanding of the information that can be encoded in ancient tidal rhythmites and rhythmite-bearing sequences. Based upon such modelling, neap-spring patterns observed within the ancient rhythmites can be directly compared to the complex and hierarchically ordered periods that exist in modern tides. Harmonic analyses of modern data indicate that neap-spring cycles can be related not only to the synodic month (related to lunar phase) but also to the tropical month (related to lunar orbital declination). In addition, close approach of the moon to the earth (lunar perigee during the anomalistic month) results in a slight but significant change in tidal height variability that is superimposed on the aforementioned cycles.

Analysis of modern data indicates that within semidiurnal systems, the principal tide-generating period is related to the synodic month, with smaller effects related to anomalistic and tropical months. In such systems, neap-spring cycles are closely related to the lunar phase. Conversely, in diurnal systems, the principal tide-generating period can be related to the tropical month; such a system would exhibit neap-spring cycles in response to lunar declination rather than lunar phase. Because the tropical month is slightly shorter than the synodic month, a diurnal system can exhibit about 27 neap-spring cycles per year rather than the expected 25 cycles produced within a semidiurnal system.

Although such neap-spring cycles are too minor to be recorded in log-based analyses, the larger scale seasonal and yearly cycles can potentially be observed in subsurface logs. A simulation of gamma-log response to these larger scale cycles is discussed, based upon assumptions involving tidal sedimentation.

RÉSUMÉ

L'élaboration d'un modèle de périodites ou séquences rythmiques de marée, basé sur des données actuelles et incluant la variation d'environnements diurnes à semi-diurnes, permet une meilleure compréhension de l'information enregistrée dans les périodites de marée anciennes et les autres types de séquences rythmiques. En se basant sur un tel modèle, les patrons de morte-eau et vive-eau observés dans les séquences rythmiques anciennes peuvent être comparés directement avec les périodes complexes et hiérachisées, déployées par les marées modernes. L'analyse harmonique des données modernes indique que les cycles de morte-eau peuvent être reliés non seulement au mois synodique (relié à la phase lunaire) mais également au mois tropique (relié à la déclination orbitale de la Lune). Lorsque la distance de la Terre à la Lune est à son minimum (périgée de la Lune durant les mois anomalistiques), un changement faible mais significatif de la variabilité de la hauteur de la marée se superpose aux cycles susmentionnés.

L'analyse de données modernes prélevées dans des systèmes semi-diurnes montre que la période générée par la marée principale est reliée au mois synodique, avec des effets mineures reliés aux mois tropique et anomalistique. Dans de tels systèmes, les cycles eau-morte et vive-eau sont étroitement reliés à la phase lunaire. À l'inverse, la période générée par la marée principale dans les systèmes diurnes peut être reliée au mois tropique. Par conséquent, un tel système pourrait exhiber des cycles morte-eau et vive-eau réagissant plutôt à la déclinaison lunaire qu'à la phase lunaire. Puisque le mois tropique est légèrement plus court que le mois synodique, le système diurne peut produire 27 cycles morte-eau et vive-eau par année, au lieu des 25 cycles prévus dans le système semi-diurne.

Bien que les cycles morte-eau et vive-eau sont trop mineurs pour être détectés par diagraphies, les cycles saisonniers à grande échelle et les cycles annuels peuvent être possiblement décelés par ce type d'analyse de sous-surface. La discussion sur la simulation de la réponse des rayons gamma à ces cycles à grande échelle se base sur des assomptions impliquant une sédimentation de marée.

[1]Currently on sabbatical leave as the F.C. Donders Visiting Professor, University of Utrecht, The Netherlands

Introduction

Modern tides exhibit a range of periodicities that potentially affect tidal sedimentation. Detailed analyses of lamina-thickness periodicities in strata interpreted as tidal deposits have indicated that subdaily, two-weekly, and even seasonal periods can be encoded in such rocks. Such rocks, which have been termed "tidal rhythmites", are potentially useful for evaluating details of ancient tidal periodicities and thus extrapolating the mechanics of the sun-earth-moon system through geological time. However, the hierarchy of modern tidal periodicities is complex, and before periods derived from ancient tidalities can be evaluated, it is imperative that the causes for the various periodicities are completely understood. To evaluate these periodicities, tidal rhythmites can be modelled using modern tidal-station data. Such modelling begins with a time series, which is the tidal-station data, and converts the information into an event series, which in the simulated tidal rhythmite records individual tidal events rather than time. In these analyses, we seek to determine if the original time series can be reconstructed from the event series.

As described herein, the origin of neap-spring cycles can be related to several modern tidal periodicities. Determination of absolute periodicities involved in neap-spring cycles of ancient tidalites is not always easy because tidal systems exhibit a complete spectrum that ranges from diurnal to semidiurnal systems. Unless tidal deposits can be conclusively determined to be the result of completely semidiurnal systems, the periodicities involved in neap-spring cycles become equivocal. In addition to neap-spring cycles, rhythmite sequences can also exhibit larger scale, seasonal and yearly cycles. Such larger scale cycles are expressed in thicknesses of strata that are potentially resolvable in log-based subsurface studies. Based upon the magnitudes of seasonal periodicities that occur in modern tidal-station data, vertical sequences produced by seasonal tidally related sedimentation can also be modelled.

Tidal Periodicities

To understand the earth-moon dynamics involved during deposition of tidal rhythmites, we have analyzed three-year-long tidal data from several modern stations. In order to best emulate the paleogeographic location of the Carboniferous tidal rhythmites that we have described previously (see Kvale *et al.*, 1989), stations were selected from tropical and near-equatorial positions. For the following analyses, three stations are discussed. These stations represent a simplified spectrum from 1) a nearly pure semidiurnal (two tides per day) system at Kolaka, Sulawesi (Fig. 1a), 2) an intermediate system representing mixed, predominantly diurnal system at Barito River, Borneo (Fig. 1b), and 3) a nearly pure diurnal (one tide per day) station at Do Son, Vietnam (Fig. 1c).

Three data sets were generated from each three-year tidal record. The first set, referred to as the "high-tide" data, consists of the heights of the high tides. This set was constructed by selecting only the high tides and excluding the intervening low tides. The second set, referred to as "flood-tide" data, was constructed by taking the difference in height between low tides that preceded high tides. Thus this set emulates a flood-dominant system. The third set, which is termed the "ebb-tide" data, was constructed by taking the difference in height between a high tide and the following low tide. This set thus roughly approximates an ebb-dominant tidal system. These are obviously the simplest and most straightforward arrangements of the data and more complicated process-oriented approaches could be taken. Nonetheless, as will be discussed, these three approaches yield generally similar results.

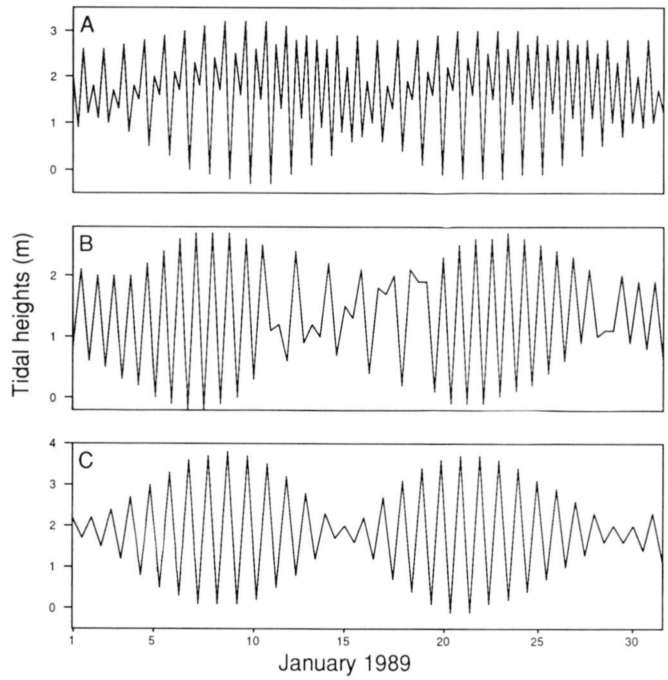

Figure 1. Predicted tidal patterns from modern equatorial stations (from data in NOAA, 1989). **A.** Kolaka, Sulawesi (lat. 4° 4′ S; long. 121° 36′ W) exhibiting a predominantly semidiurnal system; **B.** Barito River, Borneo (lat. 3° 34′ S; long. 114° 25′ E) exhibiting a mixed, predominantly diurnal system; and **C.** Do Son, Vietnam (lat. 20° 40′ N; long. 106° 49′ E) exhibiting a predominantly diurnal system.

Modern Tidal Periodicities

Before proceeding with the analyses of modern tides, a discussion detailing the various lunar periodicities is in order. Periods shorter than neap-spring periods, which include periods of two involved with semidiurnal systems, are not discussed herein (see De Boer, Oost, and Visser, 1989). Most discussions of the so-called "neap-spring" tidal periods recorded within the rock record assume that such periods are indeed related to (and forced by) the orbital position, and therefore the phases of the moon, which occur within a synodic month (full moon to full moon) with a period of about 29.53 days (Fig. 2a). (We shall designate these as "S" periods.) Modern analogs of neap-spring cycles, based upon actual sedimentation-rate measurements, have been described

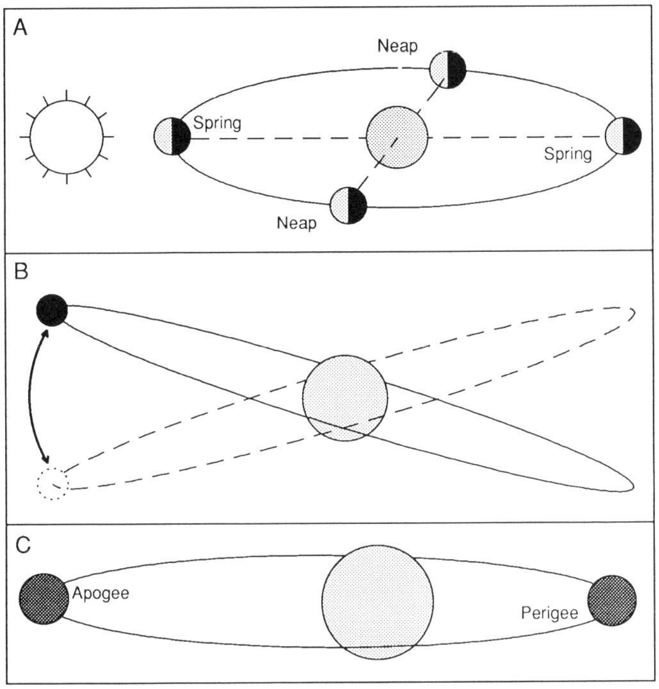

Figure 2. Main lunar periods that affect tidal heights. **A.** Synodic half-month is related to lunar phases and has a repetitive period of 14.77 days, which is the time between new and full moons. **B.** Tropical half-month is related to lunar declination and has a period of 13.66 days, which is the time required for the moon to range from a maximum northerly declination to the maximum southerly declination. **C.** Anomalistic month is related to lunar distance from the earth and has a period of 27.55 days (perigee to perigee).

by Dalrymple and Makino (1989) and Smith *et al.* (1990). Several other lunar-related tidal effects also influence tidal heights on approximately two-week and monthly periods. During a period of 27.32 days (lunar tropical month), the moon's declination ranges from its maximum northerly declination, crosses over the earth's equator to its maximum southerly declination, recrosses the equator, and then returns to its maximum northerly position (Fig. 2b). Tides controlled by the moon's declinational position relative to the earth's equator can strongly modulate or even overwhelm tidal heights related to the phases of the moon, but these tropical tides (''T'' tides) depend strongly on the latitude of the site.

Another effect arises from the changing distance of the moon from the earth during a lunar orbit (''A'' tides), the corresponding period being the anomalistic month (27.55 days), which is the time required for the moon to move from perigee to perigee (closest approach of the moon to the earth) (Fig. 2c). The various cycles have roughly the same period; however, for the following discussion, a full anomalistic month and one half of the synodic and tropical periods are of importance because in the modern areas studied, the highest tides occur coincident with lunar perigee, full and new moons, and equatorial passage of the moon. These periods will be referred to as ''A'', ''S'' and ''T'' periods respectively.

In addition to the various lunar monthly cycles, there exists a number of longer-term periodicities. These astronomical factors includes periods with wavelengths as long as thousands of years (see Macmillan, 1966; Nio and Yang, 1989); however, because our data set extends only over three years, our discussion must be limited to those periods with lengths of one year or less. Within the following discussion, these include seasonal and yearly cycles. Half-yearly tidal effects are related to the earth's axial tilt, which results in maximum solar declination during the solstices and minimum solar declination at the equinoxes. This cycle has a period of about 182.6 days; during the solstices, tidal heights as well as the diurnal inequality are maximized. Opposite effects occur during the equinoxes. Within the following discussion, these seasonal or half-yearly periods will be referred to as ''H'' periods.

Yearly variation in tidal heights is affected by the ellipticity of the earth's orbit, which results in a variation of the earth-sun distance. In addition, nonastronomical factors, including yearly variation in sea water temperature, atmospheric pressure, salinity, and direction of prevailing winds, produce long-term sea level variations, which, in turn affect tidal heights. We will refer to such periods as yearly or ''Y'' periods.

For the three modern tidal stations being analyzed, it is instructive to view the periods as described above in terms of the average number of tidal events expressed at each station. This is of course necessary when dealing with ancient tidal data. To express the cycles in terms of events, rather than in absolute time (such as hours or days), it is necessary to multiply the periods by a factor that is computed by counting the number of tidal events, such as high tides, that occurred during the three-year period, and then dividing by the number of days in the three-year period. For example, Do Son, being nearly a pure diurnal system, has a correction of 1.03, which means that over the three-year period, there were an average of 1.03 high tides per day. Conversely, the predominantly semidiurnal station at Kolaka has a correction factor of 1.93 tidal events per day. Lengths of the various periods and the periods corrected for tidal events for each station are listed in Table 1.

Table 1. Tidal periods and number of high-tide events per period

Period (code)	Length (days)	Do Son (CF=1.03)	Barito River (CF=1.38)	Kolaka (CF=1.93)
Tropical (T)	13.7	14.1	18.9	26.3
Synodic (S)	14.8	15.3	20.4	28.4
Anomalistic (A)	27.6	28.5	38.1	53.0
Seasonal (H)	182.6	189.1	252.9	352.0
Yearly (Y)	365.3	378.0	505.5	703.6

Analysis of Data

Neap-Spring Cycles

For the three selected tidal stations, harmonic spectra were generated based upon the tidal-height data. Heights of high tides were entered into computer-readable format from NOAA tidal-prediction tables (NOAA, 1987, 1988, 1989) and analyzed with the fast-Fourier program described by Horne and Baliunas (1986). Examination of the spectral analysis for the three stations (Figs. 3, 4, 5) reveals, in general, that the strongest peaks within the frequency range under discussion correspond closely to S and T periods; however, the power or magnitude of each of the periods for the diurnal systems are strikingly different from those for semidiurnal systems.

Analysis of the harmonic spectra for Do Son (Fig. 3) indicates that the T (tropical) period dominates and constitutes the primary neap-spring cycle. This is true for the high- (Fig. 3a), flood- (Fig. 3b), and ebb-dominated (Fig. 3c) data sets. The S (synodic) period is present, but is manifested only very weakly. In fact, the S period signal is barely detectable above the background "noise" of the spectra.

The spectra for the tidal station at Barito exhibit much more complexity than those from Do Son. Because Barito is a mixed, predominantly diurnal system, the spectra exhibit both T and S periodicities. In the spectrum for the high-tide data set, the T period is stronger than the S period (Fig. 4a). The flood-data spectrum (Fig. 4b) is similar; however, the ebb-data spectrum shows an inversion of the strengths of these two periods. In addition, all three data sets from Barito contain smaller scale harmonics of the S and T periodicities. These shorter-term periodicities, which may in part be produced by the processing techniques, are not readily explainable by simple astronomical parameters.

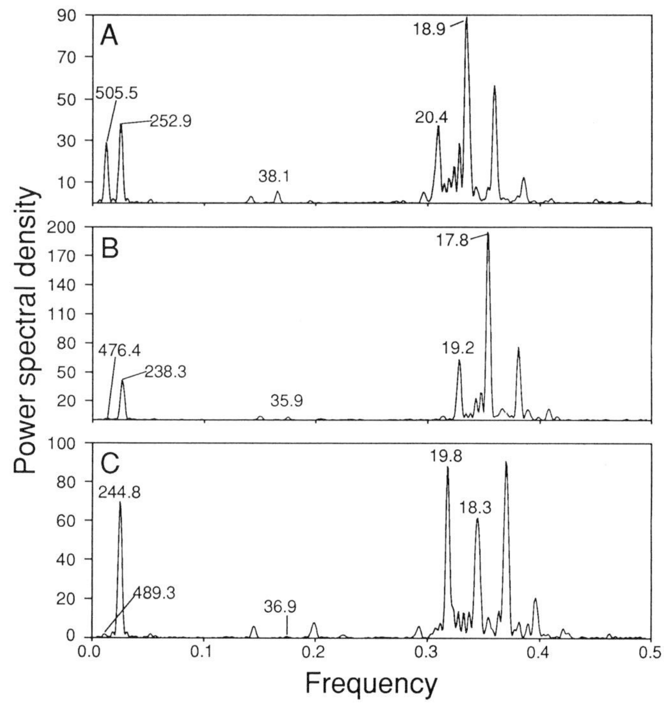

Figure 4. Power spectra from modern, mixed, predominantly diurnal tidal system at Barito River, borneo. Peaks are based upon data modelled as **A.** high tides, **B.** flood tides, and **C.** ebb tides. Spectral peaks are labeled that correspond (from left to right) to yearly, seasonal, anomalistic month, one-half synodic month, and one-half tropical month, respectively. This system is dominated by tropical periods in the high-tide (**A**) and flood-tide (**B**) models, but is dominated by synodic periods in the ebb-tidal model (**C**).

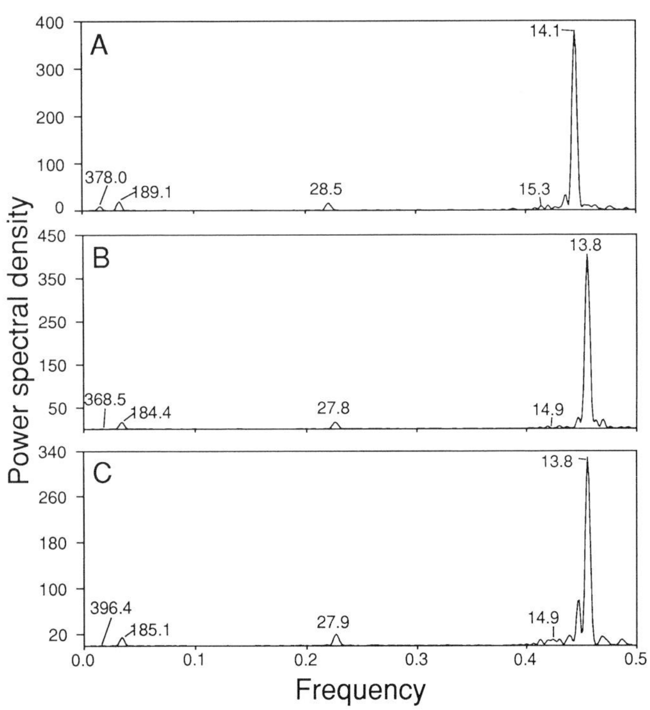

Figure 3. Power spectra from modern, predominantly diurnal tidal system at Do Son, Vietnam. Peaks are based upon data modelled as **A.** high tides, **B.** flood tides, and **C.** ebb tides. Spectral peaks are labeled that correspond (from left to right) to yearly, seasonal, anomalistic month, one-half synodic month, and one-half tropical month, respectively. This system is dominated by tropical periods, with only a subordinate expression of synodic periods.

At Kolaka, the S periods greatly dominate the harmonic spectra (Fig. 5). T periods are only weakly expressed but do constitute a significant lower-order periodicity that is well separated from the synodic peak. For the Kolaka system, difference between high-, flood-, and ebb-data sets are not significant.

Thus, Do Son and Barito (predominantly diurnal systems) suffer from mixing forces and periods are controlled by T periods (related to the declination of the moon) while Kolaka (a predominantly semidiurnal system) responds, as expected, more to the S periods (related to the phases of the moon). In addition, the T periods, which dominate the diurnal systems, are present but are of lesser magnitude within the semidiurnal system of Kolaka, Sulawesi.

Although these analyses are based upon only a few tidal stations, it is important to note that the dominant controlling period is not always the S period. In these examples, diurnal systems have more dominant T periods whereas the semidiurnal system has a more dominant S period. Systems such as Barito and Kolaka exhibit both periods; however, S periods only dominate as the systems becomes primarily semidiurnal. This difference in periods becomes especially apparent when high-tide envelopes are plotted for an entire

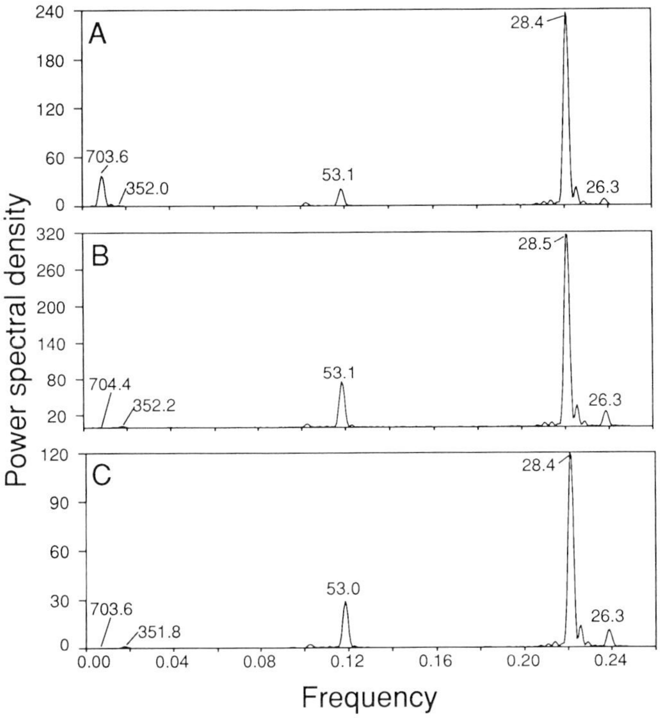

Figure 5. Power spectra from modern, predominantly semidiurnal tidal system at Kolaka, Sulawesi. Peaks are based upon data modelled as **A.** high tides, **B.** flood tides, and **C.** ebb tides. Spectral peaks are labeled that correspond (from left to right) to yearly, seasonal, anomalistic month, one-half synodic month, and one-half tropical month, respectively. This system is dominated by synodic periods with only a weak expression of the tropical period.

year (Fig. 6). The predominantly diurnal station at Do Son exhibits 27 neap-spring cycles whereas only 25 neap-spring cycles occur during the same time period at Kolaka. The somewhat shorter duration of the T period, which is controlling tides at Do Son, creates more cycles per year than areas in which S periods dominate, such as Kolaka. This difference between T and S periods is slight when a few successive neap-spring periods are plotted; however, the difference becomes highly significant when applied to ancient systems where attempts are being made to resolve earth-moon geometries.

Anomalistic-Month Periods

All three stations exhibit effects of the A period (anomalistic month) to varying degrees. At Do Son, A periods are exhibited for high-, flood-, and ebb-tidal data (Fig. 3). At Barito, these periods are most strongly expressed in the spectrum for the high-tide data (Fig. 4a). Within the spectra for the flood- and ebb-data, neither exhibits a significant A periodicity. At Kolaka, A periods are strongly expressed for all data types (Fig. 5). In fact, these periods are stronger than the tropical and most of the longer-term periods.

Longer-Term Periods

Half-yearly, or H, periods are exhibited to varying degrees at each tidal station. At Do Son, the H period effect is much more pronounced than a yearly, or Y, period (Fig. 3). Only with the high-tide data (Fig. 3a) is the yearly cycle of sufficient strength to result in a pronounced harmonic; neither flood- or ebb-data exhibit such a prominent yearly peak. For the Barito data, H periods are strongly expressed within all the harmonic spectra. In the high- (Fig. 4a), flood- (Fig. 4b), and ebb-data (Fig. 4c), the seasonal period is very strong and is much more significant than the shorter-term anomalistic period. The H period is especially strongly developed within the ebb-data set. Regarding Y periods (yearly cycle), only

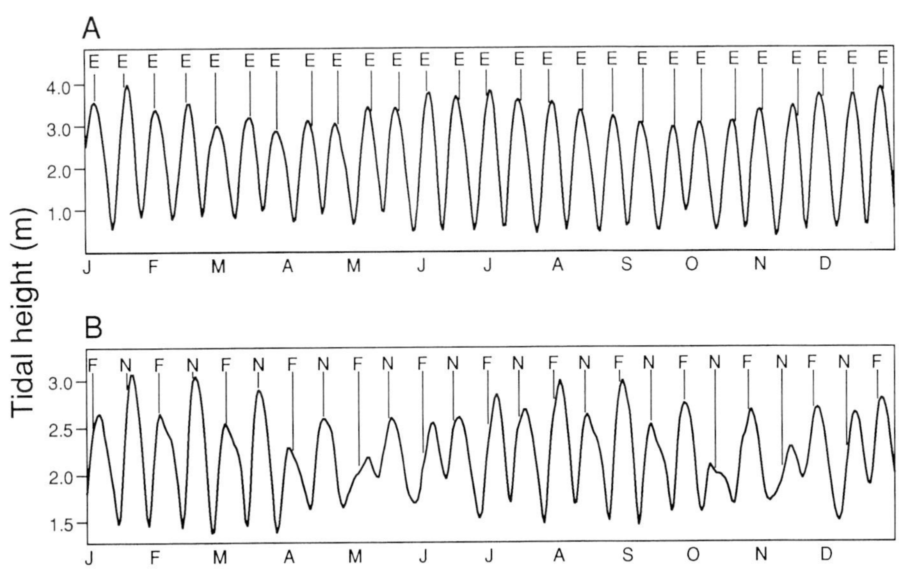

Figure 6. Envelope of highest tides per day for **A.** Do Son, Vietnam (diurnal; 377 high tides per year), and **B.** Kolaka, Sulawesi (semidiurnal, 707 high tides per year). "Spring" (highest) tides at Do Son are more in phase with equatorial (E) position of moon during tropical month. Thus, there are approximately 27 periods of highest tides during the year. Conversely, highest tides for Kolaka are more in phase with the synodic month and high tides occur during full (F) and new (N) moons. Thus, there are only about 25 periods of highest tides during the year.

the high-tide data (Fig. 4a) exhibit a strong cycle of this periodicity. Neither the flood- or ebb-data exhibit strong Y periodicities. At Kolaka, H (seasonal) periodicities are not strongly expressed for any of the various permutations of the data (Fig. 5). Similarly, Y periods are not strongly developed except for the high-tide data set (Fig. 5a).

Discussion

Neap-Spring Cycles

These analyses have a number of important implications for some of the interpretations of ancient tidal rhythmites. Of principal importance is the observation that neap-spring cycles can occur that are not directly related to the lunar phases (synodic month). In a diurnal system (such as Do Son), the neap-spring cycles are directly related to the lunar tropical month. For many geological applications, such as those which are merely attempting to document processes of tidal sedimentation, the difference between synodic and tropical periodicities is relatively unimportant. Conversely, these small differences become of paramount importance in studies (*e.g.*, Williams, 1989) that use ancient tidal rhythmites to delineate the number of lunar months or days per year in the geological past. The difficulty in differentiating between synodic and tropical periods within ancient tidal rhythmites could easily lead to wrongful assumptions regarding their utility in paleoastronomical reconstructions. For ancient systems, misinterpreting a tropical-month dominated, diurnal system as a synodic month-dominated system would lead to the wrong conclusion of more lunar months per year. Only within predominantly semidiurnal systems can the periods that produced neap-spring cycles be readily distinguished.

Long-Term Cycles

Yearly cycles, based upon counting of neap-spring cycles, have been described from rhythmic tidalites by a number of workers, including Precambrian materials (Sonett *et al.*, 1988; Williams, 1989). Similar cycles have been described by us from the Hindostan whetstones (Archer and Kvale, 1989) as well as age-equivalent strata in Indiana (Fishbaugh *et al.*, 1989). Yearly cycles, with an inferred seasonal component have also been described from a Cretaceous tidal-estuarine sequence by Rahmani (1989) and from Miocene molasse (Tessier and Gigot, 1989). This type of cycle has been described in detail previously; however, it would be instructive, using modern data, to compute how accurately the number of days per year can be inferred from tidal data.

Before such backward calculations on the number of days per year can be generated, it is necessary to know the distribution of the number of high-tide events per neap-spring cycle. Because of the various tide-producing astronomical periodicities, the number of high-tide events per cycle is not constant but exhibits a range that is more-or-less normally distributed. For the diurnal system at Do Son, Vietnam, for the three-year period under investigation, the number of

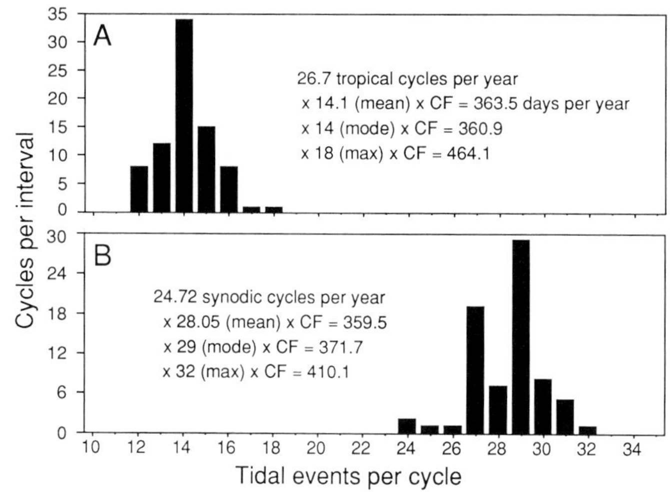

Figure 7. Histograms illustrating range in tidal events per neap-spring cycle for **A.** Do Son, Vietnam, and **B.** Kolaka, Sulawesi. Given the number of cycles per year and the correction factor (CF, or average number of tidal events per day), back computation of the number of days per year can be attempted. Use of mean value for tidal events per neap-spring cycle results in a relatively accurate prediction of days per year; conversely, use of the mode or maximum number of tidal events per cycle yields days-per-year values that are not accurate. In particular, use of the maximum number yields days-per-year estimates that are well in excess of actual values.

events ranges from 12 to 18 with a mean about 14.1 and a mode of 14 high tides per tropical period (Fig. 7a). Obviously, use of the mean value will result in the closest back-calculation; multiplying the mean by the number of tropical cycles by the correction factor (discussed previously) will result in a value of 363.5 days per year. This value is very close to the actual value. Conversely, using values other than the mean, such as the maximum value, results in days-per-year values that are much too high (464.1 days/year).

Applying the same analyses to a semidiurnal system, such as Kolaka, Sulawesi, indicates that the number of high-tide events per neap-spring cycle exhibits a range of 24 to 32 with a mode of 29 and a mean of 28.05 (Fig. 7b). Use of the mean value results in a back calculation of 359.5 days per year, whereas use of the maximum value results in an overestimation of 410.1 days per year. The overly high values for days per year by use of the maximum value in both examples is of importance because it is the technique advocated by Williams (1989) in his reconstructions on the number of days per year for the Precambrian. It is apparent that such computations are problematic when laminae counts for only a few neap-spring cycles have been generated. In fact, Williams' (1989) estimate of 400±20 days per year during the Late Precambrian, which was based upon the maximum number of laminae observed within a neap-spring cycle, is well within the range of the over-estimations described above, using the maximum number of events per neap-spring cycle, for both modern systems.

It should be noted, unfortunately, that computation of the "correction factor" described above will be extremely difficult for ancient systems unless certain special assumptions hold true. This is especially so if, as seems true for the ancient systems analyzed here, the depositional environments of the systems prevented the complete and nearly perfect preservation of each tidal event. Compression of the tidal signal would result in a perfectly semidiurnal system appearing to be mixed and for a perfectly diurnal system to appear abbreviated. Because of the compressional factor and the difficulty in resolving the degree of the compression, it will be difficult, it not impossible, to actually count the number of days per neap-spring cycle for ancient systems. Thus simple computations of the number of days per year in the geological past based on tidal rhythmites (*e.g.*, Williams, 1989), might contain inaccuracies. Conversely, more mathematically sophisticated techniques, such as those that rely upon the ratios between the various periodicities (see Sonett *et al.*, 1988) are probably more accurate in resolving ancient earth-moon geometries.

Subsurface Delineation

Although the finer scale cycles are too minor, the larger scale cycles are potentially recognizable in subsurface analyses. Seasonal and yearly cycles in tidal sedimentation, which affect not only the laminae thicknesses but also potentially the sand/clay ratios incorporated into each laminae, can be modelled and the effects on logs simulated. Based upon a five-year record for a semidiurnal tidal systems (Fig. 8a), the log response for such a sequence can be simulated once several assumptions regarding the sedimentology are made. First, we can assume that mud deposition (via clay drapes) is nearly constant, but that sand deposition is related to tidal height. In other words, there are thicker sand laminae emplaced with higher tides. Based upon such a relationship of tidal heights to sedimentation, a gamma-ray trace of such a section (assuming continuous preservation) would essentially be the inverse of the tidal-height curve. However, continuous accretion of tidal laminae cannot occur over extended periods of time because the accommodation space will become filled. This loss of accommodation space could occur either because of infilling of the site of deposition or be related to peat compaction in those rhythmites that are deposited over coals (see Kvale and Archer, 1990). Such a gradual loss of accommodation space can be simulated as a log-normal function in which there is a high potential for event preservation at the base of the section with a progressively decreasing potential toward the top of the rhythmite section (Fig. 8b). This type of selective preservation of tidal events, with high degree of preservation at the base of the sequence and low degree of preservation at the top, together with the sand/clay relationships discussed above, would result in a simulated gamma-ray log (Fig. 8c)

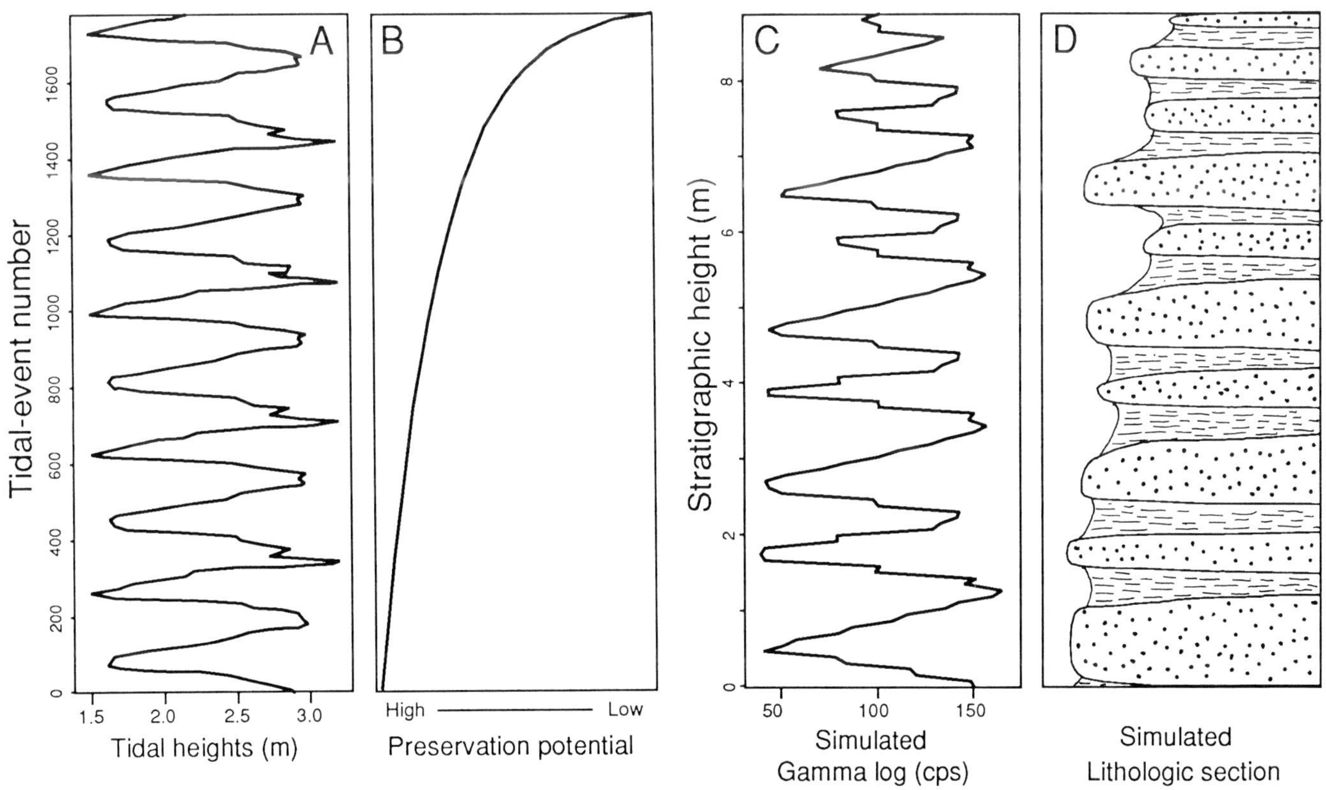

Figure 8. Simulation of subsurface, gamma-log expression of long-term tidal cycles. **A.** Filtered five-year sequence of tidal data from Kolaka, Sulawesi indicating seasonal and yearly tidal cycles. **B.** Inferred preservational potential of tidal events based upon hypothetical reduction of accommodation space by filling of channel or by underlying peat compaction (see Kvale and Archer, 1990). **C.** Resultant simulated gamma-log expression illustrating selective preservation of five-year tidal curve. This model, which produces a pattern similar to Kvale and Archer (*this volume*, Fig. 4), results in a characteristic log signature of upwardly decreasing cycle thickness and amplitude. **D.** A typical vertical sequence, consisting of alternating beds of predominantly sandstone and shale, that could be produced by this type of tidal sequence.

that delineates metre-scale cycles. The simulated vertical sequence indicates that although average grain size has remained approximately the same, the amplitude and the thickness of the cycles became reduced as the system filled the available accommodation space. This simple model is capable of producing logs that closely emulate Carboniferous tidalite sequences (see Kvale and Archer, *this volume*, Fig. 4).

Conclusions

Because of the various, nearly in-phase periods of the forces that control tidal-height fluctuations, neap-spring periods are, in predominantly diurnal systems, controlled by the lunar tropical month. Conversely, in semidiurnal systems, neap-spring periods are predominantly controlled by the more well-known lunar synodic month. These differences are relatively minor; however, tropical-month dominated systems will experience about two more neap-spring cycles per year than will synodic-month dominated systems. This difference becomes critical when time intervals and earth-moon geometries are extrapolated from ancient tidal rhythmites.

Larger scale cycles were produced in the modern systems by seasonal as well as yearly variation in tidal heights. Such cycles potentially achieve magnitudes that can be recorded in log-based subsurface analyses.

References

Archer, A. W. and Kvale, E. P. 1989. Seasonal and yearly cycles within tidally laminated sediments: an example from the Pennsylvanian of Indiana, U.S.A. *In:* Geology of the Lower Pennsylvanian in Kentucky, Indiana, and Illinois, J. C. Cobb (ed.). Illinois Basin Studies 1, Illinois Basin Consortium (Indiana Geological Survey, Illinois Geological Survey, Kentucky Geological Survey), p. 45-56.

Dalrymple, R. W. and Makino, Y. 1989. Description and genesis of tidal bedding in the Cobequid Bay — Salmon River estuary, Bay of Fundy, Canada. *In:* Sedimentary Facies in the Active Plate Margins, A. Taira and F. Masuda (eds.). Tokyo, Terra Scientific Publishing Company, p. 151-177.

De Boer, P. L., Oost, A. P. and Visser, M. J. 1989. The diurnal inequality of the tide as a parameter for recognizing tidal influences. Journal of Sedimentary Petrology, v. 59, p. 912-921.

Fishbaugh, D. A., Kvale, E. P. and Archer, A. W. 1989. Association of tidal and fluvial sediments within Lower Pennsylvanian rocks, Turkey Run State Park, Parke County, Indiana. Guidebook for 1989 AAPG Eastern Section Meeting, Bloomington, Indiana, 46 p.

Horne, J. H. and Baliunas, S. L. 1986. A prescription for period analysis of unevenly sampled time series. Astrophysical Journal, v. 302, p. 757-763.

Kvale, E. K. and Archer, A. W. 1990. Tidal deposits associated with low-sulfur coals, Brazil Fm. (Lower Pennsylvanian), Indiana. Journal of Sedimentary Petrology, v. 60, p. 563-574.

______ and ______ (*this volume*). Characteristics of two Pennsylvanian-age semidiurnal tidal deposits in the Illinois Basin, U.S.A. *In:* Clastic Tidal Sedimentology, D. G. Smith, G. E. Reinson, B. A. Zaitlin and R. A. Rahmani (eds.). Canadian Society of Petroleum Geologists, Memoir 16.

______, ______ and Johnson, H. R. 1989. Daily, monthly, and yearly tidal cycles within laminated siltstones (Mansfield Formation: Pennsylvanian) of Indiana. Geology, v. 17, p. 365-368.

MacMillan, D. H. 1966. Tides: CR Books Limited, London, 240 p.

Nio, S. D. and Yang, C. S. 1989. Recognition of tidally-influenced facies and environments. International Geoservices BV, Short Course, Note Series 1, Leiderdorp, Netherlands, 230 p.

NOAA, 1987. Tide tables, 1989, high and low water predictions, Central and Western Pacific Ocean and Indian Ocean. National Oceanic and Atmospheric Administration, U.S. Department of Commerce, 6501 Lafayette Ave., Riverdale, Maryland.

______ Tide tables, 1989, high and low water predictions, Central and Western Pacific Ocean and Indian Ocean. National Oceanic and Atmospheric Administration, U.S. Department of Commerce, 6501 Lafayette Ave., Riverdale, Maryland.

______ Tide tables, 1989, high and low water predictions, Central and Western Pacific Ocean and Indian Ocean: National Oceanic and Atmospheric Administration, U.S. Department of Commerce, 6501 Lafayette Ave., Riverdale, Maryland.

Rahmani, R. 1989. Cretaceous tidal estuarine and deltaic deposits, Drumheller, Alberta. Canadian Society of Petroleum Geologist's field guide to: Second International Research Symposium on Clastic Tidal Deposits, August 22-25, Calgary, Alberta, 55 p.

Sonett, C. P., Finney, S. A. and Williams, C. R. 1988. The lunar orbit in the late Precambrian and the Elatina sandstone laminae. Nature, v. 335, p. 806-808.

Smith, N. D., Phillips, A. C. and Powell, R. D. 1990. Tidal drawdown: a mechanism for producing cyclic sediment laminations in glaciomarine deltas. Geology, v. 18, p. 10-13.

Tessier, B. and Gigot, P. 1989. A vertical record of different tidal cyclicities: an example from the Miocene Marine Molasse of Digne (Haute Provence, France). Sedimentology, v. 36, p. 767-776.

Williams, G. E. 1989. Late Precambrian tidal rhythmites in South Australia and the history of the Earth's rotation. Journal of the Geological Society of London, v. 146, p. 97-111.

KEY WORDS: TIDAL RHYTHMITES, TIDAL PERIODICITIES, NEAP-SPRING PATTERNS, EQUATORIAL

Tidal Inlets,
Deltas and Flats

In: Clastic Tidal Sedimentology, D. G. Smith, G. E. Reinson, B. A. Zaitlin and R. A. Rahmani (eds.). Canadian Society of Petroleum Geologists. Memoir 16, p. 199-218.

Ebb-tidal delta deposits along the west Frisian Islands (The Netherlands): processes, facies architecture and preservation

L. P. SHA[1] AND P. L. DE BOER

Comparative Sedimentology Division,
Institute of Earth Sciences, University of Utrecht,
P.O. Box 80.021, 3508 TA Utrecht, The Netherlands

ABSTRACT

The morphologies of the ebb-tidal deltas of Texel, Terschelling and Ameland inlets and the resulting sand transport patterns are greatly influenced by the interaction of shore-parallel tidal currents with tidal currents through the inlets. These three inlets and their ebb-tidal deltas are similar in many aspects, and together they differ from North American Atlantic coast examples and from those of the East Frisian Islands. Sedimentary facies and sequences, and sand body geometry of these three inlets/ebb-tidal deltas are strongly controlled by this interaction of offshore tidal currents, tidal currents through the inlets, and waves. The interaction and relative strengths of both tidal current systems determine the flow patterns and orientations of the main ebb channels in the ebb-tidal deltas. Different orientations of the (asymmetrical) ebb-tidal deltas as compared to those along the East Frisian Islands and the U.S. east coast, as well as a change of the general orientation of the main ebb channel of the ebb-tidal delta of Texel Inlet since the 16th century, are related to differences and changes in the relative importance of tidal currents and waves. Waves and longshore drift tend to force the ebb-tidal deltas into a downdrift asymmetry, whereas the shore-parallel tidal currents tend to force the ebb-tidal delta into an updrift asymmetry. Under the present regime of waves and longshore drift in this area, the critical boundary between updrift and downdrift asymmetry appears to be at a tidal prism on the order of 500 x 10^6 m^3. The volume of ebb delta sand bodies is approximately 1 km^3 (Texel), a considerable part of which can be preserved in the fossil record. Holocene deposits off the West Frisian Islands indeed show that preservation of ebb-tidal delta deposits during a rise of relative sea level is possible under certain conditons. The strength of tidal influence is of great importance for determining both the basal boundary and the seaward extent of the ebb delta systems. Thus, it is inferred that preservation potential of ebb-tidal delta deposits can be positively correlated with the relative dominance of ebb-tides during their formation.

RÉSUMÉ

Les morphologies des deltas de marée de type reflux des petits bras de mer de Texel, Terschelling et Ameland, ainsi que les modèles de transport de sable associés sont grandement influencés par l'interaction des courants de marée parallèles à la côte et des courants de marée à travers les petits bras de mer. Ces trois petits bras de mer et leurs deltas de marée de type reflux sont en plusieurs points similaires entre eux, mais ils diffèrent toutefois des exemples provenant de la côte atlantique de l'Amérique du Nord et des îles Frisonnes orientales. Pour les trois exemples étudiés, les faciès et séquences sédimentaires, et la géométrie du corps sableux sont fortement contrôlés par l'interaction entre les courants de marée littoraux, les courants de marée à travers les petits bras de mer et les vagues. L'interaction et la force relative des deux systèmes de courants de marée déterminent le patron d'écoulement et l'orientation des principaux chenaux de reflux des deltas. Les différentes orientations des deltas asymétriques de marée de type reflux, comparables à celles des îles Frisonnes orientales et de la côte est des États-Unis, et le changement de l'orientation générale du principal chenal de reflux du delta du petit bras de mer Texel depuis le 16ième siècle, sont reliés aux différences et modifications de l'importance relative des courants de marée et des vagues. L'asymétrie des deltas de marée de type reflux diffèrent selon les composantes; les vagues et la dérive littorale ont tendance à produire une asymétrie dans le sens de la dérive littorale tandis que les courants de marée parallèles à la côte conduisent à une asymétrie inverse au sens de la dérive littorale. Sous le régime actuel des vagues et de la dérive littorale de cette région, un prisme de marée de l'ordre de 500 x 10 m^3 semble être la limite critique entre une asymétrie dans le sens de la dérive littorale et une asymétrie inverse au sens de celle-ci. Le volume des corps sableux du delta de marée atteint jusqu'à 1 km^3 (Texel) dont une quantité importante peut être préservée dans les archives géologiques. Dans les dépôts holocènes au large des îles Frissonnes occidentales, les dépôts deltaïques de type reflux, déposés durant une hausse relative du niveau de la mer, sont effectivement préservés sous certaines conditions. La force de l'influence des marées est importante pour déterminer tant la limite basale que l'étendue vers la mer des systèmes de delta de type reflux. Alors, il faut en déduire que le potentiel pour la préservation des dépôts de delta de marée de type reflux peut être positivement corrélée avec la dominance relative des marées descendantes durant leur formation.

[1]Present address: Geological Survey of the Netherlands, P.O. Box 157, 2000 AD Haarlem, The Netherlands

Introduction

Ebb-tidal delta deposits are accumulations of sandy sediment at the seaward side of tidal inlets (Hayes, 1975). They depend on the presence of barrier islands, which largely are the result of wave action and tidal currents through the tidal inlets. Generally micro-and mesotidal regimes are considered to favour the evolution and maintenance of tidal inlets (Hayes, 1979, 1980), but it should be pointed out that the tidal prism rather than the tidal amplitude defines the strength of tidal currents within and adjacent to the tidal inlets (Sha, 1989c).

Waves and tidal currents through the inlets have been demonstrated to be responsible for the formation and morphodynamic behaviour of ebb-tidal deltas (Bruun and Gerritsen, 1959; O'Brien, 1969; Oertel, 1972; Goldsmith *et al.*, 1975; Bruun, 1978). Thus, regional differences in the geometry of ebb-tidal deltas are logically explained on the basis of differences in wave climate and tidal range (Oertel, 1975; Hayes, 1975, 1979, 1980; Hubbard *et al.*, 1977; Nummedal *et al.*, 1977; Nummedal and Fischer, 1978). Facies and sequence models have been proposed by Oertel (1973) and Imperato *et al.*, (1988) in relation to the controlling dynamic processes (Hubbard *et al.*, 1979). It is to be expected that sand body architecture and the preservation potential of ebb delta systems are closely related to the morphodynamics. Studies about this relationship are scarce, however, and comparisons between ancient examples and modern environments are not common. This paper focusses on processes, sedimentary facies and sequences, sand body architecture, and preservation potential of ebb-tidal delta deposits in the West Frisian Islands (Figs. 1, 2).

Regional Setting of the West Frisian Islands

The tidal wave in the North Sea rotates anticlockwise around three amphidromic points. Along the Dutch coast is progrades from the southwest to the northeast, and produces nearshore reversing tidal currents parallel to the

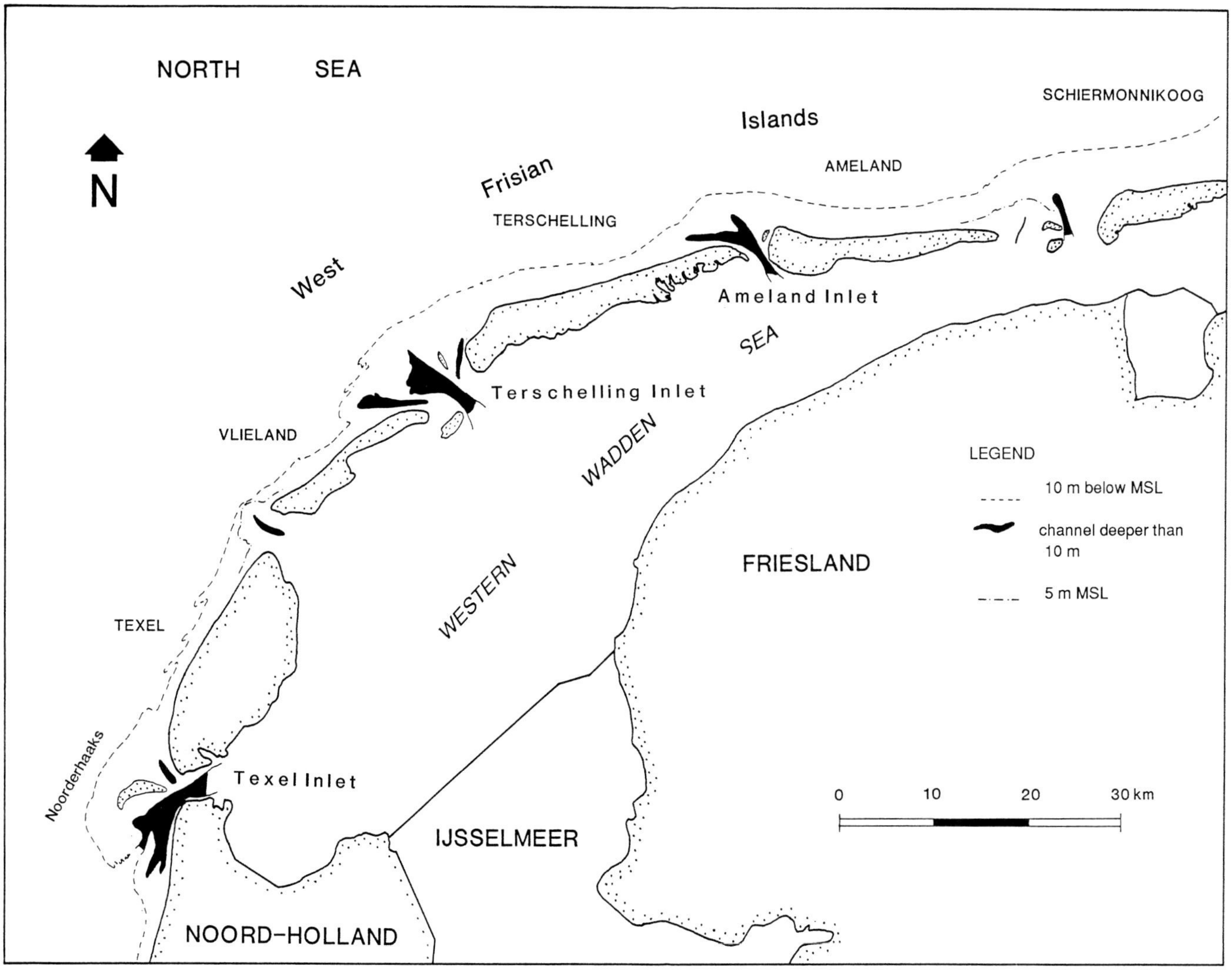

Figure 1. Location of the Texel Inlet, Terschelling Inlet and Ameland Inlet ebb-tidal deltas. Ten-metre depth lines indicate the position of the ebb-deltas; black areas are channels within the ebb-tidal deltas.

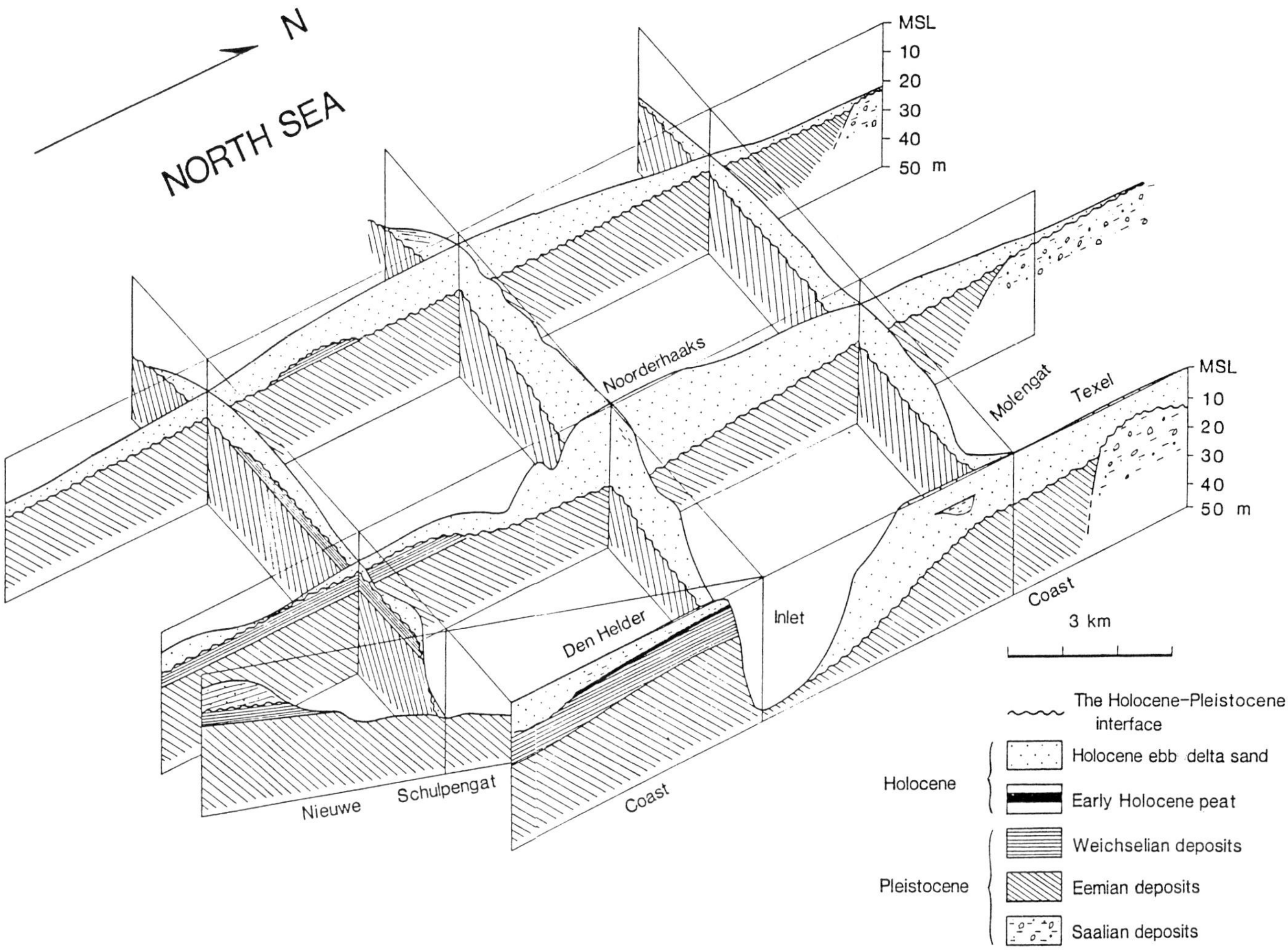

Figure 2. Fence diagram of the Texel ebb delta sand body. Five stratigraphic units are shown. The Saalian deposits at the base, found in the northern part of the ebb-tidal delta, are of glacial origin. They are poorly sorted, silty, clayey sands with gravel. The Eemian deposits above are coarse to fine grained, shell-rich sand covered by a grey, firm, clay layer. The overlying Weichselian deposits are mainly clayey silt and fine grained sand with interlayers of peat in the upper part and fine grained sand in the lower part. Peat layers at the base of the Holocene represent the early transgressive phases, and are found only onshore along the coast of Den Helder. The recent ebb delta sands consist of fine to medium grained sand with abundant marine shell fragments.

coast. The tidal range varies from 1.4 m near Texel Inlet to about 3 m in the German Bight (Dijkema *et al.*, 1980). Related to this increase in tidal amplitude toward the northeast, tidal inlets become wider and barrier islands smaller. The predominant wave direction is from the west, producing a net northeastward longshore drift. The average wave height is about 1.2 m off Texel. A net northeastward, longshore, residual current results from predominantly southwesterly winds along the Dutch North Sea coast and from deformation of the tidal wave in the nearshore shallow waters (Van Straaten, 1961; Nummedal and Fischer, 1978; Postma, 1982; Sha, 1989b). The real residual current pattern is very complicated and varies with the actual wind conditions (*cf.* Van der Giessen *et al.,* 1990).

Most of the ebb-tidal deltas along the Wadden Islands face the north and northwest, that is, they are asymmetrical to the right when seen from the land (Fig. 8, see below). However, the inlets of Texel, Terschelling, and Ameland face the southwest (Fig. 1).

Processes in the Ebb-Tidal Deltas

Sediment Transport Patterns

Sediment transport in the ebb-tidal delta of Texel Inlet was studied by means of current measurements, echo-sounding measurements of bedform asymmetry, measurements of foreset orientation and comparisons with histori cal morphological profiles. The results show mostly flood-oriented net sediment transport in the minor tidal channels and over the shoals (Fig. 3). Ebb-dominated sand transport occurs in the larger and deeper channels. Sand, which is transported to the seaward margin of the ebb-tidal delta by the ebb flow through the southwest-directed main ebb channels, is carried north by flood-dominated residual tidal currents along the ebb delta front. Landward sediment transport dominates the northern part of the ebb-tidal delta. Sand is deposited north of the ebb delta shoal 'Noorderhaaks' owing to the weak, rotational tidal currents (Fig. 3). In that area, waves transport sand shoreward, especially

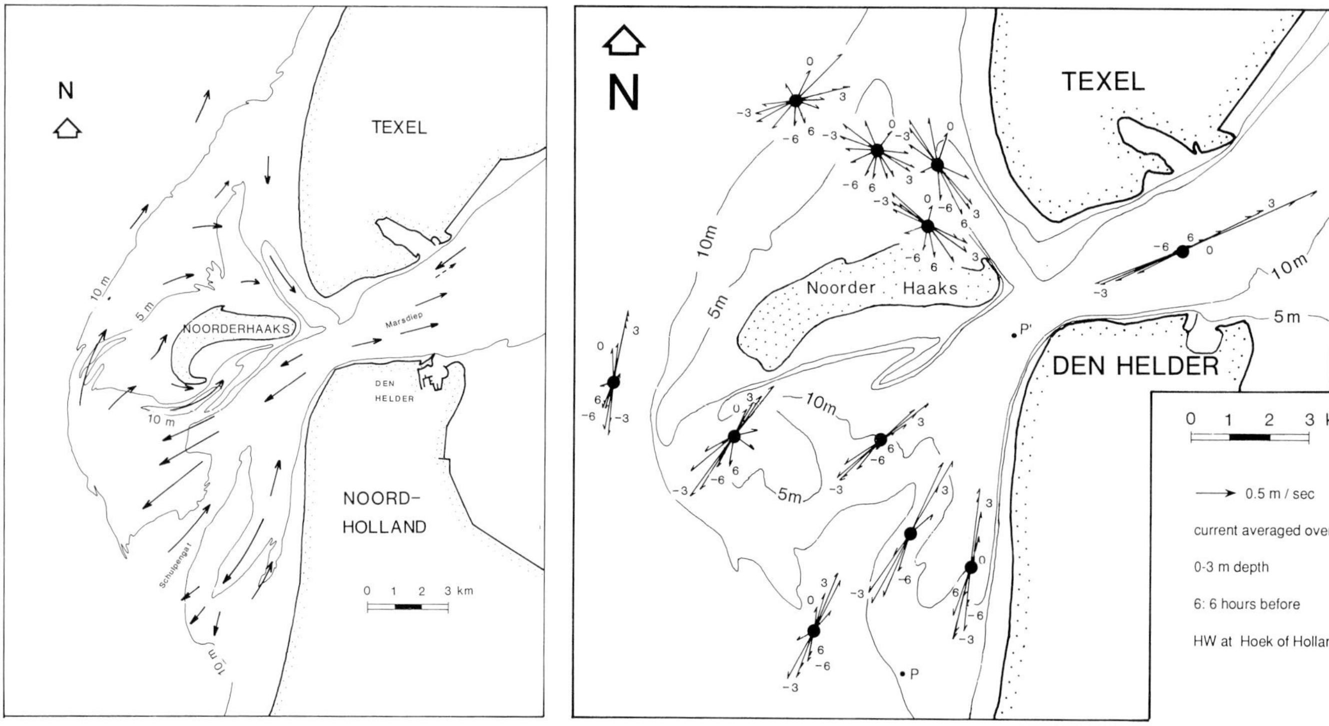

Figure 3. (**Left**) Schematic current and sand transport pattern in the ebb-tidal delta of Texel Inlet. Arrows indicate the net sand transport directions. Morphological situation as in 1985. (**Right**) Tidal current roses for a tidal cycle in the ebb-tidal delta of Texel Inlet integrated from the "Stroomatlas van Nederland" (MMAH, 1963). Figures indicate time in hours before or after high water at the Hook of Holland. Arrow length represents current velocities, averaged from the water surface to 3 m below the surface. Morphological situation as in 1963.

in the form of swash bars. Part of the sand returns to the inlet through the flood-dominated channel 'Molengat' in the north (Figs. 2, 3). The orientations of the larger bedforms in the channels (measured at different times during the tidal cycle) do not change in response to changing ebb and flood tides (Fig. 4). Orientations of foresets in the upper decimetres of the sediment generally are the same as those of the bedforms at the surface. Sand transport estimated on the basis of current velocity data (Sha, 1989b) is in accordance with the pattern shown in Figure 3. Despite the continuous rise of relative sea level, large areas of the Wadden Sea continue to be dominated by sand. Thus, there must exist a dominance of sand transport through the inlets by flood tidal currents. De Boer (1979) calculated a figure of 17 x 10^6 m^3 per year for deposition of sediment (largely sand) in the western part of the Wadden Sea. Large sand waves indeed indicate such a transport through the inlet. The North Sea and the adjacent coast are the sources of the ebb-tidal delta sands as well as of much or all of the sand in the Wadden Sea.

Sediment transport patterns similar to those in the Texel ebb-tidal delta are found in the ebb-tidal deltas of Terschelling Inlet and Ameland Inlet (Fig. 5).

Historical morphological profiles indicate erosion and deepening of the channels, and seaward progradation of the ebb delta lobe in the southern part of the ebb-tidal delta of Texel Inlet (Fig. 6). In the northern part of the ebb delta, however, the ebb-tidal delta shoal has been subjected to partial erosion since at least 1896. Such progradation and erosion patterns are caused by large-scale changes in the dynamic regime, resulting from an increase in the tidal prism as discussed below.

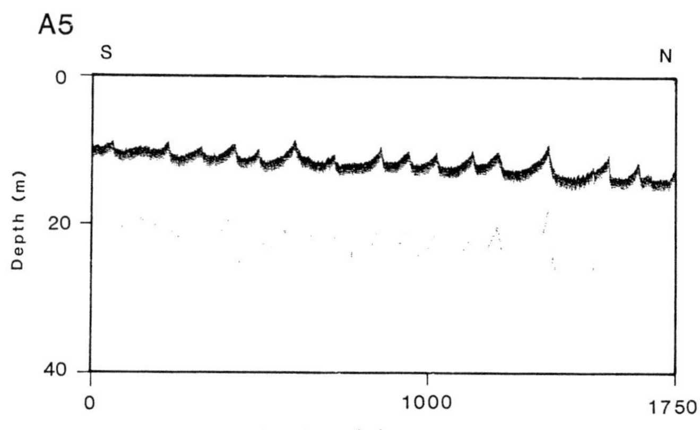

Figure 4. Echo sounding profile recorded at mid-ebb, showing the flood-oriented bedforms. Location: channel east of most southeasterly rose diagram in Figure 3b.

Effect of Tidal Current Interaction

Ebb-dominated sediment transport in all these ebb-tidal deltas is directed to the southwest. This is a result of the interaction between nearshore tidal currents parallel to the coast with tidal currents through the inlets (Fig. 7). Thus strong, bidirectional tidal currents are produced on the left side of the ebb deltas, and a weak, rotational tidal current

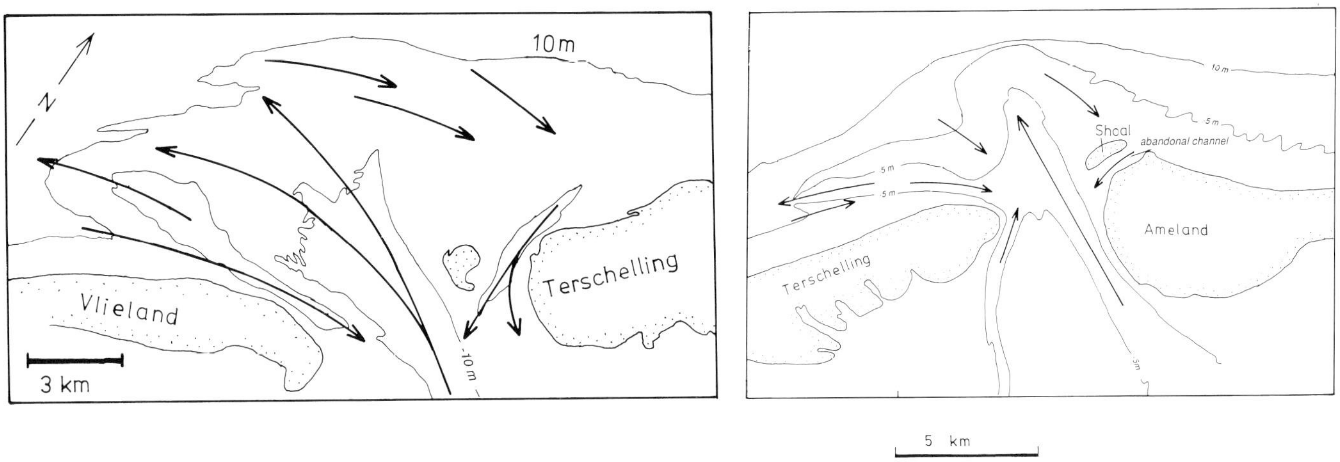

Figure 5. Schematic sand transport pattern in the ebb-tidal deltas of Terschelling Inlet (modified after Kool and de Boer; 1983) and of Ameland Inlet (modified after Lubberts and De Vos, 1987).

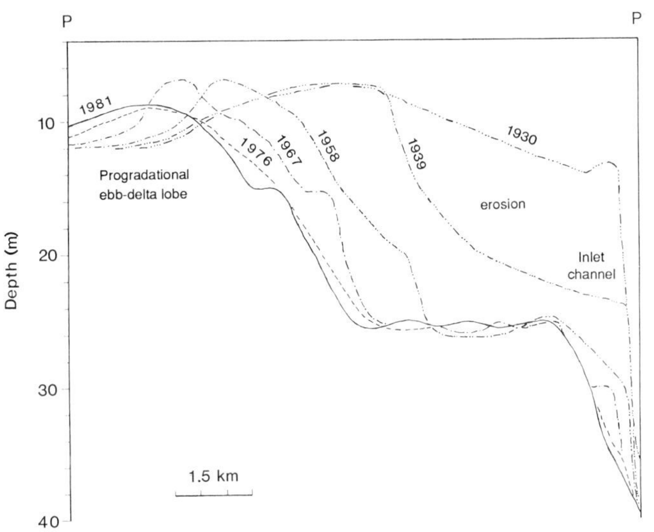

Figure 6. Geomorphological section P-P' showing the evolution of the Nieuwe Schulpengat channel since 1930. Inlet channel at the right and progradational ebb delta lobe to the left. Based on loading maps of the Ministry of Waterworks. For location see Figure 3.

pattern on the right side (Fig. 3). This tidal current pattern favours shoal development in the part of the ebb-tidal delta northeast of the inlet, and channel development to the southwest.

Morphological Patterns

Texel, Terschelling, Ameland Ebb-Tidal Deltas

The most important features of the three larger inlets (Texel, Terschelling, Ameland) are very similar. The ebb-tidal deltas and the main ebb channels are asymmetrical to the left (Figs. 3, 5), the large, ebb-dominated channels have clear seaward-prograding distal lobes (Fig. 6), and, smaller flood-dominated channels commonly develop on the landward side, adjacent to the main ebb channels.

On the north-northeast side of the inlets, sandy tidal delta shoals occupy large subtidal to supratidal areas (up to 6.5

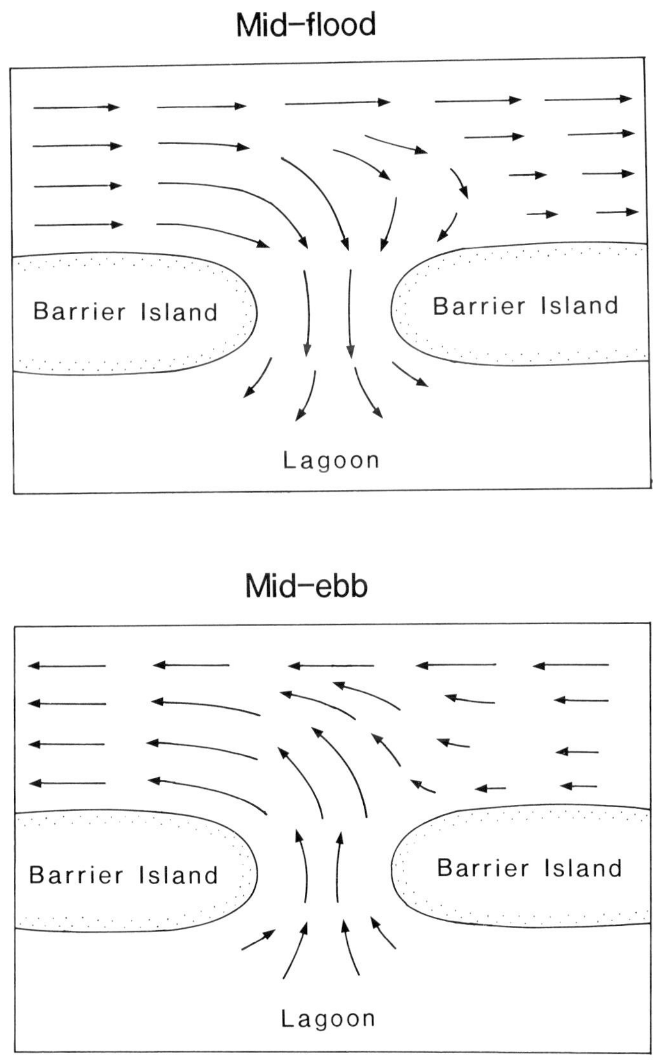

Figure 7. Schematic current patterns at mid-flood and mid-ebb showing the effect of the interaction of tidal currents parallel to the coast with the tidal currents through the inlet. Thus, strong, bidirectional currents are produced on the southwest side of the ebb-tidal deltas, and a weak, rotational tidal current pattern occurs on the northeast side. See also Figure 3b.

km²). Flood-dominated channels, which occur northeast of the shoals, tend to silt up, and become abandoned through attachment of the shoals to the island to the northeast.

Smaller Ebb-Tidal Deltas

Unlike the above three large ebb-tidal deltas, the smaller ebb deltas along the Wadden Sea coast and their main ebb channels are asymmetrically oriented to the east (Fig. 8). Explanations for such an orientation are based on the fact that wind-driven waves attack the coast predominantly from the west, producing an easterly longshore drift that forces the ebb-tidal deltas to be asymmetrical to the east (Luck, 1976; Nummedal and Penland, 1981).

Cause of Updrift and Downdrift Asymmetry

The difference between the large and the small ebb-tidal deltas seems to be controlled by the relative importance of tidal currents (offshore and through the inlet) on the one hand, and wave-generated currents on the other. Of the three larger ebb-tidal deltas with an updrift asymmetry, the asymmetry is largest in the ebb delta of Texel Inlet (tidal prism about 1050 million m³) and smallest in the ebb delta of Ameland Inlet, which has a tidal prism of only about 450 million m³. Terschelling Inlet has a tidal prism of about 850 million m³. Tidal prisms through the inlets with downdrift oriented ebb-tidal deltas are smaller. Thus, the tide-dominated ebb-tidal deltas, controlled by interaction of offshore and inlet tidal currents are updrift asymmetrical, and the ones dominated by wave-induced longshore drift are more downdrift asymmetrical.

Lynch-Blosse and Kumar (1976) interpreted changes in Brigantine Tidal Inlet, New Jersey and also attached value to the relative importance of waves/longshore drift versus tidal currents. Their example, however, concerned a much smaller system (width of the inlet about 500 m, and depth about 10 m). Lynch-Bloose and Kumar (1976) proposed that a constant tidal regime tends to produce an updrift growth of the downdrift part of the inlet system in quiet periods, whereas strong waves and storms may cause sudden shifts in the downdrift direction.

Evolution of the Channel System in the Ebb-Tidal Delta of Texel Inlet

The above mechanism also explains the morphological evolution of the ebb-tidal delta of Texel Inlet in relation to the evolution of the hydrodynamic regime. Reconstructions on the basis of historical navigation maps produced since the 16th century show that the ebb delta of Texel Inlet was asymmetrically oriented north-northwest before the middle 18th century (Fig. 9). It became symmetrical between the late 18th century and the early 19th century, and since the early 19th century it has been directed toward the southwest. The historical maps show that the maximum inlet depth has increased from 25 m in about 1583 to about 50 m at present. By combining data from modern examples in the western Wadden Sea with depth data read from old maps, a relationship between the maximum inlet depth and the tidal prism was established (Fig. 10; Sha, 1990a). The tidal prism through Texel Inlet is interpreted to have increased from about 240 x 10⁶ m³ in 1583 to 1050 x 10⁶ m³ in 1970. This increase is related primarily to sea level rise and the resulting increase of tidal amplitude. The increase in tidal volume is also due in large part to mismanagement by man when draining and exploiting extensive peat areas which previously existed in the Wadden Sea (Eisma and Wolff, 1980). Moreover, the construction of dikes since about the year 1000

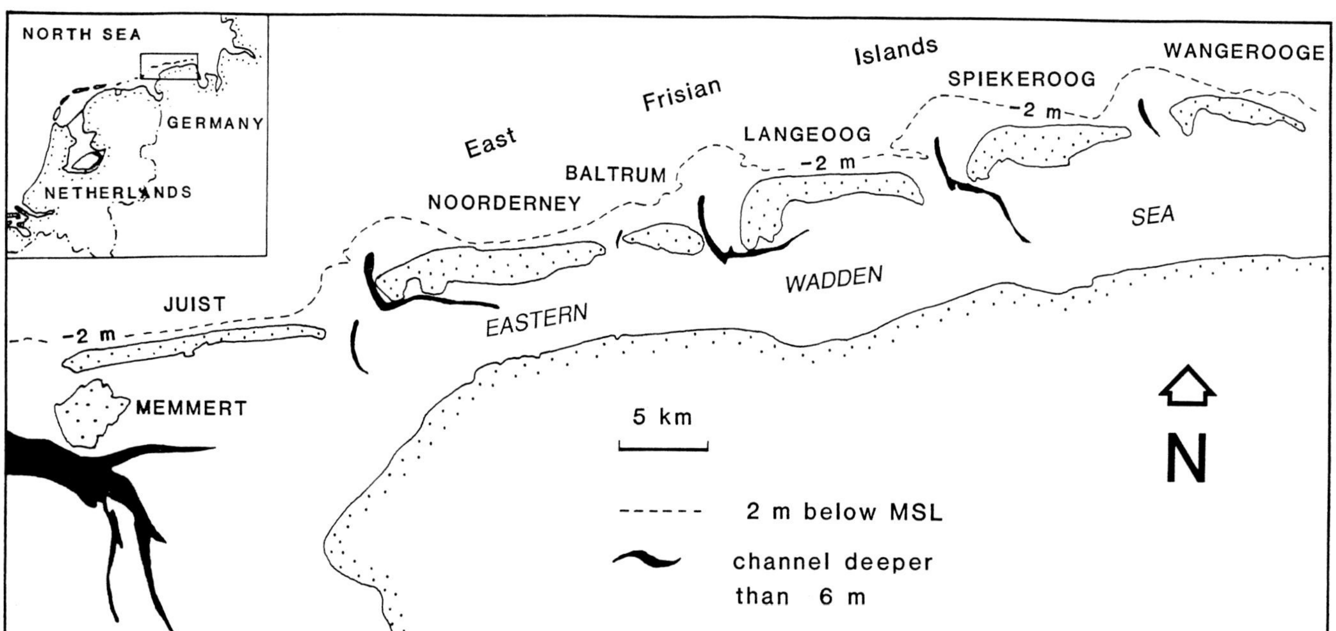

Figure 8. Morphology of the ebb-tidal deltas along the East Frisian Islands (after FitzGerald *et al.*, 1984). Note that the ebb-tidal deltas and their main ebb channels are asymmetrical to the east.

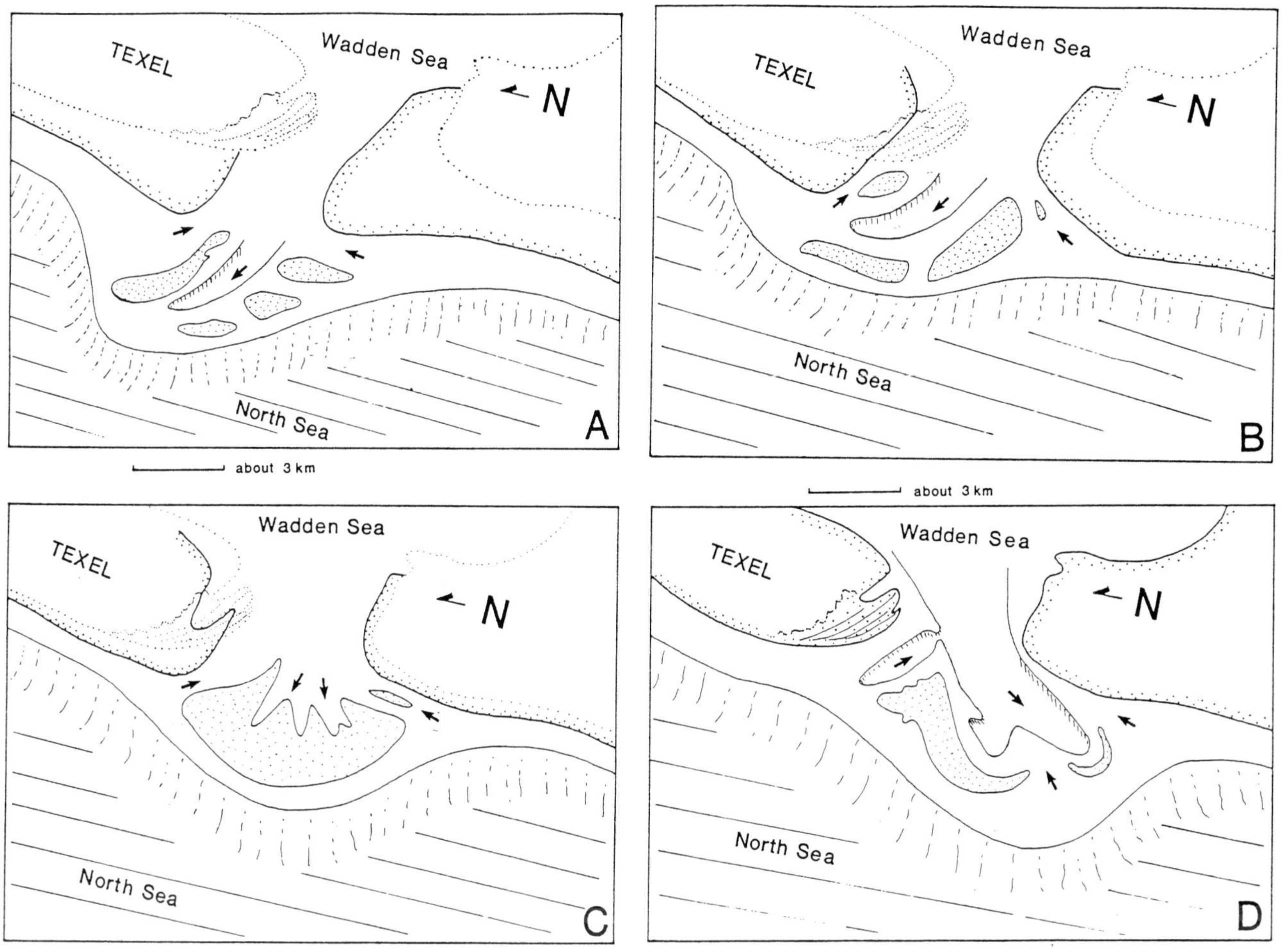

Figure 9. Schematic representation of the ebb-tidal delta of Texel Inlet, reconstructed on the basis of historical maps. **A.** 16th century; **B.** 17th to 18th century; **C.** 18th to 19th century; **D.** 19th to 20th century.

AD prevented natural subsidence to be compensated for by accumulation of fine grained sediment, so that repeated bursts of dikes during recurring severe storms caused sudden increases in volume of the Wadden Sea. In addition, the closure of tidal inlets in other areas along the Wadden Sea coast has contributed to the increase of the tidal prism through Texel Inlet. The closure of the 'Ijsselmeer' (Fig. 1) in 1932, paradoxically led to a 20 per cent increase in tidal amplitude near Texel. The increase of the tidal prism in Texel Inlet during the last few centuries may be the reason for the change of orientation (from downdrift to updrift) of the ebb-tidal delta in historical times. It appears that the change in orientation occurred when the tidal prism was about 700 x 10^6 m^3 (Figs. 9, 10).

At present the inlet of Ameland has a slight updrift asymmetry, and the tidal prism is only about 450 x 10^6 m^3. This dissimilarity must be due to differences between the deltas with respect to wave attack and tidal regime, and possibly also to changes in these factors in historical time. Morphological changes, as seen in Texel Inlet, also occurred in the inlet of Terschelling. The evolution of the historical tidal prism has not been reconstructed there, but available measurements (Ministry of Waterworks, unpubl. data) do indicate an increase in tidal prism during recent decades.

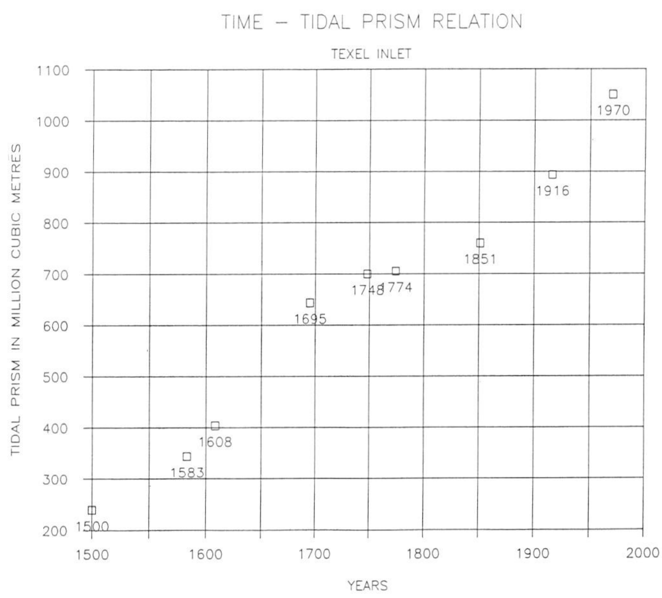

Figure 10. Estimated historical tidal prisms of Texel Inlet, based on the comparison of maximum depths of channels on historical maps with maximum depths of tidal inlets and tidal prisms of the present-day tidal inlets along the Wadden Sea.

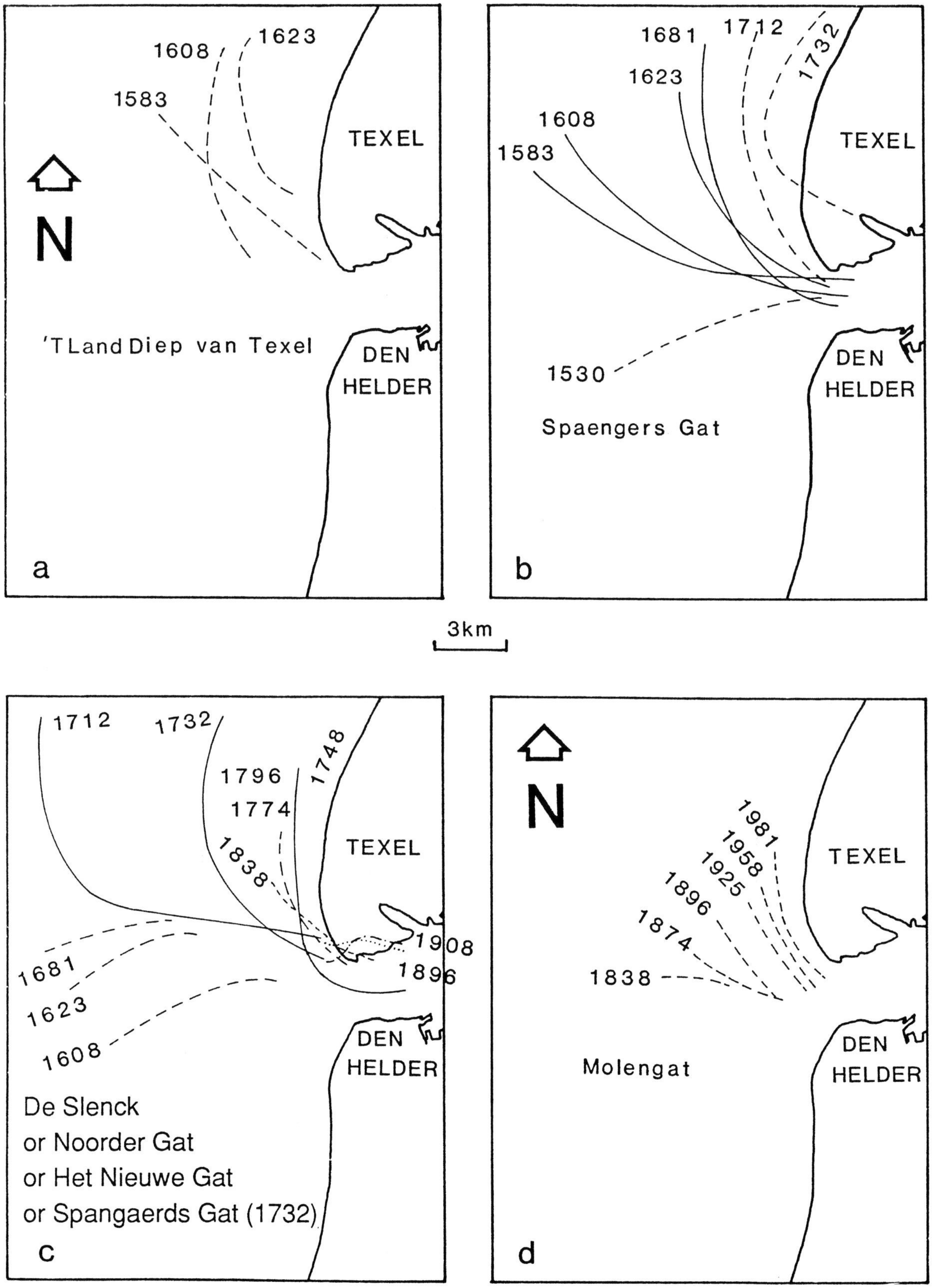

Figure 11. Rotation and migration of channels in the northern part of the ebb-tidal delta of Texel Inlet. Solid lines indicate positions and periods in which the channels acted as the main ebb channel.

ATTENTION

To take advantage of this offer, just respond by the date shown on your Personal Invitation. Even if you have other credit cards, the benefits of this Providian VISA® are hard to beat—all with no annual fee. **You've earned it, so don't let this offer pass you by!**

Z3082

ATTENTION

To take advantage of this offer, just respond by the date shown on your Personal Invitation. Even if you have other credit cards, the benefits of this Providian VISA® are hard to beat—all with no annual fee. **You've earned it, so don't let this offer pass you by!**

Z3082

Cyclical Shifting of Shoals and Channels

An important feature of the ebb-tidal delta of Texel Inlet is the clockwise rotation of the channels, together with the onshore movement of intertidal-supratidal shoals (Westenberg, 1974; Liefhebber and Berg, 1976). Such a phenomenon is cyclical and four cycles have been recognized since 1583 (Fig. 11). The cycle starts with the development of a new main (ebb) channel, which then gradually rotates to the north. The channel degenerates into a marginal flood channel, while a new main ebb channel develops farther to the southwest. Rotation of the channel occurs together with the onshore movement of the shoal (Fig. 12). The shoreward-migrating shoal buries the channel at the eastern side, and is transformed into a sand spit. The recent cycle started when the shoal 'Onrust' attached to Texel early during this century. At present the shoal 'Noorderhaaks' (Fig. 1, 2) is migrating toward the island and the 'Molengat' marginal flood-tidal channel is decreasing in importance (pers. comm., Netherlands Institute of Sea Research). Attachment of the shoal 'Noorderhaaks' to the island of Texel may be expected in the not too distant future, either by a gradual merger or, more likely, during a severe storm, which will cause filling of the 'Molengat' channel.

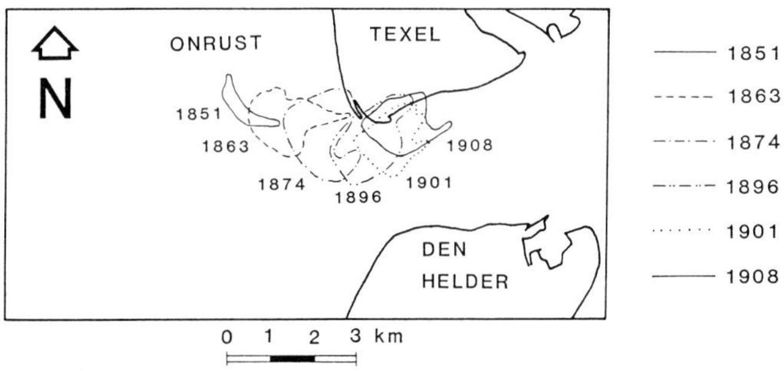

Figure 12. Migration of the inter-and supratidal shoal 'Onrust' between 1851 to 1908. The shoal attached to Texel shortly after 1908.

Similar cyclic changes also occur in the ebb-tidal deltas of Terschelling and Ameland inlets. This is evident from the clockwise rotation of the flood channel in 'Noordwest Gronden' in the ebb-tidal delta of Terschelling Inlet (Fig. 13). Also a shore-attached supratidal shoal and abandoned channel are presently seen in the ebb-tidal delta of Ameland Inlet (Figs. 5, 14).

Fitzgerald (1984) described similar cyclical patterns of growth and decay of the ebb-tidal delta at Price Inlet, South Carolina. Fitzgerald attributed this to the onshore migration of large bar complexes every 4 to 7 years.

The mechanism that causes the above-described cyclical migration is not fully clear. The most obvious explanation is that the system tends to be gradually pushed, by the action of waves and longshore currents, to a more west-northwesterly orientation. Thus the length of the main (ebb) channel increases. As a result the length of the tidal-current pathway through the main inlet channel increases (Fig. 7), and the tidal currents through this channel become weaker. At some point the distance covered by the ebb (and flood) currents (of decreasing strength) through the main ebb channel reaches a critical limit. At this stage the channel becomes ineffective and becomes a marginal, flood-dominated channel, while a channel farther to the south takes over the role of main ebb channel. Other flood-dominated channels farther north diminish in importance accordingly, and waves and storms push the larger ebb delta shoal over the marginal flood channel toward the island. By this process the channels rotate clockwise, and produce typical lateral migration patterns in one direction only.

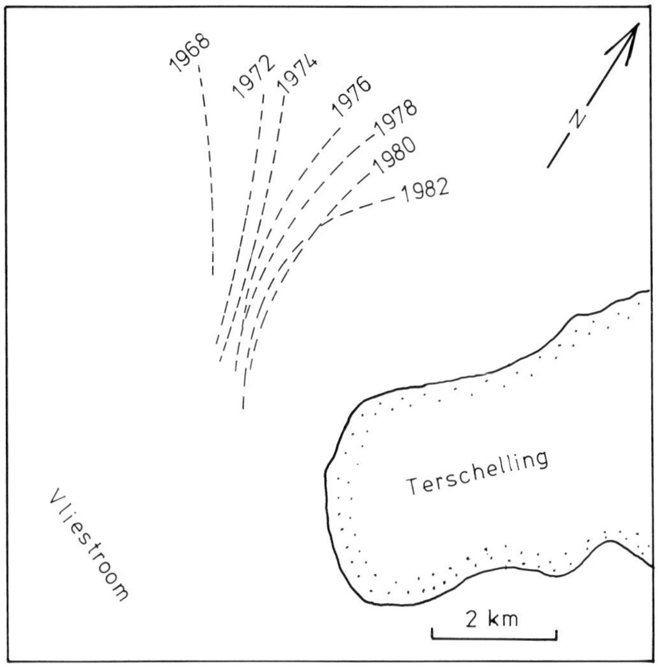

Figure 13. Channel development in the 'Noordwest Gronden' offshore Terschelling (after Kool and de Boer, 1983).

Figure 14. The shore-attached shoal on the ebb-tidal delta of Ameland Inlet (Fig. 5). The picture was taken in 1987, from the supratidal shoal which was almost attached to the island. Note the abandoned channel behind the supratidal shoal, and Ameland in the background. The now abandoned channel was 15 m deep in 1983, but only 3 m deep in 1987.

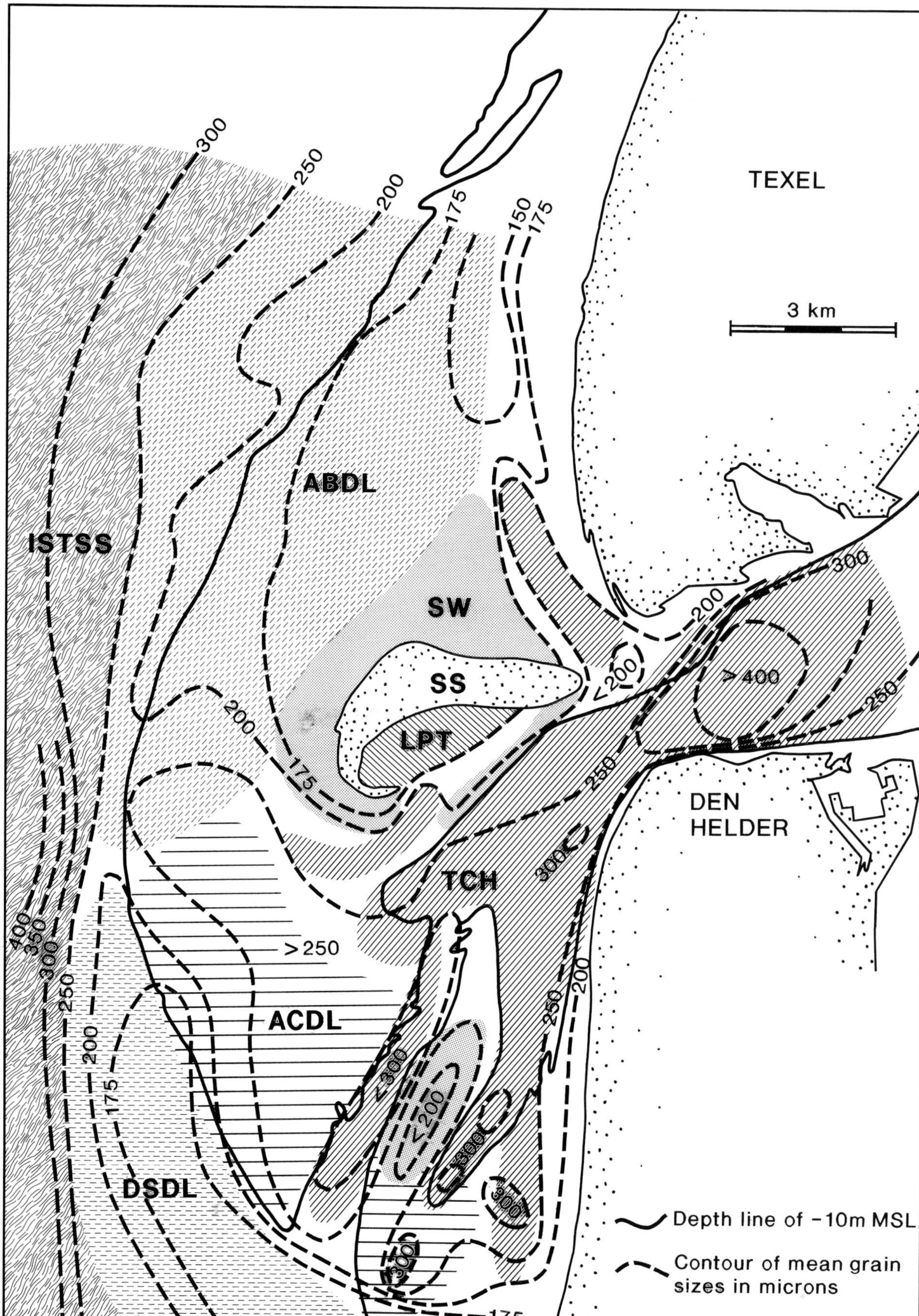

Figure 15. Facies distribution in the ebb-tidal delta of Texel Inlet. ACDL: active delta-lobe facies; DSDL: distal lobe facies; ABDL: abandoned ebb-delta lobe facies; SW: swash-bar facies; SS: supratidal shoal facies; TCH: tidal channel facies; LPT: locally protected facies; ISTSS: inner-shelf tidal sand sheet facies.

Sedimentary Facies

The distribution of surface sediments is closely related to sediment transport patterns and morphodynamics (Sha, 1990b), and is similar for the ebb-tidal deltas of the three major inlets. This is in accordance with the many morphodynamic characteristics that these systems have in common. The ebb-tidal delta deposits in the three inlets can be divided into nine facies (Fig. 15), described below.

Facies 1 (ACDL)

The active delta-lobe facies occurs seaward of the main ebb channels. It consists of prograding, lobe-like sand bodies influenced by both tidal currents and waves. Waves dominate the seaward side of the lobe, and tidal currents generally dominate the landward side. Wave-and tidal current-produced sedimentary structures are commonly interlayered (Fig. 16a).

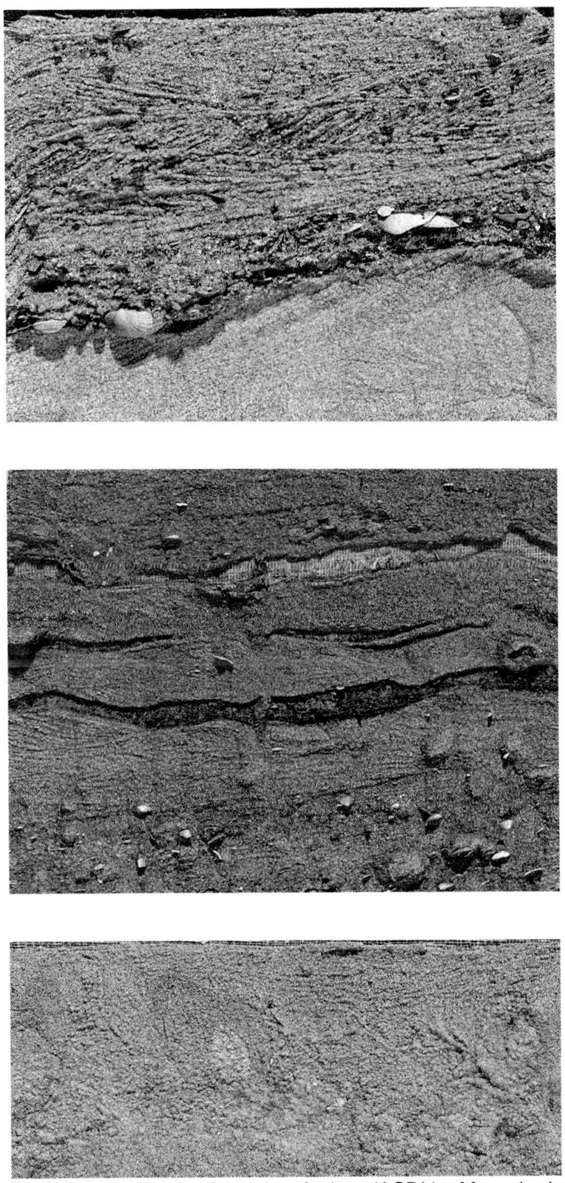

Figure 16. A. Active ebb delta lobe facies (ACDL). Megaripple crossbedded foresets with relatively coarse grained, poorly sorted, shelly sand erosionally overlying slightly inclined, parallel, finely laminated, thin, graded laminae with fine grained, well sorted, clean sand. Location: Texel Inlet. Scale in cm. **B.** Distal delta lobe facies (DSDL). Well preserved mud drapes are interlayered with ripple crosslaminated sand. An escape burrow is visible in the middle. Location: Texel Inlet. Scale in cm. **C.** Abandoned ebb delta facies (ABDL). This facies is characterized by strongly bioturbated sand. Location: Texel Inlet. Scale in cm.

Facies 2 (DSDL)

The distal lobe is situated seaward of the ebb delta lobe at a depth of >15 m below mean sea level. This area is less influenced by waves and tidal inlet currents, but longshore tidal currents do influence the bottom sediments. Fine grained sediments transported in suspension through the inlet during ebb-flow are deposited because of the deceleration of currents when arriving in the open basin. Consequently, the sediment is dominated by muddy, fine grained sand, commonly showing current and wave structures, with interbedded clay layers (Fig. 16b). Sedimentary structures are slightly bioturbated, and escape structures indicate periods of high sediment input, probably related to storms. In contrast to examples of 'storm deposits' from other settings, the result of storms within this facies is an increased input and deposition of fine grained sand and mud from the ebb-tidal channel when, during the waning phases of storms, the accumulated water in the Wadden Sea returns to the North Sea Basin.

Facies 3 (ABDL)

The abandoned ebb delta lobe facies occurs in front of the former main ebb channels, in the northeastern part of the three larger ebb-tidal deltas. The abandoned ebb-tidal delta lobe is subject to erosion, and the retreat is due to wave attack and lack of sediment supply. Parallel, hummocky or slightly inclined lamination with reworked, bioturbated, shelly sands are typical (Fig. 16c). Strong bioturbation occurs because of a low rate of sedimentation or erosion.

Facies 4 (SW)

The swash-bar facies is found in the shallow, wave-dominated zone, and develops where tidal currents are weak. It is generally shallower than 7 m below mean sea level. It consists of clean, fine grained, well-sorted sand which is regularly bedded, with thin, parallel, and finely graded laminae (Fig. 17a).

Facies 5 (SS)

The supratidal shoal facies occurs on the surface of the shoals above the normal high water line (Fig. 14). Megaripples form during storm tides, and salt pans occur within depressions between aeolian dunes. Wind ripples and wind-dune foresets are common. The sand is fine grained and well sorted. Locally, shell layers are present.

Facies 6 (TCH)

The tidal channel facies, found in active tidal channels usually shows thick, crossbedded foresets, sometimes with

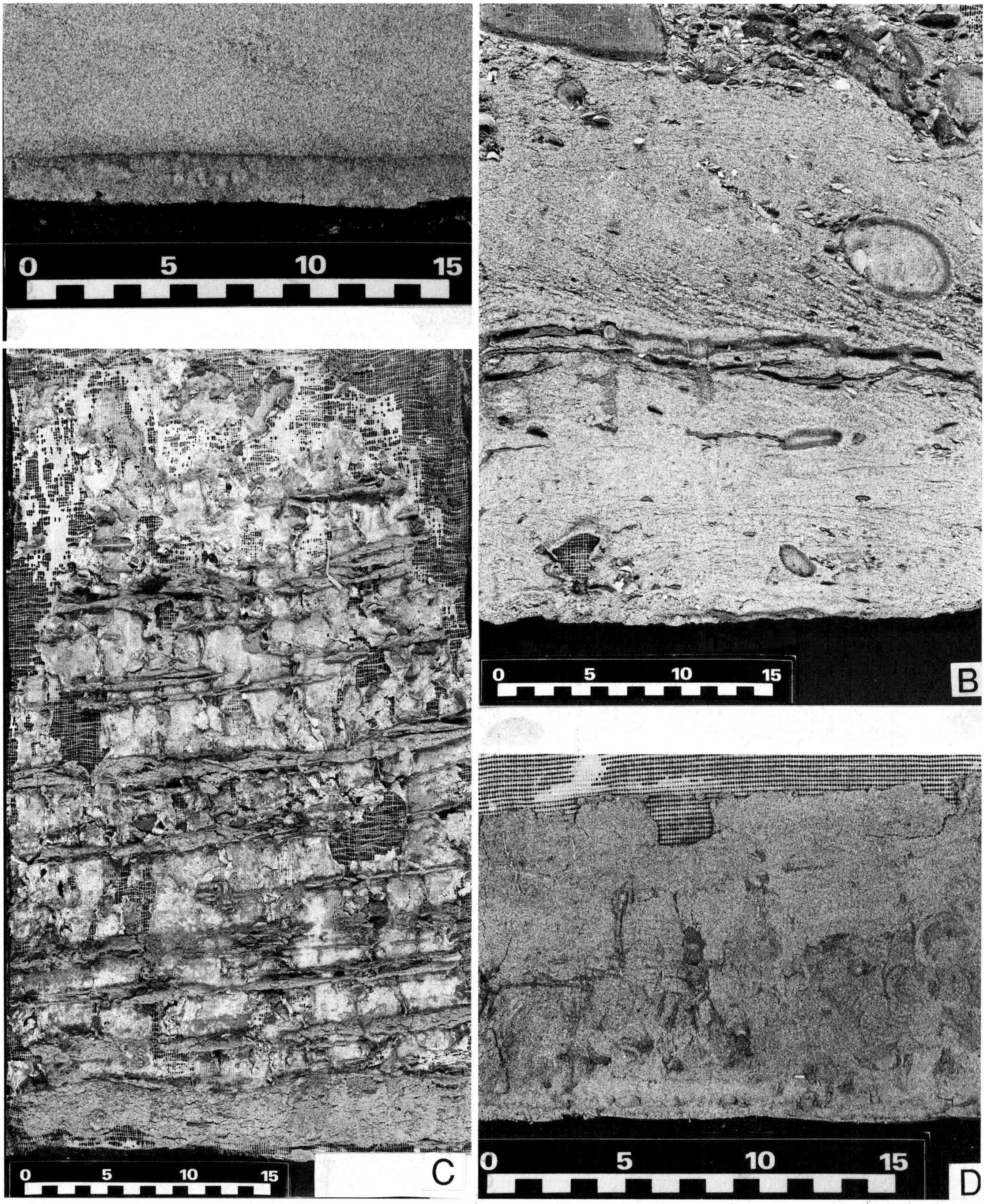

Figure 17. A. Swash bar facies (SW). Well sorted sand with graded, thin laminae. Planar inclined bedding with dark and light laminae indicate heavy mineral concentrations formed by wave action. Location: Terschelling Inlet. Scale in cm. **B.** Tidal channel facies (TCH). Coarse grained sand with abundant mud balls, shell and shell fragments. Large-scale crossbedding shows topsets, foresets and bottom sets. Location: Ameland Inlet. **C.** Abandoned channel facies (ABCH). Dominance of eutrophic organic mud, and lenticular sand ripples. Location: Ameland Inlet. Scale in cm. **D.** Partly abandoned channel facies (ABCH). Sand dominates, but is strongly bioturbates. Location: Terschelling Inlet. Scale in cm.

toe-sets, topsets, current ripple sets, reactivation surfaces, and incomplete mud drapes (Fig. 17b). Sediments consist of reworked, poorly sorted, medium to coarse grained sand with many shells and shell fragments, mud pebbles and, in the thalweg, gravels. The proportion of coarse grained sand, shells and shell fragments, mud pebbles, gravels and large foresets is higher in the deepest channels. The orientation of the large foresets depends on the dominant current direction in the channels.

Facies 7 (ABCH)

The abandoned channel facies is present in channels that are being abandoned in relation to attachment of shoals to the shoreline. The example in Figure 17c is from Ameland Inlet. The sediment consists of organic-rich mud with intercalated (lenticular) sandy ripples. This type of sediment is commonly bioturbated. In the early stage of abandonment, currents are still relatively strong, and prevent the accumulation of mud. As the tidal currents decrease, bioturbation becomes prevalent (Fig. 17c, d).

Facies 8 (LPT)

The locally protected facies was found only in Texel Inlet on the landward side of the shoal 'Noorderhaaks', and does not seem to be a common facies. Mud is deposited because of low wave energy, and intercalated, lenticular, wave-rippled sands likely produced by storms, are also present. This facies is strongly bioturbated and in many respects is similar to the abandoned channel facies.

Facies 9 (ISTSS)

The inner-shelf tidal sand-sheet facies is found seaward of the ebb-tidal deltas. Strictly speaking, it does not belong to the ebb-tidal delta setting. The sand is usually coarse grained, poorly-sorted, shelly, and reworked from underlying relict deposits. Megaripples foresets with topsets are common, but intensively bioturbated.

Sequence Models

The facies analysis, in combination with morphodynamic studies, provides a framework for constructing theoretical sedimentary sequences for ebb-tidal delta deposits. Three characteristic sequences are described, representing the effects of progradation and abandonment of ebb-tidal deltas.

Progradational Ebb Delta Lobe Sequence

The active ebb-tidal delta lobe progrades seaward over the inner-shelf tidal sand sheet, which consists mainly of reworked, underlying relict deposits. Thus a sequence of reworked inner shelf tidal sand, distal ebb delta lobe deposits and active ebb-tidal delta lobe sediments is produced. The part of the sequence produced by delta lobe progradation is upward coarsening (Fig. 18a). Bioturbation is fairly intense at the base, and diminishes toward the top.

Abandoned Ebb-Tidal Delta Lobe Sequence

In the case of deposition within a subsiding basin (*i.e.*; under conditions of relative sea level rise), abandonment of an active ebb delta lobe will be followed by a transgression of the sea. In the present-day West Frisan Islands this is especially the case, because of the glacio-eustatic rise of sea level. Marginal flood channels and swash platform/swash-bar facies form after abandonment. The marginal flood channels turn clockwise and landward, and simultaneously, shoals and swash bars migrate onshore. Thus a sequence representing the marginal flood channel facies, migrational swash bar and shoal facies is successively produced (Fig. 18b). The migrating shoal produces low angle foresets on the landward side, which are reflected in the sequence model. The sequence is truncated at the top by a ravinement surface formed during shoreface retreat. A reworked, inner-shelf tidal sand sheet covers the sequence, and if the rise of sea level continues, the series will be covered by shelf mud deposits. Below such a sequence, older deposits of the active ebb-tidal delta or of the main ebb channel may be present.

In the distal area, a marginal flood channel facies may not develop, and a ravinement surface may directly truncate ebb delta foresets.

Inlet Sequence

In the proximal part of the ebb-tidal delta, migrating channels may produce a sequence with marginal flood channel deposits on top of a channel sequence produced by lateral migration of ebb-dominated channels (Fig. 18c).

Three-Dimensional Geometry and Preservation Potential

Sand Body Geometry of Ebb Delta at Texel Inlet

A three-dimensional picture of the ebb delta sand body in Texel Inlet was constructed based on boreholes and seismic profiles (Fig. 2; Sha, 1989d). The ebb-tidal delta sand body unconformably overlies deposits of the upper Pleistocene, that is, Weichselian, Eemian and Saalian. The sand body is up to 30 m thick in the central part of the ebb-tidal delta, and radially pinches out seaward. The internal geometry of the ebb-tidal delta lobe is revealed by seismic records. The stratigraphic succession at the Pleistocene-Holocene interface and the characteristics of the lower boundary (Fig. 2) of the ebb delta sand body are closely related to the evolution of the ebb-tidal delta. The high-lying Pleistocene surface was recently incised by deep tidal-delta channels. In the southern part of the ebb-tidal delta, Weichselian deposits have been preserved (Fig. 2). This indicates that the great depth of the tidal channels is a recent phenomenon. As described above (Fig. 9), the main inlet system was situated farther north in the past, and has deepened and shifted because of the increase of tidal prism. Thus, in the northern

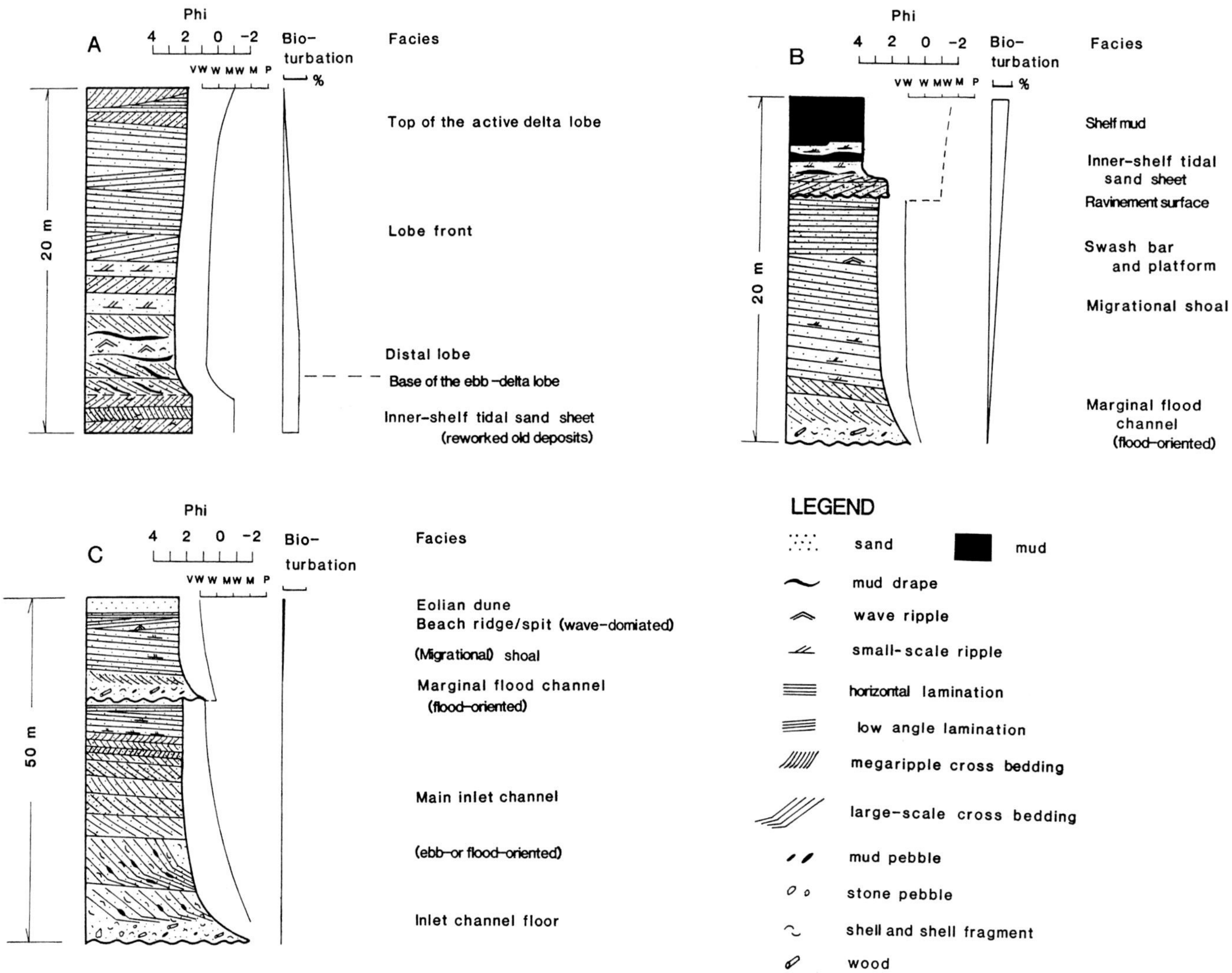

Figure 18. A. Hypothetical sedimentary sequence of a progradational active ebb delta lobe. **B.** Hypothetical sedimentary sequence of the abandoned ebb delta lobe facies. **C.** Hypothetical sedimentary sequence formed by an migrating inlet. Paleocurrent directions in the lower and upper part are more or less opposite.

part of the ebb-tidal delta, Upper Pleistocene deposits have been eroded, and the depth of the lower boundary of the ebb-tidal delta sand body, and its thickness, increase southward (Figs. 2, 19).

Lower Holocene Ebb-Tidal Delta Deposits Offshore of Terschelling and Ameland

A seismic survey offshore of Terschelling showed that ebb-tidal delta deposits can be preserved during transgression and shoreface retreat. It was shown (Sha, 1990a) that the preservation potential of ebb-tidal delta deposits depends mainly on the maximum depth of shoreface erosion and the depth of the seaward limit of the ebb-tidal deltas. These features are related to many factors, such as rate of relative sea level rise, rate of sediment supply, open-marine wave and current energy, and the tidal prism through the inlet.

The morphological map of the Terschelling coast (Fig. 20) shows the configuration of a preserved ebb-tidal delta (20 m isobath), which is not relevant to the present-day dynamic system (present-day depth of shoreface erosion is about 15 m; Postma and Kroon, 1986). Seismic lines show internal structures indicative of lateral migration of an inlet channel (Fig. 21). The angle of the accretion surfaces is similar to the slope angle of recent inlet channels. Borehole data reaveal that they must be Holocene shallow marine deposits (Sha, 1990a). A shore-transverse line (Fig. 21) shows very low-angle, seaward-dipping foresets of the ebb-tidal delta lobe and chaotic reflectors of swash platforms and ebb delta shoal deposits on the seaward side of large inlet-fill deposits (large tidal channel fill). A seismic line off Ameland even shows a sequence of large, migrational inlet channel fills (oblique reflectors) with a series of small channel fill features reflecting the cyclic migration and abandonment of small marginal flood channels (Fig. 22).

Reworking of Ebb-Tidal Delta Sands

Ebb-tidal delta sands can be preserved another way; that is, through reworking into shoreface-connected or noncon-

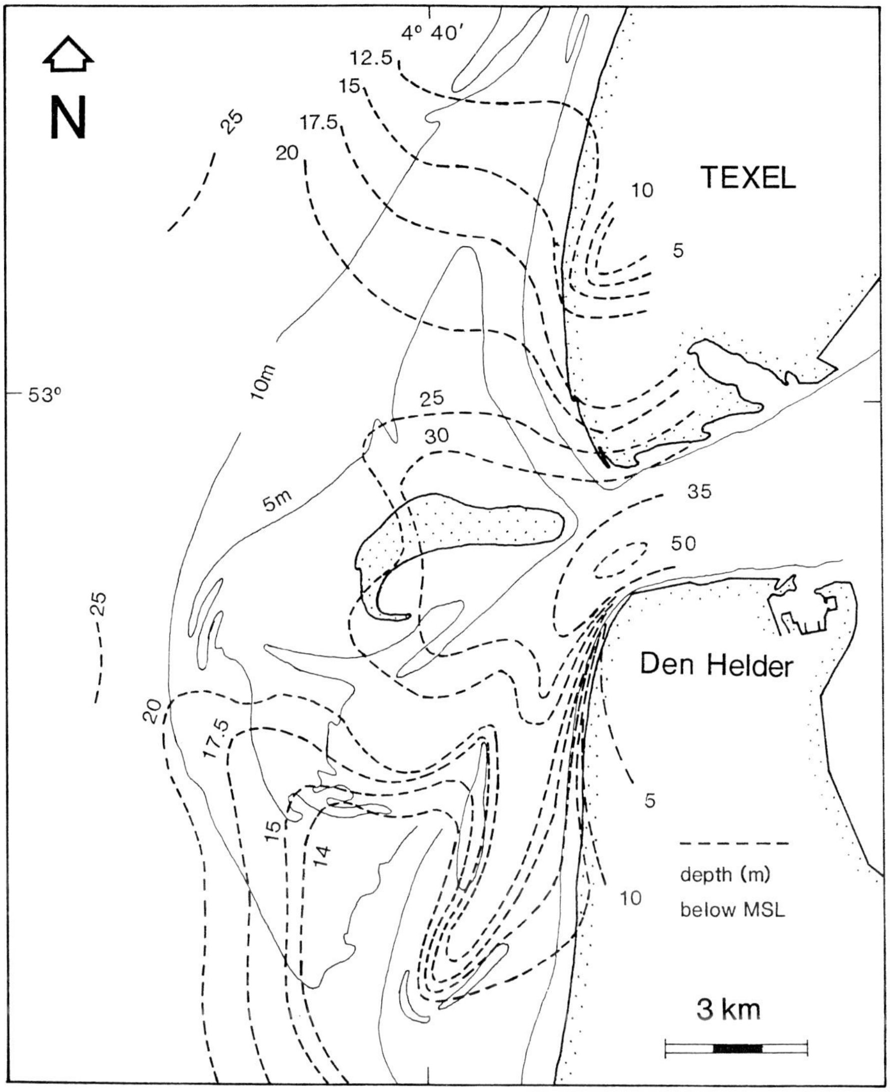

Figure 19. Depth of the lower boundary of the ebb delta sand body of Texel Inlet (contours in metres below MSL). This boundary represents the Pleistocene-Holocene interface. The thin, continuous lines show the present-day water depth in metres below MSL.

nected ridges (*cf.* Duane *et al.*, 1972). This morphodynamic process has been reported from a tidal inlet in the southwest part of The Netherlands, which was closed in 1972 (Kohsiek, 1988). After the artificial closure of the tidal inlet, the ebb-tidal delta transformed into large, longshore bars in a period of about 15 years. Small ridges, commonly present offshore of former tidal inlets along the West Frisian Islands, documented by historical maps and geological data, also are suggestive of the reworking of ebb-tidal delta sands into sand ridges.

Discussion

Comparisons With Other Examples (North America, East Frisian Islands)

Morphological models described for the North American Atlantic coast inlets and ebb-tidal deltas mostly show an interaction of inlet tidal currents and offshore waves (Hayes, 1975, 1979, 1980; Oertel, 1975). Tidal currents through the inlets form the main inlet channels, which extend seaward. Waves dominate the ebb delta shoals, form swash platforms, swash bars (Oertel, 1972), and can produce an asymmetrical geometry of the ebb-tidal deltas (Oertel, 1975; Nummedal and Fischer, 1978) and of the inlets (downdrift offset; Goldsmith *et al.*, 1975). The tidal current patterns around the inlets can also be strongly modified by waves and wave-induced longshore drift (Finley, 1978; Lynch-Blosse and Kumar, 1976; Davis and Fox, 1981). In these examples, wave influence is strong, and tidal influence, especially the interaction of offshore tidal currents and currents through the inlet, is weak compared to the West Frisian inlets and ebb-tidal deltas.

Hypothetical sedimentary sequences constructed from data collected on one of the most tide-dominated ebb-tidal deltas along the North American Atlantic coast, (Edisto Inlet, Imperato *et al.*, 1988) show a predominantly wave-dominated sedimentary facies: swash platforms, swash bars,

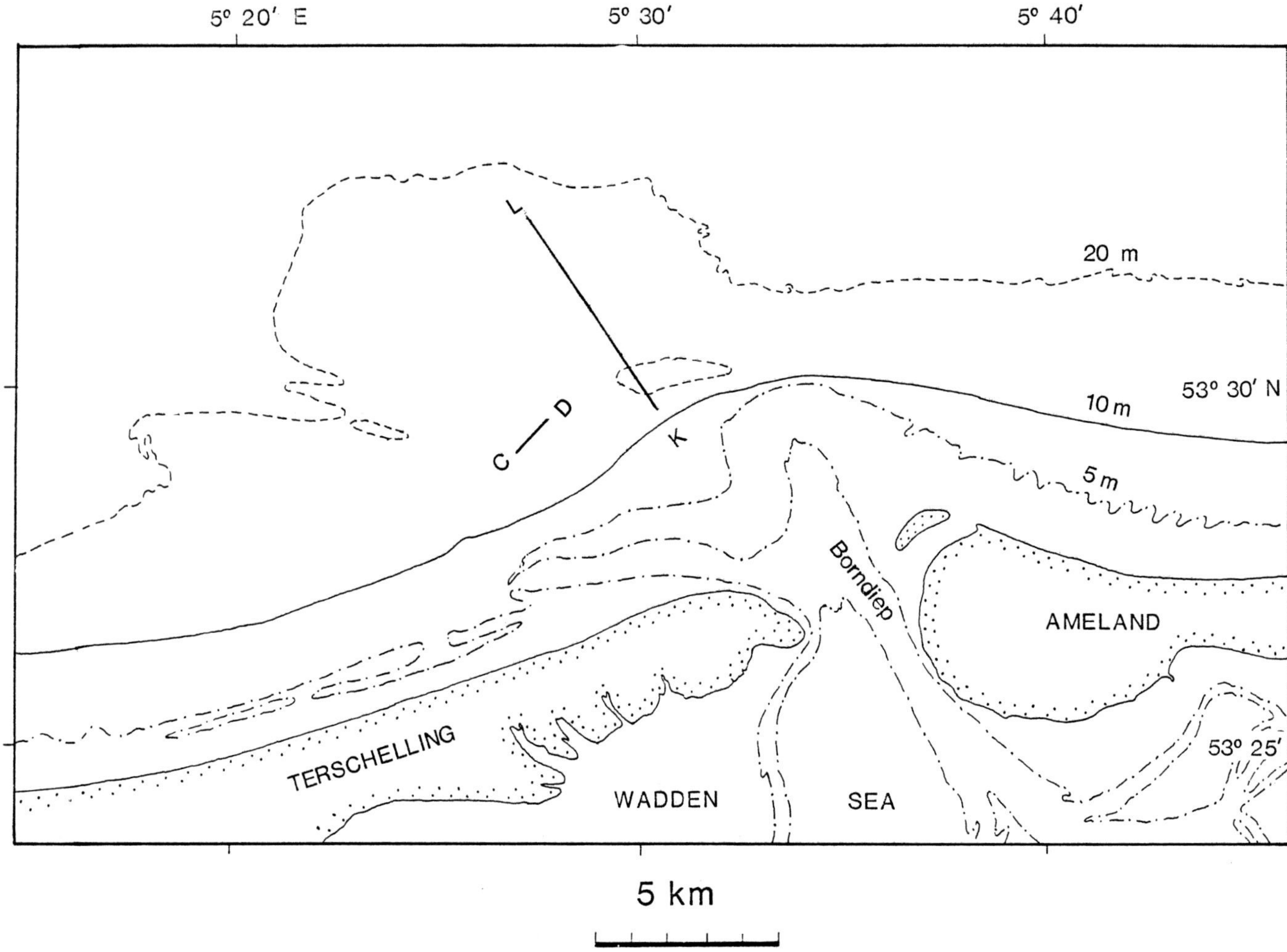

Figure 20. Morphological map of the Terschelling coast showing the locations of seismic lines of Figure 21. The ebb delta-like feature is shown by the 20 m contour (contours in metres below MSL).

abandoned channels and shoreface deposits. A tide-dominated facies was found only in the channels. Tide-influenced, ebb delta lobe and distal lobe facies, and offshore tidal sands immediately adjacent to the ebb-tidal delta, as found in the North Sea (*cf.*, Sha, 1990b), were not reported from North Edisto Inlet. Such differences again demonstrate the relative importance of the tides in the West Frisian tidal inlets.

Seismic and other surveys along the North American Atlantic coast indicate a low preservation potential for inlet channel deposits during the Holocene transgression (Field, 1980; Belknap and Kraft, 1985; Hine and Snyder, 1985; Panageotou and Leatherman, 1986). No sub-Recent ebb-tidal delta deposits have been reported there. Strikingly, the depth of shoreface scour is about 10 m in that area (Belknap and Kraft, 1985), whereas Lower Holocene inlet and ebb delta deposits have been preserved offshore off Terschelling where the depth of shoreface scour is 15 m (Postma and Kroon, 1986). Therefore, the preservation potential of inlet and ebb-tidal delta systems clearly does not only depend on the depth of shoreface erosion, but also on the depth of inlet channel scour and the seaward extension and depth of the seaward end of the ebb-tidal delta. These features, in turn, are related to the size of the tidal prism through the inlets. Therefore we speculate that the difference in tidal influence between the two areas is responsible for the difference in preservation potential. Tide-dominated inlet systems have a greater preservation potential than wave-dominated ones if other factors are similar, implying that inlet and ebb-tidal delta systems that have been preserved in the fossil record have experienced a relatively strong tide dominance.

The difference in tidal influence between the West Frisian ebb-tidal deltas and the ebb-tidal deltas of the North American Atlantic coast is attributed to two causes. Offshore tidal currents in the North Sea are relatively strong as a result of the geometry of the North Sea basin. The strength of tidal currents through the inlets is, moreover, related to the tidal prism through the inlets, and the volume of the storage basin behind. The latter is much larger in the case of the West Frisian inlets (Wadden Sea) than in the tidal inlets of the North American Atlantic coast. For instance, the tidal prism of Texel Inlet exceeds 1000 million m^3, whereas tidal

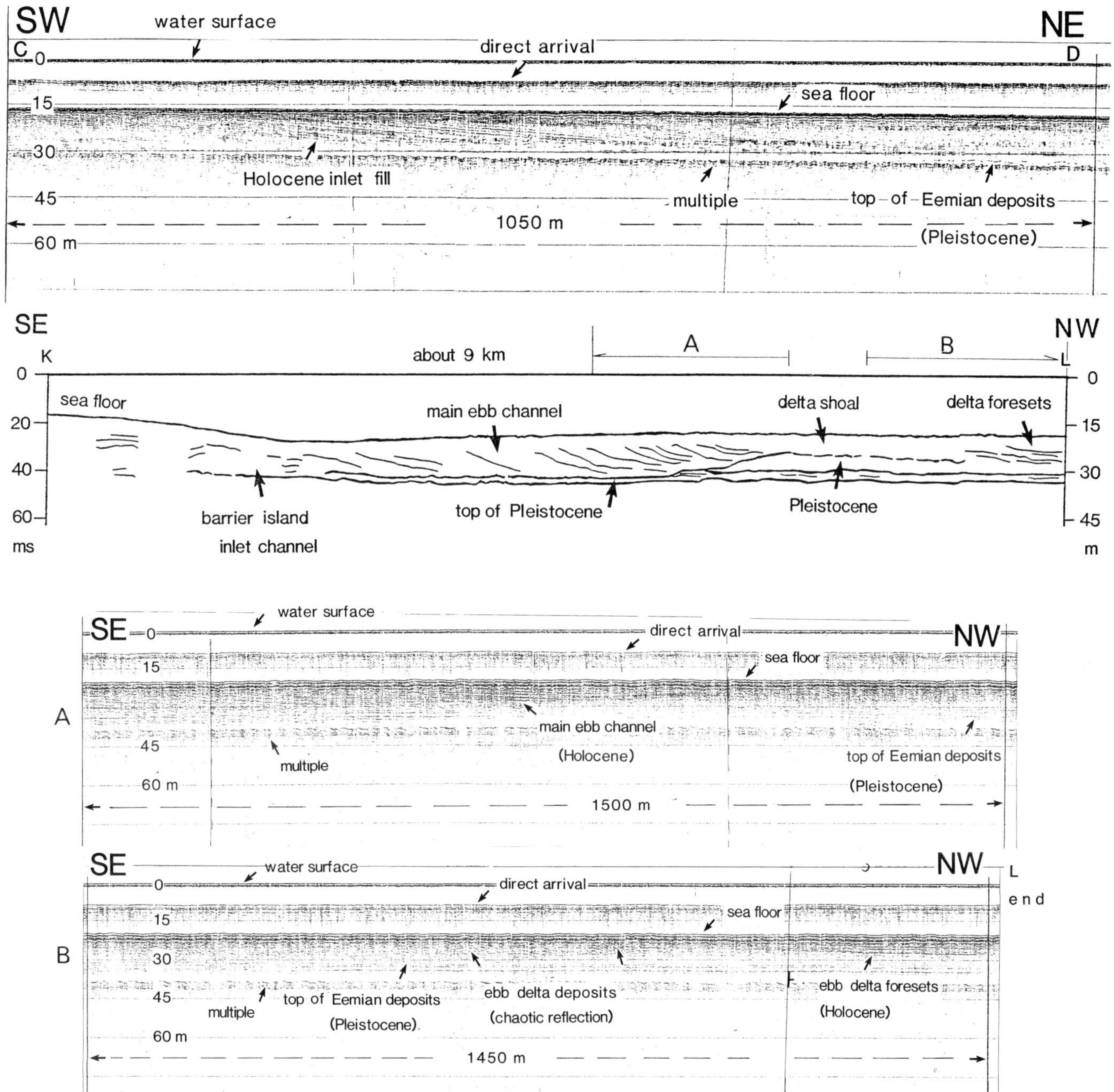

Figure 21. (**Upper**) Seismic line (C-D) parallel to the coast of Terschelling showing a low angle, clinoformal (channel fill) reflection pattern indicating lateral migration of an Early Holocene inlet (for location, see Figure 20). (**Lower**) Seismic line (K-L) perpendicular to the coast of Terschelling showing low angle seaward-dipping reflectors (shingled) and chaotic reflectors indicating relict, early Holocene ebb delta deposits (for location, see Figure 20). The inlet (channel fill reflection pattern) is landward of the preserved delta shoal deposits (chaotic reflection pattern). Compare with Figure 2. The two seismic records below correspond to the intervals indicated by horizontal lines at the right side, above the line interpretation.

prisms of the North American Atlantic inlets seldom exceed 300 million m^3 (Jarrett, 1976).

Conclusions

Similar sand transport patterns and morphologies are found in the ebb-tidal deltas of Texel, Terschelling and Ameland inlets. The sand transport patterns are strongly controlled by the interaction of tidal currents through the inlet and marine tidal currents parallel to the coast. This interaction enhances bidirectional tidal currents in the southwestern (updrift) part of the ebb-tidal deltas and results in a net seaward sediment transport there. In the northeastern (downdrift) part, it produces a rotational tidal current pattern and flood-dominated sand transport. The relatively great strength of the tidal currents through the inlets is the main reason for the aberrant morphology of the ebb-tidal deltas of the West Frisian Islands as compared to systems described from other places, for example the East Frisian Islands. The inlets

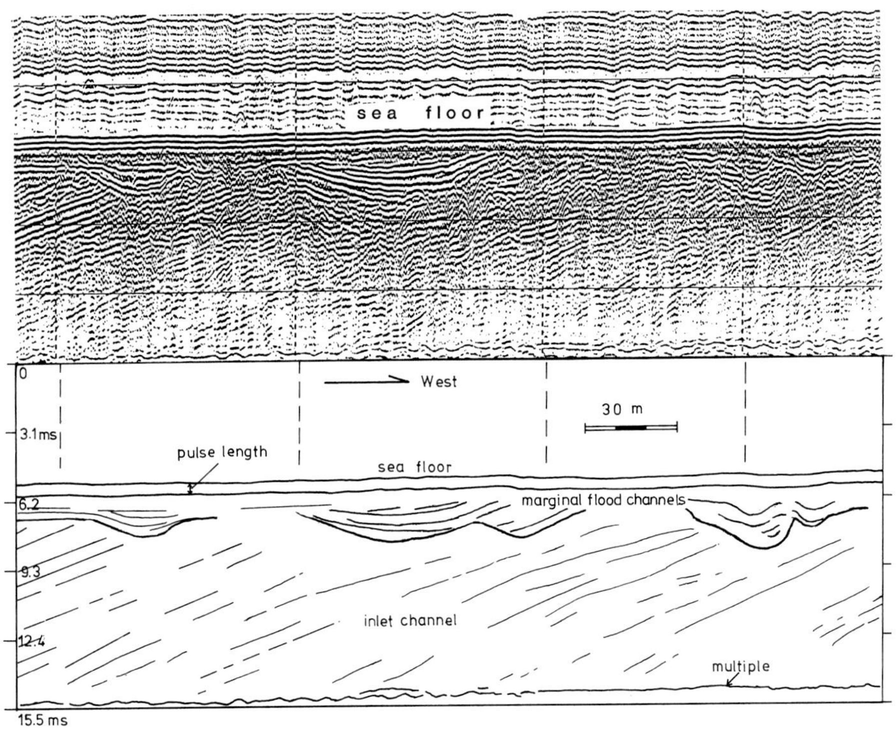

Figure 22. Seismic line off Ameland (parallel to the coast) showing a sequence of large migrational inlet channel fills (oblique reflectors) with a series of small channel fill features reflecting the cyclic migration and abandonment of small marginal flood channels. A small time delay was applied during graphic recording. Water depth is approximately 10 m.

with large tidal prisms have large ebb-tidal deltas, which are dominated by the interaction of tidal currents through the inlets with tidal currents parallel to the coast. This causes the large ebb-tidal deltas of Texel, Terschelling and Ameland inlets to be asymmetrically directed to the southwest and west (updrift). The inlets with small tidal prisms have small ebb-tidal deltas dominated by waves. Consequently wave attack and eastward-directed longshore drift drive the smaller ebb-tidal deltas of the East Frisian Islands asymmetrically to the east (downdrift).

The above model, which explains the differences in morphology between different modern ebb-tidal deltas, may also be applied to explain changes of ebb-tidal delta morphology through time. Indeed, the ebb-tidal delta of Texel Inlet was more wave- and less tide-influenced some centuries ago, and became more and more tide-dominated during the last centuries because of an increase of the tidal prism through the inlet.

With respect to their morphological evolution, the three larger ebb-tidal deltas show another common feature; that is, a cyclical migration of shoals and channels in the downdrift (north to east) part. For the ebb-tidal delta of Texel inlet, four cycles have been recognized for the period since the 16th century. The cyclical pattern in morphological development is the result of a feedback response between the morphology and the dynamic processes. In this process the main, ebb-dominated channel is pushed downdrift by wave action and longshore drift; the pathway that tides must cover in order to pass through the main channel thus becomes longer, until a certain limit is reached, at which time another, up to then subordinate channel farther to the south takes over the role of main channel. The former, main, ebb-dominated channel degenerates into a marginal flood channel. In relation to these shifts and rotations of channels, ebb delta shoals attach to the island to the northeast.

Nine sedimentary facies were recognized in the ebb-deltas of Texel, Terschelling and Ameland inlets. By combining the facies distribution of surface sediments with the morphodynamic processes, schematic models of sedimentary sequences can be constructed. These models represent hypothetical successions formed by prograding ebb-tidal delta lobes, by the abandonment of active ebb-delta lobes, and by migrating inlets. The lower bounding surface of the ebb-tidal delta body of Texel Inlet is a stratigraphic unconformity situated some 50 m below mean sea level, and formed through erosion by migrating channels. The top of the sandy body is above mean sea level locally, and the thickness of the ebb delta sands may be up to 30 m. The three dimensional architecture of the ebb-tidal delta, reconstructed from drilling and seismic lines, shows that the sand body represents a volume of more than one cubic kilometre.

Preserved ebb delta/inlet deposits are recognized on seismic profiles off Terschelling and Ameland. These profiles shows low-angle, seaward-dipping foresets of the ebb delta lobe and low-angle, inclined beds formed by the inlet channel, migrating laterally, parallel to the coast. A prerequisite for preservation is that the maximum erosional depth of the regional shoreface is less than the depth of the lower bounding surface and the base of the seaward extension of the ebb-tidal delta. In other words, the ravinement surface formed during shoreface retreat should be positioned above the lower bounding surface of the ebb-tidal delta. The lower bounding surface is generally produced by migrating channels or is formed by the pre-existing sea bottom in the more distal parts. The relative positions of these bounding surfaces depend on rate of relative sea level rise, rate of sediment supply, wave energy, and the tidal prism through the inlet.

We infer that tide-dominated inlet systems have a greater preservation potential than wave-dominated ones, if other factors are similar. Thus, there is a fair chance that inlet and ebb-tidal delta systems that have been preserved in the fossil record experienced a relatively strong tide dominance.

Acknowledgments

The Netherlands Institute of Sea Research, the Ministry of Waterworks, and the Dutch Geological Survey are acknowledged for support during various stages of this research, much of which formed part of the Ph.D. research by the first author. We thank the editors for their patience and encouragement.

References

Belknap, D. F. and Kraft, J. C. 1985. Influence of antecedent geology on stratigraphic preservation potential and evolution of Delaware's barrier systems. Marine Geology, v. 63, p. 235-262.

Bruun, P. 1978. Stability of tidal inlets. Developments in Geotechnical Engineering, Elsevier, v. 23, 510 pp.

Bruun, P. and Gerritsen, F. 1959. Natural by-passing of sand at coastal inlets. Journal of the Waterways and Harbors Division, Proceedings of American Society of Civil Engineers, v. 85, p. 75-107.

Davis, R. A. and Fox, W. T. 1981. Interaction between wave- and tide-generated processes at the mouth of a microtidal estuary, Matanzas River, Florida. Marine Geology, v. 40, p. 49-68.

De Boer, M. 1979. Morphologisch onderzoek Ameland. Verslag van het onderzoek op het Ameland wantij in 1973. Rijkswaterstaat, Studiedienst Hoorn. Nota WWKZ-79.H005, 67 p.

Dijkema, K. S., Reineck, H.-E. and Wolff, W. J. (editors) 1980. Geomorphology of the Wadden Sea area. Balkema, Rotterdam, 135 p.

Duane, D. B., Field, M. E., Meisburger, E. P., Swift, D. J. P. and Williams, S. J. 1972. Linear shoals on the Atlantic inner continental shelf, Florida to Long Island. *In:* Shelf sediment transport: process and pattern, D. J. P. Swift, D. B. Duane and O. H. Pilkey (eds.). Dowden, Hutchinson & Ross, Stroudsburg, p. 447-499.

Eisma, D. and Wolff, W. J. 1980. The development of the westernmost part of the Wadden Sea in historical time. *In:* Geomorphology of the Wadden Sea area, K. S. Dijkema, H.-E. Reineck and W. J. Wolff (eds.). Balkema, Rotterdam, p. 95-103.

Field, M. E. 1980. Sand bodies on coastal plain shelves: Holocene record of the U.S. Atlantic Inner Shelf off Maryland. Journal of Sedimentary Petrology, v. 50, p. 505-528.

Finley, R. J. 1978. Ebb-tidal delta morphology and sediment supply in relation to seasonal wave energy flux, North Inlet, South Carolina. Journal of Sedimentary Petrology, v. 48, p. 227-238.

FitzGerald, D. M. 1984. Interaction between the ebb-tidal delta and landward shoreline: Price Inlet, South Carolina. Journal of Sedimentary Petrology, v. 54, p. 1303-1318.

______, Penland, S. and Nummedal, D. 1984. Control of barrier island shape by inlet sediment bypassing: East Frisian Islands, West Germany. Marine Geology, p. 355-376.

Goldsmith, V., Bryne, R., Sallenger, A. H. and Drucker, M. D. 1975. The influence of waves on the origin and development of the offset coastal inlets of the southern Delmarva Peninsula, Virginia. *In:* Estuarine Research, Volume 2, L. E. Cronin (ed.). Academic Press, New York, p. 183-200.

Hayes, M. O. 1975. Morphology of sand accumulation in estuaries: an introduction to the symposium. *In:* Estuarine Research, Volume 2, L. E. Cronin (ed.). Academic Press, New York, p. 3-22.

______ 1979. Barrier island morphology as function of tidal and wave regime. *In:* Barrier Islands, S. P. Leatherman (ed.). Academic Press, New York, p. 1-27.

______ 1980. General morphology and sediment patterns in tidal inlets. Sedimentary Geology, v. 26, p. 139-156.

Hine, A. C. and Snyder, S. W. 1985. Coastal lithosome preservation: evidence from the shoreface and inner continental shelf off Bogue Banks, North Carolina. Marine Geology, v. 63, p. 235-262.

Hubbard, D. K., Barwis, J. H. and Nummedal, D. 1977. Sediment transport in four South Carolina inlets. American Society of Civil Engineers, Proceedings Coastal Sediments '77, p. 582-601.

Hubbard, D. K., Oertel, G. F. and Nummedal, D. 1979. The role of waves and tidal currents in the development of tidal-inlet sedimentary structures and sand body geometry: examples from North Carolina, South Carolina and Georgia. Journal of Sedimentary Petrology, v. 49, p. 1073-1092.

Imperato, D. P., Sexton, W. J. and Hayes, M. O. 1988. Stratigraphy and sediment characteristics of a meso-tidal ebb-tidal delta, North Edisto Inlet, South Carolina. Journal of Sedimentary Petrology, v. 58, p. 950-958.

Jarrett, J. T. 1976. Tidal prism-inlet area relationships. General investigation of tidal inlets. Coastal Engineering Research Centre, Ft. Belvoir, Virginia, Rep. 3, 32 p.

Kohsiek, L. H. M. 1988. Reworking of former ebb-tidal deltas into large longshore bars following the artificial closure of tidal inlets in the southwest of The Netherlands. *In:* Tide-influenced sedimentary environments and facies, P. L. de Boer, A. van Gelder and S. D. Nio (eds.). Reidel Publishing Company, p. 113-122.

Kool, G. and de Boer, M. 1983. Zeegat van het Vlie Ontwikkeling drempelgebied Thomas Smitgat Noordgat 1968-1982. Rijkswaterstaat nota Wwkz-83.H002, 24 p.

Liefhebber, P. and Berg, L. Th. 1976. Huisduinen en het oude land. Den Helder, 88 p.

Lubberts, R. K. and de Vos, F. E. M. 1987. Historical development, facies distribution and sediment dispersal, Ameland Inlet, The Netherlands. Unpublished report, Comparative Sedimentology Division, Institute of Earth Science, University of Utrecht, 53 p.

Luck, G. 1976. Inlet changes of the East Frisian Islands. Proceedings of the Conference Coastal Engineers, 15th American Society of Civil Engineers, p. 1938-1957.

Lynch-Blosse, M. A. and Kumar, N. 1976. Evolution of downdrift-offset tidal inlets: a model based on the Brigantine Inlet System of New Jersey. Journal of Geology, v. 84, 165-178.

M.M.A.H. (Ministerie van Marine, Afdeling Hydrografie) 1963. Stroomatlas Nederland: Minist. Mar., Afd. Hydrogr., The Hague, Part I.

Nummedal, D. and Fischer, I. A. 1978. Process-response models for depositional shorelines: The German and Georgia bights. American Society of Civil Engineers, Proceedings of the 16th Coastal Engineering Conference, p. 1215-1231.

______ and Penland, S. 1981. Sediment dispersal in Norderneyer Seegate, West Germany. *In:* Holocene marine sedimentation in the North Sea Basin, S. D. Nio, R. T. E. Schuttenhelm and Tj. C. E. van Weering (eds.). International Association of Sedimentologists, Special Publication, v. 5, p. 187-210.

______, Oertel, G. F., Hubbard, D. K. and Hine, A. C. 1977. Tidal inlet variability Cape Hatteras to Cape Canaveral. American Society of Civil Engineers, Proceedings, Coastal Sediment '77, p. 543-562.

O'Brien, M. P. 1969. Equilibrium flow areas in inlets on sandy coasts. Journal of Waterways and Harbors Division, Proceedings of the American Society of Civil Engineers, v. 95, p. 43-51.

Oertel, G. F. 1972. Sediment transport on estuary entrance shoals and the formation of swash platforms. Journal of Sedimentary Petrology, v. 42, p. 857-863.

______ 1973. Examination of sedimentary textures and structures of mud in layered sediments at the entrance of Georgia tidal inlets. Journal of Sedimentary Petrology, v. 43, p. 33-41.

______ 1975. Ebb-tidal deltas of Georgia estuaries. *In:* Estuarine Research, v. 2, L. E. Cronin (ed.). Academic Press, New York, p. 267-276.

Panageotou, W. and Leatherman, S. P. 1986. Holocene-Pleistocene stratigraphy of the inner shelf off Fire Island, New York: implications for barrier-island migration. Journal of Sedimentary Petrology, v. 56, p. 528-537.

Postma, H. 1982. Hydrography of the Wadden Sea: Movements and properties of water and particulate matter. Stichting Veth tot Steun aan Waddenonderzoek, Leiden, 75 p.

______ and Kroon, A. 1986. Mathematische profiel analyse van de onderzeese oever en de aansluitende zeebodem voor de Nederlandse kust. Rijkswaterstaat, Dienst Getijdewateren, notitie GWAO-86375, 63 p.

Sha, L. P. 1989a. Cyclic morphologic changes of the ebb-tidal delta, Texel Inlet, The Netherlands. Geologie en Mijnbouw, v. 68, p. 35-48.

______ 1989b. Sand transport patterns in the ebb-tidal delta off Texel Inlet, Wadden Sea, The Netherlands. Marine Geology, v. 86, p. 137-154.

______ 1989c. Variation in ebb-delta morphologies along the West and East Frisian Islands, The Netherlands and Germany. Marine Geology, v. 89, p. 11-28.

______ 1989d. Holocene-Pleistocene interface and three-dimensional geometry of the ebb-delta complex, Texel Inlet, The Netherlands. Marine Geology, v. 89, p. 207-228.

______ 1990a. Sedimentological studies of the ebb-tidal deltas along the West Frisian Islands, The Netherlands. Ph.D. Thesis Utrecht, Geol. Ultraiectina, v. 64, 160 p.

______ 1990b. Surface sediment characteristics in the ebb-tidal delta of Texel Inlet, The Netherlands. Sedimentary Geology, v. 68, p. 125-141.

Van der Giessen, A., de Ruijter, W. P. M. and Borst, J. C. (*in press*). Three dimensional current structure in the Dutch coastal zone. Netherlands Journal of Sea Research.

Van Straaten, L. M. J. U. 1961. Directional effects of winds, waves, and currents along the Dutch North Sea coastal zone. Geologie en Mijnbouw, v. 40, p. 333-346, and p. 363-391.

Westenberg, J. 1974. Kennemer dijkgeschiedenis. Verh. Kon. Ned. Akad. Wetensch. Afd. Natuurkunde, le reeks v. 27, p. 1-59.

KEY WORDS: EBB-TIDAL DELTAS, THE NETHERLANDS, WEST FRISIAN ISLANDS, SAND BODY GEOMETRY, LONGSHORE DRIFT, TIDAL CURRENTS

In: Clastic Tidal Sedimentology, D. G. Smith, G. E. Reinson, B. A. Zaitlin and R. A. Rahmani (eds.). Canadian Society of Petroleum Geologists. Memoir 16, p. 219-226.

A depositional model for a flood-tidal delta and washover sands in the late Pleistocene Paleo-Tokyo Bay, Japan

NAOMI MURAKOSHI
College of Natural Sciences, The University of Tsukuba, Ibaraki 305, Japan

FUJIO MASUDA
Institute of Earth and Planetary Science College of General Education Osaka University Toyonaka, Osaka 560, Japan

ABSTRACT

During the late Pleistocene, a flood-tidal delta landward of barrier islands developed at the eastern mouth of Paleo-Tokyo Bay, central Japan. The sediments deposited formed sand wedges tapering into lagoonal mud, and are composed of flood-tidal delta and washover facies.

The flood-tidal delta facies is composed of three major subfacies: delta-bottom facies, delta-front facies, and delta-top facies. The delta-bottom facies comprises alternations of fine to very fine grained sand and mud. Strong bioturbation is conspicuous. Sedimentary structures such as ripple lamination and mud drapes are, however, rarely preserved. The delta-bottom facies grades gradually into the delta-front facies, which displays lagoonward (landward)-inclined crossbedding with a lot of burrows. The delta-top facies consists of parallel laminated, medium to coarse grained sand, intercalated with some tabular crossbedding and trough crossbedding. A stacking of the subfacies shows a coarsening-upward sequence generated by progradation of the flood-tidal delta.

The washover facies overlies the flood-tidal delta and is distributed more lagoonward. The washover facies comprises proximal and distal subfacies. The proximal washover deposits consist of parallel-laminated, medium to fine grained sand with shell fragments at the base. Convoluted laminae are common. The distal-washover facies consists of very coarse to medium grained sand characterized by steep foreset structures with syndepositional cracks.

Enormous quantities of washover sand were deposited in a very short time interval. Some units contain features suggestive of earthquake activity during deposition. Moreover, the powerful waves responsible for the washovers may have been earthquake-generated tsunami waves.

RÉSUMÉ

Au Pléistocène tardif, un delta de marée inondé s'est développé dans la partie orientale de l'embouchure de la baie de Tokyo, Japon central, du côté terrestre d'une crête d'avant-plage émergée. Les sédiments accumulés forment un biseau de sable s'estompant dans les boues du lagon; ils sont composés de faciès de delta de marée inondé et d'un faciès de type «washover» (sédiments de la barrière relocalisés lors des intempéries).

Le delta se compose de trois faciès deltaïques majeurs: basal, frontal et sommital. Le faciès basal se compose d'une alternance de sables fins à très fins et de boues. Une intense bioturbation a affecté la préservation des structures sédimentaires; les laminations ondulés et «lentilles de boues pélagiques» («mud drapes») sont rarement observables. Le faciès basal passe graduellement au faciès frontal caractérisé par des stratifications obliques inclinées vers la lagune, i.e. vers la terre ferme. Les terriers y sont abondants. Le faciès sommital comprend des laminations parallèles de sables moyens à grossiers, interstratifiées avec des stratifications obliques tabulaires et de fosses. La succession des faciès montre une augmentation de la granulométrie vers le sommet de la séquence qui résulte de la progradation du delta de marée inondé.

Le faciès de type «washover» surmonte le delta de marée inondé et répand plus de sédiments vers le lagon. Ce faciès se compose de dépôts proximaux consistant en laminations parallèles de sables fins à moyens, avec des fragments coquilliers à la base. Les laminations convolutées sont communes. Les dépôts distaux consistent en sables moyens à grossiers caractérisés par des «structures frontales de pente» («foreset structures») très inclinées et des fissures synsédimentaires.

D'énormes quantités de sables de type «washover» se sont accumulés durant de très courts intervalles de temps. Certaines unités renferment des caractéristiques suggérant la manifestation de tremblement de terre durant la sédimentation. Les tsunamis provoqués par les tremblements de terre pourraient être responsables des puissantes vagues nécessaires à la formation de ce type de dépôts.

*Present address: Doctoral Program in Geoscience, The University of Tsukuba, Ibaraki 305, Japan

Introduction

Several stratigraphic models for barrier island systems and their constituent depositional environments are available (*e.g.*, Heron *et al.,* 1984; Moslow and Tye, 1985). Specific investigations of flood-tidal deltas have focused mainly on morphological aspects (Pierce, 1970; Hine, 1975; Boothroyd *et al.,* 1985). Only a few detailed descriptions of internal sedimentary structures and sequences of flood-tidal deltas have been published (Horne and Ferm, 1981; Hayes, 1980; Donselaar, 1984; Berelson and Heron, 1985; Israel *et al.,* 1987).

The objective of this paper is to document the vertical sequence and outline depositional processes, especially abandonment processes, for an upper Pleistocene microtidal flood-tidal delta in east-central Japan, near Tokyo.

Geological and Depositional Setting

During the last interglacial period of the Pleistocene (about 130,000-100,000 B.P.), a shallow-marine sedimentary basin called Paleo-Tokyo Bay was situated in what is now the Kanto Plain area in central Japan. Barrier islands developed on the eastern rim (the Pacific side) of Paleo-Tokyo Bay (Fig. 1; Okazaki and Masuda, 1989). The sedimentary infill of Paleo-Tokyo Bay consists, in ascending stratigraphic order, of the Narita (5-20 m thick), Ryugasaki (1-3 m thick) and Itabashi (0.5-2 m thick) formations (Aoki and Baba, 1979).

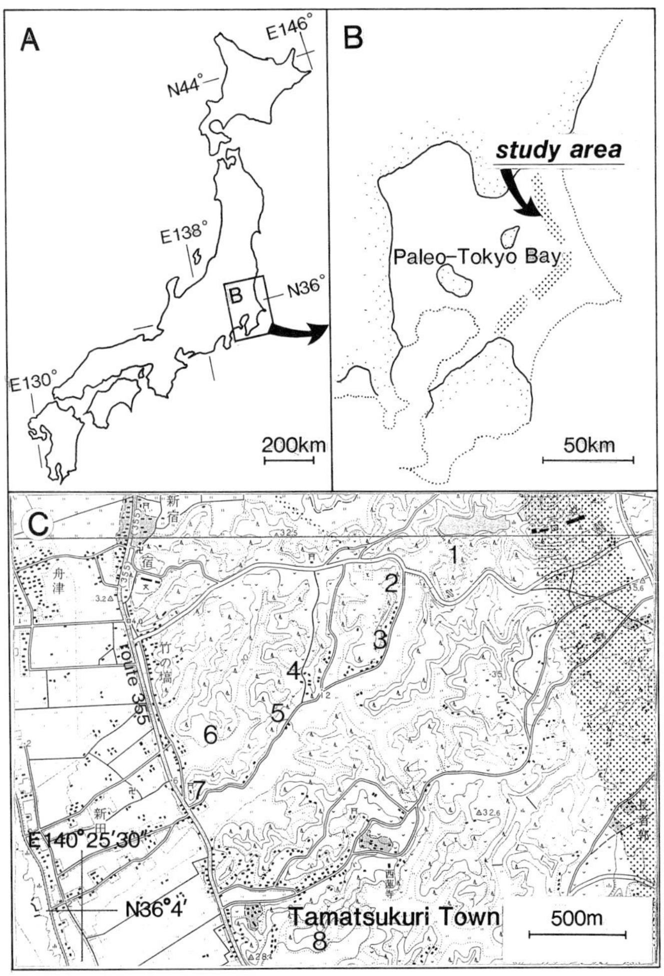

Figure 1. A. Location of Paleo-Tokyo Bay in Japan. **B.** Paleo-Tokyo Bay during the last interglacial. Dotted lines show present shoreline. Barrier islands (dark narrow area) developed on the eastern margin of the bay. **C.** Localities of studied sections. Dark belt is the inferred location of the barrier island.

On the Pacific side of the barriers, deposits are composed of alternating fine to coarse grained sands, gravels and thin mud beds. Hummocky cross-stratification and swaly cross-stratification occur in the lower strata, and trough and tabular cross-stratification occur in the upper sediments, suggesting a storm-dominated shoreface environment. The overlying unit consists of wave-ripples (0.15-0.6 m in wavelength) of granules and pebbles draped with clay. The uppermost beach facies consists of wedge-shaped horizontally-laminated, fine grained sand containing concentrations of heavy minerals. On the landward side of the barriers, deposits consists of mud and barrier associated sands. The mud is burrowed, alternating silt and clay, and is distributed over most of Paleo-Tokyo Bay. The facies suggests a calm, lagoonal environment in juxtaposition with the storm-dominant environment of the Pacific side.

It is interpreted that the tidal range in Paleo-Tokyo Bay during the Pleistocene did not exceed 2 m (microtidal), based on stratigraphic thickness of beach deposits (Masuda and Yokokawa, 1988), and height of intertidal sand waves (Masuda *et al.,* 1988). The tidal range was comparable to that in the present Tokyo Bay (1.9 m), rather than to that of the modern Pacific coast (1.4 m). Modern microtidal coasts are generally characterized by the development of active flood-tidal deltas and frequent washover sediments (Hubbard *et al.,* 1979; Heron *et al.,* 1984).

Cross-Section

The sections of the northeast-southwest trending outcrop belt are nearly normal to the west-northwest - east-southeast trending Pleistocene barrier (Figs. 1, 2). The sediments consist of mud, sand, sandy clay to clayey sand, and volcanic ash beds.

Mud resting on the underlying basement consists of alternating silt and clay. Thickness is variable, but the unit thins and disappears toward the northwest (Fig. 2). Burrows (*e.g., Rosselia* sp.) are common. The mud is interpreted as lagoonal, deposited on the floor of Paleo-Tokyo Bay.

The sand body is 1.7 km wide and consists of two major facies: flood-tidal delta (I) and washover sands (II) (Fig. 2). The flood-tidal delta facies, a southwest-wedging sand deposit, occupies the northeastern part of the sand body. The washover sand overlies the flood-tidal delta and extends into the former lagoon.

In the uppermost part of the section, sandy clay to clayey sand and/or eolian volcanic ash layers, overlie the sand facies. The sandy clay to clayey sand is rooted, indicating a

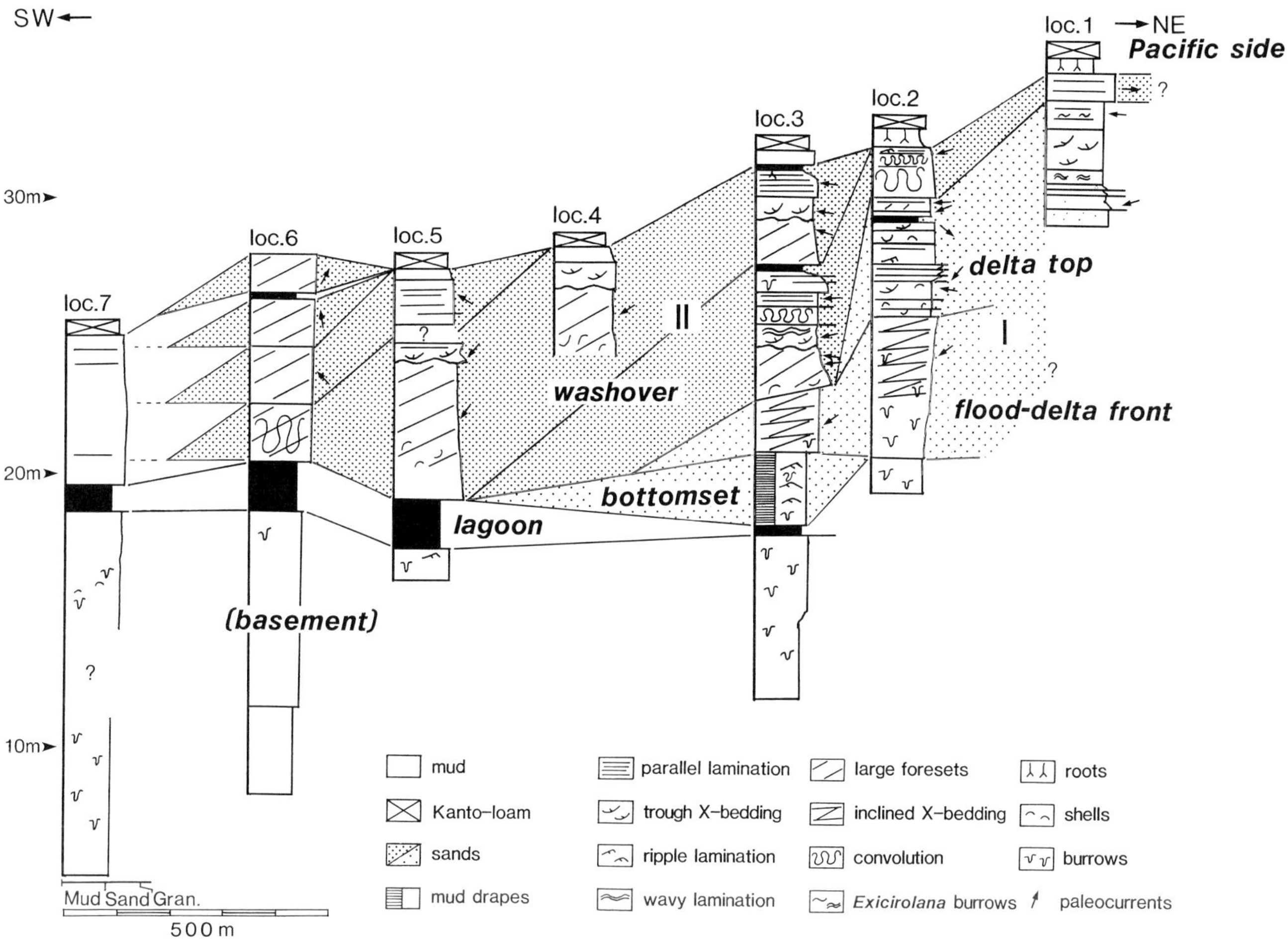

Figure 2. Stratigraphic sections along the profile of localities 1 to 7 shown in Figure 1C. The numbers on the left side indicate height above sea level.

marsh environment (Okazaki and Masuda, 1989). The volcanic ash layers are assigned to the Kanto-loam beds.

Flood-Tidal Delta

The flood-tidal delta sand body is composed of three major subfacies: a lower, very fine to fine grained sand, a middle medium grained sand, and an upper medium to coarse grained sand, stacked up to 9 m thick and wedging southwest (Fig. 2.).

The lower sand is 3 m thick, very fine to fine grained, and thins toward the southwest. Primary sedimentary structures are not well preserved because of bioturbation, but occasional ripple laminae are present. The sand contains thin layers of mud that drape ripples (Fig. 3). Some of the sands between the drapes show "thin-thick" changes, so that the mud layers form couplets. The number of mud drapes decreases upward and the facies changes gradually into the upper, medium sand facies. *Rosselia* sp. burrows are common in this facies.

The middle unit is 5 m thick, wedges southwest, and consists of inclined tabular crossbedded, medium grained sand (Fig. 4). The laminae in the crossbeds dip southwest - west-southwest. Each crossbed is 5 to 10 cm thick and the boundaries between each unit dip southwest uniformly with low

Figure 3. Distal facies of the flood-tidal delta. Ripple-laminated sand and mud drapes, some of which preserve a pattern of diurnal inequality of the tides in Paleo-Tokyo Bay (after Masuda *et al.*, 1989). The scale bar is 10 cm long; Loc. 3 (Fig. 1C).

angles (<10°). Perpendicular to the flow direction and downstream, large-scale trough cross-strata (approximately 1 m wide) can be recognized. Distinct burrow structures of *Muesteria* sp. are present in the entire facies. The contact with the upper sand is sharp to erosional.

The upper subfacies consists of oxidized, reddish brown, medium to coarse grained sand intercalated with thin (a few centimetres to one grain thick) granule layers and some rounded pebbles. Many intact shells and shell fragments are present, especially above the lower contact. Sedimentary structures such as parallel lamination, current-ripples and trough cross-stratification are dominant. Intercalated within this sand, are relatively clean, 0.2 to 0.8 m thick, south-southeast oriented, tabular crossbedded, well-sorted, medium to coarse grained sands. Some ripple-troughs are filled with thin, laterally continuous (several metres) mud layers. There are no distinct burrows in this facies.

We interpret this sand body, composed of three subunits, as a flood-tidal delta. The lower sand, including the mud drapes, is classified as the *bottomset* unit, occurring at the distal margin of the flood-tidal delta. The thin-thick changes in the sands between mud drapes (mud couplets) are interpreted as daily tidal cycles, similar to those described by Visser (1980). The thin-thick pattern is thought to indicate diurnal inequality of the tides (Masuda *et al.,* 1989). Mud content decreases upward, probably because of the relative increase of sand supply as the delta-front prograded.

The middle, inclined crossbedded sand is interpreted as *flood-tidal delta front*. It is characterized by unidirectional (flood-oriented), tabular cross-stratification with landward inclined, low angle ($<10°$) set boundaries (Fig. 4) and also

Figure 4. Delta front facies of the flood-tidal delta showing inclined sets of crossbedding with bioturbation (*e.g., Muesteria* sp.). The boundary of each cross-set generally dips lagoonward with a low angle ($<10°$). The trowel is 26 cm long; Loc. 2 (Fig. 1C).

by trough cross-stratification on the flow-normal section. The inclined boundaries were generated by 3-dimensional sand waves propagated by flood currents migrating down the delta front. Similar configurations were reported by Donselaar (1984). Dalrymple (1984) describes analogous inclined crossbedding in the sand waves from Cobequid Bay, Canada. These sand waves, however, developed in a macrotidal setting.

The upper sand unit is the *flood-tidal delta top* sediment, deposited on the delta platform. Current-ripples and the marsh mud deposit lying just above this sand suggest the platform was laid down in shallow water. Intercalated south - southeast oriented crossbeds, the paleocurrent directions of which are almost opposite to the direction of flood-flow on the delta, are interpreted as spillover lobe deposits (terminology from Hayes, 1980). The ripple-troughs filled with thin layers of mud are perhaps a result of a change of sediment-supply.

The flood-tidal delta sequence coarsens upward (Fig. 5), indicating a gradual progradation of the delta body into and over the muddy lagoonal sediment. The general coarsening-upward sequence is comparable to the sequences described by Israel *et al.* (1987) and Boothroyd *et al.* (1985).

Washover Sands

Washover sands overlie the flood-tidal delta and also the muddy lagoonal sediments in their distal portions, commonly with a sharp lower contact (Fig. 2). The sediments are composed of two major subfacies: a proximal, parallel-laminated facies, and a distal, crossbedded facies. The two facies are the same as those of the washover sand body of Schwartz (1982). The sands are generally overlain by rooted, sandy clay to clayey sand, which in interpreted as marsh deposit, and are covered by eolian volcanic ash layers (Kanto-loam).

The parallel-laminated proximal facies, occurring in the most northeasterly exposures are 2 m thick and the laminae dip slightly lagoonward. The sediments show a fining-upward trend of medium to fine grained sand. Heavy-mineral-concentrated laminae are common. Small-scale current-ripple laminae and trough crosslaminae occur within the sand. The sand in the lower portion is considerably convoluted (Fig. 6). Locally, granules and shell fragments are found at its base. No bioturbation is observed in this sediment.

Distally, the washover sands grade into large-scale (up to 3 m) steeply dipping (up to 30°) planar foresets. The distal facies overlies the delta front, bottomset, and even the lagoonal mud with a sharp lower contact. There is no mud drape in the forest even near the contact with the mud. The increase in forest height up to 3 m in a lagoonward direction matches the increased paleowater depth. The sediment consists of very coarse to medium grained sand with abundant shell fragments. The coset includes relatively coarse grained sand and shell fragments near the base and fines upward to medium grained sand.

More distally, the angle of inclination of the individual foresets increases systematically and the basal contact shows a gradual change from concave to angular (Fig. 7). Foreset dips are steepest (about 30°) and individual foresets show a normal grading with maximum thickness (up to 30 cm) at the most distal edge of the washover sand. The top of the crossbedded sand is commonly reworked and replaced by trough and tabular cross-stratified sand, which gradually changes upward into clayey sand to sandy clay and then to washover sand in some cases. There are no reactivation surfaces or bundle sequences to suggest tidally-controlled deposition of the large foreset.

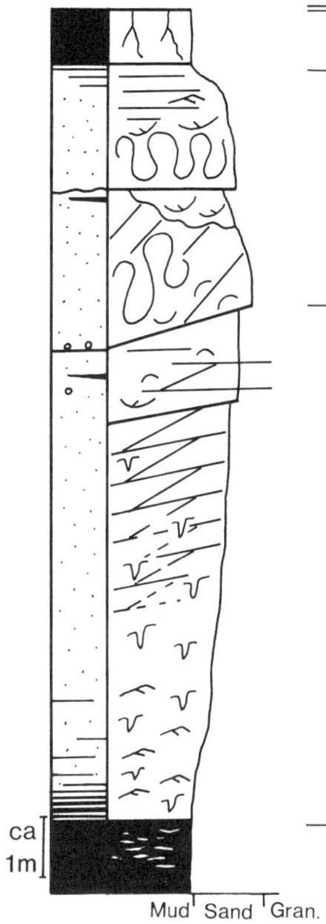

Environment		Lithofacies	Sedimentary Structures	Biofacies	Comments
marsh		sandy clay or clayey sand	structureless	abundant root traces	
washover	proximal	fine to medium grained sand shells at base	horizontal laminae with rare ripple laminae trough X-laminae convolution		sharp lower contact
	distal	coarse to medium grained sand shells at base	steep large foresets syndepositional cracks convolution		sharp lower contact
flood-tidal delta	top	medium to coarse grained sand	parallel laminae intercalating with granule layers tabular X-bedding trough X-bedding		sharp lower contact
	front	medium grained sand	inclined X-bedding	burrows *(Muesteria* sp.*)* strong bioturbation	
	bottom	fine to very fine grained sand blanketed by clay	ripple laminae with muddrapes	burrows *(Rosselia* sp.*)* strong bioturbation	upward decrease in mud content
lagoon			structureless or rare flasers	burrows *(Rosselia* sp.*)*	

Figure 5. Schematic vertical sequence and sedimentary characteristics of flood-tidal delta and washover sands.

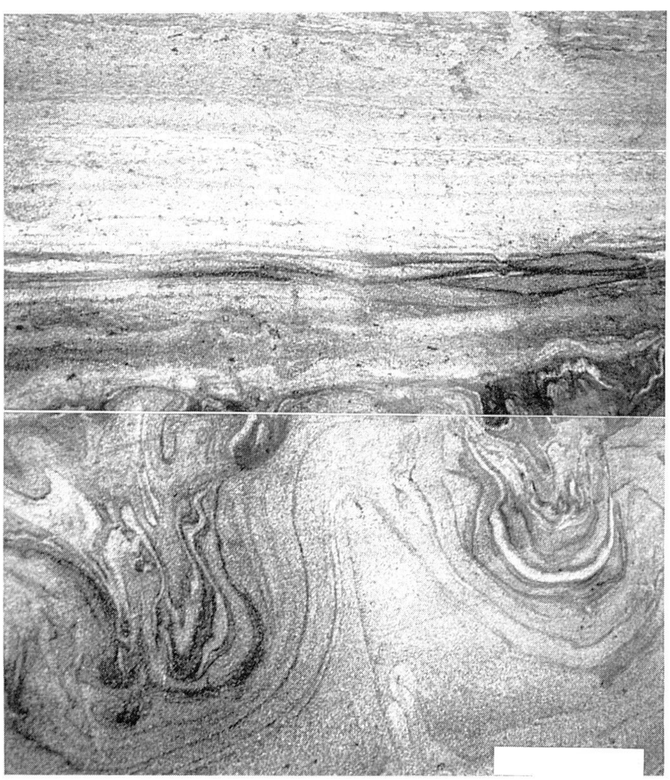

Figure 6. Proximal facies of washover sand. Small-scale ripple-lamination and trough crosslamination are visible in the middle, convolute laminae in the lower half. The scale bar is 10 cm long; Loc. 2 (Fig. 1C).

The convolution structures in the proximal parallel-laminated portion probably resulted from pore water escape. This feature is common in the rapidly deposited, poorly packed sand similar to washover sand described by Lowe (1975).

It is difficult to delimit the sediment deposited during a single washover event because the exact position of the scour surface is commonly not clear (Donselaar and Nio, 1982). In this case, however, the unit grading from a sediment rich in shell fragments and granules through a convoluted zone into a ripple- and trough-laminated scour zone may indicate a single washover event.

Although the facies of flood-tidal delta top and overlying proximal washover sand share some similar detailed structural and textural properties because both facies may

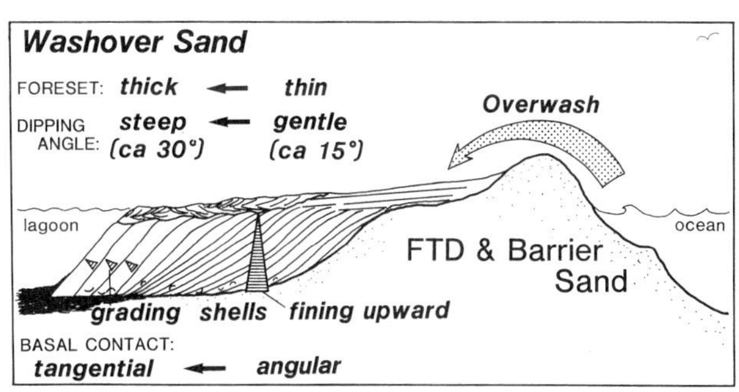

Figure 7. Schematic section of washover sand. FTD, flood-tidal delta.

have undergone some similar flow events or processes, the convenient boundary between delta-top and proximal washover facies can be defined based on the existence of spillover lobe deposits. The delta would have been active when spillover lobe sediments were being deposited, and the activity would cease during washover events.

The sediments of the distal facies are coarser than those of the proximal facies. These coarser materials would be selectively transported across the washover fan surface into the distal setting by a grain flow (shear-sorting) mechanism. This is typical of the textural trends preserved across many modern fan facies (Schwartz, 1975). The upward fining and normal grading of the foresets would be a result of the shear sorting mechanism down the avalanche face.

Each unit of distal foresets and reworked tabular and trough cross-stratified sand may represent a storm-washover event and subsequent modification by sedimentation within trenches across washover-fans.

Synsedimentary cracks characterize the distal foreset facies (Fig. 8). According to the analysis of conjugate sets (the σ_1 is almost vertical, σ_2 is parallel to the strike of the foresets, and σ_3 indicates the direction of foreset dip), the cracks would be a result of slope failure and subsequent partial collapse occurring on the surface layers of the foresets. They extend through one or several foresets, but are always cut off and buried by deposition of the next foreset. The abundant synsedimentary cracks appear to have formed before burial of the next foreset lamina, suggesting very rapid deposition of this unit.

Figure 8. Synsedimentary cracks developed in the steep foresets of distal washover facies. The cracks extend through one or several foreset units before they are covered by the next foreset. The scale is 18 cm long; Loc. 3 (Fig. 1C).

We think that the distal foresets were deposited during a decreasing stage of flow energy. Initial foresets (relatively low angle, concave) formed under higher energy conditions. Settling of suspended matter became more important with diminishing current velocity creating the concave foresets. With lessening flow energy, bedload accumulations became unstable and avalanched downward. Shear-sorting now resulted in the "avalanching" grain flow mass, and the coarsest material moved to the front and top of the downward-moving flow mass. As the result, steep, angular foresets were formed. The foresets of the final stage are the steepest (up to 30° dip) and thickest (max. 30 cm in width) and show a distinct normal grading. A single foreset grades from medium grained sand to laminated, heavy-mineral-concentrated fine grained sand. As yet, the formative mechanism of these thick, normal-grading foresets is unclear.

The overlying marsh sediment and volcanic ash layer indicate the delta and inlet system was abandoned in response to the deposition of the washover sands. After deposition of the washover sediments, this area became a supratidal marsh or upland. The inlet channel and washover throat became plugged and stabilized subaerially. A new breach or inlet probably developed elsewhere.

Tsunami Deposits

There is some evidence for tectonic instability during the deposition of washover sands. Figure 9 shows one of the out-

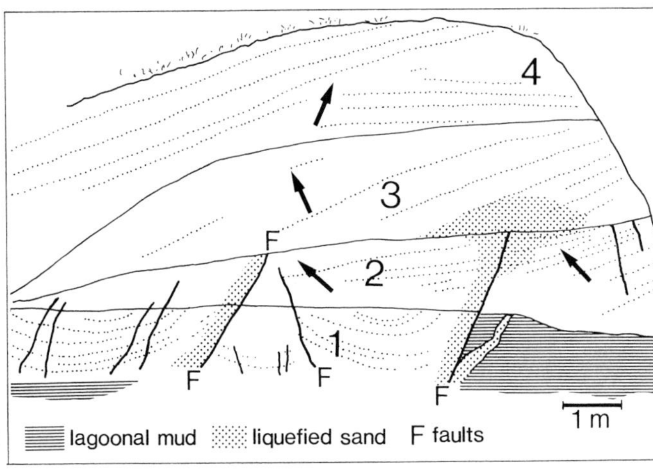

Figure 9. Field sketch of the terminal washover sands, composed of four units. Unit 1 is undulating and the other three units are distinguished by paleocurrent directions (arrows). Faults and liquefaction in units 1 and 2 precede deposition of unit 3. Loc. 6 (Fig. 1C).

crops of stacked washover foresets, which is divided into four units on the basis of their paleocurrent directions. There is a thin mud layer, suggesting a stillstand at the boundary between units 3 and 4, but no mud between boundaries 1 and 2, or 2 and 3 although it may be eroded. Many kinks and faults are observed in units 1 and 2. Unit 1 is folded and is cut by major faults. Structureless sand, attributed to liquefaction, is encountered along the faults and also appears as dikes in the lagoonal mud. The sand dikes cut the lower units (1 and 2) but are truncated by a crossbedded set of unit 3. The sand of unit 3 near the fault is partly liquefied and deformed, perhaps due to an earthquake-induced injection of sand using the fault-plane after the deposition of unit 3. This suggests that faulting and associated sand liquefaction and injection were contemporaneous with washover deposition. In other places, large-scale convolute folds and sand

Figure 10. Section of sand volcano found in the same horizon of washover sand. The scale is 85 cm long; Loc. 8 (Fig. 1C).

volcanoes are found in the same horizon of washover sand (Fig. 10).

The western Pacific Rim is characterized by widespread earthquakes (Sugimura and Uyeda, 1973). The faulting and injection occurring in washover sediments were probably caused by earthquakes. Similar earthquake-generated clastic dikes are reported by Reimnitz and Marshall (1965) from Alaska. Some earthquakes occurring near the trenches generate long-period tsunami waves that devastate coastal areas. Three metre foresets are exceptionally thick for usual washover deposits, but such large-scale sand transport would be possible by waves, such as tsunamis. Thus, some of the washover sands could have been transported by tsunami-originated overwash. Syndepositional faults and cracks in the foreset beds of the washover sands suggest a possibility that tsunamis caused rapid deposition of a large quantity of sand. Washover sediments, which plugged inlets and caused abandonment of flood-tidal delta systems, appear to have been moved to the lagoon side from the shoreface and coastal areas on the Pacific side of the barrier islands by tsunamis.

These described washover sediments may be explained by the tectonic instability of the Pacific margin in Paleo-Tokyo Bay. We can state that washover building, faulting and flood-tidal delta abandonment are closely related to the earthquakes and tsunamis, particularly sites situated near the active plate margin tectonic setting on the east coast of Japan.

Acknowledgments

J.R. Boersma of the State University of Utrecht and D.A. Leckie of the Geological Survey of Canada are thanked for critical reading and many helpful comments and suggestions. Thanks are also due to R. N. Hiscott of Memorial University of Newfoundland for critical reading and correction of our English during the preparation of this manuscript. M. Ito gave helpful advice and discussions in the field.

References

Aoki N. and Baba K. 1979. Shimosa Group of the Kasumigaura-Kitaura area, south eastern part of Ibaraki Prefecture. Tsukuba Environmental Studies, no. 4, p. 186-195 (in Japanese).

Berelson, W. M. and Heron, S. D., Jr. 1985. Correlations between Holocene flood tidal delta and barrier inlet fill sequences: Back Sound — Shackleford Banks, North Carolina. Sedimentology, no. 32, p. 215-222.

Boothroyd, J. C., Friedrich, N. E. and McGinn, S. R. 1985. Geology of microtidal coastal lagoons: Rhode Island. Marine Geology, v. 63, p. 35-76.

Dalrymple, R. W. 1984. Morphology and internal structure of sandwaves in the Bay of Fundy. Sedimentology, v. 70, p. 365-382.

Donselaar, M. E. 1984. Flood tidal delta sedimentation in the Late Cretaceous Menefee Formation (Mesaverde Group), San Juan Basin, north west New Mexico. Geologie en Mijnbouw, no. 63, p. 323-331.

______ and Nio, S.D. 1982. An Eocene tidal inlet/washover type barrier island complex in the south Pyrenean marginal basin, Spain. Geologie en Mijnbouw, no. 61, p. 343-353.

Hayes, M. O. 1980. General morphology and sedimentary patterns in tidal inlets. Sedimentary Geology, no. 26, p. 139-156.

Heron, S. D., Jr., Moslow, T. F., Berelson, W. M., Herbert, J. R., Steele, G. A. III, and Susman, K. R. 1984. Holocene sedimentation of a wave-dominated barrier island shoreline: Cape Lookout, North Carolina. Marine Geology, v. 60, p. 413-434.

Hine, A. C. 1975. Bedform distribution and migration on tidal deltas in the Chatham Harbor Estuary, Cape Cod, Massachusetts. *In:* Estuarine Research, v. 2, L. E. Cronin (ed.). Academic Press, Inc., p. 235-252.

Horne, J. C. and Ferm, J. C. 1981. Carboniferous depositional environments in the Pocahontas Basin, Eastern Kentucky and Southern West Virginia. *In:* Overwash Processes, Benchmark Papers in Geology, v. 58, S. P. Letherman (ed.). Hutchinson Ross Publishing Company, p. 304-310.

Hubbard, D. K., Oertel, G. and Nummedal, D. 1979. The role of waves and tidal currents in the development of tidal-inlet sedimentary structures and sand body geometry: examples from North Carolina, South Carolina and Georgia. Journal of Sedimentary Petrology, v. 49, p. 1073-1092.

Israel, A. M., Ethridge, F. G. and Estes, E. L. 1987. A sedimentologic description of a microtidal, flood-tidal delta, San Luis Pass, Texas. Journal of Sedimentary Petrology, v. 57, no. 2, p. 288-300.

Lowe, D. R. 1975. Water escape structure in coarse-grained sediments: Sedimentology, no. 22, p. 157-204.

Masuda, F., Nakayama, N. and Ikehara K. 1988. A cross-stratification produced by tidal current during nine days in the Pleistocene at Uchijuku in Kitaura, Ibaraki. Tsukuba Environmental Studies, no. 11, p. 91-105 (in Japanese).

_____ and Nakayama, N., Miwa, Y. and Murakoshi, N. 1989. Tidal pattern of diurnal inequality in the Pleistocene Paleo-Tokyo Bay. Annual Report of the Institute of Geoscience, the University of Tsukuba, no. 15, p. 72-77.

_____ and Yokokawa, M. 1988. Events read from Pleistocene beach deposits. Earth Monthly, v. 10, no. 8, p. 523-530 (in Japanese).

Moslow, T. F. and Tye, R. S. 1985. Recognition and characterization of Holocene tidal inlet sequences. Marine Geology, no. 63, p. 129-151.

Okazaki, H. and Masuda, F. 1989. Arcuate and bird's foot deltas in the late Pleistocene Paleo-Tokyo Bay. *In:* Deltas: Sites and Traps for Fossil Fuels, M. K. Whateley and K. T. Pickering (eds.). Geological Society Special Publication, no. 41, p. 129-138.

Pierce, J. W. 1970. Tidal inlet and washover fans. Journal of Geology, v. 78, p. 231-234.

Reimnitz, E. and Marshall, N. F. 1965. Effects of the Alaska earthquake and tsunami on recent deltaic sediments. Journal of Geophysical Research, v. 70, p. 2363-2376.

Schwartz, R. K. 1975. Nature and genesis of some storm washover deposits. Technical memorandum, no. 61, Coastal Engineering Research Center, 69 p.

_____ 1982. Bedforms and stratification characteristics of some modern small-scale washover sand bodies. Sedimentology, v. 29, p. 835-849.

Sugimura, A. and Uyeda, S. 1973. Island Arcs. Elsevier Scientific Publishing Company, 247 p.

Visser, M. J. 1980. Neap-spring cycles reflected in Holocene subtidal large-scale bedform deposits: A preliminary note. Geology, v. 8, p. 543-546.

KEY WORDS: PALEO-TOKYO BAY, FLOOD-TIDAL DELTA, WASHOVER SANDS, SYNSEDIMENTARY DEFORMATION, LAST INTERGLACIAL, TSUNAMI DEPOSITS

In: Clastic Tidal Sedimentology, D. G. Smith, G. E. Reinson, B. A. Zaitlin and R. A. Rahmani (eds.). Canadian Society of Petroleum Geologists. Memoir 16, p. 227-254.

A large flood-tidal delta and its successive spill-over apron: detailed proximal-distal facies relationships (Miocene Lignite Suite, Lower Rhine Embayment, Germany)

J. R. Boersma

Faculty of Geographical Sciences
University of Utrecht
P.O. Box 80.115, 3508 TC
Utrecht, The Netherlands

Abstract

A marine sand wedge intercalated between coastal plain sediments of an Upper Miocene Lignite Suite is described from an open pit mine in the Lower Rhine Embayment, Germany. The sand wedge has a thickness of 15 m at the inferred seaward side of the outcrop and thins landward amidst lagoonal clays over a distance of more than 2 km. Vertical and lateral facies changes were recorded in detail during the course of several years as the mine face shifted several hundreds of metres.

The wedge consists of two sand units separated by a continuous, thin clay bed. The basal interval of the lower unit, interpreted as a flood-tidal delta in its initial high energy stage, is characterized by unidirectional, tabular planer, crossbedded sets with numerous discontinuity (pause) planes reworked by subordinate wave and weak current action. The upper interval, consisting of strongly-bioturbated, essentially crosslaminated sands, developed in the sheltered platform environment behind a newly-formed spit or swash bar, part of which also was exposed. The upper unit of the sand wedge is dominated by an enigmatic type of crossbedded sets. These sets sometimes are hundreds of metres long, are consistently bundled and grade at the up- and downcurrent directions into flasered and biogenetically homogenized sands. These lateral sequences are interpreted as having been formed by aeolian sedimentation in a backbarrier wind tidal flat, which migrated into a shallow lagoon. At the boundary between the dry to damp area and the body of water, a front gradually developed, giving rise to the gradually-steepening, bundled crossbedding. Rises and falls of lagoonal water level caused the depositional system to shift back and forth. Sedimentation processes were found to be wave-dominated, the intermittency of sedimentation probably being governed by wind set-up at the shoreline. No sedimentary-structural features were found that reflected the astronomical tides. Landward from the lagoon, the flood-tidal delta bioturbated sand units join to form a terminal lobe of giant (>4 m thick) delta type crossbedding that interdigitates with lagoonal clay.

Résumé

Dans la Suite Lignite du Miocène Supérieur, un prisme de sable marin, intercalé entre les sédiments de la plaine côtière, est ici décrit à partir d'une mine à ciel ouvert dans la baie du Rhin inférieur, Allemagne. En affleurement, le prisme de sable atteint 15 m d'épaisseur du côté attribué à la mer, et s'amincit vers le continent parmi les argilites du lagon sur une distance de plus de 2 km. Les changements de faciès latéraux et verticaux ont été détaillés au cours de plusieurs années avec le recul de plusieurs centaines de mètres du front de taille de la mine.

Le prisme consiste en deux unités de sable séparées par un lit d'argile mince et continu. L'intervalle basal de l'unité inférieure, interprété comme étant un delta de marée inondé au stade initial de haute énergie, est caractérisé par des ensembles unidirectionnelles de stratifications obliques, planaires-tabulaires, avec de nombreux plans de discontinuité (arrêt) remaniés par l'action subordonné des vagues et des faibles courants. L'intervalle supérieur, composé essentiellement de grès à stratifications obliques fortement bioturbés, s'est développé dans un environnement protégé de la plate-forme, soit derrière une flèche littorale ou une barre formée par le jet de rive qui était partiellement émergée et nouvellement construite. L'unité supérieure du prisme de sable est dominée par un type énigmatique d'ensembles de stratifications obliques. Ces ensembles, parfois longs de plusieurs centaines de mètres, sont régulièrement en «faisceaux» et passent graduellement, contre et dans le sens de la direction du courant, à des grès de type «flaser» et des grès homogénéisés par une origine biologique. Ces séquences latérales résulteraient d'une sédimentation éolienne dans un estran d'arrière-barrière exposé au vent, migrant dans un lagon d'eau peu profonde. À la limite entre la région sèche à humide et la masse d'eau se développait graduellement un front aux stratifications de type «bundle» devenant graduellement de plus en plus inclinées. Dans le lagon, les montées et baisses du niveau d'eau ont provoqué l'oscillation du système de sédimentation. Les processus de sédimentation étaient dominés par les vagues, et la sédimentation intermittente était probablement gouvernée par les vents érigés le long de la ligne de rivage. Aucuns critères sédimentaires-structuraux reflétant des marées astronomiques n'ont été trouvés. Vers la partie continentale du lagon, les unités de delta de marée inondé et de sables bioturbés se rejoignent pour former un lobe terminal (>4 m d'épaisseur) avec des stratifications obliques géantes typiques des deltas, interlitées avec les argiles lagunaires.

Introduction

This paper deals with a wedge of white, apparently marine (derived) sand that lies amidst a dark suite of lignite-associated terrestrial deposits of fluviatile to deltaic origin (Fig. 1). The wedge is interpreted as representing a marine transgression of local or regional importance. Sedimentary structures, physical as well as biogenic, suggest a beach-to-nearshore environment of deposition for the clean sands that form the major part of the wedge.

Stratigraphically, wedge-shaped bodies of marine and marginal marine deposits are frequently described from continental edge settings and/or places and times of rapid relative sea level rise. Subsequent regressions preserve the record of the onlap.

Studies of Holocene transgressive deposits along the east coast of North America and the west coast of Europe provide us with many examples in our search for analogies. These studies are of two types, differing in scope as well as in resolution of sedimentary features. The first deals with barrier island depositional systems and their behaviours in response to basinal processes, coastal and submarine morphology, and sediment source. These studies clarify geometrical relationships, but tend to leave detailed sedimentary features and relationships unstudied. Among the many valuable studies of this kind are Kraft *et al.* (1979), Oertel (1985), Penland *et al.* (1988) and Carter *et al.* (1989). One of the important conclusions reached as a result of these studies is that tidal inlets, flood-tidal deltas and washover fans are intimately connected in the landward transgression of the wave-dominated barrier system, with wind transport of sand interpreted as being only of secondary importance (Kraft *et al.,* 1979). In the second type of study, process-response models are developed from aereally limited and physiographically delineated bodies, such as backbarrier associations of facies or even simple washovers. Recently, important progress has been made in the understanding of aeolian erosion and sedimentation in humid settings such as backbarrier environments (Andrews, 1970; Berelson and Heron, 1985; Boothroyd *et al.*, 1985; Deery and Howard, 1977; Fryberger *et al.*, 1990; Hennessy and Zarillo, 1987; Ritchie and Penland, 1988, and Schwartz, 1982). They provide detailed features of sedimentary (sub)facies necessary for reconstructing analogues in the interpretation of ancient deposits. As a consequence of the limitations set by their methods of investigation and the time factor, there are three shortcomings in the application of these studies to ancient strata: 1) insufficient data on the degree of postdepositional obliteration of bedding, 2) insufficient visualization of sediment arrangements at the outcrop scale of ancient successions, and 3) limited data on the preservation potential of encountered facies. The above limitations were severely felt when performing the detailed in-

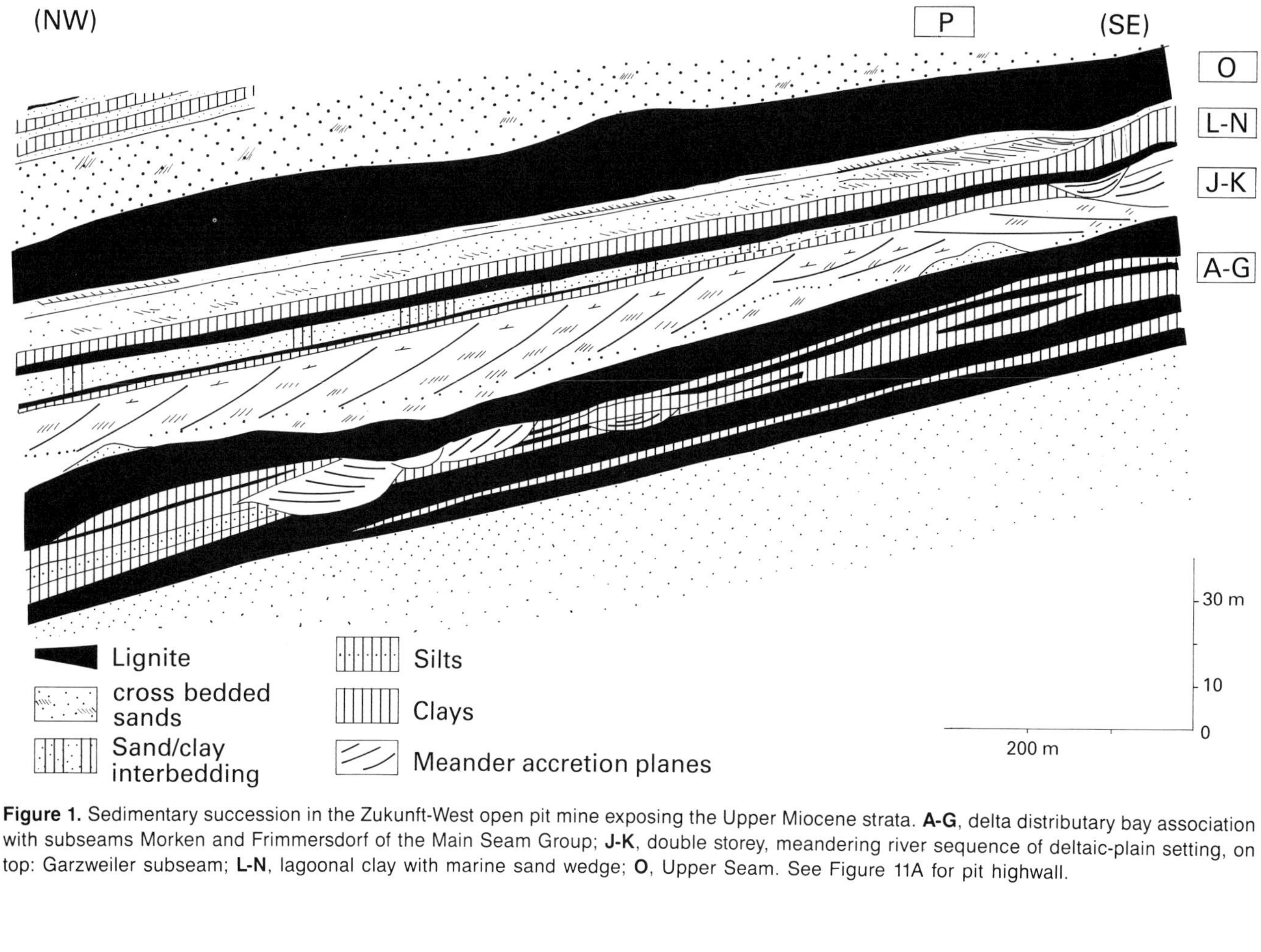

Figure 1. Sedimentary succession in the Zukunft-West open pit mine exposing the Upper Miocene strata. **A-G**, delta distributary bay association with subseams Morken and Frimmersdorf of the Main Seam Group; **J-K**, double storey, meandering river sequence of deltaic-plain setting, on top: Garzweiler subseam; **L-N**, lagoonal clay with marine sand wedge; **O**, Upper Seam. See Figure 11A for pit highwall.

vestigation of the 2.5 km long by 10 m high outcrop of marine sediments of this study. Despite the variable and rapidly changing outcrop, due to excavation, individual beds could be traced over distances of several hundred metres and their extinction or mergence into other beds established.

The objectives of this study are to give a detailed account of the different sedimentary facies and their processal implication. This is done on a wide scale, ranging from centimetres to kilometres. As such it is an attempt to bridge the gap that commonly exists between regional and outcrop studies. Furthermore, we have tried to establish the characteristics of vertical and lateral sedimentary sequences on every scale and to view them as short- and long-term reponses to processal variability and changes in the barrier-lagoon configuration.

Geological Setting

The Zukunft-West open pit mine (Fig. 2) lies on the southwestern margin of the southeast-northwest trending graben (± 60 km wide) known as the Lower Rhine Embayment (in German: Niederrheinische Bucht). The Lower Rhine Embayment forms part of the Western European Rift Belt, active since the end of the Eocene (Illies, 1977).

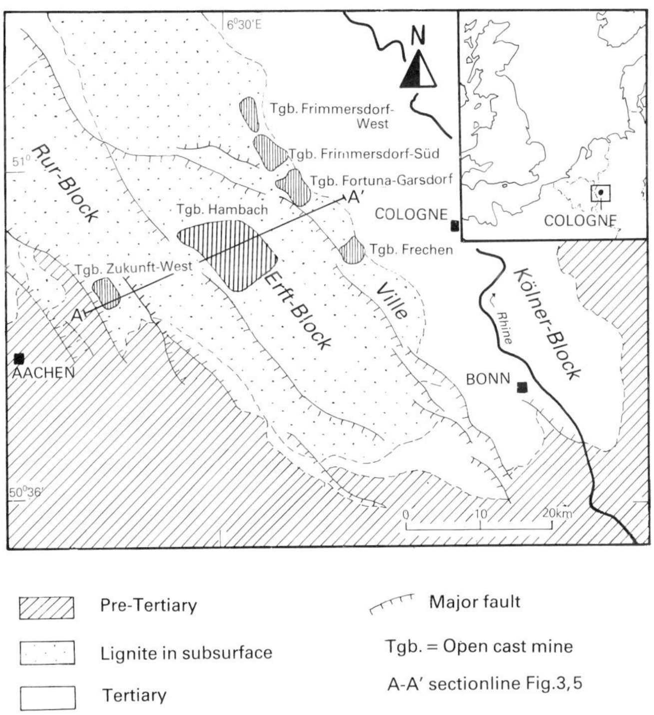

Figure 2. Location map and major geological units of the Lower Rhine Embayment.

The graben-parallel block-fault system delineates blocks that are tilted face down to the graben axis (Fig. 3). Shortly after the initial stages of rifting, the Lower Rhine Embayment was invaded by the sea, reaching its farthest inland position, Bonn, during Early to Middle Oligocene times (Fig. 2). Since then, the shoreline has retreated in a fluctuating transgressive and regressive manner northwestward, eventually vacating the basin in the Pliocene (Fig. 4).

The position of the shoreline responded to the interaction of several factors including: sediment supply, depositional processes, regional subsidence and eustatic sea level changes. At first, supply was derived from a restricted but deeply weathered hinterland, probably not extending farther south than the Slate Mountains. It was served by comparatively small pre-Rhine river systems. In the Pliocene the hinterland probably extended southward into the Upper Rhine Graben giving way to a larger primitive Rhine system that supplied higher quantities of coarser grained sediment.

Fault activity was minimal during deposition of the Main (lignite) Seam Group (Fig. 5). This permitted the uninterrupted deposition of hundreds of metres of peat in the central part of the basin (Hager, 1986). Our studies indicate that the peat accumulated largely in the backbarrier setting. Thick accumulations of sand and mud were deposited in coastal plain, shoreline and shallow marine settings, which were subject to more or less systematically shifting depocentres. Additional stratigraphic information is available in other papers (Gliese and Hager, 1978; Hager, 1968, 1986; Zagwijn and Hager, 1987; and Zagwijn and Doppert, 1978.

Setting of the Lower Rhine Embayment

Our investigations of the barrier deposits throughout the Lower Rhine Embayment indicate that shoreline processes were wave-dominated, with storms playing an important part. Longshore currents, probably directed by a longshore bar system, were at times particularly active. Severe storms created washover sheets riddled with logs and twigs, apparently eroded from nearby beach ridges. These sheet deposits were spread out "in a low-relief barrier plain comparable to the dune terrace" as described by Ritchie and Penland (1988). Eolian dunes were low or nonexistant, probably because of damp or wet substrate, attributed to high water levels in the back-barrier lagoon. Inlets were scarce, inconspicuous and probably short-lived.

In plan view, the coastline wound irregularly across the basin, reflecting deltaic depocentres and adjacent barrier systems. At one time only, during deposition of the upper part of the Main Seam Group, the coastline extended straight across the basin in a southwest-northeast direction. This coastline trend was coupled with a marine onlap, which resulted in the deposition of another wedge of marine sand (Neurather Sand) between terrestrial strata in the centre of the basin.

Sedimentary Facies and Interpretation

Most of the data were collected in three successive years from a 2.3 km long by 0.3 to 0.7 km wide segment of the continuously migrating outcrop of mine face (Fig. 6). Additional data were assembled six years later in a section 1.5 km to the east of the former segment. The wealth of information is summarized into a synoptic diagram (Fig. 7). Despite the variable exposure conditions of the rapidly mined

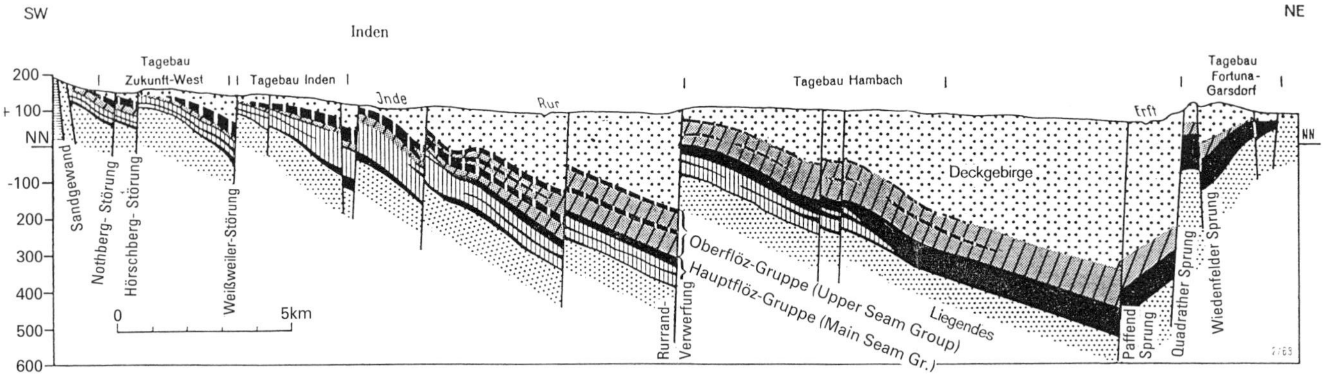

Figure 3. Generalized cross-section of the Lower Rhine Embayment showing Neogene lignite measures in a block-faulted graben structure. Open pit mine ("Tagebau") Zukunft-West is at the southern margin (after Hager, 1968).

outcrop (up to 200 m/yr.), individual beds could be traced and studied over distances of several hundred metres. This permitted detailed observations of the lateral transition of one bed type into another and wedging out patterns.

Relative positions along the profile are indicated by a metric scale along the base of Figure 7. Vertically, the succession is divided into four facies, which were termed according to their inferred genesis, from bottom to top: lagoonal clay, flood-tidal delta, spillover lobes, and a coal seam. A detailed presentation of the sand wedge lithofacies at the 1900 m location (Pt. 1900; Fig. 7) is given in Figure 8. The geometry of the sand wedge is characterized by a gradual thinning from north to south and, most prominently, by the rapid wedging-out of the flood-tidal delta facies against the lagoonal clays over a distance of a few hundred metres. This wedging is associated with a steepening of cross-beds to about 30°, producing a delta-like body of foreset sand interdigitating with clays. On the basis of paleogeographic reconstructions, the north-northwest-south-southeast strike of the synoptic section conforms to an orientation normal to the shoreline with the sea lying to the north. The terms "proximal" and "distal" will henceforth be employed to denote relative distance inland from the shoreline.

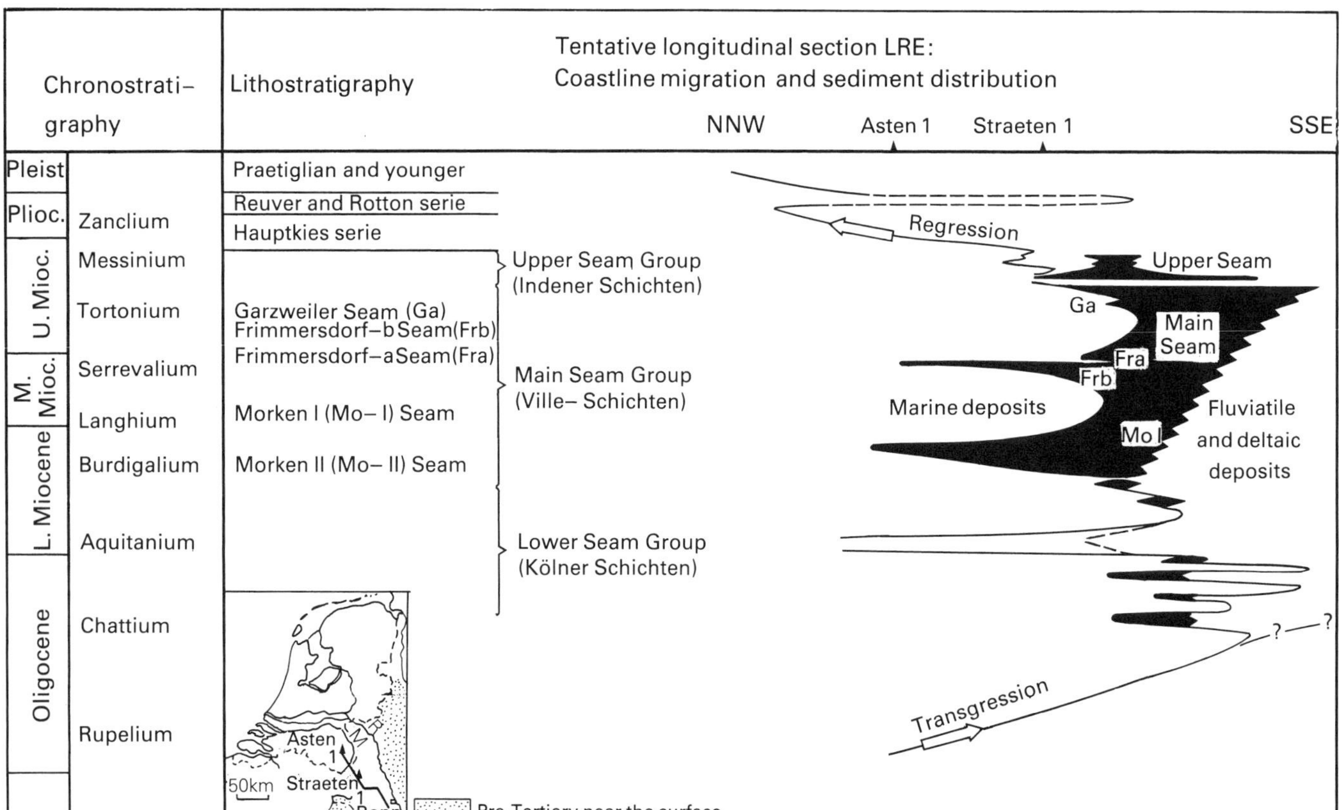

Figure 4. Stratigraphy of the Lower Rhine Embayment showing transgressional and regressional events and associated sediments (after Hager, 1981).

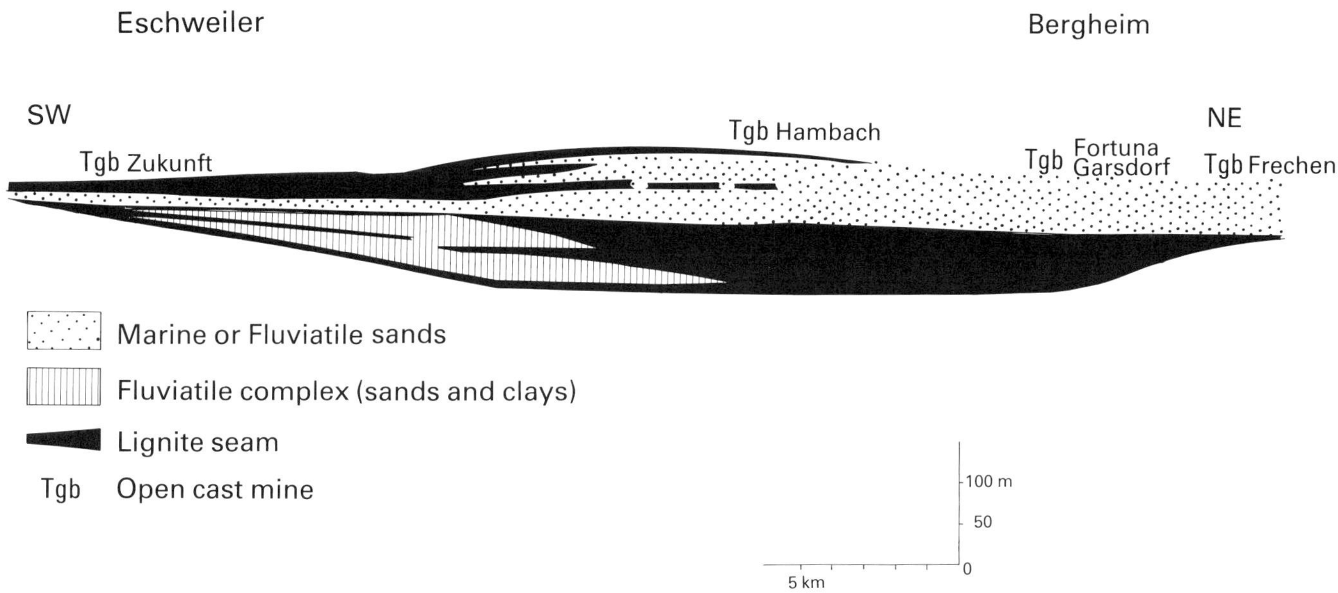

Figure 5. Schematic cross-section of the Lower Rhine Embayment disregarding the block fault displacements. Lower and Upper Seam groups shown concentrated in different parts of the basin, probably as a result of shifting fluviatile deposit centres (after Hager, 1968).

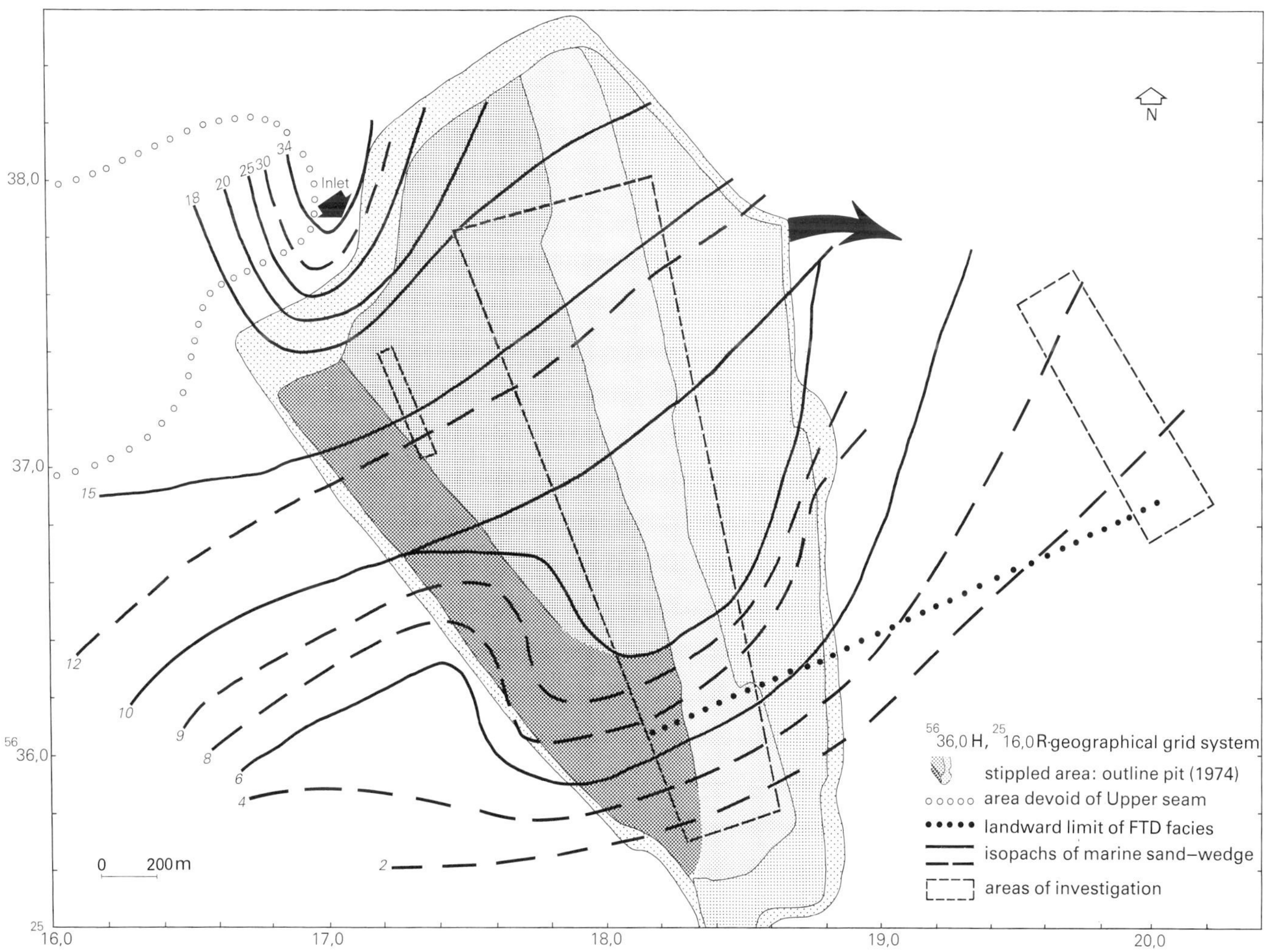

Figure 6. Isopach map of the marine sand wedge in the area of the Zukunft-West open pit mine. Thickness increases in the northwestern corner are mainly the effect of a plunging base level of the sand. This in turn is attributed to an inlet cutting the barrier. Accuracy of the isopachs is moderate since they are constructed from 1:2000 scale, southwest-northeast sections. Location of stippled area roughly conforms to that of the Zukunft-West pit in Figure 2. Diving arrow indicates direction of excavation.

Facies Descriptions

The Lagoonal Clay Facies

The clay facies has a consistent thickness of 3 to 4 m over most of the profile. Southward of Pt. 700 to about Pt. 300, the thickness gradually increases along with the gradual thinning of the overlying flood-tidal delta unit. South of Pt. 300, the clay thickens rapidly to 6 m as it completely replaces the steeply dipping delta foreset flood-tidal delta sands (*cf.* Figure 7). Close inspection in this area reveals rapidly tapering sand beds against the clay. These sand beds are distal offshoots of the ''delta'' body. Thin (mm-cm thick) laminae of sand and silt may be found throughout the clay body (Fig. 9). The thickness and density of the laminae vary vertically and are the main reason for vertical variations in gross silt/sand content of the clay. The lighter coloured layers reflect a higher silt/sand content. Biogenic structures are not easily detected and seem to be restricted to a mottling at lamina scale.

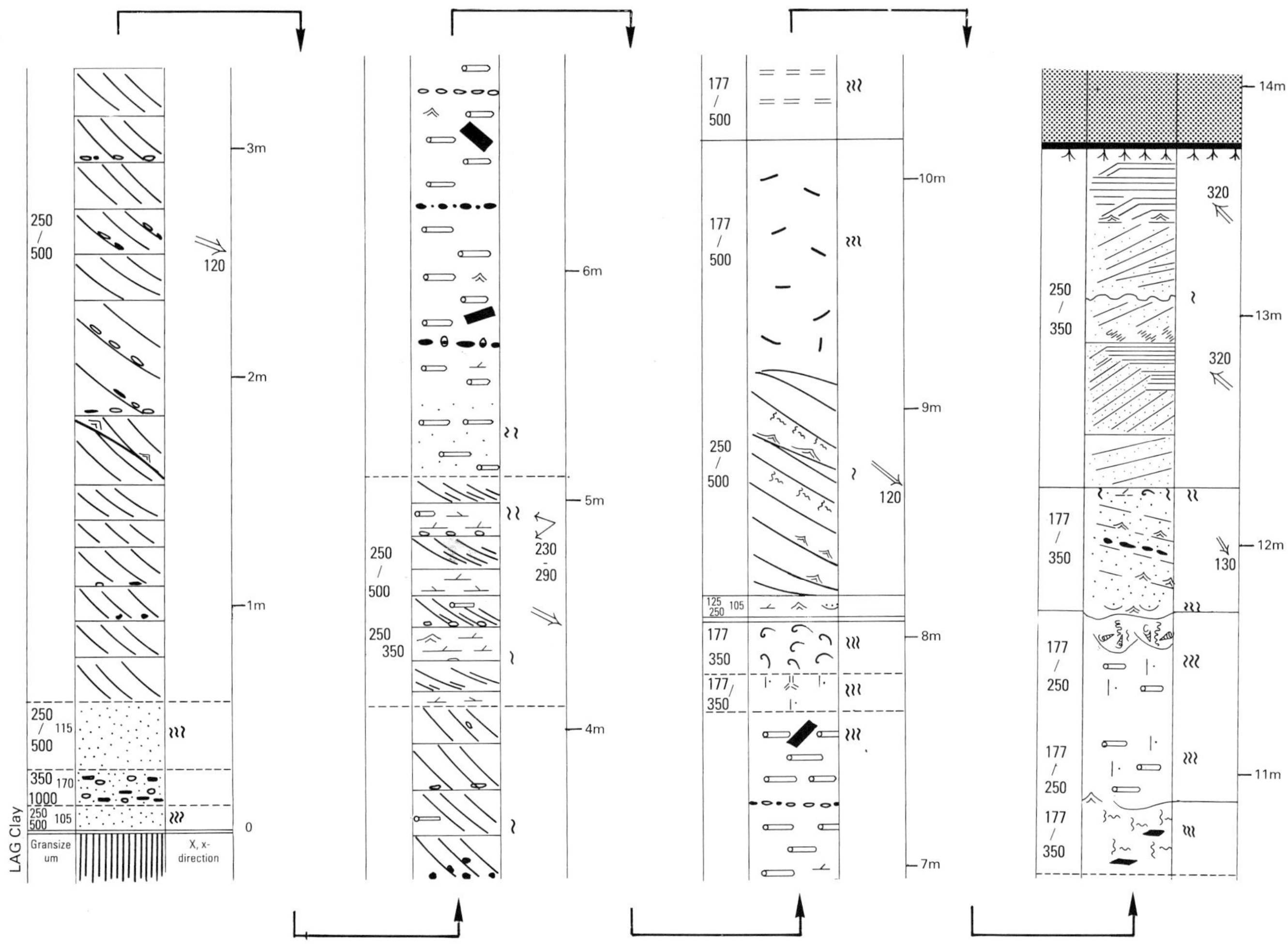

Figure 8. Detailed vertical profile of the sand wedge succession at Pt. 1900 m (see Fig. 7 for location). Flood-tidal delta and spill and blowover facies are shown together with their subunits. Toward the top of the profile the aeolian plug is represented.

- foresets, claydraped
- foresets, with organic debris
- wave ripples
- current cross-lamination
- continuous flaser
- discontinuous flaser
- cross bed | dip direction
- cross lam. | with respect to North
- wave ripple crests
- shells
- gastropods
- pod-shaped burrows
- ophiomorpha burrows
- horizontal tubes
- irregular tubes
- degree of burrowing (weak, moderate, high)
- lignite clasts
- horizontal lamination (clear, vague)
- sand, vague flasers

In the lateral interval between Pts. 350 and 100 (Fig. 7), the top few metres of the clay body contain sand beds at decimetre to centimetre scale that thin in a southerly direction. They appear to be the offshoots of the high-angle "delta" foreset of the adjacent flood-tidal delta.

The Flood-Tidal Delta Facies

Three superimposed units with contrasting physical and biogenic structures are distinguished in the flood-tidal delta facies. From bottom to top they are: the crossbedded sand unit, the bioturbated sand unit and an upper unit (Topstratum Unit) consisting of structureless sands, shell debris and clays (Fig. 7). Grain sizes fine upward and landward. Coarsest grain size is encountered between Pts. 1900 and 1500 (Fig. 7) where the sandy matrix of the large-scale crossbedded sets near the base of the unit consist of upper-medium grained sand. Well rounded pebbles of chert, clay and lignite are abundant, and although these are essentially strewn through the sets, there is a slight tendency toward their concentration along the base (Fig. 8). Large wood debris, including tree trunks, are found as well.

In a distal direction from Pt. 1500, the well rounded pebbles gradually disappear while sand grain size decreases to a lower-medium grained sand. Clays forming thin drapes or cm-thick bands become, at least visually, important. Between Pts. 500 and 250, the facies exhibits distinct heterolithic, sand-clay interbedding. Clay-clasts derived from ripped-up strata and wood debris occur commonly throughout. The colour of the sands tends to be yellowish-white except where a secondary staining from percolating groundwater has turned them to a light to dark gray-brown. This staining occurs above less permeable layers. This secondary effect may mislead the observer into overphasizing the thickness of clay layers.

Crossbedded Sand Unit

Description. The character and association of sedimentary structures vary vertically within the crossbedded unit as well as laterally along the outcrop. The basal interval, abruptly overlying the lagoonal clay, is roughly 0.5 to 1 m thick. Individual beds may attain thicknesses of a few tens of centimetres (Fig. 8). The sand beds show horizontal or low-angle inclined lamination or are massive, and are intercalated with poorly-structured mixtures of sand with clay, chert pebbles, and shell debris. Trace fossils include vertical and horizontal tubes, a few centimetres wide, of the ichnogenus *Thalassinoides.*

On top of the basal interval is a crossbedded interval with a thickness decreasing in a landward direction from about 3 m at Pt. 2000 to 0 m at Pt. 300 (Fig. 7). Crossbedding type also changes along the outcrop. Proximally, the planar-tabular sets predominate with set lengths up to a few tens of metres and set heights ranging between 20 and 100 cm. Beds of horizontal laminae also occur. They represent either offshooting bottomsets of the cross-strata or independent beds having an erosional, or at least nongradational, boundary with the adjacent crossbeds. Cherts are abundant through the sets. Small-scale cross-stratification is minimal. A few cosets, 5 to 10 cm thick, may be observed. They become more prominent distally. Dip directions in the tabular sets are unimodal with azimuths ranging between 120° and 170° (Fig. 10). The common occurrence of discontinuity or pause planes (also incorrectly from a process viewpoint called reactivation surfaces) is important. They separate bundles of cross-strata (Boersma and Terwindt, 1981). These bundles attain lengths well over one metre in the proximal, but gradually decrease to several centimetres in the distal section (Fig. 11A-E). Proximally, discontinuity planes tend to be modified by ripples built by waves and/or weak opposite flows. Distally, mm-thick clay laminae drape the discontinuity planes, while ripple sets become less conspicuous. Biogenic reworking commonly starts from the discontinuity planes (Fig. 11D), with *Skolithos* type burrows and thin, inconspicuous rootlets. *Ophiomorpha* and *Thalassinoides* ichnogenera also are common, but not bound to levels immediately below the discontinuity planes. *Chondrites* type burrows are scarce. Rootlets extending down from the discontinuity planes are thin, dispersed and commonly horizontal. Apparently they had a small biomass. They are interpreted as having belonged to small (young?) waterplants.

Figure 9. Lagoonal clay with thin silt and sand laminae, some of which have a lenticular shape resembling flat ripples.

A second type of crossbedding encountered consists of shorter, trough-shaped sets restricted to the lateral interval roughly between Pts. 1300 and 1000 (Fig. 7). Curiously, this bedding type is associated with abundant *Ophiomorpha* burrows, especially in the bioturbated sand unit above (Fig. 12).

The principle stratification plane of the crossbedded unit represented by the set-boundaries of the crossbeds remains parallel to the upper boundary of the underlying lagoonal clay between Pts. 2300 and 700. In the latter location, the sets start to incline slightly. The thickness of the crossbedded unit there is limited to 2 to 3 m. The angle of set inclination is less than 5° (Fig. 13). Vertically successive sets tends to overstep each other in their descending path. A few more hundreds of metres to the south, the crossbedded unit has shrunk to a rudimental one metre (Fig. 14A, B). Distally, this interval of descending sets ultimately leads to the steep foreset, delta-like, giant structure at Pts. 400 to 250 (Fig. 7).

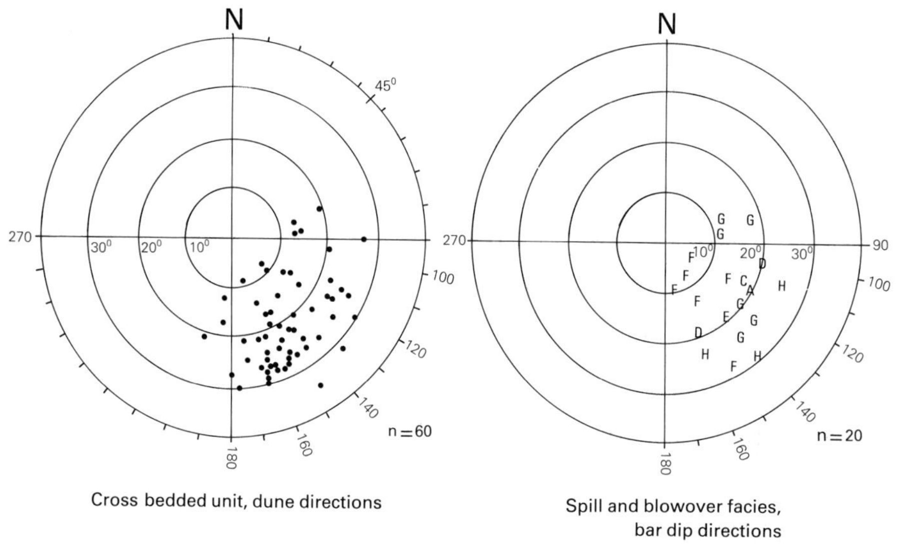

Figure 10. Cross-stratal azimuths and angles for flood-tidal delta (left) and spill and blowover facies. A to H refer to spill bars named Alpha through Hotel (see Fig. 7). Note consistency in direction.

A third and completely different type of structural association in which crossbedding plays an important part is of local occurrence (Pts. 2100 to 2300, Fig. 7). Here, crossbedded sets are highly irregular — short and trough-shaped to wide and wedge-shaped (Fig. 15). Crossbedding dips are very variable and commonly reveal a lateral transition into low angle or even horizontal bedding (Fig. 15B). Many set boundaries are transitional. Erosional boundaries sometimes are rippled and covered with slightly finer grained material. The sand is extremely well sorted. Despite the bleaching, which is a common feature in nearly all bodies of permeable sand underlying the lignite seams, ghosts of heavy mineral concentrations are visible (Fig. 15B). This structural association in many respects resembles hummocky cross-stratification. Burrows include a few *Ophiomorpha* and unlined vertical crab traces with their typical indented wall-structure produced by crawling upward. The thickness of this unit is limited to approximately 3 m.

Interpretation. Long tabular sets, unimodal crossbed directions, scattered entrained tree trunks, well-rounded chert pebbles and discontinuity planes suggest strong but intermittent currents that fed shore material into a system of long-crested dunes (sand waves). During the pauses, the sand wave fronts were subject to some erosion by waves and/or weak reverse currents. In the proximal positions, clay and silt were not permitted to settle or were subsequently removed. Distally, where the bundles are much shorter, the fines tended to build drapes or even cm-thick laminae. Wave action is interpreted to have been relatively strong in the northern (proximal) part of the outcrop. This is also indicated by the independent, horizontally-laminated intercalating beds. Between Pts. 2300 and 2100 the first interval of the crossbedded unit seems to indicate that wave effects balance the effects of currents.

It is difficult to determine the nature of the process periodicity associated with the interrupted migration of the sand waves. The common lateral recurrence of the discontinuity planes points to a high frequency of the generating process. However, the considerable degree of bioturbation starting from these planes, especially in the distal units, suggests rather long pauses (weeks, months ?), thus excluding a purely tidal genesis. This conclusion is reinforced by the absence of neap-spring tide lateral cyclicity in the proximal crossbeds (Boersma and Terwindt, 1981). The high proportion of wave-formed structures suggests that high-energy wave events of low periodicity (wave set-up) combined with the short periodicity of tides.

Bioturbated Sand Unit

Description. The crossbedded unit grades gradually into the overlying bioturbated sand unit. The crossbed sets become thinner upward and crosslamination more important. Burrowing becomes more intense upward as well. Despite the biogenic overprint, at some places (Fig. 15D), the crosslaminated physical structures may still be seen. Horizontal lamination occurs as well, usually in beds only a few centimetres thick. The burrows consist of horizontal to obliquely inclined tubes with an average diameter of 2 cm. Their net lengths are 10 cm. The walls are only slightly reinforced by organic particles, possibly agglutinated. Grain size decreases upward to that of a lower-medium to upper-fine grained sand. Sorting is good to moderate. Horizontal strings of chert pebbles and organic debris are present up to the highest levels of this unit proximally, whereas distally, the chert pebbles become less common and more dispersed. The thickness of the unit increases from about 3m at Pt. 2300 to approximately 5 m at Pt. 500. South of this position the bioturbated sand unit rapidly changes character as it merges laterally into the steeply dipping "giant" delta foresets (Fig. 7, 16). The foresets consist of a nearly pure sand. Within a short distance, however, they become a thin bedded, heterolithic alternation

Figure 11. A. General view of a 35 m high quarry wall at Pt. 1500 showing the brightly white coloured marine sand contained between fluviatile topstratum and lagoonal clay (below) and the Upper Seam. Note long, crossbedded set of the spill and blowover facies in the upper one third of the sand unit. **B.** The flood-tidal delta facies at the same location as **A.** Unidirectional, bundled crossbedded sets of the crossbedded unit pass upward into the bioturbated sand unit. Chert pebbles are abundantly strewn throughout. Bar scale is 1 m. **C.** Detailed view of crossbedded unit at the same location. The 0.7 m thick set in the centre clearly displays low angle discontinuity (pause) planes truncating cross-stratal bundles of variable length. Other, smaller crossbeds have concordant discontinuity planes only marked as draping by fines. Note, well-rounded, pebble-sized clasts of chert and wood debris. **D.** Close-up of two discontinuity planes largely build as sets or cosets of crosslamination with erosional lower set boundary. The upper one consists of wave ripples. **E.** Close-up of the basal crossbeds of **B** showing moderate burrowing partly associated with the discontinuity planes. The latter are clearly piled in the upper right hand corner. Trowel is 25 cm long.

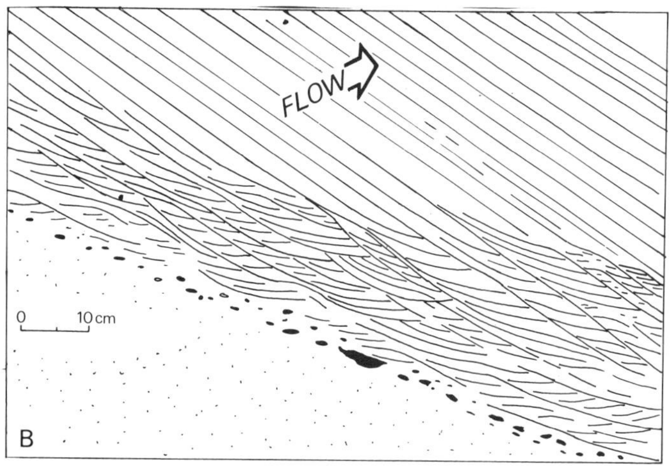

Figure 12. A. Flood-tidal delta facies, crossbedded unit at Pt. 1000. The crossbedded sets here are of the trough type with abundant wood debris. Discontinuity planes are less prominent. *Ophiomorpha* burrows are common and increase in density upward. Scale bar is 0.5 m long. **B.** Schematic drawing of one of the trough sets of **A** Crossbedding overlies and interfingers with crosslamination, which runs at right angles to it. A pebble-lag paves the base of the set. The crossbedding is interpreted as laterally filling a back-barrier, ripple-based tidal creek.

Figure 13. Flood-tidal delta facies, crossbedded unit at Pt. 700. Tabular crossbedded sets with abundant wood debris adopt an inclined position with respect to the overlying bioturbated sand unit. Note the gradual vertical transition in scale of structures. The basal 1 m of the section consists of flaser beds with the same, horizontal, tube-like burrows as in the bioturbated sand unit. Scale bar is 1 m.

of conformably dipping sand beds and fine grained, organic, mud laminae. The identity of the crosslamination is difficult to ascertain at places because of organic reworking. Current- and wave-generated crosslamination seem to be intimately associated. Bioturbation structures in the delta foreset include short, v-shaped burrows similar to those formed by pelecypods, and *Skolithos* and *Thalassinoides*-type tubes of the same sort as described for the main bioturbated sand unit. Another trace fossil is a 5 to 10 cm wide and deep, pod-like depression with walls sloping at 45° (Fig. 14F). The depression is filled with a diffuse mix of sand and disrupted clay drapes. Most burrowing seems to start from the clay covered surfaces, which apparently correspond to longer breaks in the sand supply.

Interpretation. The slight decrease in sand size in the vertical profile, together with the transition from large- to small-scale cross-stratification and the increase in density of organic structures indicate a decrease in energy levels and possibly a lower rate of sedimentation. The absence of clay drapes and occurrence of wave ripples may point to a wave-washed environment, probably of shallow depth. Unidirectional currents were still active but no longer the prime transporting agent. Chert pebbles suggest temporary, very high energy conditions (*e.g.*, a storm). The same may hold true for the parallel lamination. Unfortunately, the intense bioturbation precludes an estimation of the rate of recurrence of the high energy events. The mono-specific trace fossils, possibly from Ocypoda quadrates, suggest a brackish environment (Frey and Howard, 1969). The distal transition of the bioturbated sands into the steeply-inclined, clay-draped delta foresets are interpreted as being the result of a break in slope, where a gradually sloping platform meets

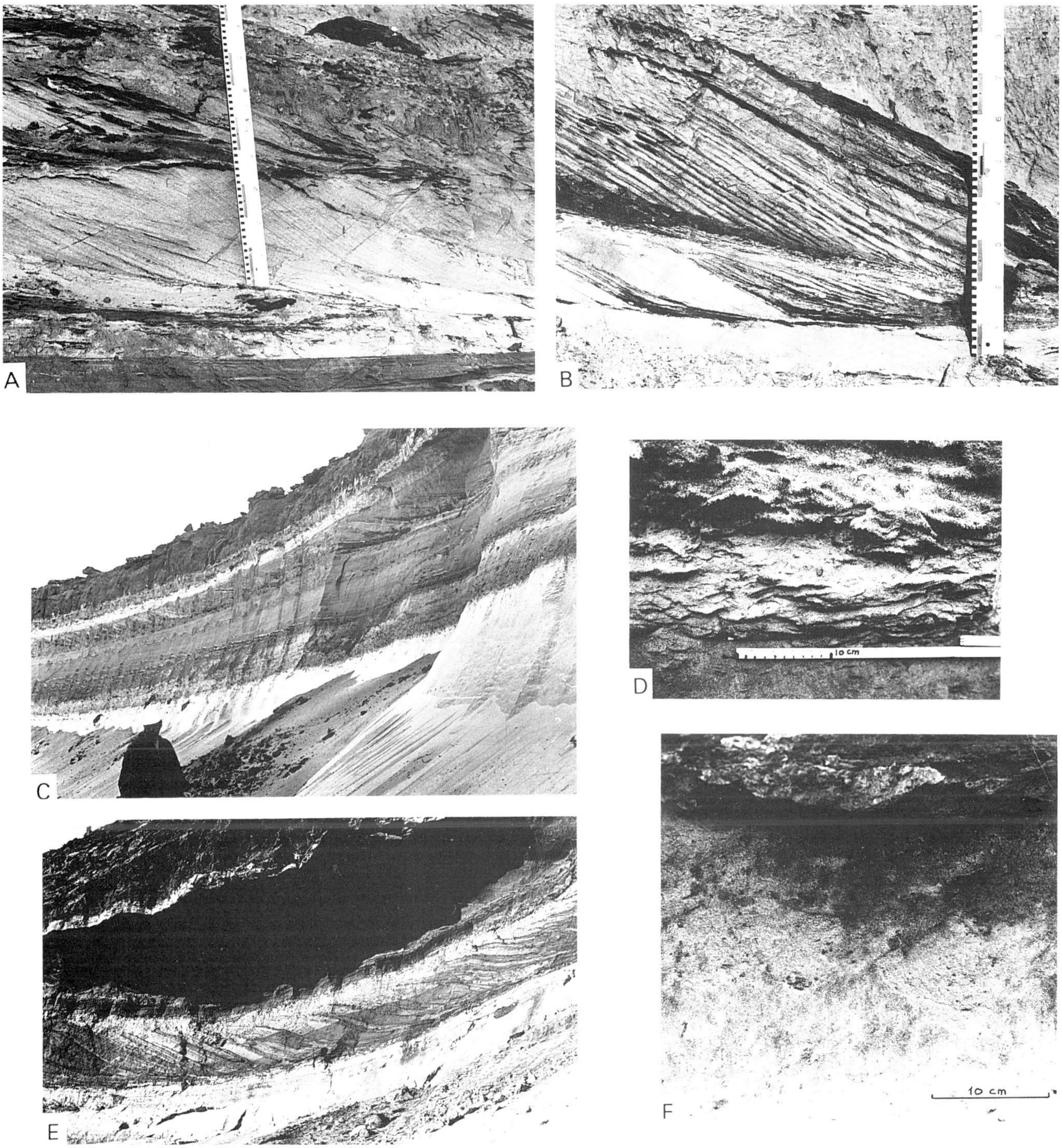

Figure 14. A, B. Flood-tidal delta facies at Pt. 500. The crossbedded unit, wedged out to 1 m or less, rapidly passes upward into the bioturbated sand unit. Individual crossbeds overstep and tend to descend. Discontinuity planes are commonly closely spaced and draped by mud or organic muck while oppositely-directed, crosslaminated cosets (see **B**) may be intercalated. **C.** Spill and blowover facies overlying the light-coloured flood-tidal delta facies at Pt. 1500. Along the top, part of the extensive crossbed "Delta" is visible. It overlies a faintly-laminated bioturbated sand and the low-angle dipping offshoot of "Echo". **D.** Top of the flood-tidal delta facies at Pt. 1900 made up of a conglomeratic sand bed with shells, shell hash, wood debris and flat, well-rounded chert pebbles. It overlies the bioturbated sand unit. **E.** Close-up of "Delta" with a succession of sigmoidal bundles separated by multiple, mud-draped discontinuity planes. See lithofacies log of Figure 8. **F.** Damp aeolian deposits of the spill and blowover facies immediately below the aeolian plug at Pt. 1900 (*cf.*, Figure 8). Note pod-like depressions filled with homogenized or distorted sands. Dark fragments are wood debris and gastropod shells. Structures are interpreted as hoof imprints of large mammals.

a body of standing water. The break in slope probably was initiated by successive, over-stepping bars as is apparent from the inclined sets of the crossbedded unit between Pts. 700 and 500.

There are no objective criteria available to interpret the depth of deposition and its change with time. Since there is only a slight change in grain size and burrowing activity up-profile, while the ichnofacies remains the same, and the sand remains well sorted and thus deposited in well agitated shallow (<0.5 m) waters, it seems likely that water depth increased only slightly in the course of deposition of the bioturbated sand unit. However, in order to create room for the sedimentary body, a relative subsidence of 2 to 3 m for the proximal to distal parts of the unit, respectively, is required. Although eustatic sea level rise may have played an important part, it is believed that compaction of the thick clays and peats in the shallow underground is likely to have substantially contributed. This compaction may have been accelerated by the burden of the crossbedded unit.

Topstratum Unit

Description. This unit is rather thin, fine grained and overlies the bioturbated sand unit (Fig. 7). It increases in thickness from less than 0.5 m at Pt. 2000, to slightly over 1 m at Pt. 300 (Fig. 7). It consists of sand and clay beds. In a southerly direction the sandy beds become progressively clay-rich (flaser-bedding) and thin out. The topmost bed in this unit is an organic clay with abundant wood and plant debris. It constitutes the "middle-clay" that separates the flood-tidal delta facies from the spill and blowover facies all along the outcrop. Underlying the "middle-clay" is a 20 to 50 cm clayey bed containing a mixture of fine sand, pelecypod shells and shell-hash, chert and clay pebbles, and disrupted, clayey filaments resembling algal mats (Fig. 15E, 22A). Usually this mixed bed is strongly bioturbated. Locally, some flaser-structures are encountered in sands that seem to have escaped the churning. It is difficult to decide whether the commonly irregular contact of the clayey sand bed with the underlying sand bed is due to erosion or burrowing activity. The sand beds in this unit have thicknesses of about 1 dm. They are of two types: 1) vaguely parallel laminated, pure sands, and 2) completely bioturbated or homogeneous sands with some organic debris and, possibly, some shell-hash. Some dispersed chert pebbles are present. Trace fossils include abundant horizontal tubes of the same type as described for the bioturbated sand unit. A few *Ophiomorphia*-like reinforced tubes (Fig. 15E) occur as well. The sand-fill of the tubes stands out colour-wise.

Interpretation. The discrete bedding of this unit suggests deposition may have taken place in separate events. The mixture of chert and clay clasts, as well as organic debris of various sizes, point to intermittent, high energy events dumping a mixture of material. The horizontally laminated chert-mixed sands probably reflect deposition under an upper flow regime, and the pure clays times of very quiet conditions. This reconstruction contrasts with that for the underlying bioturbated sand unit, where a more steady type of deposition is envisaged. If the interval between successive depositional events were long enough, organic reworking would have created a completely churned deposit. Considering its intense biogenic reworking, the shell-mix bed probably reflects a prolonged time of virtual nondeposition of fines and a dense clam settlement. Some sporadic storm events merely touched the environment, leaving some pebbles, while the fines were swept away. The overall tranquility of the environment was either the result of an increased water depth or a sheltered position. The latter interpretation is preferred, however, as the successive spill and blowover unit clearly reflects a nearly subaerial environment.

Spill and blowover facies

Description. A striking feature of this facies is the occurrence of laterally very extensive crossbedded sets in association with beds ranging from clays to pure, commonly homogenized sands (Fig. 7). Individual bed thicknesses range between 0.1 to slightly over 1 m. A continuous lateral and vertical transition occurs between these bed types. Five main lithotypes are distinguished: 1) heterolithic and bundled crossbeds, 2) low angle-inclined laminated flaser beds, 3) homogeneous to vaguely laminated or flasered sands, 4) clay or sand/clay mixed beds, and 5) megaflaser crossbeds. Types 1 to 3 characteristically appear transitional to each other in the extensive beds referred to above. The spill and blowover facies is contained between the topstratum unit, below, and the clay and upper coal seam, above (Fig. 7). Part of this interval, in the far northern part of the outcrop, is occupied by a different but, apparently, genetically related subfacies referred to as the aeolian plug. It contains crossbedded, laminated, and horizontally bedded, pure sands. The detailed vertical and lateral distribution of units is depicted in Figure 7. For purposes of quick reference, the extensive crossbeds are called Alpha through Hotel.

Heterolithic and Bundled Crossbeds. The crossbeds usually occur as isolated, planar-tubular sets that can be traced over tens to a few hundred metres (Fig. 14C, E). Set thicknesses are 0.5 to 1 m. The cross-stratal bundles consist of pure medium grained sand or upper-fine sands contaminated with fine organic particles (Fig. 17A, B, 18A, C). The bundles are separated by discontinuity (pause) planes lined with clay or organic debris. Typically, the clay drapes are multiple and cover or enclose reverse sets of ripples, which tend to be more abundant toward the toe of the discontinuity plane. The cross-lamination is identified as being of wave and current origin. Though a 30° dip may be achieved, cross-stratal dip commonly is less than the natural angle of repose and varies considerably along the set, giving the impression of temporary degeneration and rejuvenation of the sediment body during migration (Fig. 18A, C). The ideal lateral sequence of sedimentary structures encountered in these very extensive sets is shown in Figure 20.

The magnitudes of the cross-stratal bundles vary along

Figure 15. A-C. Stratification types in the plug of nearshore sands, flood-tidal delta facies, crossbedded unit at Pt. 2100. Note irregular association of horizontal, low-angle and crossbedding types. Dark laminae probably are ghosts after heavy mineral concentrations. Set boundaries are gradational or erosional. Crab burrows with typical indented wall structures are common (**C**). **D-F.** Flood-tidal delta facies, bioturbated sand unit. Horizontal tubes of the *Thalassinoides suevicus* type in vertical **(E)** and almost horizontal **(F)** sections. Note the different colours of the sandy fill of the tubes. At places where the sediment was not completely churned, physical structures including horizontal and crosslamination can be distinguished **(D)**. In **(E)** the topstratum unit, which covers the bioturbated sand unit is shown. It consists of a lower light-coloured "sand mix" bed and an upper "clay mix" bed. The "sand mix" usually contains shell hash, plant debris and some chert pebbles. It is homogenized by bioturbation. The "clay mix" in addition has torn up clay drapes and disrupted organic filaments possibly derived from algal mats. Both bed types are attributed to washover events. Compare also Figure 22A.

Figure 16. Large foreset delta structure of the flood-tidal delta facies, bioturbated sand unit at Pts. 500 to 300. **A.** View of quarry wall. From left to right a steepening and successive flattening of the strata may be seen. Along the base of the unit the sandy foresets interdigitate with and are replaced by lagoonal clay (Fig. 7). **B.** Initially the steep foresets are sandy without much mud draping. **C.** Less than 50 m south of **(B)** the foresets become strongly heterolithic. The sand beds between the drapes display parallel and crosslamination of wave and current types. Current directions vary strongly between downdip and parallel strike. Bioturbation structures descend from individual mud drapes. V-shaped traces are attributed to clams.

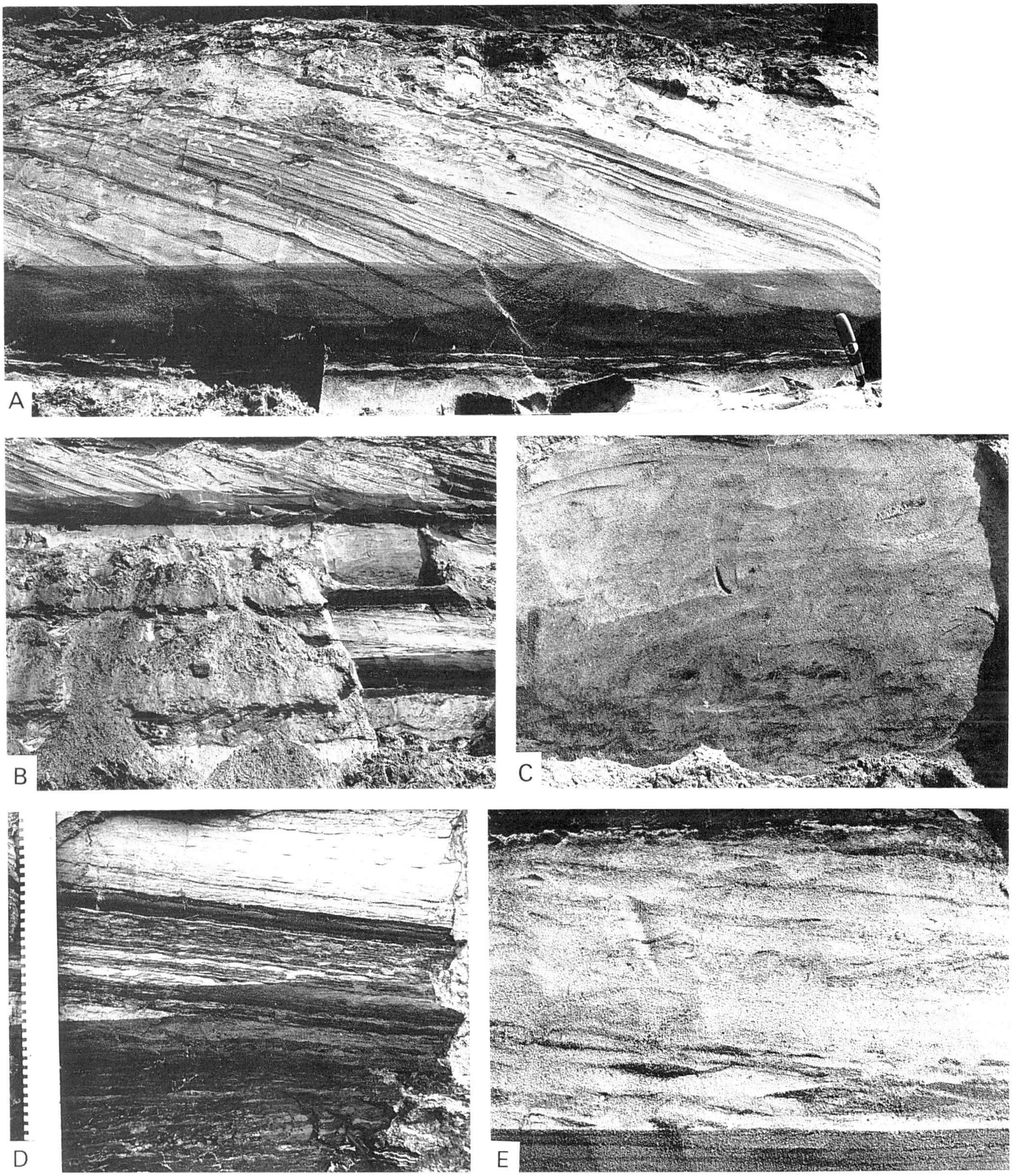

Figure 17. A-E. Spill and blowover facies at Pt. 1000. **B.** Three sand beds in superposition show different types of sedimentary structures that potentially may occur in juxtaposition. Height of exposure 2.5 m. **A.** Close-up view of the crossbedded set "Golf" 70 cm thick (see Fig. 7). Bundles of crossbedded sand are separated by single or multiple drapes of clayey or organic material. Enclosed by multiple drapes, incised into the sand bundles but especially encountered along the base of the set, are crosslaminated sets. The crossbed is contained between lenticular sand-clay bedding. The upper lenticular bed contains wave ripples and *Skolithos*-type burrows. Horizontal, tube-like burrows become more numerous going upward in the crossbed. They apparently are genetically related to the upper set boundary or even the overlying lenticular bedding. **C.** Homogenized, faintly flasered, fine to medium sand bed. Going landward (S) this bed changes into the crossbedded set "Hotel" 50 cm thick. **E.** Close-up of low angle flaser bed, the offshoot of the crossbed "Fox-2". Bundles of clean, parallel or crosslaminated sand are separated by pause planes of crosslaminated sand or single mud drapes. Thickness of light-coloured bed: 40 cm. **D.** Low-angle inclined lenticular bedding at Pt. 900, the final stage of "Fox-2". The overlying low angle sandy flaser bed becomes "Hotel" farther downstream.

Figure 18. Spill and blowover facies at Pt. 1750. **A.** General view of the 2.5 m thick exposure with crossbed "Charlie" overlying a sand-clay lenticular bed at the base. **C.** Close-up view of "Charlie" with cross-stratal bundles of variable thickness and structure. Toward the toe, crosslamination is more prominant. The direction of crosslamination direction may be up and down cross-stratal dip as well as strike parallel. **E.** A strike section. **(F, D and B)** are superimposed close-up views of the beds on top of "Charlie". **F.** A regularly-flasered homogeneous sand followed by a wave-rippled, lenticular, sand-clay alternation. **D.** The same lenticular bedding overlain by low-angle flaser bedding, the offshoot of "Bravo". **B.** Two sand beds homogenized by bioturbation largely of the common, horizontal-tube type. Note ghost after crossbedding in the lower bed. Partitions on bars indicate decimetres.

Figure 19. Stratification types of the aeolian plug, spill and blowover facies at Pts. 2200 to 1700. **A.** Northern flank of mound-shaped sand ridge seen in distance. Sedimentary structures include elegantly curving crossbeds and low-angle bedding. **B.** and close-up view **D**, S-shaped aeolian cross-strata rest on homogenized bed with fine rooting attributed to grasses in the damp interdune or high aeolian flat area. **C.** Top of ridge with stems and root system of redwood tree. **E.** Climbing adhesion pseudo crossbeds and wavy ripple-lamination. **F.** Densely-rooted homogenized sand beds with soil formation at top, lying at the base of the aeolian plug (*cf.* Fig. 8).

the set. Greatest thickness occurs where cross-stratal dip is steepest. No systematic variation in bundle thickness, as is sometimes encountered in tidal deposits, was noticed. The clay drapes between the cross-stratal bundles thicken toward the toe of the set and shoot off, asymptotically approaching the top of the underlying bed.

Burrowing in the isolated crossbeds is generally weak to moderate and of two types. The first type is physically related to the upper set boundary and the thin clayey bed that overlies it. The burrows are of *Skolithos* and *Thalassinoides suevicus* ichnogenera. The latter consists of the same horizontal tubes (Fig. 17A) as are encountered in the bioturbated sand unit. The second type of burrowing is related to the pause planes separating the cross-stratal bundles. Here the burrows either run parallel to the multiple drapes or penetrate the underlying bundle. The result is a fine to coarse mottling (Fig. 18C). Thin, irregular, commonly horizontal roots also contribute to the organic reworking. They occur in the same two positions as do the faunal traces. Their relationship with the pause planes seems most remarkable, as it suggests that pauses were of considerable duration.

The upper set boundary of the sand beds, crossbedded or otherwise, is erosional and irregular, irrespective of grain size and structure of the overlying bed (Fig. 17A, 18A, F). If the overlying bed is clayey, the irregularities may be attributed to bioturbation penetrating downward. Where sand overlies, the irregularities are pod-like structures similar to those described by Fryberger *et al.* (1990) in periodically flooded interdune areas. They may also be a result of differential resistance of the cross-strata to scouring action of the wind.

Low Angle-Inclined Laminated Flaser Beds. Laminated flaser beds are the longitudinal counterpart of the crossbeds, in the "upstream" as well as in the "downstream" sections (Figs. 17D, E, 18D, 20F). The longest stretches of laminated flaser beds, however, tend to occur upsection, and may show a bundled bedding similar to the crossbeds, although, it is usually thinner and less well delineated. The bundles 1 to 10 cm thick, consist of distinctly to vaguely parallel-laminated, pure, fine to medium grained sand. The pause planes may appear at 1 to 5 cm thick cosets of rather irregular, flaser crosslamination, the wave or current origin of which is difficult to establish. Wave action seems to play an important part, as is indicated by the irregularly scoured lower boundary of these cosets, a feature characteristic of wave ripples.

In other instances, the laminated flaser beds are delicate and wavering, and the drapes are discontinuous and almost flat. The drapes may consist dominantly of sand-sized or smaller organic particles, probably macerated plant debris or reworked peat. Sets of crosslamination can hardly be distinguished in this type of laminated flaser bed.

Organic reworking of these beds tends to be weak, and results in mottling. Similar to the high-angle crossbeds, burrowing and sparse rootlets may either start from the upper set boundary and pause planes or occur dispersed throughout. Horizontal tubes are encountered as well.

Homogeneous to Vaguely Laminated or Flasered Sands. These sands are the same grain size as the two bed types described above: upper-fine to lower-medium grained sand. This bed type constitutes a transitional series of subtypes that have in common a high degree of bioturbation. They differ in the rate of intercalation of fine grained laminae. The pure sand end member may show a ghost after horizontal or crossbedding structures without any fine grained draping. Upon close inspection, an apparently homogeneous sand bed may be seen to consist of two or more subbeds (Fig. 18B).

The other end member of this bed type constitutes a faintly visible alternation between 1 to 5 cm thick sand layers and traces of fine, usually organic, material (Fig. 18F). These fine grained traces may combine to form bands, 1 to 5 cm thick, which resemble the crosslaminated pause planes in the laminated inclined flaser beds. The subtype intermediate between the aforementioned end members is characterized by a homogenized sand throughout with some disrupted flasers in the lower one third of the beds (Fig. 17C).

Remains of whole or broken pelecypod and gastropod shells may be found in some beds. The shells are almost completely dissolved preventing sampling and identification. Pothole-like, semicircular depressions filled with homogeneous or contorted sands are encountered along the tops of some homogeneous sand beds (Fig. 14F). They are 10 to 15 cm wide and deep. In horizontal section they appear to be

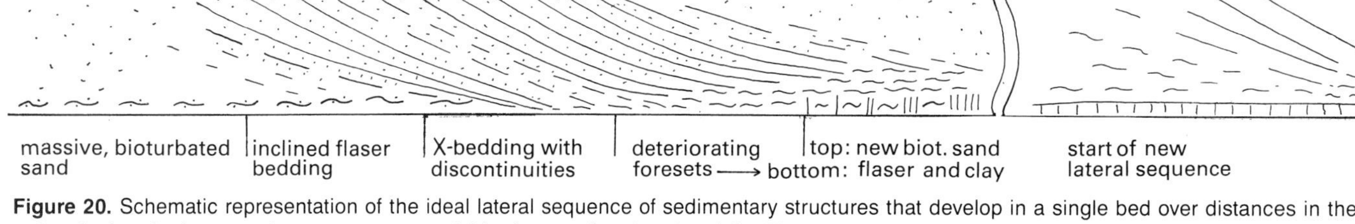

Figure 20. Schematic representation of the ideal lateral sequence of sedimentary structures that develop in a single bed over distances in the order of a hundred metres, in response to a variable degree of surface humidity. The massive sands represent the relatively dry (damp) stages; the low-angle, slightly flasered intervals form on a humid, adhering surface; the clay-draped high-angle foresets build at a water's edge. As the backbarrier water level drops, the lateral sequence reverses from high to progressively lower angles of deposition. Bed thicknesses range between 30 cm and 1 m.

circular or oval in outline and sometimes crosscut each other. Gastropod shells are commonly mixed with the distorted sand-fill. They bear some resemblance to mammal hoof-prints (Van der Lingen and Andrews, 1969; Lewis and Titheridge, 1978). Most common types of biogenic structures include diffuse fine mottling and the horizontal tubes described previously.

Clay or Sand-Clay Mixed Beds. These beds are usually less than 0.1 m thick in the proximal portion of the outcrop, however, landward of Pt. 800, they increase to thicknesses of decimetres up to one metre. Also, the clay beds in the top one metre of the spill and blowover facies are thicker than those lower down. Most of these beds can be traced over several hundred metres. They act as markers separating the sand beds.

The clay bed may differ quite markedly in appearance. Its composition ranges between a sandy to silty clay, and lenticular or flaser bedding may be present (Fig. 18B, F). Physical structures range from crosslamination through interrupted sand laminae-in-mud to massive. Organic reworking likewise varies between undisturbed to completely churned. The lower boundary usually is irregular or scoured into the underlying bed (Fig. 18F). The scouring is most visible where crosslamination occurs along the base of the clayey bed. An irregular lower boundary is characterized by erratically-shaped protrusions into the underlying bed. These are attributed to macroorganisms, probably clams. As stated before, thin vertical tubes of the *Skolithos* ichnogenus, and rootlets may be encountered (Fig. 17A).

Megaflaser Crossbeds. Megaflasers are restricted to the far southern part of the outcrop in the area where the flood-tidal delta facies wedges out. Between about Pts. 500 and 300, a single unit, up to 2 m thick, is exposed, which fills a slight depression probably formed by differential compaction in the subsurface. In a northern as well as southern direction, the mega-flaser unit grades into small-scale flaser beds. In the north, the small-scale flasers change into the regular crossbed-dominated spill and blowover facies. Both types of flaser bed consist of very light coloured, fine to medium grained sand with abundant organic debris (Fig. 22B). Some of the crossbeds have distinct basal lags of wood and plant debris with gastropod shells. The megaflaser unit fines upward, as a result of a proportional decrease of crossbeds and increase of crosslaminated cosets as well as interset clay drapes and biogenic reworking. The unit is topped by sandy clay beds, which in turn are abruptly overlain by a massive,

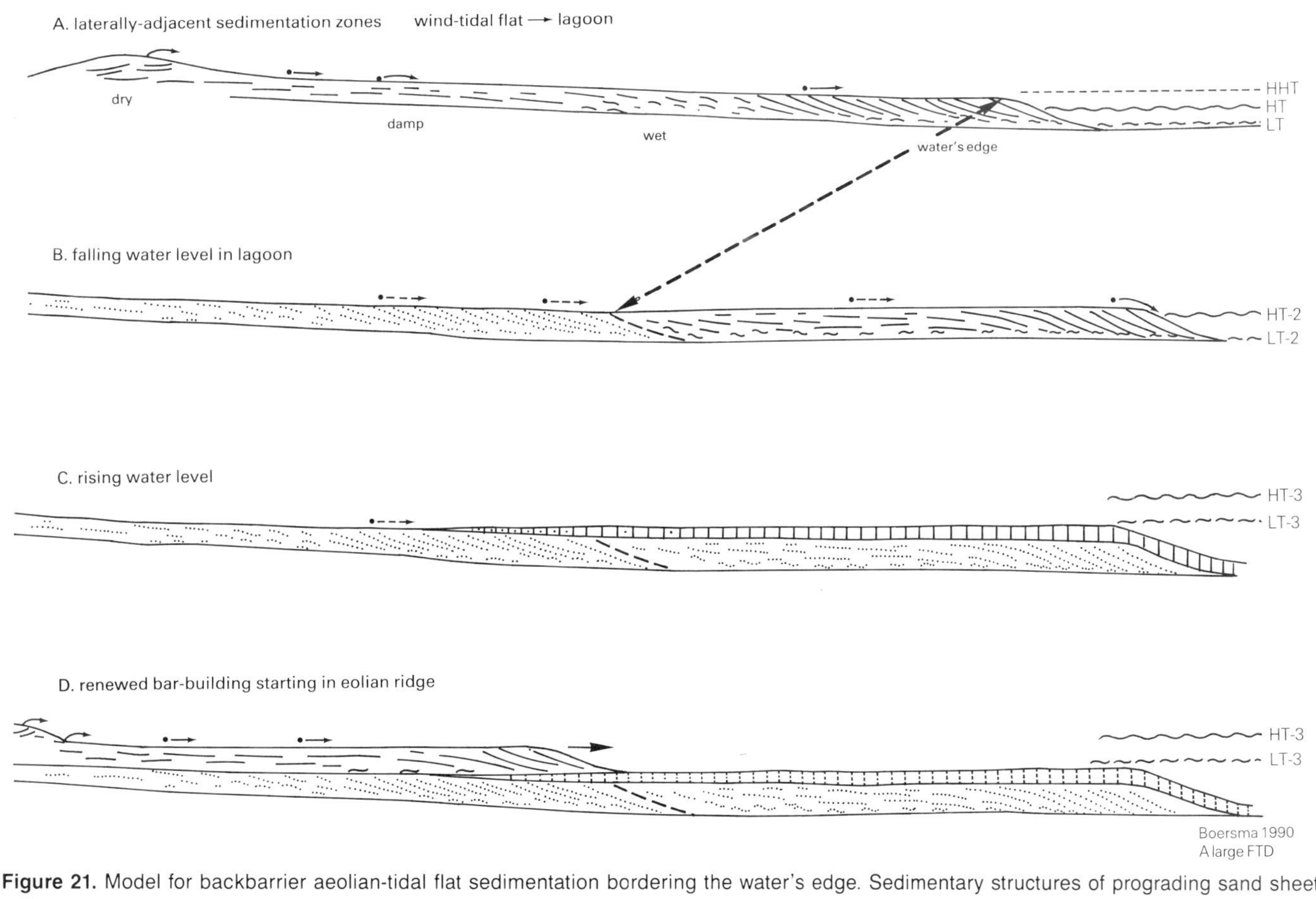

Figure 21. Model for backbarrier aeolian-tidal flat sedimentation bordering the water's edge. Sedimentary structures of prograding sand sheet differ according to relative position of water table (**A**). As the water level falls (**B**), the site of the high-angle water's edge shifts lagoon-inward. At the former water's edge, the topographical step is evened out by damp aeolian deposition in the lee area, creating subhorizontal stratification. A major rise in lagoonal water level (**C**) results in an extensive though thin clay layer covering the initial sand sheet and a more duneward reconstruction of the water's edge (**D**). Stages **A-B** explain the lateral sequence of structures (Fig. 20).

Figure 22. A. Topstratum unit of flood-tidal delta facies with separate beds, some of them containing sand rich in wood debris, shells and chert pebbles. Scale: ruler 0.5 m. **B.** Megaflaser sequence filling the slight depression between Pts. 500 and 300. Spill and blowover facies. Note energy-decreasing and fining-upward sequence. Sands with abundant wood debris. Scale bar is 1 m. These deposits are interpreted as having resulted from a tidal channel sweeping a water's edge.

silty clay. The sandy clay beds can be traced northward into the bed overlying crossbed "Hotel" (Fig. 7). Cross-stratification, large- as well as small-scale, is of current origin. Cross-stratal discontinuity planes are common. They are marked by organic particles and are essentially concordant; that is, lacking erosional modification. Longitudinally, individual crossbeds commonly grade into crosslaminated cosets. No systematic repetition of discontinuity planes is observed.

Aeolian Plug

In the far northern end of the outcrop between Pts. 2200 and 1700 m (Fig. 7), part of the spill and blowover facies merges into and is partly overlain by a lenticular, 400 m wide body of white, well sorted sand (Fig. 19). In its core, the aeolian plug measures slightly over 3 m in thickness.

Sedimentary structures include horizontal, low angle and high angle crossbedding of the gentle and elegantly curving type so typical of aeolian deposits. Accessory structures are climbing adhesion ripples and low-amplitude wavy ripple lamination. Toward the base of the body, homogeneous sand beds strongly marked by root horizons become increasingly common. Laterally (southward) the gentle type of crossbed intercalates between homogeneous beds and, with distance, merges into the heterolithic crossbeds of the spill and blowover facies (see "Bravo" and "Charlies", Fig. 7). Along the top of the mound-shaped body, the stem and root system of a Miocene Redwood (*Sequoia sempervirens*?) is present (Fig. 19C).

Interpretation of the Spill and Blowover Facies

The bundled structure of the isolated, extensive crossbeds suggests the repetitive supply of sand to a bar front with low-to high-angle dipping foresets. The thicknesses of the sand bundles vary laterally in a nonsystematic way, which means that the transport capacity of the current was irregular. Grain size and sorting suggest a coastal provenance. Periods of sand transport are separated by times of lower basinal energy. Lowest energy conditions existed during deposition of the drapes of clay or organic material. The multiple characteristics of many drapes, commonly enclosing ripples, indicate comparatively long pauses relative to the next influx of sand. Current ripples are common along the bases of the sets and show directions at right angles to or obliquely up the foreset dip of the bar. Pause planes suggest an aquatic setting. This genesis may be extended to the sand bundles as long as they dip at the natural angle of repose for the material. The poor foreset sorting and especially the admixture of fines do not seem to correspond with a subaquatic slipface. Even less consistent with the idea of aquatic deposition, is the low-angle inclination of the cross-stratal bundles in the long sets that lie up- and downsection from the truly crossbedded, high angle parts. The low angle bundles apparently were not deposited through normal avalanching of the grains. A process of grain fall and adherence, as is encountered in humid, aeolian settings, seems a more likely mechanism. An analogous reasoning applies to the laminated flaser beds.

The homogeneous to vaguely laminated to flasered sands consist of slightly finer sand than the crossbedded and laminated beds. The admixture mainly silt-sized particles, making them dirty is attributed to bioturbation. Current ripples that could indicate an aquatic genesis are scarce. The ghosts after crossbedding are clearly devoid of clay draping. This, as well as the absence of pebbles suggests they may have been formed by aeolian processes. The flat, gently curved drapes of fine, commonly organic material that occurs in some of the flaser troughs are another indication of aeolian sedimentation. They show hardly any crosslamination. These features are analogous to aeolian ripple beds on which organic dust adheres as a rising groundwater level produces a damp surface. An important clue as to the probability of an aeolian origin of the homogeneous and vaguely flaser beds is their lateral transition into the strata of the aeolian plug.

Biogenic structures, notably the horizontal tubes classified as *Thalassinoides suevicus*, are rather difficult to interpret as they do not seem to fit precisely those of the modern genera. Best fit may be the slanting, obtuse L-shaped burrow of the fiddler crab *Uca pugilator* (Frey and Howard, 1969). *Uca* is known to occur locally in backbarrier environments so abundantly that it masks or even destroys any physical structures (Deery and Howard, 1977). Other options include *Ocypode quadrata*, and a beetle.

The clay and sandy clay mixed beds, traceable over large distances while fining and thickening landward, probably record conditions of inundation. The intercalation of interrupted sand laminae and wave ripples may point to aeolian-supplied sand subsequently reworked by waves. The clay-covered surface was vegetated, displaying the diffuse root system produced by small water plants. Furthermore, clams and possibly polychaete worms were present.

The megaflaser unit, as encountered in a depression at the base of the "Spill" between Pts. 500 and 300, bears no resemblance to any other part of this facies. It does resemble, however, the beds included in the onset of the large delta foresets of the underlying flood-tidal delta facies. The megaflaser unit clearly is the product of a weak, bimodal current. The prevalent direction was N 120°.

The major depositional activity of this facies took place at the migrating boundary between the aeolian and the aquatic domain. Currents of a bimodal nature were largely active at the foot of the bar front that gave rise to the extensive bedding units. Their considerable length, built up as a result of many hundreds of progradational pause events, indicates a high degree of environmental stability.

Environments of Deposition

Depositional Environment of the Flood-Tidal Delta Facies

The composition and sorting of the flood-tidal delta facies strongly indicates a coastal setting. The crossbedded unit was deposited by dominant, intermittent, unidirectional currents, which transported sand waves (long-crested megaripples) landward. During the pauses, the surface was reworked by waves and oppositely-directed, highly subordinate currents. The dominant current system decreased in strength lagoonward (south), which permitted the subordinate system to become more important distally. Wave action decreased in a landward direction. The pauses, especially in the more distal position of the outcrop, were long enough to permit bioturbation starting from the pause planes.

With time, the crossbedded unit graded vertically into the bioturbated sand unit, current strength decreased considerably and wave energy slightly. The lower energy environment was still agitated by waves, but permitted the settling of a monospecific burrowing fauna, possibly a kind of Ocypode.

The upper one metre of the flood-tidal delta facies reflects a marked change in depositional conditions with the occurrence of a laterally continuous clay bed, which thickens lagoonward. Although the increasing amounts of mud indicate lower-energy conditions, interbeds of sand, 0.2 to 0.5 m thick, point to sporadic, high-energy fluxes.

Geometry and stratigraphic position of the entire sand wedge further restricts the interpretation of depositional environment to that of a backbarrier association which includes inlet, estuary, tidal or aeolian flat, dunes, washover and flood-tidal delta subenvironments. Washover events, as described by numerous authors including Andrew (1970), Schwartz (1975, 1982), Deery and Howard (1977) and Hennessy and Zarillo (1987), tend to deposit moderately-sorted sand with commonly abundant debris, and their basal contacts are sharp, with pebbles or shell fragments. Individual washover sand units may have thicknesses of between less than 5 cm to about 1 m. Physical structures are dominated by horizontal and lów angle lamination. Crossbedding may grade laterally away from these structures, but is of a restricted occurrence. Many of these primary structures tend to fall victim to intense bioturbation and rooting during the prolonged passive phases that came between individual washover acts (active phases).

The crossbedded unit and a large part of the bioturbated sand unit, however, seem to fit better into an inlet-to-flood-tidal delta interpretation. Despite the numerous studies on flood-tidal deltas, there is not enough knowledge about diagnostic sedimentological features to support a definitive interpretation. The flood-tidal delta model as adopted by Hayes (1975) recognizes five, distinct morphological elements: flood ramp, flood channels, ebb shield, ebb spit and ebb channels. This model applies predominantly to the mesotidal settings of Northern New England, which exhibit limited wave action, and where prominent ebb return flows modify the flanks and leading edge of the flood-tidal delta. As correctly stated by Boothroyd (1985), the flood-tidal delta characteristics may vary considerably from this model as ebb influence is replaced by wave action. Ebb shield, ebb spit and ebb channels then tend to not apply to the model, a situation examplified by The Chatham Harbour Estuary (Hine, 1975), the Ninigret flood-tidal delta (Boothroyd, 1985) and the Thyborøn Kanals Delta (Flyfoto-atlas Danmark, 1980).

Factors favouring the wave influence reportedly include a wide inlet opening, a position of the flood-tidal delta immediately adjacent to the inlet and a wide lagoon having a long wave fetch (Boothroyd, 1985). Factors that would reduce ebb return flow are a microtidal setting and the existence of additional inlets to the lagoon through which flood water may drain. Microtidal lagoons are classified as open, leaky or restricted (Kjerfve, 1986). As such, the flood-tidal delta at Chatham Harbour Estuary, which extends into Nantucket sound, has a smaller tidal range and a thirty minute offset of the tidal phase compared to the sea beyond the inlet. The Thyborøn Kanals Delta builds into a wide lagoon, which is connected crosscountry with the Baltic Sea.

Modern flood-tidal deltas, especially those having a marked velocity and time asymmetry, commonly show sand wave covered surfaces arranged in a system of partly overlapping lobes. In lower energy situations, unidirectional sand waves seem to be restricted to the axial parts of the subtidal floodramps. Laterally this area merges into rippled areas, which may become increasingly muddy away from the zone of maximum velocity (Hennessy and Zarillo, 1987). Stronger evidence for the occurrence of events producing unimodal crossing in flood-tidal deltas come from ancient examples (Carter, 1978; Kirschbaum, 1989).

Reconstruction of Process Events

The sequence of events leading to the deposition of the flood-tidal delta facies in this study may have been as follows (*cf.* Fig. 23). A long, low and narrow barrier insufficiently keeping pace with a rising sea level was washed over several times before a major break took place. The lower 0.5 to 1 m of our synoptic section (Fig. 7) clearly represents washovers, which were largely homogenized after deposition. Scarce *Ophiomorpha* burrows point to an intertidal to shallow subtidal environment. The major breach seems to have eroded a rather narrow and deep inlet channel, the position and azimuth of which can be seen on Figure 6. As the inlet widened, steep-fronted lobes of shoreline-derived sand migrated into the lagoon, forming a pile of crossbed sets. This initial phase in the construction of the flood-tidal delta is illustrated in Figure 23, Stage 1. The shallow depth of deposition is inferred from: 1) the common occurrence of chert pebbles, which probably were derived from longshore transport; 2) the common occurrence of wave ripples covering the bar fronts and adjacent, deeper water; and 3) the temporary colonization by small plants (seagrass?) during pauses in the migration of the bars. Additional evidence comes from the inferred wave-dominated character of the flood currents. It can be expected that a deeper subtidal position of the bars would have resulted in a tidal cyclicity of the cross-strata, at least during some periods. On the other hand, a shallow position would favour currents irregularly released by wind set-up. Another argument supporting the apron-like deposition is the absence of seaward-dipping, master bounding surfaces in the crossbedded unit, as would be expected with the floodramp of a flood-tidal delta ascending from the deeper inlet waters.

Sufficient space for the prolonged piling of crossbedded units at approximately the same water depth in the backbarrier lagoon probably was caused largely by compaction of the underlying lagoonal and fluviatile clays and the peat (*cf.* Fig. 1). As the crossbedded unit was deposited, current strengths decreased as is indicated by the decreasing bedding height and grain size. This probably was due to the building up of the plug of nearshore sands likely representing a swashbar or spit. Initially, the currents could still go around the obstacle, as is demonstrated by the occurrence of crossbedding at the level of the plug at Pts. 1900-1500 (Fig. 7). This second phase of flood-tidal delta construction is demonstrated in Figure 23, Stage 2.

The different type of trough crossbedding as encountered between Pts. 1300 and 1000 (Fig. 7), which is associated with numerous *Ophiomorpha* burrows and a high proportion of organic debris (Fig. 12), probably represents a tidal channel. The reason for not finding discrete channel margins may have to do with the trend of the channel, which probably was at a slight angle with respect to the outcrop. Also, the occasional lateral filling of the channel by sediment transported across the flood-tidal delta and subsequent platform (= bioturbated sand unit) surfaces may have obscured the channel outline (Fig. 12).

As the spit or swash bar developed, the backbarrier area became more sheltered and turned into a lower energy platform. At a later stage, the platform extended seaward across the plug of nearshore sand, probably as a result of the final healing and vertical deposition of the barrier. The backbarrier platform apparently had a shallow depth and was wave- and current-agitated during most of its existence. Some analogy seems to exist with the shallow tidal lagoons, described by Carter *et al.* (1989). These reportedly develop as the lagoon becomes filled up by the flood-tidal delta. Water depths

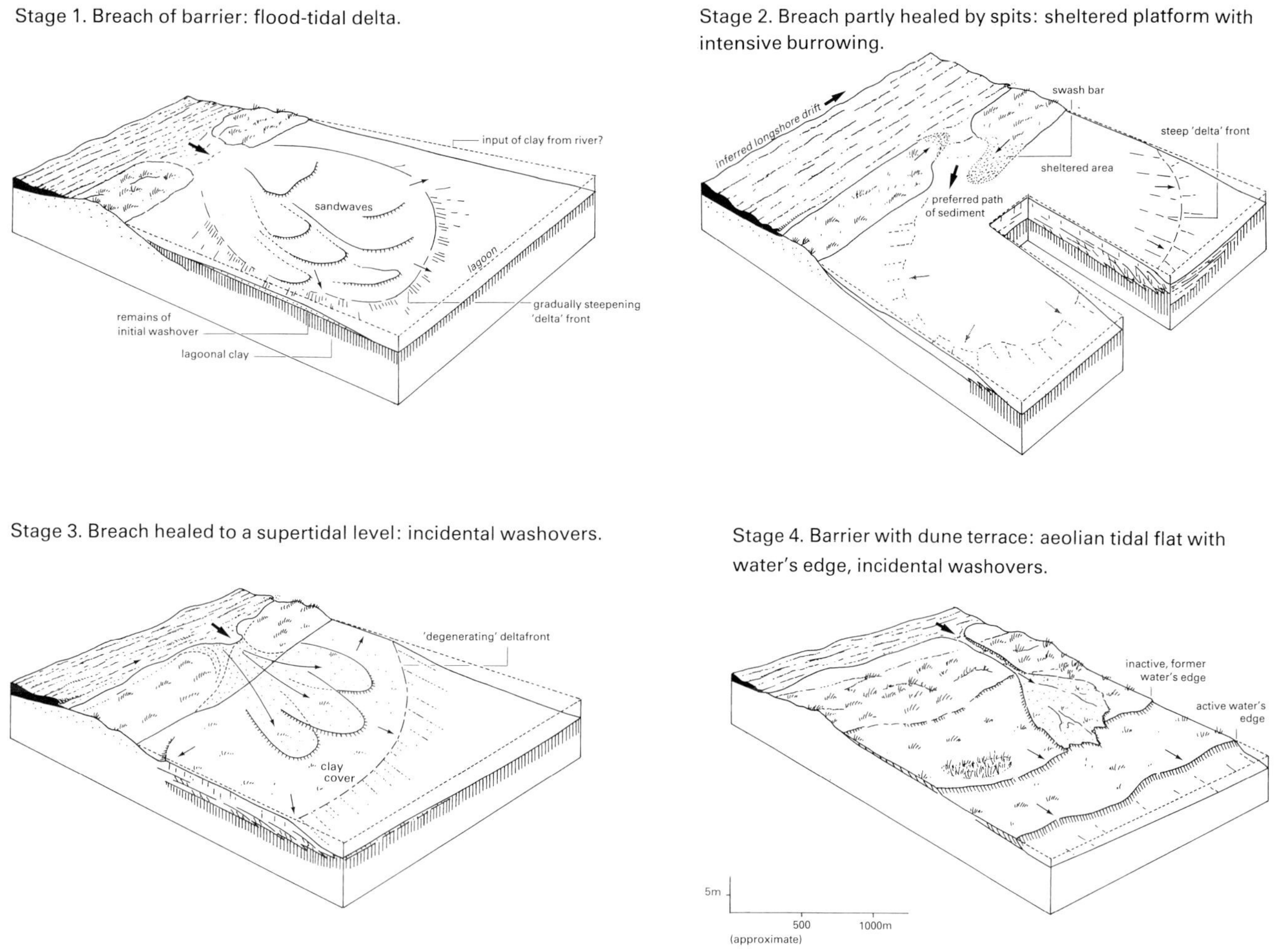

Figure 23. Diagrammatic interpretation of the four main stages in the barrier-backbarrier evolution and related sediment types, eventually resulting in the flood-tidal delta and spill and blowover marine sand wedge.

then range between 1 and 2 m. During deposition of the bioturbated sand unit in the platform environment, there were occasional high energy events that produced parallel-laminated, thin sand beds, sometimes associated with chert pebbles and shell hash. The exact nature of these events is difficult to establish. They may represent lagoonal storm conditions or distal washovers. Other analogues of the bioturbated sand unit may be found in the low-energy flood-tidal delta and sandy flat as described by Hennessy and Zarillo (1987), Andrew's (1970) stagnant fan, and the flood-tidal deltas described by Heron *et al.* (1984). Most authors agree on the fining-upward character of the inlet flood-tidal delta lagoon complex.

Landward, the bioturbated sand unit merges into the steeply inclined, large delta foresets. They form a kind of terminal lobe (Boothroyd *et al.*, 1985) which gradually came into existence as the waves crossing the platform dumped sand over the gently sloping frontal edge of the crossbedded unit.

The discrete bundles of sand in the large crossbeds seem to have been deposited at intervals marked by higher energy storm events. Between these intervals, clays were deposited and burrowing by clams was permitted. This suggests a mean recurrence interval in the order of years.

The topstratum unit reflects a major change in the mode of sediment transport and sedimentation. From being a more or less continuously agitated environment, it changed into one in which washover events spread pure or debris laden sand onto a tranquil, clay covered surface. This change of condition may be explained by a final restoration of the barrier to a level where open ocean waves, even under high wind setup could no longer spill across the barrier.

The medial clay that forms the continuous boundary between the flood-tidal delta and the spill and blowover facies is testament that once again the lagoon deepened behind the stabilized and possibly seaward-prograding barrier.

Depositional Environment of the Spill And Blowover Facies

The interpretation of this facies, with its enigmatic, long crossbeds showing a bundled structure over tens or hundreds of metres, presents problems. Most puzzling is the laterally symmetrical sequence within a single bed: homogenized sand, slightly flasered sand, low-angle inclined flaser sand, high-angle bundled crossbedding, and low angle inclined flaser sand (Fig. 20).

The stratigraphic position of the spill and blowover facies, contained between a flood-tidal delta and a lignite, would very well fit a washover complex. However, it is very difficult to envisage how successive washover events could maintain lee faces of the bars that formed the extensive crossbedded sets. Successive washovers may be expected not to have a constant strength and to dump most of their material at irregular distances inland from the shoreline. High-energy events would then very likely cause the bar to be levelled and thus no longer be the site of cross-stratal deposition. At only one place (Pts. 2100-1800), 1 m above the base of the spill and blowover unit, a single bed containing ripped-up peat clasts and shell debris is encountered, which could be explained as a washover. Even so, the bed is replaced at distance by pure, homogenized and eventually flasered and low-angle crossbedded sand. As argued earlier, it is likely that the sands, including those connected with the single washover bed, were distributed by aeolian processes. More support for the aeolian interpretation comes from the identification of an aeolian plug at Pts. 2300 to 1800 and its lateral interdigitation with the regular spill and blowover facies (Fig. 7).

The question of extensive, bundled crossbedded sets contained in a lateral sequence (Fig. 20) can be related to the observations of Fryberger *et al.* (1990) who found comparable sets forming where a dunefield moves into a body of water. These so-called water's edge structures, in our case would have been formed at the edge of the gradually sloping backbarrier aeolian flat where it met the high tide mark. It may be envisaged that sands derived from the dry parts of the barrier were blown across areas that were increasingly damp or even periodically wet. In the damp areas the sand accumulated through adherence, whereas in the wet areas, sand transport was checked by the clay covered surfaces, thus building a gradually increasing depositional slope. This eventually led to a high-angle bar front subject to high tide aquatic modification and clay draping, and low-tide accumulations of aeolian-transported sand.

The vertical accumulation of sand in the damp areas probably responded to groundwater levels. If a bed is accreted sufficiently high to escape the capillary action, most of the additional sand is transferred to the wet area and the water's edge. This situation is depicted in Figure 21A. Lowering of the lagoonal water level may cause a shift of the different aeolian flat environments farther into the lagoon. This will have a pronounced effect on the steep bar front at the water's edge. The adherence of wind-blown sand at the now relatively dry lee side of the bar will then even out the relief. A new relief and water's edge will be created downwind (Fig. 20B). This may explain the lateral sequence at the base of the spill and blowover unit, where four, successive bar-building phases (Alpha, Echo, Fox 1, Fox 2) can be identified separated by homogeneous and flasered sand tracts (Fig. 7).

A rise in lagoonal water level will have the opposite effect on the water's edge, which will be pushed back toward the aeolian domain. The inundation of the lower portion of the sand flat will cause deposition of a clay layer, which in the subsequent act of reworking may become a sand-clay mixture (Fig. 21C).

The question why in this case sand deposition in the damp and humid areas did not leave a more definite structural record than a vague and crinkly lamination must be addressed. This question becomes all the more important as many authors (*e.g.*, Andrews, 1970; Deery and Howard, Fryberger *et al.*, 1990) stress the frequent occurrence of adhesion structures, like warts and ripples. Two factors may account for this apparent discrepancy. First, deposition in most places took place on grass-covered surfaces, preventing the construction of ripples and continuous horizontal lamination. Second, and this is recognized in many a study, post-depositional processes such as bioturbation and the building of air vesicles by a rising water table very likely will obliterate any structures formed in the barren areas.

The spill and blowover unit apparently records several rises and falls of water level as is demonstrated by the different clay layers and the proximal-distal shifts of the crossbedded set (Fig. 21D). There is a weak indication that rises in water level may be associated, at least in part, with storm events at the shoreline. This may be concluded from the washover bed discussed above. This washover bed situated 1 m above the base of the unit merges into a crossbedded tract ("Hotel-1"), which in turn moves across a clay bed.

It is tantalizing to hypothesize from the lateral succession in Figure 7 that the rather elevated and seaward position of "Hotel-1" has to do with a high lagoonal water level corresponding to the washover event. Then, as the water level dropped, a new water's edge developed more distally and lower into the lagoon ("Hotel-2"), the intervening tract to "Hotel-1" being bridged by humid-aeolian sedimentation.

The topographic configuration of the backbarrier landscape also contained tidal channel elements. This may be inferred from the bar "Hotel-2", which is flanked by megaflaser deposits at Pts. 500 to 300, filling a slight depression (Fig. 22B).

Apparently, tidal channels locally swept the front of the aeolian flat and sometimes were pushed laterally by the flat's extension lagoonward. As the lagoon was filled with aeolian sediments, the barrier that acted as the source area for the sand apparently shifted seaward, giving way to the deposition of the top clays of the spill and blowover unit, which were followed by a peat marsh that became the Upper Lignite seam.

Discussion and Conclusions

Magnificent exposures in a rapidly backward-shifting face of the open lignite mine Zukunft-W permitted a detailed description and facies interpretation of the more than 2 km long and up to 15 m thick wedge of marine sands resting on and distally interdigitating with a lagoonal clay body.

The sand wedge comprises a lower and upper interval, the flood-tidal delta and spill and blowover facies respectively, separated by a thin but continuous clay layer that marks a pronounced break in the supply of beach-derived sands and chert pebbles.

Distally, that is, in a mainland direction, the flood-tidal delta facies displays high-angle delta foresets before finally wedging between the lagoonal clays. The spill and blowover facies also tapers distally, but seems to extend beyond the outcrop as a triple, m-thick layer of highly bioturbated sands separated by thin clays.

The flood-tidal delta facies consists of three, superimposed units that reflect a decrease in energy and a change in sedimentary processes with time. From bottom to top: 1) the crossbedded unit, 2) bioturbated sand unit, and 3) the topstratum unit.

The crossbedded unit starts with a roughly one metre thick subunit that is dominantly horizontal or low-angle bedded. Clay, chert and lignite pebbles are mixed with poorly sorted sands. This subunit probably represents the initial breach of the barrier. Sediment was shed into the lagoon, probably as a kind of subaquous "washover", with density or debris flow characteristics.

A major interval of tabular-planar crossbedded sets characterised by cross-stratal bundles of variable length overlies this subunit. Bundles are separated by pause planes, which show reworking by waves and weak currents and draping by clay, silt or organic material. The draping becomes more prominent distally. This, together with decreasing grain size, set size, bundle size and increasing biogenic reworking of the sets in the same direction, points to decelerating currents away from the barrier, probably as a result of flow expansion into the lagoonal waters. The absence in the crossbeds of lateral sequences of regularly extending and contracting cross-stratal bundles as are commonly encountered in inshore tidal settings and which are connected with neap-spring tide cycles seems to indicate the (at least) subordinate influence of the astronomical tides. It is more likely that wind or wave set-up at the coastline was responsible for the intermittency of flow. The crossbedded unit was apparently deposited as a slightly lagoonward-dipping subaquous fan, which in its proximal position may have been constricted by the inlet and the flaring extensions of the barrier through which it broke. Water depth probably was in the order of a few metres. The greater thickness of the present unit may be explained by the penecontemporanous compaction of the underlying lagoonal clays.

Going upward, the crossbedded unit grades into the bioturbated sand unit with a reduction in set size and grain size, while an increasing density of a mono-specific, tube-like ichnofossil causes a progressive obliteration of sedimentary structures. As far as biogenic homogenization permits observation, small current- and wave-formed crosslaminated sets and occasional horizontal lamination seem to predominate in the upper portion of the unit. This, together with the relatively clean character of the sands, indicates a sheltered backbarrier platform environment that was almost constantly wave-agitated. The waves, however, were rather small and the water depth limited (<0.5 m). The present relatively great (3-5 m) thickness of the bioturbated sands may be attributed to compaction in the shallow subsurface, just like the crossbedded unit.

The roughly 1 m thick topstratum unit that overlies the bioturbated sands and tops the flood-tidal delta facies consists of dirty, strongly bioturbated sands, commonly containing clay chips, shell hash, chert pebbles and plant debris. These sands alternate with even more clayey mixtures. Roots, probably from seagrasses, and erosional features may mark the top of some beds. The topstratum unit is attributed to washover processes, which brought sands into an intertidal, clay-covered, backbarrier flat. During transport and deposition, some mixing between newly introduced and resident material took place, and subsequent biogenic reworking completed the churning.

The decreasing energy going upward in the flood-tidal delta facies probably was a result of decreasing inlet efficiency as the tidal prism reduced along with the partial infilling of the lagoon (*cf.* Hine 1975). At a later stage the inlet seems to have been obstructed considerably by the growth of swash bars, possibly in the form of spits extending from the barrier nucleus. The sedimentary product of such a bar/spit, characterized by the high and low angle crossbedding typical for the upper nearshore interval of shoreline sequences encountered in other exposures of the present basin, was actually found in the extreme northern part of the out crop ('Plug of nearshore sands').

The thin but continuous layer of clay ('middle-clay') that extensively covers the flood-tidal delta facies records the final closure of the inlet (*cf.* Kraft *et al.*, 1979).

The spill and blowover facies that follows shows an enigmatic set of sedimentary features, including grain size, sedimentary and biogenic structures, and most notably, a lateral sequential arrangement, which presented many difficulties in the interpretation, as modern analogues are not yet known. Characteristically, sand beds wholly or partly reworked by biogenic activity (roots, clams and decapods) tend to grade distally into low angle-inclined flaser beds, which in turn gradually steepen to form closely-bundled crossbeds. The bundles are commonly draped by mud or organic muck. These crossbedded sets may distally grade back into the homogenized sands from which they proximally stem. A single bed containing one or more of these lateral sequences can typically be followed over several hundred metres.

Significantly, some of these beds in their 'up-current' direction were found to originate in a nucleus of aeolian strata with ridge-shaped geometry ('Plug of back-shore aeolian sands'). This led to the interpretation that the spill and blowover facies was largely deposited in a humid-to-wet back-barrier wind-tidal flat, and at its steep edge facing the adjoining body of lagoonal waters.

At this 'water's edge', crossbedding developed as aeolian

sand avalanched into the water. At one place, a 2 m deep tidal creek, characterized by large- and small-scale flaser bedding, was found to flank the water's edge. Incidental washover events were recognized in the northern, seaward outcrop of the spill and blowover facies; these washovers were partly reworked by aeolian activity.

Coastal paleogeography during deposition of the flood-tidal delta facies was probably characterized by a narrow barrier backed by a wide lagoon as is reflected by washovers extending over thick lagoonal muds instead of being restricted to the sandy nucleus of the barrier (Hennessy and Zarillo, 1987).

Long, narrow, linear barriers, backed by wide, shallow, possibly open lagoons are typical for wave-dominated microtidal environments (Heron *et al.*, 1984). Moslow and Tye (1985), among others, point to the fact that wave-dominated inlets show a dominance of flood direction. This should be the case, especially where the time asymmetry of the tidal currents is enhanced by an offset between the tidal rhythms of the open sea and the part of the lagoon lying close to the inlet (*cf.* Hine, 1975).

The open character of the lagoon, which resulted in it receiving large quantities of mud from the nearby river-mouth, is reflected in the thick accumulation of lagoonal muds (a minimum of 3 m remained after compaction by overburden). Basinal reconstructions (Hager, 1968, 1981, 1986; Hager, Kothen and Spann, 1981; Hager and Prüfert, 1988) indicate the likelihood of a major river system with a gradually increasing capacity and rapidly extending delta system debouching not more than 10 km to the northeast of the area of sand wedge accumulation.

The question of what caused the onlap of the marine sand wedge and associated lagoonal clays over the essentially fluviatile suite of the upper Main Seam group has to be addressed.

Moslow and Heron (1979) found in the barrier system of the Core Banks area that overwash was the primary mode of barrier migration in the face of a rapidly rising sea level; as sea level rise decreased, inlet formation and migration accompanied by flood-tidal deposition became the primary mode of landward migration. According to Kraft *et al.* (1979), factors determining the evolution of barrier systems are: nearshore surface and submarine morphology; source and size of sediment available; wind, wave and tidal processes impinging on the barrier system; climate; and variations in relative sea level change. They consider wave-dominated processes to be the only factor in determining the volume of sediment in motion. Hager *et al.* (1981) stressed the importance of high local compaction rates caused by thick peats and clays in the subsurface for patterns of sediment accumulation in the Lower Rhine Embayment. According to Hager and Prüfert (1986), the deposition of the clastic wedge discussed in the present contribution can be correlated with a rising tract in the sea level curve. At the same time two other factors reportedly played an important part in sediment distribution: increased block tilting and a considerable increase in the capacity of the main river system that crossed the embayment less than 10 km to the east of the present outcrop. This river system eventually, in Upper Seam time, came to feed a delta lying 15 to 20 km to the north. This rapid northward shift of the depocentre may have deprived the present area from the sands necessary to keep the barrier sufficiently nourished in the face of the rising sea level.

Clays ejected by the river system and caught by basinal circulation patterns, did accumulate in the area, however, entering the open or restricted lagoon either directly through smaller distributaries or indirectly from the seaside through inlets. A major role in the above reasoning is assigned to the direction of longshore transport. It seems reasonable to assume that the essentially east-west trending coastline of the Lower Rhine Embayment in Upper Miocene time was subject to a similar westerly meteorological circulation pattern as that which exists nowadays in the North Sea area. This would have created a strongly predominant easterly longshore drift, which probably was affected in only a minor way by tides in the essentially shallow sea beyond. The above considerations suggest that the onlap of the marine sand wedge and the associated lagoonal clays constitutes a local transgression of a wave-dominated coast determined primarily by starvation of this part of the coastline together with high local compaction rates, an increased subsidence by block tilting, and, possibly, a rise in eustatic sea level.

Acknowledgments

I gratefully remember the friendly ambiance and scientific enthusiasm with which J. Gliese (formerly Geologisches Landesamt, GLA, Nordrhein-Westfalen, Krefeld) shared the early part of the investigations. His enforced abandonment of the profession was a great loss. Sincere thanks and appreciations are extended to H. Hager (GLA) for his guidance and stimulating ideas, especially with respect to the stratigraphic relationships of the basin. Rheinbraun AG, concessionaire of the lignite mines, not only gave permission to enter the quarries but also supplied generous support in many respects, not in the least by furnishing free housing to our group through several years. Among the numerous people who contributed far beyond their duty, I'd like to mention Dr. Stahl (Zukunft-W), and Dr. Nehring and Mr. Kothen (Köln). It is to be hoped that Rheinbraun will continue to give scientific access to the unique but temporary geological inheritance of which it is the trustee.

Part of the dataset was collected by the graduate students C. Groen, B.J. Wolters, L. D. Alblas, and H. A. J. M. van der Meijden. The manuscript benefitted from discussions with and critical remarks by C. J. Dabrio (Madrid), L. Krystinik (Ft. Worth), J. H. J. Terwindt (Utrecht) and two unknown referees. D. G. Smith (Calgary) showed a remarkable forbearance and a continued interest in my manuscript.

References

Andrews, P. B. 1970. Facies and genesis of a hurricane-washover fan, St. Joseph Island, Central Texas Coast. Bureau of Economic Geology, University of Texas, Austin, Report of Investigation, n. 67, 147 p.

Berelson, W. M. and Heron, S. D., Jr. 1985. Correlations between Holocene flood-tidal delta and barrier island inlet fill sequences, Back Sound-Shackleford Banks, North Carolina. Sedimentology, v. 32, p. 215-222.

Boersma, J. R. and Terwindt, J. H. J. 1981. Neap-spring tide sequences of intertidal shoal deposits in a meso tidal estuary. Sedimentology, v. 28, p. 151-170.

Boothroyd, J. C. 1985. Tidal inlets and tidal deltas. *In:* Coastal Sedimentary Environments, R. A. Davis (ed.). Springer Verlag, New York, p. 445-532.

______ , Friedrich, N. E. and McGinn, S. R. 1985. Geology of microtidal coastal lagoons, Rhode Island, Marine Geology, v. 63, p. 35-76.

Carter, C. H. 1978. A regressive barrier-protected deposit: depositional environments and geographic setting of the Late Tertiary Cohansey Sand. Journal of Sedimentary Petrology, v. 48, p. 933-950.

Carter, R. W. G., Forbes, D. L., Jennings, S. C. and Orford, J. D. 1989. Barrier and lagoon coast evolution under differing relative sea-level regimes. Examples from Ireland and Nova Scotia. Marine Geology, v. 8, p. 221-242.

Deery, J. R. and Howard, J. D. 1977. Origin and character of washover fans on the Georgia coast, U.S.A. Gulf Coast Association, Geological Society Transactions, v. 27, p. 259-271.

Flyfoto-atlas Danmark 1980, Bygd Geografforlaget I/S, Esbjerg.

Frey, R. W. and Howard, J. D. 1969. A profile of biogenic sedimentary structures in a Holocene barrier island saltmarsh complex. Gulf Coast Association, Geological Society Transactions, v. 19, p. 427-444.

Fryberger, S. G., Krystinik, L. F. and Schenk, C. J. 1990. Tidally flooded back-barrier dunefield Guerrero Negro area, Baja California, Mexico. Sedimentology, v. 37, p. 23-44.

Gliese, J. 1971. Fazies und Genese der Kölner Schichten (Tertiär) in der Sudlichen Niederrheinischen Bucht. Sonderveröffentlichung no. 19, Geologisches Institut der Universität, Köln, n. 19, 91 p.

______ and Hager, H. 1978. On brown coal resources in the Lower Rhine Embayment (West Germany). Geologie en Mijnbouw, v. 57, p. 517-525.

Hager, H. 1968. Zur Gleichstellung und Genese der Flöze im Rheinischen Braunkohlenrevier. Fortschritte in der Geologie von Rheinland und Westfalen, Krefeld, v. 16, p. 73-84.

______ 1981. Das Tertiär des Rheinischen Braunkohlenreviers – Ergebnisse und Probleme. Fortschritte in der Geologie von Rheinland und Westfalen, Krefeld, v. 29, p. 529-563.

______ 1986. Peat accumulation and sygenetic clastic sedimentation in the Tertiary of the Lower Rhine basin (F.R. Germany). Mémoire du Société Géologique de la France, n. 149, p. 51-56.

Hager, H., Kothen, H. and Spann, R. 1981. Zur Setzung der rheinischen Braunkohle und ihrer klastischen Begleitschichten Fortschritte in der Geologie von Rheinland und Westfalen, Krefeld, v. 29, p. 319-352.

Hager, H. and Prüfert, J. 1988. Tertiär. *In:* Geologie am Niederrhein, H. D. Hilden (ed.). Geologisches Landesamt Nordrhein-Westfalen, Krefeld.

Hayes, M. O. 1975. Morphology of sand accumulation in estuaries: an introduction to the symposium. *In:* Estuarine Research, L. E. Cronin (ed.). Academic Press, London, p. 3-22.

Hennessy, J. T. and Zarillo, G. A. 1987. The interrelation and distinction between flood-tidal delta and washover deposits in a transgressive barrier island. Marine Geology, v. 78, p. 35-56.

Heron, S. D., Jr., Moslow, F. G., Berelson, W. M., Herbert, J. R., Steele, C. A. and Susman, K. R. 1984. Holocene sedimentation of a wave-dominated barrier shoreline: Cape Lookout, North Carolina. Marine Geology, v. 60, p. 413-434.

Hine, A. C. 1975. Bedform distribution and migration patterns on tidal deltas in the Chatham Harbor Estuary, Cape Cod, Massachusetts. *In:* Estuarine Research, L. E. Cronin (ed.). Academic Press, New York, v. 2, p. 235-252.

Illies, H. J. 1977. Ancient and recent rifting in the Rhine Graben. Geologie en Mijnbouw, v. 56(4), p. 329-350.

Kirschbaum, M. A. 1989. Lagoonal deposits in the Upper Cretaceous Rock Springs Formation (Mesaverda Group), Southwest Wyoming, Marine Geology, v. 88, p. 349-364.

Kjerfve, B. 1986. Comparative oceanography of coastal lagoons. *In:* Estuarine Variability, D. A. Wolfe (ed.). Academic Press, New York, p. 235-252.

Kraft, J. C., Allen, E. A., Belknap, D. F., John, C. J. and Maurmeyer, E. M. 1979. Processes and morphologic evolution of an estuarine and coastal barrier system. *In:* Barrier Islands from the Gulf of St. Lawrence to the Gulf of Mexico, S. P. Leatherman (ed.). Academic Press, New York, p. 148-183.

Lewis, D. W. and Titheridge, D. G. 1978. Small scale sedimentary structures resulting from foot impressions in dune sands. Journal of Sedimentary Petrology, v. 48, p. 835-838.

Lingen, G. J. van der and Andrews, P. B. 1969. Hoof-printed structures in beach sand. Journal of Sedimentary Petrology, v. 39, p. 350-357.

Oertel, G. F. 1985. The barrier island system. Marine Geology, v. 63, p. 1-18.

Penland, S., Boyd, R. and Suter, J. R. 1988. Transgressive depositional systems of the Mississippi Delta Plain. A model for barrier shoreline and shelf sand development. Journal of Sedimentary Petrology, v. 58, p. 932-949.

Ritchie, W. and Penland, S. 1988. Rapid dune changes associated with overwash processes on the deltaic coast of South Louisiana. Marine Geology, v. 81, p. 97-122.

Schwartz, R. K. 1975. Nature and genesis of some storm washover deposits. Coastal Engineers Research Centre, Technical Memoir 61, 69 p.

______ 1982. Bedform and stratification characteristics of some modern small-scale washover sand bodies. Sedimentology, v. 29, p. 835-850.

Zagwijn, W. H. and Doppert, J. W. Chrs. 1978. Upper Cenozoic of the Southern North Sea Basin, paleoclimatic and paleogeographic evolution. Geologie en Mijnbouw, v. 47, p. 577-588.

______ and Hager, H. 1987. Correlations of continental and marine Neogene deposits in the south-eastern Netherlands and the Lower Rhine district. Mededelingen Werkgroep Tertiar en Kwartaire Geologie, v. 24(1-2), p. 59-78.

KEY WORDS: FLOOD-TIDAL DELTA, SAND WEDGE, LOWER RHINE EMBAYMENT, PROXIMAL-DISTAL FACIES RELATIONSHIPS SEDIMENTARY STRUCTURES AND PROCESSES

In: Clastic Tidal Sedimentology, D. G. Smith, G. E. Reinson, B. A. Zaitlin and R. A. Rahmani (eds.). Canadian Society of Petroleum Geologists. Memoir 16, p. 255-274.

Tidal and seasonal cycles in a Lower Cambrian shallow marine sandstone (Hardeberga Fm.) Scania, Southern Sweden

LARS HAMBERG

Institute of Geology
University of Copenhagen, Öster Voldgade 10
1350 Copenhagen, Denmark

ABSTRACT

The Lower Cambrian Hardeberga Formation in Scania (southern Sweden) is 120 m thick and rests directly on the weathered and peneplained crystalline Proterozoic basement. The formation comprises a basal, 15 to 25 m thick, arkosic sandstone unit, overlain by three, 30 to 50 m thick, vertically stacked marine sequences. The basal unit consists of an immature, conglomeratic sandstone grading into crossbedded, arkosic sandstone. Paleocurrents were consistently toward the south. The basal unit consists of a fluvio-deltaic deposit, truncated by a marine erosion surface (a ravinement).

The three marine sequences are separated by ravinements and were deposited as a result of a northerly directed regional transgression over the southern part of the Baltic Shield. Each sequence contains three major facies associations that document the progradation of a mixed tide-and wave-influenced barrier island shoreline. Each ravinement surface is mantled by a 10 m thick unit, initiated by muddy heterolithic strata overlain by a hummocky cross-stratified and coarse grained wave-rippled quartzose sandstone. This succession represents the transition from an inner shelf to shoreface environment. The paleoshoreline was oriented approximately east-west, parallel to the orientation of the coarse grained wave ripples. The inner shelf-to-shoreface association is truncated by a 10 m thick tabular sandstone body, which shows giant-scale epsilon crossbedding, occasionally interrupted by wave ripples. This deposit is interpreted to have been formed by lateral migration of a tidal inlet delta-barrier complex. The tidal inlet sandstone is erosionally overlain by spit shoreface bar-trough crossbedded deposits or a 1.5 to 2 m thick wedge of epsilon crossbedded sandstone that represents an on-barrier creek deposit. The inlet, spit shoreface and back-barrier creek deposits constitute the barrier island facies association. This association is erosionally overlain by a backbarrier tidal channel facies association consisting of stacked vertical sequences of A and B units. The A units are 2 to 7 m thick, crossbedded sandstone beds displaying regular mud drapes, acceleration-deceleration structures, and a bipolar, east-west paleocurrent pattern, parallel to the paleocoastline. The A units reflect deposition in active tidal channels. The B units typically comprise 2 to 10, stacked 0.2 to 0.8 m thick microsequences of crossbedded sandstone passing upward into bioturbated or tidal-bedded and burrowed sandstone. This cyclic alternation between bedform migration and dense bioturbation is interpreted as indicating deposition during winter storms alternating with longer periods of low-energy deposition in abandoned tidal channels, during the calm summer season.

RÉSUMÉ

Dans la province de Scanie (Suède méridionale), la Formation Hardebergera du Cambrien Inférieur atteint 120 m de puissance et repose directement sur le socle cristallin protérozoïque, altéré et presqu'aplani. La base de la formation comprend une unité de grès arkosique de 15 à 25 m d'épaisseur, surmontée par une succession verticale de trois séquences marines de 30 à 50 m d'épaisseur. L'unité basale se compose d'un grès conglomératique immature passant graduellement à un grès arkosique à stratifications obliques. Les paléocourants étaient constamment dirigés vers le sud. L'unité basale consiste en dépôts fluvio-deltaïques tronqués par une surface de ravinement.

Les trois séquences marines se sont déposées durant une trangression marine régionale, en progression vers le nord sur la partie méridionale du Bouclier baltique; elles sont séparées par des discontinuités de type ravinement. Chaque séquence contient des associations de trois faciès majeurs qui reflètent la progradation d'une crête d'avant-plage émergée, influencée par l'action des vagues et marées. Chaque surface de ravinement est recouverte par dix mètres de faciès composés de couches hétérolithiques argileuses, surmontés par des grès quartzeux à stratifications obliques avec creux et bosses («hummocky»), et des grès quartzeux à grains grossiers avec rides de vague. Cette succession représente la transition d'un environnement de plate-forme interne à un environnement de la zone infratidale. L'ancienne ligne de rivage était orientée approximativement est-ouest, parallèle à l'orientation des rides de vague à grains grossiers. L'association plate-forme interne et zone intertidale est tronquée par une masse de grès tabulaire de 10 m d'épaisseur, avec des stratifications obliques de type epsilon de grandes dimensions, occasionnellement interrompues par des rides de vague. Ce dépôt est interprété comme le résultat de la migration latérale d'un petit bras de mer de marée d'un complexe delta-barrière à stratifications obliques. Le grès du petit bras de mer de marée est surmonté en discordance d'érosion par des dépôts de barres et fosses d'un cordon littoral, ou par un prisme de 1.5 à 2 mètres d'épaisseur, de grès à stratifications obliques de type epsilon résultant de la sédimentation d'un ruisseau développé sur la barrière. Les dépôts du petit bras de mer, du cordon littoral et du ruisseau développé sur la barrière, constituent l'association des faciès d'une crête d'avant-plage émergée. Cette association est surmontée par un faciès de chenal de marée d'arrière-barrière constitué par des séquences verticales des unités A et B.

Les unités A, de 2 à 7 m d'épaisseur, contiennent des couches de grès à stratifications obliques montrant des «lentilles de boues pélagiques» («mud drapes»), des structures d'accélération-décélération, et un patron bipolaire est-ouest des paléocourants parallèles à l'ancienne ligne de rivage. Les unités A représentent des dépôts dans des chenaux de marée. Les unités B typiques contiennent de 2 à 10 microséquences de 0,2-0,8 m d'épaisseur, de grès à stratifications obliques passant verticalement à des grès bioturbés ou lités sous l'action des marées, et des grès avec terriers. Cette alternance cyclique entre une migration des «morphologies de fond» («bedforms») et une intense bioturbation est interprétée comme le résultat d'une sédimentation saisonnière, i.e. durant les tempêtes hivernales, et la saison calme estivale où la sédimentation se caractérisait par un niveau de faible énergie dans les chenaux de marée abandonnés.

Introduction

More than 10 per cent of the world's present coastline consists of barrier islands, displaying a geomorphology characteristic of micro- and mesotidal, wave-influenced shorelines and associated wide, low gradient shelves (Cromwell, 1971; Heward, 1981).

Early Cambrian eustatic sea level rise caused the development of extensive epeiric seas that developed worldwide over the existing cratonic continents (*cf.* Klein, 1977a; Surlyk, 1980; Dott *et al.*, 1986). The broad shelf areas probably were highly influenced by tides (Klein and Ryer, 1978; Cram, 1979). Many Cambrian epeiric coastlines may therefore have analogues in the transgressive, low gradient and tide-influenced barrier island environments of the modern epicontinental North Sea coasts (*cf.* Heward, 1981; Oertel, 1985; Niedoroda *et al.*, 1985).

The current picture of the backbarrier environment is commonly referred to as a dendritic drainage pattern developed symmetrically around tidal inlets. This is true for the pattern of marsh islands and channels common in mid-latitude backbarrier lagoons (Hayes, 1975). In contrast, the backbarrier region of the Wadden Sea possesses only fringing marshes and is exposed to relatively high wave and wind energy (Fitzgerald and Penland, 1987). As a consequence, the Wadden Sea backbarriers are generally more sandy, and sand-dominated subenvironments are found at the tidal inlet mouths and adjacent intertidal channels (Reineck, 1967; Fitzgerald and Penland, 1987). Because of this sand dominance and reduced vegetation, the Wadden Sea geomorphology probably represents more closely than other modern shorelines the nonvegetated Early Cambrian tidal shoreline environment of southern Sweden.

This paper describes a tidally-influenced, backbarrier, sand-dominated system with coast-parallel channels and shoals situated just behind the barrier island, with an example from the Lower Cambrian Hardeberga Formation in Sweden and Denmark (Fig. 1).

The objectives of this paper are to demonstrate that fossil (especially Precambrian and Early Paleozoic), sandy, tidal deposits that bear a pronounced coast-parallel paleocurrent pattern are best explained as having been generated behind coastal barrier islands in a backbarrier tidal channel system, and to describe this system in terms of its possible modern analogue, the Wadden Sea, by means of dynamics and depositional successions.

The recurring problems of the exceptionally mature texture and widespread, sheet-like geometry of many Lower

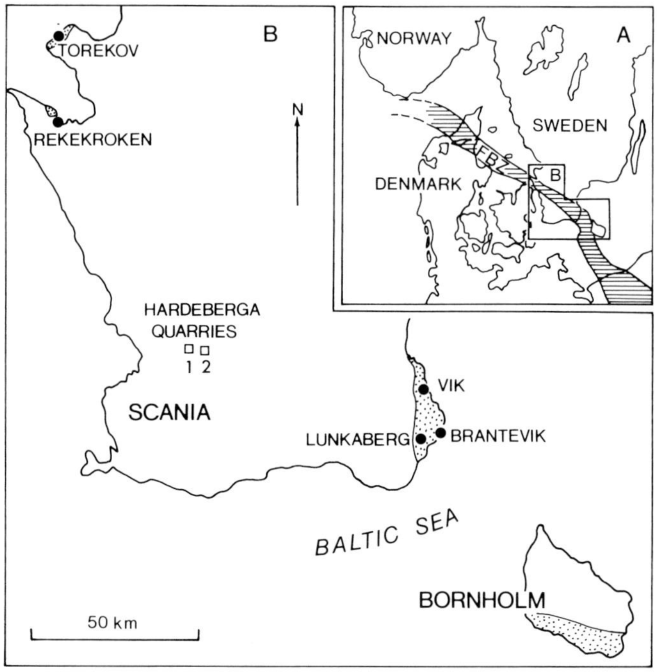

Figure 1. Location map showing the distribution of the outcrops of the Hardeberga Formation in Scania (southern Sweden) and on Bornholm (Denmark). The demarcation of the Fennoscandian Border Zone (FBZ) is inferred from EUGENO-S Working Group (1988).

Paleozoic marine quartzose sandstones together with the apparent paucity of shale and muddy tidal-flat deposits have been previously discussed by several authors (Klein, 1977b; Dott and Byers, 1981; Dott *et al.*, 1986). These problems are also considered here.

Geological Setting and Stratigraphy

The Lower Cambrian Hardeberga Formation forms a widespread, sheet-like, quartzose sandstone in southern Scandinavia (Poulsen, 1966; Bergström and Gee, 1985). The formation marks the onset of the marine transgression of the peneplained, Proterozoic, crystalline basement of the Baltic Shield. In the latest Precambrian and Early Cambrian, this craton comprised all of Scandinavia and the western part of Russia in a micro-continent, Baltica (Surlyk, 1980; Bergström and Gee, 1985). The marine transgression is related to the latest Proterozoic-earliest Cambrian openings of the Iapetus Ocean, and was probably the result of an increased sea floor spreading rate (Matthews and Cowie, 1979; Bergström and Gee, 1985; Piper, 1985). Cocks and Fortey (1982) suggested that a large ocean, the Tornquist Sea, bordered southern Scandinavia in the Early Cambrian. Paleomagnetic evidence suggests that the Baltic Shield was located in the southern hemisphere at mid latitudes (45°) (Scotese *et al.*, 1979; Piper, 1985).

On the island of Bornholm in the Baltic Sea (Fig. 1), the 60 m thick marine Hardeberga Formation rests on the 100 m thick continental fluvio-aeolian sandstone of the Neksø Formation (Dam and Clemmensen, 1988) (Fig. 2). The fluvial deposits show a consistent drainage pattern toward the south. Northward into Scania (Figs. 1, 2) these continental deposits are replaced by thin deltaic deposits (10-20 m) and farther north at Rekekroken and Torekov, the marine Hardeberga Formation rests directly on weathered Proterozoic basement (Hamberg, 1988), indicating that the marine deposits are increasingly younger toward the north (Hadding, 1929; Bergström, 1981). This suggests a northerly directed, marine transgressive onlap.

In Scania, the Hardeberga Formation is approximately 120 m thick (Lindström and Staude, 1971; Bergström, 1981) and has yielded acritarchs that indicate a Lower Cambrian *Holmia A* level (Vidal, 1981). The formation is subdivided into four members (from base to top: the Lunkaberg, Vik, Brantevik and Tobisvik members (Fig. 3). The Hardeberga Formation is succeeded by the Norretorp Formation which consists of glauconitic shelf sandstones (Bergström, 1981). These shallow marine sediments are overlain by marine shale and carbonates that mark the maximum advance of the transgression (Surlyk, 1980; Bergström *et al.*, 1982) (Fig. 2).

Today the Hardeberga Formation in Scania is situated within the Fennoscandian Border Zone (Fig. 1), between exposed Proterozoic basement to the north and a thick Mesozoic and Tertiary succession in the south, under which it is buried (Bergström, 1981). In the Mesozoic, the Fennoscandian Border Zone was characterized by block faulting and strike-slip movements (Surlyk, 1980; EUGENO-S Working Group, 1988), but paleomagnetic evidence suggests that no large lateral movements have taken place since the Sweconowegian (Bylund, 1984). This conclusion is supported by the similar trends of paleocoastlines and paleocurrents

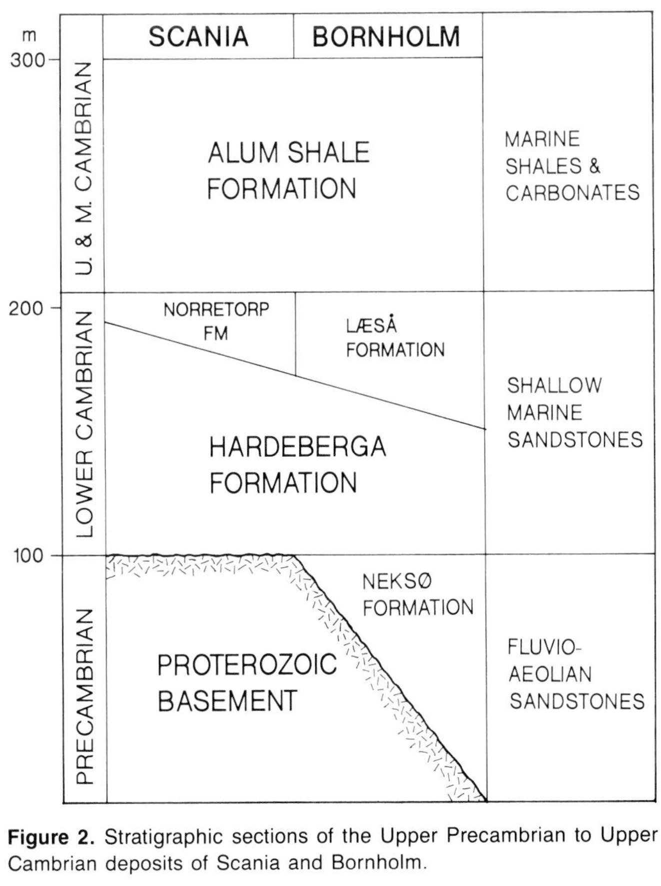

Figure 2. Stratigraphic sections of the Upper Precambrian to Upper Cambrian deposits of Scania and Bornholm.

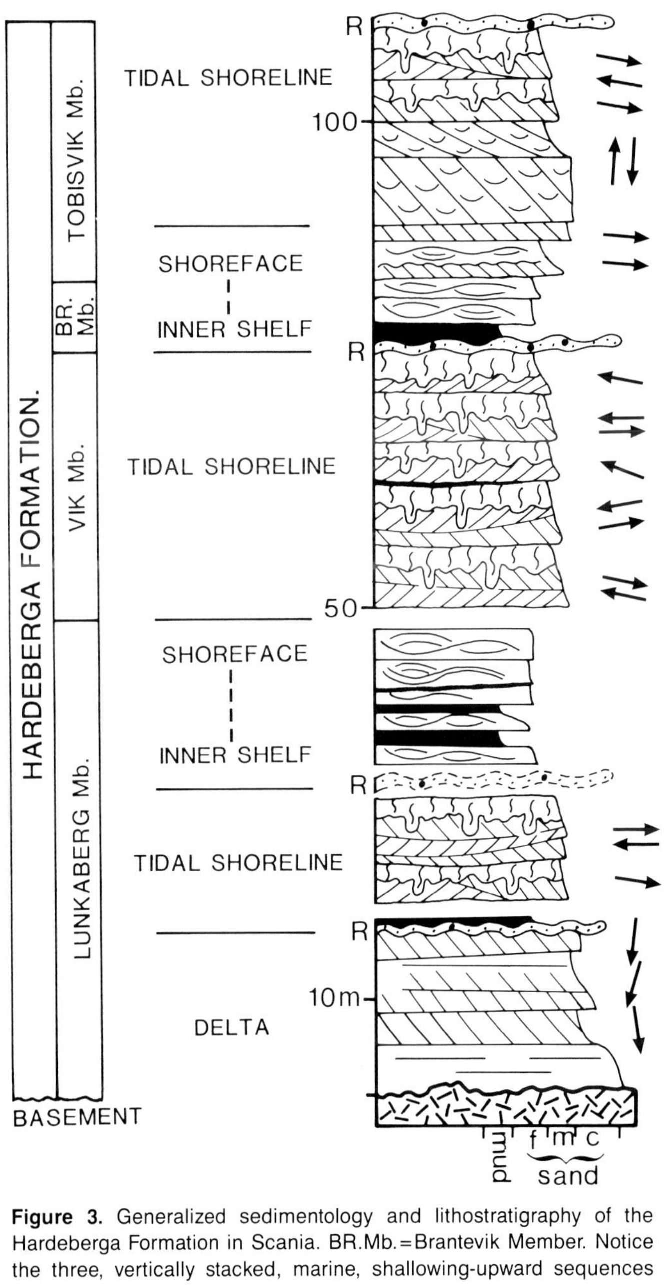

Figure 3. Generalized sedimentology and lithostratigraphy of the Hardeberga Formation in Scania. BR.Mb.=Brantevik Member. Notice the three, vertically stacked, marine, shallowing-upward sequences separated by ravinement surfaces (R). (For legend, see Fig. 10).

of the Hardeberga Formation in Sweden and on Bornholm, across the border zone (Hamberg, 1988), (Fig. 1).

Sedimentology

The Hardeberga Formation in Scania comprises a basal, nonmarine, 15 to 25 m thick, arkosic sandstone unit succeeded by three, 30 to 50 m thick, vertically stacked, marine facies sequences. The overlying Norretorp Formation is made up of a similar, fourth, marine sequence. Each facies sequence is separated by a marine erosional surface or ravinement.

The basal unit includes a lowermost 5 to 7 m, thick, immature, pebbly to conglomeratic and structureless arkosic sandstone, which passes gradationally into a coarse grained arkosic and crossbedded sandstone (Fig. 3). The crossbedded sets range in thickness from decimetres to metres and are arranged in broadly wedge-shaped beds, divided by horizontally laminated pebbly horizons. Paleocurrent direction was consistently toward the south (Fig. 3). The basal unit is truncated by a ravinement included in the lowermost marine facies sequence (Fig. 3).

Broad, wedge-shaped beds in an immature, arkosic sandstone with a strongly unimodal paleocurrent distribution pattern suggest deposition in wide and shallow fluvial channels. This interpretation is consistent with the southerly paleocurrents, which are similar to the paleocurrent pattern obtained in the corresponding fluvial deposits of the Nekso Formation on Bornholm.

The marine truncation of the fluvial deposits suggests that they were most probably deposited in the coastal zone as a braid-delta. The character of this nonvegetated Early Cambrian delta is probably comparable to the tidally-influenced sandur shorelines on Iceland, described by Hine and Boothroyd (1978).

The following focuses on the marine facies sequences of the Hardeberga Formation. Each marine facies sequence contains five facies associations named after their respective interpreted environments.

Inner Shelf to Shoreface Facies Association

This sand-dominated association occurs in the upper Lunkaberg Member, the entire Brantevik Member, the lower part of the overlying Tobisvik Member, and the Norretorp Formation (Fig. 3). The only coherent facies sequence is observed in the Brantevik and Tobisvik members at Hardeberga (Fig. 4) and is discussed below in terms of the dynamics and orientation of the paleoshoreline.

The inner shelf to shoreface facies association is recognized as a 10 m thick, but widespread and tabular sandstone unit above the ravinement surfaces (Fig. 4). The ravinement is a low-relief erosional surface mantled by pebbly gravel, typically wave rippled, and with oblate, intraformational sandstone clasts.

The lowermost metre of the association comprises the muddiest deposits of the Hardeberga Formation. It consists

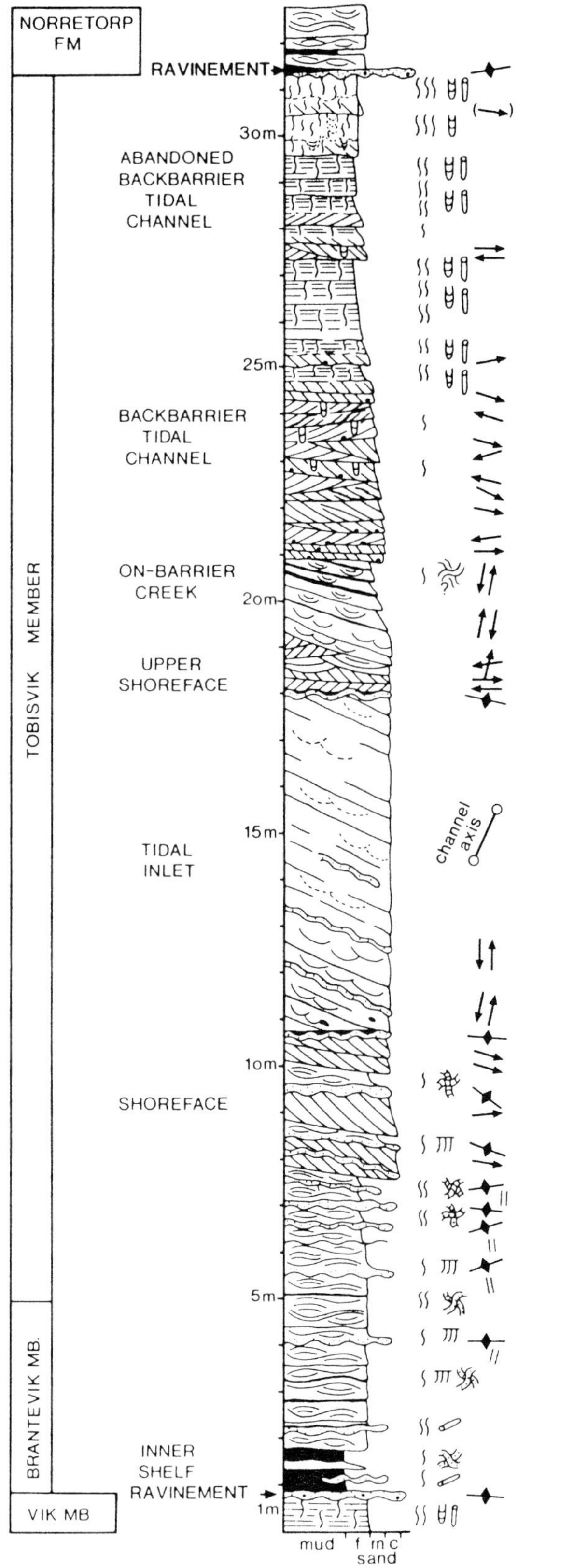

Figure 4. Vertical section of the uppermost Hardeberga Formation in Scania, which represents a prograded barrier island shoreline preserved between two ravinement surfaces. Notice the consistent east-west orientation of the coarse grained wave ripple crestlines. Tobisvik Member, Hardeberga Quarries, Scania. (For legend, see Fig. 10).

of a glauconitic mudstone interbedded with 2 to 10 cm thick, fine to coarse grained, sharp based sand layers (Fig. 4). Overlying this facies is a series of fine grained, alternating hummocky cross-stratified and burrowed beds, dominated by the horizontal trace fossils *Planolites, Palaeophycus* and *Didymaulichnus*.

The next 4 to 5 m of the sequence consist of 20 to 50 cm thick, amalgamated, hummocky cross-stratified to parallel beds with an increasing frequency of interbedded, 10 to 15 cm thick, but laterally extensive, coarse grained and wave rippled layers (Figs. 4, 5a). This unit is succeeded by a 3 to 4 m thick, coarse grained sandstone unit, composed of a 0.5 to 1 m thick, trough crossbedded set or coset displaying coarse grained wave ripples at the top, and draped by a fine grained, hummocky to swaley cross-stratified sandstone (Fig. 4). Similar wave rippled erosional surfaces, draped by finer grained sand containing sporadic trace fossils, are recognized as discontinuity surfaces within the crossbedded sets (Figs. 4, 5b).

The paleocurrent directions of the crossbedded part are tightly grouped around the east and southeast and the orientations of the crestlines of the coarse grained wave ripples, both from the lower interbedded part and the upper coarse grained part, are grouped around a vector mean of 90° (Fig. 6a, b). Parting lineation in the hummocky beds is uniformly orientated perpendicular to the crestlines of the associated coarse grained wave ripples (Fig. 4).

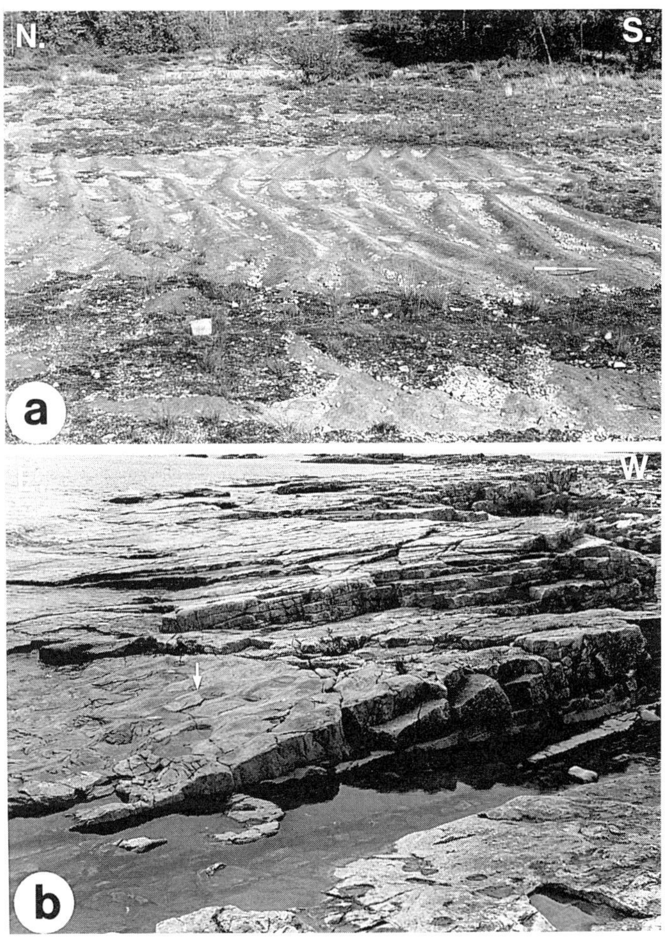

Figure 5. Lower shoreface deposits: **a.** surface exposure of coarse grained wave ripples; measuring stick is 0.6 m. **b.** 0.5 m high, large-scale crossbedded set formed by storm-enhanced longshore current. The exposed wave rippled surface (arrow) represents an internal, wave-ripple-modified discontinuity surface related to waning storm phase oscillatory movements. Brantevik, Scania.

Interpretation

The sand-dominated association of hummocky cross-stratification and coarse grained wave ripples most likely represents deposition on the shoreface (Dott and Bourgeois, 1982; Hunter and Clifton, 1982; Leithold and Bourgeois, 1984; Leckie, 1988; Hunter *et al.*, 1988; Leckie and Krystinik, 1989).

The presence of sharp-based layers and the lack of wave-produced sedimentary structures in the lower muddy section suggests that sedimentation took place below storm wave base. The coarsening-upward facies sequence is interpreted as having been produced by progradation of the shoreface zone over the inner shelf zone.

The co-existence of thin but widespread, coarse grained, wave rippled layers interbedded with fine grained, parallel to hummocky cross-stratified layers may indicate a shoreface setting similar to that observed by Hunter *et al.* (1988). They described a comparable lithological relationship of exposed, sheetlike bands of coast-parallel wave ripples bordered by fine grained sand developed in water depths of 10 to 20 m.

According to the orientation of both fossil coarse grained wave ripples (Wright and Walker, 1981; Leckie and Walker, 1982; Leckie, 1988), and similar modern examples (Yorath *et al.*, 1979; Hunter *et al.*, 1988), the well defined, east-west trending wave ripple crestlines suggest a broad, east-west orientation of the paleocoastline. The southerly direction of fluvial paleocurrents suggest that the Early Cambrian sea lay to the south. The large-scale crossbedding showing consistent east to southeast peleocurrent directions probably was generated by unidirectional and apparently uniform, coast-parallel currents. Moreover, the regular appearance of internal wave rippled discontinuity surfaces and the vertical development of the crossbedded sets followed by wave ripples and hummocky beds suggest shoreface deposition from

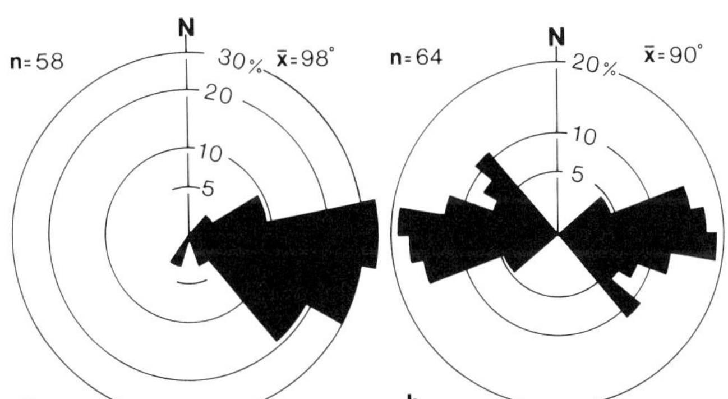

Figure 6. Paleocurrent data from the inner shelf to shoreface deposits in Scania: **a.** coarse grained crossbedded sandstone, and **b.** crestline of coarse grained wave ripples with ripple heights greater than 5 cm.

sistent east to southeast paleocurrent directions probably was generated by unidirectional and apparently uniform, coast-parallel currents. Moreover, the regular appearance of internal wave rippled discontinuity surfaces and the vertical development of the crossbedded sets followed by wave ripples and hummocky beds suggest shoreface deposition from storm enhanced, easterly, shoreparallel currents followed by oscillatory movements within the same storm event. The oscillatory movements probably reflect the waning phase of the storm (Niedoroda *et al.*, 1985; Swift and Nummedal, 1987; Leckie, 1988). A shoreface sub-environment subjected to dynamic behavior similar to that interpreted above, undergoing episodic storm conditions with shoreparallel to shore-oblique transport of coarse grained sand, may be that producing the storm controlled, oblique-running ridge and swale morphodynamic rip-channel features observed by Flemming and Antia (1989).

Tidal Inlet Facies Association

This association is best observed in the Tobisvik Member at Hardeberga (Fig. 4). It forms a 7 to 8 m thick, tabular sandstone body that erosionally overlies the coarse, crossbedded and wave rippled shoreface sediments (Figs. 4, 7). The sandstone body is medium to very coarse grained with abundant smaller pebbles and, together with the shoreface sediments, constitutes the coarsest of the marine deposits. No pronounced fining-or coarsening-upward in the sandstone body were recognized (Fig. 4).

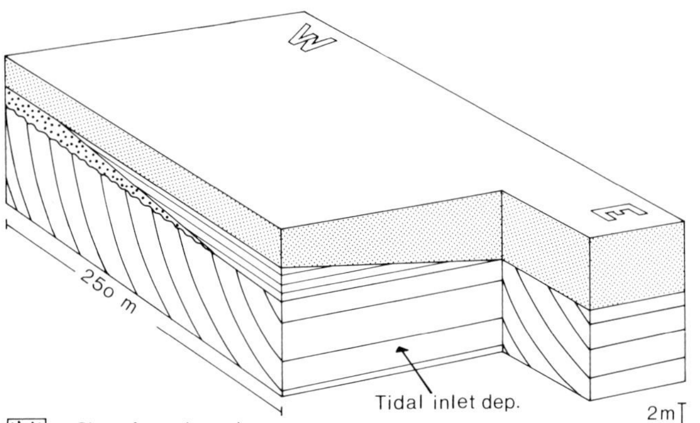

Figure 7. Diagrammatic outline of the vertical and lateral relationships between the tidal inlet deposits and the overlying upper shoreface, on-barrier creek and backbarrier tidal channel deposits. Notice the wedge-out of the backbarrier deposits toward the south, paleo-seaward. This truncation surface appears to be horizontal, but it actually dips 1 to 2° north (Fig. 7). Tobisvik Member, Hardeberga Quarry No. 2. Slight vertical exaggeration.

The lower erosional surface may be covered by large mud clasts, and it occasionally truncates shallow, 1.5 to 4 m deep and 50 to 80 m wide, channelized scours cut into the shoreface sediments, with trough axes directed north-south (Fig. 8a).

In general the tabular sandstone body appears structureless, but in quarry number 2 at Hardeberga (Fig. 1), it contains beds 7 to 8 m high and 20 to 80 cm thick, dipping 15 to 23° toward the east (Fig. 8b). The inclined beds are crossbedded, and to some extent wave rippled. The crossbedded intrasets show foreset dips arranged perpendicular to the east dipping inclined beds with a paleocurrent pattern directed north-south (Fig. 4). The thicker, inclined beds, or two to three amalgamated, thinner beds may be wave rippled in their upper parts and separated by a mud drape. On the basis of the present condition of the outcrops, no identification of any lateral development of the inclined beds or wave rippled and mud draped surfaces was possible.

Interpretation

The 7 to 8 m high inclined beds (Fig. 4) with perpendicularly arranged intrasets may be regarded as large forms of Allen's (1965) epsilon cross-stratification and are here interpreted as having been deposited by lateral accretion in a migrating tidal inlet, consistent with the sedimentological observations of the migrating slope of the Doboy Sound Inlet (Hoyt and Henry, 1967).

Tidal inlets migrate shoreparallel in the direction of the longshore current by spit accretion on the updrift side of the inlet and erosion of the downdrift side (Hoyt and Henry, 1967; Kumar and Sanders, 1974). In agreement with this, the uniform easterly dip of the inclined strata indicates migration parallel to the east-west orientation of the paleocoastline and suggests the presence of an easterly directed longshore current.

The bipolar paleocurrent patterns of the intrasets probably reflect tidal currents in the inlet channel. The restriction of the coarsest sediment at the shoreface and inlet suggests that sand was supplied by longshore drift, reworked and redistributed by tidal currents in the inlet in accordance with modern inlet dynamics (*cf.* Kumar and Sanders, 1974; Reddering, 1983). Recently, Nielsen *et al.* (1988, their Fig. 7) have described comparable intrasets bound by lateral accretion surfaces from a spit-platform deposit developed in the tidally influenced part of a terminating spit system.

It seems that generally, inlets influenced by longshore currents possess more than one channel, developing a deeper main (ebb) channel and margins occupied by minor (flood) channels (Kumar and Sanders, 1974; Hayes, 1980; Hanisch, 1981; Nummedal and Penland, 1981). However, as observed by Hanisch (1981), the larger ebb-channel is displaced in a downdrift direction during inlet migration.

The inclined beds described here seem to have occupied the full part of the inlet margin and migration probably took place, at least locally, as one, large, accretionary face. This was most likely located in a relatively permanent position next to the marginal (flood) channel, and was subjected to a lower and more uniform current strength. The occasional occurrence of shallow, channelized scours beneath the inclined strata may indicate scouring of a channel, probably the ebb channel, in shorter periods of diminished migration.

The lateral termination of the larger inclined beds or set

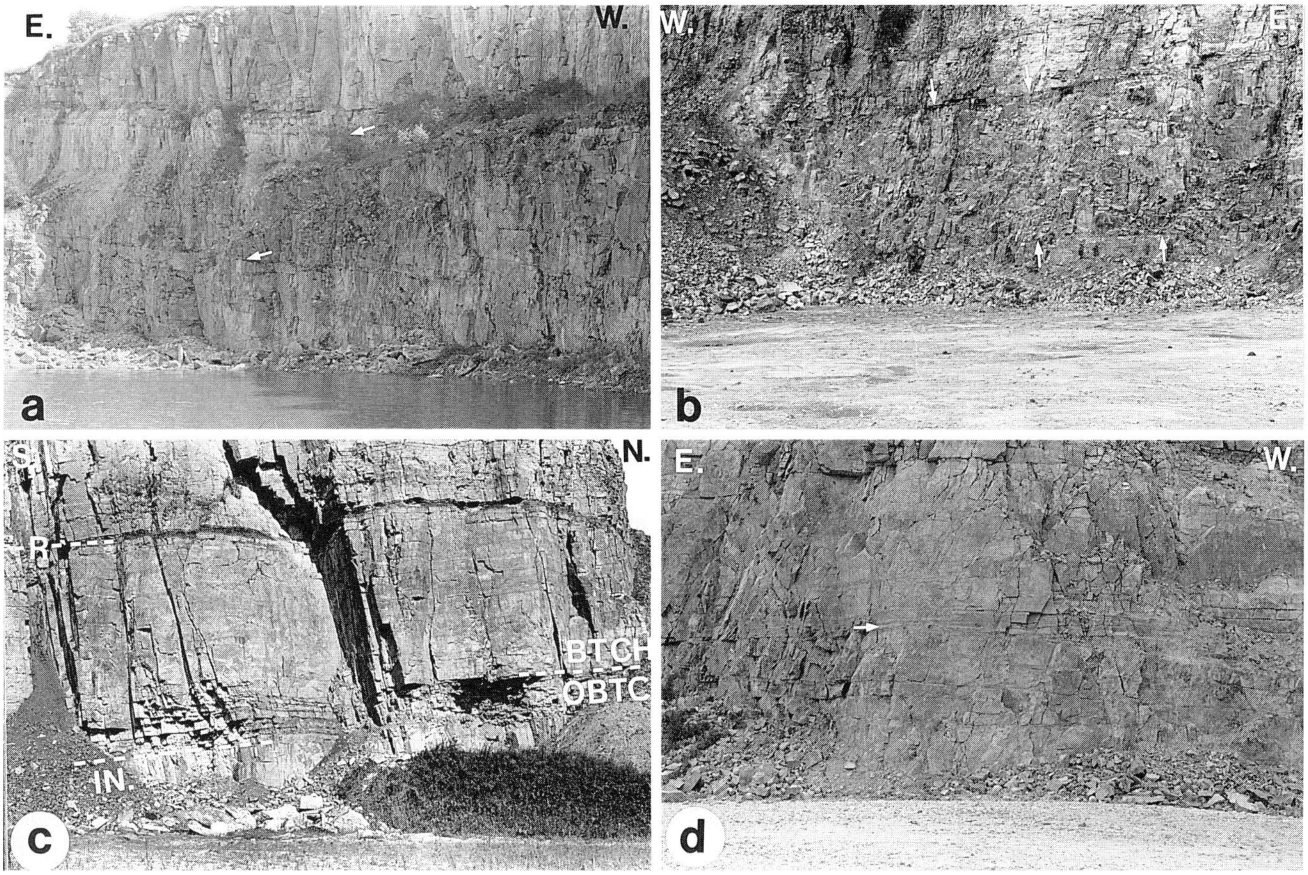

Figure 8. Barrier island deposits: **a.** Typical structureless appearance of the 7 to 8 m thick tidal inlet deposits (between arrows), which truncate a north-south directed and 50 m broad, shallow scour (just below the lower arrow). Hardeberga Quarry No. 1. **b.** Exposure of the tidal inlet deposits (between arrows) revealing the internal architecture of inclined beds dipping 15 to 23°. Notice the sharp truncation (upper arrows) of the top of the inlet deposits by the backbarrier tidal channel deposits. Person for scale, Hardeberga Quarry No. 2. **c.** Section from the top of Tobisvik Member exposing the muddier part of the on-barrier tidal creek deposits (OBTC) sharply truncated by the sandy backbarrier tidal channel deposits (BTCH). The tidal inlet sediments (IN) are exposed below the on-barrier creek deposits, and R marks the ravinement surface related to the overlying Norretorp Formation. Hardeberga Quarry No. 1. **d.** A 1.5 m thick unit of inclined strata within the coarse grained proximal part of the on-barrier creek deposit, sharply truncated by backbarrier tidal channel deposits (arrow). Hardeberga Quarry No. 2.

of amalgamated beds by erosional wave ripple bedding is probably related to storm agitation. Storm erosion and migration of inlets indeed is a common observation (Reddering, 1983), and it is possible that the larger inclined strata represent accelerated migration during storm conditions as a function of the increased sediment supply.

Upper Shoreface Facies Association

The upper shoreface facies association consists of a 2.5 m thick, coarse to medium grained crossbedded sandstone with an average set thickness of less than 20 cm. The paleocurrent directions are widespread, but show a tendency for a bidirectional east-west arrangement, and a few northerly directions (Fig. 4). Where crossbedded sandstone overlies the inlet deposit, the latter is persistently truncated by patchy, coarse grained strata displaying wave ripples with east-west trending crestlines and interbedded with a muddy heterolith, or directly by the crossbedded sandstone (Figs. 4, 7).

Interpretation

The sharp truncation of the inlet facies is similar to the sharp erosion of the spit-platform deposits produced by storm waves and migrating bar-trough systems in the upper shoreface zone described by Nielsen *et al.* (1988). The coarse grained, trough crossbedded sandstone units with bidirectional paleocurrents parallel to the paleocoastline, are interpreted as having been produced by the migration of small, subaqueous dunes in the troughs of upper shoreface bars, similar to the bar-trough systems described by Carter (1978), Hayes (1980) and Nielsen *et al.* (1988).

The persistent truncation of the inclined beds of the inlet channel facies by bar-trough deposits most likely reflects wave reworking and recurved spit formation on the inlet platform in the updrift end of the barrier island (*cf.* also Hayes, 1980). The coarse grained, wave ripple truncation and associated heterolithic units probably represent shoreface, storm generated deposition on the inlet-platform.

Barrier Tidal Creek Facies Association

These deposits form 1.5 to 2 m thick, wedge-shaped sandstone bodies within 3-dimensional channel forms. Usually these deposits sharply truncate tidal inlet deposits, but may laterally cut into and wedge-out over the upper shoreface deposits (Fig. 7). The barrier tidal creek facies is truncated by the backbarrier tidal channel facies association in all the sections (Figs. 4, 7, 8c). The sandstone body fines upward and consists of 10 to 25 cm thick inclined beds, dipping 7 to 18°, with internal trough crossbedding arranged perpendicular to the dip of the inclined strata (Figs. 4, 8d). The dip directions of the inclined strata are uniform for each specific sandstone body, either toward the east or west, with axes of the channels perpendicular to the paleoshoreline.

The sandstone bodies commonly show lateral development within a distance of 50 m from coarse grained, 20 to 25 cm thick, inclined sets (Fig. 8d), to an increasingly muddy and fine grained heterolithic facies, possessing thinner inclined beds with a sigmoidal shape (Fig. 9). Most of the inclined heterolithic strata are graded, have a sharp lower contact and are draped by a mm thick mud layer, which is slightly eroded by the succeeding inclined bed. The muddy layers commonly reveal an undifferentiated repichnia ichnofauna and a few vertical burrows.

Lateral sequences or sets of both the sandy and the heterolithic inclined strata, 0.2 to 1 m thick, are separated by erosional, convex-upward discontinuity surfaces. The number of discontinuity surfaces increases toward the distal heterolithic part (Fig. 9).

Interpretation

The thin and low-angle inclined strata with orthogonally arranged, trough crossbedded intrasets and abundant biogenic traces are interpreted as having been deposited by lateral accretion of a marine point bar in a 1.5 to 2 m deep, high sinuosity channel, resembling Allen's (1965) point bar system. Small, high sinuosity channels occur in the coastal environment predominantly as tidal creeks, although both modern and fossil examples of tidal creeks are considerably muddier than the present example (Reineck, 1958; Bridges and Leeder, 1976; Klein, 1977b; De Mowbray, 1983; Smith, 1987).

The thicker and more proximal parts of the channel sandstone are similar to the underlying and truncated inlet and shoreface deposits in grain size. This, together with the observed position directly on top of the tidal inlet sediments indicates that the original channel probably: 1) eroded into, redistributed and migrated through the unconsolidated shoreface and upper inlet sediments and, 2) represents an on-barrier meandering drainage channel (Fig. 4).

Modern examples of comparable sandy on-barrier channels are described from mature, prograding and migrating, mesotidal barrier islands (Bernard and Leblanc, 1965; Duc and Tye, 1987). Duc and Tye (1987) document the development of a tidal creek drainage network on the expanded barrier island, which almost completely reworked dune, foreshore and shoreface sandy sediments during lateral migration. An on-barrier location of the Early Cambrian channels may likewise account for their relatively sandy character.

The dominance of north-south channel axes directions suggests the existence of a drainage pattern at right angles to the original east-west orientated barrier island, consistent with the preferred orientation of modern tidal creeks on mature, Holocene barrier islands (Bernard and Leblanc, 1965).

The internal bipolar crossbedding of both the sandy and heterolithic inclined strata argues for a possible tidal influence. On the other hand, the clearly graded, inclined, mud draped beds that dominate the middle and distal heterolithic part may have been generated by one, steadily declining flow phase or event as demonstrated for similar beds in fluvial point bars (Allen, 1970; Bridge and Diemer, 1983). Storm-agitated waters are likely to provide conditions for the generation of sediment-loaded flows and subsequent graded beds.

The larger discontinuity surfaces within the point bar deposits probably indicate erosion caused by an increased discharge as described from fluvial point bars by Bridges and Leeder (1976), Van Der Meulen (1982), and from tidal flat point bars by De Mowbray (1983). They are here interpreted as being related to abnormal discharge during storms, because they increase in number, and the spaces between them decrease from the sandy toward the heterolithic part along with an increasing presence of the graded beds (Fig. 9).

The lateral wedging-out, together with the increasing mud content and presence of storm related erosion and deposition, points to channel abandonment with reduced tidal influence and increasing storm-dominated suspension load.

The inlet, spit-shoreface and on-barrier tidal creek

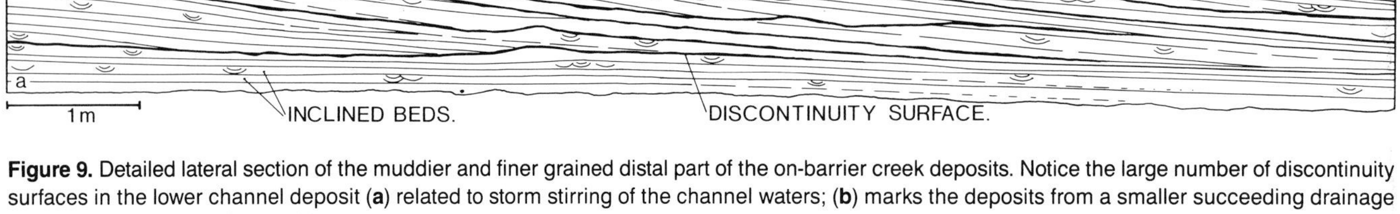

Figure 9. Detailed lateral section of the muddier and finer grained distal part of the on-barrier creek deposits. Notice the large number of discontinuity surfaces in the lower channel deposit (**a**) related to storm stirring of the channel waters; (**b**) marks the deposits from a smaller succeeding drainage channel. Hardeberga Quarry No. 2.

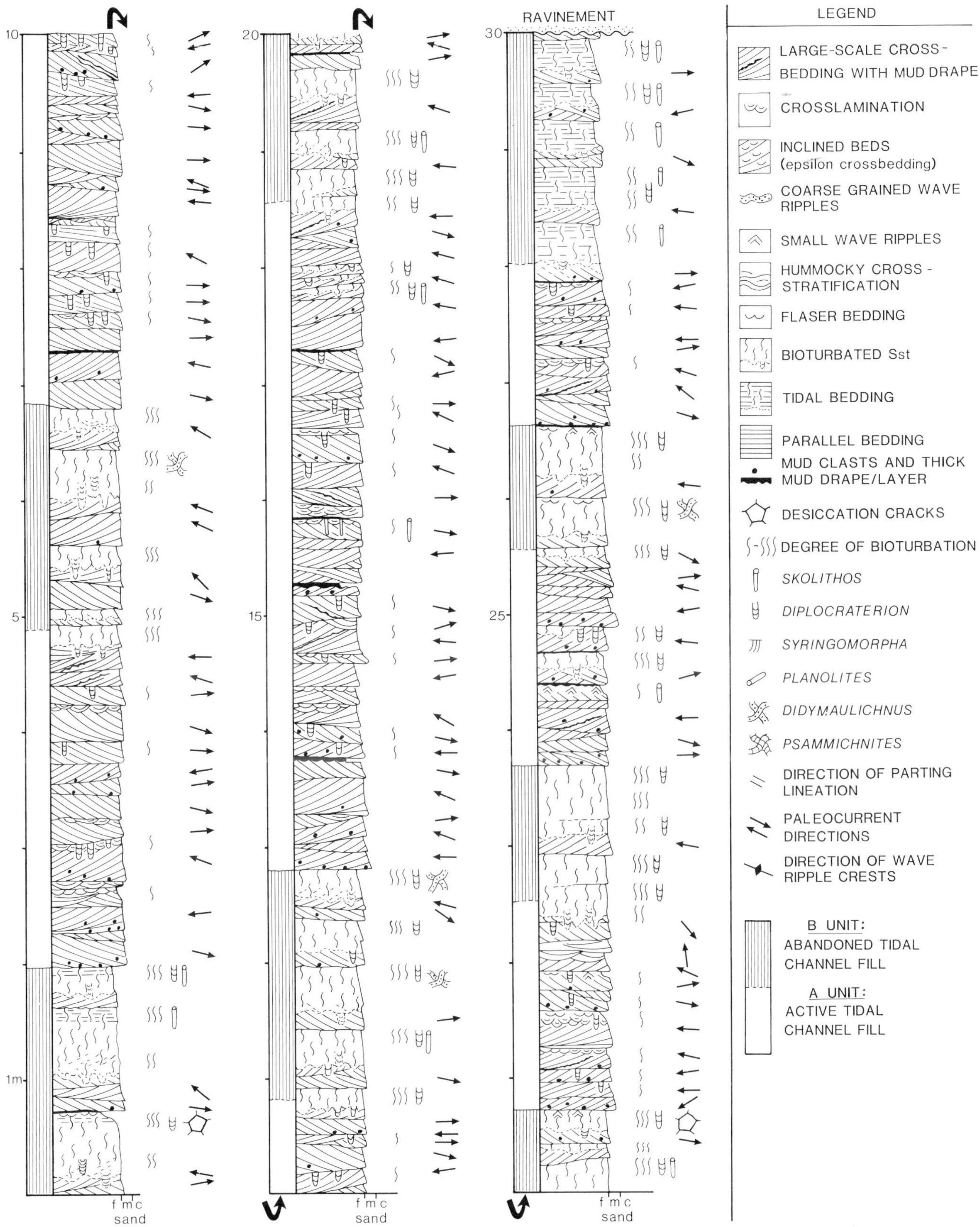

Figure 10. Detailed section through the backbarrier tidal channel deposits. Notice the general lack of vertical development of the facies, except for the stacking of alternating active and abandoned channel fill, and the concentration of tidal-bedded layers just below the ravinement surface. Notice also the consistent east-west and east-southeast-west-northwest paleocurrent directions, both within the active and the abandoned fills. Vik Member, Hardeberga Quarries.

deposits constitute what is left of a former barrier island. Figure 7 sketches the locally observed intricate architecture of the barrier island deposits and the increasing importance of the backbarrier tidal deposits in a landward direction toward the north.

Backbarrier Tidal Channel Facies Association

Throughout the Hardeberga Formation, the backbarrier tidal deposits are characterized by stacked, facies sequences consisting of a 2 to 6 m thick lower unit of large-scale, crossbedded sandstone (A unit), overlain by a 1 to 4 m thick upper unit of alternating crossbedded and burrowed sandstone layers (B unit). The crossbedded A units represent the active channel fill and the B units, the infill during abandonment (Fig. 10). The thickest amalgamated sequence recorded, the Vik Member, is approximately 30 to 40 m thick and comprises more than 7 stacked A-B unit couplets (Fig. 10). The vertical stacking is the only facies arrangement recorded (Fig. 10). The association forms the middle part of the Lunkaberg Member, the entire Vik Member and the upper half part of the Tobisvik Member (Fig. 3).

Active Channel Fill: A units

The A units all possess a sharp lower boundary and are gradationally overlain by the burrowed B unit (Fig. 10). They are recognized as laterally consistent tabular sandstone bodies, and can be followed for at least 500 m. The sandstone is medium grained, quartzose, and is dominated by trough crossbedded sets, 0.15 to 0.35 m thick containing a few tabular crossbedded sets, 0.3 to 0.5 m thick. Dispersed vertical burrows, predominantly *Diplocraterion parallelum* and occasionally *Skolithos linearis*, together with escape traces, occur throughout the A units (Figs. 10, 12). The burrows in general increase in number upward toward the overlying B-units without any change in their physical structures (Figs. 10, 16d).

The trough crossbedded co-sets display a large suite of both characteristic and diagnostic tidal sedimentary structures, including numerous reactivation surfaces, herringbone crossbedded structures, reactivation and deceleration structures, mud drapes, and mud layer couplets, described and discussed below.

Mud Layer Couplets

In the A units of the Hardeberga Formation only a few examples of this particular structure are observed. They appear as regular, sigmoidal, discontinuity bands in the trough crossbedded sets (Fig. 11a). The mud layer couplets consist of mm thick mud layers (mostly fine grained silt and minor clay), separated by a 0.5 cm thick sandstone layer (Fig. 11b). The sandstone layer commonly possesses a faint ripple morphology with crosslamination directed up the major foreset (Fig. 11b).

The mud layer couplets enclose tidal bundles (*cf.* Visser, 1980) and the thickness of the bundles varies from 3 to 12 cm (Fig. 11a, b). Unfortunately no observation of a continuous development of the bundles exceeds 2.5 m in length and they are therefore not suitable for any statistical analysis.

Mud layer couplets are generated during the slack water periods between the dominant and subordinate currents (Visser, 1980). The trough crossbedding of the tidal bundles was produced by the migration of 3-dimensional, small to medium dunes in the direction of the dominant current. The presence of mud layer couplets suggests deposition during slack water under the influence of a semidiurnal tidal regime (Nio and Yang, 1989).

Reactivation and Decleration Structures

Besides the numerous reactivation surfaces, two examples of reactivation and deceleration structures are recognized. Single mud drapes in the bottomsets of the large-scale crossbedded sets are commonly surrounded by opposite-directed, climbing ripple laminae that continue erosionally up the back side of the major foreset strata (Figs. 11c, d, 12).

Crossbedding is clearly related to the dominant current, whereas the reverse-directed climbing ripples represent deposition of the subordinate current (Fig. 12). The overlying ripple laminae are related to the initial reactivation of the dominant current, similar to the acceleration structures described by Boersma (1969) and Boersma and Terwindt (1981a).

Thicker crossbedded sets are commonly erosionally overlain by obliquely orientated crosslaminae or smaller trough crossbedding (Fig. 12). Similar relationships are commonly observed in modern intertidal environments produced by late stage runoff spreading the flow under the influence of the topography of the larger bedforms (Klein, 1970; Boersma and Terwindt, 1981b; Elliott and Gardiner, 1981). A similar spreading of the flow and cannibalizing of the larger bedforms may also occur subtidally (De Mowbray and Visser, 1984; Reddering, 1987). The obliquely orientated crossbedding is interpreted as representing decay of major bedforms during neap tide with generally lower water depth in shallow subtidal or intertidal settings.

Paleocurrent Pattern

A plot of all paleocurrent data from the backbarrier tidal channel association shows a polymodal distribution. Studies of modern tidal environments by Klein (1970) and Boersma and Terwindt (1981a) demonstrated that this represents the interaction of several different current systems within the tidal sub-environment, and that only a careful discrimination of the bedform to paleocurrent relationships reveals the subtle tidal paleocurrent pattern.

The paleocurrents related to the crossbedded sets with mud drapes and reactivation structures represent the dominant tidal paleocurrent system and reveal a conspicious overall bipolar west-northwest-east-southeast directed pattern (Fig. 13a, b). This paleocurrent system was established throughout southern Sweden and on Bornholm (Hamberg, 1988). Within a single A unit, the paleocurrent system may

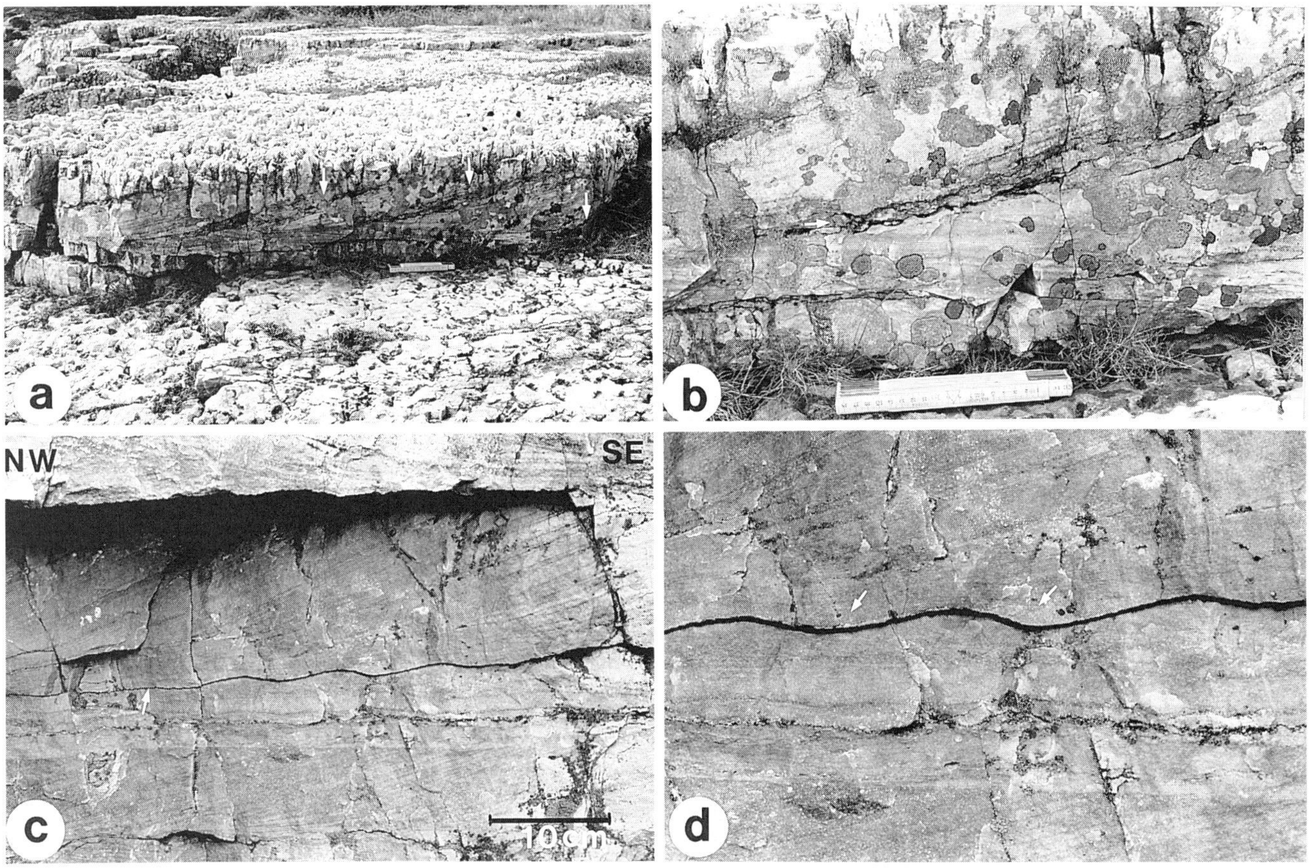

Figure 11. Tide generated sedimentary structures from the active channel fill of the backbarrier deposits: **a.** Mud-layer couplets within crossbedded sandstone (arrows, ruler is 25 cm long). The vertical burrows and irregular upper surface of the crossbedded unit represents the former presence of a bioturbated layer, developed during channel abandonment. **b.** Enlargement of mud-layer couplets. Notice the faint ripple forms in the thin sandstone layer enclosed by the mud drapes (arrow). The mud drapes are preferentially weathered. **c.** Single mud drape (arrow), now weathered, in the bottomset of a crossbedded set representing deposition by the dominant tidal current (northwest), **d.** Enlargement of the mud drape surrounded by underlying climbing ripple laminae related to the subordinate current (southeast) and overlying ripple laminae (below arrows) representing the reactivation of the dominant tide (northwest); see Figure 12. Vik Member at Vik.

diverge locally from the overall pattern and show a bimodality directed west-northeast (Fig. 13d).

The paleocurrents related to the minor crossbedded sets orientated obliquely to the larger crossbedded sets reveal a bipolar distribution directed approximately north-south (Fig. 13c).

Interpretation of the A Units

The thin but widespread crossbedded A units with regularly occurring mud drapes and mud layer couplets suggest deposition in subtidal channels (*cf.* Visser, 1980; Terwindt, 1988). The observed overall rectilinear pattern of the paleocurrent directions in relation to the dominant tide emphasizes a lateral restriction of the flow, supporting the interpretation of channelized deposition.

The characteristic sparsity of mud drapes, apparently high migration rates and strikingly high degree of paleocurrent reversal, deduced from the larger crossbedded sets, may be regarded as an effect of the original small to medium bedform size, (*cf.* Boothroyd and Hubbard, 1975). The rarity

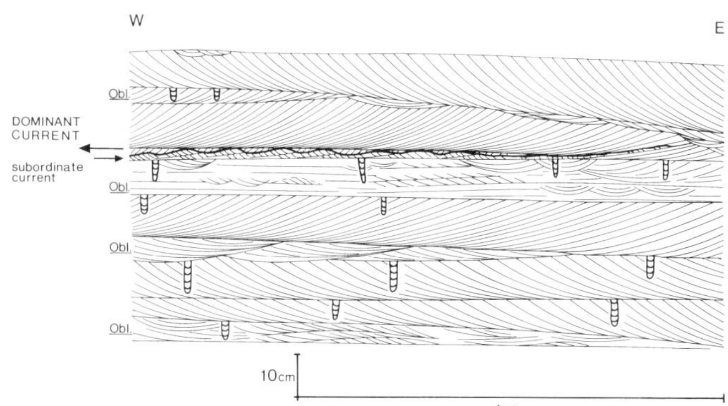

Figure 12. Sketch of crossbedded section in the active channel fill of the backbarrier tidal deposits. Vik Member at Vik. Notice the single mud drape enclosed by subordinate and dominant current ripple laminae described in Figure 11, the uniform small thickness of the crossbedded sets, and the dispersed vertical burrows that emphasize longer interruptions of the sedimentation. Obl=obliquely oriented crosslaminae and bedding representing decaying of the major bedforms during neap tide. The vertical arrangement does not necessarily imply any persistent neap-spring cyclicity but probably reflects interfingering of adjacent flood or ebb dominated bedform fields, most likely within the tidal shoals.

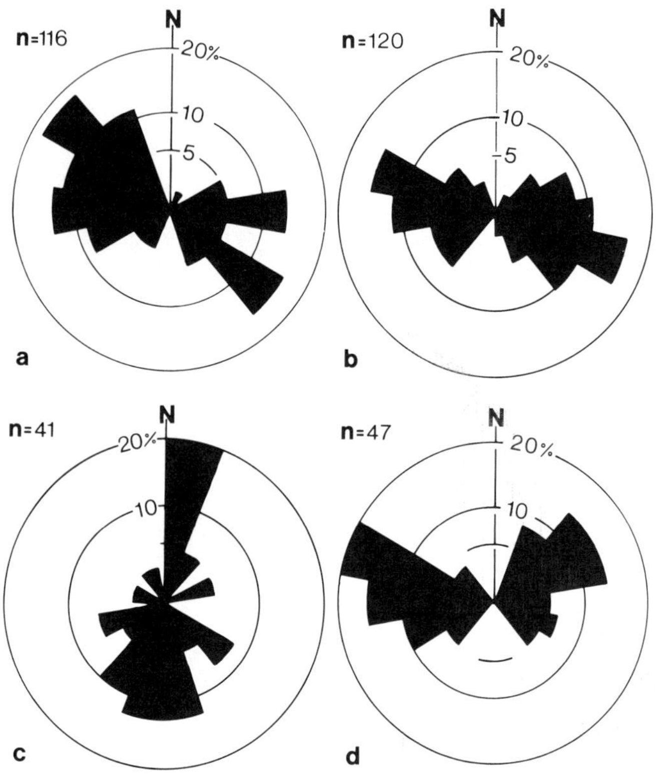

Figure 13. Paleocurrent data from the active channel fill (A units) of the backbarrier deposits. **a** and **b.** Crossbedding representing the dominant tide: (**a**) measured in the Hardeberga Quarries and (**b**) measured at Vik (Fig. 1). Notice the consistent east-southeast-west-northwest bipolarity although the measurements are separated by 70 km. **c.** Directions of the obliquely orientated crossbedding. **d.** Paleocurrents measured within a single A unit at Vik. Notice the diverging pattern compared to the overall east-southeast–west-northwest orientations.

of burrows is also an indication of rapid migrations and high sedimentation rates.

Modern, sandy tidal channel environments generally form a complex of channels divided by shallow water shoals characterized by rapid lateral migration of the channels and frequent interfingering of shoals and channels (Klein, 1970; Nio *et al.*, 1980; Boersma and Terwindt, 1981a; Van Den Berg, 1982).

The apparent random occurrence of true subtidal deposits in close proximity to shallow sub- or intertidal, decaying bedform deposits suggests interfingering of adjacent sub-environments of different water depths. This supports the interpretation of an original channel and shoal complex morphology, and individual A units are interpreted as representing deposition within active tidal channels and on tidal shoals.

In addition, the observed recurrent change of direction of the dominant paleocurrent up through most of the active channel fills may be explained as being caused by migration and interfingering of minor, ebb- or flood-dominated bedform fields (Klein, 1977b; Boersma and Terwindt, 1981a, b).

The paleocurrent directions of the dominant current indicate a west-northwest-east-southeast orientation of the original channel and shoal complex, parallel to the paleocoastline (Figs. 6b, 13a, b). The local occurrence of diverging bimodal paleocurrent directions probably reflects the effect of the tidal shoal topography on the tidal currents as observed in modern shoal complexes (Klein, 1970; Nio *et al.*, 1980). Modern, sandy tidal channel and shoal environments are described from estuaries and sandy backbarrier environments in the Wadden Sea (Nio *et al.*, 1980; Boersma and Terwindt, 1981a, b; Van Den Berg, 1982; Nummedal and Penland, 1981; Fitzgerald and Penland, 1987). The assumed shallowness of the channels, together with the coast-parallel orientation of the dominant tidal current, suggests an original backbarrier position of the channel and shoal complex. The stratigraphic position of the tidal channel deposits, on top of and wedging out over the barrier island deposits in a southerly direction (seaward) likewise suggests a backbarrier setting. A similar, coast-parallel orientation of sandy, fossil, backbarrier channel deposits is described by Carter (1978).

Abandoned Channel Fill: B units

The alternating crossbedded and bioturbated layers of the B units can be subdivided into two types of 0.3 to 1 m thick microsequences according to the type of the burrowed facies (Figs. 14, 15). In both types the lower part consists of a large-scale crossbedded set or coset that commonly shows bipolar foreset dips directed east-west or east-southeast-west-northwest, reactivation surfaces, and pause planes with mud drapes (Fig. 10).

Type 1 of the burrowed facies is a bioturbated layer with a remarkably consistent thickness of 20 to 30 cm (Fig. 14a, b). Within a lateral distance of 1 to 5 m, two or three stacked microsequences may be transformed into one, thick, amalgamated, fully bioturbated horizon by the obliteration of the dividing cross-stratified layers (Fig. 14c), explaining the development of apparently thicker bioturbated layers.

Trace fossils in the bioturbated layers are dominated by vertical domichnia burrows, *Diplocraterion parallelum* and *Skolithos linearis*. Both types were probably formed by suspension-feeding worms (Fürsich, 1974). The *Diplocraterion* burrows are always protrusive and commonly possess a bell-shaped outline related to animal growth (*cf.* Fürsich, 1974). The burrows may be as deep as 25 to 30 cm but seldom deeper. The intensity of biogenic reworking decreases downward, so that well defined trace fossils together with faint cross-strata are only seen at the transition to the underlying crossbedded layer (Fig. 14b). Exposed surfaces of this transition show a density of *D. parallelum* sometimes exceeding 600 U-tubes per m^2.

Small, thin J to broad U-shaped burrows are observed in very few examples in the upper, thoroughly bioturbated parts. They probably represent burrows from juvenile *"Diplocraterion"* animals (worms).

When preserved, the uppermost parts of the bioturbated layers reveal horizontal trace fossils, *Didymaulichnus* and

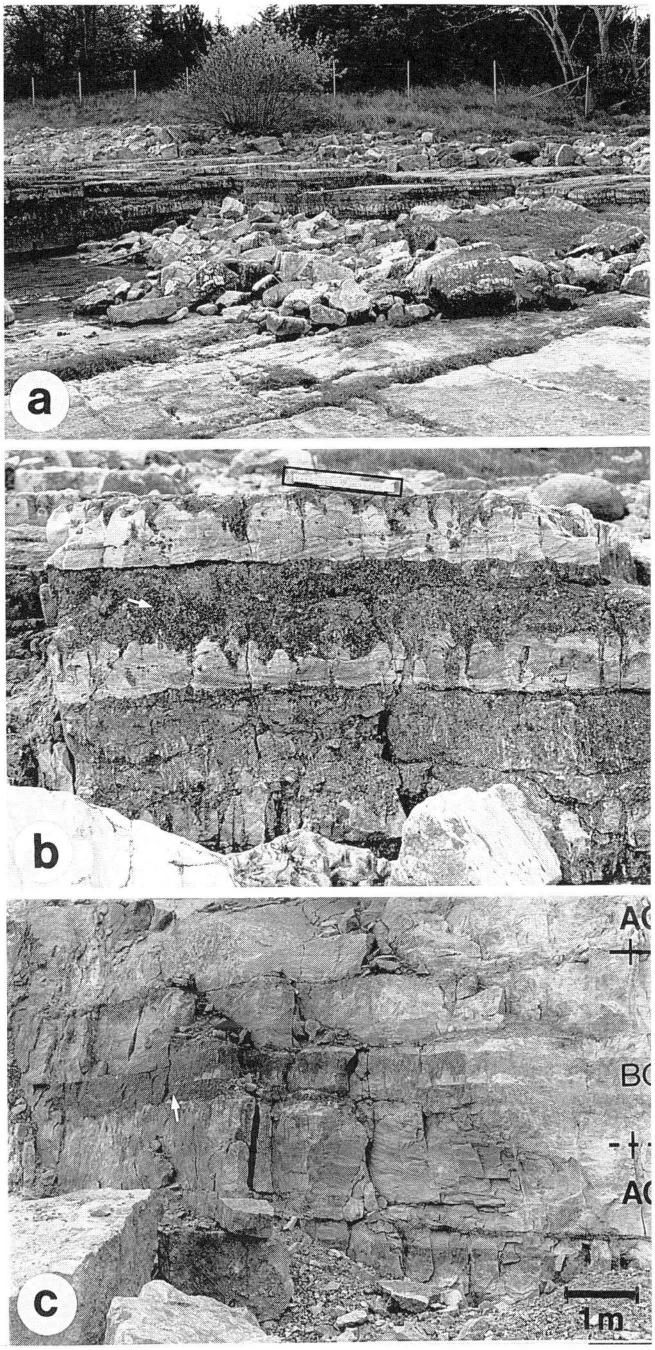

Figure 14. Abandoned backbarrier channel fill. **a** and **b.** Continuous stacking of alternating physically structured and bioturbated layers, type 1. Notice the consistent thickness of 20 to 30 cm of the bioturbated layers and faint traces of cross-strata (arrow). **c.** Lateral obliteration of the dividing, crossbedded layers (arrow) and resulting amalgamation of the bioturbated layers in abandoned channel fill (BC), enclosed by active channel fill (AC).

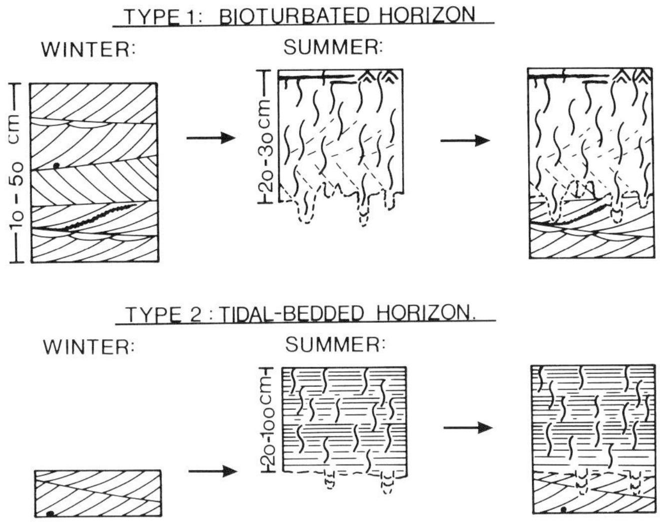

Figure 15. Seasonal sedimentation patterns in abandoned, backbarrier tidal channels. The crossbedded lower part of both types reflects recurrent migration of bedform fields during several winter storm set-ups within the backbarrier abandoned channels. The overprinting type 1 bioturbated horizon represents colonization and biogenic reworking through the calm summer season; the type 2 tidal-bedded horizons record active daily deposition during several spring-neap tide periods in the summer season. (For legend, see Fig. 10).

occasionally *Planolites*, and exhibit flaser laminae and small wave ripples, sometimes flat-crested, double-crested or as interference ripples (Fig. 10). Furthermore, cm thick polygonal desiccation cracks are observed in the tops of some B units (Fig. 10).

Interpretation of the Type 1 Burrowed Facies

Typical modern examples of bioturbated profiles start out as a result of colonization by larvae, with subsequent growth and "bioturbation" (Rhoads, 1967; Carey, 1987). Larval colonization in the type 1 facies is supported by the presence of possible juvenile burrows in the top, by the presence of bell-shaped growth structures in the adult burrows, and also by the extent of the bioturbated horizons, providing an area too large to permit any possible invasion by adult worms from adjacent populations.

The faintly visible crossbedding and upward increase in bioturbation seem to indicate that the type 1 bioturbated layers were originally fully crossbedded and probably reflect a period of non-migration and colonization by larvae.

Any speculation as to the time it took for colonization and bioturbation can only be tentative, since no modern examples of true *Diplocraterion* burrows are known. However, Rhoads (1967) and Carey (1987), measured the time it took selected species of polychaetes to rework a 10 to 30 cm thick profile, and found it to be several months to years or from the early spring to the late summer. In other words, the biogene reworking is expected to be slow, probably requiring at least half a year, to be accomplished.

Modern examples of tide-influenced alternating physically structured and biogenically reworked layers comparable to the microsequences of the B units are described from abandoned estuarine subtidal channels (Van Den Berg, 1981, 1982) and from a construction excavation facing the North Sea (Unsöld, 1974), which may be regarded as a semi-enclosed and abandoned intertidal channel. In both settings, deposition and bioturbation is completely governed by seasonal fluctuations of energy, temperature, salinity etc. Raised energy within the abandoned tidal channels, mostly during winter storms, is responsible for the formation of the structured sections, whereas the following calm summer period gives time for animal colonization and bioturbation (Unsöld, 1974; Van Den Berg, 1981; *cf.* also Carey, 1987).

The microsequences of the B units most likely represent deposition controlled by similar seasonal energy fluctuations. The crossbedded layer can likely be attributed to the migration of bedforms during several storm setups through a period of rough conditions, probably winter. An extended period of calm conditions, most likely the summer, resulted in colonization and a bioturbated profile (Fig. 15). The paleocurrents of the lower crossbedded parts are identical to the ones recorded in the active channel fills (A units), and the B units are interpreted as representing infilling of abandoned, backbarrier, coast-parallel tidal channels.

Storms are known to raise energy and water levels within backbarrier environments for several days. The widespread presence of tidal structures in the crossbeds illustrates that storms lasted long enough to be overprinted by tidal activity. The observed wedge-out of the crossbedded layers seems to emphasize that the original bedform field generated during storms was insignificant and presumably sporadically developed.

Small wave ripples and desiccation cracks on the tops of the bioturbated horizons and in the tops of the preserved abandoned channel fills indicate shallow water deposition, not necessarily intertidal, but with temporary exposure.

The type 2 burrowed facies of the B units is characterized by the presence of tidal bedding, disturbed by penetrating vertical burrows but never fully bioturbated (Fig. 16). The tidal bedded layers are usually 30 to 35 cm thick but amalgamated layers may be as thick as 1 m (Figs. 4, 16d). The tidal bedding consists of rhythmic couplets comprising a 2 to 6 mm thick sandstone layer draped by a mud layer less than 1 mm thick, succeeded by a thinner, 1 to 2 mm thick sandstone layer draped by a similar mud veneer (Fig. 16a-c). The sandstone layers all possess sharp lower and upper boundaries and normally appear ungraded, but are sometimes crosslaminated.

Tidal bedding is arranged in 10 to 20 cm thick cycles characterized by an upward thinning of the individual sandstone layers, accentuated by upward terminations of the larger vertical burrows (Fig. 16a-c). In addition, the thin bedded top parts of the sections commonly contain sandstone layers of almost equal thickness, although this part normally is highly bioturbated (Fig. 16a-c). Any precise and detailed counting of the sandstone layer couplets is prevented by the burrowing, but counting of eleven fining-upward cycles results in a minimum number of 22 to 31 thick-thin sandstone layer couplets per section. The 30 to 35 cm thick tidal bedded layers contain 2 to 3 thinning-upward cycles (Fig. 16a), whereas the 1 m thick layers seems to contain at least 6 thinning-upward cycles.

Interpretation of the Type 2 Burrowed Facies

The alternating sandstone and mudstone layer couplets resemble the tidal bedding originally defined by Reineck and Wunderlich (1969). From their description it is evident that the tidal current responsible for deposition of the thicker sandstone layer represents the dominant tide, whereas the current that deposited the upper, thin sandstone layer is the subordinate tide. Mud drapes represent the slackwater periods during the shifting of the tide.

The observed upward thinning and equalized thickness of the tidal bedding, together with an increasing burrowing within each cycle, most likely reflect deposition from continuously decreasing current velocities, accompanied by a transition to dominant and subordinate tides of equal magnitude. Each cycle containing approximately 22 to 31 couplets is interpreted as reflecting the development from spring to neap tide within a semidiurnal tidal regime (Fig. 16a-c). This interpretation is in agreement with the tidal cyclicity of the exceptionally continuous Carboniferous example of non-burrowed tidal bedding described by Kvale *et al.* (1989) and the inferred overall semidiurnal tidal regime indicated by the presence of mud layer couplets.

In modern tidal environments, tidal bedding is generated on tidal flats and in shallow channels, but owing to frequent storm erosion the vertical end product never fully equals the actual depositional tidal cycle (Reineck and Wunderlich, 1969). The apparent continuity and regularity of each thinning-upward tidal cycle shown here suggests a minimum influence from storm erosion, and implies daily sedimentation within a sheltered and most likely shallow subtidal environment. Some two to three spring-neap tidal cycles within the thinner, 30 to 35 cm tichk type 2 burrowed layers, and perhaps as many as six tidal cycles within the 1 m thick amalgamated layers suggest that each burrowed and tidal bedded layer represents one to three months of daily deposition. This time perspective of the type 2 burrowed layers is consistent with a half yearly and seasonal development of the type 1 bioturbated layers. The microsequences containing the type 2 burrowed layers are correspondingly interpreted as representing a seasonal sedimentation from winter storms followed by a summer period of daily tidal and spring-neap tidal cycle accumulation in an abandoned tidal channel (Fig. 15).

Discussion of Backbarrier Dynamics

The first appearance of a bioturbated or tidal-bedded layer is here used as the demarcation of the active versus the abandoned channel fill, but an initial decrease of the energy of the tidal channel is suggested by the increasing density of burrows toward the top of the active channel fill (Figs. 10, 16d). The thicknesses of the corresponding A and B units indicate that the depth of the open channels probably never exceeded 7 to 8 m (Fig. 10).

The above interpretations are used to explain the driving mechanisms responsible for the development of a

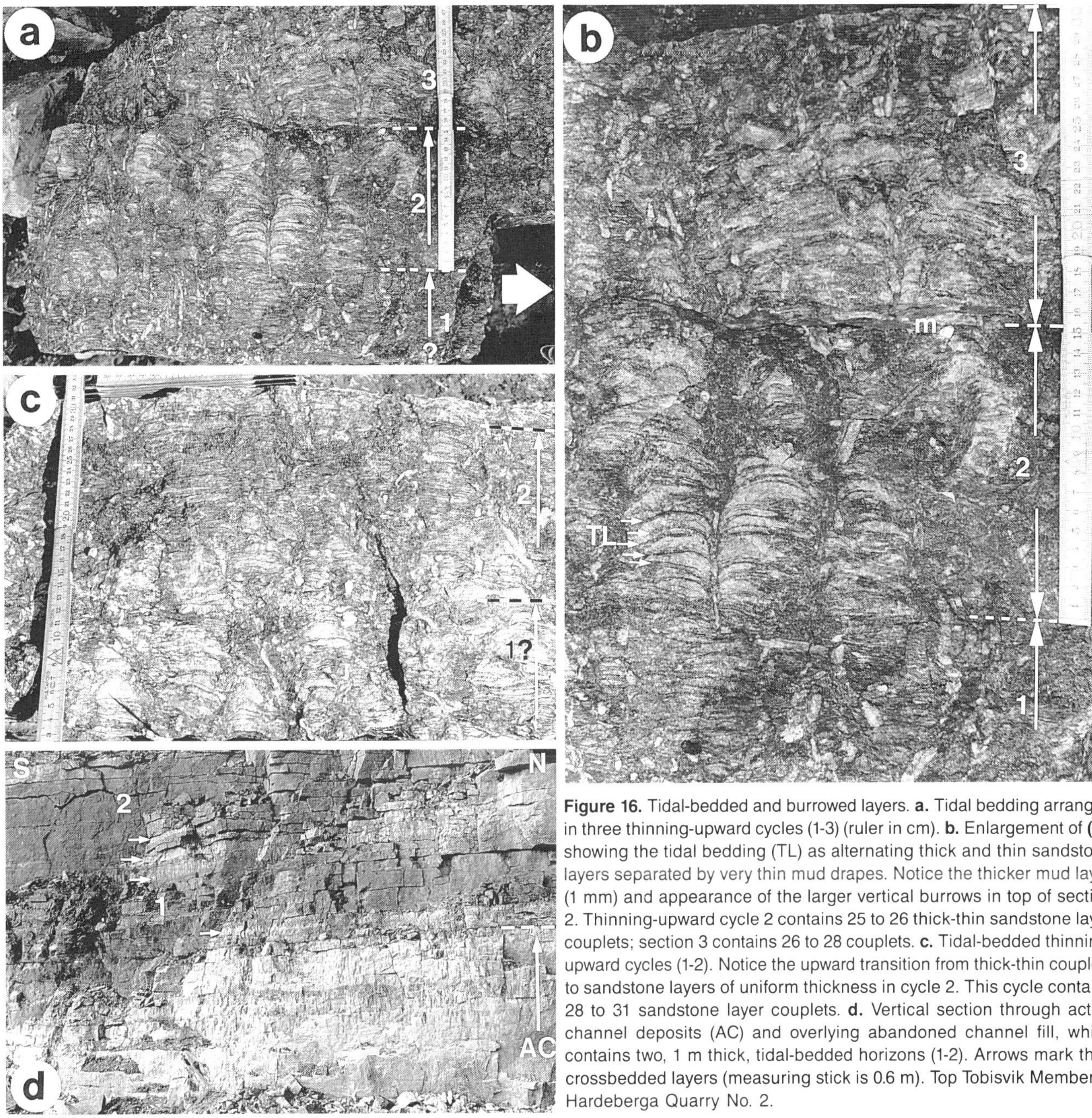

Figure 16. Tidal-bedded and burrowed layers. **a.** Tidal bedding arranged in three thinning-upward cycles (1-3) (ruler in cm). **b.** Enlargement of **(a)**, showing the tidal bedding (TL) as alternating thick and thin sandstone layers separated by very thin mud drapes. Notice the thicker mud layer (1 mm) and appearance of the larger vertical burrows in top of section 2. Thinning-upward cycle 2 contains 25 to 26 thick-thin sandstone layer couplets; section 3 contains 26 to 28 couplets. **c.** Tidal-bedded thinning-upward cycles (1-2). Notice the upward transition from thick-thin couplets to sandstone layers of uniform thickness in cycle 2. This cycle contains 28 to 31 sandstone layer couplets. **d.** Vertical section through active channel deposits (AC) and overlying abandoned channel fill, which contains two, 1 m thick, tidal-bedded horizons (1-2). Arrows mark thin, crossbedded layers (measuring stick is 0.6 m). Top Tobisvik Member at Hardeberga Quarry No. 2.

pronounced, coast-parallel, backbarrier drainage pattern, such that 1) the tidal inlet migrated actively, and 2) since the backbarrier sediments in general are finer grained than the fluvially derived deltaic sediments, the backbarrier region did not form part of the lower, marine-influenced stretch of a bent river. The supply by run-off from the mainland may have been rather small or absent, as it is along the Wadden Sea today (Veenstra, 1983).

There seem to exist at least two dynamic situations in which modern backbarrier drainage channels evolve into a pronounced coast-parallel pattern. One dynamic situation occurs when the backbarrier channels fall behind the faster migrating inlet, causing an updrift asymmetry of the drainage pattern and consequently a tendency towards coast parallelism (*cf.* Fitzgerald and Penland, 1987). Another dynamic situation has been described recently by Fitzgerald and Penland (1987) from the backbarrier region behind the East Friesian Islands, where prevailing unidirectional and shore-parallel winds cause migration of the drainage channels. The winds are responsible for the longshore drift and inlet migration. They also force the backbarrier tidal drainage pattern to evolve asymmetrically and coast-parallel in a downdrift direction, particularly in cases where barrier inlets migrate more slowly than backbarrier channels (Fitzgerald and

Penland, 1987). Where inlets migrate through sandy shoreface deposits, as in this Cambrian example, the migration occurs relatively quickly, but in addition, an increased content of sand in the backbarrier environment correspondingly enhances the migration of the backbarrier channels (Fitzgerald and Penland, 1987). In the Early Cambrian, no stabilizing vegetation of marsh existed and the backbarrier region seems to have been more sandy than modern counterparts. Therefore it is likely that the Cambrian backbarrier channels fully exposed to the prevailing wind would migrate faster than the tidal inlet.

According to Fitzgerald and Penland (1987) the latter mechanism seems to indicate the development of several modern sandy and nonvegetated backbarrier regions. This mechanism may therefore apply to the Precambrian and early Paleozoic settings as well, and is preferred as the main reason for the uniform, coast-parallel paleocurrents of the sub- to intertidal backbarrier channels of the Cambrian barrier system of the Hardeberga Formation (Fig. 17).

Fitzgerald and Penland (1987) demonstrated that the coast-parallel drift of the backbarrier channels is caused by headward erosion of the channels into the drainage divides and subsequent infilling of the adjacent channels on the opposite sides of the drainage divides. As a result, fine grained material is deposited on the drainage divides and in associated downdrift channels. This depositional mechanism may explain the presence of tidal-bedded abandoned channel fills of the Cambrian backbarrier deposits. Deposition probably took place subtidally on the drainage divides or in the sheltered heads of the downdrift channels. In addition, the confinement of the thicker tidal-bedded deposits to the tops of the backbarrier sediments, just below the ravinement surfaces (Fig. 10), may be explained as a "frozen picture" of the actual sediments of the drainage divides and infilled

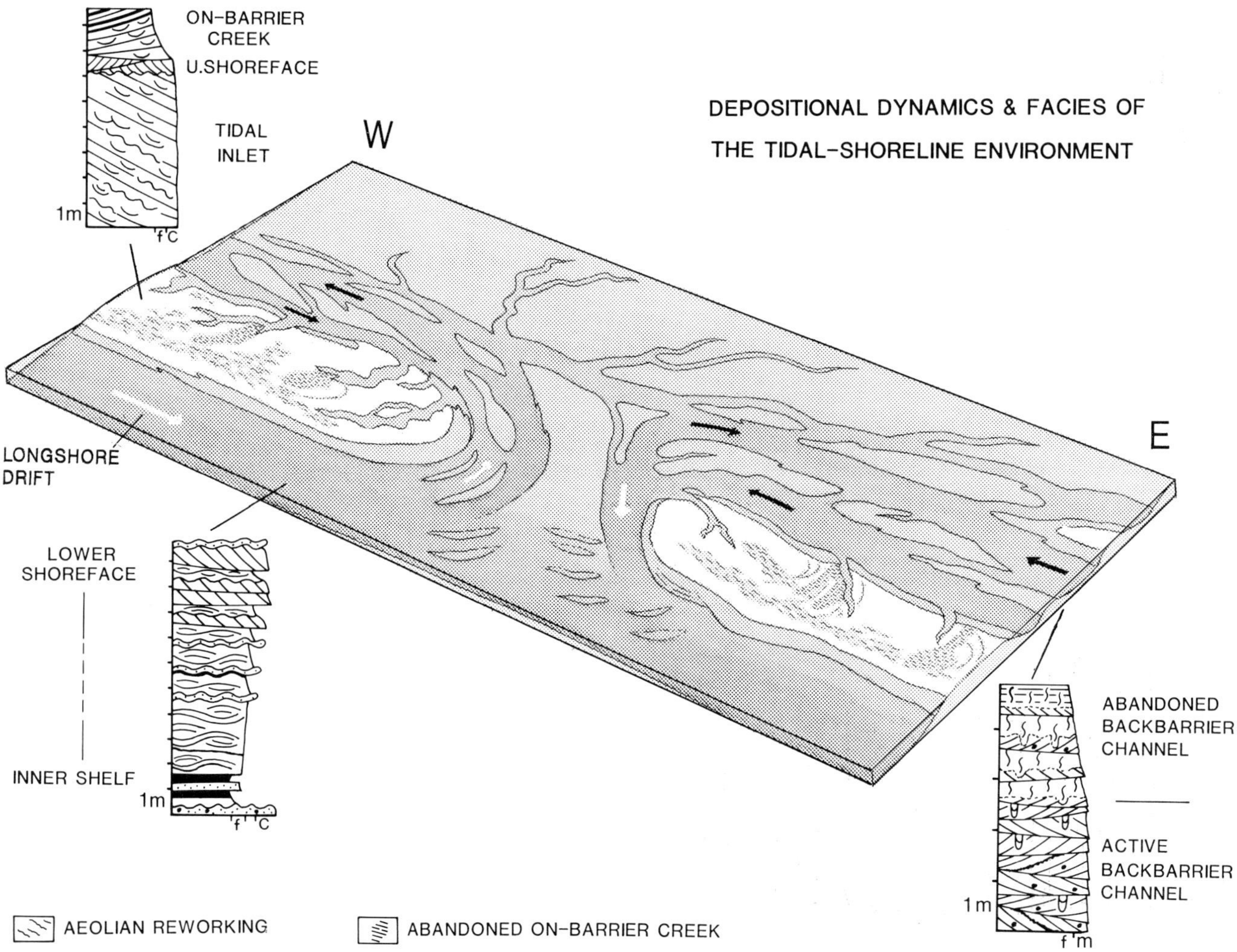

Figure 17. Depositional model of the Hardeberga Formation. The tide-influenced Early Cambrian coastline was developed as a barrier island shoreline characterized by an easterly longshore drift and tidal inlet migration. The unvegetated surfaces of the barrier islands were reworked by aeolian processes and laterally migrating on-barrier creeks, probably initiated in subaerial spit-bars. The barriers enclosed a sandy backbarrier system with coast-parallel tidal channels and shoals, which also eroded into sandy tidal flats. Persistent westerly winds resulted in active downdrift migration of the backbarrier channel system and abandonment of the updrift channels. The barrier islands may have been approximately 2 km wide. The generalized sedimentological logs represent the recorded facies and facies relationships of the shoreface, barrier island and backbarrier depositional environment (for legend, see Fig. 10).

channels. Elsewhere, these were eroded by the continuous coast-parallel migration of the sandy, backbarrier channel system. Fully bioturbated abandoned channel fills are clearly the most frequently abandoned channel facies and probably represent abandonment of the larger exposed channels within the up-drift tidal channel system. Here, abandonment was favoured by narrowing, reduced activity and sparse sedimentation, as the entire channel system migrated downwind and downdrift (Fig. 17). The monotonously repeated stacking of active and abandoned channel deposits may consequently be regarded as deposition from several generations of downdrift-migrating backbarrier channel systems during continuous sea level rise. This recurrent downdrift migration may also account for the sheet-like geometry of the backbarrier deposits and the absence of any flood-tidal delta deposits.

The ubiquitous mud drapes, mud clasts and tidal bedding suggest that mud was present in the backbarrier water column, and that muddy tidal flats should form. The lack of any muddy tidal-flat component may therefore be explained as a function of the generally higher wind and wave energy of the backbarrier region and absence of stabilizing vegetation known to be of importance in capturing the mud in modern barrier systems (J. Bartholdy, pers. comm.).

The mature and commonly well rounded appearance of sand grains from the Cambrian craton (Klein, 1977b; Dott *et al.*, 1986), is recognized in the Hardeberga Formation. Dott *et al.* (1986) clearly documented evidence indicating a marine redistribution of former aeolian deposits. This also seems to have been the situation during the initial transgression of the fluvio-aeolian deposits of the Neksø Formation on Bornholm. The former existence of likewise aeolian deposits on the Swedish parts of the Baltic Shield is verified by the presence of wind faceted clasts in the basal deposits (Hadding, 1929). The entire marine facies development of the Hardeberga Formation (Fig. 3) and the maturing of the sand grains outside the beach zone may be achieved by prolonged abrasive processes in the tidal zone (Klein, 1977b) and by the aeolian action on the barren, nonvegetated Cambrian barrier island surface.

Depositional Model of the Hardeberga Formation

The depositional environment of the Lower Cambrian Hardeberga Formation is summarized in Figure 17. The facies sequence, with a vertical development from inner shelf deposits to backbarrier tidal channels, represents progradation of a barrier island shoreline through longshore drift and accumulation of sand in the shallow shoreface and offshore zones (Fig. 17).

The Early Cambrian paleoshoreline of southern Sweden and Bornholm was orientated approximately east-west and most likely developed as a linear barrier island shoreline, intersected by rivers. The shoreline was characterized by uniform, eastward, longshore transport and continuous eastward lateral migration of tidal inlets (Fig. 17). The barren and nonvegetated barrier islands were reworked by laterally migrating creeks and also by aeolian processes (Fig. 17). Consequently, the barrier island deposits are distinguished as coarse grained units of crossbedded and wave rippled, inclined beds, 7 to 8 m high, representing the tidal inlet. These are erosionally overlain by upper shoreface bar and trough deposits of the barrier spit and a 1 to 2 m thick, fining-upward unit of epsilon cross-stratified, sandy to muddy deposits representing on-barrier creeks (Fig. 17).

The barrier islands enclosed a sandy backbarrier tidal environment with coast-parallel tidal channels divided by tidal shoals (Fig. 17). The tidal channel system may have been bordered landward by sandy tidal flats, frequently eroded by the laterally migrating tidal channels (Fig. 17). The backbarrier deposits are recognized as stacked, 3 to 10 m thick facies sequences of crossbedded, tidal channel fills overlain by several cycles of alternating crossbedded and bioturbated layers, representing seasonal sedimentation in abandoned tidal channels (Fig. 17). The seasonal sedimentation occurred as migration of minor bedform fields within the abandoned channels, activated by storm set-ups during the winter, and followed by colonization and subsequent bioturbation during calm summer periods (Fig. 15). Alternating crossbedded and burrowed tidal bedded layers represent a similar deposition governed by seasonal fluctuations of energy on the drainage divides or in sheltered abandoned tidal channels.

A record of seasonally controlled sedimentation is consistent with the inferred paleogeographic location on the higher latitudes of the southern hemisphere suggested from paleomagnetic data (Scotese *et al.*, 1979; Piper, 1985).

The overall sheet-like geometry of the Hardeberga Formation may be explained as being a result of the suggested predominant longshore drift and migration of both the tidal inlets and backbarrier channel systems, as well as an effect of the flat morphology of the underlying peneplained crystalline basement.

The progradation of the barrier islands took place within an overall transgressive phase, and the Hardeberga Formation contains three, stacked, progradational facies sequences (Figs. 3, 4). According to Davies and Clifton (1987), similar stacked progradational facies sequences indicate conditions of slow sea level rise and rapid sedimentation. The termination of each progradation by a ravinement surface (Figs. 3, 4), suggests a phase of reduced sediment supply and erosional retreat of the barrier island shoreface in response to the rising sea level (Van Straaten, 1965; Davies and Clifton, 1987).

The original presence of barrier islands, reworked by onbarrier creeks and enclosing a well developed, coast-parallel tidal drainage system similar to the Wadden Sea, suggests deposition under mesotidal or microtidal conditions (Hayes, 1975; Oertel, 1985).

This barrier island and tidal backbarrier facies model may be applicable to Upper Precambrian and Cambrian sandy tidalites, which possess a coast-parallel tidal paleocurrent pattern.

Conclusions

It may be concluded that:

1. The Hardeberga Formation represents stacked deposits of barrier islands and backbarrier tidal systems.

2. The backbarrier tidal deposits show a pronounced coast-parallel paleocurrent pattern and were developed as a system of coast-parallel tidal channels divided by sand bars.

3. Each channel succession consists of an active phase influenced by tidal currents and an abandoned phase influenced by seasonal deposition.

4. The channelized system was separated by drainage dividing areas represented by tidal-bedded deposits that document deposition from seasonal fluctuations of the energy, neap-spring sedimentation and finally, a daily tidal sedimentation in a semidiurnal system not very different from the present one.

Acknowledgments

This paper is based on M.Sc. thesis work carried out by the author at the University of Copenhagen. I am especially indebted to G. K. Pedersen (who supervised the study) and to H. Tirsgaard, J. Bartholdy, C. F. Hansen and F. Surlyk for valuable discussion and critical comments on an earlier version of this paper. I am grateful to O. B. Berthelsen for the photographic illustrations and R. Madsen for the drawings.

References

Allen, J. R. L. 1965. Fining upwards cycles in alluvial successions. Geological Journal, v. 4, p. 229-246.

______ 1970 Studies in fluviatile sedimentation: a comparison of fining-upwards cyclothems, with special reference to coarse-member composition and interpretation. Journal of Sedimentary Petrology, v. 40, p. 298-323.

Berg, J. H. Van Den 1981. Rhythmic seasonal layering in a mesotidal channel fill sequence, Oosterschelde Mouth, the Netherlands. International Association of Sedimentologists, Special Publication 5, p. 147-159.

______ 1982. Migration of large-scale bedforms and preservation of cross-bedded sets in highly accretional parts of tidal channels in the Oosterschelde, SW Netherlands. Geologie en Mijnbouw, v. 61, p. 253-263.

Bergström, J. 1981. Lower Cambrian shelly faunas and biostratigraphy in Scandinavia. *In:* Short papers for the 2nd International Symposium on the Cambrian System, M. E. Taylor (ed.). p. 22-25.

______ and Gee, D. G. 1985. The Cambrian in Scandinavia. *In:* The Caledonide Orogen Scandinavia and Related Areas, D. G. Gee and B. A. Sturt (eds.). Wiley, p. 247-271.

______ , Holland, B., Larsson, K., Norlyng, E. and Sivhed, U. 1982. Guide to excursions in Scania. Sveriges geologiska Undersökning, serie Ca. 54, p. 1-95.

Bernard, H. A. and Leblanc, R. J. 1965. Resume of the Quarternary geology of the northwestern Gulf of Mexico Province. *In:* The Quarternary of the United States, H. E. Wright, Jr. and D. G. Frey (eds.). Princeton, N.J., Princeton University Press, p. 137-185.

Boersma, J. R. 1969. Internal structures of some tidal megaripples on a shoal in the Westerschelde Estuary, The Netherlands. Geologie en Mijnbouw, v. 48, p. 409-414.

______ and Terwindt, J. H. J. 1981a. Neap-spring tide sequences of intertidal shoal deposits in a mesotidal estuary. Sedimentology, v. 28, p. 151-170.

______ and ______ 1981b. Berms on an intertidal shoal: Shape and internal structure. International Association of Sedimentologists, Special Publication 5, p. 39-49.

Boothroyd, I. C. and Hubbard, D. K. 1975. Genesis of bedforms in mesotidal estuaries. *In:* Estuarine research II, L. E. Cronin (ed.). p. 217-234, London.

Bridge, J. S. and Diemer, J. A. 1983. Quantitative interpretation of an evolving ancient river system. Sedimentology, v. 30, p. 599-623.

Bridges, P. H. and Leeder, M. R. 1976. Sedimentary model for intertidal mudflat channels, with examples from the Solway Firth, Scotland. Sedimentology, v. 23, p. 535-552.

Bylund, G. 1984. Paleomagnetic aspects on the Tornquist Line, in summaries of talks. Geologiska Freningens i Stockholm Forhandlingar, v. 106, p. 379-397.

Carey, D. A. 1987. Sedimentological effects and paleoecological implications of the tube-building polychaete *Lanice conchilega* Pallas. Sedimentology, v. 34, p. 49-66.

Carter, C. H. 1978. A regressive barrier and barrier-protected deposits: depositional environments and geographic setting of the Late Tertiary Cohansey sand. Journal of Sedimentary Petrology, v. 48, p. 933-950.

Cocks, L. R. M. and Fortey, R. A. 1982. Faunal evidence for oceanic separations in the Paleozoic of Britain. Journal of the Geological Society of London, v. 139, p. 467-480.

Cram, J. M. 1979. The influence of continental shelf width on tidal range: paleoceanographic implications. Journal of Geology, v. 87, p. 441-447.

Cromwell, J. E. 1971. Barrier coast distribution: a world-wide survey. *In:* Barrier Islands: Benchmark Papers in Geology, M. J. Schwartz (ed.). Dowden, Hutchinson and Ross, Stroudsburg, Pennsylvania, 408 p.

Dam, G. and Clemmensen, L. B. 1988. Nekso Sandsten fra Nedre Kambrium. Varv, v. 3, p. 84-88.

Davies, R. A., Jr. and Clifton, H. E. 1987. Sea-level change and the preservation potential of wave-dominated and tide-dominated coastal sequences.

De Mowbray, T. 1983. The genesis of lateral accretion deposits in Recent intertidal mudflat channels, Solway Firth, Scotland. Sedimentology, v. 30, p. 425-435.

______ and Visser, M. J. 1984. Reactivation surfaces in subtidal channel deposits, Oosterschelde, southwest Netherlands. Journal of Sedimentary Petrology, v. 54, p. 811-824.

Dott, R. H., Jr. and Byers, C. W. 1981. SEPM research conference on modern shelf and ancient cratonic sedimentation, the orthoquartzite-carbonate suite revisited. Journal of Sedimentary Petrology, v. 51, p. 329-347.

______ and Bourgeois, J. 1982. Hummocky stratification: Significance of its variable bedding sequences. The Geological Society of America, Bulletin, v. 93, p. 663-680.

______ , ______, Fielder, G. W., Stenzel, S. R. and Winfree, K. E. 1986. Aeolian to marine transition in Cambro-Ordovician cratonic sheet sandstones of the northern Mississippi valley, U.S.A. Sedimentology, v. 33, p. 345-367.

Duc, A. W. and Tye, R. S. 1987. Evolution and stratigraphy of a regressive barrier-backbarrier complex: Kiawah Island, South Carolina. Sedimentology, v. 34, p. 237-251.

Elliott, T. and Gardiner, A. R. 1981. Ripple, megaripple and sandwave bedforms in the macrotidal Loughor Estuary, South Wales, U.K. The International Association of Sedimentologists, Special Publication 5, p. 51-64.

EUGENO-S Working Group, 1988. Crustual structure and tectonic evolution of the transition between the Baltic Shield and the North German Caledonides (the EUGENO-S Project). Tectonophysics, v. 150, p. 253-348.

Fitzgerald, D. M. and Penland, S. 1987. Backbarrier dynamics of the East Friesian Islands. Journal of Sedimentary Petrology, v. 57, p. 746-754.

Fleming, B. W. and Antia, E. E. 1989. On the origin of nearshore ridge and swale topography along the southern mesotidal North Sea coast (abstract): Second International Research Symposium on clastic tidal deposits, Calgary, Canada, p. 34.

Fürsich, F. T. 1974. On *Diplocraterion* Torell 1870 and the significance of morphological features in vertical, spreiten-bearing, U-shaped trace fossils. Journal of Paleontology, v. 48, p. 952-962.

Hadding, A. 1929. The Pre-Quarternary sedimentary rocks of Sweden III. The Paleozoic and Mesozoic sandstones of Sweden. Meddelande Lunds Universitetet Geologisk-Mineralogisk Institution, v. 41, 287 p.

Hamberg, L. 1988. Tidal and storm dominated shallow marine depositional systems related to the transgression of the Lower Cambrian Hardeberga Formation, Scania. (in Danish). Unpublished M.Sc. thesis, University of Copenhagen, 127 p.

Hanisch, J. 1981. Sand transport in the tidal inlet between Wangerooge and Spiekeroog (W. Germany). The International Association of Sedimentologists, Special Publication 5, p. 175-185.

Hayes, M. O. 1975. Morphology of sand accumulation in estuaries: an introduction to the symposium. *In:* Estuarine Research, L. G. Cronin. Geology and Engineering Academic Press, Inc., v. 2, p. 3-22.

______ 1980. General morphology and sediment patterns in tidal inlets. Sedimentary Geology, v. 26, p. 139-156.

Heward, A. P. 1981. A review of wave-dominated clastic shoreline deposits. Earth Science Review, v. 17, p. 223-276.

Hine, A. C. and Boothroyd, J. C. 1978. Morphology, processes and recent sedimentary features of a glacial-outwash plain shoreline, southern Iceland. Journal of Sedimentary Petrology, v. 48, p. 901-920.

Hoyt, J. H. and Henry, V. J. 1967. Influence of island migration on barrier island sedimentation. The Geological Society of America, Bulletin, v. 78, p. 77-86.

Hunter, R. E. and Clifton, H. E. 1982. Cyclic deposits and hummocky cross-stratification of probable storm origin in Upper Cretaceous rocks of the Cape Sebastian area, southwestern Oregon. Journal of Sedimentary Petrology, v. 52, p. 127-143.

______ , Dingler, J. R., Anima, R. J. and Richmond, B. M. 1988. Coarse-sediment bands on the inner shelf of southern Monterey Bay, California. Marine Geology, v. 80, p. 81-98.

Klein, G. De V. 1970. Tidal origin of a Precambrian quartzite the lower fine-grained quartzite (Dalradian) of Islay, Scotland. Journal of Sedimentary Petrology, v. 40, p. 973-985.

______ 1977a. Tidal circulation model for deposition of clastic sediments in epeiric and mioclinal shelf seas. Sedimentary Geology, v. 18, p. 1-12.

______ 1977b. Clastic Tidal Facies. Continuing Education Publication Company, Champaign, Illinois, 149 p.

______ and Ryer, T. A. 1978. Tidal circulation patterns in Precambrian, Paleozoic and Cretaceous epeiric and mioclinal shelf seas. Geological Society of America, Bulletin, v. 89, p. 1050-1058.

Kumar, N. and Sanders, J. E. 1974. Inlet sequence: a vertical succession of sedimentary structures and textures created by lateral migration of tidal inlets. Sedimentology, v. 21, p. 491-532.

Kvale, E. P., Archer, A. W. and Johnson, H. R. 1989. Daily, monthly, and yearly tidal cycles within laminated siltstones of the Mansfield Formation (Pennsylvanian) of Indiana. Geology, v. 17, p. 365-368.

Leckie, D. A. 1988. Wave-formed, coarse-grained ripples and their relationship to hummocky cross-stratification. Journal of Sedimentary Petrology, v. 58, p. 607-622.

______ and Krystinik, L. F. 1989. Is there evidence for geostrophic currents preserved in the sedimentary record of inner to middle-shelf deposits? Journal of Sedimentary Petrology, v. 59, p. 862-870.

______ and Walker, R. G. 1982. Storm and tide-dominated shorelines in Cretaceous Moosebar, Lower Gates Interval outcrop equivalents of Deep Basin gas trap in western Canada: The American Association of Petroleum Geologists, Bulletin, v. 66, p. 138-157.

Leithold, E. L. and Bourgeois, J. 1984. Characteristics of coarse grained sequences deposited in nearshore, wave-dominated environments - examples from the Miocene of southwest Oregon. Sedimentology, v. 31, p. 749-775.

Lundström, M. and Staude, H. 1971. Beitrag zur Stratigraphie den Underkambrischen Sandsteine des sdlichsten Skandinavien. Geologica et Palontologica, v. 5, p. 1-8.

Matthews, S. C. and Cowie, J. W. 1979. Early Cambrian transgressions. Journal of the Geological Society of London, v. 136, p. 133-135.

Meulen, S. Van Der 1982. The sedimentary facies and setting of Eocene point bar deposits, Monllobat Formation, Southern Pyrenees, Spain. Geologie en Mijnbouw, v. 61, p. 217-227.

Niedoroda, A. W., Swift, D. J. P. and Hopkins, T. S. 1985. The shoreface. *In:* Coastal Sedimentary Environments, R. A. Davies (ed.). Springer-Verlag, Berlin, p. 533-624.

Nielsen, L. H., Johannesen, P. N. and Surlyk, F. 1988. A Late Pleistocene coarse-grained spit-platform sequence in northern Jylland, Denmark. Sedimentology, v. 35, p. 915-937.

Nio, S-D. and Yang, C. S. 1989. Recognition of tidally-influenced facies and environments. International Geoservices, Short Course Note Series 01, 200 p.

Nio, S-D., Van Den Berg, J. H. Goesten, M. and Smulders, F. 1980. Dynamics and sequential analysis of a mesotidal shoal and intershoal channel complex in the eastern Scheldt (Southwestern Netherlands). Sedimentary Geology, v. 26, p. 263-279.

Nummedal, D. and Penland, S. 1981. Sediment dispersal in Norderneyer Seegat, West Germany. International Association of Sedimentologists, Special Publication 5, p. 187-210.

Oertel, G. F. 1985. The barrier island system. Marine Geology, v. 63, p. 1-18.

Piper, J. D. A. 1985. Continental movements and breakup in Late Precambrian-Cambrian times: prelude to Caledonian orogenesis. *In:* The Caledonide Orogen -Scandinavia and related Areas, D. G. Gee and B. A. Sturt (eds.). Wiley, New York, p. 19-34.

Poulsen, C. 1966. Cambro-Silurian stratigraphy of Bornholm: Meddelelser Dansk Geologisk Forening, v. 16, p. 117-137.

Reddering, J. S. V. 1983. An inlet sequence by migration of a small microtidal inlet against longshore drift: the Keurbooms Inlet, South Africa. Sedimentology, v. 30, p. 201-218.

______ 1967. Subtidal occurrences of ladder-back ripples: their significance in palaeo-environmental reconstruction. Sedimentology, v. 34, p. 253-257.

Reineck, H-E. 1958. Longitudinale Schrgschicken im Watt. Geologische Rundschau, v. 47, p. 73-82.

______ 1967. Layered sediments of tidal flats, beaches and shelf bottoms of the North Sea. *In:* Estuaries, G. E. Lauff (ed.). The American Association of Advanced Scientists, Publications, v. 83, p. 191-206.

______ and Wunderlich, F. 1969. Die Entstehung von Schichten im Watt. Senchenbergiana Maritima, v. 1, p. 85-106.

Rhoads, D. C. 1967. Biogenic reworking of intertidal and subtidal sediments in Barnstable Harbor and Buzzards Bay, Massachusetts. Journal of Geology, v. 75, p. 461-476.

Scotese, C. Bambach, R. K., Barton, C., Voo, R. Van Der and Ziegler, A., 1979. Paleozoic base maps. Journal of Geology, v. 87, p. 217-278.

Smith, D. G. 1987. Meandering river point bar lithofacies models: modern and ancient examples compared. The Society of Economic Paleontologists and Mineralogists, Special Publication 39, p. 75-91.

Straaten, L. M. J. U. Van 1965. Coastal barrier deposits in South-and North-Holland. *In:* Barrier islands: Benchmark Papers in Geology, M. L. Schwartz (ed.). Doweden, Hutchinson and Ross, Inc., 1973, v. 9, p. 171-217.

Surlyk, F. 1980. Denmark. *In:* Geology of the European countries, Denmark, Finland, Iceland, Norway, Sweden, 1-50. Dunod. Published in cooperation with the Comité National Français de Géologie (C.N.F.G.) on the occasion of the 26th International Geological congress.

Swift, D. P. 1988. Palaeo-tidal reconstructions of inshore tidal depositional environments. *In:* Tide-influenced sedimentary Environments and Facies: Sedimentology and Petroleum Geology, P. L. Boer, A. Van Gelder and S-D. Nio (eds.). Reidel, p. 233-263.

______ and Nummedal, D. 1987. Hummocky cross-stratification, tropical hurricanes and intense water storms (Discussion). Sedimentology, v. 34, p. 338-344.

Unsöld, G. 1974. Jahreslagen und Aufwachsraten in Schlicksedimenten eines künstlichen, gezeitenoffenen Sedimentationsbeckens (Wattgebiet südlich Nordstrand/Nordfriesland). Meyniana, v. 26, p. 103-111.

Veenstra, H. J. 1983. Introduction to the geomorphology of the Wadden Sea area. *In:* Ecology of the Wadden Sea, W. J. Wolff (ed.). A. A. Balkema, Rotterdam, report 1, p. 8-19.

Vidal, G. 1981. Micropaleontology and biostratigraphy of the Lower Cambrian sequences in Scandinavia. *In:* Short papers for the 2nd International Symposium on the Cambrian system, M. E. Taylor (ed.). The United States Geological Survey, open-file report 81-743, p. 232-235.

Visser, M. J. 1980. Neap-spring cycles reflected in Holocene subtidal large-scale bedform deposits: a preliminary note. Geology, v. 8, p. 543-546.

Wright, M. E. and Walker, R. G. 1981. Cardium Formation (U. Cretaceous) at Seebe, Alberta: storm-transported sandstones and conglomerates in shallow marine depositional environments below fair weather wave base. Canadian Journal of Earth Science, v. 18, p. 795-809.

Yorath, C. J., Bornhold, B. D. and Thomson, R. E. 1979. Oscillation ripples on the northeast Pacific continental shelf. Marine Geology, v. 31, p. 45-58.

KEY WORDS: LOWER CAMBRIAN HARDEBERGA FORMATION, SOUTHERN SWEDEN, BALTIC SHIELD, SEASONAL CYCLES, TIDAL CYCLES

In: Clastic Tidal Sedimentology, D. G. Smith, G. E. Reinson, B. A. Zaitlin and R. A. Rahmani (eds.). Canadian Society of Petroleum Geologists. Memoir 16, p. 275-282.

Time-series study of mesoscale tidal bedforms, Martens Plate, Wadden Sea, Germany

RICHARD A. DAVIS, JR.

Department of Geology, University of South Florida Tampa 33620

B. W. FLEMMING

Senckenburg Institute Wilhelmshaven 2940, Federal Republic of Germany

ABSTRACT

Continuous bathymetric profiles, side-scan sonar surveys, and current velocity measurements were taken over a complete lunar tidal cycle through a tidal channel system in the German Wadden Sea near the islands of Spiekeroog and Wangeroog. Data were collected at low, mid-flood, high and mid-ebb stages under neap, mean and spring tidal conditions. A broad spectrum of medium- and large-scale bedforms (dunes) displays a range in size, shape and orientation that is related to specific tidal conditions. The nature of bottom configuration, time-velocity relationships and tidal stage show that although there are patterns displayed at individual locations, there is great geographic and temporal variation.

Irregularities in the channel floor caused by various scales of bedforms provide shielding from tidal currents during given phases of the tidal cycle. This shielding inhibits deterioration and may prevent reorientation of bedforms during tidal reversals. Locations only hundreds of metres apart may display both ebb- and flood-dominated bedforms. Height of bedforms ranges from tens of centimetres to greater than two metres at a single location between neap and spring tide conditions.

Velocity profiles show a log-normal distribution, with the accelerating phase displaying a steeper slope than the decelerating phase. Greatest shear velocities occur on the decelerating phase of the tidal cycle and bed roughness values peak near the end of the decelerating tidal currents.

RÉSUMÉ

Dans un système de chenaux de marée près des îles Spiekeroog et Wangeroog dans la mer Wadden, Allemagne, des profils bathymétriques continus, des relevés de sonar à balayage latéral et des mesures de vélocité de courant ont été pris durant un cycle lunaire de marée complet. Les données ont été prélevées à marée basse, moyenne et haute, et à marée descendante en position médiane, ceci sous des conditions de faibles, moyennes et grandes marées. Une grande gamme de «morphologies de fond» (dunes) à moyenne et grande échelle montre une variation en dimensions, formes et orientations qui sont reliées à des conditions spécifiques de marée. Malgré les modèles obtenus à des localités individuelles, la nature de la configuration du fond, la relation temps-vélocité et les stades de la marée démontrent une grande variation géographique et temporelle.

Les irrégularités du fond des chenaux, causées par les différentes dimensions des «morphologies de fond», fournissent une protection contre les courants de marée durant certaines phases du cycle de la marée. Ces écrans de protection empêchent la détérioration et peut-être la réorientation des «morphologies de fond» durant les inversions de marée. À une même localité, la hauteur des lits varie de quelques dizaines de centimètre pour des faibles marées, à plus de deux mètres pour des conditions de syzygie.

Les profils de vélocité donnent une distribution log-normale avec une pente plus forte pour la phase d'accélération que pour celle de décélération. Les plus grandes vitesses de cisaillement s'enregistrent durant la phase de décélération du cycle de marée, et les valeurs de rugosité des lits culminent près de la fin de la décélération des courants de marée.

INTRODUCTION

Time-series studies of bedforms in the natural tidal environment represent the best way to understand fundamental process-response mechanisms and their relative contributions to the local depositional record. Whereas numerous studies deal with momentary or time-averaged sediment transport phenomena in tide-dominated environments (*e.g.*, Collins *et al.*, 1981; Yang, 1986), comparatively few studies have concentrated on temporal aspects of depositional processes within the time frame of the diurnal and/or lunar tidal cyclicity. In this context various studies have investigated bedform migration or intertidal sand bodies (*e.g.*, Dalrymple *et al.*, 1978; Lambiase, 1980) which in a few cases may have been continued over several tidal cycles (*e.g.*, Boothroyd, 1985).

To date, no comprehensive time-series studies on bedform development and behavior have been published from Holocene subtidal environments. Interpretation and recognition of distinct tidal signatures in the stratigraphic record over the past ten years (Visser, 1980; van den Berg, 1982; Nio *et al.*, 1983; Allen and Homewood, 1984; Yang and Nio, 1985; Kreisa and Moiola, 1986) have demonstrated the need for the study of comparable modern analogues. In response to this demand, the present study concentrates on the time-dependent nature of mesoscale bedforms in a subtidal channel system of the German Wadden Sea (Fig. 1). Incorporated

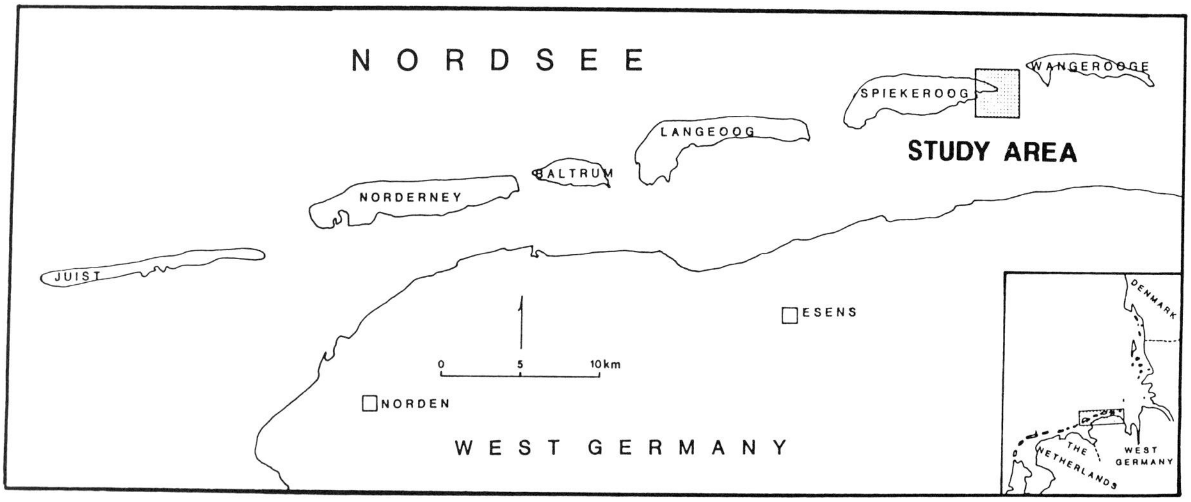

Figure 1. Location map of study area in the East Friesian Islands and the German Wadden Sea.

are coordinated foathometer and side-scan surveys, which have been supplemented by velocity profile measurements and sediment sampling at several locations in the channel system. The 16-day field program included the primary constituents of a complete lunar tidal cycle: spring, mean and neap conditions.

The main objectives of this study were: 1) to evaluate bedform scale and orientation, and hence the mode of cross-stratification, in order to aid in the interpretation of cores recovered from adjacent intertidal sand bodies, and 2) to obtain a clearer picture of the process-response mechanisms responsible for bedform variability observed over the course of both diurnal and lunar tidal cycles.

Location and Regional Setting

The southern North Sea shows a progressive transition from a microtidal, wave-dominated coast at Den Helder in The Netherlands, through a mixed energy, mesotidal coast along the Friesian Islands (Fig. 1), to a tide-dominated coast at the apex of the German Bight. The same trend is repeated from there to the Danish coast to the north. The East Friesian Islands of the German coast represent a high and mixed energy, mesotidal coast characterized by short drumstick barrier islands and wide inlets. Wave energy along the North Sea coast is high with a mean annual significant wave height of 1.6 m off the island of Norderney and storm waves up to 4.8 m annually (Dette, 1977). The net littoral drift along this reach is from west to east (Nummedal and Penland, 1981). Spring tidal range is above 3 m (Fig. 2) and tidal prisms are huge because of the large area of the Wadden Sea. The approximate tidal prism for the Harle Inlet is 1.8×10^8 m^3.

The island of Spiekeroog is the second island from the east in the East Friesian Islands (Fig. 1). It is separated from the adjacent island to the east, Wangeroog, by the Harle Inlet. The site of the study reported herein is just landward of the inlet and the eastern portion of Spiekeroog (Fig. 3). This general area is the focus of an ongoing coastal morphodynamics project at the Senckenberg Institute. The aspect of the overall project that is of concern in this report is the process-response conditions which develop bedforms and the time-series character of these bedforms in the tidal channels of Martens Plate and Muschelbank (Fig. 4).

Historical data show that Martens Plate has grown substantially in a seaward direction over the past few centuries and that the tidal channels have also moved over a broad area of the Wadden Sea. In fact, the channels exhibit movement of hundreds of metres in only a few decades (Reineck, 1958).

Methods

The study period was chosen to span a complete lunar tidal cycle. Within this time span a quasi-synoptic sampling and measurement program was carried out, which included current profile measurements, side-scan sonar and fathometer surveys, as well as sediment sampling at spring, mean and neap tidal conditions. For logistical reasons the velocity profiles were measured one or two days removed from the fathometer and side-scan surveys.

Six stations located in two tidal channels were occupied over the appropriate tidal cycles during June, 1988. These

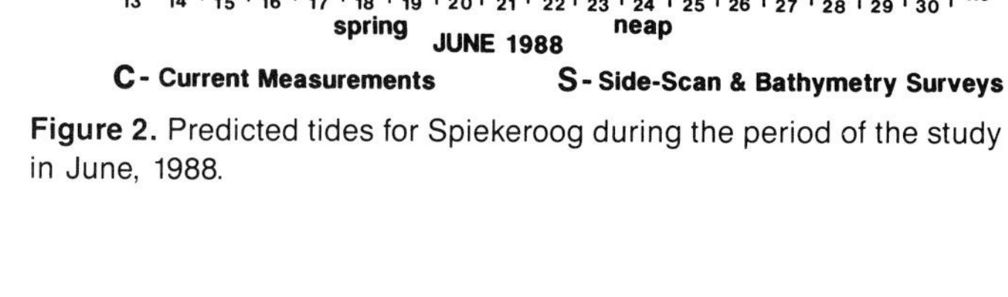

Figure 2. Predicted tides for Spiekeroog during the period of the study in June, 1988.

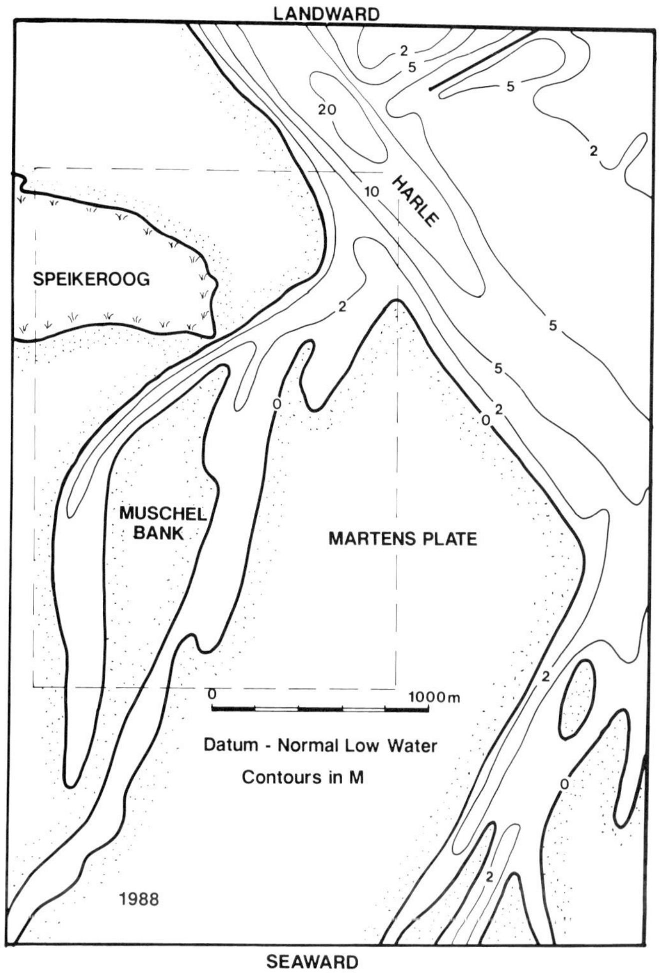

Figure 3. General morphology of study area near the island of Spiekeroog. Dashed rectangle indicates area of bathymetry shown in Figure 4.

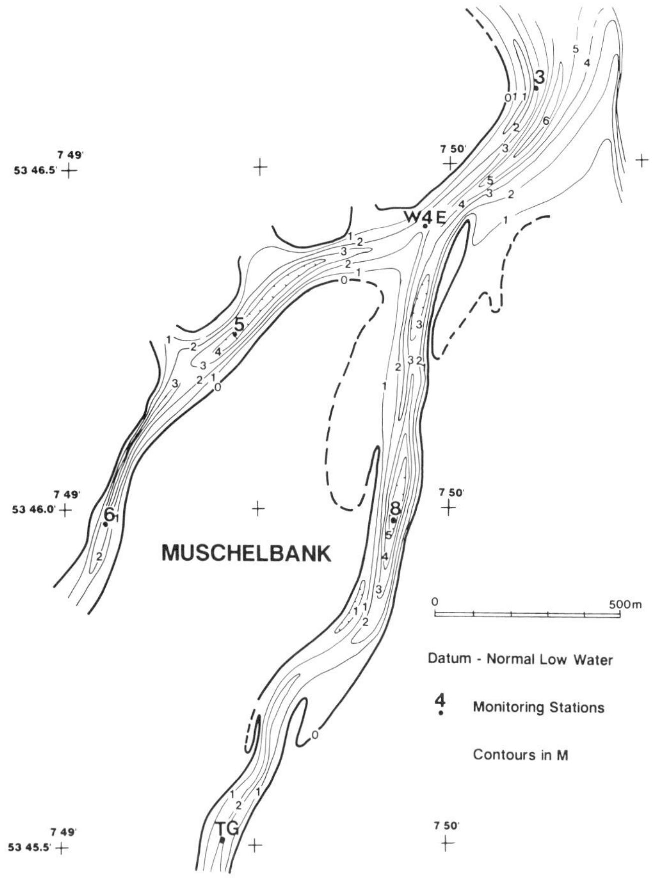

Figure 4. Bathymetry of tidal channels in the study area showing locations where velocity profiles were measured and sediment samples were collected.

included one in each channel (W and E) near station 4 (Fig. 4). Fathometer traces were recorded continuously along the channels at high, mid-ebb, low and mid-flood conditions during spring, mean and neap conditions. Each of the six sites was occupied as often as possible during a 12.5 hour cycle. This typically produced 12 to 14 current measurements at each site over the tidal cycle.

Velocity profiles were measured at each station using a General Oceanics direct readout flowmeter with readings taken at 0.5 m above the bed and at 1.0 m intervals to the surface. The position of the lowest reading relative to the crest or trough of a bedform could not be determined. In addition, one bottom grab sample of sediment was collected during the flood and ebb stage during spring tidal conditions. These samples were separated into gravel, sand and mud fractions and the sand fraction was analyzed in the settling tube at the Senckenberg Institute.

The Furuno Model FE-6200 recording fathometer was used limiting bedform resolution to those greater than about 3 m in spacing and wave heights of 20 cm. The tidal channels were traversed along a course marked by buoys in an effort to reoccupy the same position on successive tidal stages. Side-scan traces indicate that the bedforms are rather two-dimensional, which permits some deviation from the course without significant change in bedform parameters.

Results

Although all locations were surveyed, only two adjacent sites will be considered in detail in this report (#3 and #4W, (Fig. 4) because the observations at these sites can be considered as representative of the general patterns at all locations.

Bedforms

Overall bedform dimensions in the study area range from below resolution limits of the equipment to spacing over 40 m and heights over 2 m (Fig. 5). The large feature on Figures 5 and 6, which extends landward from location #3 and upon which bedforms are superimposed, is probably also a bedform, a huge sand wave. In most cases the bedforms exhibit sufficient asymmetry to identify the direction of flow responsible for their development and migration.

Fathometer traces were recorded during spring, mean and neap tidal stages in order to consider the differences in bedforms as a response to variation in tidal prism and therefore

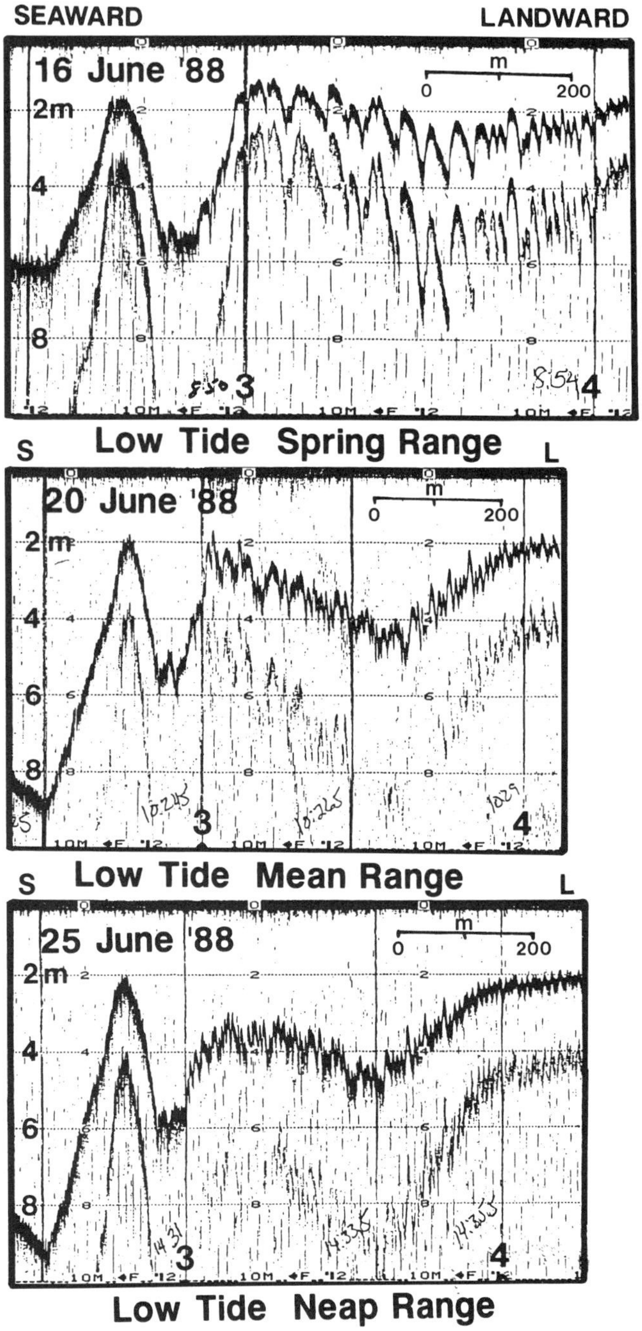

Figure 5. Fathometer traces between locations 3 and 4W at low tide during various conditions of the lunar cycle.

current velocity. The channels are distinctly ebb-dominated and therefore an example was chosen at low tide in order to consider the probable best bedform development. The three traces shown in Figure 5 display marked changes through this lunar cycle with a direct correlation between tidal prism and size of bedform. Spring conditions produced bedforms with a maximum size near 2 m in height and 40 m in length. The same site during neap conditions showed a maximum height of 0.8 m and length of 12 m. Mean values fall in between (Fig. 5).

The variation displayed at these sites over only a single tidal cycle is also considerable. The example of one cycle during spring conditions shows a range from 0.30 m in wave height and 5 m in length during high tide to 2.0 m in height and 40 m length during low tide.

Almost without exception, the bedforms display an ebb orientation (Fig. 6). The ebbing cycle, culminating with low tide conditions, produces the largest and best developed bedforms whereas the flood tide cycle results in much smaller bedforms. This is to be expected because of the ebb-domination of the tidal channels and the strong channelization of ebbing tidal currents. Flood orientation is only achieved at location 3 during high tide, and at location 4 the bedforms are nearly symmetrical on the fathometer trace (Fig. 6).

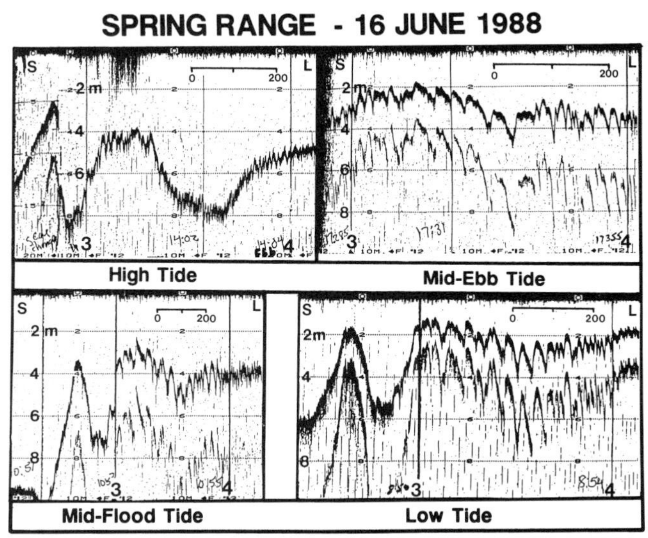

Figure 6. Fathometer traces between locations 3 and 4W at various tidal stages during spring conditions. Seaward (S) and landward (L) directions are indicated.

Bedform measurements from the fathometer traces at each location are summarized on a time-series format that includes tidal stage, bedform orientation, spacing height and index (Figs. 7, 8). Regression lines for each of the measured parameters are also included to display overall trends.

The trends are similar at both locations 3 and 4W. Water depth fluctuates with tidal stage and the bedform size responds but does so in a different manner at each site. Orientation remains in the ebb direction throughout the cycle at location 4W but does reverse at high tide only at location 3.

The wavelengths of the bedforms show considerable variation over the lunar cycle between the two locations; wavelength decreases greatly at location 3 but is constant at location 4W. This is a response to the greater tidal prism and greater current velocities at location 3. There is more variation in bedform height, but at station 4W this variation is limited to spring conditions (Figs. 7, 8).

All bedform parameters, except the bedform index at lo-

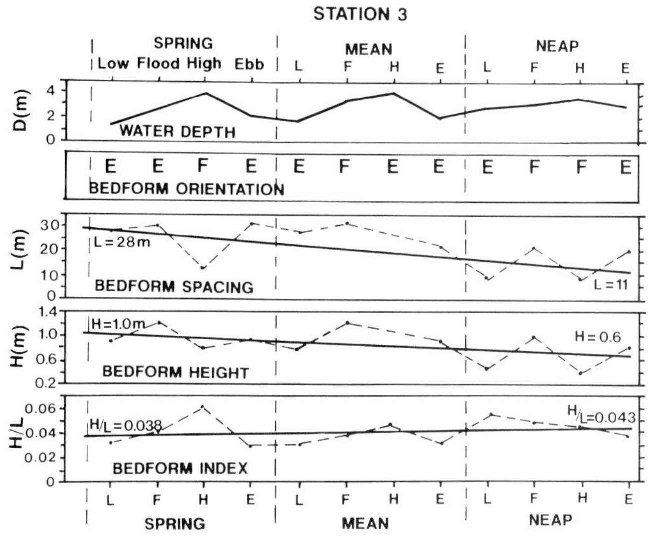

Figure 7. Summary data of bedform parameters at location 3 over various lunar tidal conditions.

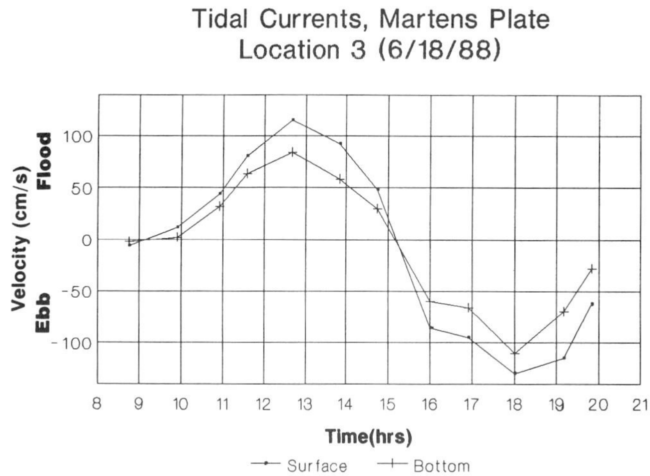

Figure 9. Time-velocity curves at location 3 during spring conditions on June 18, 1988.

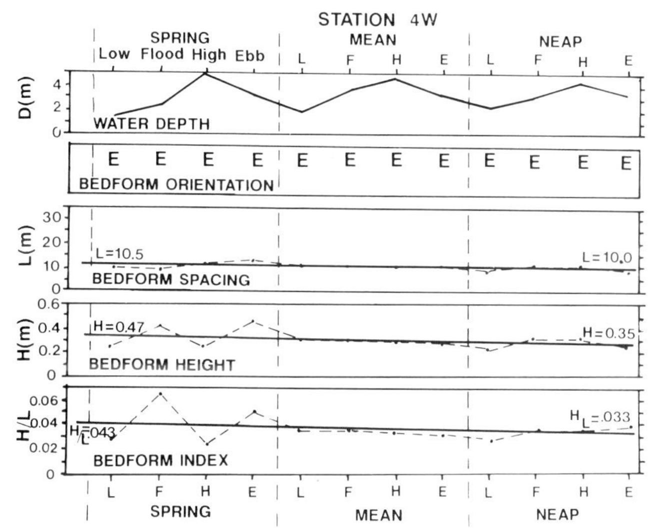

Figure 8. Summary data of bedform parameters at location 4W over various lunar tidal conditions.

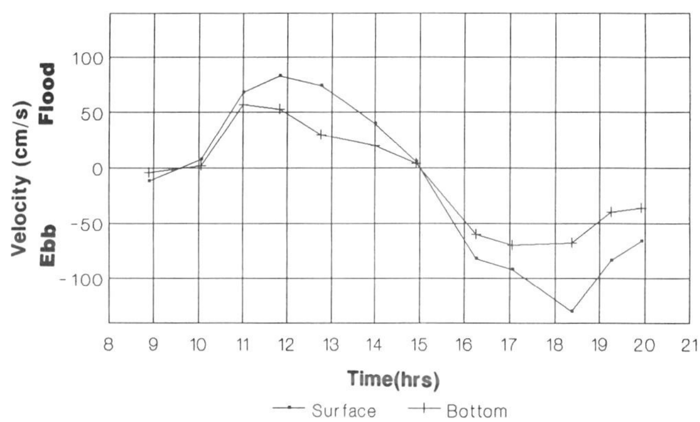

Figure 10. Time-velocity curves at location 4W during spring conditions on June 18, 1988.

cation 3, show a general and expected decrease in scale when decreasing tidal range over the lunar cycle (Fig. 7). Although the wavelength decreased greatly at this site, the height did not decrease accordingly.

Tidal Currents

Time-velocity curves of tidal currents were plotted for the bottom (0.5 m above channel floor) and the uppermost (depth of 1.0 m) readings over each measured tidal cycle. Differences between near-surface and bottom velocities were about 20 to 30 cm/s at peak conditions (Figs. 9, 10). Comparison of spring conditions with neap conditions shows only modest differences in maximum velocity at the bed; typically about 10 cm/s. This reflects the small variation in tidal range between spring and neap conditions (Fig. 2) and therefore the nominal change in tidal prism from spring to neap.

The best data sets are available for the near-spring conditions of June 18, 1988 which also represent the most energetic situation as a result of the high tidal prism and related high current velocities. At both locations 3 and 4W, the relative hydraulic conditions exhibit a very similar trend in the course of the tidal cycle under consideration in this report. All of the velocity profiles have an essentially logarithmic shape, with some deviations being observed in the upper water column during the decelerating phase of the flooding tide (Figs. 11, 12). It is significant that the velocity profiles are always steeper during the accelerating phase of the tide. This applies to both the ebbing and the flooding phase of the tidal cycle. In correspondence the roughness lengths (*Zo*) are consistently larger during the decelerating phases of each tide, reaching values of over 10 cm in some cases (*e.g.*, Fig. 11; 4W, ebbing tide).

The well-defined logarithmic shape of most velocity profiles, and hence, the reliable determination of roughness lengths allowed the calculation of shear velocities (U*) on

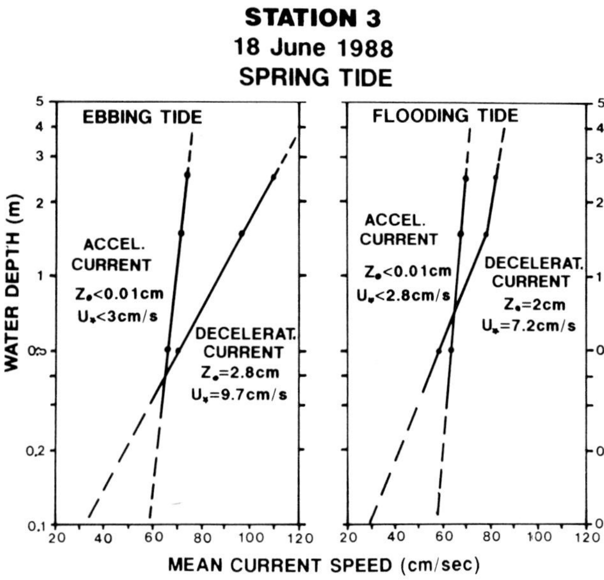

Figure 11. Logarithmic plots of velocity profiles at location 3.

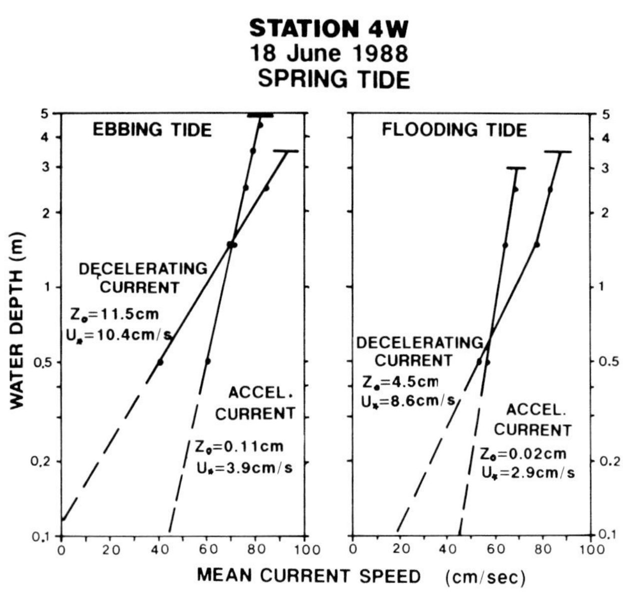

Figure 12. Logarithmic plots of velocity profiles at location 4W.

the basis of the rearranged Karman-Prandtl equation (*cf.* Sternberg, 1972; Dyer, 1986):

$$U^*=Uz/5.75 \log (Z+Zo/Zo)$$

Because shear velocities are directly proportional to the bed shear stresses by the relationship $\tau = gU^{*2}$, they provide a better criterion for the sediment transport capacity of the flow than a velocity measured some distance above the bed.

As is clearly evident from the time-series data in Figures 13 and 14, the shear velocities document an ebb dominance at both locations. Furthermore, in both cases, the shear velocities show a stronger time-velocity asymmetry toward the decelerating phase of each tide than is evident from the U_{100} values. This feature is even more pronounced in the time of progression of *Zo* values. Bed roughness in general is lower at location 3 as compared to location 4W, especially

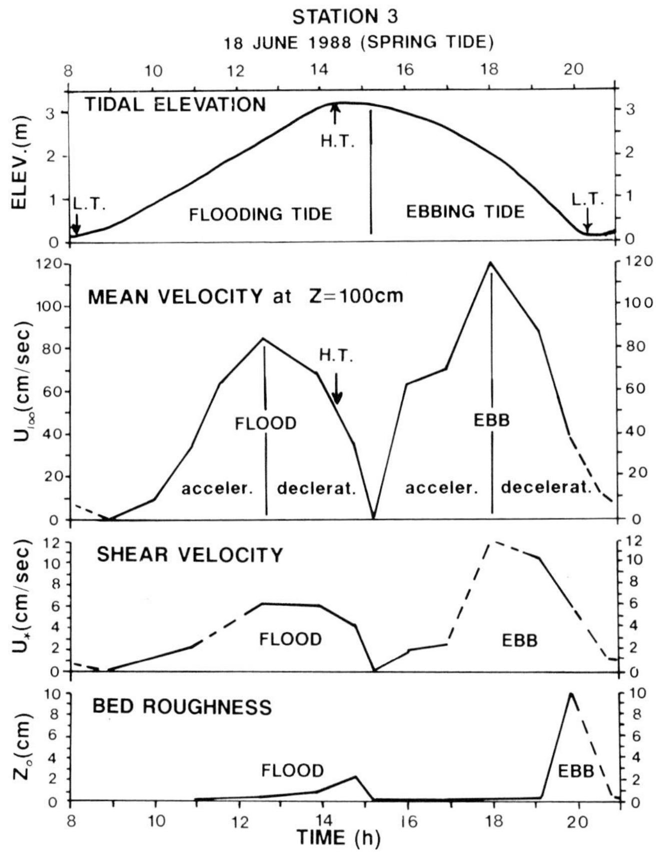

Figure 13. Hydraulic parameters at location 3 during spring tidal conditions.

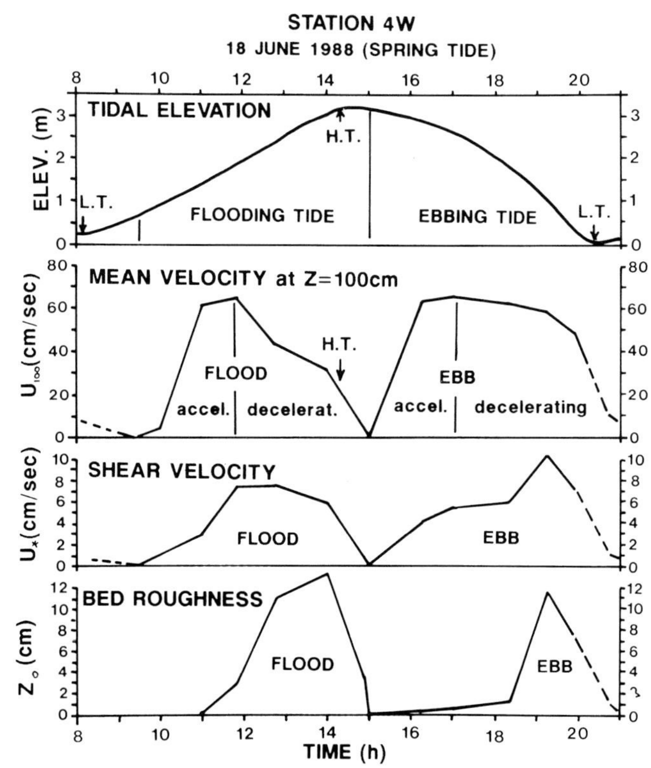

Figure 14. Hydraulic parameters at location 4W during spring tidal conditions.

during the decelerating phase of the flood cycle, in spite of the larger bedforms recorded at location 3. Peak roughness is coincident with high and low tidal stages, respectively, at each site. A marked difference in the roughness trends at the two stations, however, is evident in the extraordinary high *Zo* values observed during the flood cycle at station 4W, as opposed to those at location 3. This may point to a significant difference in bedform response during the flooding tide at the two locations. The reasons for this difference may be the greater tidal prism at location 3 and the distinctly ebb-dominated conditions at 4W as compared to 3. The balanced ebb and flood cycle is likely to reverse bedform orientation and thereby minimize roughness as the next cycle begins. The strong ebb-dominance at 4W does not permit reversal of bedforms. Thus the flooding currents are flowing over ebb-oriented bedforms, which may be the reason for the high roughness values.

On the other hand, it must be remembered that the Karman-Prandtl equation is based on measurements in steady, equilibrium flows. The inherently unsteady flow conditions in this study never reach an equilibrium state. As a result, it may be appropriate to use the steady-state equation in spite of the logarithmic shape of the velocity profiles above 0.5 m from the bed. There is some evidence in the literature that under such conditions the observed roughness lengths are more apparent than real (McLean, 1983).

Discussion

The data obtained by a time-series study of tidal flow conditions and bedform responses over a lunar tidal cycle have provided sufficient information to satisfy the main objectives as defined in the introduction. Thus, although the study area shows only small changes in tidal range between spring and neap conditions, the bedform response is quite striking. Even small changes in peak current velocities over a lunar cycle are sufficient to cause substantial adjustments in bedform dimensions. These changes are independent of grain size, which varies little, and they are only marginally affected by small fluctuations in water depth for a given tidal stage. Such information will be of great value in the interpretation of sediment cores from adjacent tidal sand bodies. It can be used to assist in the reconstruction of the environmental conditions on the basis of the scale, orientation and angle of inclination of cross-stratification observed in the stratigraphic record of tide-dominated environments.

Repeated side-scan sonar and fathometer surveys have revealed that although the tidal channels overall are ebb-dominated, there is local flood or ebb domination as is evident from bedform orientation throughout tidal cycles. On the basis of such information it was possible to optimize the choice of current metering stations. The current data in turn have demonstrated that single point measurements are inadequate for the assessment of sediment transport phenomena. Thus, without velocity profile measurements it would not have been possible to detect the significant differences in profile steepness, roughness length, and hence, shear velocities, between accelerating and decelerating phases of the tidal flow. All of these are important variables in bedform formation and maintenance.

Roughness lengths, similar to shear velocities, achieve their greatest values during declerating phases of the tidal flow where bedforms would be expected to be best developed. On the other hand, the unusually large *Zo* values (*cf.* Soulsby, 1983; Dyer, 1986) suggest that the scale of roughness elements at the bed is not the only parameter defining the roughness length. It would appear that bedform shape, bedform orientation and possibly even the thickness of the moving bed layer may be involved in determining the observed roughness lengths.

Conclusions

Lunar time-series data on bedforms and the currents that produce them provide information that gives direction to proper analysis of stratigraphic data from sandy, tide-dominated channel environments. These data indicate that:

1. Bedforms in tidal channels may not reverse direction during the tidal cycle. This negates the possibily of forming and preserving reversing cross-stratification, a criterion called upon by most investigators to indicate tidal conditions.

2. Bedform size and shape does change systematically through the tidal cycle. This produces a rhythmic sequence of changes in scale and angle of dip of cross-stratification that is tidal in origin.

These observations have obvious and important relevance to interpretations from the ancient stratigraphic record.

Acknowledgments

This study was supported primarily by the Senckenberg Institute. Davis was funded by a summer research fellowship from Deutscher Akademischer Austauschdienst (DAAD). The authors are indebted to the numerous personnel and students of the Senckenberg Institute, Wilhelmshaven for their assistance with the fieldwork.

References

Allen, P. A. and Homewood, P. 1984. Evolution and mechanics of a Miocene tidal sandwave. Sedimentology, v. 31, p. 63-81.

Berg, J. H. van den, 1982. Migration of large-scale bedforms and preservation of crossbedded sets in highly accretional parts of tidal channels in the Olsterschelde, S.W. Netherlands, Geologie en Mijnbouw, v. 61, p. 253-263.

Boothroyd, J. C. 1985. Tidal inlets and tidal inlet sandbodies. *In:* Coastal Sedimentary Environments, R. A. Davis (ed.). 2nd edition, New York, Springer-Verlag, p. 445-532.

Collins, M. B., Amos, C. L. and Evans, G. 1981. Observations of some sediment transport over intertidal flats, The Wash, U.K. *In:* Holocene Marine Sedimentation in the North Sea Basin, S. D. Nio, *et al.*, (eds.). International Association of Sedimentologists, No. 5, p. 81-98.

Dalrymple, R. W., Knight, R. J. and Lambiase, J. J. 1978. Bedforms and their hydraulic stability relationships in a tidal environment, Bay of Fundy, Canada. Nature, v. 275, p. 100-104.

Dette, H. H. 1977. Ein Vorschlag zur Analyse eines Wellenklimas. Die Kuste, v. 31, p. 166-180.

Dyer, K. R. 1986. Coastal and estuarine sediment dynamics. New York, John Wiley & Sons, 342 p.

Kreisa, R. D. and Moiola, R. J. 1986. Sigmoidal tidal bundles and other tide-generated sedimentary structures of the Curtis Formation, Utah. Geological Society of America, Bulletin, v. 97, p. 381-387.

Lambiase, J. J. 1980. Sediment dynamics in the macrotidal Avon River estuary, Bay of Fundy, Nova Scotia. Canadian Journal of Earth Sciences, v. 17, p. 1628-1641.

McLean, S. R. 1983. Turbulence and sediment transport measurements in a North Sea tidal inlet (The Jade). *In:* North Sea Dynamics, J. Sündermann and W. Lenz (eds.). Springer-Verlag, Berlin, p. 436-452.

Nio, S. D., Siegenthaler, C. and Yang, C. S. 1983. Megaripple cross-bedding as a tool for the reconstruction of paleohydraulics in a Holocene subtidal environment, S.W. Netherlands. Geologie en Mijnbouw, v. 62, p. 499-510.

Nummedal, D. and Penland, S. 1981. Sediment dispersal in Norderneyer Seegat, West Germany. *In:* Holocene Marine Sedimentation in the North Sea Basin, S. D. Nio *et al.* (eds.). International Association of Sedimentologists Special Publication, No. 5, p. 187-210.

Reineck, H. E. 1958. Longitudinale Schragschicht im Watt. Geologie Rundschau, v. 47, p. 73-82.

Soulsby, R. L. 1983. The bottom boundary layer of shelf seas. *In:* Physical Oceanography of Coastal and Shelf Seas, B. Johns (ed.). Amsterdam, Elsevier, Chapter 5.

Sternberg, R. W. 1972. Predicting initial motion and bedload transport of sediment particles in the shallow marine environment. *In:* Shelf Sediment Transport: Process and Pattern, D. J. P. Swift, D. B. Duane and O. H. Pilkey (eds.). Stroudsburg, Hutchinson & Ross, p. 61-82.

Visser, M. J. 1980. Neap-spring cycles reflected in Holocene sub-tidal large-scale bedform deposits: a preliminary note. Geology, v. 8, p. 543-546.

Yang, C. S. 1986. Estimate of sand transport in the Oosterschelde tidal basin using current velocity measurements. Marine Geology, v. 72, p. 143-170.

Yang, C. S. and Nio, S. D. 1985. The estimation of palaeohydrodynamic processes from subtidal deposits using time series analysis methods. Sedimentology, v. 32, p. 41-57.

KEY WORDS: LUNAR TIDAL CYCLES, TIDAL CHANNELS, WADDEN SEA, DUNES, TIME-VELOCITY RELATIONSHIPS

In: Clastic Tidal Sedimentology, D. G. Smith, G. E. Reinson, B. A. Zaitlin and R. A. Rahmani (eds.). Canadian Society of Petroleum Geologists. Memoir 16, p. 283-290.

The influence of flocculation on cohesive sediment transport in a microtidal estuary

Morten Pejrup

University of Copenhagen
Øster Voldgade 10
1350 Copenhagen K
Denmark

Abstract

Salt flocculation causes small clay- and silt-sized particles in estuarine waters to coagulate, resulting in much higher settling velocities than that undergone by the original single particles. Thus, the flocculation process changes the behavior of suspended cohesive sediment as it enters the estuarine environment. In the Grådyb tidal area of western Denmark, it can be shown that suspended concentration of cohesive sediment controls about 80 per cent of the variation in median settling velocity of cohesive suspended sediment; higher concentrations give increasingly higher settling velocities. In addition, wind speed influences the concentration level of suspended sediment. As a rule, settling velocities will generally be higher during periods of high wind speed and this will enhance the efficiency of settling and scour-lag processes. Over a three week monitoring period, measurements show an ebb oriented net transport during calm weather periods and flood oriented net transport during periods of high wind speed. This shift is thought to be due to the formation of larger sediment flocs during adverse weather periods. Such a mechanism may be an important controlling factor on the transport of cohesive sediments in shallow, microtidal areas with fairly high suspended concentrations of cohesive sediment.

Résumé

Dans les eaux estuariennes, les sels précipités par floculation provoquent la coagulation des particules de la taille des silts et argiles, d'où une plus grande vitesse de sédimentation pour les agglomérats que pour les particules individuelles originales. Donc, les processus de floculation modifient le comportement du sédiment cohésif en suspension introduit dans l'environnement estuarien. Dans la région de marée de Grådyb, partie occidentale du Danemark, il est démontré que la concentration des sédiments cohésifs en suspension contrôle, à environ 80 pourcent, la variation de la médiane de la vitesse de sédimentation du sédiment cohésif en suspension. Par conséquent, les plus grandes concentrations correspondent à de plus grandes vitesses de sédimentation. Aussi, la vitesse du vent influence le niveau de concentration du sédiment en suspension. En règle générale, les vitesses de sédimentation seront généralement plus élevées lors de vents violents, ce qui a pour effet d'augmenter l'efficacité de la sédimentation, des processus d'érosion et d'accumulation des particules grossières. Les mesures recueillies au cours d'un monitorage, d'une durée de trois semaines, démontrent que le transport net s'oriente avec le reflux pendant les périodes de temps calme et avec le flux pendant les périodes de vents violents. Ce renversement s'explique peut-être par la formation d'agglomérats ou flocons plus grossiers durant les périodes de mauvais temps. Un tel mécanisme peut être un facteur de contrôle important sur le transport du sédiment cohésif dans les régions d'eaux peu profondes et microtidales, avec une assez forte concentration de sédiment cohésif en suspension.

Introduction

The influence that salt flocculation may have on cohesive sediment transport in estuaries has been discussed extensively in the literature, (Meade, 1972; Zabawa, 1978; Schubel *et al.*, 1978; Krone, 1978; Gibbs, 1985; Kranck, 1984, 1986; Eisma, 1986; Owen, 1977; Dyer, 1988, 1989; Mehta, 1989; and Sternberg, 1989). Some of these authors almost disregard flocculation as an important factor for sediment dynamics in estuaries, whereas others claim it to be a dominant factor in controlling cohesive sediment transport. One reason for this dispute may be that different methods have been used for analyzing settling velocity and grain size of suspended sediment. Commonly used methods include coulter counter techniques (*e.g.*, Kranck, 1975), the settling tube bottom withdrawal method (*e.g.*, Owen, 1971; Burt, 1986), photography (*e.g.*, Kranck, 1984; Eisma, 1986), and laser diffraction techniques (*e.g.*, Bale and Morris, 1987). Finally, the Rouse formula may be used to compute indirectly the settling velocity of suspended sediment. However, another major reason for the diverging conclusions on the effect of flocculation on sediment dynamics in estuaries may be that investigations have been carried out in differing estuarine environments. Therefore, part of the problem may be that, in most studies, no information is given about the specific kind of estuarine environment under investigation.

A number of papers conclude that suspended sediment concentration strongly controls the flocculation process (*e.g.*, Migniot, 1968; Owen, 1971; Hunter and Liss, 1979; Kranck, 1981; Gibbs, 1983; Burt, 1986; Pejrup, 1988a; Puls *et al.*, 1988; and van Leussen, 1988). One can therefore hardly expect that investigations carried out in tidal areas with low suspended sediment concentrations will show flocculation to be important. Furthermore, if flocculation has any significant influence on the sediment dynamics, it is because the flocs have high enough settling velocities to settle to the bottom in periods of slack water. Therefore, investigations from deep estuarine environments with moderate or low suspended sediment concentrations; for example, fjords, cannot be expected to show that flocculation is a dominant factor controlling cohesive sediment dynamics (*e.g.*, Syvitski, 1981). In such environments, the estuarine net circulation of water will of course have a much stronger bearing on cohesive sediment transport.

Figure 1 exhibits the relationship between concentration and depth, and the relative importance of flocculation within an estuary. Flocculation has been shown to increase exponentially with concentration (Krone 1962; Burt 1986; Mehta, 1989). At the same time, however, its importance decreases linearly with increasing water depth because the sediment flocs must have time to be deposited, especially if the flocculation of small particles into larger aggregates is to have a significant impact on cohesive sediment transport.

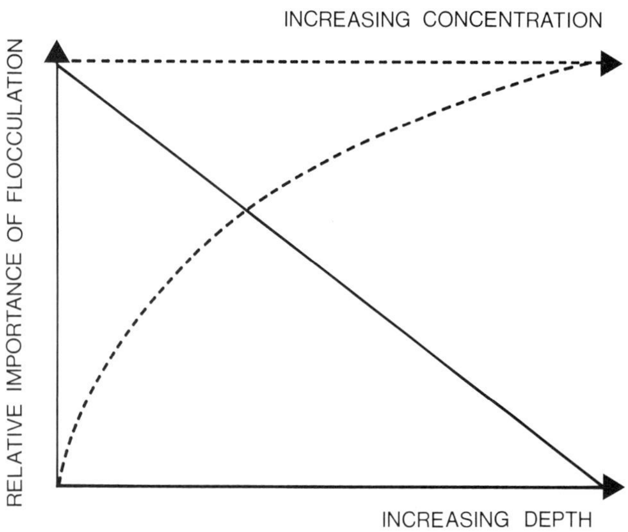

Fig. 1. Schematic presentation of the relative importance of flocculation in an estuary as a function of suspended concentration (dotted line) and depth (solid line).

In order to measure the impact of flocculation on cohesive sediment transport it is necessary to choose an estuarine environment with high concentrations of fine grained suspended sediment and moderate depths. Most parts of the European Wadden Sea fulfill both these criteria. Mean depths at high water are within 2 to 4 m, and from many areas, such as Dollard, (Postma, 1967), Jadebusen, the Lister Dyb tidal area (Pejrup, 1988a) and the Grådyb tidal area (Gry, 1942; Pejrup, 1986), very high concentrations of cohesive suspended sediment have been measured, especially during periods of stormy weather. Extensive intertidal areas consist of mud flats, indicating a heavy net accumulation of fine grained sediment. This can only be achieved in shallow estuarine environments if the settling and scour-lag processes are effective in net landward transport of fine particles (Postma 1954, 1967; van Straaten and Kuenen, 1958), because depth and salinity gradients are too small to create an estuarine circulation of any significance in the lagoons behind the barrier islands.

This study deals with the Grådyb tidal area in the Danish part of the Wadden Sea (Fig. 2). Mean tidal range is 1.5 m and mean salinities vary from about 5‰ in the northern part of the area to about 30‰ near the tidal inlet. Suspended concentrations are low at the estuary mouth but increase rapidly toward the north and south. Maximum concentrations in the inner parts of the area are in the order of 1000 to 2000 mg/l. A more detailed description of the study area has been given elsewhere (*e.g.*, Pejrup, 1986, 1988a, 1988b; Bartholdy and Pheiffer Madsen, 1985).

Results and Discussion

Regional Variation in Suspended Sediment Concentrations

In shallow estuarine environments the processes of settling and scour-lag will cause the concentration of fine grained suspended sediment to increase from the tidal inlet toward the inner parts of the estuary because of the decreasing mean velocity of the tidal flow (Postma, 1967). This is also the case for the estuarine area studied here (Fig. 3). A corresponding increase in mean suspended sediment concentrations has been described for the northern part of the area by Pejrup (1983). Tidal variation in concentration was described at six locations evenly distributed over an area of about 40 km^2. In Table 1 mean values for four different tidal periods are given for the six localities together with mean wind speed throughout the period; the geographical variation is shown in Figure 4. The measurements show that mean concentration increases from south to north in both adverse and calm weather; this compares well with earlier work presented by Gry (1942). Furthermore, mean concentration at all six localities increases with increasing wind speed; inflow from river and sea is assumed constant. This is because of resuspension of bottom sediment by wind-generated waves.

Parameters Controlling Suspended Sediment Concentrations

The parameters governing the concentration levels and tidal variations of fine grained suspended sediment in shallow microtidal areas are: 1) wind speed, 2) wind direction, 3) bottom shear stress (current velocity), 4) water depth, 5) temperature, and 6) salinity.

A conceptual mathematical model relating suspended sediment concentration to the most important of these parameters was developed by Pejrup (1986). The following three model parameters were computed:

1. $W = V^2 S^{-4}$
2. $D = Sin(winddir - \phi)$
3. $\Delta U^2 = (U_{0.6}^2 - U_c^2)$

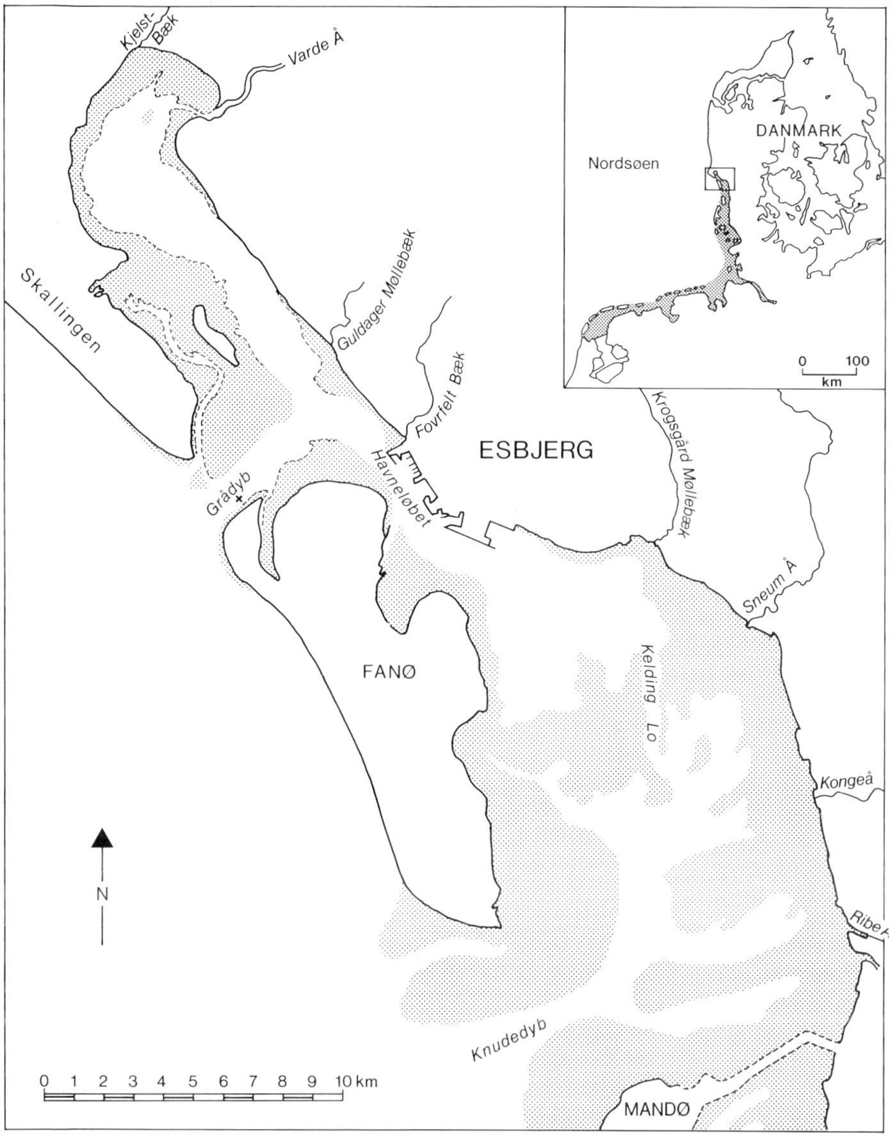

Fig. 2. Location of study area in the northernmost part of the European Wadden Sea. Shaded area indicates the intertidal flats.

$$V^2 = \frac{\sum_{i=1}^{37} V^2_i\,(38-i)}{\sum_{i=1}^{37} i}$$

where:

V_{37} is the mean wind speed measurement taken from 17.5 to 18 hours before observation

V_{36} is the mean wind speed measurement taken from 17 to 17.5 hours before observation

V_i is the mean wind speed measurement taken from 0 to 0.5 hour before observation

S is the salinity ‰

ϕ is a constant

$U_{0.6}$ is the current velocity 0.6 m above the bed, m/s

U_c is the critical erosion velocity 0.6 m above the bed, m/s

The model parameter W relates to wind conditions and the origins of the water masses, because the V^2 parameter characterizes wind speed during the preceding 18 hours based on half-hourly mean values, while the parameter S^{-4} expresses the origin of the water because salinity decreases from tidal inlet to estuary head. D expresses the influence of the wind direction and ΔU^2 expresses the influence of the excessive bottom shear stress on suspended sediment concentrations.

The model predicting the suspended sediment concentrations (SSC) at site B (Fig. 4) takes the following form: $SSC = 3.7 + 44176.5 \cdot W - 26000.2 \cdot W \cdot D + 4.5 \cdot W \cdot \Delta U^2$

This model can account for 85 per cent of the variance of suspended sediment concentrations and acts in a physically correct way. The parameter W alone accounts for approximately 74 per cent of the variance of suspended sediment concentrations (Pejrup, 1986). An important con-

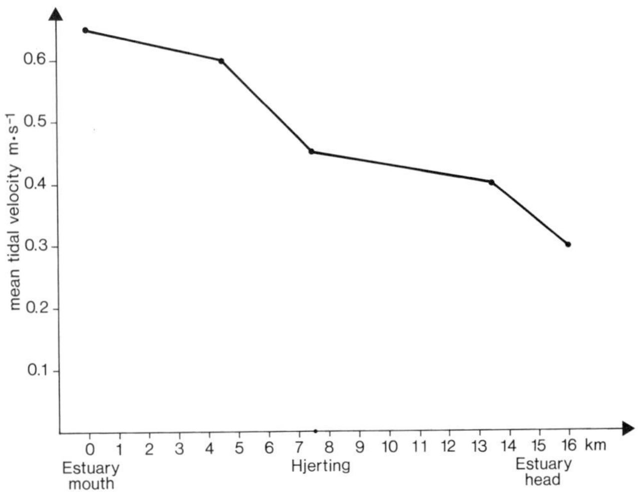

Figure 3 Mean tidal velocity along the northern estuary axis of Grådyb tidal area.

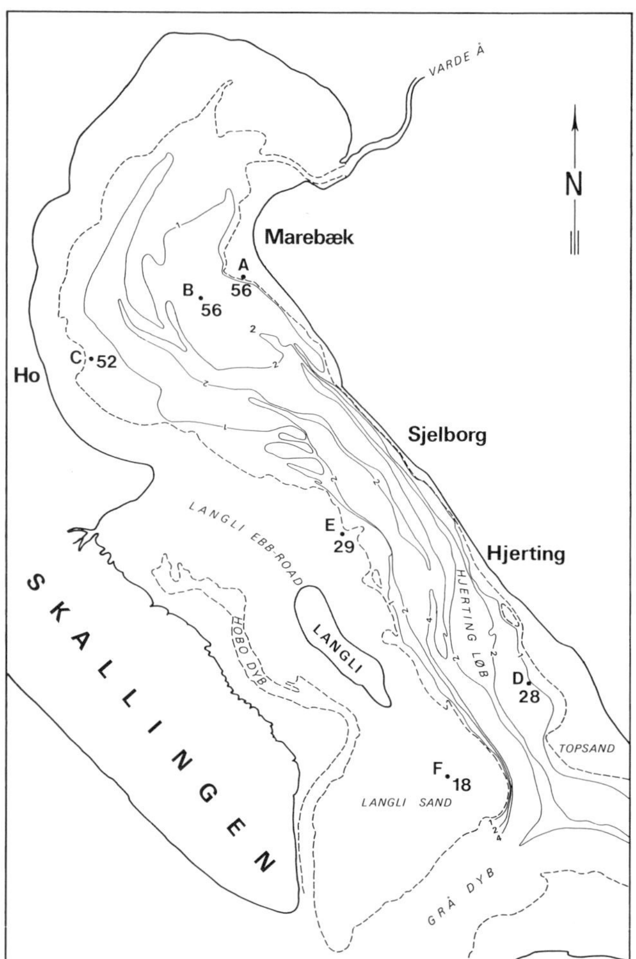

Fig. 4. Regional variation in mean concentration of suspended sediment (mg/l) measured over four tidal periods. (see Table 1).

clusion of this investigation is that wind speed is the major determinant for concentration level of fine grained suspended sediment in this area.

Flocculation of the Suspended Sediment

In situ settling tube analysis of suspended sediment at two sites within the Danish Wadden Sea (Pejrup, 1988a) indicated that suspended sediment concentration controls about 80 per cent of the variance in median settling velocity of the suspended flocs. These findings are consistent with results presented by Burt (1986), Puls *et al.* (1988) and Dyer (1989). The estimated settling velocities were described by the equation:

$$W_{50} = 2.80 \cdot 10^{-5} \cdot C^{0.51} \qquad r^2 = 0.79$$

where:

W_{50} is the median settling velocity in m/s

C is suspended concentration in mg/l

r^2 is the variance

This means that maximum median settling velocities are in the order of 10^{-3} m/s, corresponding to concentrations of approximately 1000 mg/l.

When the results from these two investigations are combined (Pejrup 1986, 1988a), it is found that the highest settling velocities of the cohesive suspended sediment can be expected during periods of maximum concentration caused by elevated wind speeds. This is because high concentration levels enhance flocculation, producing larger flocs with higher settling velocities.

Deposition of Suspended Sediment on Tidal Flats

Mean depth in the study area during high tide is 2.3 m, suggesting that on average, it will take 30 to 60 minutes for the large flocs to be deposited. The larger the suspended concentration, the greater the median settling velocity. Increased settling velocities cause the time for minimum concentration to be displaced into the ebb period, and the higher the settling velocity, the longer the time is that deposition persists into the ebb period. Furthermore, the scour-lag will be more pronounced in such periods because when a large amount of sediment is deposited on the bottom, it may consolidate so that a larger bottom shear stress is needed for erosion. This delays the time of the resuspension peak. The final result is that mean concentration of suspended sediment is higher during the flood than during the ebb period, when suspended sediment concentrations are high.

Two examples of this tidal variation in suspended sediment concentration are shown in Figure 5, where two periods are compared. The tidal period on August 3 has a low concentration level, whereas the period on June 6 has a high level of suspended concentrations (Table 1). It is clearly demonstrated that concentration minimum is displaced about 1 hour into the ebb period in the tidal period with high mean concentrations and, furthermore, that the resuspension peak is strongly delayed on June 6.

These findings imply a net headward transport of cohesive sediment in the estuary during periods with high suspended sediment concentrations caused primarily by high wind speeds. Settling velocities within the range 0.5×10^{-3} to 1.5×10^{-3} m/s correspond to quartz grains with a fall diameter of 25 to 45 μm. This is within the 8 to 100 μm range for fall diameters for which Postma (1967) states the settling- and scour-lag mechanisms to be efficient. If this reasoning is correct it should be possible to measure enhanced headward transport of suspended cohesive sediment during periods of high wind speeds and high suspended concentrations. On the other hand, if such an enhanced landward net transport of cohesive sediment can be demonstrated, it will stress the importance of salt flocculation as a major determinant of the suspended transport of cohesive sediment in this shallow, microtidal estuary.

Net Transport of Cohesive Sediment

To investigate this theory, transport of suspended cohesive sediment was measured at a tidal flat station situated

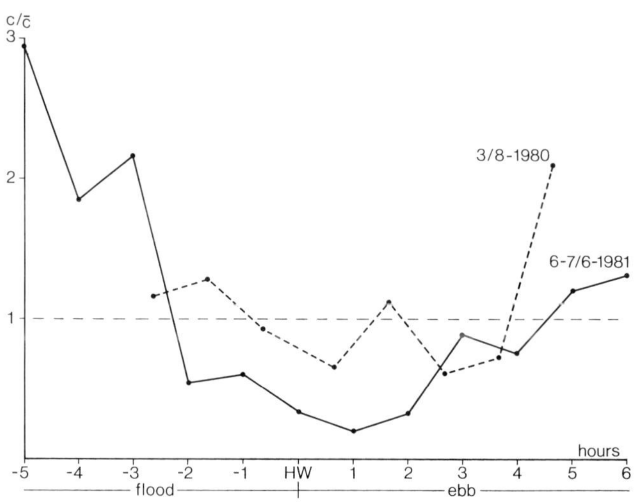

Fig. 5. Tidal variation of suspended sediment concentration at site B for a tidal period with low concentration level (3-8-80) and high concentration level (6-7/6-81). Mean concentrations are given in Table 1. The Y-axis shows measured concentrations, C, made dimensionless by division with the mean concentration of the tidal period $\overline{C}$.

at position B during a three week period (Fig. 4). Current velocity and direction were measured with an Aanderaa RCM 4 self-recording current meter every 10 minutes. These records were used to compute mean velocity. Water level was measured continuously with an Ott Bandschreiber, and suspended sediment concentration with a Partech Suspended Solid Monitor (type 7000 3 RP, equipped with a twin probe type S100, Gippel, (1989)). Suspended sediment transport was calculated for 10-minuted intervals during the three week investigation period. The cumulated values for flood and ebb periods, respectively, can be seen in Figure 6 together with the cumulated net transport.

Roughly, the investigation period can be divided into two halves. One from September 1st to September 11th, with low transport values due to low suspended concentration; and another from September 11th to September 22nd, exhibiting high transport values as a result of high suspended concentrations. The difference in suspended mean concentration for the two periods was caused by higher peak values in wind velocity in the latter part of the three week period.

Net transport of suspended sediment in the first period is directed out of the estuary. However, this net transport direction shifts in the second part of the period toward the estuary head. This shift in net transport direction is a result of the enhanced effect of the two lag-processes caused by the increase in settling velocity of the suspended sediment.

The annual mean net transport computed on the basis of these data, (Pejrup, 1988c) corresponds well to a sediment budget for the area set up by Bartholdy and Pheiffer Madsen (1985).

Settling Velocity of the Suspended Sediment

The increased settling velocities may be due either to the formation of increasingly larger sediment flocs or to the suspension of larger single, silt-sized particles. From a number of in situ settling tube analyses of settling velocities of suspended sediment, Pejrup (1988b) showed that it is primarily the particles $< 4.5\phi$ in diameter that take part in the flocculation process. This conclusion was made on the basis of breaks in the log-probability grain size curves (equivalent fall diameter) on almost all the analyzed samples. These findings are consistent with results presented by Carson *et al.*

Table 1. Mean concentration of suspended sediment in mg/l for six different tidal flat localities in the northern part of the Grådyb tidal area over four tidal periods. The numbers are based on hourly sampling for that part of the tidal period when water covers the tidal flat stations. Last column gives the mean wind speed within the tidal period. Locations are shown in Figure 4.

Date	A	B	C	D	E	F	Wind Speed m/s
3-8-80	32	22	23	7	17	3	2.5
22-9-80	29	29	29	6	18	8	3.4
13-5-81	64	60	120	37	44	25	5.0
6-6-81	99	113	36	61	38	36	9.3
Mean	56	56	52	28	29	18	

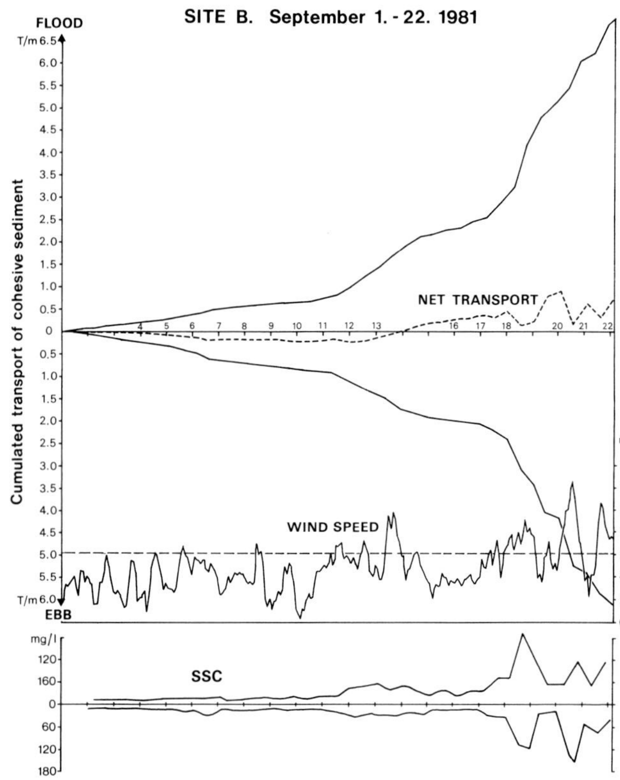

Fig. 6. Cumulated values of gross- and net transport values of cohesive sediment at site B together with variation in mean wind speed and mean suspended sediment concentration (SSC).

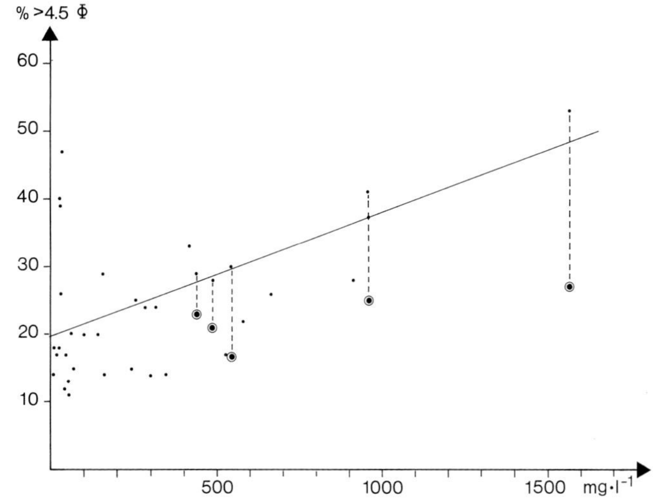

Fig. 7. The fraction of suspended sediment samples coarser than 4.5 ϕ plotted versus suspended concentration. Large solid circles indicate this fraction after the suspended sediment has been dispersed in an ultrasonic bath for 15 minutes.

(1988) who found that suspended organic-mineral aggregates in The Great Sound, New Jersey, consist of particles $< 5\phi$ in diameter. The flocs described in this study, with equivalent diameters between 4.5 to 7.0ϕ are formed as a result of physiochemical aggregation (salt flocculation). These large flocs may be made from much smaller, and more resistant aggregates owing their existence to biogenic activity. However, these aggregates are too small to be described accurately with the in situ settling tube analysis.

If it is the suspension of larger, single mineral grains that is responsible for the increase in median settling velocity in the Grådyb tidal area, one would expect that the percentage of the suspended sample $> 4.5\phi$ in diameter should increase with increasing suspended concentration. In Figure 7 this fraction is plotted against suspended concentration for 35 suspended sediment samples analyzed with the settling tube. There is poor correlation between an increase in grain size fraction and increasing suspended concentration, with a correlation of:

$$\% > 4.5\phi = 19.3 + 0.018\ C\ mg/l \qquad r^2 = 0.30$$

At least part of this correlation is, however, explained by the formation of flocs larger than 4.5ϕ at very high concentration levels. This is illustrated in Figure 8, where two samples of very high suspended sediment concentrations have been analyzed in situ as well as in the laboratory after reflushing of the sediment from the filters and treatment in an ultra sonic bath. It appears that the grain size curves begin to

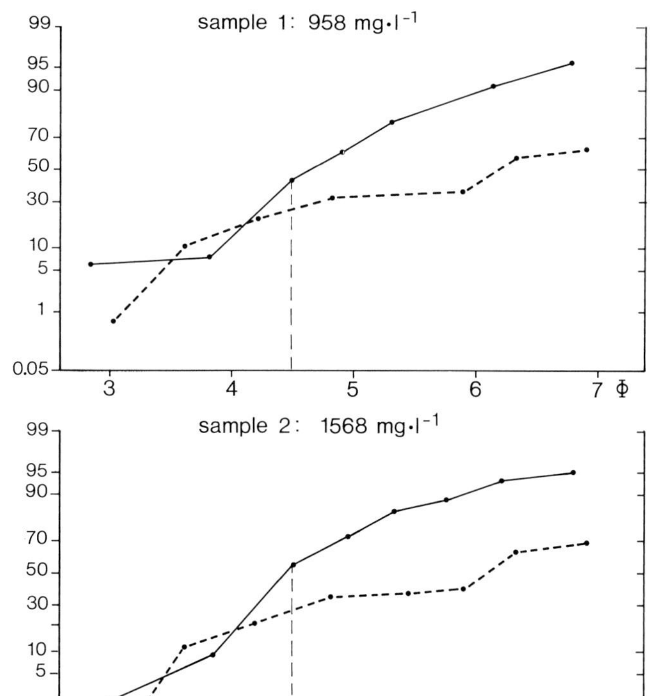

Fig. 8. Two examples of equivalent grain size curves based on settling tube analysis. Solid line indicates in situ analysis, whereas dotted line indicates analysis in the laboratory after treatment of the samples in an ultrasonic bath. The curves begin to diverge between 4.0 and 4.5 ϕ indicating that flocs in this grain-size interval were present during the in situ analysis.

diverge between diameters of 4.0 and 4.5ϕ. Five such double analyses have been plotted in Figure 7, where the dispersion of the flocculated suspension suggests an almost constant amount of 20 per cent for the grain size fraction coarser than 4.5ϕ in diameter. It can therefore be concluded that the increase in median settling velocity is almost exclusively due to the formation of larger sediment flocs. Consequently, it is the formation of sediment flocs that is responsible for the shift in settling velocity and thereby for the shift in net sediment transport from a seaward to a headward direction.

Conclusions

1. The combined influences of settling velocity and bathymetry control the transport and deposition of cohesive sediment within estuaries, but wind conditions are the major determinant for concentration of cohesive sediment in microtidal estuaries such as the Grådyb tidal area in Denmark.

2. Increasing concentrations of cohesive sediment will lead to higher settling velocities, primarily caused by salt flocculation of the fine grained, cohesive sediment and not by erosion of larger, single mineral particles.

3. The settling and scour-lag processes will be more efficient when settling velocity of the suspended sediment increases, causing the time of minimum concentration and peak of resuspension to be displaced into the ebb period. This asymmetry in suspended sediment concentration causes the highest mean concentration during the flood period resulting in a greater headward transport of sediment within the estuary.

4. Flocculation, together with wind speed, are the major factors controlling cohesive sediment transport in the Grådyb tidal area. This is probably true for most of the European Wadden Sea, where hydrographic and bathymetric conditions are very similar to those in the investigated area.

Acknowledgments

The funds for this study were provided by the Danish Natural Science Research Council. I am indebted to A. Jensen for participating in the fieldwork, to J. Bartholdy for valuable discussions, and to J. Ulrich for drafting.

References

Bartholdy, J. and Pheiffer Madsen, P. 1985. Accumulation of fine-grained material in a Danish tidal area. Marine Geology, v. 67, p. 121-137.

Bale, A. J. and Morris, A. W. 1987. In situ measurements of particle size in estuarine waters. Estuarine, Coastal and Shelf Science, v. 24, p. 253-263.

Burt, T. L. 1986. Field settling velocity of estuary muds. *In:* Estuarine Cohesive Sediment Dynamics, A. J. Metha (ed.). Lecture Notes on Coastal and Estuarine Studies 14: Springer, Berlin, p. 126-150.

Carson, B., Carney, K. F. and Meglis, A. J. 1988. Sediment aggregation in a salt-marsh complex, Great Sound, New Jersey. Marine Geology, v. 82, p. 83-96.

Dyer, K. R. 1988. Fine sediment particle transport in estuaries. *In:* Physical Processes in Estuaries, J. Dronkers and W. van Leussen (eds.). Springer, Berlin, p. 295-310.

______ 1989. Sediment processes in Estuaries: Future requirements. Journal of Geophysical Research, v. 94, p. 14,327-14,339.

Eisma, D. 1986. Flocculation and de-flocculation of suspended matter in estuaries: Netherlands Journal of Sea Research, v. 20, p. 183-199.

Gibbs, R. J. 1983. Coagulation rates of clay minerals and natural sediments. Journal of Sedimentary Petrology, v. 53, p. 1193-1203.

______ 1985. Estuarine flocs: Their size, settling velocity and density. Journal of Geophysical Research, v. 90, p. 3249-3251.

Gippel, C. J. 1989. The use of turbidity instruments to measure stream water suspended sediment concentration. Monograph Series no. 4, Departments of Geography and Oceanography, University of New South Wales, 204 p.

Gry, H. 1942. Das Wattenmeer bei Skallingen. Folia Geographica Danica, v. 2, 137 p.

Hunter, K. A. and Liss, P. S. 1979. The surface charge of suspended particles in estuarine and coastal waters. Nature, v. 282, p. 823-825.

Kranck, K. 1975. Sediment deposition from flocculated suspensions. Sedimentology, v. 22, p. 111-123.

______ 1981. Particulate matter grain-size characteristics and flocculation in a partially mixed estuary. Sedimentology, v. 28, p. 107-114.

______ 1984. The role of flocculation in the filtering of particulate matter in estuaries. *In:* The Estuary as a Filter, V. S. Kennedy (ed.). Academic Press, p. 159-175.

______ 1986. Settling behavior of cohesive sediment. *In:* Estuarine Cohesive Sediment Dynamics, A. J. Mehta (ed.). Lecture notes on Coastal and Estuarine Studies, v. 14, Springer, Berlin, p. 151-169.

Krone, R. B. 1962. Flume studies of the transport of sediment. *In:* Estuarial shoaling processes: Final Report; Hydrographic Engineering Sanitary Engineering Research Laboratory, University of California, Berkeley, 110 p.

______ 1978. Aggregation of suspended particles in estuaries. *In:* Estuarine Transport Processes, B. Kjerfve (ed.). University of South Carolina Press, Columbia, South Carolina, p. 177-190.

Meade, R. H. 1972. Transport and deposition of sediments in estuaries. Geology of America, Memoir 133, p. 91-120.

Mehta, A. J. 1989. On estuarine cohesive sediment suspension behavior. Journal of Geophysical Research, v. 94, p. 14,303-14,314.

Migniot, C. 1968. Etude des propriétés physiques de différents sédiments très fins et de leur comportement sous des actions hydrodynamiques. La Houille Blanche, v. 23, p. 591-620.

Owen, M. W. 1971. The effect of turbulence on the settling velocity of silt flocs. Proceedings of the Congress of the International Association of Hydraulic Research 14th, Paris, Paper D4, p. 1-6.

______ 1977. Problems in the modeling of transport, erosion, and deposition of cohesive sediments. *In:* The Sea VI, Marine modelling, E. D. Goldberg, I. N. McCave, J. J. O'Brien and J. H. Steele (eds.). Wiley-Interscience, New York, p. 515-537.

Pejrup, M. 1983. Et metodestudium af den suspenderede sedimenttransport i et vadehavsmiljo, Ph.D. thesis Unv. Copenhagen, (English summary and figure texts). Medd. fra Skallinglaboratoriet., v. 29, 178 p., Reitzel, Copenhagen.

______ 1986. Parameters affecting fine-grained suspended sediment concentrations in a shallow micro-tidal estuary, Ho Bugt, Denmark. Estuarine, Coastal and Shelf Science, v. 22, p. 241-254.

______ 1988a. Flocculated suspended sediment in a micro-tidal environment. Sedimentary Geology, v. 57, p. 249-256.

______ 1988b. The triangular diagram used for classification of estuarine sediments: A new approach. *In:* Tide-influenced Sedimentary Environments and Facies, P. L. de Boer, A. van Gelder and S. D. Nio (eds.). Reidel Publishing Company, p. 289-300.

______ 1988c. Suspended sediment transport across a tidal flat. Marine Geology, v. 82, p. 187-198.

Postma, H. 1954. Hydrography of the Dutch Wadden Sea. Archives Néerlandaises de Zoologie, v. 10, p. 405-511.

______ 1967. Sediment transport and sedimentation in the estuarine environment. *In:* Estuaries, G. H. Lauff (ed.). American Association for the Advancement of Science, Publ. no. 83, p. 158-179.

Puls, W., Kuehl, H. and Heymann, K. 1988. Settling velocity of mud flocs: Results of field measurements in the Elbe and the Weser estuary. *In:* Physical Processes in Estuaries, J. Dronkers and W. van Leussen (eds.). Springer-Verlag, Berlin, p. 404-424.

Schubel, J. R., Wilson, R. E. and Okubo A. 1978. Vertical transport of suspended sediment in upper Chesapeake Bay. *In:* Estuarine Transport Processes, B. Kjerfve (ed.). University of South Carolina Press, Columbia, South Carolina, p. 161-175.

Sternberg, R. W. 1989. Instrumentation for estuarine research. Journal of Geophysical Research, v. 94, p. 14,289-14,301.
Syvitski, J. P. M. 1981. Particle interaction in fjord suspended sediment. Marine Geology, v. 39, p. 215-242.
Van Leussen, W. 1988. Aggregation of particles, settling velocity of mud flocs. A review. *In:* Physical Processes in Estuaries, J. Dronkers and N. van Leussen (eds.). Springer-Verlag Berlin, p. 347-403.
Van Straaten, L. M. J. U. and Kuenen, Ph. H. 1958. Tidal action as a cause of clay accumulation. Journal of Sedimentary Petrology, v. 28, p. 406-413.
Zabawa, C. F. 1978. Microstructure of agglomerated suspended sediments in Northern Chesapeake Bay estuary. Science, v. 202, p. 49-51.

KEY WORDS: FLOCCULATION, WADDEN SEA, GRÅDYB, COHESIVE SUSPENDED SEDIMENT, MICROTIDAL ESTUARY

In: Clastic Tidal Sedimentology, D. G. Smith, G. E. Reinson, B. A. Zaitlin and R. A. Rahmani (eds.). Canadian Society of Petroleum Geologists. Memoir 16, p. 291-300.

Effects of seagrass beds on tidal flat sedimentation, Corner Inlet, southeast Australia

WU-YI ZHUANG

Department of Geography
Institute Building
University of Sydney
Sydney, Australia
2006

JOHN CHAPPELL

Department of Biogeography and Geomorphology
The Australian National University
GPO Box 4, Canberra, ACT 2601 Australia

ABSTRACT

Seagrass beds have been found to effectively baffle current flow and to increase sedimentation in shallow water environments, but few quantitative studies of this process have been carried out on tidal flats. Field monitoring of current velocity and suspended sediment concentrations (SSC) over the seagrass beds and adjacent sand flats (without seagrass cover) in intertidal environments of Corner Inlet, southeast Australia, indicates that the seagrass beds (*Zostera muelleri, Posidonia australia*) can reduce current velocity and suspended sediment concentrations near the bed by as much as 40 per cent and 25 per cent respectively. This effect is particularly evident near the sediment substrate, and it decreases toward the water surface. The general tendency of a shoreward decrease in wave and tidal current energy across intertidal flats is thus modified.

Because of the reduced bottom current velocity and the sediment trapping effect, seagrass has a significant influence on intertidal sedimentation. This is clearly shown by the distinct differences in sediment texture and sedimentary structures between sand flats with and without seagrass beds in the intertidal environments of Corner Inlet. Flats with seagrass generally accumulate a higher percentage of silt and clay than those without. Consequently, sediments accumulating on intertidal flats of Corner Inlet display finer and coarser sediment alternations across the intertidal zone, a characteristic apparently related to the presence or absence of seagrass beds, respectively.

The depositional effects of seagrass on intertidal sedimentation are also revealed by the internal sedimentary structures of tidal deposits. Seagrass beds tend to produce layers of highly bioturbated mud or sandy mud that overlap structureless, uniform sand or shelly sand deposited in areas free of seagrass. An upward fining stratigraphic sequence can be produced by the migration of seagrass. Thus, it is suggested that the temporal and spatial variations of seagrass density may leave a signature in the sedimentary record.

RÉSUMÉ

Dans les environnements d'eau peu profonde, il a été démontré que les herbiers filtrent avec efficacité les courants et qu'ils accroissent la sédimentation. Cependant, peu d'études quantitatives ont porté sur ce processus agissant sur les estrans. Au sud-est de l'Australie, dans la zone intertidale du petit bras de mer de Corner au-dessus des herbiers et des estrans sableux adjacents (sans couverture d'herbes marines), un monitorage sur le terrain a relevé la vélocité des courants et des concentrations du sédiment en suspension (CSS). Il indique que les herbiers (*Zostera muelleri, Posidonia australia*) peuvent réduire près du lit la vélocité des courants ainsi que les concentrations du sédiment en suspension par des pourcentages respectifs, de 10 et 25. Cet effet, d'autant plus évident près de la surface des sédiments, décroît vers la surface de l'eau. Dans la zone intertidale, la tendance générale d'une diminution vers la côte de l'énergie des courants de marée et des vagues en est alors modifiée.

À cause de la vélocité réduite du courant de fond et du piégeage de sédiments, les herbes marines ont une grande influence sur la sédimentation de la zone intertidale. Dans la zone intertidale du petit bras de mer de Corner, ce fait est clairement démontré par les différences dans la texture du sédiment et les structures sédimentaires entre les estrans sableux, avec ou sans végétation, et les herbiers. L'estran avec les herbes marines accumule un plus grand pourcentage de silt et d'argile comparativement à celui sans herbes. Conséquemment, l'accumulation de sédiments à travers la zone intertidale du petit bras de mer de Corner montre une alternance de sédiments plus fins et plus grossiers, apparamment caractéristique de la présence et l'absence d'herbiers.

Les effets des herbes marines sur la sédimentation de la zone intertidale sont également révélés par les structures sédimentaires internes des dépôts. Les herbiers tendent à produire des couches de boues ou de sables boueux fortement bioturbées qui se superposent aux sables homogènes ou coquilliers accumulés dans les aires libres de ces herbes. Une séquence stratigraphique «virtuelle», i.e. avec une granulométrie décroissante de la base vers le sommet, peut résulter de la migration des herbiers. Ainsi donc, il est probable que les variations, temporelle et spaciale, de la densité des herbes marines laissent une signature dans les archives sédimentaires.

Introduction

Seagrasses are widely distributed in shallow coastal water throughout temperate and tropical areas, and have been the subject of many ecological and environmental studies. With their roots and rhizome systems established in the sediments, and their high growth rates and dense, leafy foliage, seagrasses have been found to have considerable influence on their environments. Wood *et al.* (1969) suggested that seagrasses act in several ways to control or modify the ecosystems of coastal water. The influence of seagrass on depositional environments was previously discussed by Ginsburg and Lowenstam (1958), who recognized that dense carpets of seagrass (*Thalassia*) in Florida Bay have a profound stabilizing influence on sand and mud substrate. Studies of carbonate sedimentation in Shark Bay of Western Australia (Davies, 1970), indicate that seagrasses are especially important in controlling sedimentary processes, and this was reflected by the sediments and carbonate accumulation within the seagrass communities, promoting the formation of carbonate banks. Scoffin (1970) working in the Bimini area, Bahamas, observed that a dense cover of seagrass can protect the substrate from tidal currents of up to 70 cm/sec. Wayne (1976) suggested that wave energy crossing dense seagrass beds can be decreased as much as 20 per cent because of the baffling effect of seagrass blades. In terms of deposition, Lynts (1966) and Patriquin (1975) indicated a positive associative between grain size characteristics and seagrass cover. Wanless (1981) documented an fining-upward sedimentary sequence generated by the migration of seagrass beds in southeast Florida.

Some recent studies have attempted to evaluate the effect of density of seagrass cover on current flow (Fonseca *et al.*, 1982; Fonseca, 1989), or to examine the impact of seagrass on coastal ecosystems (Short and Short, 1984; Castel *et al.*, 1989). To date, there is a general acceptance that seagrass can effectively baffle current flow, stabilize sediment substrate and thereby have an important influence on sedimentation. However, there are very few detailed studies that have documented the specific influence of seagrass beds on intertidal sedimentation.

Extensive seagrass beds (*Zostera muelleri, Heterocostera tasmanica* and *Posidonia australis*) exist in Corner Inlet, a large back-barrier embayment on the eastern coast of Victoria, southeast Australia (Fig. 1). Seagrass beds occur in both intertidal flats (typically *Zostera*) and shallow subtidal environments (typically *Posidonia*). The area provides an ideal field laboratory in which to investigate the depositional effects of seagrass on intertidal sedimentation. This paper is concerned specifically with the effect of seagrass beds on the intertidal dynamic regime and sediment distribution in this environment.

Study Area

Corner Inlet, situated in the southeast corner of Australia, is a shallow back-barrier embayment covering approximately 700 km^2 (Fig. 1) in a warm, temperate environment

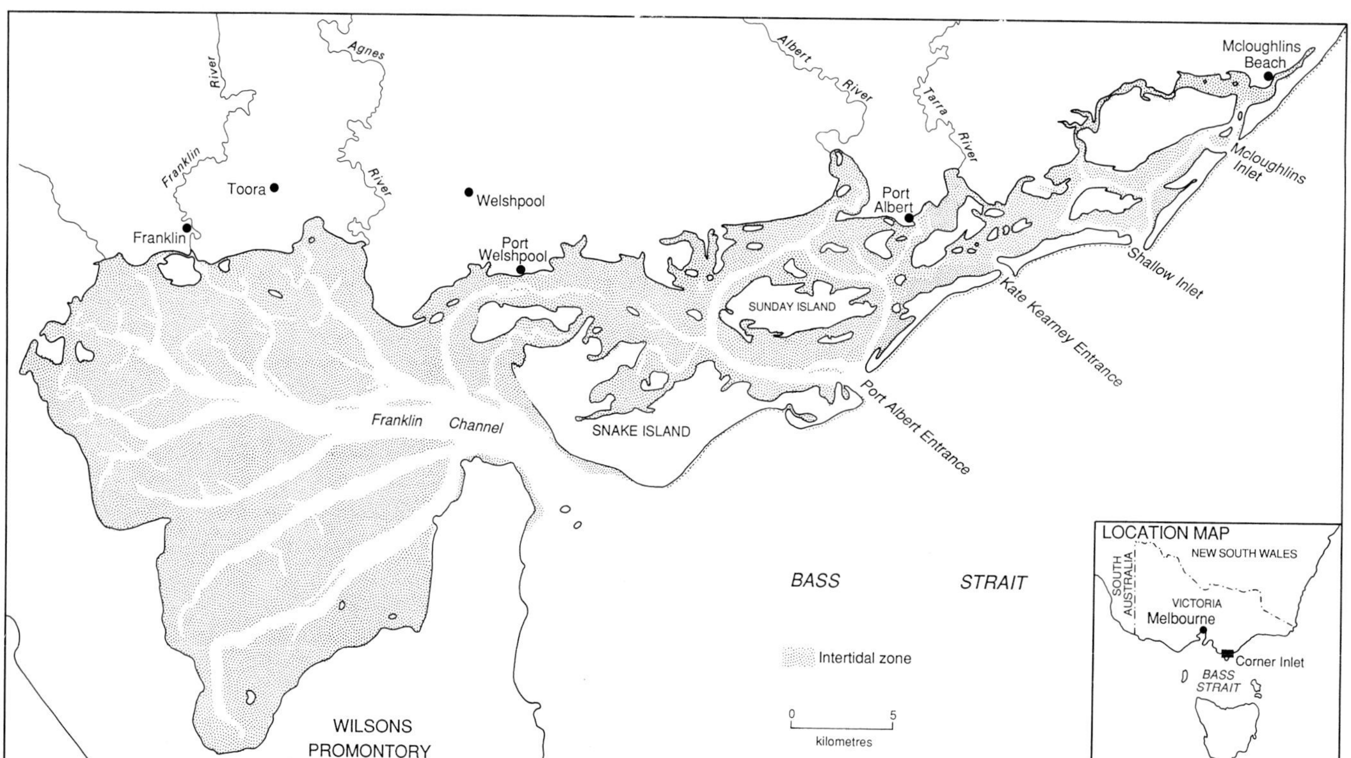

Figure 1. Map of Corner Inlet, southeast Australia, showing sites of field surveys at McLoughlans Beach, Port Albert, Port Welshpool, Toora, and Port Franklin.

with strong oceanic influence. Five entrances connect this embayment with the open sea (Fig. 1). The inlet experiences a strong tide-dominated hydrodynamic regime with a mean tidal range of approximately 2 m. Fluvial influence is minor. Both the mainland and lee shores of the barriers are bordered by well developed tidal flats (Fig. 1). Three basic tidal depositional environments have been recognized (Zhuang, 1988) according to their morphodynamic characteristics: 1) low energy protected tidal flats, represented at McLouglans Beach; 2) high energy elongated tidal flats, such as at Port Welshpool; and 3) extensive open tidal flats, which are well developed near Toora Beach and Port Franklin.

Grain size analysis shows that all depositional tidal flats in Corner Inlet are sand dominated. The intertidal environments of Corner Inlet can be subdivided into six geomorphic/sedimentological units (Fig. 2A). From high tide level to the subtidal channel, they are: 1) tidal salt marsh, which is the highest of the intertidal zones, and is colonized by salt marsh vegetation (*Sarcocornia spp. Arthrocnemum spp.*); 2) mangrove mud flat, which encompasses the region between mean high spring tides (MHST) and mean high neap tides (MHNT) and is colonized by the white mangrove (*Avicennia marina*); 3) upper sand flat, which lies just in front of the mangrove-fringed mud flat; 4) upper seagrass muddy sand flat, which occurs approximately at mid tide level (MTL) and is colonized by seagrass (*Zostera muelleri*); 5) lower sand flat, which is composed of rippled sand or shelly sand; and 6) lower seagrass mud flat, between low water neap tides (MLNT) and mean low spring tide (MLST), is covered by dense seagrasses (dominantly *Zostera muelleri*) and sandy mud (Fig. 2A). Each of these units has its distinctive, small-scale physiography, types of sedimentation and biological processes (Zhuang, 1988), and each usually occupies a specific shore-parallel zone of intertidal elevation (Fig. 2B). Some tidal flats do not contain all six units, in which case the vertical range of those present may be expanded.

Two major species of seagrass occur along the intertidal flats and shallow subtidal zone in Corner Inlet. *Posidonia australis* is dominant subtidally, and *Zostera muelleri* and *Heterozostera tasmanica* occur intertidally. In particular, the *Zostera* beds are very extensive on both upper intertidal and lower intertidal muddy sand flats (Fig. 2).

Methods

The depositional tidal flats were surveyed along transects normal to shore, and observations were made in both areas with seagrass beds and adjacent sand flats without seagrass cover. Experiments were designed to simultaneously measure

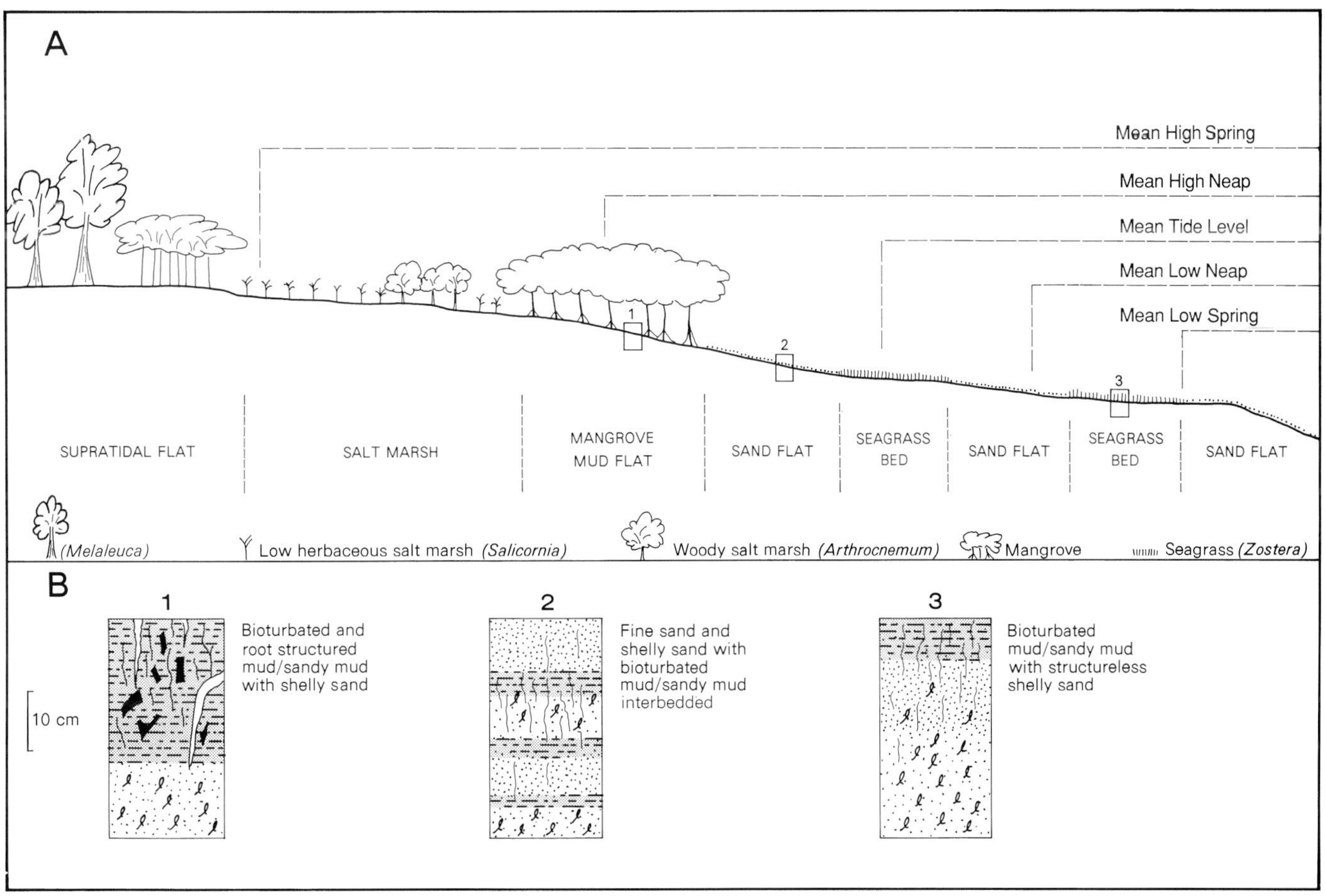

Figure 2. Typical zonation of the intertidal environments at Corner Inlet. **A.** diagrammatic cross-section of intertidal zonation showing geomorphic units of the depositional tidal flat in relation to tide level in Corner Inlet; mean spring and neap tidal ranges are shown on the right relative to mean tidal ranges (MTL). **B.** generalized sedimentary sequence occurring in mangrove, sand flat and seagrass beds in Corner Inlet.

the tidal current velocity and suspended sediment concentrations. Five measurement stations were established. Each station consisted of two sites less than 2 m apart — one on the seagrass bed and one on the adjacent sand flat. Surface sediment samples from these sites were collected extensively during the winter and summer months (July 1987-March 1988) when the hydrodynamics surveys were conducted.

Current velocities were measured using a bidirectional ducted flow meter, and an OTT current meter of the type used for streamflow measurements. Operation of the ducted flow meter is described by Bradshaw *et al.* (1978) and Short and Wright (1984). With the OTT current meter, readings were taken at 15 minute intervals with durations of 100 seconds. Simultaneously bottom velocity was measured at 15 cm above the bed (which is normally within the seagrass's canopy) and surface velocity was measured 5 cm below water surface. Readings were taken simultaneously over the seagrass bed and adjacent sand flat. During each experiment, measurements were carried out through a complete tidal cycle. Current velocity and suspended sediment concentration information presented in this paper were based on measurements carried out over four complete tidal cycles at McLoughlans Beach and Port Welshpool, respectively, during spring tide periods of January, 1988.

Suspended sediment concentrations (SSC) was determined gravimetrically from water samples collected using a set of evacuated bottles. Each was connected with two thick-wall plastic tubes: one for air outlet and the other for water inlet; the apparatus was supported by an aluminum pole. The position of the water inlet was adjusted to the chosen water depth. This technique is based on the principal of suction and is modified from Nielsen *et al.* (1982). Water samples were taken from near bottom (20 cm above bed) and the surface (15 cm below water level), over both seagrass bed and sand flat (without seagrass) at hourly intervals through a complete tidal cycle. Water samples were filtered in the laboratory using pre-weighed Sartorius filters (pore size 0.45 m), and dried at 110°C. The SSC of each sample was calculated from the dried weight of the filter papers.

In order to sample the very top layer of surface sediment, which is believed to directly reflect local depositional processes, a spatula was carefully used to collect the top several millimetres of sediment. Some 240 surface sediment samples were collected from the various intertidal units at the five field sites indicated in Figure 1. The bulk samples were processed to determine contents of carbonate and organic matter, and particle size distribution. A Rapid Sediment Analyser (RSA) based on a 2 m settling tube with computerized balance at its base and Centrifugal Automatic Particle Analyser (CAPA-300) were used for the sand fraction and mud fractions respectively. Additionally, 45 box core samples were taken from seagrass beds and sand flats using a modified method of Short and Wright (1984). The box cores were impregnated with Ciba-Geigy Araldite (K79 kit) to give a 40×25 cm peel of near surface sedimentary structures. Sediment samples were taken from box cores for grain size analysis.

Results

Tidal Current Velocity

Measurement of current velocity in the water column above seagrass beds and in areas devoid of seagrass is the basis for determining the effect of seagrass on current flow. Velocity data in Figure 3, chosen to be the representative of the study stations, indicate that the bottom current velocity over the seagrass bed is much lower (as much as 40-45%) than that on the adjacent sand flat (Fig. 3A), while the difference near the water surface is less than 5 per cent (Fig. 3B). These differences are sustained throughout the complete tidal cycle. The difference in bottom velocity is greatest during mid-stages of both flood and ebb tides, when the tidal current reaches its maximum. The bottom velocity over seagrass beds rarely exceeded 20 cm/sec during our observations. Results from McLoughlans Beach and Port Welshpool show a very similar current velocity profile. These differences in bottom and near-surface current velocities are sufficient to affect tidal sediment transport and deposition (Wayne, 1976; Fonseca *et al.*, 1982).

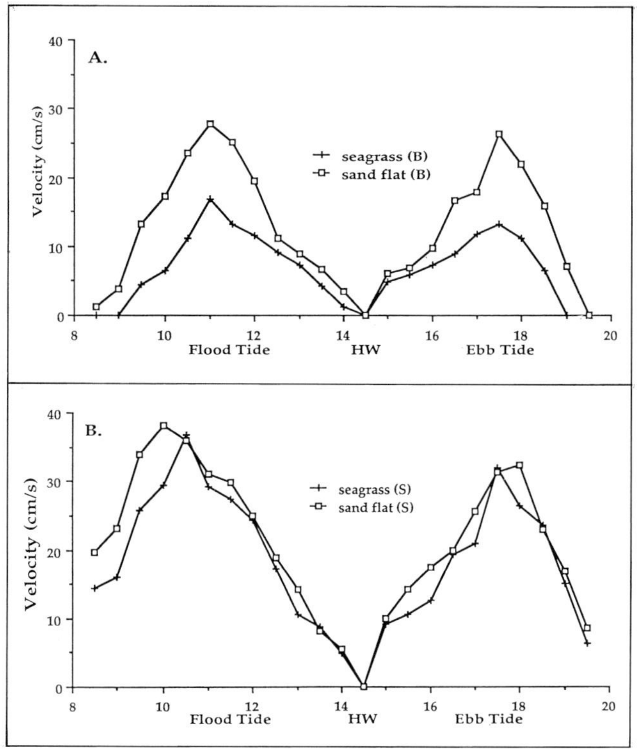

Figure 3. Variations in both bottom and surface current velocities over the seagrass bed and sand flat at the Port Welshpool site during one tidal cycle on 20 March, 1988. **A.** comparison of bottom current velocity measured 15 cm above seagrass and sand flat beds. **B.** comparison of surface current velocity measured at 5 cm below water surface over seagrass and sand flat beds.

Suspended Sediment Concentration

The suspended sediment concentrations in water samples from the seagrass bed system and the adjacent sand flat (Fig. 4) increase with current velocity in both cases, implying resuspension of fine sediment from the bed. Surprisingly, suspended sediment concentration values near the bottom are lower over the seagrass bed than over the adjacent sand flat (Fig. 4A). Suspended sediment concentration values decrease downward over the seagrass, but are much the same for both near-surface and the bottom on the sand flat (Fig. 4B). Moreover, the SSC profile (value) on the seagrass bed is consistently lower than over the sand flat.

There are two possible sources of resuspended sediment: either it comes from the underlying bed, or it is advected in suspension by tidal flow. The suspended sediment profile normally shows an increase in concentration with depth (Stanley and Swift, 1976; Friedman and Sanders, 1978), and this is the case over the sand flat sites in our study. However, our results indicate that the reverse is true over the seagrass beds. The SSC value over the seagrass beds is lower toward the bottom, suggesting that suspended sediment advected into the seagrass bed is removed in the zone of low current velocity. This process would transfer fine/clay size sediment preferentially into the seagrass bed. This interpretation was supported by the results of particle size analyses, from transects across the tidal flats, showing a higher percentage of fine grain (silt and clay) concentrations (15-20%) in the seagrass beds than in the adjacent sand flats (Fig. 5).

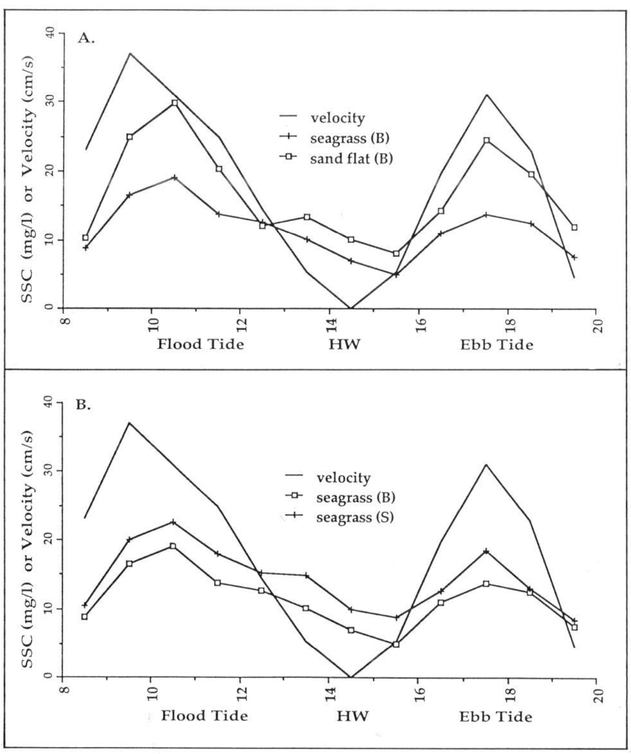

Figure 4. Variations in surface current velocity and suspended sediment concentrations of seagrass and sand flat measured at the Port Welshpool site during one tidal cycle on 20 March, 1988. **A.** bottom suspended sediment concentration of seagrass compared with that for sand flat; surface current velocity (sand flat) indicated. **B.** comparison of suspended sediment concentration value over the seagrass measured at surface and bottom; surface current velocity (sand flat) indicated.

Surface Sediment

The surface sediments from both seagrass bed and sand flat are predominantly sand, though the sand fractions in seagrass beds are smaller than on the adjacent sand flats (Fig. 5). Most of the sand-sized fraction, which normally constitutes 75 to 95 per cent of each sample, is mature, fine to very fine quartz sand with a minor feldspar and heavy mineral component. The remainder of the sand fraction is carbonate skeletal material. Much of the silt and clay fraction is organically produced. The gravel fraction, when present, consists mainly of shell fragments (Zhuang, 1988). As shown in Figure 5, the seagrass bed is characterized by a higher percentage of silt and clay (15-20%) and a higher content of organic matter (3.6-4.8%) (Fig. 6) than the adjacent sand flat. This clearly indicates that the seagrass can enhance the

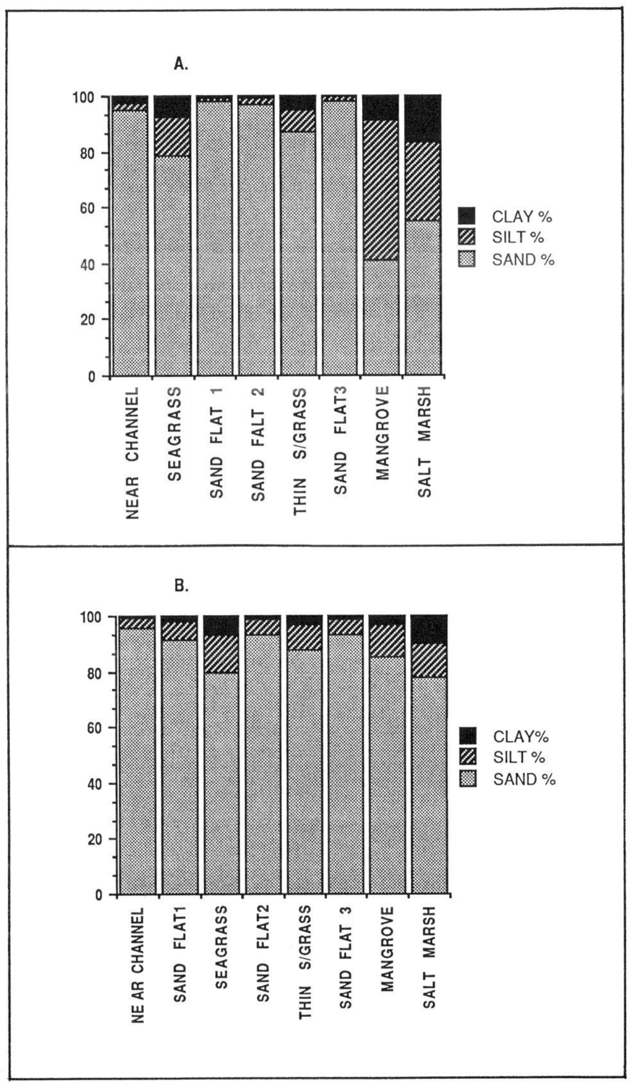

Figure 5. Textural variation (sand: silt/clay ratio) across the intertidal zone at Port Welshpool (**A**) and Toora Beach (**B**), Corner Inlet.

Table 1. Quantitative observations on the relationship of seagrass beds to intertidal sedimentary development in Corner Inlet, southeast Australia

Location	Seagrass Bed				Sand Flat			
	Mud (%)	Grain size (mean)	Skewness	Organic matter (%)	Mud (%)	Grain size (mean)	Skewness	Organic matter (%)
McLoughlan B.	20.5	2.65	0.65	2.85	7.40	1.96	−0.40	0.41
Port Albert	22.1	2.70	0.66	3.40	12.1	2.04	−0.23	1.15
Port Welshpool	18.6	2.48	0.55	2.01	5.30	1.78	−0.10	0.88
Toora Beach	19.6	2.51	0.51	4.21	10.6	1.94	0.13	1.44
Foster Beach	21.3	2.75	0.45	4.30	7.85	1.89	−0.05	2.76
Yanakie Beach	23.4	2.60	0.47	3.55	4.10	1.73	−0.36	1.09

Note:
1. the numbers presented here are the mean values of 15 surface samples from each field location
2. mean grain size value is calculated by the moment method and is presented in phi scale
3. the value of skewness is calculated by the moment method

accumulation of fine grained sediment and organic detritus. These results agree well with the previous studies which directly or indirectly suggest a positive correlation between seagrass cover and fine sediment accumulation (Lynts, 1966; Scoffin, 1970; Marshall and Lukas, 1970; Orth, 1977; Kenworthy *et al.*, 1982). A quantitative summary of the relationship between seagrass beds and intertidal sedimentary development is given in Table 1.

In terms of mechanisms involving sediment transport, the proportions of bedload, saltation and suspension transport of seagrass sediments are quite different from sand flat sediments. The sediment from seagrass beds has a smaller saltation fraction than the sediment from the adjacent sand flat (Table 2). Seagrass sediments are found to contain more finer grained material than those from the sand flat. The texture of seagrass sediment is charaterized by a strong fine-skewness and finer grain size (higher mud fraction), and sand flat sediments display a coarse-skewness, as is evidenced by the intertidal sediment analysis from the Welshpool site (Fig. 7 and Table 2). This difference in sediment texture suggests a different in hydrodynamic regime and sediment transport over the seagrass beds and sand flats.

Internal Sedimentary Structure

Results from box cores indicate that the seagrass in intertidal environments influences sedimentary structures (Zhuang, 1988). The seagrass bed tends to leave a layer of fine grained, highly organic sediment in the tidal sediment bodies (Fig. 8). A fining-upward stratigraphic sequence occurs when the seagrass bed invades areas previously free of seagrass cover (*e.g.*, sand flat). Figure 8 is a box core photo of seagrass bed sediments at McLoughlan Beach. The sediment on the top 15 cm has a higher content of finer sediment (20% of mud mean grain size 2.45ϕ), organic matter of 3.06%, is well sorted and strongly fine-skewed. In contrast, the lower sediments of the box core show a lower per-

Table 2. Grain size characteristics and transport dynamics of intertidal sediments at Port Welshpool, Corner Inlet, southeast Australia

Location	Transported Weight (%)			Sand (%)	Mud (%)	Mean	Skewness	Samples
	Bedload	Saltation	Suspension					
Tidal channel	40%	59.5%	0.5%	97.3	2.7	1.35	−0.75	13
Sand flat	9%	89.0%	2.0%	94.8	5.2	1.78	−0.14	16
Seagrass bed	2%	88.0%	10.0%	81.1	8.9	2.53	0.59	17
Mangrove	1.2%	83.8%	15.0%	59.3	40.7	2.49	0.63	15
Salt marsh	0.0	64.2%	35.8%	65.6	34.4	2.56	0.55	15

Note:
1. the transported weight (%) is estimated from the log-probability plots of each sample and the grain size range is between 0.0 phi and 4.5 phi
2. mean grain size value is calculated by the moment method and is presented in phi scale
3. the value of skewness is calculated by the moment method
4. samples refer to the number of observation samples for a particular depositional environment, and the value presented here is the mean of observed samples

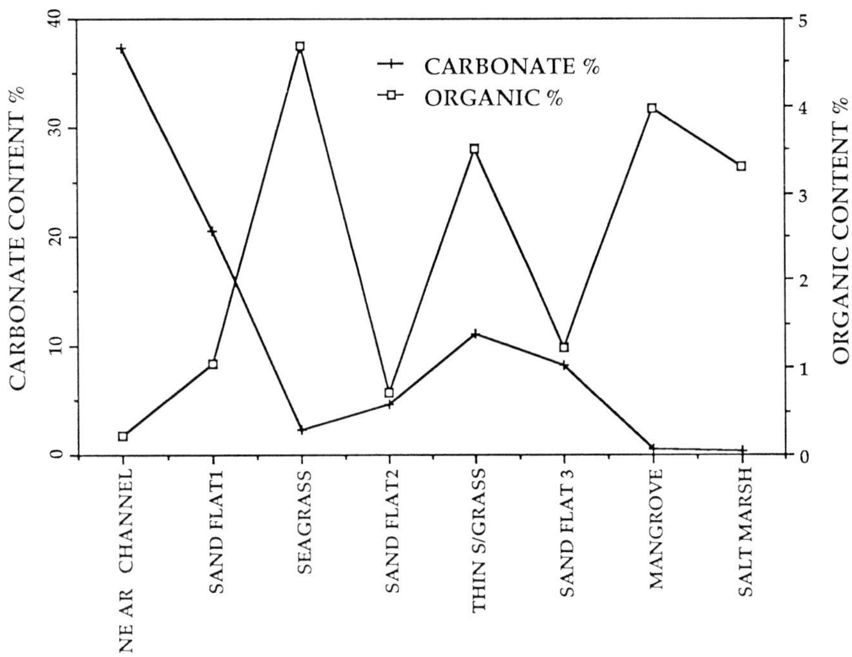

Figure 6. Variations in organic matter and carbonate contents across the intertidal zone at Toora Beach, Corner Inlet.

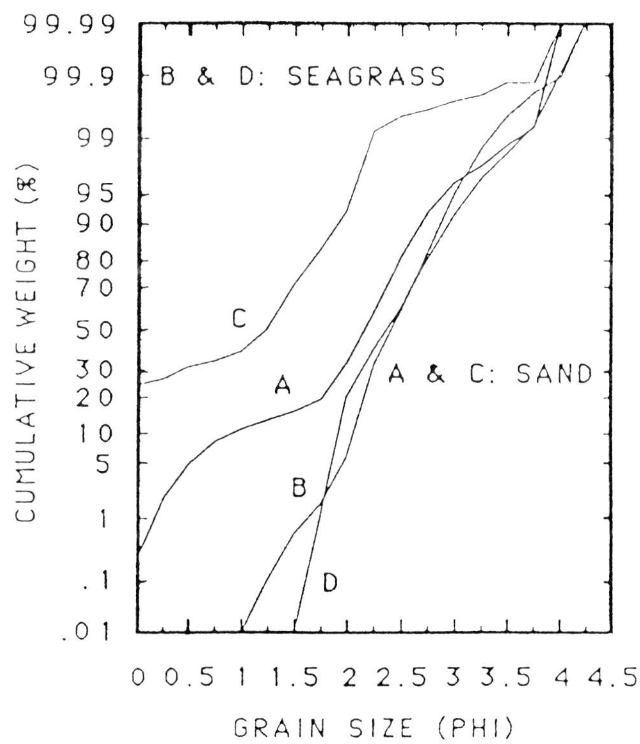

Figure 7. A comparison of grain size distributions (log-probability scale) of seagrass and non-seagrass sediments in Corner Inlet.

centage of mud content (7.2%), are strongly coarse-skewed (−0.90) and moderately well sorted, and thus are believed to represent a former sand flat environment. In other cases, a multiple fining-upward sequence can be preserved in the vertical sequence (Fig. 9), indicating a temporal and spatial change of seagrass cover in a particular intertidal environment.

Discussion and Conclusions

Depositional Effects of Seagrass Beds

Our study of seagrass bed and sediment interaction effects in intertidal environments suggests that seagrass has at least three distinct effects on tidal sedimentary processes.

The Baffle Effect

It has been suggested by many researchers that seagrass can substantially baffle current velocity both in shallow coastal water (Ginsberg and Lowenstam, 1958) and in laboratory experiments (Fonseca *et al.*, 1982). Because the seagrass is pliable and semipermeable, its stems and leaves become a physical obstacle to current flow. Above the seagrass canopy, tidal flow is unobstructed (Fig. 3). Results of observed current velocities over seagrass beds and adjacent sand flats in the intertidal environment of Corner Inlet indicate that seagrass beds can significantly reduce tidal current velocity by as much as 40 per cent near the sediment-water interface (Fig. 3a), but have little effect on the surface velocity (Fig. 3b). Obviously, this reduction of near-bottom current velocity is a function of seagrass baffling. It is clearly demonstrated by the velocity profiles (Fig. 3) and is considered here to be the principle cause of deposition in seagrass beds.

The Trapping Effect

The lower values of suspended sediment concentration over seagrass beds and the higher proportion of fine grained sediment in the beds themselves suggests that the reduction

Figure 8. A fining-upward sequence in the peel of a box core taken from seagrass substrate, intertidal zone at Mcloughlans Beach, Corner Inlet. Layer A is seagrass sediments and layer B is sand flat deposits.

of bottom current velocity by the seagrass traps fine grained sediment. Fine grained sediment may also be trapped directly on the seagrass leaves. Once the fine grained particles are stuck to the seagrass leaves, they no longer behave as discrete grains (Scoffin, 1970), and thus are not easily resuspended.

The Binding Effect

This occurs as a result of seagrass roots penetrating the sediment substrate and holding the sediment together, making it resistant to erosion.

Seagrass-Sediment Interactions

It is evident that seagrass beds affect (modify) the pattern of tidal deposition expected from sole physical sedimentation in intertidal environments. The net results of the depositional influence of seagrass in Corner Inlet is clearly demonstrated by the pattern of grain size distribution across the intertidal zone (Fig. 5).

Tidal flow advection and deposition of fine sediment in the seagrass bed implies that in areas of seagrass, sediments will accumulate at a faster rate than on the adjacent sand flat. However, the seagrass bed cannot act as a perfect sink for accumulating fine sediment (Kennedy, 1984) for long periods. Interactions between tidal currents, wave processes and seagrass will change as sedimentation continues, and may eventually stop vertical accretion. If the seagrass zone should become elevated above neighbouring sand flats, wave action and tidal velocities over the seagrass may increase. Resulting increased turbidity may kill the seagrass and the area may then become a sand flat (*cf.* Kemp *et al.*, 1983; Orth and Moore, 1983). Our box core samples show that sand flats at some localities are immediately underlain by seagrass deposits, which contain higher percentages of mud and organic matter. An example is shown in Figure 9, where sandy deposits (layers A and C) overlie organic muddy sands (layers B and D). In other cases, a fining-upward sequence was produced by the onlap (migration) of seagrass onto a previous sand flat. This is clearly shown in the vertical texture change of Figure 8. The top layer is a highly organic mud believed to be the seagrass deposits; the bottom section is a layer of a fine to medium grained, sand flat sediment.

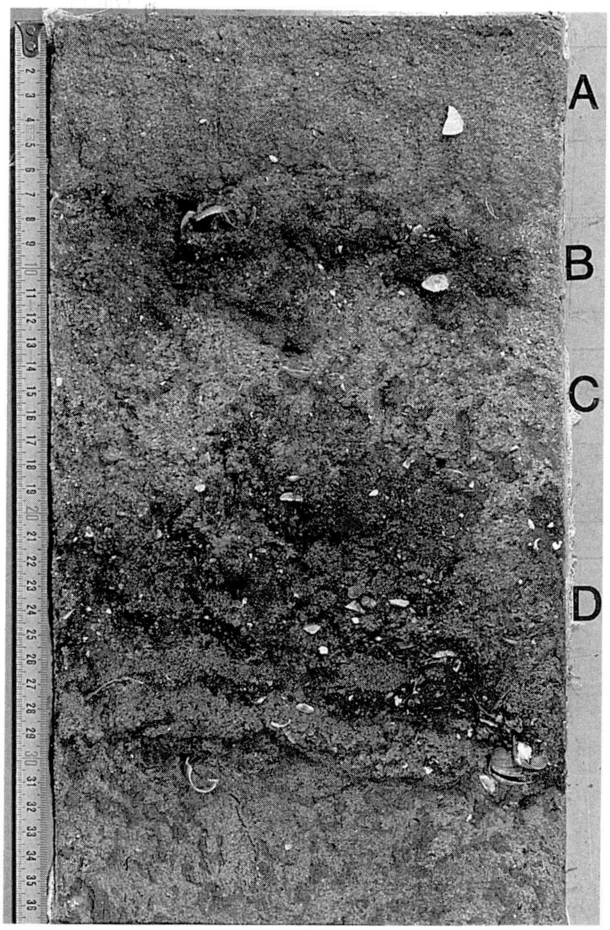

Figure 9. Cyclic fining-upward sequence observed in the peel of a box core taken from the sand flat in the intertidal zone at Mcloughlans Beach, Corner Inlet. Layers A and C are sand flat sediments and layers B and D are seagrass sediments.

Alternatively, sediment advection may decrease if the seagrass bed is relatively elevated. Finally, storm waves may disperse the fine sediment accumulated on the seagrass bed during normal conditions and then reverse the direction of fine sediment transport (it is believed here that fine sediment is transported from sand flat to seagrass during normal conditions), while at the same time damaging the seagrass bed. We have no direct observations of these processes, but variations from the intertidal zonation of Figure 2 and box core data indicate that seagrass beds are not static features in Corner Inlet.

Tidal Sediment Distribution

The general shoreward fining of intertidal sediment, suggesting a tendency of shoreward-decreasing tidal energy as reported by many previous workers (Klein, 1984; Li and Li, 1982; Reineck and Singh, 1980; van Straaten, 1961; Wang, 1983; Weimer *et al.*, 1982), is not generally applicable in Corner Inlet. Our study shows that an alternation of greater and lesser sand: silt/clay ratios is related to seagrass beds trapping fine sediment. The accumulation of fine grained sediments is a result of three influencial factors as discussed above: 1) current reduction by seagrass leaves; 2) the entrapment of suspended sediment particles carried over the seagrass beds from adjacent sand flats; and 3) the preservation of silt- and clay-sized particle produced in situ from the seagrass biota by organic, biochemical and mechanical breakdown. The extent of these influences appears to vary through time, reflecting the interaction between seagrass-induced sedimentation and local hydrodynamics.

Our study also suggests that the existence of seagrass beds in the intertidal zone will exert structural and functional effects (*e.g.*, reducing current velocity, trapping sediment and preventing resuspension of finer sediment) on tidal depositional environments. Because of these effects, the natural processes of tidal deposition will be modified, alternating in space and time between seagrass and sand flat regimes. As a result, normal tidal sediment distribution in Corner Inlet shows an alternation of fine and coarse grained sediment related to the presence or absence of seagrasses (Figs. 2A, 5) instead of a progressive change in grain size.

The alternation of seagrass beds and sand flats across the intertidal zone in Corner Inlet is interesting, and not fully explained by our study. It may reflect the wave energy distribution, averaged through tidal cycles, in this low energy shoaling environment, and major storms. Integrated current monitoring is required to explore this.

The role played by seagrass in trapping fine grained sediments and stabilizing the substrate in intertidal environments is thus geologically significant. It is also clear that seagrass beds leave a signature in the sediment body, being preserved as layers of highly organic matter within a finer sediment matrix (Fig. 2B). Further investigations of the upper Holocene sedimentary record, using sites with contrasting aggradational histories, may suggest mechanisms that cause spatial variation of seagrass beds, as well as improving our understanding of sedimentary processes in tidal flat environments.

Acknowledgments

Financial and logistical support for this work was provided by the Australian National University. We would like to thank D. Kelleher and the Department of Conservation of Forests and Lands in Yarram for their field assistance, and members of the Cartographic Unit of the Research School of Pacific Studies (ANU) for their assistance with some diagrams. We are very grateful to A. D. Short, Li Congxian, D. Neil, B. G. Thom and H. R. Wanless, for their review comments.

References

Bradshaw, M. P., Chappell, J., Hales, R. S. and Wright, L. D. 1978. Field monitoring and analysis of beach and nearshore hydrodynamics. 4th Australian conference of Coastal and Oceanographic Engineering, p. 171-175.

Castle, J., Labourg, P-J., Escaravage, V., Auby, I. and Garcia, M. E. 1989. Influence of seagrass beds and oyster parks on the abundance and biomass patterns of meso- and macrobenthos in tidal flats. Estuarine, Coastal and Shelf Science, v. 28, p. 71-85.

Davies, G. R. 1970. Carbonate bank sedimentation, Eastern Shark Bay, Western Australia. *In:* Carbonate Sedimentation and Environments, Shark Bay Western Australia, B. W. Logan, G. R. Davies, J. F. Read and D. E. Cebulski (eds.). American Association of Petroleum Geologists, Memoir 13, Oklahoma, U.S.A., p. 85-168.

Fonseca, M. S. 1989. Sediment stabilization of *Halophila decipiens* in comparison to other seagrasses. Estuarine, Coastal and Shelf Science, v. 29, p. 501-507.

______ , Fisher, J. S., Zieman, J. C. and Thayer, G. W. 1982. Influence of the seagrass, *Zosteramarina* L., on current flow. Estuarine, Coastal and Shelf Science, v. 15, p. 351-364.

Friedman, G. M. and Sanders, J. E. 1978. Principles of Sedimentology. Wiley, New York, 792 p.

Ginsberg, R. N. and Lowenstam, H. A. 1958. The influence of marine bottom communities on the depositional environment of sediments. Journal of Geology, v. 66, p. 310-318.

Kemp, W. M., Boynton, W. R., Twilley, R. R., Stevenson, J. C. and Means, J. C. 1983. The decline of submerged vascular plants in Upper Chesapeake Bay: summary of results concerning possible causes. Journal of Marine Science and Technology, v. 17, p. 78-89.

Kennedy, V. S. 1984. The Estuary as a Filter. Academic Press, New York, 514 p.

Kenworthy, W. J., Zieman, J. C. and Thayer, G. W. 1982. Evidence for the influence of seagrasses on the benthic nitrogen cycle in a coastal plain estuary near Beaufort, North Carolina (USA). Oecologia, v. 54, p. 152-158.

Klein, G. D. 1984. Intertidal flat and intertidal sand bodies. *In:* Coastal Sedimentary Environments, R. A. Davis (eds.). 2nd edition, Springer-Verlag, New York, p. 187-224.

Li, C. and Li, P. 1982. Sediments and sand bodies on the tidal flats. Oceanologia et Limnologia Sinica, v. 13, p. 48-59.

Lynts, G. W. 1966. Relationship of sediment-sized distribution to ecologic factors in Buttonwood Sound, Florida Bay. Journal of Sedimentary Petrology, v. 36, p. 66-74.

Marshall, N. and Lukas, K. 1970. Preliminary observations on the properties of bottom sediments with and without eelgrass, *Zostera marina* cover. Proceedings of National Shellfisheries Association, v. 60, p. 107-111.

Nielsen, P., Green, M. O. and Coffey, F. 1982. Suspended sediment under waves. Coastal Studies Unit, Technical Report 82/6, Coastal Studies Unit, University of Sydney, 157 p.

Orth, R. J. and Moore, K. A. 1983. Chesapeake Bay: an unprecedented decline in submerged aquatic vegetation. Science, v. 222, p. 51-53.

______ 1977. The importance of sediment stability in seagrass communities. *In*: Ecology of Marine Benthos, B. Coull (ed.). University of South Carolina Press, Columbia, p. 281-300.

Patriquin, D. G. 1975. Migration of blowouts in seagrass beds at Barbados and Carrocou, West Indies and its ecological and geological implications. Aquatic Botany, v. 76, p. 710-717.

Reineck, H. E. and Singh, I. B. 1980. Depositional Sedimentary Environments. 2nd edition, Springer-Verlag, New York, 549 p.

Scoffin, T. P. 1970. The trapping and binding of subtidal carbonate sediments by marine vegetation in Bimini Lagoon, Bahamas. Journal of Sedimentary Petrology, v. 40, p. 249-273.

Short, A. D. and Wright, L. D. 1984. Field methods in wave dominated surfzone and nearshore environments. *In:* Divers, Submersibles and Marine Science, N. C. Flemming (ed.). Occasional Paper in Biology, No. 9, Memorial University of Newfoundland, p. 86-99.

Short, F. T. and Short, C. A. 1984. The seagrass filter: purification of estuarine and coastal water. *In:* The Estuary as a Filter, V. S. Kennedy (ed.). Academic Press, New York, p. 395-413.

Stanley, D. J. and Swift, D. J. P. 1976. Marine sediment transport and environment management. John Wiley & Sons, New York, 602 p.

Van Straaten, L. M. J. U. 1961. Sedimentation in tidal flat areas: Journal of the Alberta Society of Petroleum Geology, v. 9, p. 203-226.

Wang, Y. 1983. The mud flat system of China. Canadian Journal of Fisheries and Aquatic Sciences, v. 40, p. 160-171.

Wanless, H. R. 1981. Fining-upward sedimentary sequences generated in seagrass beds. Journal of Sedimentary Petrology, v. 51, p. 445-454.

Wayne, C. 1976. The effects of sea and marsh grass on wave energy. Coastal Research Notes, v. 4, p. 6-8.

Weimer, R. J., Howard, J. D. and Lindsay, D. R. 1982. Tidal flats and associated tidal channels. *In:* Sandstone Depositional Environments, P. A. Scholle and D. Spearing (eds.). American Association of Petroleum Geologists, Wisconsin, p. 191-246.

Wood, E. J. F., Odum, W. E. and Zieman, J. C. 1969. Influence of sea grasses on the productivity of coastal lagoons. *In:* Coastal Lagoons, a Symposium, C. A. Ayala and F. G. Phleger (eds.). Universidad Nacional Autonoma de Mexico, Ciudad Universitaria, p. 495-502.

Zhuang, W. Y. 1988. The sedimentary character of tidal flat system in Corner Inlet, Southeast Australia. Abstract. 26th Congress of the International Geographic Union. Sydney, p. 671.

KEY WORDS: SEAGRASS, TIDAL FLATS, CORNER INLET, SOUTHEAST AUSTRALIA, SAND FLATS

In: Clastic Tidal Sedimentology, D. G. Smith, G. E. Reinson, B. A. Zaitlin and R. A. Rahmani (eds.). Canadian Society of Petroleum Geologists. Memoir 16, p. 301-312.

Sedimentology of subarctic tidal flats of western James Bay and Hudson Bay, Ontario, Canada

I. P. Martini

Department of Land Resource Science
University of Guelph
Guelph, Ontario, Canada, N1G 2W1

Abstract

The 3 to 5 m thick muddy and sandy tidal flats on the western and southwestern shores of marine, mesotidal, subarctic, locally brackish, intracratonic basins of James Bay and Hudson Bay, resemble those of other basins, except they do not contain major tidal channels and they display characteristic features that reflect their cold setting. These features include the following: 1) surficial scours caused by ice floes or ice-pushed boulders and filled with poorly sorted, polymictic sand and gravel; 2) scattered, frost-resistant, subrounded, igneous and metamorphic glacial erratics locally reworked by sea ice; 3) frost-shattered, angular pebbles of limestone and dolostone; and 4) coarse, ice-rafted material enclosed in fine grained, water-laid deposits. The macro-infauna is restricted in numbers of species (primarily *Macoma balthica* and *Hydrobia minuta*) and locally in numbers of individuals (along brackish coasts).

The coastal marshes are dominated by *Puccinellia phryganodes, Carex subspathacea*, and *Hyppuris tetraphilla*, and the marsh deposits develop diagnostic laminae reflecting either mat root structure of *Puccinellia phryganodes*, or buried, discretely spaced hollow stems of *Hippuris tetraphilla*. The marshes grade landward into extensive emerging peatlands (fens and bogs), where, 100 km inland, 3 to 4 m of peat have accumulated in the last 3500 to 4000 years. Cryoturbation and other pedogenic processes imprint emerged suquences with convolutions and gleysolic characteristics.

Résumé

Les bassins intracratoniques des baies James et d'Hudson sont des bassins subarctiques marins, localement saumâtres, à régime mésotidal. À leurs marges sud-est, les estrans boueux et sableux, de 3 à 5 m d'épaisseur, ressemblent à ceux de d'autres bassins, mais ils s'en différencient par l'absence de chenaux de marée majeurs et la présence de caractéristiques reflétant un environnement froid. Ces caractéristiques sont: 1. des «figures en relief» (surficial scours) produites par des banquises ou de gros blocs transportés par la glace, remplies par des sables et graviers polygéniques mal triés; 2. des roches erratiques dispersées, d'origine ignée ou métamorphique, subarrondies, résistantes à la gelée et localement remaniées par la glace de mer; 3. des galets anguleux de calcaires et dolomies, éclatés par le gel; 4. du matériel grossier transporté par des radeaux de glace et déposé dans des sédiments fins en milieu submergé. L'endofaune macroscopique contient un nombre restreint d'espèces (principalement *Macoma balthica et Hydrobia minuta*), et localement, un nombre restreint d'individus le long des côtes saumâtres.

Les marais côtiers sont dominés par *Puccinellia phryganodes, Cares subspathacea* et *Hyppuris tetraphilla*. Les laminations diagnostiques de ces dépôts reflètent soit des tapis de structures de racine appartenant à *Puccinellia phryganodes*, soit l'enfouissement d'*Hippuris tetraphilla* dont les tiges creuses sont légèrement espacées. Les marais passent graduellement vers l'intérieur à des tourbières («fens» et «bogs», respectivement traduits de l'anglais par tourbières minérotrophes et tourbières ombrotrophes). Durant les derniers 3500 à 4000 ans, de 3 à 4 m de tourbes se sont accumulés à une distance d'environ 100 km à l'intérieur du continent. Les séquences émergées sont marquées par des convolutions et des caractéristiques spécifiques au gley, formées par la cryoturbation et autres processus pédologiques.

Introduction

The tidal flats of James Bay and Hudson Bay differ from other cold climate tidal flats in that they have developed in a combination of semidiurnal mesotides (3 m), and relatively low energy conditions on very flat shores. These flats can be strongly affected by freshwater fluvial plumes.

The subarctic area studied is underlain by flat-lying Paleozoic carbonates and shales, and the surface was glaciated during the Pleistocene. The area is still emerging (0.7 m/century) as a result of isostatic rebound (Norris and Sanford, 1968; Craig, 1968; Webber *et al.*, 1970).

The sea is covered by ice for six months of the year, is affected by severe storms in the fall, and by a persistent (approximately 1 knot) geostrophic, counterclockwise current (Prinsenberg, 1983). Depositional tidal flats several kilometres wide have developed in embayments, updrift from low (5 m), narrow (30-50 m), long (2-4 km) transverse ridges that act as natural groins, and in protected areas behind sand barriers or bouldery shoals (Martini and Protz, 1978; Martini, 1981b, 1986). Some of the tidal flats experience changes in salinity ranging from brackish conditions, attributed to a high, fluvial influx in spring and as a result of in situ melting of ice blocks, to high salinity during local winter freezeup of the intertidal zone. High salinities are also caused by summer evaporation of storm surge waters, and by desalination of underlying Pleistocene, marine sediments (Glooschenko and Martini, 1987; Price and Woo, 1987, 1988).

Objectives

The objective of this paper is to present the characteristic depositional features of subarctic tidal flats of western James Bay and southwestern Hudson Bay (referred to hereafter as tidal flats of James Bay and Hudson Bay) (Fig. 1). The paper emphasizes aspects of tidal flat deposits associated with cold climatic conditions.

Sampling Design and Analytical Methods

Representative coastal transects were sampled in summer from the low tide mark to upper marshes (Fig. 1). Surficial grab samples, 24 cm deep Shenkenberg box cores, and samples from hand excavated pits (in drier areas) were collected every 200 to 400 m depending on the width of the flat. Additional samples were taken from areas showing rapid lateral variation in surficial sedimentary features and/or vegetation type. The fresh box-core samples were photographed, then epoxy peels (mixture of equal amounts of FAUMO 1000 resin and FAUMO CURE 62, backed with cheesecloth) were taken. The remaining sediments were used for textural, mineralogical (modal analyses, heavy minerals, clay mineralogy), and chemical (pH, calcium carbonate equivalent, organic matter, conductivity, and organically-bound iron and aluminum) analyses (Folk, 1974; McKeague, 1978; Martini and Protz, 1978, 1980, 1981; Martini *et al.*, 1979).

Samples for infaunal studies consisted of three round cores (11 cm in diameter by 15 cm long) collected from the vertices of an equilateral triangle randomly placed near each sediment sample location on the unvegetated intertidal flats. Macro-organisms were retrieved by passing the sample through a 200 micron sieve, and were preserved in an alcohol solution for identification and counting. Small polychaetes and oligochaetes were separated by floatation in a sugar solution (Anderson, 1959). The ages of *Macoma balthica* specimens were estimated by counting annual growth rings and length of the shells (Clarke, 1980; Martini and Morrison, 1987).

Results

Eight major zones have been recognized across typical low-lying, sandy and muddy tidal shores of James Bay and Hudson Bay: shelf, shallow subtidal pools, lower tidal flats, upper tidal flats, high tidal flats, lower marsh, upper marsh, and inland wetlands and peatlands.

Shelf

Little information is available on modern shelf deposits of James Bay and Hudson Bay, but some parts of them can be observed where uplifted by postglacial rebound and exposed along river banks (Pelletier, 1968, 1986; Kranck and Ruffman, 1982; Martini and Grinham, 1983). The exposed modern shelf deposits vary from local, ice-rafted gravel and washed, sandy gravels on shoals to silt and silty clays, up

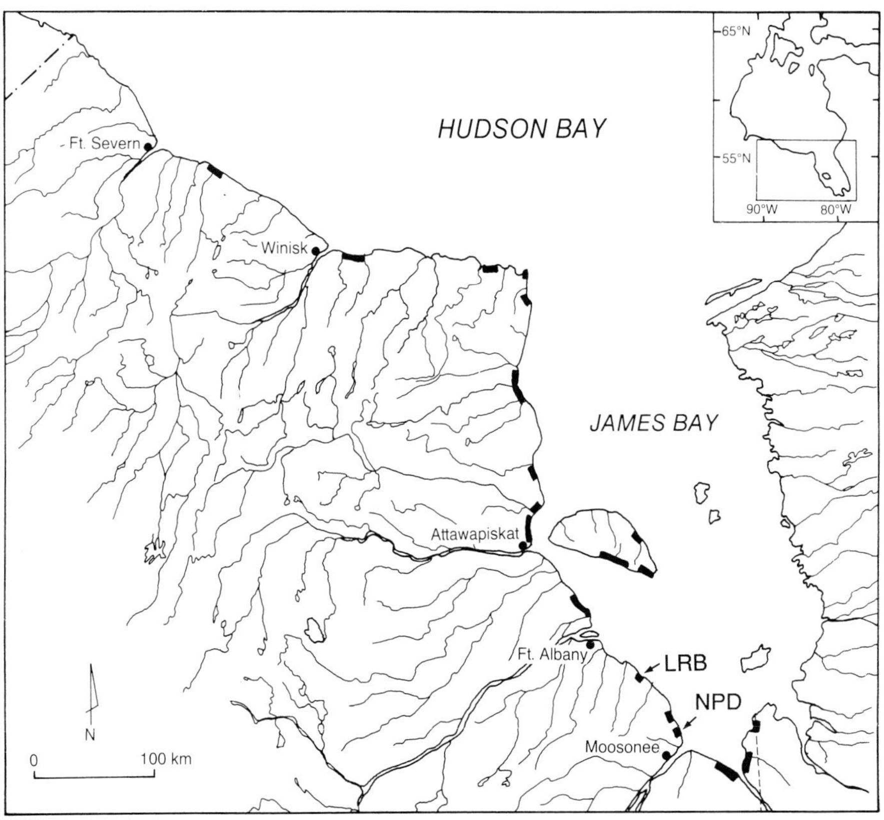

Figure 1. Location map of extensive tidal flats studied along the Ontario coast. NPD and LRB: coastal transects mentioned in this paper.

to 60 m thick, deposited in deeper depressions (Martini and Grinham, 1983). The fine grained deposits are sparsely fossiliferous, massive to faintly laminated, with thin interlayers of crosslaminated sand and occasional "lonestones" (Fig. 2A). The silty component suggests a glaciomarine origin with some lonestone contributions from recent ice rafting.

Shallow Subtidal Pools

This zone has been studied most extensively at North Point (NPD) where an outer, bouldery shoal protects a shallow (0.5-1.5 m), wide (500-700 m), subtidal pool on the seaward side of the wide (2 km) tidal flats (Fig. 1) (Martini and Protz, 1978; Clarke *et al.*, 1982). The pool contains extensive growth of *Zostera marina*, which traps silty sand. Several bald and planed-off hummocks (less than 1 m high) are present, some rising nearly to low tide levels. A restricted macrobenthic fauna is present with numerous individuals. *Macoma balthica* has concentrations ranging from about 8600/m^2 for the whole pool to about 12,800/m^2 in the heavily vegetated area. *Hydrobia minuta* has concentrations of 8300/^{2}, locally as high as 33,600/m^2 in the vegetated areas.

The internal sedimentary sequence is dominated by wavy bedding (Fig. 2B). Hairline rootlets, thin worm burrows and larger *Macoma* ones are abundant, but they never obliterate primary sedimentary structures.

The major sedimentological processes acting in this environment are related to tidal flows, wave action and ice scouring and rafting. Tidal flows carry considerable suspended material into the pools, and waves resuspend large amounts of fine sediment. Ice pressure ridges, which form on the sea side of the outer shoals, effectively protect the subtidal pools from ice scour; however, grounding of shore ice generates a hummocky bottom topography by squeezing sediments in fissures (similar to the "inverted mud cracks" of Parker (1963), and by concentrating material into cracks on the ice surface during melting. Ice floes floated by high tides and currents during spring breakup are responsible for truncating the crests of the hummocks.

Lower Tidal Flat

This zone extends from the mean low tide level to where a slight break in slope and conspicuous change in surficial features occurs in the unvegetated intertidal zone (Fig. 3). No significant tidal channels are present on the tidal flat. Sediments are composed of medium grained sand with scattered pebbles and subrounded, isolated igneous and metamorphic boulders. The surface of the flats is covered by tidal-flood-generated ripples commonly showing truncated crests. Ice gouges are readily observed in the spring after breakup, and are preserved throughout the subsequent summer. The infauna, primarily *Macoma balthica*, comprises both mature and immature individuals in concentrations of up to 6600/m^2 (August) (Fig. 4, Table 1). *Hydrobia* is also

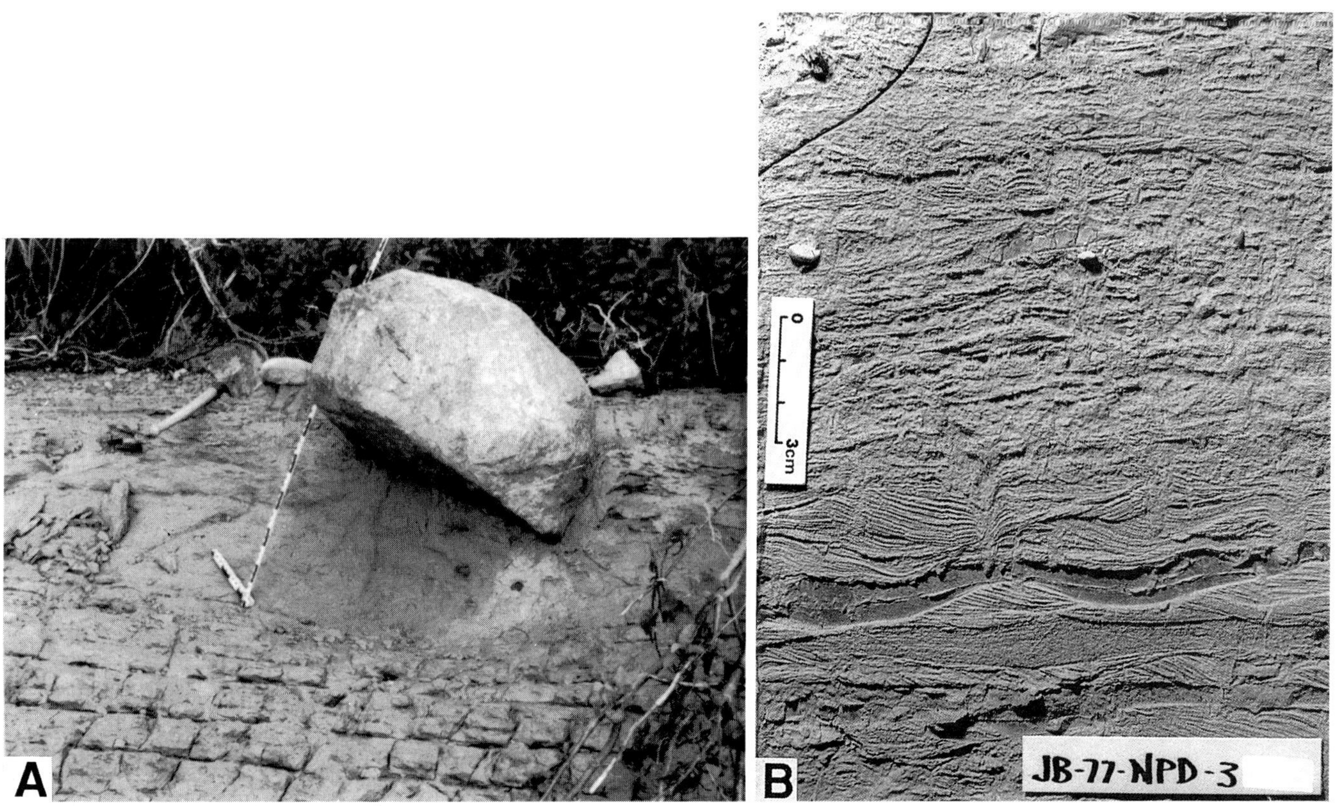

Figure 2. Characteristic features of sublittoral deposit and of lower tidal flat. **A.** Large dropstone in fossiliferous, marine, sublittoral silty clay; **B.** Box core taken from subtidal pool. Note ripple crosslaminae, flasers and burrows.

present, but not in as large numbers (3370/m^2 in the peak period of mid-August) as *Macoma* (Clarke, 1980).

The lower tidal flat has a characteristic internal sequence exhibiting ripple crosslaminae, alternating with plane beds, flaser bedding, numerous *Macoma* burrows, rare deformation structures, and rare, isolated pebbles and lenses of granules (Fig. 5A). "Diffused lenses" are also present, and are characterized by a gradual fining from a central core (coarse sand and granules in fine grained matrix) toward the surrounding better sorted, fine grained, silty sand deposits (Grinham, 1980; Grinham and Martini, 1983).

Tides and waves rework most of the granular sediments, including part of the ice-rafted material, except for the "diffuse lenses" which retain a core of rafted material intact. Strong winds perennially blowing over the exposed flats, accelerate the thin water sheets of ebbing tides resulting in supercritical flows capable of truncating ripple marks and creating plane beds. In addition, ice scours, boulder pushing, and rafting are directed primarily alongshore and landward, indicating the persistance of offshore ice cover and the importance of onshore ice-push ridges.

Upper Tidal Flat

The transition between the lower and upper tidal flats is gradational. The upper tidal flat is characterized by slightly steeper slopes, less uniformity, and smoother sandy surfaces than the lower flat, and may contain strewn pebbles and boulders (Fig. 5B). Ripple marks are still the dominant surficial feature. Considerable macrobenthos occur, primarily juvenile *Macoma balthica*, numerous *Hydrobia minuta*, and polychaetes and oligochaetes (Table 1).

Internal sedimentary sequences show crosslaminae alternating with plane beds, "cut and fill" features composed of interbedded polymictic, poorly sorted fill, isolated ice-rafted pebbles, and *Macoma* burrows.

Ice plays an important role in moulding this zone. Various types of scour develop and may persist. Some are long (80 m), straight and deep (up to 25 cm in depth), and are formed by the landward bulldozing of boulders pushed by pack ice. Another scour type is both shorter and shallower and presents characteristic, small, terminal hooks directed seaward. These scours are interpreted as having been formed by boulders and pebbles carried on the underside of ice floes (Martini, 1981a), the hooks being generated during the tidal reversal. Other types of scour are characterized by shallow and very wide, brush-like marks. They are formed by wide ice floes grounding in the uppermost part of the tidal flats as they are transported shoreward. Other ice-related features include rings of rafted pebbles; low (20-30 cm), long (several tens of metres) ridges composed of pebbles; sand and silt washed from the surface of the ice into cracks between ice blocks; and other erosional and depositional structures described elsewhere (Martini, 1981a; Dionne, 1985). Some

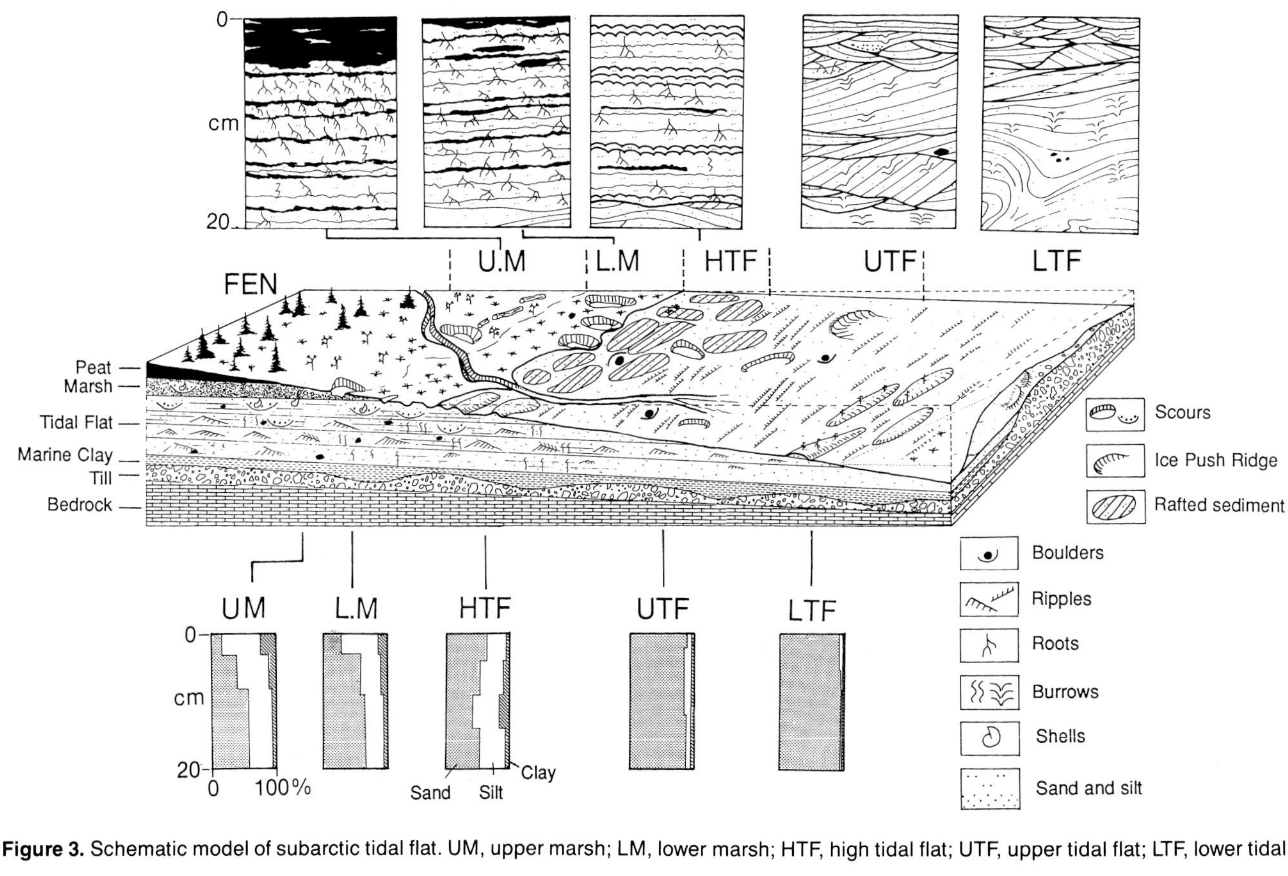

Figure 3. Schematic model of subarctic tidal flat. UM, upper marsh; LM, lower marsh; HTF, high tidal flat; UTF, upper tidal flat; LTF, lower tidal flat.

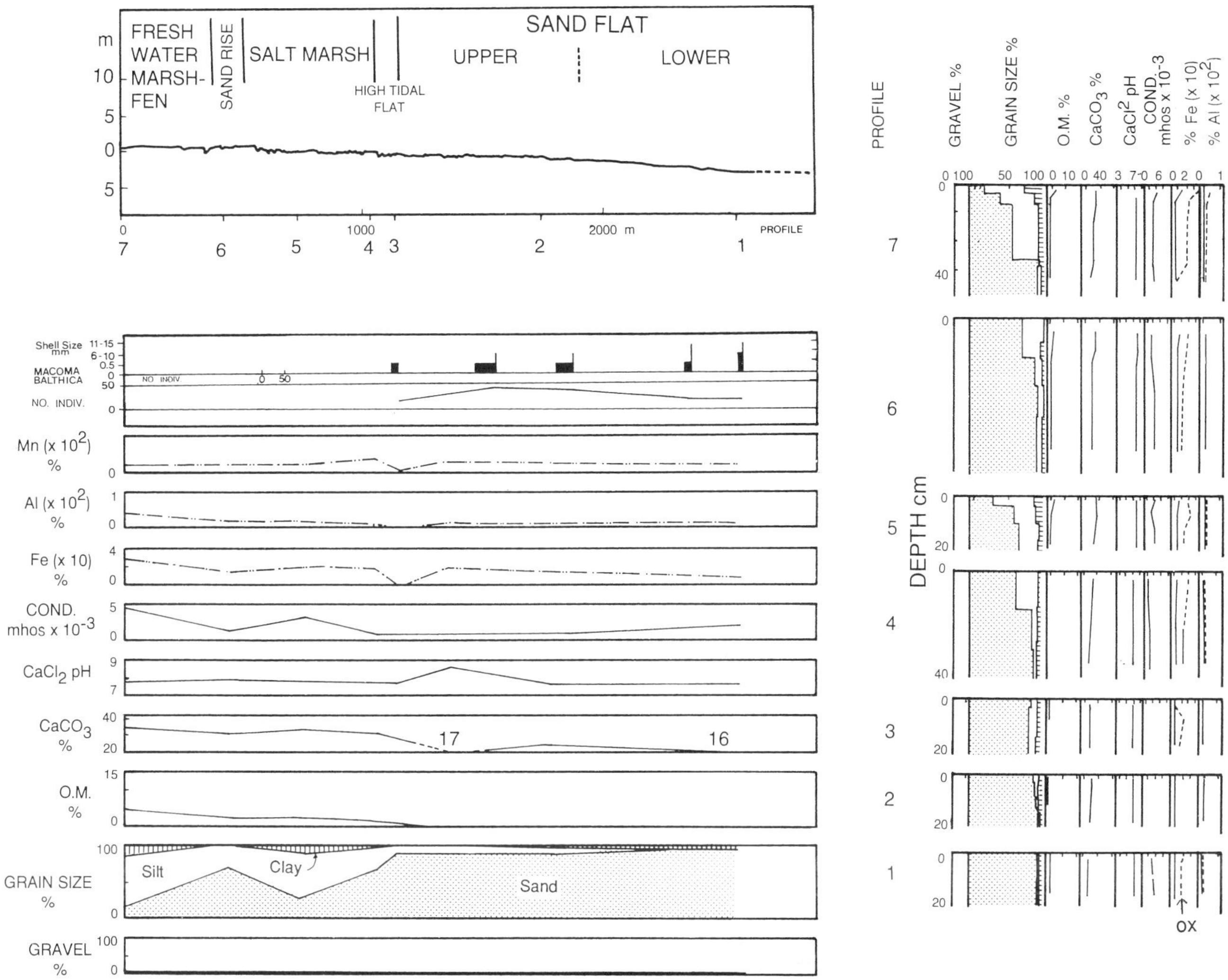

Figure 4. Cross-section of a tidal flat (LRB, Fig. 1) with morphological divisions and lateral variations in grain size, chemical composition and concentration of *Macoma balthica*. The right part of the figure illustrates vertical variations in grain size and chemical composition in seven box core profiles. This is an example of the data summary available for each one of more than 50 transects along the Ontario coasts of James Bay and Hudson Bay.

of these features are disrupted during the summer, but many remain visible in areas that experience relatively low sedimentation rates. Ice scouring moulds the surficial morphology of the flats, which generates intrastratal surfaces marked by burrow truncations in cross-section.

High Tidal Flat

In the vast, unobstructed upper tidal flats a relatively narrow, sandy, silty zone marks the transition between the unvegetated parts of the tidal flats and the marshes. It may constitute almost the whole intertidal zone in flats developed behind barriers. Principal characteristics of the high tidal flat are the development of widespread algal mats and incipient colonization by clumps of *Puccinellia phryganodes* on slightly higher hummocks (5-15 cm high) near its landward edge. Surficial sedimentary features include ripple marks stabilized by algal mats, disseminated patches of pebbles, unsorted coarse sand, and isolated boulders (Fig. 6A). Thin, organic-rich, silt drapes cover everything during summer. In spring, numerous, large (10-15 cm in diameter) gas bubbles form under the algal mat, and locally burst. Mudcracks may occur, but they are pervasive. Sparse infauna is present: juvenile *Macoma balthica* are found along the banks of draining creeks, *Hydrobia minuta* are stranded near clumps of growing vegetation, and small polychaetes are present throughout (Table 1; Clarke *et al.*, 1982). Numerous tracks and trails are left on these flats by seabirds, geese, and by caribou and polar bears in the more northern areas.

Internal sequences display wavy, organic-rich, silty laminae alternating with thin, irregular sand and fine gravel beds (Fig. 6B). Ripple crosslaminae are mantled with organic-rich, silty drapes. Isolated pebbles occur. Relatively small synsedimentary deformations (loadings and folds) are common. Very few small burrows are present.

In addition to tidal and wave activities associated with the high tidal flat, growth of algal mats and ice reworking of sediments are important processes acting in this tidal zone. The sustained growth of algal mats, deposition of fine

Table 1. Characteristic features of tidal flats in the study area. (OM, organic matter)

FACIES	FEN	MARSH UPPER	MARSH LOWER	TIDAL FLAT HIGH	TIDAL FLAT UPPER	TIDAL FLAT LOWER	SHALLOW SUBTIDAL
LITHOLOGY	OM	Sandy silt		Sandy silt	Sandy silt	Sand	Silty sand
SORTING		Fair		Fair	Good-fair	Good	Good-fair
COLOUR	Drk. gray	M. gray	Lt. brown	M./drk. gray	Lt./M. gray	Lt./drk. gray	M. gray
CONDUCTIVITY (mS/cm)	0.3-4.4	0.6-6.5	0.4-7.5	0.8-18	1.0-6.0	1.7-4.2	2.1-2.2
pH	6.4-7.8	7.3-8.0	7.4-7.9	7.7-8.2	7.6-8.6	7.3-8.1	7.7-7.8
$CaCO_3$ %	1.0-34	9.1-44	18-45	6-42	17-40	16-39	23-30
OM %	3-7.6	1.1-8.8	0.3-4.8	0.1-2.1	0-7.0	0-0.7	0.1-0.6
SILT %	3-60	26-72	31-74	2-72	0-69	0-59	2.7-27
GRAVEL %	0-5	0-2.3	0-1.4	0-1.0	0-5	0-52	0-15
SEDIMENTARY STRUCTURES	A-OM	A-Laminae C-Scours R-Logs C-Cut and fill		A-Algal lam. C-Ripples C-Scours, Rafted sed. C-Cut and fill	A-Ripples R-Dunes R-silty clay lenses C-Scours, Pebbles R-Boulders	A-Ripples R-Herringbone C-Deformation structure C-Scours	A-Ripples C-Laminae A-Silty, clay lenses R-Pebbles
BIOTURBATION	A-#	A-# C-&		C-# R-&	A-&	A-&	A-& C-#
SHELLS		Tr	R	Tr	Tr		
MACOMA/m^2				0-4800	650-4800	200-6600	8600-12800
HYDROBIA/m^2					1400-7500	400-3370	8300-33600
PLANTS	Graminaceous *Larix*	*Puccinellia* *Carex* *Hippuris*		Algal mat	R. Algae	R. Algae	*Zostera marina*

A = abundant
C = common
R = rare
& = burrows
# = roots

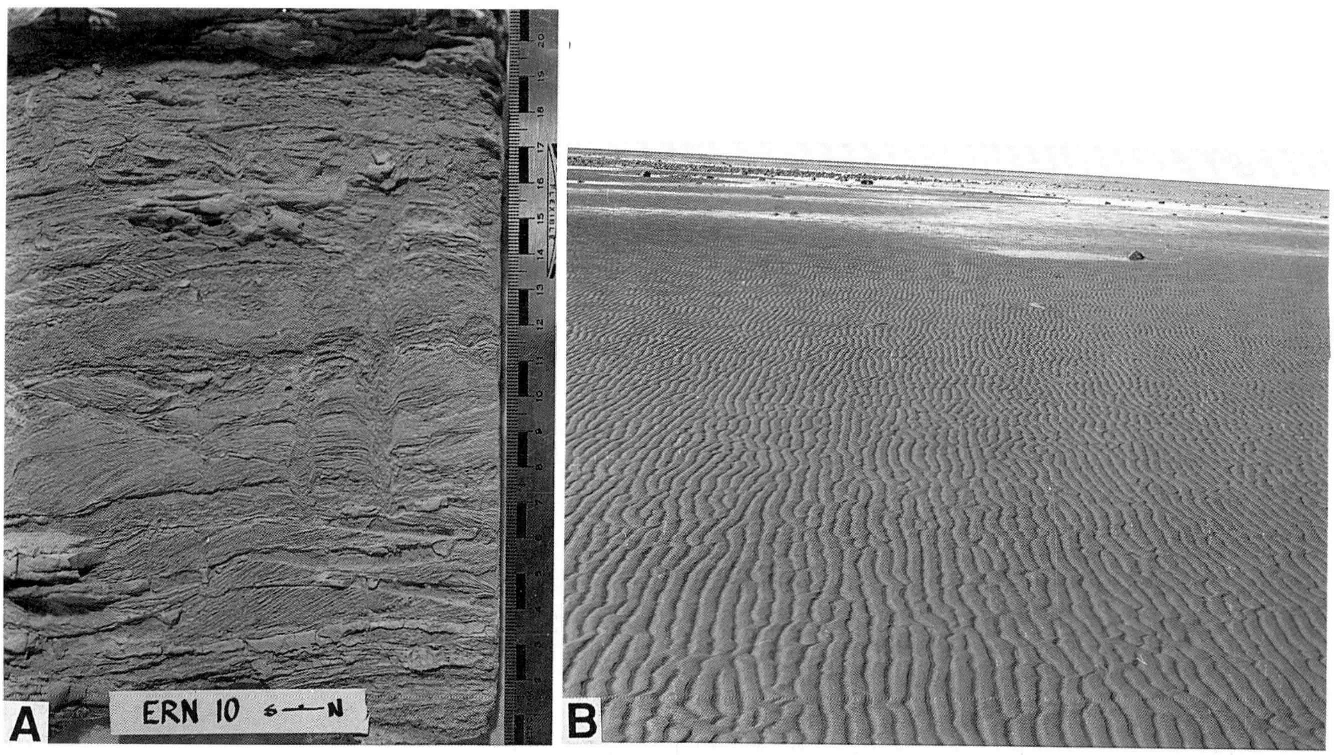

Figure 5. Features of unvegetated tidal flats. **A)** Box core from a lower tidal flat showing ripple crossbeds alternating with plane beds. Note burrows of *Macoma balthica*. **B)** Ripple marks in lower tidal flat (foreground) and boulder-strewn upper tidal flat in the background.

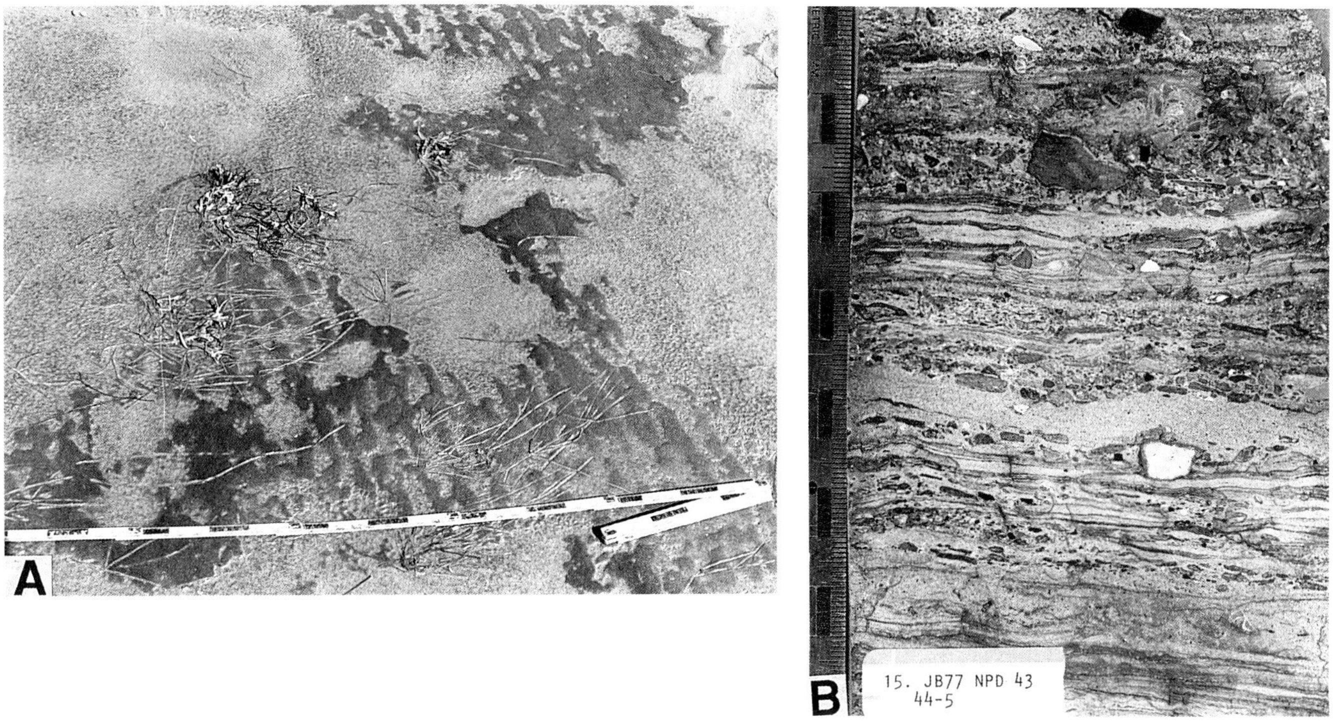

Figure 6. Characteristic features of high tidal flats. **A.** Algal mats protecting sediments and allowing colonization of *Puccinellia phryganodes* on drier hummocks; **B.** Shallow box core showing fine, algal-rich laminae alternating with ice-rafted material locally reworked by waves and tidal currents.

grained sediments during tidal cycles, and recurrent injection of coarse material by ice rafting have led to the development of clastic stromatolitc-like laminae alternating with very thin, coarse grained, gravelly, sandy layers.

Lower Marshes

The floral assemblages of lower marshes is dominated by *Puccinellia phryganodes* (Fig. 7A), except along the brackish coasts affected by fluvial fresh water plumes where *Hippuris tetraphilla* associations develop (Fig. 7B). Salinity inversion occurs in certain brackish shores where a narrow *Puccinellia phryganodes* zone develops inland from the *Hyppuris tetraphilla* zone (Martini and Protz, 1978; Glooschenko and Clarke, 1982; Glooschenko and Martini, 1987). These saline inland zones are associated with evaporating sea waters stranded during storm surges, and/or to upwelling of interstitial waters from underlying marine clays (Price and Woo, 1987, 1988).

The *Puccinellia*-dominated marsh commonly exhibits numerous interconnected pools that form "jig-saw puzzle" patterns. Development of these patterns is fostered by intense ice scouring in the lowermost part of the marsh, and by the activities of geese (flightless during the breeding season), which can number in the hundreds of thousands (Dionne, 1972; Jefferies *et al.*, 1979; Martini, 1981a).

Much of the suspended load in tidal flood waters is trapped at the seaward edge of the mat-structured *Puccinellia phryganodes* salt marsh, where reworked disarticulated and articulated *Macoma* shells also concentrate. This is one reason for landward changes in the salt marsh deposits, such as fining in grain size, increase in pH, increase of organic matter, high concentrations of organically-bound iron and aluminum, decrease in calcium carbonate equivalent, and variable changes in conductivity (Fig. 4; Table 1).

Marsh sedimentary sequences are characterized by irregular, thin, organic-rich laminae, alternating with lenses and thin layers of silt with minor amounts of sand and granules. The sedimentation of coarse grained layers, which alternate with the organic rich laminae, is little affected by pre-existing vegetation. This is probably a result of major storms in the fall and spring, when vegetation is either flattened or not yet grown, such that the marsh presents a quasi-smooth depositional surface. The contacts between the laminae are irregular, and crenulated by rootlets. Whereas the *Puccinellia phryganodes* marsh is characterized by laminae intensely microturbated by mat-like roots (Fig. 7C), the *Hyppuris tetraphilla* marsh develops laminae (irregular rhythmites) of well sorted silt and clay alternating with algal-bonded laminae, pinching out and deformed around the isolated growing stems (Fig. 7D). The laminae of the *Hippuris tetraphilla* marsh reflect the less restricted distribution of the suspended load as a result of the more open vegetation structure, and they may record tidal flood cycles. However, laminae are deformed by plants and modified by storms, and do not provide accurate, continuous data for studying diurnal and spring tidal cycles.

Figure 7. Characteristic features of lower marshes. **A.** Lower marsh showing the *Puccinellia phryganodes* mat, and ice-scoured ponds; **B.** Looking diagonally down onto lower marsh of brackish areas with *Hippuris tetraphilla*; **C.** Cross-section of *Puccinellia phryganodes* lower marsh showing alternating laminae of organic-rich mud (dark bands) and sand and silt with local pebbles. Note irregular "crinkly" boundaries of the laminae resulting from roots; **D.** Shallow box core in a *Hippuris tetraphilla* marsh showing laminae (rhythmites) deformed by plant stems.

Upper Marsh

The upper marsh is characterized by an association of *Carex subspathacea*, with pools colonized by small aquatic ferns, locally by sparse *Hippuris tetraphilla* and, farther inland, by *Juncus balticus* and *Salix* plant associations. Creeks are deeply entrenched (1-3 m). Driftwood lines of twigs, logs, sand and shells transported onto the marsh during storm surges are common. They constitute drier strandlines readily recognizable in aerial photographs.

Accumulation of organic matter (up to 20-25 cm thick) is due to increased organic production in the more stable, less saline marsh, and lower rates of decomposition under waterlogged, cold conditions. The fresh water marsh cannot be classified as a "peatland", which by definition, must contain at least 30 to 40 cm of peat (Gore, 1983; National Wetlands Working Group, 1988). Desalination and partial decalcification of underlying sediments leads to progressively more acid conditions inland (Fig. 4, Table 1; Martini and Protz, 1978; Martini *et al.*, 1979; Martini and Protz, 1980, 1981; Protz, 1982).

The internal sequence of the upper marsh displays crenulated interstratification of organic-rich laminae, and silty and sandy storm layers with the organic content (O.M.) increasing landward (Fig. 4). Root disturbance is intense. Deposits of iron oxide (ferrans) form around plant roots in drier marsh sediments. No larval insect burrows have been recognized in the samples analyzed, although they undoubtedly occur.

The upper marsh is affected by frost conditions and by discontinuous permafrost in the northernmost shores (Martini and Protz, 1981). Major effects of ground ice include deformation (convolutions) of the organic and finer mineral sediments, and generation of horizontal, locally reticulate voids (Fig. 8; Mackay, 1974; Van Vliet-Lanoe, 1988). In northern regions, these processes affect layers of tidal sequences down to a depth of 2 m.

Figure 8. Convolutions affecting subarctic tidal sequences under a thin peat cover.

Inland Wetlands and Peatlands

Inland from the coastal marshes, permafrost development varies depending on whether the location is in the northern continuous permafrost area (inland from Hudson Bay), or in the south (inland from James Bay), where sporadic permafrost exists (Brown, 1968). In the north, the landscape is hummocky because of differential growth of ice lenses. These areas are covered by tundra near the coast and by boreal forest farther inland. To the south, the marshes grade landward into unfrozen fens. These minerotrophic peatlands resemble large, monotonous graminaceous prairies and locally exhibit sparse *Larix laricina* (Tamarack) forests (Martini *et al.*, 1980; Sims *et al.*, 1983; Martini and Glooschenko, 1985). The minerotrophic, thus nutrient-rich conditions of the fens leads to a relatively rapid transformation and homogenization of the organic matter into amorphous peat sequences. Where the growth of *Sphagnum* is favored, ombrotrophic bogs form and contain thick peat. For instance, about 100 km inland from the present shoreline, 3 to 4 m of peat has developed in the 3000 to 4000 years since emergence (Protz *et al.*, 1984).

Comparison with other Tidal Flats

The subarctic tidal flats of western James Bay and southwestern Hudson Bay are similar to those of other subarctic coasts, but differ both in terms of their surficial structure and sedimentary sequences. This section will first compare the tidal flats of James Bay and Hudson Bay with other cold climate flats, and then with other examples of temperate and warm tidal flats.

Dionne (1988) attempted a classification of cold tidal flats based on the relative importance of ice effects and ground frost. He also compared major geomorphological differences between tidal flats in cold regions and those in temperate and warm regions. He did, however, point out that cold climate tidal flats have considerable variability and that data on them is generally limited. For instance, it is correct to state that erosional action by ice floes is more intense in cold-temperature and subarctic areas than in the high arctic, but in addition, the protective action of ice is more important in northern arctic areas than farther south. Similarly, in high latitude arctic flats, continuous permafrost is present but in subarctic areas, where more numerous freeze and thaw cycles occur annually, the effect of frost shattering is more important.

From another perspective, ice cover and permafrost can make contributions, but do not uniquely determine the type of sedimentary facies developed in the flats. Many other variables affect sedimentation, such as tidal range, wave conditions during the open season, availability of various types of sediments, bedrock conditions, vegetation type, regional slope of the land, presence of barriers, rate of uplift or submergence, and salinity of the waters. Depending on the interaction of these parameters, intertidal zones like those of western James Bay and southwestern Hudson Bay coasts vary considerably. Some coasts are steep and predominantly erosional, with the deposition of sand and gravel ridges occurring only in the subtidal zone and at the high tide level. Other, flatter coasts exhibit low sedimentation rates, with sand deposition in the subtidal zone, sand waves in the lower part of the intertidal zone, patchy sandy and bouldery deposits in the unvegetated intertidal area, beach ridges at the high tide levels and muddy deposits in marshes behind the beach ridges. Still other coasts have normal to high sedimentation rates and exhibit the characteristics that have been described in this paper (also see Martini, 1981b, 1982). Only this latter type generates sequences that have all the features encountered in what are considered classical tidal flat deposits; that is, those showing fining landward and upward sequences, prevalence of ripple crosslamination, rare herringbone cross-beds, plane beds, locally well defined infauna and associated burrow zones and typical salt marsh laminae (Reineck and Singh, 1975). Therefore, when comparing tidal flats to establish the effects of cold conditions, similar depositional environments (that is, similar tidal flats) should be considered. This is done in the following paragraphs.

Aitken *et al.*, (1988) compared the macrotidal sandy flats of the Pangnirtung Fiord (mid-arctic) in southeastern Baffin Island to those of the Bay of Fundy (cold-temperate). They found the structures to be rather different since the arctic flats were protected by glacial erratics, which were mainly absent in the Bay of Fundy. They also found a significant difference in the faunal compositions; for instance the absence of *Corophium volutator* in the north. However, they established that in the rock record there would be an overall similarity in ichnofauna developed in the two tidal flats, which for both cases would resemble a *Skolithus* ichnofacies.

In comparing these eastern seaboard macrotidal flats with mesotidal subarctic flats of western James Bay and southwestern Hudson Bay, some similarities are encountered, but also very major differences. First, the substratum in the James and Hudson Bay flats do not allow development of bounding boulder barricades such as those of the subarctic and arctic eastern coasts of Canada (Rosen, 1979; McCann *et al.*, 1981; Aitkin *et al.*, 1988). Instead those barricades are replaced by bedrock or diamict shoals at the low tide level in the James Bay Hudson Bay area, which constitute preferential grounding sites for ice floes and ice pressure ridge formation, thus developing bouldery surfaces. This is consistent with the development of boulder-strewn flats inland from the shoals. A second major difference occurs in the composition and distribution of infauna. The subarctic tidal flats of James Bay and Hudson Bay contain well defined *Macoma balthica* zones, but lack *Corophium volutator*, and they have few occurrences of *Arenicola marina*. *Macoma balthica* is the principal burrowing organism in these subarctic tidal flats and is locally very abundant, but nowhere is animal bioturbation so intense that primary sedimentary structures are totally obliterated. Third, whereas extensive coastal salt marshes occur in James Bay and Hudson Bay, there is almost no marsh development in Pangnirtung Fiord, and totally different, *Spartina* marshes occur in the Bay of Fundy (Glooschenko *et al.*, 1988).

There are also strong differences between subarctic tidal flats of southwestern James and Hudson bays and the arctic flats of Foxe Basin, located farther north. Throughout the Foxe Basin, intertidal flats are essentially barren of macrofauna or contain few worms (Martini, unpublished data). This low intertidal production of Foxe Basin contrasts with the richness of the subtidal zones (Grainger, 1959). Lack or scarcity of macro-infauna has been reported from several other locations, such as river-dominated coasts of James and Hudson bays, and intertidal zones in Cook Inlet, Alaska (Murphy *et al.*, 1972; Bartsch-Winkler and Ovenshine, 1984; Martini and Morrison, 1987). Reworking of the intertidal zone by ice and seasonal frost are suggested as causes for the scarcity of bioturbating macrofauna (Dionne, 1988). Other suggested causes include strong currents and scouring, washing out of food-supporting fine particles from coarse sediments, local effects of pollution, and variation in salinity (Bartsch-Winkler and Ovenshine, 1984).

In Hudson and James bays the variation in salinity has a great effect on the restricted intertidal macrofauna, more so than limited changes in sand granulometry of the substratum, intensity of ice scouring, and predation by migratory shorebirds (Martini and Morrison, 1987). The latter two variables cause local mass mortalities, which affect the size distribution of *Macoma balthica* on the intertidal flats, but cause only temporary changes in the overall annual number of individuals. Those areas scoured in the spring by ice and subject to predation for parts of the summer, are rapidly recolonized by juvenile spats derived from larger and more mature *Macoma balthica* stocks occupying the shallow subtidal zone. The mature subtidal population is, however, placed under stress and spats do not survive in intertidal areas where freshened by fluvial plumes (D'Anglejan, 1980; Grenon, 1982; Martini and Morrison, 1987). Depletion of intertidal macrofauna in Foxe Basin is also in part associated with strong salinity variations caused by heavy spring runoff, by in situ melting of ice, which covers the lower part of the flats until mid-summer, and by drainage from the melting active layer of coastal areas throughout the summer (Martini, unpublished data).

For all the variation that may exist between the subarctic and arctic shores, cold tidal flats such as those of Hudson and James bays generate clastic deposits unmistakably distinct from those of cold-temperate and warmer shores (Kellerhals and Murray, 1969; Swinbanks and Murray, 1981; Ginsburg, 1975). Some differences are due to modern sedimentary processes, whereas others, such as the extensive polymictic source material, are related to the Pleistocene glaciation. Typical characteristic features of the cold tidal flats of James Bay and Hudson Bay include the following: a) poor sorting and bimodal grain size distributions because of ice rafting of coarse material onto reworked fine grained material; b) other textural-compositional inversions where boulders are composed exclusively of frost-resistant igneous and metamorphic rocks and retain the roundness they acquired during glacial transport, while pebbles and granules are angular and are composed primarily of frost-susceptible carbonates derived from local bedrock; c) restricted faunal assemblages locally rich in individuals, with prevalence of *Macoma balthica* and *Hydrobia minuta* (Clarke, 1980; Clarke *et al.*, 1982; Grinham and Martini, 1983; Martini and Morrison, 1987). As in other tidal flats, burrows persist, but most infauna shells are destroyed or transported and preserved in sand ridges in the lower parts of salt marshes (Yeo and Risk, 1981; Martini and Morrison, 1987); d) intense bioturbation by roots of *Puccinellia phryganodes* occurs and delicate, partially deformed lensing tidal laminae of the *Hippuris tetraphilla* are found in the brackish marshes; e) thick peat forms in the unconfined inland fens and bogs (Martini and Glooschenko, 1985); and f) well developed convolutions and gleyed soils with vertical, iron-rich, calcareous rhyzoconcretions are common.

The usual questions posed at this point are whether cold climate features in modern environments can be preserved, and even if they are, how important would they be in the geological record. The first question is a rhetorical one, because every modern sedimentary feature can be preserved and most of them have been recognized in the rock record. Preservation of tidal flats and other Holocene coastal deposits of the Canadian cratonic inland seas requires a relatively rapid submergence, as could be provided by an eustatic sea level rise due to atmosphere warming and collapse of major ice sheets. If instead, atmospheric cooling were to occur, a newly-developed glaciation would not necessarily wipe out the coastal sedimentary record, as demonstrated by a similar offlap interglacial sequence preserved under diamicts of the ultimate glaciation in parts of the Hudson Bay Lowland (Skinner, 1973; Shilts, 1986). The presence of a thick peat cap, resistant to erosion, enhances preservation of the coastal sequences. Therefore, cold climate tidal flats have the potential for preservation and recognition comparable to that of tidal flats in temperate and warm climates.

As for the second question, the geological record is biased toward warm climate deposits, but there is increasing evidence that worldwide unconformities exist, many associated with eustatic sea level variations triggered by glaciation. Nonglacial, cold climates may have occurred frequently in the geological past, and should have led to the formation and preservation of cold climate marine and coastal deposits (many still to be recognized). Many such deposits formed in nonglaciated areas may be thinner than those of warmer areas because of lower sedimentation rates, and may lack convenient lithological tracers like those of the Quaternary. Thus they may be difficult to recognize unless found in the proper stratigraphic sequence. For instance, in the Lower Permian of eastern Australia, glaciomarine deposits bearing ice-rafted erratics, cold-water fauna and carbonates are strong indicators of cool to cold settings. These characteristic glaciomarine deposits intertongue with more uniform, less bioturbated, sandy coastal and fluvial deposits, which lack any clear indication of cold climate origin. It is only because of the conformable contact between the two units that one can confidently identify the latter as having cold climate origin (Rao, 1981; Rao and Green, 1982; Martini and Banks, 1989).

Synthesis

The tidal flat sequences of western James Bay and southwestern Hudson Bay contain features common to many other tidal shorelines. They differ because they lack tidal channels in the lower tidal flats; they have restricted infauna, and lack intensely bioturbated facies, although characteristic burrows occur; their marshes differ in plant species and in certain types of sediment laminae from others described in the geological literature; and extensive, virgin, subarctic to boreal peatlands develop inland.

Ice and frost conditions have left a strong imprint on the James and Hudson Bay tidal deposits. Although seasonal freezing and local permafrost conditions are Holocene features of the emergent land, the overall cold climatic setting is a continuation of Pleistocene environments. Thus, the present-day sedimentary characteristics associated with cold climates form a continuum that is diagnostic of a glacial and postglacial, emergent, coastal sequence. Such a sequence, in general, is characterized by frost-shattered bedrock, overlain by thin tills overlain locally by residual, fine grained shelf sediments, which are, in turn, overlain by thin (2-10 m), recent coastal deposits and peat.

The modern coastal deposits reflect the nature of their multiple sources and the effect of various sedimentary processes. They contain widespread, fairly well sorted, ripple crosslaminated silty sands typical of tidal flats. Disseminated rounded crystalline boulders, and polymictic lenses and layers of ice-rafted material with a predominance of angular carbonate pebbles variously shaped depending on their frost susceptibility are also common. These are capped by thinly laminated, organic-rich, root-disturbed, locally concretionary (rhyzoconcretions) marsh deposits with sparse, ice-rafted silt and sand lenses and isolated pebbles and boulders.

Acknowledgments

I wish to thank colleagues, assistants and helicopter pilots who participated in the Hudson Bay Lowlands (HBL) research project and withstood severe field conditions and my company for several seasons. Special thanks to pilot H. Dale, whose skill and dedication allowed completion of the work under "quasi-impossible" flying conditions along the Hudson Bay coast. Comments of reviewers (J. C. Dionne and D. L. Forbes) and editors helped make the manuscript more readable. Financial support was provided by Environment Canada and the National Science and Engineering Research Council (Grant A-7371). Helicopter support was provided by Ocean and Aquatic Sciences, Central Region.

References

Aitken, A. E., Risk, M. J. and Haward, J. D. 1988. Animal-sediment relationships on a subarctic intertidal flat, Pangnirtung Fiord, Baffin Island, Canada. Journal of Sedimentary Petrology, v. 58, p. 969-978.

Anderson, R. O. 1959. A modified flotation technique for sorting bottom fauna samples. Limnology and oceanography, v. 4, p. 223-225.

Bartsch-Winkler, S. and Ovenshine, A. T. 1984. Macrotidal subarctic environment of Turnagain and Knik Arms, upper Cook Inlet, Alaska: sedimentology of the intertidal zone. Journal of Sedimentary Petrology, v. 54, p. 1221-1238.

Brown, R. J. E. 1968. Permafrost investigations in Northern Ontario and Northeastern Manitoba. National Research Council. Division of Building Research, Ottawa, 75 p.

Clarke, K. E. 1980. The ecology of a subarctic coastal system, North Point, James Bay, Ontario. Unpublished M.Sc. thesis, University of Guelph, Ontario 232 p.

______ , Martini, I. P. and Glooschenko, W. A. 1982. Sedimentary characteristics of the coastal environment of North Point, Ontario. Naturaliste canadien, v. 109, p. 385-397.

Craig, B. G. 1968. Late-glacial and post-glacial history of the Hudson Bay region. *In:* Earth Science Symposium on Hudson Bay, P. J. Hood (ed.). Geological Survey of Canada, Paper 68-53, p. 63-77.

D'Anglejan, B. 1980. Effects of seasonal changes on the sedimentary regime of a subarctic estuary, Rupert Bay (Canada). Sedimentary Geology, v. 26, p. 51-68.

Dionne, J. C. 1972. Characteristiques des schorres des regions froides, en particulier de l'estuarie du Saint-Laurent. Z. Geomorphology NF. Supplbd. 13, p. 131-162.

______ 1985. Formes, figures et facies sedimentaires glaciels des estrans vaseaux des regions froides. Palaeogeography, Palaeoclimatology, Palaeoecology, v. 51, p. 415-451.

______ 1988. Characteristic features of modern tidal flats in cold regions. *In:* Tide-influenced sedimentary environments and facies, P. L. De Boer *et al.* (eds.). D. Reidel Publishing Company, p. 301-332.

Folk, R. L. 1974. Petrology of Sedimentary Rocks. Hemphill Company, Austin, 182 p.
Ginsburg, R. N. (editor) 1975. Tidal deposits. Springer-Verlag, New York, 428 p.
Glooschenko, W. A. and Clarke, K. 1982. The salinity cycle of a subarctic salt marsh. Naturaliste canadien, v. 109, p. 483-490.
______ and Martini, I. P. 1987. Vegetation of river-influenced coastal marshes of the southwestern end of James Bay, Ontario. Wetlands, v. 7, p. 71-84.
______ , ______ and Clarke-Whistler, K. 1988. Salt marshes of Canada. *In:* Wetlands of Canada, National Wetlands Working Group. Ecological Land Classification Series, No. 24. Environment Canada, Ottawa, p. 347-377.
Gore, A. J. P. 1983. Mires: swamp, bog, fen and moor (2 vols.). Elsevier, Amsterdam.
Grainger, E. H. 1959. The annual oceanographic cycle at Igloolik in the Canadian Arctic. 1. The zooplankton and physical and chemical observations. Journal of Fisheries Research Board Canada, v. 16, p. 453-501.
Grenon, J. R. 1982. The macrobenthic fauna of Estmain Estuary (James Bay, Quebec) before the diversion. Naturaliste canadien, v. 109, p. 793-802.
Grinham, D. F. 1980. Intertidal sedimentation in Akimiski Strait, James Bay, Canada. Unpublished M.Sc. Thesis, University of Guelph, Ontario. 178 p.
______ and Martini, I. P. 1983. Sedimentology of the Ekwan Shoal, Akimiski Strait, James Bay, Canada. Sedimentary Geology, v. 37, p. 273-294.
Kellerhals, P. and Murray, J. W. 1969. Tidal flats at Boundary Bay, Fraser River Delta, British Columbia. Bulletin of Canadian Petroleum Geology, v. 17, p. 67-91.
Kranck K. and Ruffman, A. 1982. Sedimentation in James Bay. Naturaliste canadien, v. 109, p. 353-361.
Jefferies, R. L., Jensen A. and Abraham, K. F. 1979. Vegetation development and the effect of geese on vegetation at La Perouse Bay, Manitoba. Canadian Journal of Botany, v. 57, p. 1439-1450.
McCann, S. B., Dale, J. E. and Hale, P. B. 1981. Subarctic tidal flats in areas of large tidal range, southern Baffin Island, eastern Canada. Geographie physique et Quaternaire, v. 35, p. 183-204.
MacKay, J. R. 1974. Reticulate ice veins in permafrost, Northern Canada. Canadian Geotechnical Journal, v. 11, p. 230-237.
McKegue, J. A. 1978. Manual on soil sampling methods of analysis. Soil Research Institute, Ottawa. 212 p.
Martini, I. P. 1981a. Ice effect on erosion and sedimentation on the Ontario shores of James Bay, Canada. Z. Geomorphology, N.F., v. 25, p. 1-16.
______ 1981b. Morphology and sediments of the emergent Ontario coast of James Bay, Canada. Geografiska Annaler, v. 63A, p. 81-94.
______ 1982. Geomorphological features of the Ontario coast of Hudson Bay. Naturaliste canadien, v. 109, p. 415-429.
______ 1986. Canadian Inland Seas. Elsevier, Amsterdam. 494 p.
______ and Banks, M. R. 1989. Sedimentology of the cold-climate, coal-bearing, Lower Permian "Lower Freshwater Sequence" of Tasmania. Sedimentary Geology, v. 64, p. 24-41.
______ Cowell, D. W. and Wickware, G.M. 1980. Geomorphology of southwestern James Bay: a low energy, emergent coast. *In:* The Coastline of Canada, S.B. McCann (ed.). Geological Survey of Canada, Paper 80-10, p. 293-301.
______ and Glooschenko, W. A. 1985. Cold climate peat formation in Canada, and its relevance to Lower Permian Coal Measures of Australia. Earth-Science Review, v. 22, p. 107-140.
______ and Grinham, D. F. 1983. Sedimentary conditions and deposits of Akimiski Strait, James Bay, Canada. Sedimentary Geology, v. 37, p. 251-272.
______ and Morrison, R. I. G. 1987. Regional distribution of *Macoma balthica* and *Hydrobia minuta* on the subarctic coasts of Hudson Bay and James Bay, Ontario, Canada. Estuarine, Coastal and Shelf Science, v. 24, p. 47-68.
______ and Protz, R. 1978. Coastal geomorphology, sedimentology and pedology of southern James Bay, Ontario, Canada. Land Resource Science, Technical Memo. 78-1, University of Guelph, Ontario, 207 p.
______ and ______ 1980. Coastal geomorphology, sedimentology and pedology of northern James Bay, Ontario, Canada. Land Resource Science, Technical Memo 80-1, University of Guelph, Ontario, 117 p.
______ and ______ 1981. Coastal geomorphology, sedimentology and pedology of Hudson Bay, Ontario, Canada. Land Resource Science, Technical Memo. 81-1, University of Guelph, Ontario, 322 p.
______ , ______ , Grinham, D., King, W. A. and Clarke, K. E. 1979. Studies of coastal sediments, soil, and biota, James Bay, Ontario, Canada. Land Resource Science, Technical Memo. 79-1, University of Guelph, Ontario, 169 p.
Murphy, R. S., Carlson, R. F., Nyquist, D. and Britch, R. 1972. Effect of waste discharges into a silt-laden estuary; a case study of Cook Inlet, Alaska. Institute of Water Resource Report No. IWR-26, Fairbanks, University of Alaska, p. 6-25.
National Wetlands Working Group, 1988. Wetlands of Canada. Ecological Land Classification Series, No. 24. Sustainable Development Branch, Environment Canada, Ottawa, and Polyscience Publication Inc., Montreal, Quebec, 452 p.
Norris, A. W. and Sanford, B. W. 1968. Paleozoic and Mesozoic Geology of the Hudson Bay Lowlands. *In:* Earth Science Symposium on Hudson Bay, P. J. Hood (ed.). Geological Survey of Canada, Paper 68-53, p. 169-206.
Parker, J. 1963. Inverted mud cracks. Journal of Sedimentary Petrology, v. 33, p. 230-231.
Pelletier, B. R. 1968. Submarine physiography, bottom sediments, and models of sediment transport in Hudson Bay. *In*: Earth Science Symposium on Hudson Bay, P. J. Hood (ed.). Geological Survey of Canada, Paper 68-53, p. 100-135.
______ 1986. Seafloor morphology and sediments. *In*: Canadian Inland Seas, I. P. Martini (ed.). Elsevier, Amsterdam. p. 143-162.
Price, J. S. and Woo Mong-Ko 1987. Origin of salt in coastal marshes of Hudson and James bays. Canadian Journal of Earth Sciences, v. 25, p. 145-147.
______ and ______ 1988. Studies of a subarctic coastal marsh, II. Salinity. Journal of Hydrology, v. 103, p. 293-307.
Prinsenberg, S. J. 1983. Present and future circulation and salinity in James Bay. Naturaliste canadien, v. 109, p. 827-841.
Protz, R. 1982. Development of gleysolic soils in the Hudson Bay and James Bay coastal zones, Ontario. Naturaliste canadien, v. 109, p. 491-500.
______ , Ross, G. J., Martini, I. P. and Terasmae, J. 1984. Rate of Podzolic soil formation near Hudson Bay, Ontario. Canadian Journal of Soil Science, v. 54, p. 31-49.
Rao, C. P. 1981. Criteria for recognition of cold-water carbonate sedimentation: Berriedale Limestone (Lower Permian), Tasmania, Australia. Journal of Sedimentary Petrology, v. 51, p. 491-506.
______ and Green, D. C. 1982. Oxygen and carbon isotopes of Early Permian cold-water carbonates, Tasmania, Australia. Journal of Sedimentary Petrology, v. 52, p. 1111-1125.
Reineck, H. E. and Singh, I. B. 1975. Depositional sedimentary environments. Springer-Verlag, New York, 430 p.
Rosen, P. S. 1979. Boulder barricades in central Labrador. Journal of Sedimentary Petrology, v. 49, p. 1113-1124.
Sims, R. A., Cowell, D. W. and Wickware, G. M. 1983. Classification of fens near southern James Bay, Ontario using vegetational physiognomy. Canadian Journal of Botany, v. 50, p. 2608-2623.
Shilts, W. W. 1986. Glaciation of the Hudson Bay region. *In:* Canadian Inland Seas, I. P. Martini (ed.). Elsevier, Amsterdam. p. 55-78.
Skinner, R. G. 1973. Quaternary stratigraphy of the Moose River basin, Ontario. Geological Survey of Canada, Bulletin 225, 77 p.
Swinbanks, D. D. and Murray, J. W. 1981. Biosedimentological zonation of Boundary Bay tidal flats, Fraser River Delta, British Columbia. Sedimentology, v. 28, p. 201-237.
Van Vliet-Lanoe, B. 1988. Le role de la glace de segregation dans les formations superficielles de l'Europe de l'ouest. These de Doctorat d'Etat. Unversite de Paris I — Sorbonne, Centre de Geomorphologie du C.N.R.S., 2 v., 854 p.
Webber, P. J., Richardson, J. W. and Andrews, J. T. 1970. Post glacial uplift and substrate age at Cape Henrietta Maria, southeastern Hudson Bay, Canada. Canadian Journal of Earth Sciences, v. 7, p. 317-325.
Yeo, R. K. and Risk, M. J. 1981. The sedimentology, stratigraphy and preservation potential of intertidal deposits in the Minas Basin system, Bay of Fundy. Journal of Sedimentary Petrology, v. 51, p. 245-260.

KEY WORDS: SUBARTIC, TIDAL FLATS, JAMES BAY, HUDSON BAY, MESOTIDAL BASINS

Nearshore
and
Shelf Systems

In: Clastic Tidal Sedimentology, D. G. Smith, G. E. Reinson, B. A. Zaitlin and R. A. Rahmani (eds.). Canadian Society of Petroleum Geologists. Memoir 16, p. 315-320.

Tidal influence on foreshore deposits, Pacific coast of Japan

MIWA YOKOKAWA AND FUJIO MASUDA

Institute of Earth and Planetary Sciences
College of General Education
Osaka University
Toyonaka, Osaka 560, Japan

ABSTRACT

Grain fabric and the depositional volume of modern foreshore deposits from the Pacific coast of southern Ibaraki Prefecture, Japan, are influenced by tides. Reversals of imbrication dip were found that may result from predominance of uprush transport during the flood stage and backwash transport during the ebb stage. The thickness of sediment deposited during both rising and falling water stages varies directly with tidal range for that cycle, so that diurnal inequality leads to alternations of thick (3-4 cm) and thin (1-2 cm) layers on the foreshore.

Pleistocene foreshore deposits exposed near the Pacific coast are also characterized by reversals of grain imbrication and may have been controlled by tidally-influenced uprush-backwash processes. The Pleistocene example also possesses alternating thin and thick layers, suggesting a diurnal variability in tidal pattern similar to the modern tidal pattern along the Pacific coast.

RÉSUMÉ

Dans la zone infratidale de la côte pacifique du Japon, au sud de la préfecture Ibaraki, les marées influencent l'orientation des grains et la quantité de sédiments modernes accumulés. Les inversions de pendage des imbrications indiquent peut-être une prédominance du transport de jet de rive durant le flux, et du transport par le retour du courant des vagues durant le reflux. L'épaisseur du dépôt sédimentaire d'un cycle flux-reflux varie directement avec l'ampleur de ce cycle. Par conséquent, les inégalités diurnes engendrent dans la zone d'avant-plage une alternance de couches épaisses (3-4 cm) et minces (1-2 cm).

Près de la côte du Pacifique affleurent des dépôts pléistocènes de la zone infratidale également caractérisés par une imbrication inverse des grains; ils ont pu être contrôlés par les processus de transport influencés par les marées, i.e. transport de jet de rive et par retour de courant. Cet exemple pléistocène se caractérise aussi par une alternance de couches minces et épaisses, ce qui suggère une variation diurne du cycle des marées semblable à celle observée actuellement sur la côte du Pacifique.

INTRODUCTION

Foreshore deposits are characterized by parallel lamination (Thompson, 1937) formed by the combined processes of wave uprush and backwash (Clifton, 1969; Harms, 1975). Tidal influence on deposition in the swash-backwash zone has been discussed by Duncan (1964) and Otvos (1965). In this study, we examine tidal influence on foreshore deposition by analyzing grain fabric within laminae.

Studies of grain orientation on the surface of foreshore deposits (*e.g.*, Nanz, 1955; Curray, 1956; Taira, 1989) have shown that elongate grains orient themselves normal to the coastline. Taira (1989) predicted that landward-imbricated grains should dominate foreshore deposits. However, seaward-dipping grains have been found in ancient beach sandstone (Shelton and Mack, 1970).

We studied grain imbrication of successive vertical sections in Japanese foreshore deposits, and found that: 1) the dominant dip direction of sand grain imbrication alternates in response to tidal cycles; and, 2) changes in depositional thickness correlate with the pattern of tidal diurnal inequality.

The ebb and flood depositional model from a modern beach was then applied to nearby Pleistocene foreshore deposits, where ebb and flood deposits could also be distinguished, and a tidal pattern of diurnal inequality similar to the modern one was found.

STUDY AREA

We studied the beach of Kouya Village in Ibaraki Prefecture, Pacific coast, central Japan (Loc. 1, Fig. 1). The setting is a wave-dominated (mean wave height and period are 1.0 m and 7.5 sec., respectively), microtidal, open coast. The spring tidal range is 1.6 m, and neap tidal range is 20 cm. The tidal curve shows strong diurnal inequality, particularly in the low tide height (Fig. 2).

The sand comprising the beach is fine to medium grained, and arenitic. Porosity averages 25 per cent. Parallel laminae dip slightly seaward at about 2 to 6° in wedge-shaped sets.

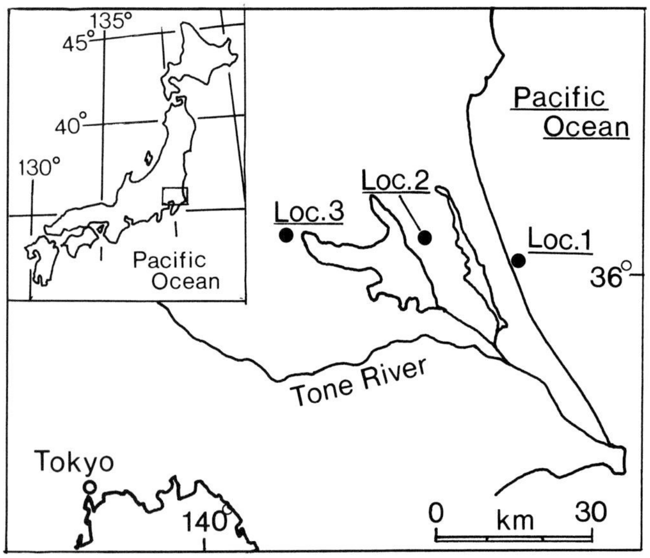

Figure 1. Locations of modern foreshore deposits (Loc. 1) and Pleistocene foreshore deposits (Locs. 2 and 3).

Sampling and Analytical Methods

Monitoring of deposition on the beach face was initiated just after a storm event, because renewed sedimentation and beach accretion is most pronounced following storm erosion. Sand can accumulate on the beach at a high rate during several post-storm calm days (Allen, 1982; Fig. 3a). Thin metal rods 3 mm in diameter and 54 cm long were inserted into the beach face. Sedimentation was then monitored by slipping aluminum rings over the rods at high water and at low water. The light weight of the rings minimized the potential for sinking into the sediment due to weight alone. With the placement of each ring, the distance between the top of the rod and the beach surface was measured. A ring could be buried when deposition occurred (Fig. 3b), or exposed and displaced downward from the top of the rod when the foreshore was eroded. Deposition did not always occur during both flood and ebb periods, but we attempted to collect the samples in which only continuous deposits were represented.

After continuous deposition had occurred, a vertical section normal to the coastline and adjacent to the rod was quarried with a knife. Quick-setting, low-viscosity glue was carefully sprayed obliquely on the vertical quarried surface to preserve the grain fabric. The glue impregnated a 0.5 to 1.0 cm surface layer. The impregnated sand layer was then removed from quarried wall (Fig. 3c) and brought to the laboratory without disturbing the sand-grain fabric. The sand layers were completely impregnated with resin in the laboratory. A vertical section of the impregnated surface was then cut out and polished to observe grain fabric, grain size, and sedimentary structures.

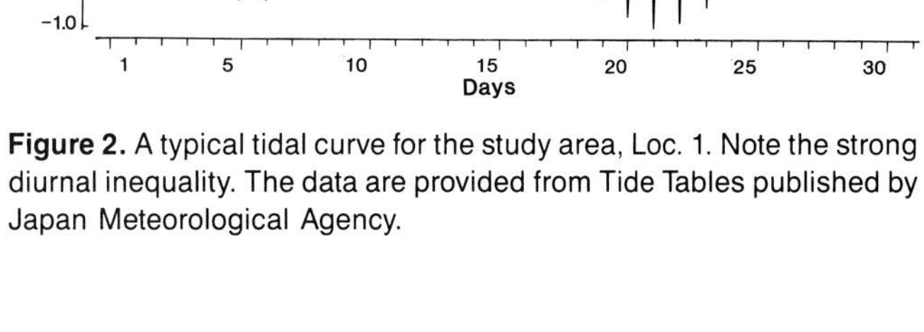

Figure 2. A typical tidal curve for the study area, Loc. 1. Note the strong diurnal inequality. The data are provided from Tide Tables published by Japan Meteorological Agency.

The number of grains with long axes dipping either seaward or landward was counted under the microscope or on photomicrographs; 100 to 300 grains were measured along traverses parallel to the beach surface (Fig. 4). Grains that were not imbricated either landward or seaward and spherical grains, amount to about 13 per cent of the total grain population and were not counted. The landward-dipping proportion, the so called ''imbrication ratio'' in this paper, which is the number of landward-dipping grains divided by the sum of the number of seaward-dipping and landward-dipping grains, was calculated for each traverse. When the value is more than 50 per cent, landward-dipping grains predominate, and when less than 50 per cent, seaward-dipping grains predominate.

Ebb and Flood Deposits

For several calm days after a storm, parallel-laminated, fine to medium grained sand (0.5-2.5ϕ) was deposited on Kouya Beach. The sand was moderately well sorted to well sorted (Folk and Ward, 1957).

By noting the positions of rings emplaced at successive stages of the tide, specific layers could be assigned to either rising or falling tide. Imbrication ratios (Fig. 4) within these layers show that seaward-dipping grains predominate in the deposit during the flood half-cycle, whereas landward-dipping grains predominate during the ebb half-cycle. This suggests that during the flood half cycle, uprush processes control the grain fabric, producing a mean imbrication with seaward dip. Conversely, during the ebb half-cycle, backwash processes leave the strongest imprint and landward-dipping grains predominate (Fig. 5). Both landward-dipping and seaward-dipping imbrication are present, consistent with the findings of Shelton and Mack (1970) and Taira (1989). The stippled zone in Figure 4 represents small-scale, tabular cross-laminae with a height of about 1 cm, deposited during the flood half-cycle. The foresets dip landward, meaning that the crosslaminae were produced by uprush. This supports the view that uprush processes are more effective during flooding tide. Grain size can change between flood and ebb deposition, such that laminar sets can be differentiated into flood or ebb deposits. Conversely, it was observed that the influence of the tide is not so great on the very coarse grained to granular foreshore sand deposits. Such deposits formed under high wave energy just after the storm, and show remarkable dominance of seaward-dipping imbrication

Figure 3. a. A trench through the foreshore, Kouya Beach, normal to the coastline. The gravelly layer in the middle (P, post-storm deposits) was deposited on an erosional surface during the waning phase of a storm. After storm waning, parallel-laminated, fine to medium grained sand was deposited (F, fair-weather deposit) over the gravelly layer. **b.** Trench wall normal to the coastline, Kouya Beach. Note the buried monitoring ring (white arrow), placed 24 hours previously. Since that time, 8 cm of sand has been deposited. **c.** An impregnated sample. Scale is 18 cm long. White arrow points to a buried ring. Other monitoring rings are buried in this deposit.

(imbrication ratios are 56-80%), independent of the tidal cycle. In this case, the backwash was ineffective for removing grains, because most of uprush water infiltrated into the very coarse grained sand and/or granule-size gravel.

Although flood and ebb deposits in the foreshore record are distinguishable by the analysis of grain fabric, the mechanisms of deposition and hydrodynamic interpretation are not yet clear. Several studies have been published on the hydrodynamics of uprush-backwash processes. The Froude number observed on recent beaches is generally greater than 1 in the backwash, but less than 1 in the uprush (Tanner, 1977). Uprush velocities of 0.7 to 1.9 m/sec and backwash velocities of 0.3 to 1.2 m/sec, were obtained from experimental measurements on a rigid surface sloping 34° (Machemehl and Herbich, 1971). However, the relationship between the water (tidal) level and hydraulic condition of uprush-backwash is not yet clear.

An additional complication is the infiltration capacity of foreshore sand, which significantly differs between the flood half-cycle and the ebb half-cycle (Duncan, 1964). High infiltration capacity may cause decreasing water volume in the backwash relative to that in the uprush, intensifying the uprush influence during the flood half-cycle. On the other hand, the water volume varies little between the uprush and backwash on wet sand during the ebb half-cycle (Clifton, pers. comm.). As mentioned previously, infiltration capacity has a pronounced influence in very coarse grained to granular sand. In addition, it is well known that beach gravels are imbricated in a seaward dipping direction (Reineck and Singh, 1980), probably for the same reason.

Depositional Thickness and Tidal Influence

The tide in the study area shows strong diurnal inequality (Fig. 2). The tidal pattern is reflected in the depositional thickness of the foreshore. The depositional thickness of paired ebb and flood deposits during fair-weather conditions shows an alternation of thick and thin layers (Fig. 6), each layer consisting of an ebb-flood pair. The thick composite layers are the products of deposition during the high range for the ebb half-cycle down to the lower low water, and the approximately equal high range for the next flood half-cycle. On the other hand, the thin composite layers correlate with the small range for the ebb half-cycle down to the higher low water, and the associated small range for the following flood half-cycle. In other words, the thick layers are deposited during cycles of higher tidal range and the thin layers are deposited during cycles of smaller tidal range.

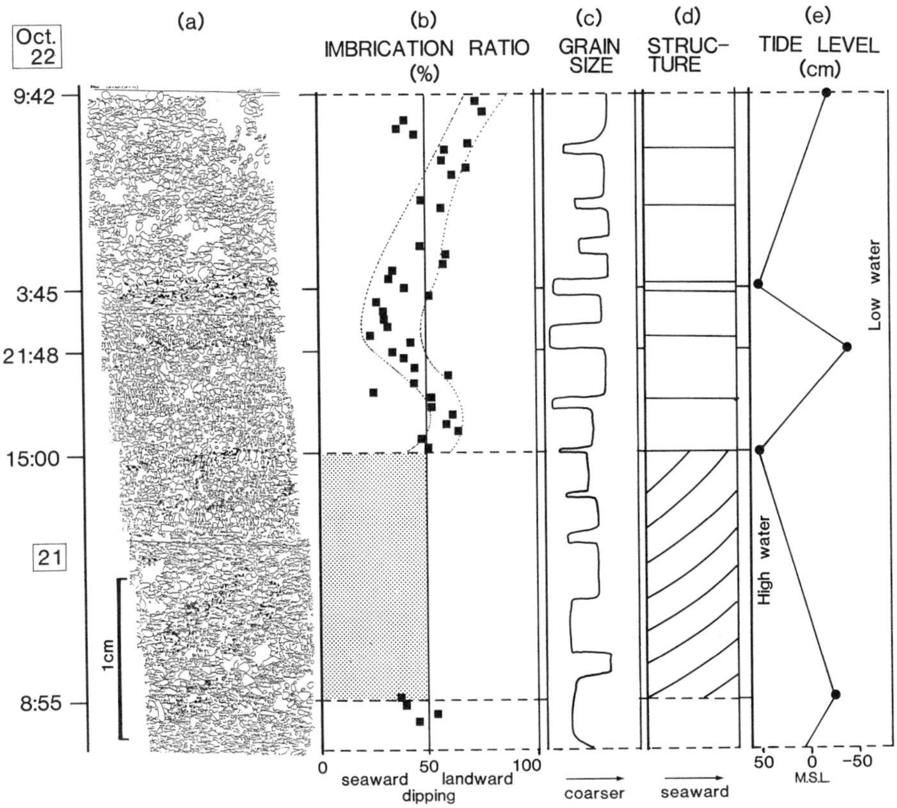

Figure 4. Analysis of grain fabric of the foreshore deposit. Sampling dates and times of high and low water are shown on the left. **a.** A trace of sand grains on the vertical section normal to the coastline. **b.** Imbrication ratio measured from **a**. The stippled zone has small tabular crosslaminae with foresets that dip landward. The imbrication ratio shows landward-dipping predominance in the ebb half cycle, and seaward-dipping predominance in the flood half-cycle. Three aomalous points in the uppermost part of the last ebb half-cycle are interpreted as recording final swash uprush before sampling. **c.** Apparent grain size. **d.** Sedimentary structures consisting mostly of parallel-lamination, but some small, 1 cm high crosslaminated beds are present. **e.** Tide levels for comparison with the imbrication ratio **b**. During the flood half-cycle, seaward-dipping grains predominate, whereas during the ebb half-cycle, landward-dipping grains predominate.

Application to Pleistocene Foreshore Deposits

We applied the results obtained from the modern foreshore deposits to Pleistocene foreshore deposits to determine possible indicators of paleotidal patterns. The samples were taken from the Narita Formation of the Shimosa Group, which is found in the southern part of Ibaraki Prefecture and the northern part of Chiba Prefecture. The Narita Formation consists mainly of shallow marine deposits, including foreshore deposits, and accumulated during the last interglacial period, in response to the expansion of Paleo-Tokyo Bay (Masuda, 1989). These Pleistocene deposits are generally horizontally bedded and undeformed. In the upper part of the Narita Formation, beach deposits are well developed, with parallel laminae comprising low-angle, wedge-shaped sets containing heavy mineral concentrations. The trace fossil ''*Exicirolana chiltoni japonica*'' (Kikuchi, 1972), which is similar to *Macraronichus segregatis* (Clifton and Thompson, 1978), is commonly present in the beach deposits.

Samples of the Pleistocene foreshore deposits were taken normal to the paleocoastline using quick-setting glue. The shoreline trend is defined by the regional paleogeography, and dip direction of foreshore bedsets. The samples were taken in a part of the sediment that appears to have been deposited without a significant break. Several layers, more than 1 cm thick and rich in heavy minerals, are intercalated in the foreshore deposits. These layers may correspond to surfaces of erosion or reworking, and thus samples were taken between such layers.

The results of our analysis of the Pleistocene foreshore deposits are illustrated in Figure 7. The flood and ebb deposits can be distinguished based on the imbrication ratio, although the deviations are smaller than those of modern deposits. Pairs of ebb and overlying flood deposits form alternations of thick and thin layers. These results indicate that deposition on the Pleistocene beach was strongly influenced by tides, and that the tides had a significant diurnal inequality, similar to the modern example from the Pacific coast of central Japan.

Acknowledgments

We would like to express our appreciation to T. Sunamura, The University of Tsukuba, for many helpful suggestions during the course of this work, and to J. B.

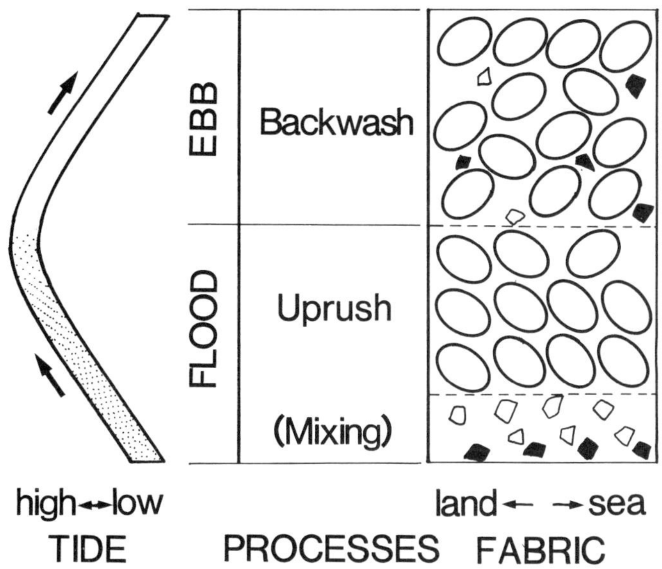

Figure 5. A model summarizing the fabrics of the fine to medium grained foreshore sands deposited during flood and ebb half-cycles under fair-weather conditions. Uprush processes predominate during the flood half-cycle, and generate seaward-dipping imbrication. At the base of the flood deposit, remobilized, fine grained heavy minerals (black angular particles) from the underlying surface are commonly present. Conversely, backwash processes predominate and produce landward-dipping imbrication during the ebb half-cycle.

Southard, Massachusetts Institute of Technology, H. E. Clifton, United States Geological Survey, R. N. Hiscott, Memorial University of Newfoundland, R. J. Cheel, Brock University, M. R. Gibling, Dalhousie University, and M. Ito, Chiba University for critical reading of an earlier version of this manuscript and suggested improvements.

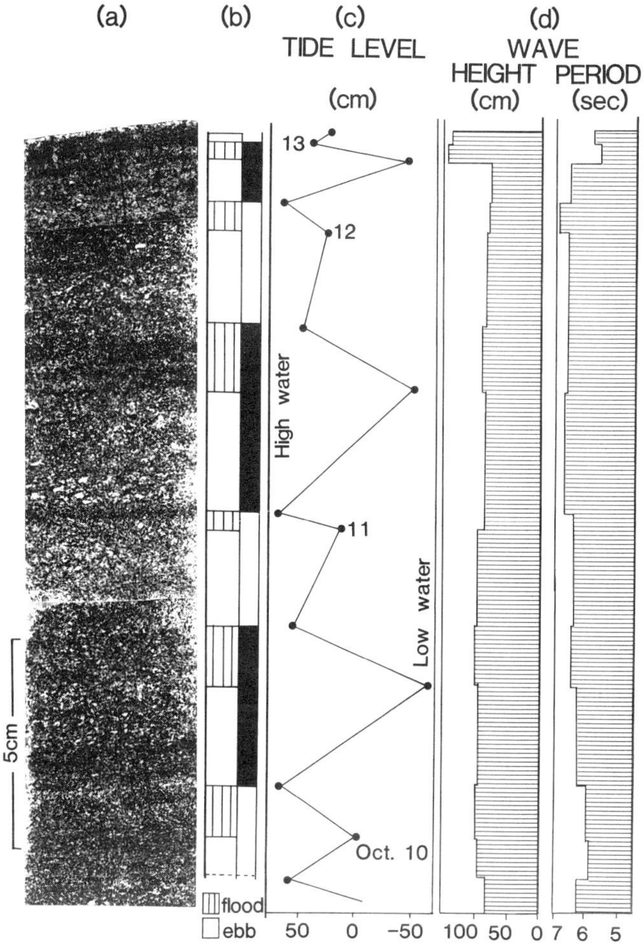

Figure 6. a. Fine grained foreshore sands deposited on calm days (October 10-13, 1987, in Kouya Beach). **b.** Paired ebb (white area) and flood (striped pattern) deposits show an alternation of thick and thin beds (black and white bars, respectively) that appear to correlate with the tidal pattern **c**. **c.** Tidal pattern of diurnal inequality. **d.** Wave conditions, which were calm and consistent over the observation period. (Wave data provided by Construction Office of Kashima Port, Ministry of Transport.)

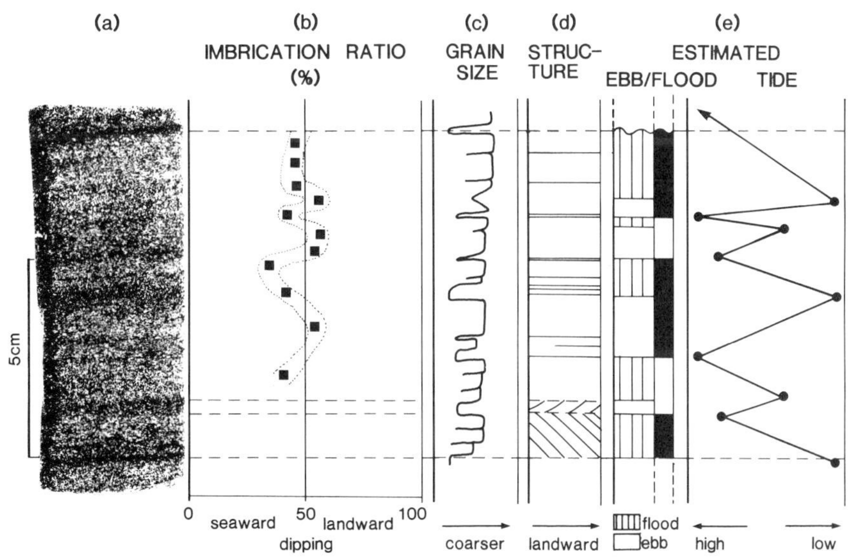

Figure 7. Analysis of grain fabric of Pleistocene foreshore deposits in Kozayama Tamatsukuri-Town, Ibaraki Prefecture. **a.** Photograph of the impregnated section. **b.** Imbrication ratio that shifts gradually upward from seaward dip dominance to landward dip dominance. **c.** Apparent grain size, showing slight changes. **d.** Sedimentary structures: mostly parallel-lamination, with small crosslaminae in the lower part used to estimate the phases of the tidal cycle. **e.** Ebb and flood deposits inferred from **b** and **c**. Paired ebb and overlying flood deposits show alternation of thick and thin layers (black and white bars, respectively), suggesting that the Pleistocene beach was influenced by tides with diurnal inequality.

REFERENCES

Allen, J. R. L. 1982. Sedimentary structures: Their character and physical basis, v. 21, Elsevier Scientific Publishing Company, 663 p.

Clifton, H. E. 1969. Beach lamination: nature and origin. Marine Geology, v. 7, p. 553-559.

Clifton, H. E. and Thompson, J. K. 1978. *Macraronichus segregatis*: a feeding structure of shallow marine polychaetes. Journal of Sedimentary Petrology, v. 48, p. 1293-1302.

Curray, J. R. 1956. Dimensional grain orientation studies of recent coastal sands. American Association of Petroleum Geologists, Bulletin, v. 40, p. 2440-2456.

Duncan, J. R. Jr. 1964. The effects of water table and tide cycle on swash-backwash sediment distribution and beach profile development. Marine Geology, v. 2, p. 186-197.

Folk, R. L. and Ward, W. C. 1957. Brazos river bar: A study in the significance of grain size parameters. Journal of Sedimentary Petrology, v. 27, p. 3-26.

Harms, J. C. 1975. Stratification and sequence in prograding shoreline deposits. *In:* Depositional Environments as Interpreted from Primary Sedimentary Structures and Stratification Sequences, J. C. Harms, J. B. Southard, D. R. Spearing and R. G. Walker (eds.). Society of Economic Paleontologists and Mineralogists, Short Course No. 2, p. 84-115.

Kikuchi, T. 1972. A characteristic trace fossil in the Narita Formation and its paleogeographical significance. Journal of Geological Society of Japan, v. 78, p. 137-144.

Machemehl, J. L. and Herbich, J. B. 1971. Non-steady flow on sloping beach with large roughness elements. Proceedings of the 14th Congress, International Association for Hydraulic Research (Paris), v. 2, p. 177-186.

Masuda, F. 1989. Tidal deltas and barrier islands. *In:* Barrier islands in Paleo-Tokyo Bay: Field trip guide to 96th annual meeting of Geological Society of Japan, Yasuhiko Makino and Fujio Masuda (eds.). Geological Society of Japan, Tokyo, p. 188-196.

Nanz, R. H. 1955. Grain orientation in beach sands: A possible means for predicting reservoir trends. Journal of Sedimentary Petrology, v. 25, p. 130 (Abstract).

Otvos, E. G. Jr. 1965. Sedimentation-erosion cycles of single tidal periods on Long Island Sound beaches. Journal of Sedimentary Petrology, v. 35, p. 604-609.

Reineck, H. E. and Singh, I. B. 1980. Depositional Sedimentary Environments, 2nd edition, Springer-Verlag, Berlin, p. 145-147.

Shelton, J. W. and Mack, D. E. 1970. Grain orientation in determination of paleocurrents and sandstone trends. American Association of Petroleum Geologists, Bulletin, v. 54, p. 1108-1119.

Taira, A. 1989. Magnetic fabrics and depositional processes. *In:* Sedimentary facies in the active plate margin, Asahiko Taira and Fujio Masuda (eds.). Terra Scientific Publishing Company, Tokyo, p. 43-77.

Tanner, W. F. 1977. Froude regimes in the swash zone. *In:* Coastal Sedimentology, W. F. Tanner (ed.). Geology Department, Florida State University, Florida, p. 289-297.

Thompson, W. O. 1937. Original structures of beaches, bars, and dunes. Geological Society of America, Bulletin, v. 46, p. 723-752.

KEY WORDS: DIURNAL INEQUALITY, EBB AND FLOOD DEPOSITS, FORESHORE DEPOSITS, GRAIN IMBRICATION, MODERN AND PLEISTOCENE FORESHORE DEPOSITS IN JAPAN, VARIATION OF DEPOSITIONAL VOLUME

In: Clastic Tidal Sedimentology, D. G. Smith, G. E. Reinson, B. A. Zaitlin and R. A. Rahmani (eds.). Canadian Society of Petroleum Geologists. Memoir 16, p. 321-334.

Lateral variations in tidal deposits of the Upper Cretaceous Neuquén Basin, western Argentina

CLAUDIO A. BARRIO

Shell C.A.P.S.A.
Exploration and Production
Avda Roque Saenz Peña 788
Buenos Aires (1383), Argentina

ABSTRACT

Shallow marine to coastal, clastic and evaporitic sediments of the Allen (Loncoche) Formation were deposited during the late Campanian-early Maastrichtian in the Neuquén foreland basin (western Argentina). On the basis of several stratigraphic sections, four facies associations representing different tidal subenvironments have been recognized: A) crossbedded sandstones; B) massive mudstones with dolomites and volcanic ashes; C) sandstones and mudstones, and D) evaporites. Facies association A is composed of quartz-felspathic crossbedded sandstones, with epsilon and sigmoidal crossbedding, and clay drapes being the principal tide-related sedimentary structures. A subtidal to intertidal depositional environment is inferred for facies association A. Facies association B is composed of montmorillonitic mudstones with intercalations of volcanic ash, stromatolites and dolomites, containing a brackish to marine fauna. Facies association B was deposited in an intertidal setting. Facies association C, characterized by mixed montmorillonite-rich mudstones and feldspathic sandstones with lenticular, wavy, and flaser bedding, is considered to be a subtidal to intertidal lagoonal deposit. Facies association D is composed of evaporites and stromatolites formed in an upper intertidal to supratidal environment.

The arrangement of facies associations on the east side of the foreland basin resembles the classical tidal-flat examples of the North Sea, while toward the northwest, paleoenvironments correspond to modern, tide-dominated estuarine environments. Differences in the depositional paleoenvironments are attributed to differences in the tidal regimes, which in turn are related to the basin's geometry. A microtidal regime affected the depositional environments of the low-relief coast to the southeast, while a macrotidal regime influenced the depositional environments of the funnel-shaped basin toward the northwest. Additional factors affecting the distribution of facies associations include paleoslope and sediment input. Gentle paleoslopes and reduced sediment input affected deposition on the eastern, cratonic side of the foreland basin, whereas increased sediment input and steeper paleoslopes affected deposition in the western, retro-arc side of the basin.

RÉSUMÉ

Durant le Campanien tardif et le Maastrichtien précoce, les sédiments clastiques et évaporitiques de la Formation Allen (Loncoche) ont été déposés dans le bassin d'avant-pays Neuquén (Argentine occidentale). Ils représentent des dépôts marins d'eau peu profonde passant à des dépôts littoraux. L'examen de plusieurs sections stratigraphiques a permis de reconnaître quatre associations de faciès représentant des sous-environnements de marée différents: (A) grès à stratifications obliques; (B) mudstones massifs avec des dolomites et cendres volcaniques; (C) grès et mudstones, et (D) évaporites. L'association A se compose de grès feldspathiques à stratifications obliques avec des structures sédimentaires reliées aux marées dont les principales sont des stratifications obliques de type epsilon, des «lentilles d'argiles pélagiques» («clay drapes») et des structures sigmoïdes. Un milieu de sédimentation de la zone subtidale à intertidale est inféré à cette association de faciès A. L'association B se compose de mudstones à montmorillonite, avec des intercalations de cendres volcaniques, stromatolithes et dolomites; elle contient une faune marine à saumâtre. Les faciès de l'association B se sont accumulés dans une zone intertidale. L'association C se caractérise par un mélange de mudstones riches en montmorillonite et de grès feldspathiques avec des lits lenticulaires, onduleux et de type «flaser». Elle est considérée comme appartenant à un milieu de sédimentation de lagon variant de la zone intertidale à subtidale. L'association D se compose d'évaporites et stromatolithes, formés dans la partie supérieure de la zone intertidale et dans la zone supratidale.

L'arrangement des ces quatre associations de faciès, de la partie orientale de ce bassin d'avant-pays, ressemble aux exemples classiques d'estran de la mer du Nord. Toutefois, vers le nord-est du bassin, les paléoenvironnements correspondent aux environnements estuariens modernes prédominés par les marées. Les différences dans la sédimentation des paléoenvironnements sont attribuées aux différences des régimes de marées, lesquelles sont reliées à la géométrie du bassin. Un régime microtidal a affecté la partie sud-est du bassin où les milieux de sédimentation sont reliés à une côte de faible relief. Un régime macrotidal a influencé les milieux de sédimentation vers le nord-ouest du bassin où il devient en forme d'entonnoir. Les autres facteurs influençant la distribution des associations de faciès comprennent la paléopente et l'apport de sédiments. La sédimentation a été affectée, du côté est du bassin d'avant-pays face au craton, par des paléopentes faibles et un apport réduit de sédiments, et du côté ouest face au bassin rétro-arc, par des paléopentes plus fortes et un apport accru de sédiments.

Introduction

The Neuquén Basin is a sedimentary basin in western Argentina (Fig. 1A) with deposits ranging in age from Late Triassic to Tertiary. It is bordered to the northeast by the Sierra Pintada massif and to the south by the North Patagonian massif (Fig. 1A), which define the original depositional limits of the Jurassic and Cretaceous marine rocks. The present western flank is formed by the fold and thrust belt of the Andean (Principal) Cordillera (Digregorio *et al.*, 1984).

Three major sedimentary supercycles are represented by the strata of the basin: Jurásico (Late Triassic-Late Jurassic), Andico (Late Jurassic-Early Cretaceous), and Riograndico (Late Cretaceous-early Tertiary) (Groeber, 1929, 1946, 1953). This paper focuses on the upper part of the Riograndico cycle, in the sedimentary sequence known as the Malargüe Group (late Campanian-Paleocene), which comprises four formations (Fig. 2). To the southeast the group includes (from base to top) the Allen, Jagüel, Roca and Carrizo formations (Digregorio and Uliana, 1980; Uliana, 1979). Toward the northwest, the Malargüe Group is composed of (from base to top) the Loncoche, Jagüel, Roca and Pircala formations (Ramos, 1981; Barrio, 1989) (Fig. 2). The basal Allen (Loncoche) Formation is the subject of this paper.

The study area is located in the Neuquén Province of western Argentina between 37°00′ and 39°00′ S latitude and 70°00′ and 67°30′ W longitude, where the Malargüe Group is well exposed in three areas, shown in Figure 1. A total of nine sections of the Allen (Loncoche) Formation were

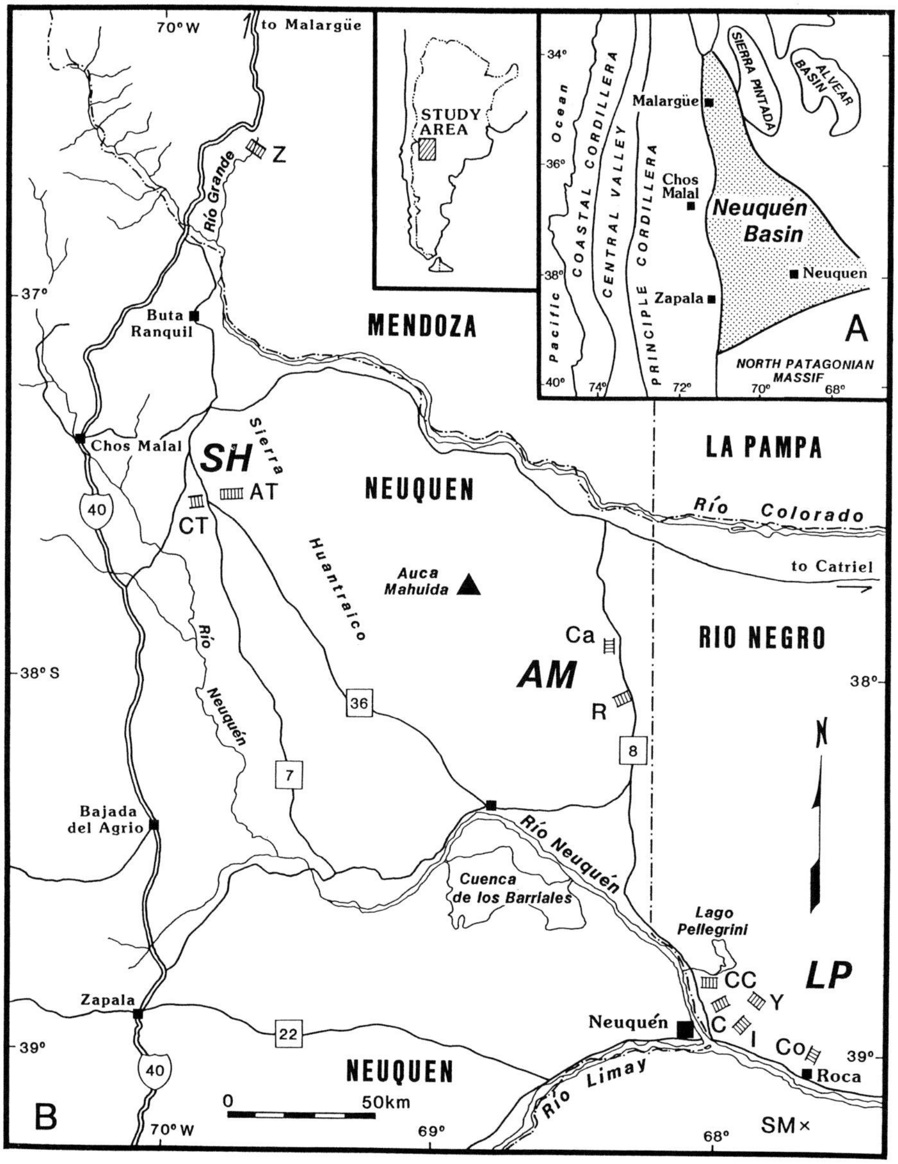

Figure 1. Location map of the study area, in western Argentina. **a)** regional background of the Neuquén Basin. **b)** the three areas with good exposures of Upper Cretaceous-lower Tertiary strata in the northern Neuquén Basin. Letters indicate study areas: LP, Lago Pellegrini (Co, Cantera Corral section; I, Indupa section; C, Cementerio de Cinco Saltos section; CC, Contralmirante Cordero section; Y, La Yesera quarry section). AM, Auca Mahuida (Ca, El Caracol section; R, Jagüel de Los Rosauros section). SH, Sierra de Huantraico (CT, Cerro Tormenta section; AT, Aguada del Tuco section). Z, Zampal (after Uliana and Musacchio, 1978).

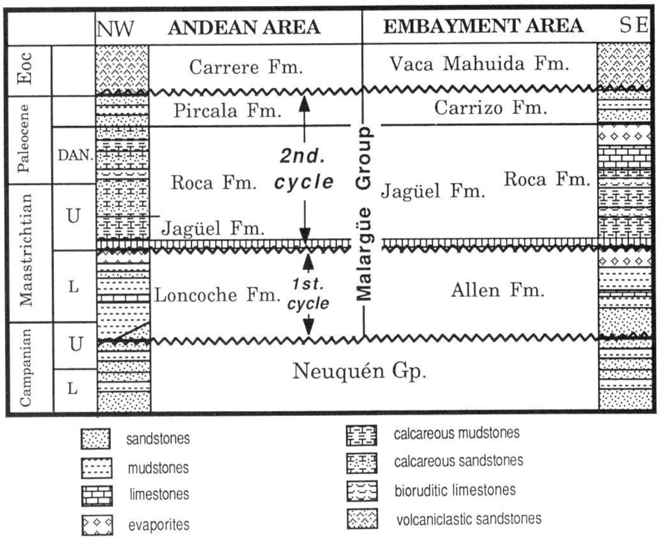

Figure 2. Stratigraphy of the Malargüe Group.

measured and described in three areas. They will be referred to in the text by capital letters taken from the name of the measured sections, as follows (see Fig. 1 for location):

1. The southern Lago Pellegrini-General Roca area (LP), a continuous outcrop belt of Upper Cretaceous-lower Tertiary strata (five sections: Indupa (I), La Yesera (Y), Cementerio de Cinco Saltos (C), Contralmirante Cordero (CC) and Cantera Corral (Co)).
2. The central Auca Mahuida area (AM), 80 km northwest of the first area, and in general similar to the Lago Pellegrini-General Roca area (two sections: El Caracol (Ca) and Jagüel de los Rosauros (R)).
3. The northern Sierra de Huantraico area (SH), 105 km northwest of the Auca Mahuida area, where excellent exposures of the Malargüe Group contain sections very different from those in the south (two sections: Cerro Tormenta (CT) and Aguada del Tuco (AT)).

The objectives of this paper are two-fold: 1) to provide a detailed facies analysis and interpretation of the Allen Formation; and 2) to document the arrangement of lateral facies variations within equivalent units and to relate these variations to an overall, regional, depositional model.

Allen (Loncoche) Formation

Stratigraphy

The Allen (Loncoche) Formation was referred to in the past as "Senoniano Salobre", that is the "Brackish Senonian" because of the fresh water and brackish fauna in the unit (Wichmann, 1924; Keidel, 1925; Frenguelli, 1930). The Allen Formation (Roll, 1939) is the name used today for the basal section of the Malargüe Group on the eastern side of the Neuquén Basin. Three members were described for the Allen Formation by Andreis *et al.* (1974) and Uliana and Dellapé (1981); a basal sandy unit interpreted as a sand flat, a middle pelitic section (intertidal), and sabkha evaporites at the top.

The Loncoche Formation is the accepted name for the equivalent unit toward the western side of the basin (Groeber, 1946). In the Malargüe area (Mendoza province) Groeber (1946), Uliana and Musacchio (1978), and Volkheimer (1978) described the Loncoche Formation as being composed of more than 100 m of crossbedded, volcanic-rich sandstone, mudstone and tuff with continental microfossils (Uliana and Musacchio, 1978), continental flora (Papu, 1987) and paleosols (Volkheimer, 1978). To the south, in the Sierra de Huantraico area, Bertels (1968), Marcón (1975) and Ramos (1981) described the Loncoche Formation as comprising greenish siltstone and mudstone with intercalated limestone containing fresh water ostracodes (Bertels, 1968). Based on the fresh water fauna and the considerable thickness of the sequence, Bertels (1969b) assigned a deltaic environment of deposition to the sediments of the Loncoche Formation. Ramos (1981) considered this unit to have been deposited in a lacustrine environment.

Age

Bertels (1964, 1969a) and Ballent (1980) assigned an early to middle Maastrichtian age for the Allen (Loncoche) Formation. A late Campanian age, based on sequence position, was proposed for the basal section of the Loncoche Formation by Ramos (1981) for the Sierra de Huantraico area (Fig. 2). In the southern Mendoza area, Uliana and Musacchio (1978) gave a Maastrichtian age to the Loncoche Formation, whereas a lower Campanian age (80 Ma) was assigned to this formation by Legarreta and others (1989).

Sequence

The Allen (Loncoche) Formation includes a sequence (Mitchum and others, 1977; Van Wagoner *et al.*, 1988) bounded by unconformities at the base and top. It is bounded at the base by an erosional disconformity (Andreis *et al.*, 1974) over the Anacleto Member of the Rio Colorado Formation (Neuquén Group; De Ferrariis, 1968). In the field this disconformity is marked in the Lago Pellegrini area by large rip-up clasts of the Anacleto Member (Uliana and Dellapé, 1981). In Mendoza province the unconformity is indicated by 10 cm volcanic clasts at the base of the sequence (Legarreta *et al.*, 1989). In the other areas this erosive surface is obscure and can easily go unrecognized.

The top of this depositional sequence is an erosional disconformity between evaporites of the Allen (Loncoche) Formation and open marine shales of the Jagüel Formation, which constitutes the base of the second depositional sequence of the Malargüe Group.

Sedimentology

The Allen (Loncoche) Formation consists of four facies associations, defined by Reading (1986) as a group of facies that are genetically or environmentally related:

A. crossbedded sandstone
B. mudstone with dolomite and volcanic ash
C. sandstone and mudstone
D. evaporites

All four facies associations were recognized in two of the three study areas. Development of individual facies associations varies from one location to another (Table 1). Facies association A is thickest (18 m) in the Lago Pellegrini area, and thins toward the west. In the Sierra de Huantraico area, facies association A varies in thickness from 8 m in the Aguada del Tuco section to 0 m in the Cerro Tormenta section. The thickness of facies association B increases significantly from east to west. Facies association C is well represented in the Lago Pellegrini area, is less developed in the Sierra de Huantraico area, and is not present at all in the Auca Mahuida area. Facies association D was encountered in all the three areas, with better development toward the east side of the basin (Table 1).

Table 1. Regional distribution of the facies associations in the three study areas with outcrops of the Allen (Loncoche) Formation.

	Facies Associations	E LP	Regional Distribution AM sections	W SH
A	crossbedded sandstone	I:18 m CC:12 m C:4 m	Ca:3.7 m R:1 m	AT:8 m CT:0 m
B	mudstone with dolomite and volcanic ash	CC:25 m C:15 m	Ca:53.4 m R:43 m	AT:110 m CT:104 m
C	sandstone and mudstone	Y:20.5 m CC:12 m	—	CT:4.75 m
D	evaporites	Y:17 m Co:7.32 m	Ca:6 m R:1.3 m	CT:6.5 m

Facies Association A: Crossbedded Sandstone

Regional Distribution

In the Lago Pellegrini area (see Fig. 1 for location of measured sections), facies association A was recognized in sections CC, I and C (Fig. 3). In the Auca Mahuida area, facies association A was found in the Ca (3.7 m) and R sections (1 m) (Fig. 4). Eight metres of facies association A were

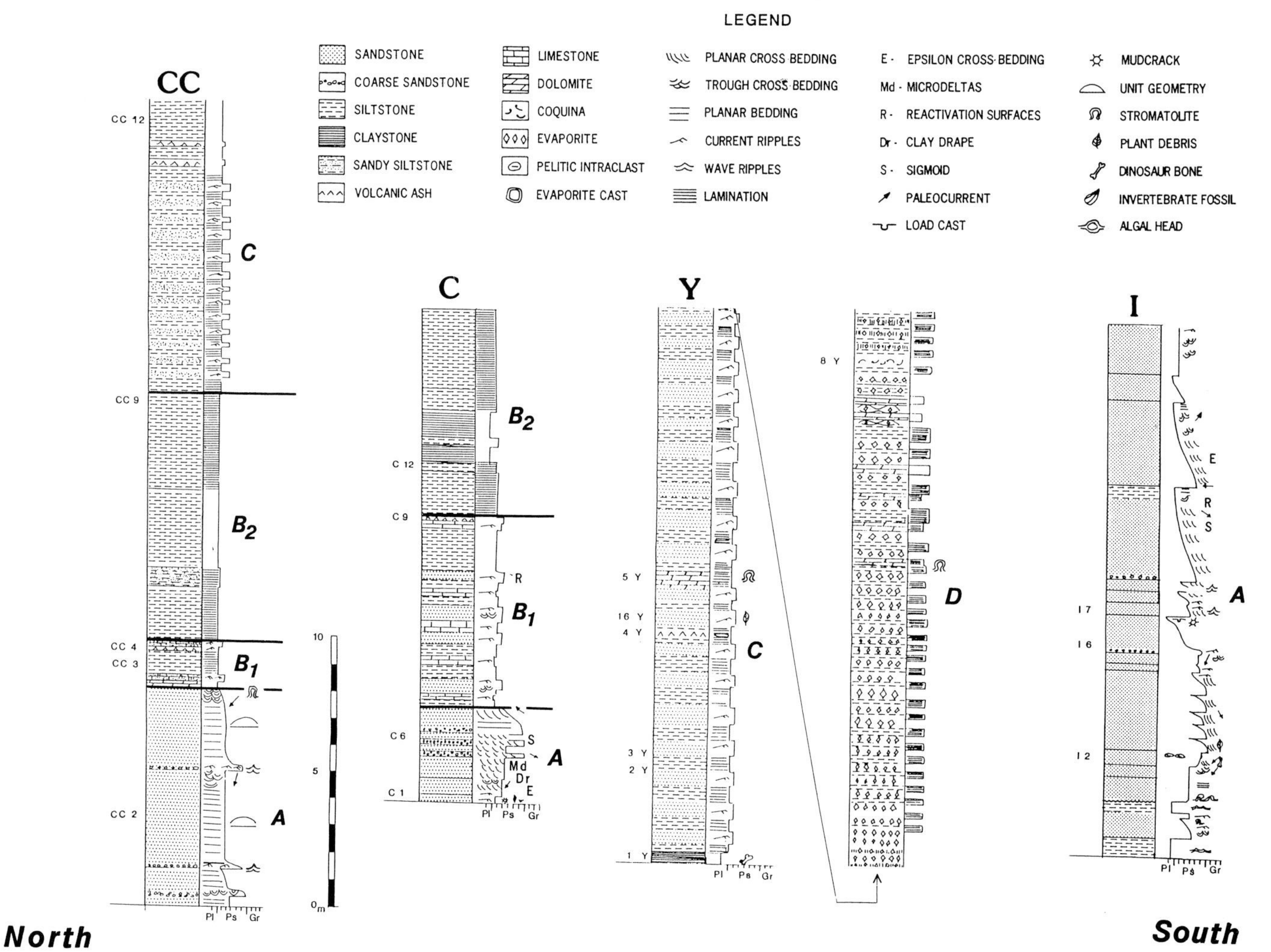

Figure 3. Measured sections for the Lago Pellegrini area. Contralmirante Cordero (CC), Cementerio (C), Indupa (I) and La Yesera (Y). See Figure 1 for location of sections.

measured in the AT section (Fig. 5) in the Sierra Huantraico area (Table 1).

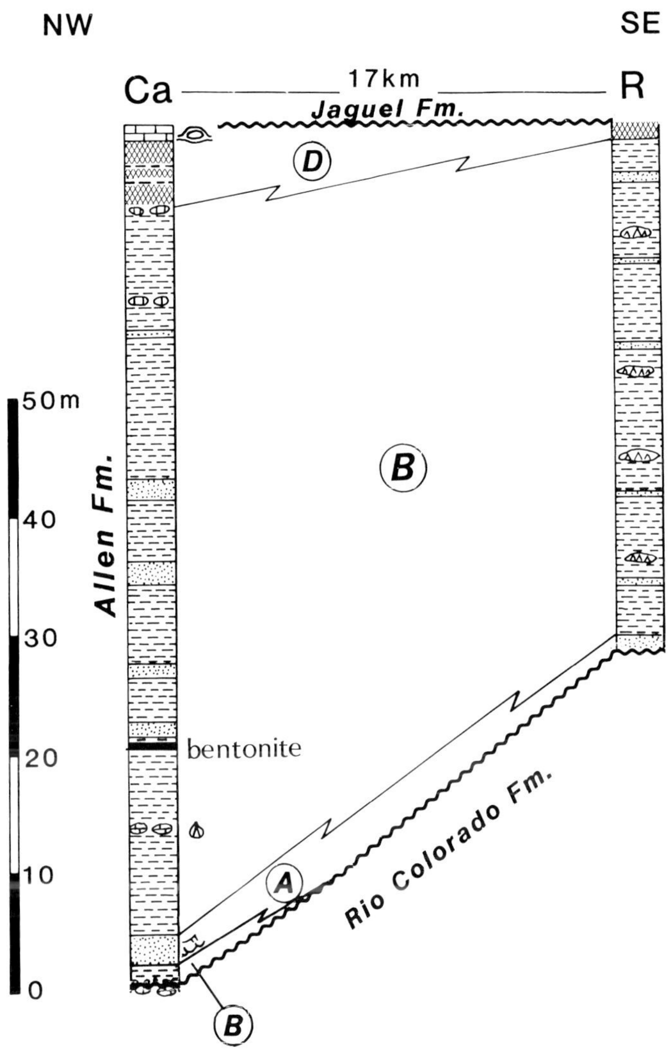

Figure 4. Measured sections for the Auca Mahuida area. El Caracol section (Ca) and Jagüel de Los Rosauros section (R). Legend same as in Figure 3.

Lithology and Sedimentary Structures

The idenitification of this facies association in the field is relatively easy because of changes in grain size and colour relative to the underlying red beds of the Neuquén Group. The main characteristic of facies association A is the predominance of grey and yellow sandstones, which form a distinctive ledge. The thickness varies over short distances, from 14 m in section CC to 4 m in section C (6.5 km to the south) and 18 m, a maximum value, in section I (4 km south of section C, Fig. 3). The contact with the mudstone of overlying facies association C is concordant and well marked.

Strata range from medium to thick bedded and are both tabular and lenticular. Tabular and trough crossbedding is abundant in this facies. Associated with the primary sets are sets of epsilon crossbedding or inclined heterolithic stratification (IHS) (Thomas *et al.*, 1987), characterized by large-scale foresets and low angle dips (Allen, 1963), with foresets fining upward from coarse sandstone at the base to fine grained sandstone or siltstone at the top.

Reactivation surfaces (Collinson, 1970) are common and in some cases associated with "microdelta cross-bedding" (Reineck and Singh, 1975; Dalrymple, 1984) (Fig. 6). Clay drapes in the foresets of the crossbedding were also seen in this area. Medium- to small-scale sigmoidal crossbedding characterized by S-shaped foresets (Kreisa and Moiola, 1986) were described in section I (Fig. 7).

In the upper parts of sections I and CC (Fig. 3) a fining-upward cycle 3 to 3.5 m thick is present, beginning with wave-rippled bedding in a coarse grained sandstone, followed by planar crossbedding and trough crossbedding in a medium grained sandstone, and ending with massive shale or laminated, fine grained sandstone. The three-dimensional geometry of these sets is evident in the CC section, where flat set bottoms and convex set tops outline large bars.

Ripples in sections I and CC show the internal lamination of "wave-rippled bedding" (Reineck and Singh, 1975). In plan view they are symmetrical with straight crests or asymmetrical with linguoid crests. In some of the ripples the trough is filled with fine grained lime mudstone. Climbing ripples are common in the Auca Mahuida sections. Wavy and lenticular bedding is also common. Some ripples are present in the foresets of the large-scale channelized structures, a characteristic typical of epsilon crossbedding, which indicates flow subparallel to the strike of the inclined beds (Collinson and Thompson, 1982). Other structures include rill-marks, rain prints and mud cracks (desiccation cracks), all indicating subaerial exposure of the sediments. Burrowing is very rare.

The paleocurrent measurements in sections I and CC show a flow direction predominantly to the southeast (Andreis *et al.*, 1974), with subordinate directions to the northwest and southwest (Barrio, 1989).

Fossils

Dinosaur bones and plant debris are both found at the bottom of the sequence in the LP area. A carnosaur (*Abelisarius comahuensis*) was discovered in the Lago Pellegrini area (Bonaparte and Novas, 1985). A hadrosaur was described by Powell (1987) from Salitral Moreno (SM in Fig. 1), associated with titanosaurs, theropods, birds, turtles, gastropods and plant fragments.

Depositional Environment

Many of the sedimentary structures described for facies association A are not diagnostic of a marine or fluvial environment. However, it is considered, (based mainly on the most complete sections on the eastern side of the basin), that facies association A probably represents subtidal to intertidal deposits as suggested by the conjunction of: epsilon crossbedding or inclined heterolithic stratification, which may be

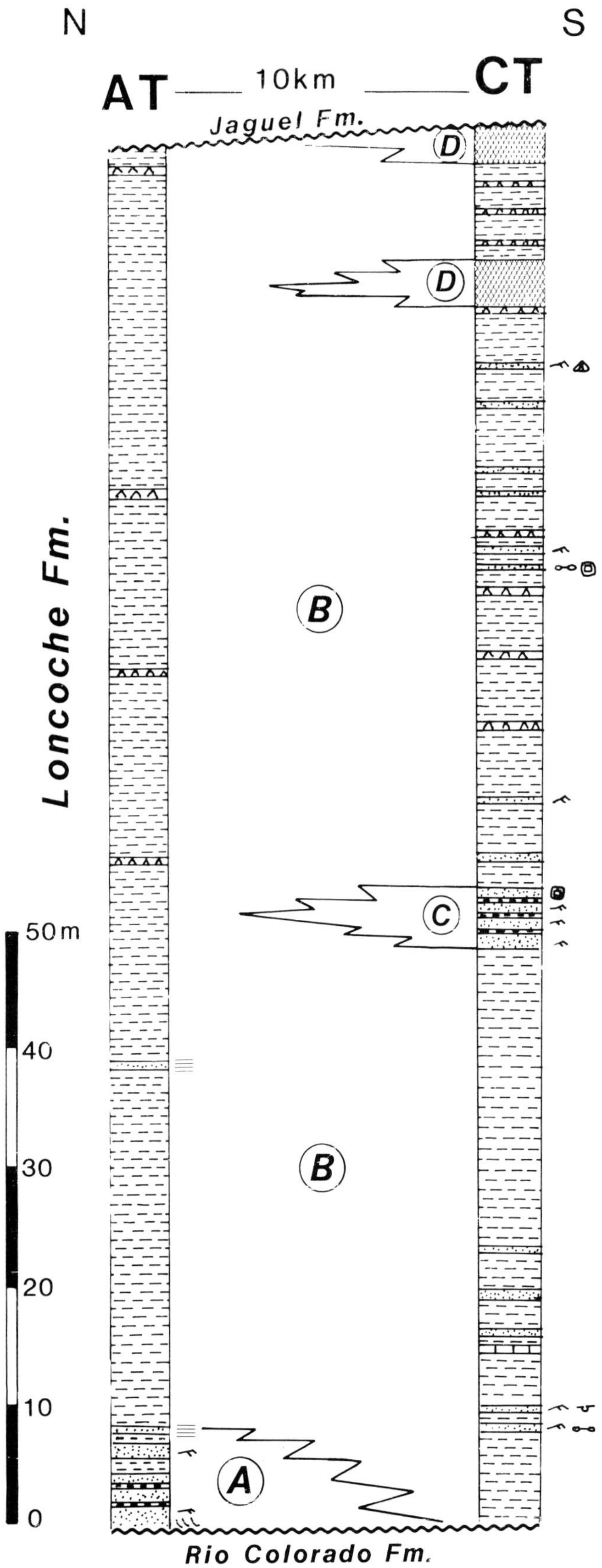

Figure 5. Measured sections for the Sierra de Huantraico area. Cerro Tormenta section (CT) and Aguada del Tuco section (AT). Legend same as in Figure 3.

Figure 6. Reactivation surfaces associated with microdelta crossbedding in facies association A. Rock hammer is 33 cm long.

found in subtidal channels (Evans, 1965; Reineck, 1972; Allen, 1963; Klein, 1977; Thomas *et al.*, 1987); reactivation surfaces related to time-velocity asymmetry of the currents (Klein, 1970); clay drapes in the foresets commonly associated with alternating ebb and flood tidal currents (Visser, 1980; Terwindt, 1988); and sigmoidal foresets related to fluctuations in current velocity that are characteristic of tidal environments (Allen, 1984). In addition, the occurrence of dinosaur bones and plant debris shows the proximity of land. Additional evidence is provided by the northwest-southeast linear trend of the outcrops of facies association A, which are limited to the westernmost locations (see fence diagram, Fig. 8).

An intertidal setting for parts of facies association A in the Lago Pellegrini area is indicated by the presence of desiccation cracks and rain drop impressions (Reineck and Singh, 1975). Rill marks are verification of very shallow waters. The microdeltas associated with reactivation surfaces are indicative of runoff currents and emergence (Dalrymple, 1984).

Figure 7. Sigmoidal crossbedding in the I section, facies association A. Rock hammer is 33 cm long.

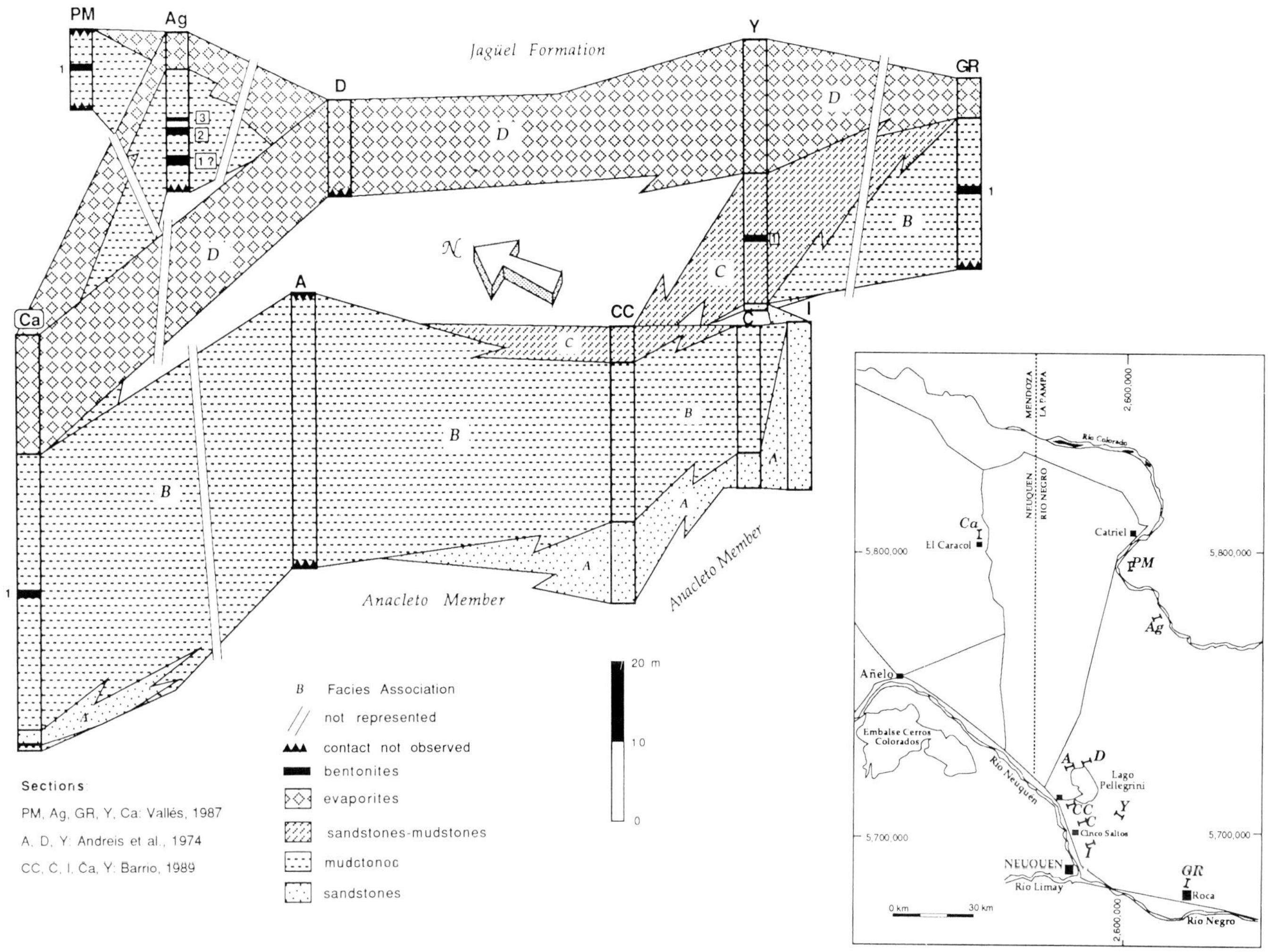

Figure 8. Fence diagram for the facies associations of the Allen (Loncoche) Formation in the Lago Pellegrini and Auca Mahuida areas.

Wave ripple troughs filled with fine grained material as described are found in intertidal deposits (Clifton, 1983).

Facies association A in the northern part of the basin (SH and AM areas) lacks diagnostic features for paleoenvironmental interpretation. The depositional environment of facies association A in these areas will be considered under the discussion heading.

Facies Association B: Mudstone with Dolomite and Volcanic Ash

Regional Distribution

Facies association B has been recognized in the study area in all the locations where sections of the Allen Formation were measured (Table 1). The contact with underlying facies association A is sharp, and paraconformable with the Anacleto member of the Neuquén Group (Cazau and Uliana, 1972) in the CT section, while the contact with overlying facies association C or D is transitional. The total thickness of facies association B in the Lago Pellegrini area is 25 m in section CC and 15 m in section C (Fig. 3). Facies association B in the Auca Mahuida area (Ca and R sections, Fig. 4) increases in thickness to 53.4 m. Facies association B was recognized in the Sierra de Huantraico area in the CT section (104 m) and the AT section (110 + m; Fig. 5).

Lithology and Sedimentary Structures

In some of the measured sections it was possible to distinguish two facies subassociations. The B-1 facies subassociation is characterized by mudstone intercalated with calcareous sandstone, dolomite and volcanic ash in very thin beds. The B-2 facies subassociation is composed of claystone and siltstone in a monotonous sequence (Fig. 9).

The B-1 facies subassociation comprises a greenish, thick to medium bedded (from 4.4 m to 0.4 m), laminated or massive mudstone with intercalations of thin bedded, pale orange, calcareous sandstone and dolomite in beds 0.1 to 0.2 m thick. This facies subassociation is 7 m thick in section CC, 2 m thick in section C in the LP area, and 74 m thick in CT section in the SH area. The calcareous sandstone contains trough crosslamination and symmetrical and climbing ripples. Pelitic intraclasts, small load casts and evaporite casts were observed in the SH area. Clasts of the underlying san-

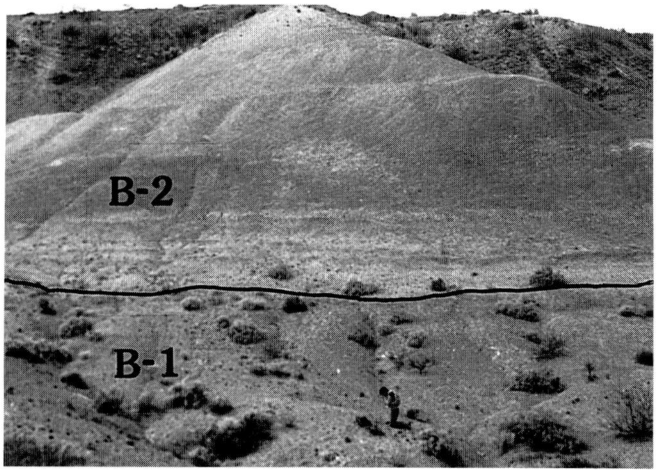

Figure 9. Outcrop view of facies association B showing the two facies subassociations described in the text. In the bottom part of the photograph the B-1 facies subassociation and in the top, the B-2 facies subassociation. See person for scale.

dy facies association A are common at the base of facies association B.

The dolomite is fine grained, greyish orange with LLH (laterally linked hemispheroids; Logan and others, 1964), and stromatolites, which in plan view are circular, with diameters of 2 cm. Secondary gypsum is present as thin veins 1 to 2 cm thick. The volcanic ash is very fine grained, well consolidated, silicified, and shows symmetrical ripple-bedding.

The B-2 facies subassociation consists of a monotonous sequence of intercalated, medium to thick bedded, laminated or massive, grey and green claystone and siltstone. Thin carbonate beds are common in facies subassociation B-2, and thin horizons of volcanic ash and gypsum are present. The thickness of this facies subassociation is 10 m in CC section, 8 m in C section, and 30 m in CT section.

Carbonate concretions with marine invertebrates (Wichmann, 1927) form a continous layer in the middle part of the Ca section (AM area). At the top of the Ca section massive gypsum nodules surrounded by fibrous gypsum are present, as well as abundant, small, discoidal carbonate concretions. A bentonite level in facies association B (Ca section) was located by Vallés (1987) (Fig. 4).

Fossils

Benthic foraminifera (Kielbowicz, 1980) and brackish water ostracodes (Angelozzi, 1980) were described from facies association B in the LP area. No burrows were observed in the field.

Wichmann (1924, 1927) discovered fresh water ostracodes, plant debris, fresh water and marine bivalves, crocodile bones, turtles plaques, and fish teeth in the AM area. Bertels (1969a, 1970) classified fresh water ostracodes from facies association B in the Auca Mahuida area. More recently, Vallés (1987) has recovered marine bivalves (*Isognomon*) from this facies association.

Bertels (1968, 1972) described lower Maastrichtian fresh water ostracodes and charophytes from the SH area. However, analysis of thin sections revealed marine fauna in facies B for this area (brachiopods, mollusks, forams; Barrio, 1989).

Depositional Environment

An intertidal environment is inferred for facies association B for the eastern sections, based on the presence in the mudstone of evaporite casts, stromatolites and dolomites. The evaporitic casts represent desiccation in very shallow waters and were reported for intertidal deposits in the Persian Gulf (Illing *et al.*, 1965). Stromatolites usually occur in intertidal environments (Awarik *et al.*, 1976). The dolomite found in this facies association was probably formed by the evaporation of sea water by reflux or evaporite pumping in an intertidal setting (Hsü and Siegenthaler, 1969; Blatt, 1982). The brackish water fauna (Wichmann, 1924; Bertels, 1964; Ballent, 1980), and the marine fauna (Kielbowicz, 1980; Vallés, 1987) reported for this facies association on the eastern side of the basin suggests an estuarine ecological environment, which could be related to a coastal environment partially protected from the open ocean by a barrier deposit (facies association A?).

To the west, in the Sierra de Huantraico area, thickening of facies association B and the repetition of evaporite casts in the section point to a vertical aggradation of the deposits with cycles about 30 to 50 m thick (section CT, Fig. 5). In this area, facies association B seems to have been deposited under a wider range of tidal conditions, probably from subtidal to upper intertidal.

One particular feature of facies association B is the absence of burrowing and the lack of vegetation. This may be related to excessive sediment mobility (Frey and Howard, 1987).

Facies Association C: Sandstone and Mudstone

Regional Distribution

Facies association C was recognized in the Lago Pellegrini and the Sierra de Huantraico areas, but is absent in the Auca Mahuida area (Table 1). It is very well exposed in the Lago Pellegrini area in the "La Yesera" quarry (section 20.5 m thick) (Fig. 3) and at the top of the CC section (12 m thick) (Fig. 3). Facies association C in the Sierra de Huantraico area was identified in the CT section, where it is 4.75 m thick (Fig. 5).

Lithology and Sedimentary Structures

Facies association C is characterized by the alternation of laminated, greenish olive mudstone and thin lenticular beds of light green, feldspathic sandstone (Fig. 10). The contacts with the underlying and overlying facies associations are gradational.

Facies association C is characterized by lenticular, wavy, and flaser bedding (Fig. 11) (Reineck and Singh, 1975). Also

Figure 10. Outcrop view of facies association C showing the interbedding of sandstones (light) and mudstones (dark) typical of this facies association. See person for scale.

present are lenticular beds (0.2 m long, 0.1 m thick) of very fine grained, white, pyroclastic rocks. Associated with the heterolithic strata is a lenticular stromatolite bed at the top of the Y section, up to 40 cm thick. Some vertical clastic dikes composed of mudstones disrupt the bedding of this facies association.

In the SH area, the calcareous sandstones contain intraclasts of the underlying pelitic layer, associated with small load and flute casts. Evaporite casts were also found (Fig. 5).

Fossils

In the LP area, a specimen of *Plesiosauria sp.* was described by Andreis *et al.* (1974); brackish water ostracodes were described by Bertels (1964). No burrows were observed.

Depositional Environment

Facies association C is characterized by interbedded, ripple crossbedded sandstone and mudstone, which represent alternating periods of traction and decantation (Evans, 1965; Reineck and Singh, 1975; Mack and James, 1986). These structures are typical of but not exclusive to mixed tidal flats (Nio and Yang, 1989).

Supporting the tidal flat interpretation is the occurrence of the stromatolites found in facies association C (Awarik *et al.*, 1976). Evaporite casts described in the CT section indicate temporary subaerial exposure of the deposits.

The fine grained volcanic ash beds represent aerial volcanic input to areas without strong currents such as lagoons, where preservation of discrete beds could occur (Andreis *et al.*, 1974; Vallés, 1987). The presence of a plesiosaur and brackish water ostracodes in facies association C also supports the interpretation of a protected lagoon. In the fence diagram for the Lago Pellegrini area (Fig. 8) facies association C is present landward of the thicker, facies association A, which may have been a barrier enclosing the lagoon.

Facies Association D: Evaporites

Regional Distribution

The greatest thickness measured for facies association D is in the Lago Pellegrini area (Table 1). Facies association D was measured in LP area in sections Y (17 m) and Co (7.32 m) (Fig. 3). In the AM area, facies association D was measured in sections Ca (6 m) and R (1.3 m) (Fig. 4); 6.5 m of facies D were measured in the CT section (Fig. 5) in the SH area. The contact with underlying facies association C or B is transitional.

Lithology and Sedimentary Structures

Each of the measured sections is characterized by a different arrangement of the evaporitic structures, but nodular or "chicken-wire" structure is the most common (Fig. 12). Enterolithic folding, laminar and massive gypsum were also described. A lenticular coquina bed was seen intercalated with the shales and evaporites at the top of section Y.

The top of the Ca section in the AM area contains

Figure 11. Lenticular bedding in facies association C (Fig. 10), La Yesera section. Lens cap is 5.5 cm in diameter.

Figure 12. Nodular or ''chicken-wire'' structure in facies association D, El Caracol section. Black bar scale is 5 cm long.

lenticular stromatolites (''algal heads'') surrounded by fibrous gypsum, denoting extremely arid conditions (Halley, 1976).

Fossils

In the shale intercalated with the evaporite beds (Y section), Ballent (1980) discovered lower Maastrichtian, nonmarine ostracodes associated with forams, bryozoans and bivalves. No burrows were seen.

Depositional Environment

The nodular or ''chicken-wire'' structure is the most common structure in facies association D, interpreted by many authors as indicating a supratidal setting (Shearman, 1966; Bosellini and Hardie, 1973). However, some authors have claimed that the nodular structure is related more to early diagenesis in the hosting sediments than to depositional conditions (Murray, 1964).

It is difficult to attribute a primary shallow water environment to structures in evaporites, which in many cases have a diagenetic origin. Warren and Kendall (1985) pointed out that the existence of a sabkha environment can be demonstrated by a ''trinity'' of subtidal, intertidal and supratidal deposits, including algal mats and an upper evaporitic bed eroded at the top. In the depositional sequence that constitutes the Allen (Loncoche) Formation, the supratidal position of facies association D is demonstrated by its occurrence as the culmination of a fining-upward sequence, starting with the basal sandstones deposited in a subtidal to intertidal setting (facies association A), moving into intertidal muddy deposits (facies association B), and the heterolithic mudstone-sandstone of an intertidal environment (facies association C). In some of the sections, facies association D represents an intertidal deposit as is evident from the intercalation of the evaporites with shale and stromatolites containing marine and continental fauna (Ballent, 1980), and from the thickness of the evaporitic section (Warren and Kendall, 1985).

The presence of coquina beds in the La Yesera section and marine skeletal fragments in the Cerro Tormenta section points to a recharge of marine waters by storms in a partially restricted environment that only allowed the passage of high-energy currents.

Discussion

A block diagram representing the paleogeography of the Neuquén Basin in the late Campanian-early Maastrichtian was constructed based on the individual facies association interpretations (Fig. 13). It is assumed that geomorphological and basin tectonics factors have affected sedimentation processes.

Geomorphology

The Neuquén Basin is characterized by its closure to the north and west giving it a triangular geometry (Fig. 14A). This triangular embayment geometry probably controlled coastal sedimentary processes with an enhancement of the tidal regime to the north (Hayes, 1975, 1979; Fig. 14B). Similar conclusions regarding the influence of basin geomorphology on tidal range were reached by Horne (1982) for the Carboniferous in Northeastern Kentucky and by Wescott (1982) for the Tar Springs Sandstone (Upper Missisippian) in southern Illinois.

The depositional sequence of the Allen (Loncoche) Formation in the eastern part of the study area (Lago Pellegrini and Auca Mahuida areas) is similar to the lateral arrangement of the tidal flats of the North Sea described by Evans (1965), Reineck (1972) and Klein (1971). The tidal deposits of the North Sea are characterized by a basal sandy facies association, a mixed shale-sand flat, and a supratidal flat. In the North Sea, the supratidal flat is represented by a tidal marsh, rather than by an evaporitic deposit as is the case for the Allen sequence (Fig. 13). This sequence, with a presence of a barrier (facies association A; Fig. 13) resembles more a micro- to mesotidal setting. This scheme is in agreement with the theoretical model (Fig. 14B), which is characterized by a microtidal range at the open side of the embayment (Hayes, 1975).

Toward the north (SH area), the abundant, fine grained sediments of the sequence, the mixing of fresh water and marine fauna, the passage landward (toward the north) from a shallow marine to a fluvial facies (Groeber, 1946; Uliana and Musacchio, 1978; Volkheimer, 1978), and the relative limited geographic extent of these deposits, suggest an estuarine environment (Hayes, 1975; Clifton, 1983; Brownridge and Moslow, 1989). Based on Hayes' model and the presence of an estuary, a macrotidal range can be inferred in the Sierra de Huantraico area (Fig. 14A).

The fauna described for most of the facies associations is characterized by a mixture of brackish water, marine and fresh water species, indicating a coastline with partially protected environments such as lagoons that were better developed at the sides of the estuary (Fig. 14B). Toward the north, in the Sierra de Huantraico area, a mainly fresh water

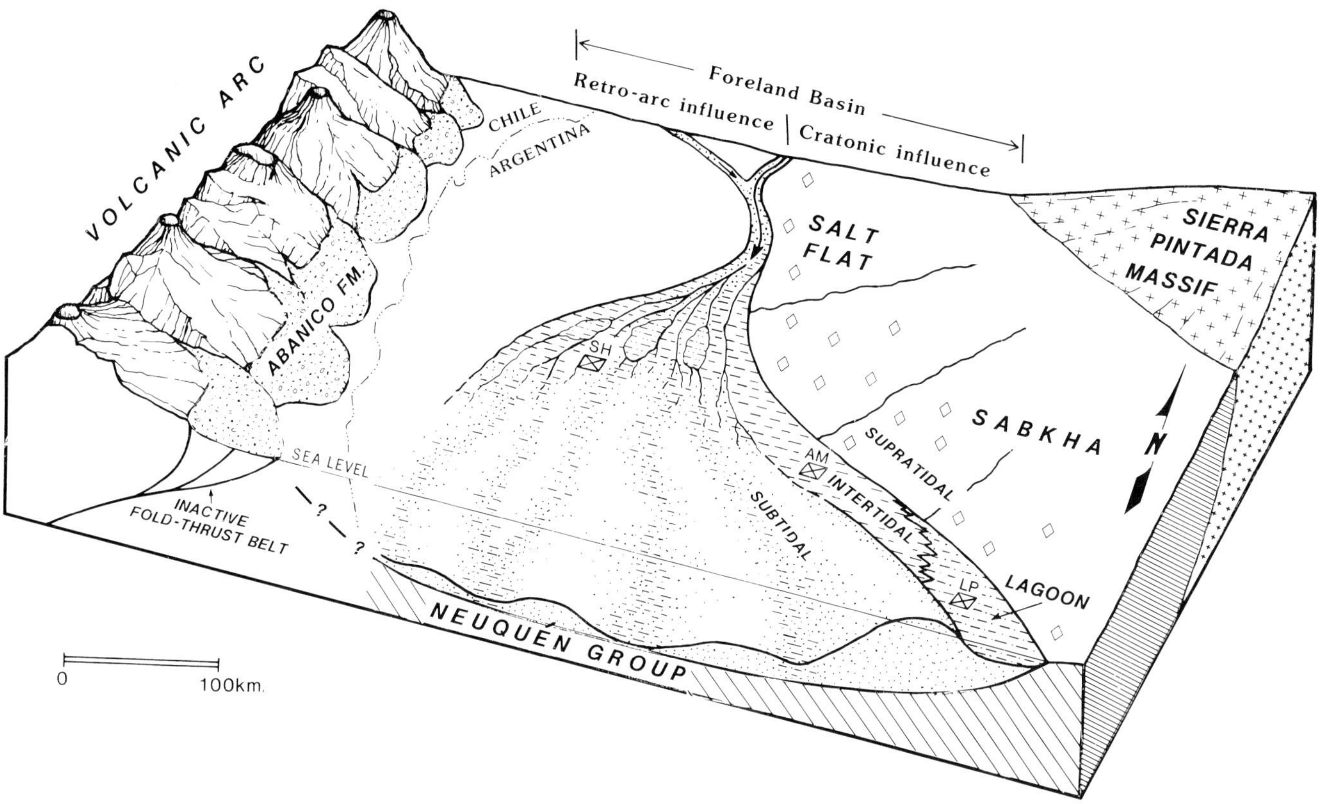

Figure 13. Paleoenvironmental and paleogeographic reconstruction of the depositional environment of the Allen (Loncoche) Formation in the northern Neuquén Basin during the late Campanian-early Maastrichtian. LP, Lago Pellegrini; AM, Auca Mahuida; SH, Sierra de Huantraico.

fauna was found (Bertels, 1968). However, in the Zampal section (Z, Fig. 1) forams were described by Uliana and Musacchio (1978), probably representing short-lived marine ingressions (*cf.* Frey and Howard, 1987).

Basin Tectonics

The western side of the Neuquén Basin corresponds to a retro-arc-influenced basin limited to the west by a volcanic arc (Digregorio *et al.*, 1984). The eastern side can be characterized as a cratonic-influenced basin (Fig. 13). As the consequence of the dissimilarity in tectonic settings between the sides and thus paleoslopes of the basin, sediment input and source rocks were different.

On the northern side of the foreland basin, the development of the tide-dominated estuary environment is associated with a steeper paleoslope and a higher sediment input. The materials were carried by rivers running south (shown by the fining trend of the grain size of the Loncoche Formation toward the south) parallel to the foreland basin axis. The sediment was deposited on a macrotidal coast (Fig. 14A). The volcanic arc provided large volumes of volcanics, mainly feldspars and other highly erodible minerals to the south in the estuary area, which explains the abundance of fine sediments there. The source is very evident in the fine grained deposits of facies association B, which to the north in the Sierra de Huantraico area show abundant feldspars and smectites and some levels with zeolites (Barrio, 1989). These minerals are closely related to an acid volcanic source, probably transported as volcanic ash by westerly winds. The sandstone composition (facies association A) also shows an enrichment in volcanic lithics on the western side of the basin (Barrio, 1989).

Well described estuaries are the Colorado Estuary in southern California (Thompson, 1968, 1975; Meckel, 1975) and the Ord River Estuary located in North Australia (Coleman, 1976), both modern deposits developed in extremely arid areas with variable fluvial inputs carrying high suspended clay-silt loads, under somewhat similar conditions as those inferred here for the Neuquén Basin in the Late Cretaceous.

On the cratonic side of the basin, as a result of the gentle paleslope, the drainage system was poorly developed and the amount of sediment carried to the basin was much less than to the west. The sediment input on this side of the basin possibly came from the Sierra Pintada Massif, located in a peneplain area rich in meso-siliceous volcanics (Permian-Triassic Choiyoi Formation; Rapela and Kay, 1988) that provided quartz and feldspars. The development of thick sequences of evaporites in a supratidal environment points to a tectonic quiescence in this area, accompanied by periods of detrital sediment starvation. The presence of an inferred barrier coast on the eastern side of the foreland basin also points to a more stable platform (Tankard, 1986).

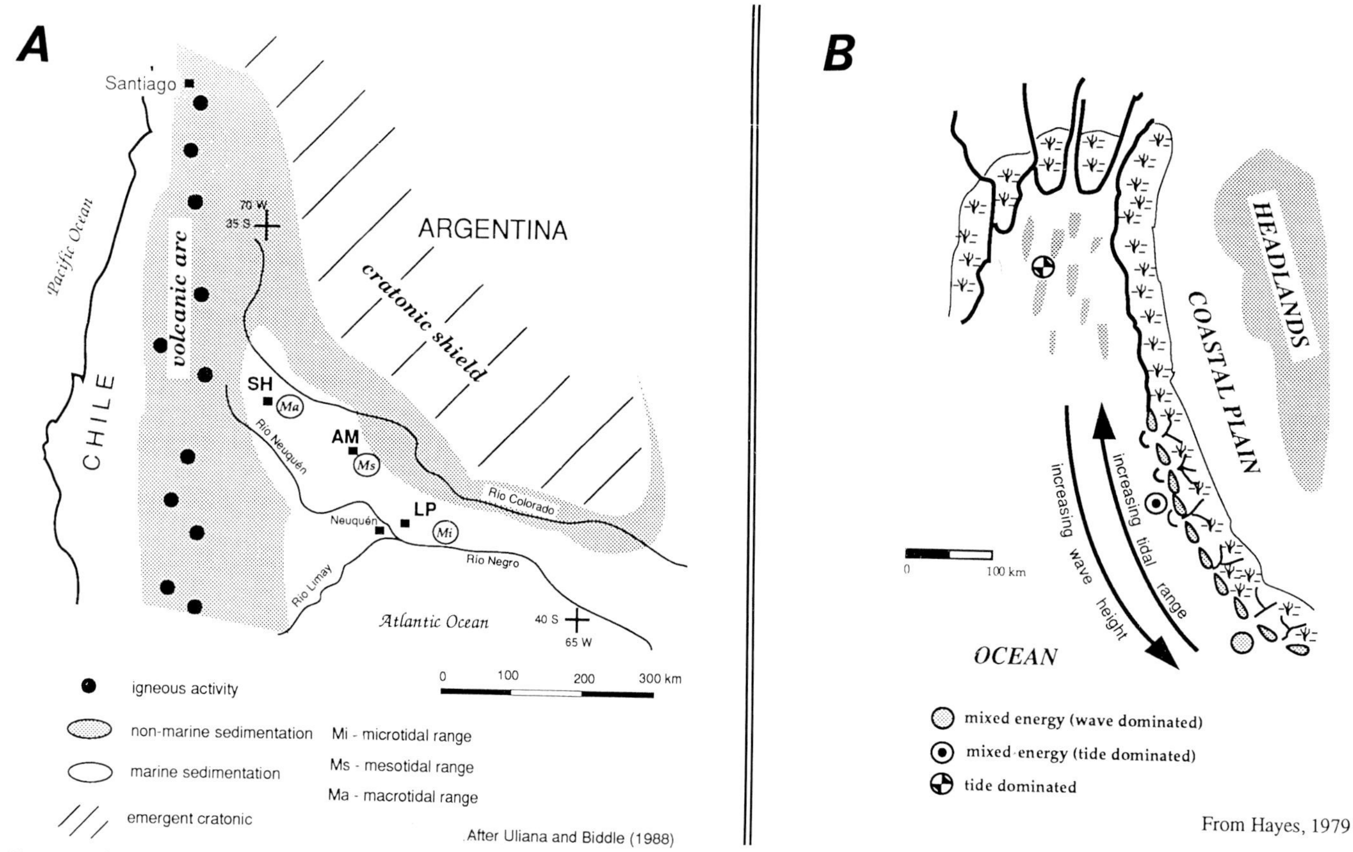

Figure 14. Neuquén Basin geometry during the Late Cretaceous, **A,** (after Uliana and Biddle, 1988) compared to the theoretical embayment model, **B,** (after Hayes, 1979). In the Neuquén Basin (**A**): LP, Lago Pellegrini area; AM, Auca Mahuida area; SH, Sierra de Huantraico area.

Lateral Variation of the Facies Association

Facies association A is thickest on the east side of the basin, probably as a result of the local development of microtidal barriers (Fig. 13). To the west, where a tide-dominated estuary is thought to have existed, facies association A is absent in some sections. In the estuary this facies association probably represents parts of tidal bar deposits that are usually concentrated at the center of the estuary (Hayes, 1975). The low supply of sand in the deltaic area may have restricted the development of tidal ridges.

Facies association B is thickest on the west side of the basin, a consequence of a high sediment supply in a tide-dominated estuary associated with thickening by flexural loading and vertical aggradation of deposits. The wave energy in this area was low enough to permit the preservation of fine grained deposits without reworking (Thompson, 1975). The abundance of mud deposits in the SH area probably represents the middle, muddy portion of the estuary (*cf.* Rahmani, 1988; Clifton, 1983).

Facies association C is better developed on the eastern side of the basin (Fig. 3). This is probably related to its development in a microtidal setting protected by a large barrier (facies association A). To the west, where estuaries occur, this facies association is restricted to ponds or minor lagoons in the tidal plain (Fig. 13).

Changes in the structures of the evaporites of facies association D from one section to another could be related to the size and setting of the sabkhas, the nature of the host sediment, the nature of the offshore water body, diastrophic control, and/or topographic control (Kendall, 1984). Paleogeography appears to be the most important factor controlling the deposition of facies association D. The greatest thickness of this facies association was described for the eastern side of the basin, implying a greater stability in the area and a lack of major streams carrying coarse grained continental sediments into the basin. To the west, lesser development of the evaporites could be related to more active deposition. Evaporites in the estuary were probably deposited in salt pans in the upper estuary plain (Fig. 13).

Conclusions

1. Regional sedimentological studies allowed the differentiation of four facies associations in the upper Campanian-lower Maastrichtian Allen (Loncoche) Formation outcropping in the Neuquén Basin. The basal, sandy facies association A is considered to have been deposited in a subtidal to intertidal setting with local development of barriers. For the predominantly pelitic facies association B, a subtidal to intertidal setting is inferred. Facies association C, characterized by the intercalation of sandstone and mudstone in thin beds, is considered to be a typical lagoonal deposit. The evaporitic facies association D was probably deposited in an upper intertidal to supratidal sabkha environment.

2. The arrangement of facies associations changes from east to west across the basin, reflecting different tidal paleoenvironments. On the east side of the foreland basin,

paleoenvironments resemble the classical tidal-flat examples of the North Sea, while toward the west, paleoenvironments correspond to modern, tide-dominated, estuarine environments. Differences in the depositional paleoenvironments are attributed to differences in the tidal regimes, which were related to the basin geometry. A microtidal range controlled the depositional environments of the low-relief coast to the east, while a macrotidal regime influenced the depositional environments of the funnel-shaped basin toward the north. Additional factors affecting facies association distributions include paleoslope, sediment input and climate. Gentle paleoslopes and reduced sediment input affected deposition on the eastern cratonic side of the foreland basin, whereas increased input and steeper paleoslopes affected deposition in the western, retro-arc side of the basin.

Acknowledgments

The author acknowledges financial assistance from CONICET (National Research Council of Argentina; PID 390450285 to L. A. Spalletti). Centro de Investigaciones Geológicas (La Plata, Argentina) provided research facilities at the beginning of this investigation. C. E. Gazzera and J. C. Giannattassio assisted in the fieldwork. At the Earth Sciences and Resources Institute, W. H. Kanes encouraged my studies in the Neuquén Basin and allowed the time to complete the study. Comments and discussions with C. E. Macellari, L. A. Spalletti, T. DeVries and P. Allen enhanced this contribution. My wife, Cindy helped with the illustrations. Reviews by G. Plint (Univ. of Western Ontario, Canada) and K. M. Bohacs (Exxon Prod. Research, USA) clearly improved the final manuscript.

References

Allen, J. R. L. 1963. The classification of cross-stratified units with notes on their origin. Sedimentology, v. 2, p. 93-114.

______ 1984. Sedimentary Structures. Their Character and Physical Basis. Developments in Sedimentology 30, Elsevier, Amsterdam, 663 p.

Andreis, R. A., Iñiguez Rodriguez, A. M., Lluch, J. J. and Sabio, D. A. 1974. Estudio sedimentológico de las Formaciones del Cretácico Superior del área del Lago Pellegrini (Prov. de Río Negro, Rep. Argentina). Asociación Geológica Argentina Revista, v. 29, p. 85-104.

Angelozzi, G. N. 1980. Dos nuevas especies de *Ilyocypris* (Ostracoda) de ambiente salobre del Cretácico superior en la Cuenca del Neuquén, República Argentina. Ameghiniana, v. 17, p. 163-167.

Awarik, S. M., Margulis, L. and Barghoorn, E. S. 1976. Evolutionary processes in the formation of Stromatolites. *In:* Stromatolites: Developments in Sedimentology 20, M. R. Walter (ed.). Elsevier, Amsterdam, p. 149-162.

Ballent, S. C. 1980. Ostracódos de ambiente salobre de la Formación Allen (Cretácico Superior) en la provincia de Río Negro (República Argentina). Ameghiniana, v. 27, p. 67-82.

Barrio, C. A. 1989. Sedimentology of the Malargüe Group (Late Cretaceous-early Tertiary), Neuquén Basin, western Argentina. (Unpublished Ph.D. dissertation), University of South Carolina, 221 p.

Bertels, M. 1964. Micropaleontología del Paleoceno de General Roca (Prov. de Río Negro). Revista Museo de La Plata, Nueva Serie 4, Paleontología, v. 23, La Plata, Argentina, p. 125-184.

______ 1968. Micropaleontología y estratigrafía del Límite Cretácico-Terciario en Huantraico (Pcia del Neuquén). Ameghiniana, v. 5, p. 279-295.

______ 1969a. Estratigrafía del límite Cretácico-Terciario en Patagonia Septentrional. Asociación Geológica Argentina Revista, v. 24, p. 41-54.

______ 1969b. Micropaleontologia y Estratigrafía del límite Cretácico-Terciario en Huantraico, Provincia del Neuquén, Parte I. Ameghiniana, v. 5, p. 279-298.

______ 1970. Los foraminíferos planctónicos de la Cuenca Cretácico-Terciaria en Patagonia septentrional (Argentina), con consideraciones sobre la estratigrafía del Fortín Gral. Roca (Prov. de Río Negro). Ameghiniana, v. 7, p. 1-56.

______ 1972. Ostrácodos de agua dulce del miembro inferior de la Formación Huantraico (Maastrichtiano inferior). Provincia de Neuquén, República Argentina. Ameghiniana, v. 9, p. 173-182.

Blatt, H. 1982. Sedimentary Petrology. H. Freeman and Company, San Francisco, 564 p.

Bonaparte, J. F. and Novas, F. E. 1985. Abelisaurus comahuensis, N.G., N.Sp., Carnosauria del Cretácico Tardío de Patagonia. Ameghiniana, v. 21, p. 259-265.

Bosellini, A. and Hardie, L. A. 1973. Depositional theme of a marginal marine evaporite. Sedimentology, v. 20, p. 5-27.

Brownridge, S. and Moslow, T. F. 1989. Mesotidal estuarine facies of the Glauconitic Member, Drayton Valley, Alberta. *In:* Modern and Ancient examples of clastic tidal deposits-A core and peel workshop, G. E. Reinson (ed.). Canadian Society of Petroleum Geologists, Second International Research Symposium on Clastic Tidal Deposits, Calgary, Alberta, p. 80-88.

Cazau, L. B. and Uliana, M. A. 1973. El Cretácico superior continental de la Cuenca Neuquina. V Congreso Geológico Argentino, Actas III, p. 131-163.

Clifton, H. E. 1983. Estuarine deposits. *In:* Sandstone depositional environments, P. A. Scholle and D. Spearing (eds.). American Association of Petroleum Geologists, Memoir 31, p. 179-189.

Coleman, J. M. 1976. Deltas: Processes of Deposition and Models for Exploration. Continuing Education Publication Company, Champaign, Illinois, 102 p.

Collinson, J. D. 1970. Bedforms of the Tana river, Norway. Geology Analer, v. 52 A, p. 31-55.

______ and Thompson, D. B. 1982. Sedimentary structures. George Allen and Unwin, London, 194 p.

Dalrymple, R. W. 1984. Runoff microdeltas: a potential emergence indicator in crossbedded sandstones. Journal of Sedimentary Petrology, v. 54, p. 825-830.

De Ferrariis, C. I. C. 1968. El Cretácico del norte de la Patagonia. III Jornadas Geológicas Argentinas, Buenos Aires, Actas 3, p. 121-144.

Digregorio, J. H. and Uliana, M. A. 1980. Cuenca Neuquina. *In:* Geología Regional Argentina, J. C. M. Turner (ed.). Academia Nacional de Ciencias, Córdoba, Argentina, II, p. 985-1032.

______, Gulisano, C. A., Gutierrez Pleimling, A. R. and Minniti, S. A. 1984. Esquema de la evolución geodinámica de la Cuenca Neuquina y sus implicancias paleogeográficas. IX Congreso Geológico Argentino, S. C. de Bariloche, Actas 2, p. 147-162.

Evans, G. 1965. Intertidal flat sediments and their environments of deposition in the Wash. Quarterly Journal Geological Society of London, v. 121, p. 209-245.

Frenguelli, J. 1930. Nomenclatura estratigráfica Patagónica. Anales de la Sociedad Científica de Santa Fe, Tomo III.

Frey, R. W. and Howard, J. D. 1987. Mesotidal Estuarine Sequences: A perspective from the Georgia Bight. Journal of Sedimentary Petrology, v. 56, p. 911-924.

Groeber, P. 1929. Líneas fundamentales de la geología del Neuquén, sur de Mendoza y regiones adyacentes. Dirección Minas, Geología, Hidrología, Publicaciones 58.

______ 1946. Observaciones geológicas a la largo del meridiano 70°. I. Hoja Chos Malal: Asociación Geológica Argentina Revista, v. 1, p. 177-208.

______ 1953. Andico. *In:* Geografía de la República Argentina. Sociedad Argentina Estudios Geográficos, v. 2, p. 349-510.

Halley, R. B. 1976. Textural variation within Great Salt Lake algal mounds. *In:* Stromatolites: Developments in Sedimentology 20, M. R. Walker (ed.). Elsevier, p. 435-446.

Hayes, M. O. 1975. Morphology of sand accumulation in estuaries: An introduction to the Symposium. *In:* Estuarine Research, E. L. Cronin (ed.). Academic Press Inc., New York, p. 3-22.

______ 1979. Barrier Island morphology as a function of tidal and wave regime. *In:* Barrier Islands - from the Gulf of St. Lawrence to the Gulf of Mexico, S. P. Leatherman (ed.). Academic Press New York, p. 1-27.

Horne, J. C. 1982. Effects of Carboniferous shoreline geometry on paleocurrent distribution. *In:* Clastic Depositional Systems (Short Course prepared by R.P.I.); Lecture 14, Denver, Colorado, p. 509-516.

Hsü, K. J. and Siegenthaler, C. 1969. Preliminary experiments on hydrodynamic movement induced by evaporation and their bearing on the dolomite problem. Sedimentology, v. 12, p. 11-25.

Illing, L. V., Wells, A. V. and Taylor, J. C. M. 1965. Penecontemporaneous dolomite in the Persian Gulf. *In:* Dolomitization and limestone diagenesis a symposium, L. C. Pray and R. C. Murray (eds.). Society of Economic Paleontologists and Mineralogists, Special Publication 13, p. 89-111.

Keidel, J. 1925. Sobre la estructura tectónica de las capas petrolíferas en el oriente del territorio del Neuquén. Dirección General de Minas, Geología e Hidrología, Publicación Especial 8, Buenos Aires.

Kendall, A. C. 1984. Evaporites. *In:* Facies Models, R. G. Walker (ed.). Geoscience Canada, Second Edition, Canada, p. 259-298.

Kielbowicz, A. 1980. Estudio de la microfauna de sedimentitas del Cretácico superior de la zona del Lago Pellegrini, Pcia. de Río Negro. (Trabajo Final Licenciatura, Unpublished), Universidad Nacional de Buenos Aires.

Klein, G. De Vries, 1970. Tidal origin of a Precambrian quartzite, the lower fine grained quartzite (Middle Dalradian) of Islay, Scotland. Journal of Sedimentary Petrology, v. 40, p. 973-985.

______ 1971. A sedimentary model for determining paleotidal range. Geological Society of America, Bulletin, v. 82, p. 2582-2592.

______ 1977. Tidal circulation model for deposition of clastic sediment. *In:* Epeiric and Mioclinal shelf seas; Sedimentary Geology, v. 18, p. 1-12.

Kreisa, R. D. and Moiola, R. J. 1986. Sigmoidal tidal bundles and other tide-generated sedimentary structures of the Curtis Formation, Utah. Geological Society of America, Bulletin, v. 97, p. 381-387.

Legarreta, L., Kokogian, D. A. and Boggetti, D. A. 1989. Depositional Sequences of the Malargüe Group (Upper Cretaceous-lower Tertiary), Neuquén Basin, Argentina. Cretaceous Research, v. 10, p. 337-356.

Logan, B. R., Rezak, R. and Ginsburg, R. N. 1964. Classification and environmental significance of algal stromatolites. Journal of Geology, v. 72, p. 68-83.

Mack, G. H. and James, W. C. 1986. Cyclic sedimentation in the mixed siliciclastic-carbonate Abo-Hueco transitional zone (Lower Permian), South Western New Mexico. Journal of Sedimentary Petrology, v. 56, p. 635-647.

Marcon, V. 1975. Geología del Cerro Villegas (Prov. de Neuquén) con especial mención del Terciario. Segundo Congreso Ibero-Americano de Geología Económica, Actas 4, p. 209-226.

Meckel, L. D. 1975. Holocene sand bodies in the Colorado Delta area, Northern Gulf of California. *In:* Deltas, M. L. Broussard (ed.). Houston Geological Society. Houston, USA, p. 87-98.

Mitchum, R. M., Vail, P. R. and Thompson III, S. 1977. Seismic stratigraphy and global changes of sea level, Part 2: The Depositional Sequence as a Basic Unit for Stratigraphic Analysis. *In:* Seismic Stratigraph: Applications to hydrocarbon exploration, C. E. Payton (ed.). American Association of Petroleum Geologists, Memoir 26, p. 53-62.

Murray, R. C. 1964. Origin and Diagenesis of Gypsum and Anhydrite: Journal of Sedimentary Petrology, v. 34, p. 512-523.

Nio, S. D. and Yang, C. S. 1989. Recognition of tidally-influenced facies and environments. Short Course Notes #01, International Geoservices BV, The Netherlands, 230 p.

Papu, O. H. 1987. Palinología y Paleoambiente de la Formación Malargüe (Cretácico Superior), Mendoza, Argentina. X Congreso Geológico Argentino, Simposio Cuencas Sedimentarias.

Powell, J. E. 1987. Hallazgo de un dinosaurio Hadrosaurido (Ornithischia, Ornithopoda) en la Formación Allen (Cretácico Superior) de Salitral Moreno, Prov. de Río Negro, Argentina. X Congreso Geológico Argentino, Actas III, p. 149-152. San Miguel de Tucumán.

Rahmani, R. A. 1988. Estuarine tidal channel and nearshore sedimentation of a Late Cretaceous epicontinental sea. Drumheller, Alberta, Canada. *In:* Tide-influenced sedimentary environments and facies, P. L. de Boer, A. van Gelder and S. D. Nio (eds.). D. Reidel Publishing Company, Dordrecht, p. 433-474.

Ramos, V. A. 1981. Descripción Geológica de la hoja 33 c. Los Chihuidos Norte, Pcia del Neuquén. Servicio Geológico Nacional, Boletín 182, p. 1-103. Buenos Aires.

Rapela, C. W. and Kay, S. M. 1988. Late Paleozoic to Recent Magmatic evolution of Northern Patagonia. Episodes, v. 11, p. 175-182.

Reading, H. G. 1986. Facies. *In:* Sedimentary Environments and Facies, H. G. Reading (ed.). Blackwell Scientific Publications, Oxford, p. 4-19.

Reineck, H. E. 1972. Tidal Flats. *In:* Recognition of Ancient sedimentary environments, J. K. Rigby and W. K. Hamblin (eds.). Society of Economic Paleontologists and Mineralogists, Special Publication 16, Tulsa, p. 146-159.

______ and Singh, I. B. 1975. Depositional Sedimentary Environments. New York. Springer Verlag, 439 p.

Roll, A. 1939. La Cuenca de los estratos con Dinosaurios al sur del río Neuquén (Unpublished) Y.P.F., Argentina.

Shearman, D. J. 1966. Origin of marine evaporites by diagenesis. Transactions Institute Mineralogy and Metallogeny, B, v. 75, p. 208-215.

Tankard, A. J. 1986. Depositional Response to Foreland Deformation in the Carboniferous of Eastern Kentucky. American Association of Petroleum Geologists, Bulletin, v. 70, p. 853-868.

Terwindt, J. 1988. Paleo-tidal reconstructions of inshore tidal depositional environments. *In:* Tide-Influenced Sedimentary Environments and Facies, P. L. de Boer, A. van Gelder and S. D. Nio (eds.). Reidel Publishing Company, Dordrecht, The Netherlands, p. 233-264.

Thomas, R. G., Smith, D. G., Wood, J. M., Visser, J., Calverley-Range, E. and Koster, E. 1987. Inclined heterolithic stratification-terminology, description, interpretation and significance. Sedimentary Geology, v. 53, p. 123-179.

Thompson, R. W. 1968. Tidal flat sedimentation on the Colorado River Delta, northwestern Gulf of California. Geological Society of America, Memoir 107, 133 p.

______ 1975. Tidal flat sediments of the Colorado Delta, northwestern Gulf of California. *In:* Tidal Deposits, a Casebook of Recent Examples and Fossil Counterparts, R. N. Ginsburg (ed.). Springer Verlag, New York, p. 57-65.

Uliana, M. A. 1979. Geología de la región comprendida entre los ríos Colorado y Negro. Pcias de Neuquén y Río Negro. (Unpublished Ph.D. Thesis), Universidad Nacional de La Plata, Argentina.

______ and Musacchio, E. A. 1978. Microfósiles calcáreos no-marinos del Cretácico superior en Zampal. Prov. de Mendoza, Argentina. Ameghiniana, v. 15, p. 111-135.

______ and Dellapé, D. A. 1981. Estratigrafía y evolución paleoambiental de la sucesión Maastrichtiano-Eoterciaria del engolfamiento Neuquino (Patagonia septentrional). VIII Congreso Geológico Argentino, San Luis, Actas 3, p. 673-711.

______ and Biddle, K. T. 1988. Mesozoic-Cenozoic Paleogeographic and Geodynamic Evolution of Southern South America. Revista Brasileira de Geocências, v. 18, p. 172-190.

Vallés, J. M. 1987. Posición estratigráfica y distribución de los horizontes de bentonita en Rio Negro, Neuquén y La Pampa. Rep. Argentina. X Congreso Geológico Argentino, Actas II, p. 33-37. San Miguel de Tucumán.

Van Wagoner, J. C., Posamentier, E. J., Mitchum, R. M., Vail, P. R., Sarg, J. F., Loutit, T. S. and Hardenbol, J. 1988. An overview of the fundamentals of sequence stratigraphy and key definitions. *In:* Sea-level Changes: An Integrated Approach, C. K. Wilgus, H. Posamentier, C. G. St. C. Kendall and C. A. Ross (eds.). Society of Economic Paleontologists and Mineralogists, Special Publ. 42, p. 39-46.

Visser, M. J. 1980. Neap-spring cycles reflected in Holocene subtidal large scale bedform deposits: a preliminary note. Geology, v. 8, p. 543-546.

Volkheimer, W. 1978. Descripción Geológica de la Hoja 27b, Cerro Sosneado. Provincia de Mendoza. Secretaría de Estado de Minería. Boletín N° 151.

Warren, J. K. and Kendall, C. G. St. C. 1985. Comparison of sequences formed in marine sabkha (subaerial) and salina (subaqueous) settings. Modern and Ancient. American Association of Petroleum Geologists, Bulletin, v. 69, p. 1013-1023.

Wescott, W. A. 1982. Depositional setting and history of the Tar Springs Sandstone (Upper Mississippian), Southern Illinois. Journal of Sedimentary Petrology, v. 52, p. 353-366.

Wichmann, R. 1924. Nuevas observaciones en la parte oriental del Neuquén y en el territorio de Río Negro. Dirección General de Minas, Publicaciones 2. Buenos Aires.

______ 1927. Sobre la facies senoniana de los estratos con dinosaurios y su fauna. Boletín Academia Nacional de Ciencias, Córdoba (Argentina), v. 30, p. 383-406.

KEY WORDS: ARGENTINA, NEUQUÉN BASIN, CRETACEOUS, ESTUARY, BARRIER, LAGOON, SABKHA, FORELAND

In: Clastic Tidal Sedimentology, D. G. Smith, G. E. Reinson, B. A. Zaitlin and R. A. Rahmani (eds.). Canadian Society of Petroleum Geologists. Memoir 16, p. 335-348.

Tidal sand sheets of the Late Albian Joli Fou-Kiowa-Skull Creek marine transgression, Western Interior Seaway of North America

INDRANIL BANERJEE

Geological Survey of Canada
Institute of Sedimentary and Petroleum Geology
3303-33nd Street N.W.
Calgary, Alberta
T2L 2A7

ABSTRACT

Subsurface facies analysis of the Lower Cretaceous hydrocarbon-bearing Basal Colorado Sandstone in the Cessford and Countess fields of southern Alberta, Canada reveals a tidal sand sheet facies (100 km × 30 km × 10 m) displaying cross-stratification, mud couplets, tidal bundles and reactivation surfaces. The Basal Colorado Sandstone was deposited in the Western Interior Seaway as a basal transgressive unit during the Joli Fou transgression (Kiowa-Skull Creek equivalent, earliest Late Albian). This sandstone overlies tidal flat-lagoon-coastal swamp sediments of the Mannville Group and underlies open marine shales of the Joli Fou Formation.

Other stratigraphically equivalent sandstone units studied in outcrop (the Sunkay Member in the Ghost River section in southern Alberta, and the Lower Flood Member of the Blackleaf Formation in Montana, U.S.A.) and in the subsurface (Spinney Hill Sandstone in southwestern Saskatchewan) also reveal characters indicative of a tidal sand sheet facies.

A regional compilation of isopach maps of these sandstones, the available paleocurrent data and a previously compiled paleotidal regime map were used to reconstruct the paleogeography of the Lower Cretaceous tidal sea. Although the Cretaceous Western Interior Seaway was wave dominant, it seems that during periods of transgression, tidal currents did transport and deposit large amounts of sand.

RÉSUMÉ

Le Grès Basal Colorado des champs pétroliers Cessford et Countess dans le sud de l'Alberta, Canada, contient des hydrocarbures. L'analyse de ses faciès en sous-surface révèle des faciès de marée composés de sables en feuillets (100 km x 30 m x 10 m), caractérisés par des stratifications obliques, des couples d'argile, des «faisceaux» («bundles») de marée et des surfaces de réactivation. Le Grès Basal Colorado, déposé dans la partie occidentale du détroit des plaines intérieures de l'Ouest canadien, constitue l'unité basale de la séquence transgressive de Joli Fou (Kiowa-Skull Creek équivalant du début de l'Albien tardif). Ce grès surmonte les sédiments d'estran, de lagon et de marais côtier du Groupe Mannville, et est sous-jacent aux «shales» de mer ouverte de la Formation Joli Fou.

D'autres unités de grès stratigraphiquement équivalentes ont été étudiées en affleurements (Membre Sunkay de la section Ghost River dans le sud de l'Alberta) et en sous-surface (Grès Spiney Hill dans le sud-ouest de la Saskatchewan); elles montrent également des caractères indicatifs de faciès de marée formés de sables en feuillets.

La reconstitution paléogéographique de la mer au Crétacé Inférieur se base sur la compilation régionale de cartes de courbes isopaques de ces grès, des données de paléocourants disponibles, et une carte préexistante réunissant les données du paléorégime de marée. Bien qu'au Crétacé le détroit des plaines intérieures de l'Ouest canadien était dominé par l'action des vagues, il semble que durant les périodes de la transgression, les courants de marée ont transporté et déposé d'énormes quantités de sable.

INTRODUCTION

The Joli Fou-Kiowa-Skull Creek marine transgression, a Late Albian marine cycle recorded within the Cretaceous of the Western Interior Basin is, according to Kauffman (1977) and Caldwell (1984), an eustatically controlled event. The northern "boreal" sea and the southern "gulfian" sea joined together to form one continuous Interior Seaway across the North American continent as a consequence of this transgression. In the southern Alberta Plains, the Joli Fou Formation (Fig. 1), consisting of 50 to 60 m of black marine shale, records this transgression. The Joli Fou shale rests unconformably on a paralic sequence of coal-shale-sandstone of the Mannville Group. The actual transgressive surface is marked locally by erosional scour, thin chert-pebble conglomerates, and concentrations of phosphate and glauconite, which mark a depositional hiatus (Banerjee, 1989a;

Geological Survey of Canada Contribution Number 58590

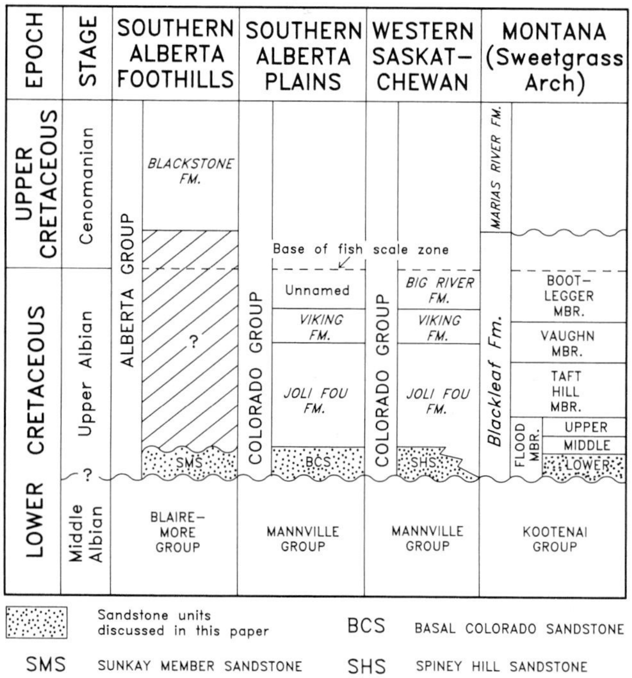

Figure 1. Lower Cretaceous units in Alberta, Saskatchewan and Montana.

Caldwell, 1984; Caldwell *et al.*, 1978). The sandstone units to be discussed in this paper were deposited as basal beds during this transgressive event (Fig. 2).

This study is the consequence of a detailed sedimentological study of a 10 m thick subsurface reservoir sandstone in an oilfield in southern Alberta, Canada (Banerjee, 1986; 1989a, b). During the study it was found that this reservoir sandstone, called the Basal Colorado Sandstone (BCS), has many features of a tidally-deposited sand sheet. The BCS occurs above a marine transgressive surface that overlies estuarine sediments, and it is overlain by a 50 m thick sequence of open marine shales (Fig. 3, 4). A regional search, both in the outcrop and in the subsurface, led to the identification of four more sandstone bodies that display features characteristic of tidal sand sheets (Figs. 1, 2). The objective of this paper is to describe the relevant sedimentary features of these sandstone units and to put them in a regional perspective within the paleotidal regime of the Lower Cretaceous Western Interior Seaway of North America.

Methods and Objectives of Study

The subsurface study of the Basal Colorado Sandstone in the Cessford Field is based on geophysical logs from 510 wells, and 49 cores. The subsurface study in the Countess Field is based on 502 geophysical logs and 18 cores. The isopach map of the BCS has been extended beyond the Cessford and Countess fields on the basis of examination of surrounding geophysical well logs. The Mill Creek and Ghost River outcrops in the southern Alberta Foothills, the Marias Pass section in Montana, and three sections near Great Falls, Montana were measured for this study. Paleocurrent data were collected from two measured sections (Mill Creek and Sun River, Great Falls). Isopach maps, measured cross-sections and detailed core log descriptions from published literature (Cobban *et al.*, 1976; Lang and McGugan, 1988; Simpson, 1982) were also used to supplement the database.

In the Cessford Field, twenty-two samples were analyzed for micropaleontological and palynological data, thirty-five for carbon/sulphur ratios and fourteen for glauconite content. Fifteen samples from outcrops were also analyzed for palynomorphs. Mostly palynological data were used for subsurface-to-outcrop correlation.

The objectives of this study are threefold:

1. To show that the sandstones discussed lie on the same unconformity surface.
2. To show that the sandstones are equivalent in age.
3. To suggest that they are deposited by tidal currents on a shallow shelf as discontinuous sand sheets.

Beyond the objectives stated above, speculations about the paleotidal regimes involved (*e.g.*, tidal paleo-currents and paleotidal ranges) are made.

Facies analysis of the BCS in the Cessford and Countess fields will serve as a model for comparison with other sandstone units. The detailed sedimentology of the BCS has been published elsewhere (Banerjee, 1989a) and only a summary will be included here.

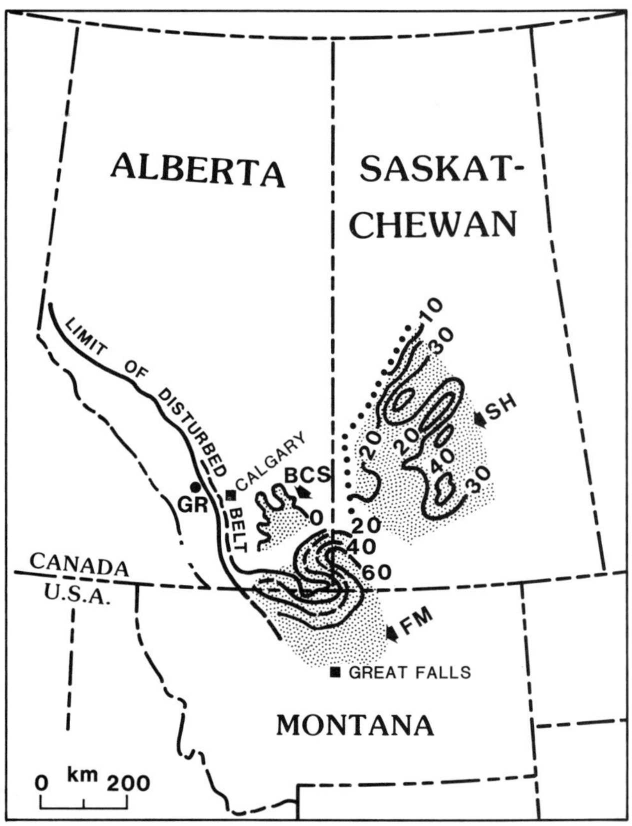

Figure 2. Location map of the sandstone sheets discussed in the paper. Isopach data of some sandstones shown. BCS, Basal Colorado Sandstone in Cessford and Countess fields; GR, sandstone in Ghost River sections; SH, Spinney Hill sandstone (isopach compiled from Simpson, 1982); FM, Flood Member (isopach compiled from Lang and McGugan, 1988).

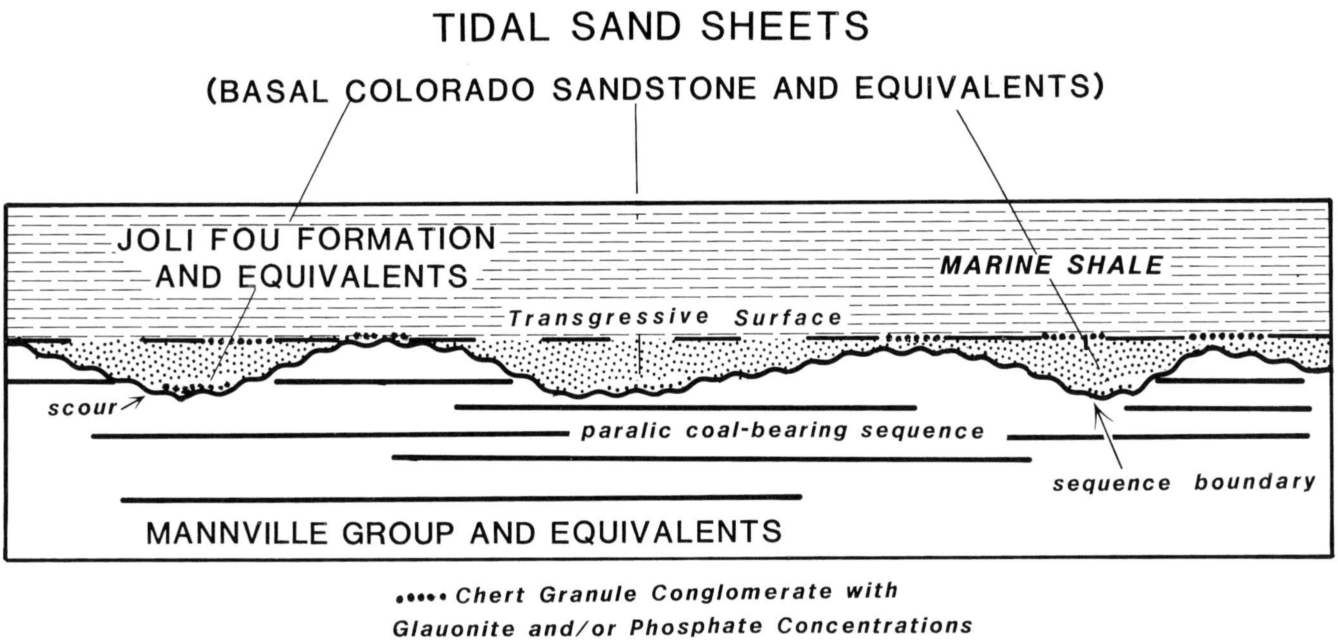

Figure 3. A schematic sequence stratigraphic framework for the Basal Colorado Sandstone and equivalent units in the Western Interior Seaway of North America.

Tidal Deposition During Marine Transgression

There are several reasons why tidal deposition of sand may be enhanced during marine transgression (Houbolt, 1968; Nio, 1976). First, more sand becomes available during transgression because rivers that discharged their loads farther offshore during the previous lowstand become drowned and fluvial sands can then be reworked by tidal currents, and enhanced coastal erosion during transgression produces more sand. Second, because the area of the newly-formed sea is continuously increasing during transgression, more and more sands are exposed to the action of tidal currents. This increase in the inundated area also results in an increase in the width of the continental shelf, which in turn, increases the tidal range and produces a tide-dominated shelf (Cram, 1979; Klein de Vries, 1982). Third, as the transgression continues there is a gradual waning of tidal currents on the shelf, resulting in the cessation of growth of the sand bodies and ultimate burial by marine mud. This ensures preservation of sands (Nio, 1976; Stride, 1988). In fact, Nio (1976) presented a facies model illustrating the initiation, growth, stagnation and burial of tidal sand wave complexes during marine transgression. Some of the sandstone facies, to be described later, show some similarity to this model.

Studies of recent offshore tidal sands reveal the presence of two contrasting facies (Stride, 1982; Belderson, 1986): tidal sand bank facies and tidal sand sheet facies. Differences in the size and shape of the sand bodies, in the grain size of their constituent sand, and in their internal sedimentary structures and resulting vertical sequences, have been tabulated in Table 1. According to Stride (1982) and Belderson (1986), tidal sand sheets are a lower-energy facies than tidal sand banks. The Lower Cretaceous stratigraphic units described here are believed to belong to the former facies, and were initiated during the early phases of the Late Albian Joli Fou transgression. They ceased to grow during the peak of the transgression, and were ultimately buried by marine shales.

All of the sandstone units described in this paper are marine, as is indicated by the occurrence of: 1) an open marine dinoflagellate assemblage consisting of 3 to 19 species (E.H. Davies, pers. comm.), 2) a marine trace fossil assemblage including *Arenicolites, Chondrites, Paleophycus, Planolites, Skolithos* and *Thalassinoides*, and 3) a high glauconite content. As well, all the units lie on the same transgressive unconformity surface, which overlies nonmarine or marginal marine sediments, are overlain by marine shales, and have sheet-like geometries.

The stratigraphic setting and the fossil content suggest an inner shelf environment in which storms and tides were the two main agents of transportation and deposition of sand. A general dominance of tidal deposition has been inferred from: 1) the presence of 0.3 to 0.4 m thick, high-angle ($>20°$) cross-stratified sets signifying strong (0.6-1.0 m/s) currents; 2) the presence of flaser, wavy and lenticular bedding indicating fluctuations in current velocity; and 3) the presence of thin mud partings suggesting slack water periods. Diagnostic tidal features such as mud-draped foresets and truncated interference ripples are rare.

Sequence Stratigraphic Framework

The Joli Fou marine transgression discussed above is recorded in southern Alberta by the Basal Colorado Sandstone (BCS) which occupies broad scours cut during a sea

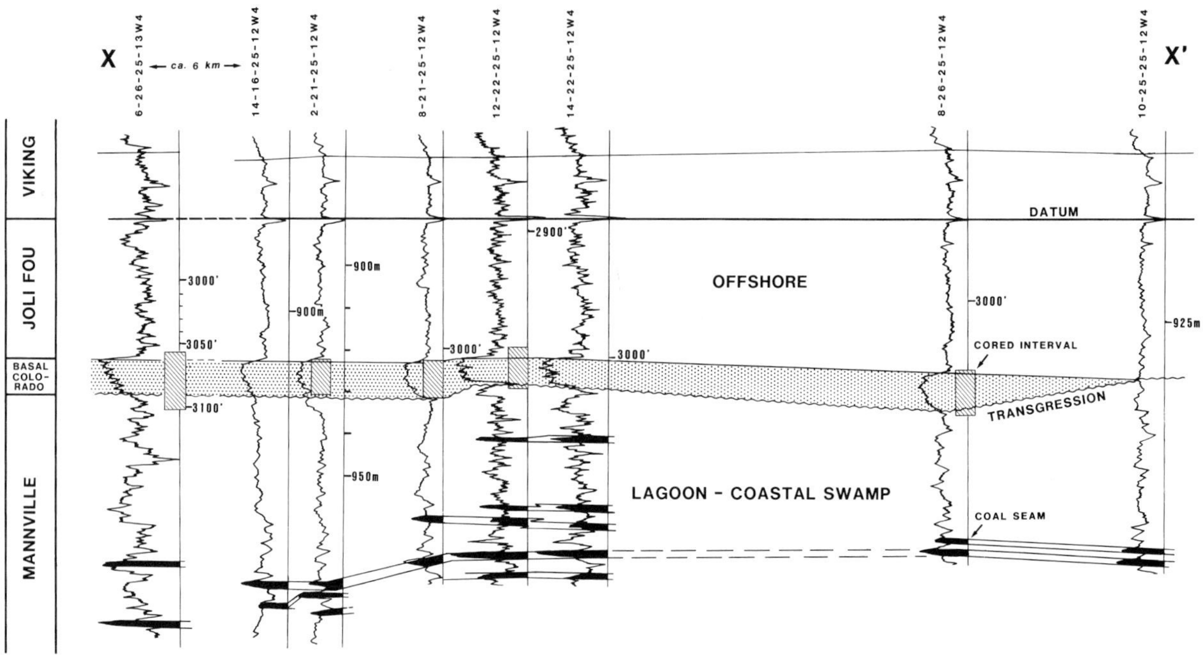

Figure 4. Stratigraphic cross-section of the Basal Colorado Sandstone (BCS) in the Cessford Field, Alberta. Location of line XX′ is shown in Figure 6. All geophysical logs are gamma ray.

level lowstand (Banerjee, 1989; Figs. 3, 4). The scoured surface is situated on the coal-bearing coastal sedimentary sequences of the Mannville Group, forming a lowstand surface of erosion. This unconformable surface can be recognized regionally. The equivalent Skull Creek-Kiowa marine transgression in the U.S.A. is also recorded by scour-based, discontinuous sandstone sheets such as the Fall River Sandstone and Plainview Sandstone (Weimer and Land, 1972), which have been interpreted as incised valley fills located on a sequence boundary. Thus the Basal Colorado Sandstone-Mannville Group unconformity is thought to be a regionally identifiable sequence boundary (Fig. 3).

The top of the Basal Colorado Sandstone exhibits several features that suggest a hiatal surface caused by increasing water depth:

1. Coarse-grained gravelly wave ripples on top of the sandstone suggesting slow deposition and winnowing.

2. Concentration of glauconite and/or phosphatic material suggesting low sedimentation rates.

3. Presence of thin (5-10 cm) limestone beds indicating cessation of clastic input.

This hiatal surface has been interpreted in this study as a transgressive surface. This interpretation is supported by paleontological, geochemical and mineralogical data as shown in Figure 5. Where the discontinuous tidal sand sheets are absent, the sequence boundary and the transgressive surface coincide and only one surface can be recognized.

To summarize, the base of the BCS is a sequence boundary containing scours formed during a sea level lowstand, and the top of the BCS is a hiatal or a transgressive surface. Where the sandstone is absent, these two surfaces coincide.

Basal Colorado Sandstone: Cessford Field

Facies analysis of the Basal Colorado Sandstone (BCS) in the Cessford Field revealed that:

Table 1. Diagnostic features of the two contrasting facies of offshore tidal sands (sand bank, sand sheet; after Stride, 1982, p. 124, Table 5.2) and of the Basal Colorado Sandstone, Cessford Field, Alberta, Canada.

	Sand Bank Facies	Sand Sheet Facies	BCS in Cessford Field
Dimensions (Average)	50 km × 3 km × 40 m	400 km × 50 km × 10 m	100 km × 30 km × 10 km
Location	Embayment or open sea	Open sea	Open marine (open marine flora and fauna)
Grain size	Well-sorted sand	Well-sorted gravel to silt	well-sorted gravel to very fine grained sand
Structures	Commonly high-angle crossbeds separated by master bedding (<6°). Up to 3 orders of bounding surfaces and very thick (1-5 m) high-angle crossbeds may occur. Mud drapes rare.	High-angle (>20°), megaripple-scale (0.3-0.4 m thick) crossbeds. Master bedding (3° to 15°) separating crossbeds. Mud drapes present.	High-angle, megaripple-scale crossbeds in the lower part and crosslamination in the upper part (5-8 cm). Mud drapes, tidal bundles and mud couplets present.

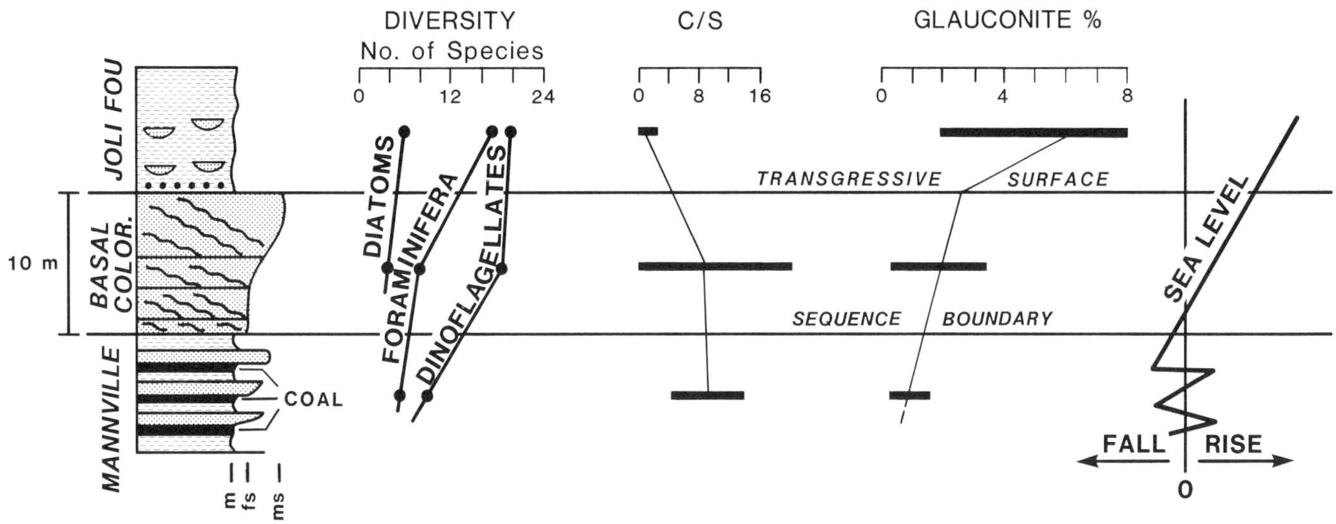

Figure 5. Summary of micropaleontological, geochemical and mineralogical data of the Mannville-Basal Colorado-Joli Fou interval in the Cessford Field of southern Alberta. Micropaleontological data are derived from 22 samples from 15 wells, C/S (Carbon /Sulphur) ratio from 35 samples from 7 wells and glauconite data from 14 samples from 10 wells. Values plotted on the micropaleontological scale are averages for each stratigraphic unit. The C/S ratio and glauconite percentage scales show range of values represented by the horizontal bars, and the average values plotted on the bars have been taken as representative of each stratigraphic unit. Each parameter shown is an indicator of paleosalinity, and increasing marine influence from the Mannville upward is reflected by the data. Inferred trends of sea level rise are also shown.

1. The BCS unconformably overlies the shale-coal-sandstone sequence (lagoon-tidal flat-coastal swamp) of the Mannville Group, and underlies the thick (>50 m) black shales (offshore marine) of the Joli Fou Formation (Figs. 3, 4).

2. The BCS occupies a paleotopographic low on the transgressive surface. Lenses of matrix-supported, gravelly, coarse-grained sandstone representing a transgressive lag lie at its base.

3. The BCS is generally medium grained and well-sorted (Banerjee, 1989b), contains 2 to 4 per cent glauconite and has an average thickness of 8 m (maximum 16 m). The lower half of the BCS is crossbedded and the upper half is cross-laminated and bioturbated.

4. The isopach map of the BCS (Fig. 6) defines a northwest elongated, thin sheet (100 km × 30 km × 10 m).

5. Four sedimentary facies have been identified within the BCS (Fig. 7a, b, c, d): A) crossbedded, gravelly, coarse grained sandstone (tidal); B) crossbedded (with mud drapes), medium grained sandstone (tidal); C) graded bedded, rippled or parallel-laminated, coarse to medium grained sandstone (storm); and D) crosslaminated, bioturbated, fine grained sandstone interbedded with shale (low-energy tidal). In Facies A and B, diagnostic tidal features (Terwindt, 1988) such as mud-draped foresets, mud couplets, and reactivation surfaces have been identified (Banerjee, 1989a).

6. The facies map of the BCS (Fig. 6) shows a southeasterly decrease in grain size, from gravelly, coarse grained sand to very fine grained sand, suggesting a south-southeast-flowing tidal current.

7. The interbedded shales in the BCS have yielded an open marine fossil assemblage consisting of 8 species of foraminifera and 19 species of dinoflagellates (J. H. Wall and E. H. Davies, pers. comm.).

8. The overall sedimentary features of the BCS are similar to those of the tidal sand sheet facies (Table 1).

9. The vertical succession within the BCS illustrates the initiation (beginning of transgression), growth (slow transgression) and decay (peak of transgression) of the BCS tidal sand sheet (see Banerjee, 1989a, Fig. 14b, c).

10. The tidal sequence is punctuated by a number of 30 to 40 cm thick, sharp-based, graded sandstones (Fig. 7c) presumably of storm origin. "Storm" beds such as these, also containing features such as gravel ripples (see Fig. 13a, Banerjee, 1989a) are concentrated toward the base. Areally, as the tidal sandstones thin out southeastward, storm beds encased in marine shale become pronounced.

In the Countess Field, located southwest of the Cessford Field, the Basal Colorado Sandstone is a separate, thinner (maximum thickness ≤ 12 m) but similar-sized sandstone sheet with the same kind of sedimentary facies of tidal origin. Toward the southeast, the Countess and Cessford tidal sand sheets coalesce and widen into a broad (>110 km), thin (≃8 m) sheet (Fig. 6) the limit of which has not yet been mapped.

Sunkay Member Sandstone: Ghost River Outcrop

At Ghost River, approximately 160 km west of the Cessford Field, a 13 m thick sandstone sequence rests with a sharp, erosional base on top of the Beaver Mines Formation of the Blairmore Group, and is overlain by black marine shales of the Blackstone Formation. A regional cross-section (Fig. 8) drawn from geophysical logs shows the

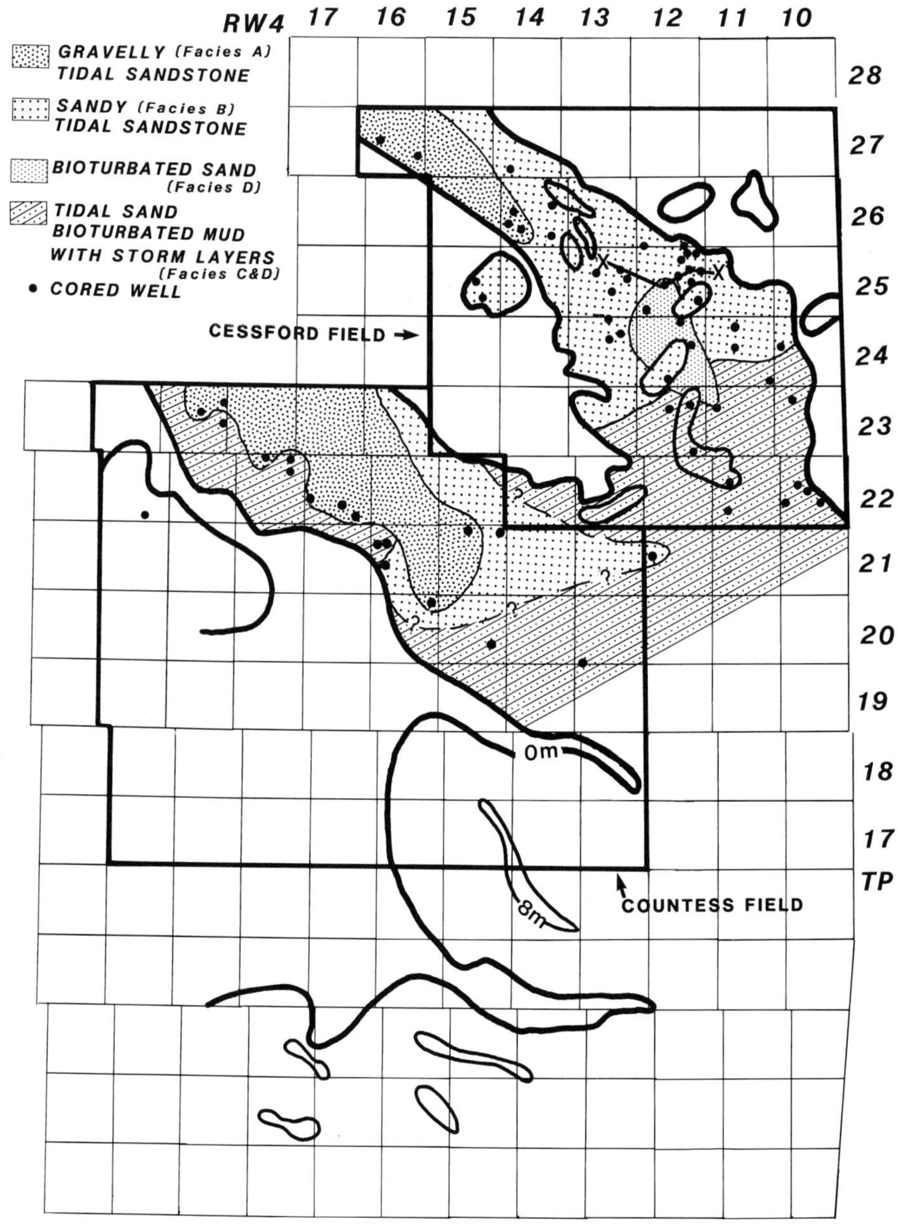

Figure 6. Facies map (with isopachs) of the Basal Colorado Sandstone in the Cessford and Countess fields. Line of cross-section XX′ (Fig. 4) is shown. Note southeasterly decrease of overall grain size.

interpreted correlation of strata in the Alberta plains with those of the Ghost River section. From foraminifera obtained from the shales overlying the sandstones, Wall (1967) and Caldwell *et al.* (1978) ascribed a Cenomanian age to the Blackstone Formation, which is believed to be much younger than the Joli Fou Formation overlying the BCS in the Cessford Field. The dinoflagellate assemblage found in the sandstone studied here indicates an Albian age for the basal sandstone of the Sunkay Member in the Ghost River outcrop and yielded four species common in the BCS (E. H. Davies, pers. comm.). On the basis of physical correlation and fossil evidence, it is assumed that the basal sandstone of the Sunkay Member in the Ghost River outcrop is age equivalent to the BCS. To reconcile the conflicting age data from foraminifers and dinoflagellates, an unconformity on top of the Sunkay Member sandstone may be assumed (Figs. 1, 8) although there is no physical evidence for it.

The vertical facies sequence of the sandstone units is shown in Figure 9. Like the BCS, this unit also has a coarser grained, crossbedded lower part and a finer grained crosslaminated (rarely bioturbated) upper part. Tidal features found in the sequence are:

1. Cross-stratification (>20° and 30-40 cm thick) with a few exhumed megaripples (Fig. 11a), which have shale layers (mud drapes) adhering to the surfaces.

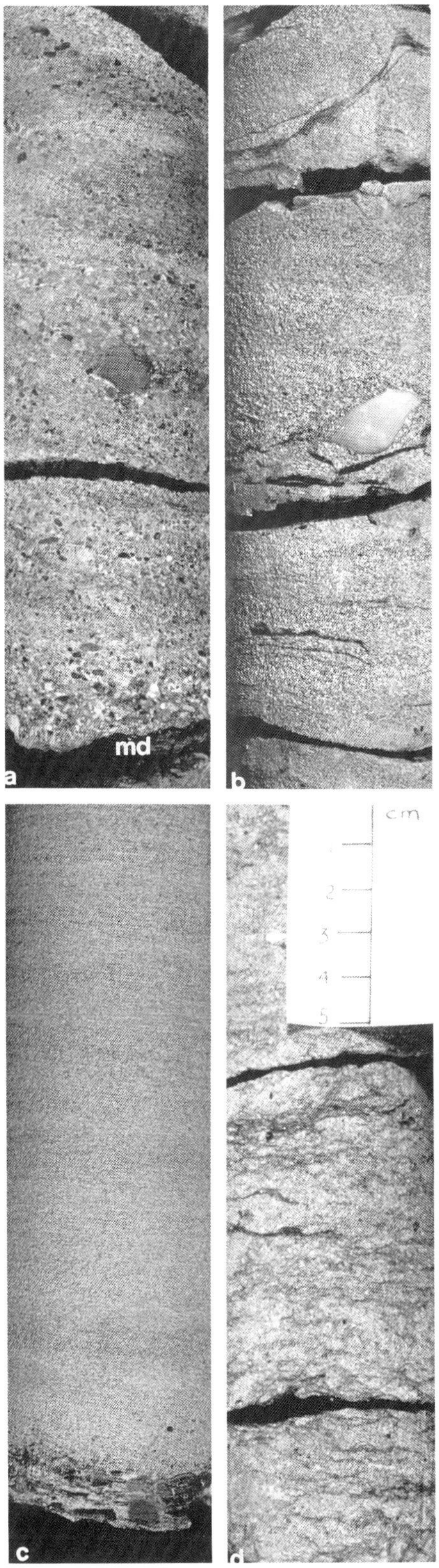

Figure 7. Core photographs of four representative facies of the Basal Colorado Sandstone (Cessford Field). Scale in **d** applies to all photographs: **a.** Gravelly, crossbedded, coarse grained sandstone with mud drape (md) (Facies A); **b.** Crossbedded, medium grained sandstone with mud drapes and siderite clasts (Facies B); **c.** Graded and parallel-laminated, medium gained sandstones (Facies C); **d.** Bioturbated and flaser bedded, fine grained sandstone (Facies D).

2. Flaser-, wavy-, and lenticular-bedding in the upper part (Fig. 11b).

As in the Cessford occurrence, this sandstone unit is interrupted by massive to graded 30 to 50 cm thick beds of conglomerate and low-angle crossbedded, gravelly, coarse grained sandstones of presumed storm origin (Fig. 9a). The areal extent and three dimensional shape of the sand body is not known because of the effects of folding and thrusting.

Lower Flood Member: Blackleaf Formation Outcrop, Montana

Stratigraphically equivalent rocks have also been studied in Montana, U.S.A. where the Lower Flood Member (Cobban *et al.*, 1976; Lang and McGugan, 1988) of the Blackleaf Formation occupies the same stratigraphic position as the Basal Colorado Sandstone and rests on the same transgressive surface overlying nonmarine rocks (Fig. 12). The correlation is based on:

1. The equivalence of the Mannville and Kootenai groups.
2. The recognition of the *Haplophragmoides Gigas* Zone indicative of the Joli Fou Formation in the Flood Member.
3. The occurrence of the *Inoceramus comancheanus* in the Flood Member.
4. The occurrence of a dinoflagellate assemblage in the Lower Flood Member, which is similar to that found in the BCS (E. H. Davies, pers. comm.).

The measured section at Sun River Valley, Great Falls Montana (Fig. 13) shows a 5 m thick vertical sequence of wavy bedded, sandstone-mudstone alternations overlain by a 1 m crossbedded (megaripple scale), medium grained sandstone. The trace fossil assemblage and dinoflagellates within the sequence suggest considerable marine influence. Sigmoidal, mud-draped foresets (Fig. 14a) occur within the crossbedded sandstone on the top of the section (Fig. 13). Both current and wave ripples occur within the wavy-bedded, lenticular bedded part of the section (Figs. 13, 14b). A few interference and flat-topped ripples also suggest that the water depth intermittently reached very shallow intertidal depths. Paleocurrents measured from the crossbeds indicate (Fig. 13) a dominant south-southwest current, which lies within a 45° arc with the BCS paleocurrent, interpreted to be south-southeast. The diagnostic tidal features in this sandstone are: 1) the sigmoidal mud-draped foresets (Fig. 14a); and 2) flat-topped interference ripples (not shown).

The isopach map of the Flood Member drawn by Lang and McGugan (1988), is included in the regional compilation map (Fig. 2) indicating its general relationship with other tidal sand sheets. It should be recognized however, that the Flood Member consists of two sandstones separated by a shale (Fig. 12). Only the Lower Flood Member represents the tidal sand sheet. On the assumption that both sandstones are tidal in origin, the overall shape of the Flood Member may reflect the shape individual sand sheets or the stacked package of the two sheets.

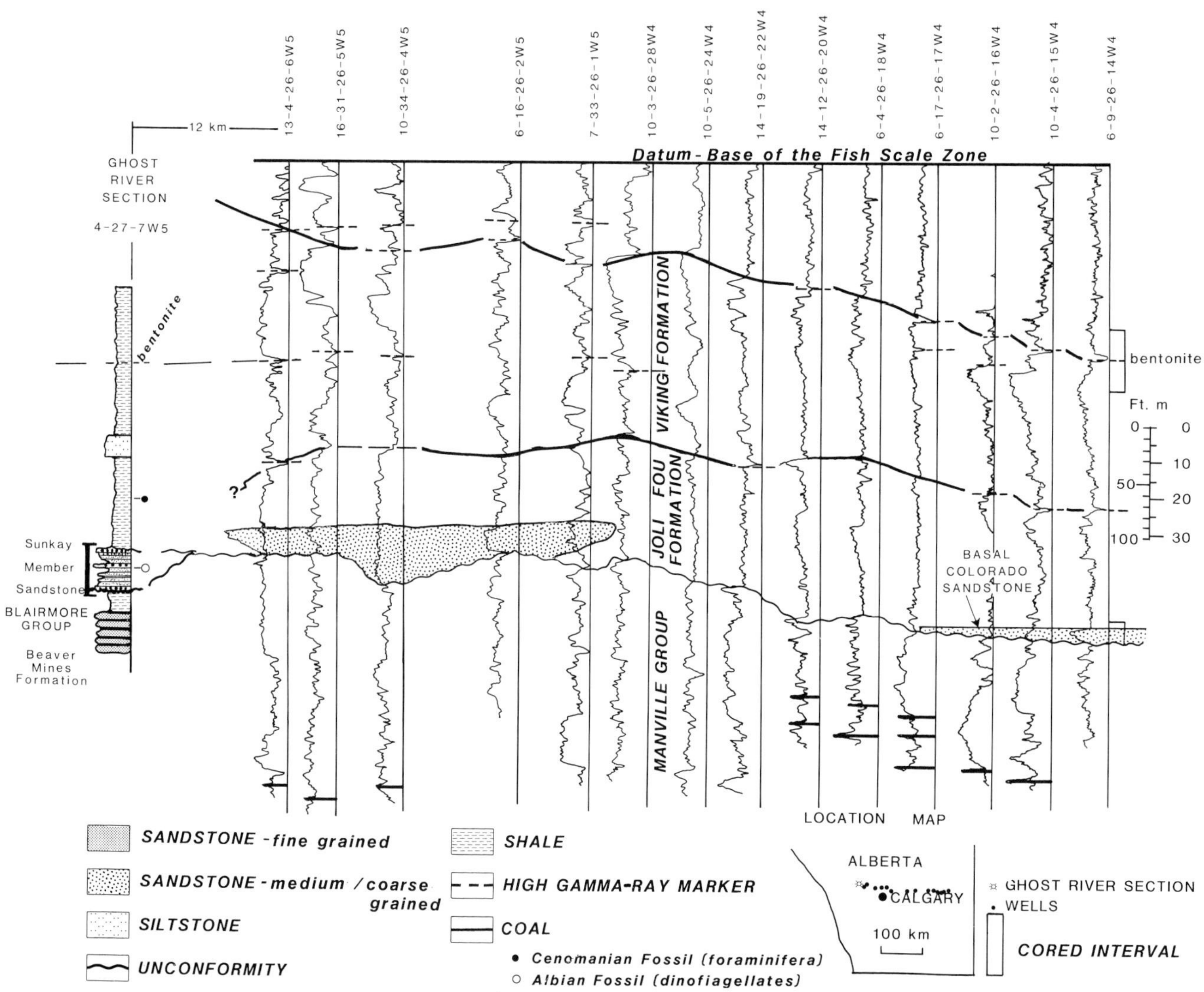

Figure 8. Stratigraphic cross-section based on subsurface correlation of 14 wells and one outcrop section. All geophysical logs shown are gamma ray. Discontinuous sand sheets lie on the Joli-Fou-Mannville unconformity, which rises stratigraphically toward the west. Conflicting ages deduced from dinoflagellate and foraminiferal studies can be reconciled by an unconformity on top of the Sunkay Member Sandstone, but there is little evidence for it.

oturbated and crosslaminated, fine grained sandstone interbedded with shales.

Although the tidal origin of Spinney Hill cannot be established beyond doubt, the presence of flaser bedding and megaripple cross-stratification may suggest tidal action. Flaser bedding is very common throughout and megaripple-scale (with an average thickness of 0.3 m and maximum of 1.0 m) crossbeds with or without mudstone partings occur particularly toward the east (Simpson, 1982, p. 74). The Spinney Hill sandstone lies on the same transgressive surface as the other sandstones described earlier, and exhibits a sheet-like form with a south-southwest orientation (Fig. 2).

Spinney Hill Sandstone, Saskatchewan, Canada (Subsurface)

The Spinney Hill Sandstone in Saskatchewan occupies a similar stratigraphic position as does the Basal Colorado Sandstone (BCS) in Alberta (Fig. 15). This sandstone has not been studied by the author but the following description has been compiled from the descriptions of cores by Simpson (1979, 1982). In one of the type wells, the Spinney Hill (Fig. 16) consists of 16.5 m of dominantly sandstone with numerous, thin shale layers. The presence of glauconite, trace fossils such as *Chondrites* and *Teichichnus* and body fossils such as *Ophiuroides* strongly suggest a marine influence. Typically, the Spinney Hill Sandstone overlies an erosional surface and consists of a crosslaminated, coarse grained sandstone at its base overlain by a sequence of bi-

Regional Paleogeography

From the previous discussion it is apparent that there are four sandstone sheets that lie above a major transgressive unconformity overlying the Blairmore-Mannville Group and underlying the Joli Fou Formation. The sheets have two major orientations, south-southeast and south-southwest (Fig. 2), range in size from 100 km $\times$ 150 km $\times$ 10 m in

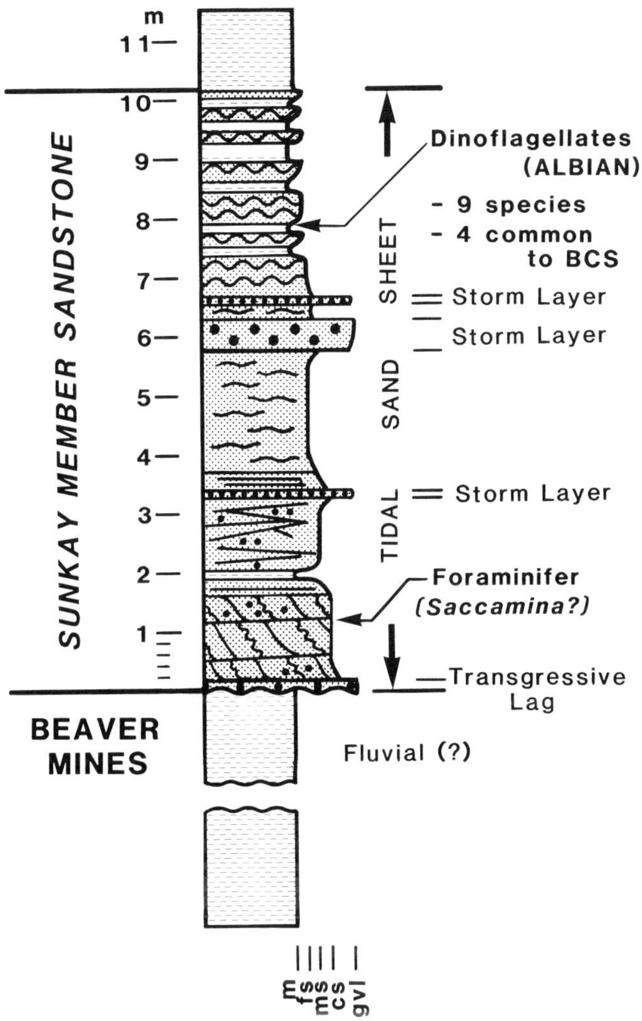

Figure 9. Vertical sequence in the Ghost River section. Transgressive lag at the bottom and storm layers shown in the middle are structureless conglomerates. The tidal sand sheet facies fines upward and is overlain by presumably marine shales.

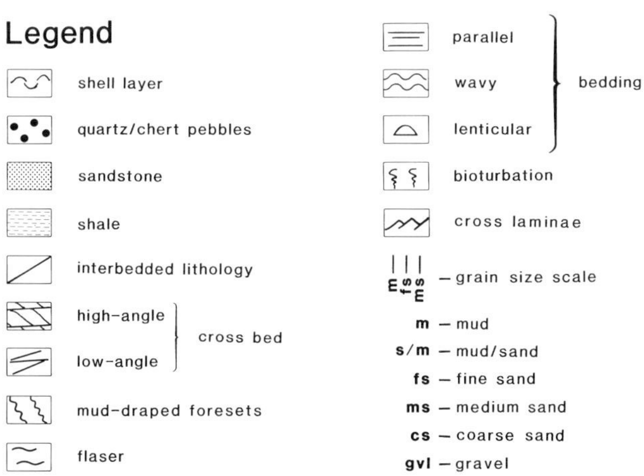

Figure 10. Legend for Figures 9, 12, 13 and 16.

Figure 11. Photographs of tidal structures in the sandstone, Ghost River section. a. Exhumed megaripples (arrows), with wavelengths of approximately 1.5 m. b. Wavy bedded sandstone-shale sequence near the top of the sandstone sequence.

the Cessford and Countess fields, Alberta, to 500 km × 300 km × 40 m in southwestern Saskatchewan. Evidence from the sandstone indicates paleocurrents varying from south-southeast to south-southwest. All sheets underlie offshore marine shales and overlie fluvial (in the southern Foothills of Alberta and Montana) and lagoon-tidal flat-coastal swamp sediments (in the southern Alberta Plains). With the proposed tidal sand sheet origin as a working hypothesis, an attempt is made to place these sand sheets within the over-

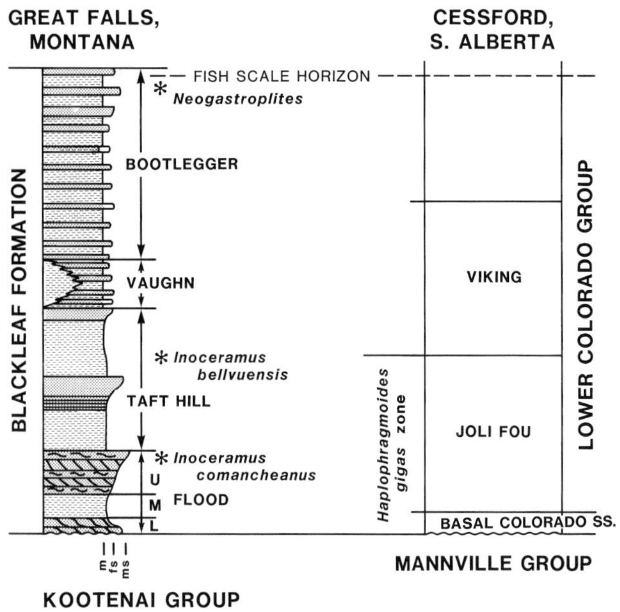

Figure 12. Correlation diagram showing stratigraphic equivalence of the Basal Colorado Sandstone in southern Alberta and the Lower Flood Member, Blackleaf Formation, Montana. Not to scale. Legend in Figure 10.

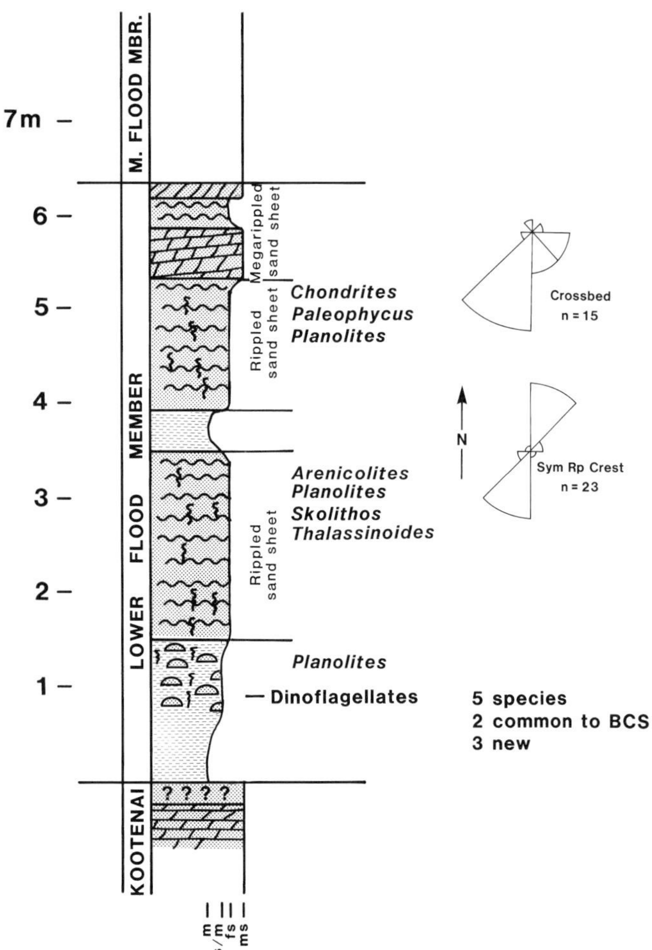

Figure 13. Measured section of the Lower Flood Member, Blackleaf Formation, Sun River Valley, Great Falls, Montana. For legend see Figure 10.

Figure 14. Photographs of tidal structures in the Lower Flood Member, Sun River section, Great Falls, Montana. Pen 15 cm long. **a.** Sigmoidal foresets (arrow) converging towards bottomsets. Gaps in the sandstone indicate mudstone layers that have been eroded out. **b.** Wavy bedded alternation of fine grained sandstone and black shale.

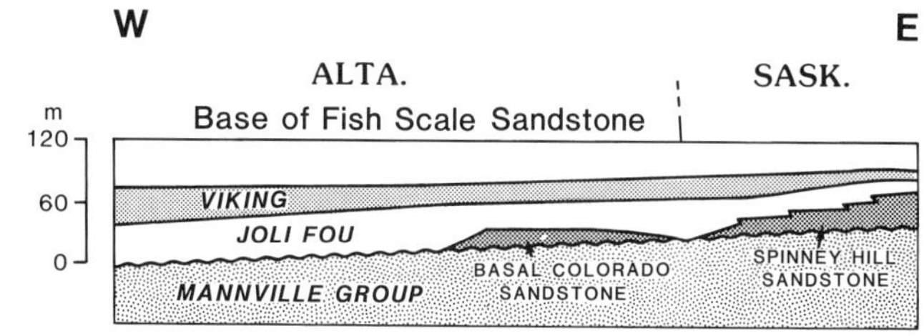

Figure 15. Stratigraphic correlation of the Spinney Hill Sandstone showing its stratigraphic equivalence to the Basal Colorado Sandstone. The line of cross-section is approximately 500 km long across southern Saskatchewan.

all paleotidal regime of the Cretaceous Western Interior Seaway. The map in Figure 17 is incomplete, because all the sandstone sheets that formed during the Joli Fou-Kiowa-Skull Creek transgression have not been located and plotted. But this map presents a picture of numerous thin, disconnected sandstone sheets probably located in isolated, shallow, paleotopographic depressions like the one occupied by the BCS in the Cessford Field (Banerjee, 1989a). As the transgression progressed, tidal currents waned and these sheets ceased to grow, stagnated and were ultimately buried by offshore muds (Joli Fou Shale). A model of aggradation

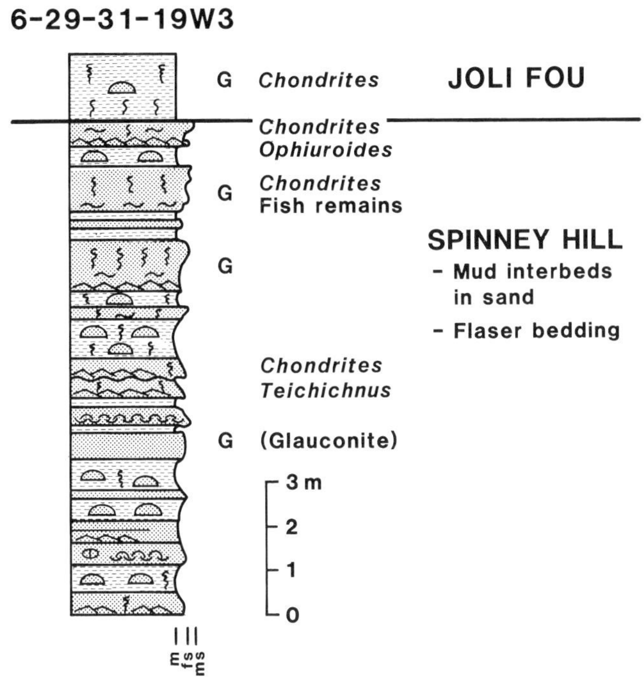

Figure 16. Core log of the Spinney Hill Sandstone in its type well (after Simpson, 1979). Two sedimentary structures suggestive of tidal deposition are shown. Occurrences of trace fossils, body fossils and glauconite are also indicated. For legend see Figure 10.

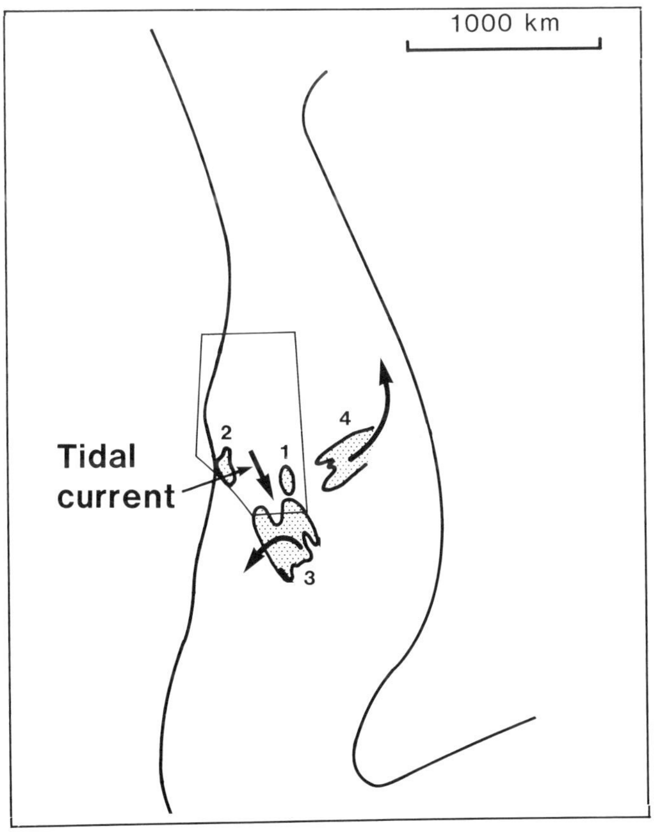

Figure 17. Postulated paleotidal regime of the Western Interior Seaway during the Joli Fou transgression. Four sandstone sheets are indicated: 1. Basal Colorado Sandstone (Cessford and Countess), 2. Sunkay Member Sandstone (Ghost River), 3. Lower Flood Member and 4. Spinney Hill. Directions of sediment-moving tidal currents (arrows) are speculative.

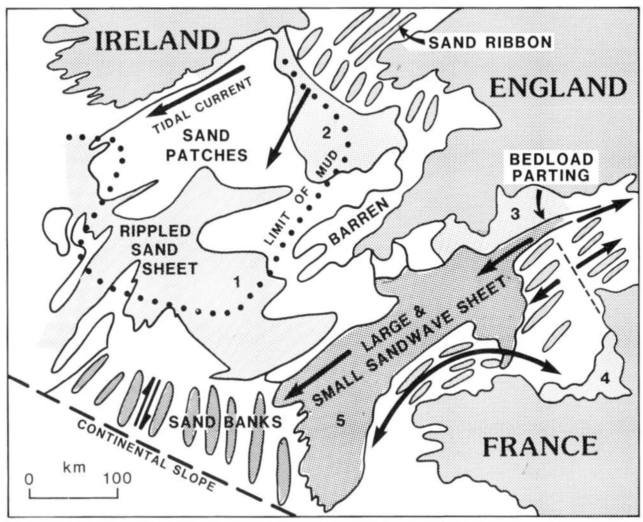

Figure 18. Possible modern analogue of the Lower Cretaceous tidal sand sheets in the Celtic Sea. Likely analogues are the rippled sand sheets numbered 1 to 4 and the (large and) small sandwave sheet numbered 5. Note that the limit of mud deposition extends at least 100 km offshore (after Stride, 1982).

of offshore sand wave complexes has been presented by Nio and Yang (1989). According to them, aggradation takes place during sea level rise and continues as long as sea level conditions result in tidal currents strong enough to generate megaripples. At higher sea levels, a more symmetrical tidal current pattern occurs which initiates a degradational stage, when the sand waves cease to grow and become moribund.

According to the model presented here, the tidal sand sheets were deposited in the inner shelf. It may be argued that features like mud-draped foresets can only be produced and preserved in sheltered estuaries (Terwindt, 1981), but the difficulty of postulating enough estuaries opening simultaneously at various points in the Seaway to produce so many tidal sand sheets is a real problem. Any mud deposition depends primarily on the concentration of suspended mud in the seawater, and according to Swift and Rice (1984), the Western Seaway was a muddy, epeiric sea.

A comparison may be made between the BCS and the recent deposits of the Celtic Sea, bordered by Ireland, England and France (Fig. 18), a shallow (<100 m deep) sea in which a large amount of sand is being transported and deposited by tidal currents operating above an eroded surface of Pleistocene glacial sediments. With tidal ranges between 4.5 and 2.0 m, the average maximum near-surface tidal currents during spring vary approximately between 0.5 to 1.5 m/s, transporting all sizes, from fine sand to gravel. Apart from the barren patches of rocky areas, the sea floor is covered with gravel sheets, sand sheets and sand banks which, in turn, contain different hierarchies of bedforms such as ripples, small sand waves (megaripples) and large sand waves. The shallow depth, nearness to coastline and tidal current transport during marine transgression, make the Celtic Sea a good recent analogue to the environment of the Cretaceous

Western Interior Seaway during Basal Colorado Sandstone deposition.

Conclusions

Probable examples of modern offshore tidal sand sheet facies from the Cretaceous Western Interior Seaway in Alberta, Saskatchewan and Montana have been documented in this paper. All the sandstone sheets were formed during the Late Albian Joli Fou-Kiowa-Skull Creek transgression, when tidal deposition of sands was enhanced. The sheets unconformably overlie either nonmarine or marginal marine sediments (Mannville, Blairmore and Kootenai groups) and are overlain by marine shales (Joli Fou-Blackstone-Blackleaf Formation).

A marine origin for the sandstones is indicated by their microfossil and trace fossil assemblages and glauconite content. The sandstones also contain high-angle, megaripple-scale (30-40 cm) crossbeds and a few mud-draped foresets suggestive of tidal-current depositional controls.

The paleocurrent and isopach data of the sandstone sheets appear to be consistent with the previously reconstructed paleotidal regime of the Cretaceous Western Interior Seaway. The recent depositional setting of the Celtic Sea offshore England, Ireland and France, can be taken as a modern analogue for the tidal sand sheets.

Acknowledgments

The author is indebted to D. A. Leckie and A. P. Hamblin of the Institute of Sedimentary and Petroleum Geology for critically reading the first draft of the manuscript and offering many helpful suggestions.

References

Banerjee, I. 1986. The Basal Colorado Sandstone (Lower Cretaceous) in the Cessford Field, Alberta (tide versus storm origin of a basal transgressive sandstone). 1986 Core Conference, Canadian Society of Petroleum Geologists, p. B-1–B-7.

_____ 1989a. Tidal sand sheet origin of the Basal Colorado Sandstone (Albian): A subsurface study in the Cessford Field, southern Alberta, Canada. Bulletin of Canadian Petroleum Geology, v. 37, p. 1-17.

_____ 1989b. Source rock potential and maturation levels of shales within the Basal Colorado Sandstone and adjacent strata, Cessford Field, Southern Alberta. Bulletin of Canadian Petroleum Geology, v. 37, p. 113-116.

Belderson, R. H. 1986. Offshore tidal and non-tidal sand ridges and sheets: differences in morphology and hydrodynamic setting. *In:* Shelf Sands and Sandstones, R. J. Knight and J. R. McLean (eds.). Canadian Society of Petroleum Geologists, Memoir 11, p. 293-301.

Bridges, P. H. 1982. Ancient offshore tidal deposits. *In:* Offshore Tidal Sands, Process and Deposits, A. H. Stride (ed.). London, Chapman and Hall, p. 172-189.

Caldwell, W. G. 1984. Early Cretaceous transgressions and regressions in the southern interior plains. *In:* The Mesozoic of Middle North America, D. F. Stott and D. J. Glass (eds.). Canadian Society of Petroleum Geologists, Memoir 9, p. 173-203.

_____ North, B. R., Stelck, C. R. and Wall, J. H. 1978. A foraminiferal zonal scheme for the Cretaceous System in the Interior Plains of Canada. *In:* Western and Arctic Canadian Biostratigraphy, C. R. Stelck and B. D. Chatterton (eds.). Geological Association of Canada, Special Paper 18, p. 495-575.

Cobban, W. A., Erdmann, C. E., Lemke, R. W. and Maughan, E. K. 1976. Type sections and members of the Blackleaf and Marias River formations (Cretaceous) of the Sweetgrass Arch, Montana. United States Geological Survey, Professional Paper 974.

Cram, J. M. 1979. The influence of continental shelf width on tidal range: Paleoceanographic implications. Journal of Geology, v. 87, p. 441-447.

Donselaar, M. E. 1984. Flood tidal delta sedimentation in the Late Cretaceous Menefee Formation (Mesaverde Group), San Juan Basin, NW New Mexico. Geologie en Mijnbouw, v. 63, p. 323-331.

_____ 1989. The Cliff House Sandstone. San Juan Basin, New Mexico: model for the stacking of 'transgressive' barrier complexes. Journal of Sedimentary Petrology, v. 59, p. 13-27.

Elliott, T. 1987. Tidally-produced sigmoidal cross-bedding in Cretaceous shelf sandstone bodies, Western Interior Basin, U.S.A. (Abs.). Abstracts volume, Society of Economic Paleontologists and Mineralogists, Annual Midyear Meeting, Austin, Texas, p. 25.

Houboult, J. J. 1968. Recent sediments in the Southern Bight of the North Sea. Geologie en Mijnbouw, v. 47, p. 245-273.

Kauffman, E. G. 1977. Geological and Biological Overview: Western Interior Cretaceous Basin. Mountain Geologists, v. 14, p. 75-99.

Klein de Vries, G. 1982. Probable sequential arrangement of depositional systems on cratons. Geology, v. 10, p. 17-22.

Kreisa, R. D. and Moiola, R. J. 1984. Sigmoidal bundles and other tide-generated sedimentary structures of the Curtis Formation, Utah. American Association of Petroleum Geologists, v. 97, p. 381-387.

Lang, H. L. and McGugan, A. 1988. Cretaceous (Albian-Turonian) foraminiferal biostratigraphy and paleogeography of northern Montana and southern Alberta. Canadian Journal of Earth Sciences, v. 25, p. 316-342.

Leckie, D. A. 1986. Tidally influenced, transgressive shelf sediments in the Viking Formation, Caroline, Alberta. Bulletin of Canadian Petroleum Geology, v. 34, p. 111-125.

_____ and Cheel, R. J. 1986. Tidal channel facies of the Virgelle member (Cretaceous Milk River Formation), southern Alberta. *In:* Current Research, Part B, Geological Survey of Canada, Paper 86-1B, p. 637-645.

McCrory, V. L. and Walker, R. G. 1986. A storm and tidally-influenced prograding shoreline — Upper Cretaceous Milk River Formation of southern Alberta, Canada. Sedimentology, v. 33.

Nio, S. D. 1976. Marine transgressions as a factor in the formation of sandwave complexes. Geologie en Mijnbouw, v. 55, p. 18-40.

_____ and Yang, C. S. 1989. Recognition of tidally-influenced facies and environments. Short course note series #01. International Geoservices, Leiderdorp, 230 p.

Simpson, F. 1979. Lithologic descriptions of selected cored sections from Lower Colorado Group (Cretaceous) of west-central Saskatchewan. Saskatchewan Mineral Resources, Report 160, 121 p.

_____ 1982. Sedimentology, Palaeoecology and Economic Geology of Lower Colorado (Cretaceous) Strata of West-Central Saskatchewan. Saskatchewan Department of Mineral Resources, Report no. 150.

Slater, R. D. 1985. A numerical model of tides in the Cretaceous Seaway of North America. Journal of Geology, v. 93, p. 333-345.

Stott, D. F. 1963. The Cretaceous Alberta Group and Equivalent Rocks, Rocky Mountain Foothills, Alberta. Geological Survey of Canada, Memoir 317, 306 p.

Stride, A. H. (editor) 1982. Offshore tidal sands. London, Chapman and Hall, 222 p.

_____ 1988. Preservation of marine sandwave structures. *In:* Tide-influenced Sedimentary Environments and Facies, P. L. Boer *et al.* (eds.). D. Reidel Publishing Company, p. 13-22.

Swift, D. J. and Rice, D. D. 1984. Sand bodies on muddy shelves: a model for sedimentation in the Western Interior Cretaceous Seaway, North America. *In:* Siliciclastic Shelf Sedimentation, R. W. Tillman and C. T. Siemers (eds.). Society of Economic Paleontologists and Mineralogists, Special Publication, n. 34, p. 43-62.

Terwindt, J. H. 1981. Origin and sequences of structures in inshore mesotidal deposits of the North Sea. *In:* Holocene Marine Sedimentation in the North Sea Basin, S. D. Nio, R. T. E. Shuttenhelm and Tj. C. van Weering (eds.). Special Publication no. 5, International Association of Sedimentology, p. 4-26.

_____ 1988. Paleo-tidal reconstructions of inshore tidal depositional environments. *In:* Tide-influenced sedimentary environments and facies, P. L. de Boer, A. van Gelder and S. D. Nio (eds.). Dordrecht, Reidel, p. 233-263.

Wall, J. H. 1967. Cretaceous foraminifera of the Rocky Mountain Foothills, Alberta. Research Council of Alberta, Bulletin 20, 185 p.

Weimer, R. J. and Land, C. B. Jr. 1972. Field guide to Dakota Group (Cretaceous) stratigraphy. Golden-Morrison area, Colorado. Mountain Geologist, v. 9, p. 241-267.

KEY WORDS: SAND SHEETS, ALBIAN, JOLI FOU, SKULL CREEK, KIOWA, WESTERN INTERIOR SEAWAY

In: Clastic Tidal Sedimentology, D. G. Smith, G. E. Reinson, B. A. Zaitlin and R. A. Rahmani (eds.). Canadian Society of Petroleum Geologists. Memoir 16, p. 349-370.

Storm- and tide-influenced prograding shoreline sequences in the Middle Devonian Mahantango Formation, Pennsylvania

WILLIAM L. DUKE

Department of Geosciences
The Pennsylvania State University
University Park
Pennsylvania 16802

ANTHONY R. PRAVE

Department of Earth and Planetary Sciences
The City College of New York
New York
New York 10031

*and PennStrat**

ABSTRACT

Thickening- and coarsening-upward depositional cycles are present in the Middle Devonian Mahantango Formation, Pennsylvania. They are found in fossiliferous, shallow marine deposits (terrigenous shale and mudstone, quartzose siltstone and sandstone, and quartz-pebble conglomerate) displaying various physical and biogenic structures. These cycles record shoaling associated with shoreline progradation; they are bounded by erosional disconformities (transgression surfaces) overlain by reworked lags. Three orders of superimposed coarsening-upward cycles are developed: large cycles (30-80 m thick), typically are composed of several small cycles (2-20 m thick) and a third order of cycles is recognized in proximal exposures.

An idealized summary sequence for Mahantango facies records an upward transition from: 1) abundantly fossiliferous mudstone with thin sandstone beds, interpreted as distal offshore storm deposits; to 2) similar mudstone interbedded with hummocky cross-stratified sandstones, interpreted as proximal offshore, subtidal storm deposits; to 3) amalgamated, hummocky sandstone bodies, interpreted as lower shoreface, subtidal storm deposits; to 4) swaley sandstone, intepreted as wave-dominated nearshore deposits from the upper shoreface; to 5) thin, flat-bedded sandstone exhibiting a wide variety of physical and biogenic structures, interpreted as shallow subtidal channel-mouth shoals or ebb-tidal deltas from the upper shoreface; to 6) trough crossbedded, medium to very coarse grained sandstone and pebbly sandstone displaying reversing flow directions, reactivation surfaces, and lateral accretion bedding, interpreted as shallow subtidal to possibly intertidal deposits of tide- and river-influenced estuaries/distributaries; to 7) relatively thin and laterally extensive, sandy conglomerates with erosional bases and abundant trace and body fossils, interpreted as reworked "transgressive lags".

This succession records shoaling from a shallow, muddy, storm-dominated sea to a sandy, barrierless shoreface. The lower shoreface was storm-dominated; the upper shoreface was tide-dominated and wave-influenced. Paleocurrent data reflect a straight shoreline (oriented roughly northeast-southwest to north-northeast - south-southwest) dissected by sinuous, shore-normal tidal channels that probably received coarse sediment from braided streams on the lower coastal plain. A mesotidal range is inferred, with channels incised to subtidal depths of about 10 m. The coast and shelf of the central German Bay represents a partial modern analogue for this system.

Episodic progradation to the northwest produced the superimposed cyclicity. Thinner cycles can be traced only a few tens of kilometres; they likely resulted from autocyclic processes or tectonic events of restricted spatial and temporal scale. Thicker cycles, traceable for many tens to hundreds of kilometres, are best attributed to major tectonic events in the adjacent Acadian highlands.

RÉSUMÉ

Des cycles sédimentaires négatifs (i.e devenant plus épais et plus grossiers vers le sommet) sont présents dans la Formation de Mahantango du Dévonien Moyen de la Pennsylvanie. Ils sont rencontrés dans des dépôts marins fossilifères d'eau peu profonde (shales et mudstones terrigènes, siltites et grès quartzeux, et conglomérats à galets de quartz) caractérisés par différentes structures physiques et d'origine biologique. Ces cycles enregistrent une diminution de la bathymétrie associée à la progradation de la ligne de rivage; ils sont limités par des discontinuités d'érosion (surfaces de transgression) surmontées par des «dépôts résiduels» («lags») remaniés. Trois ordres de cycles sédimentaires négatifs se superposent: des grands cycles (30-80 m d'épaisseur) typiquement composés de plusieurs petits cycles (2-20 m d'épaisseur), et des cycles de troisième ordre reconnues dans la zone proximale.

La séquence idéalisée des faciès de la Formation Mahantango montre la transition verticale entre: 1) des mudstones très fossilifères, avec des lits de grès minces interprétés comme des tempestites distales déposées au large du rivage; et 2) des mudstones similaires aux précédants, interstratifiés avec des grès à stratifications obliques avec creux et bosses («hummocky»), interprétés comme des tempestites proximales et subtidales déposées au large du rivage; et 3) des corps sableux amalgamés avec creux et bosses («hummocky»), interprétés comme des tempestites subtidales déposées dans la partie inférieure de la zone infratidale; et 4) des grès avec des laminations de type «swaley» (laminations obliques à faible

* The Penn State Student Stratigraphy Group. Members (and co-authors of this paper) are Said Al-Hajri, Clifford A. Cuffey, Marc C. Ericksen, Diana K. Fisler, Michael D. Guebert, James C. Matthews, Seth E. Narehood, John B. Ritter, and Koen R. Vogel.

pendage formées dans des faibles dépressions locales), interprétés comme des sédiments littoraux dominés par les vagues, déposés dans la partie supérieure de la zone infratidale; et 5) des grès minces et tabulaires montrant une grande variété de structures physiques et d'origine biologique, interprétés comme des hauts-fonds à l'embouchure du chenal subtidal d'eau peu profonde, ou comme des deltas de marée de type reflux de la partie supérieure de la zone infratidale; et 6) des grès à grains moyens à très grossiers et des grès à cailloux, chacun déployant des laminations obliques de fosses, avec des inversions de directions d'écoulement, des surfaces de réactivation, et des stratifications d'accrétion latérale; et 7) des conglomérats sableux relativement minces et latéralement étendues, avec des bases en contact d'érosion, et des traces fossiles et des fossiles en abondance, interprétés comme étant des «dépôts résiduels de transgression» («transgressive lags»).

Cette succession enregistre une décroissance de l'épaisseur de la couche d'eau marine, soit des sédiments boueux d'une mer dominée par les tempêtes passant à des dépôts infratidaux sableux sans barrière. La partie inférieure de la zone infratidale était dominée par les tempêtes tandis que la partie supérieure était dominée par les marées et influencée par les vagues. Les données de paléocourant reflètent une ligne de rivage rectiligne (orientée approximativement NE-SO à NE-SSO) découpée par des chenaux sinueux de marée normale qui recevaient probablement le sédiment grossier provenant des ruisseaux anastomosés de la plaine côtière inférieure. Un marnage de type mesotidal est inféré à ce système, et les chenaux sont entaillés à une profondeur subtidale d'environ 10 m. La côte et la plate-forme de la partie centrale de la baie Allemande représente, en partie, un analogue moderne de ce système.

La progradation périodique vers le nord-ouest a produit la superposition des cycles. Les cycles les plus minces peuvent être tracés seulement sur quelques dizaines de kilomètres; ils résultent probablement de processus autocycliques ou d'événements tectoniques à échelles, spatiale et temporelle, restreintes. Les cycles plus épais, continus sur plusieurs centaines de kilomètres, sont attribués aux événements tectoniques majeurs affectant les hautes-terres adjacentes de l'Acadien.

Introduction

As part of a larger investigation of tectonic and eustatic influences in the development of sequence-scale stratal geometries in the central portion of the Appalachian Acadian foreland basin, we have examined depositional cyclicity in the Middle Devonian (Givetian) Mahantango Formation of the Hamilton Group in Pennsylvania (Fig. 1). Mahantango exposures exhibit a hierarchically arranged assemblage of thickening- and coarsening-upward depositional cycles 2 to 80 m thick. These cycles are found in terrigenous, shallow marine deposits (shale and mudstone, mudstone interbedded with quartzose siltstone and sandstone, and thick sandstone bodies and quartz-pebble conglomerate) displaying a diverse assemblage of body fossils and primary physical and biogenic structures. Contacts between vertically adjacent cycles typically are sharp and erosional, and they commonly are overlain by thin accumulations of relatively coarse, fossiliferous debris. We have measured ten Mahantango exposures exhibiting some thirty large-scale coarsening-upward cycles, each 20 to 80 m thick (Fig. 1). The major objective of this paper is to document the sedimentology and the nature of regional depositional cyclicity in the Mahantango.

Herein we present: 1) descriptions of the facies defined in our work; 2) two detailed measured sections showing vertical facies successions and paleoflow data; 3) a correlation diagram of depositional cycles and formal lithostratigraphic members; and 4) an idealized vertical facies succession

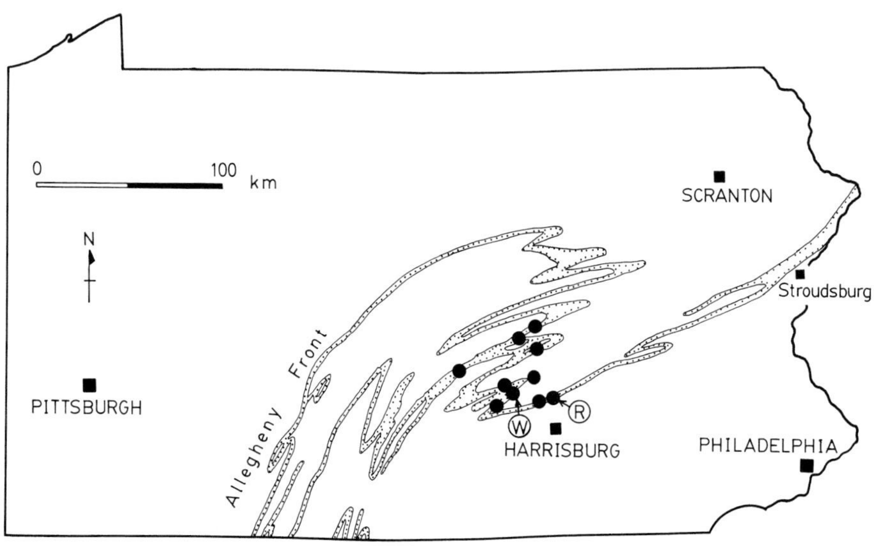

Figure 1. Locations of exposures for this study (filled circles) superimposed on a map of Middle Devonian outcrop in Pennsylvania. "W" and "R" refer to the Watts Exit and Rockville exposures, respectively.

defined from our complete descriptive database. This standard Mahantango facies succession reflects a vertical transition from storm-dominated to tide-dominated sedimentation in Mahantango cycles, which we interpret as shoaling successions produced by the episodic progradation of a straight, tide-dominated shoreline into a storm-dominated foreland basin. This succession provides a basis for the recognition and interpretation of major morphological elements in the "mixed-influenced" deposits of the Mahantango shoreline system. Together with our cycle correlations, the standard facies succession provides important constraints for the interpretation of "incomplete" shoaling cycles elsewhere in the stratigraphic record. Some results from our work have been presented previously in abstracts by Prave and Duke (1989, 1990 a,b) and Duke and Prave (1989).

Stratigraphy of the Mahantango Formation in Pennsylvania

The Mahantango Formation constitutes the bulk of the Middle Devonian Hamilton Group in Pennsylvania (Fig. 2). Since the 1930's, numerous investigators have conducted stratigraphic studies of this unit, and the internal subdivision of the formation has been repeatedly revised (see summary of Faill *et al.*, 1978). Various members have been defined in different areas of the state (Fig. 2). The salient features of the present subdivision are: 1) the limited geographic extent of the coarse lithologies dominating the Montebello Member in central Pennsylvania; and 2) the vertical alternations between shale and siltstone/sandstone in various parts of the Mahantango in both southwestern and northeastern Pennsylvania.

Stratigraphic relations between these lithostratigraphic members are uncertain, due largely to pronounced variation in lithologies and facies over relatively short distances laterally. For example, the dominant sandstones, pebbly sandstone, and conglomerates of the Montebello Member in the Harrisburg area are replaced by the dominant shales, siltstones, and fine sandstones of the Chaneysville Member in southwest-central Pennsylvania, over a distance of only 100 km (Fig. 2). Intertonguing and lensing between these various lithologies are common, leading to complicated (and differing) correlations involving the formal members in Pennsylvania (for further discussion, see Faill and Wells, 1974, p. 81, Faill *et al.*, 1978, and Dennison and Hasson, 1976). Correlations are even less certain between the Mahantango and equivalent units in adjacent states.

Kaiser (1972), Goldring and Bridges (1973), Faill and Wells (1974), and Hoskins (1976) apparently were the first workers to recognize the well developed coarsening-upward cycles in the Mahantango. Faill *et al.* (1978) discussed the superimposition of coarsening-upward cycles of different scale; this hierarchical arrangement was described as a "nested" cyclicity. Faill and Wells (1974) presented correlations of large-scale cycles in exposures measured in the Millersburg 15-minute quadrangle; their work demonstrated that cycle correlations in the Mahantango are easily defined, in marked contrast to the formal member correlations. Previously, Kaiser (1972) had presented excellent, detailed measured sections showing large- and small-scale cycles in exposures throughout Pennsylvania, but he made no attempt to rank or correlate them. Despite the early recognition of the usefulness of cycle correlations in this unit, the formal

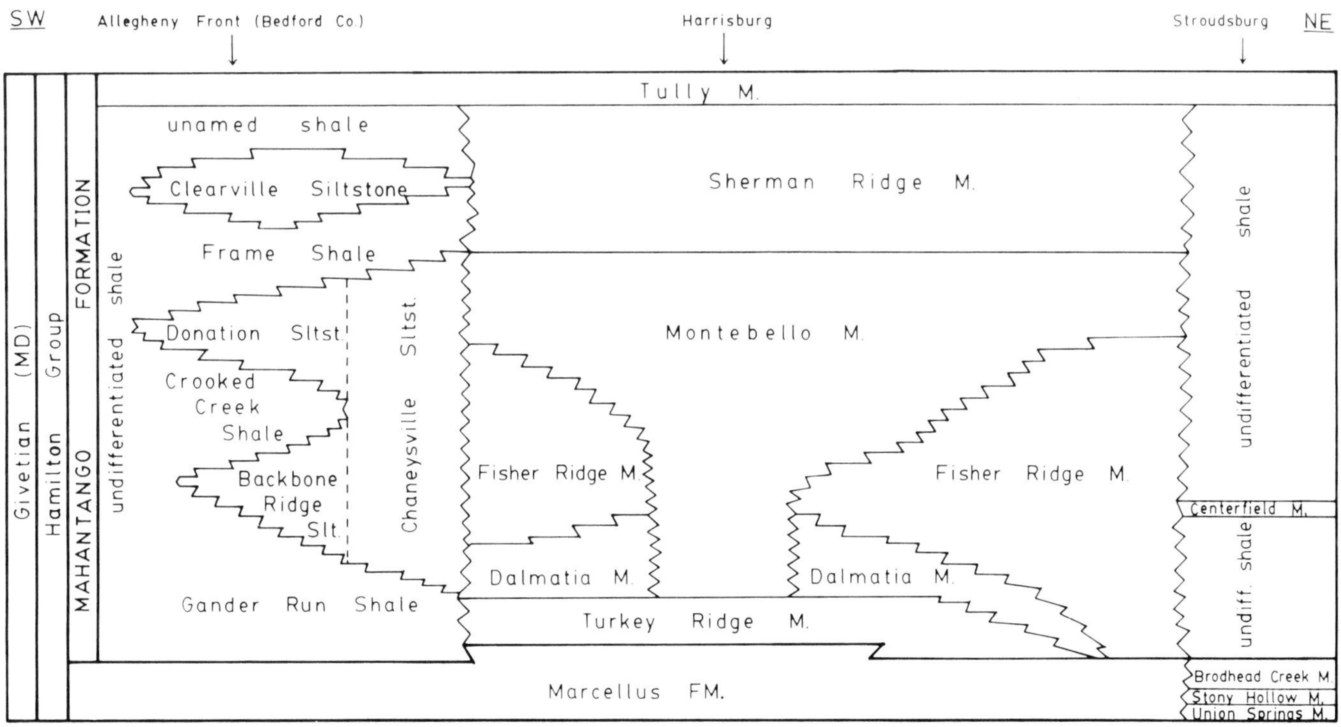

Figure 2. Lithostratigraphic subdivisions of the Hamilton Group in Pennsylvania, after various authors (see text).

lithostratigraphic member subdivisions of the Mahantango have persisted. In the present paper, we shall focus on the genetically significant depositional cycles in the Mahantango, disregarding the formal member subdivisions of this unit.

Previous Sedimentological Interpretations

Recent investigations of Mahantango sedimentology have produced widely varying interpretations of the same exposures. The spectrum of interpretations includes a fluvial-dominated delta (Kaiser, 1972), a storm-dominated prograding shoreline (Goldring and Bridges, 1973), and a system of storm-generated shelf sand ridges (Sarwar and Smoot, 1983; Sarwar, 1984). Our inference of a regionally straight storm- and tide-influenced prograding shoreline system dissected by channels adds a new element to this spectrum. The following descriptions will help provide the reader with a basis for discriminating between these varied alternatives.

Facies Defined in this Study

We have measured ten Mahantango exposures for this study (Fig. 1), preparing detailed vertical graphic logs (measured at a scale of 1:40) of selected exposures. Both physical and biological attributes of the rocks were recorded, and paleoflow indicators were measured wherever possible. We have defined 19 distinctive facies from our descriptive database, but for the purpose of this paper these have been consolidated into seven generalized facies.

Facies 1: Intervals Dominated by Shale And Mudstone

Description

Ranging in thickness from 1 m to a few tens of metres, occurrences of Facies 1 are composed mainly of shale or

Figure 4. Shale with thin sandstone beds from an occurrence of Facies 1 in overturned strata from Rockville (198 m level on Fig. 27). Up is to left. This occurrence overlies a multibed lag from Facies 7, which in turn overlies a tidal channel sandstone from Facies 6. Note thickening-and coarsening-upward trend in Facies 1.

Figure 3. Fossiliferous shale from an occurrence of Facies 1 at Watts (51 m level on Fig. 26).

mudstone (Fig. 3). Locally, these fine grained deposits are interbedded with very thin beds of coarse siltstone or very fine to fine grained sandstone (Fig. 4). Where such interbedded lithologies are present, they typically define well developed coarsening-upward trends. Occurrences of Facies 1 are either drab (olive-green to grey) or, rarely, red. Colour variation is probably diagenetic in origin and is not discussed further in this report.

The mudstones and shales of Facies 1 commonly are abundantly fossiliferous. A variety of brachipods, gastropods, pelecypods, trilobites, bryozoans, crinoids, rugose and tabulate corals, and rare cephalopods and fish bones have been encountered. Various trace fossils have been identified; these derive mainly from the *Cruziana* ichnofacies of Frey and Pemberton (1984) and include *Teichichnus, Chondrites, Paleophycus*, and *Planolites*. Many occurrences include intensely bioturbated mudstone or sandy mudstone. In some intervals, all primary physical sedimentary structures have been obliterated, leaving only a ropey or knotted texture of crosscutting, poorly defined burrows. More commonly, however, individual burrows can be recognized; most of these appear to be *Teichichnus* or *Planolites*.

The very thin siltstone and sandstone beds locally constitute about 10 to 40 per cent of the thickness of an occurrence of this facies (Figs. 4, 5). These beds have sharp, flat bases and typically are a few millimetres to centimetres thick; rarely, sandstones are as thick as 15 cm. These beds generally are continuous across the outcrop, but some pinch out locally. Many are graded, with gradational upper contacts with overlying mudstone; others have sharp upper contacts, some of which clearly are erosional. Internally, these beds are apparently structureless or are flat-laminated, wavy laminated, and/or trough crosslaminated; many resemble B, BC, and C-Type Bouma sequences. Symmetrical wave ripples (Fig. 6) are rare, and some of the thickest sandstone beds exhibit decimetre-scale hummocky cross-stratification (*i.e.*, hummocky cross-strata with undulatory laminae possessing spacings less than 1 m; Fig. 7).

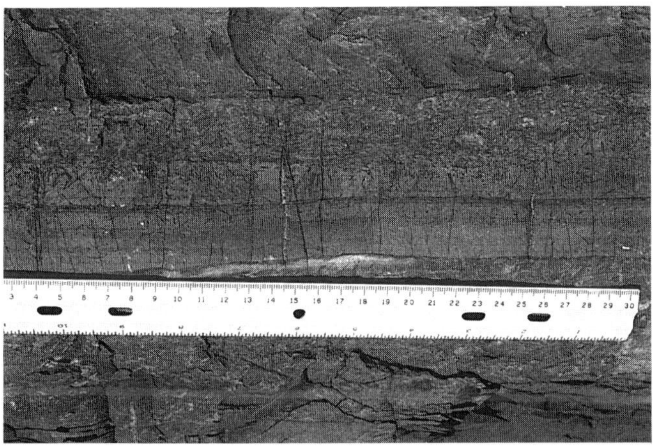

Figure 5. A thin, sharp-based sandstone bed from an occurrence of Facies 1 at Rockville (12 m level on Fig. 27). Basal interval of flat lamination is succeeded by an intensely bioturbated horizon of barely discernible trough crosslamination. Scale in centimetres and inches.

Figure 6. A thin, sharp-based, flat-laminated sandstone bed from an occurrence of Facies 1 at Rockville (14 m level on Fig. 27). Note symmetrical vortex ripples capping bed. Scale in centimetres and inches.

Figure 7. An unusual occurrence of low-angle (hummocky?) cross-stratification in a thin, sharp-based, flat-laminated sandstone bed from an occurrence of Facies 1 at Rockville (16 m level on Fig. 27). Note crosslaminae indicating lateral migration (to the left) of symmetrical vortex ripples capping bed. Also note biogenic structures penetrating the upper portion of bed. Scale in centimetres and inches.

Interpretation

The fossiliferous shales and mudstones of Facies 1 are interpreted as fairweather hemipelagic deposits from the subtidal offshore zone of an open shallow sea. The siltstones and sandstones locally interbedded with this finer material are interpreted as distal, wave-formed storm deposits derived from an adjacent shoreline. This interpretation is based on the abundance of open marine body fossils and the rare occurrence of vortex ripples and decimetre-scale hummocky cross-stratification in the thicker sandstone beds. It is supported by the assemblage of trace fossils present in many examples, indicative of deposition in an open, shallow marine environment below fairweather wave base but above storm wave base. See Duke (1990) for a further discussion of the formation of storm-generated Bouma sequences above wave base.

Facies 2: Sharp-Based Hummocky Cross-Stratified Sandstone Beds

Description

Occurrences of Facies 2 (Fig. 8) generally are a few metres to over 10 m thick and consist of sharp-based sandstone beds 10 to 150 cm thick interbedded with fine grained deposits similar to those of Facies 1. Shale and mudstone typically represent about half of the facies thickness; locally, however, sandstone constitutes as much as 80 per cent of an interval. Sandstones in this facies typically form very well developed thickening- and coarsening-upward trends.

Siltstone and sandstone beds in this facies have sharp, flat bases that locally display oriented sole marks. Internally, the thinner of these beds (<10 cm thick) display flat-lamination or normal grading. Typically, these beds have flat,

Figure 8. Upper part of a thickening-and coarsening-upward cycle at Watts (30-33 m level on Fig. 26). Facies 2 is gradationally succeeded by Facies 3, which is erosionally overlain by a bioturbated lag from Facies 7. Note well developed hummocky cross-stratification in Facies 2 and 3. Field book is 12x18 cm.

sharp to gradational upper surfaces; less commonly, they are capped with symmetrical vortex ripples, asymmetrical ripples, transitional "combined-flow" ripples, or "interference" ripples. Thicker sandstone beds typically pinch and swell over a few metres laterally; they are hummocky cross-stratified and/or flat-laminated (Fig. 9). Grain size of hummocky sandstones is consistently very fine to lower fine (< 0.15 mm). Hummocky undulations exhibit spacings of 1 to 3 m. Tops of hummocky beds commonly are sharp, displaying surficial truncation of laminae; less commonly, they are mantled with the various ripple forms that cap some of the thinner beds. Locally, upper fine and medium grained sandstone beds included in this facies exhibit high-angle (trough?) cross-bedding, but these are very rare. Certain of the thicker sandstone beds have locally foundered into underlying muds; these display soft sediment deformation structures such as load casts and ball-and-pillow structures. Thin layers of scattered quartz granules and/or small pebbles are rarely encountered.

Figure 9. Hummocky cross-stratified sandstone beds from an occurrence of Facies 2 at Watts (46 m level on Fig. 26). Hammer (upper center) provides scale.

The siltstones and sandstones in this facies typically display a wide variety of trace fossils, including forms from both the *Cruziana* and *Skolithos* ichnofacies. *Planolites, Teichichnus, Rosselia, Skolithos*, and *Chondrites* are most common. Disarticulated and reworked brachiopod shells and other fossil fragments are present in many sandstones; they commonly are concentrated just above the sharp bases of beds. Intensely bioturbated intervals are present locally; as in Facies 1, *Teichichnus* burrows are most abundant. These intervals may be several decimetres to a few metres thick and are composed of very fine to fine grained, argillaceous sandstone and sandy mudstone. They exhibit isolated remnants of hummocky cross-stratified and flat-laminated sandstone beds locally preserved from complete bioturbation (Fig. 10). These remnant portions of hummocky beds are about 10 to 30 cm thick and are penetrated by numerous *Teichichnus* burrows.

Interpretation

The sharp-based siltstone and sandstone beds of Facies

Figure 10. Preserved "remnants" of laminated sandstone in an intensely bioturbated interval of argillaceous sandstone from an occurrence of Facies 2 at Rockville (115 m level on Fig. 27). Scale in centimetres and inches.

2 are interpreted as discrete storm deposits emplaced well above effective storm wave base in a proximal offshore, subtidal setting normally characterized (during fairweather conditions) by the slow deposition of hemipelagic muds. This interpretation is based on the abundance of open marine body fossils and the ubiquitous hummocky cross-stratification and other wave-formed structures in the sandstone beds. It is supported by the presence of a mixed assemblage of trace fossils characteristic of the opportunistic invasion of storm deposits in proximal offshore settings (Pemberton and Frey, 1984; Pemberton, 1987).

Facies 3: Amalgamated Hummocky Cross-Stratified Sandstone Bodies

Description

Facies 3 consists of thick (2 - 10 m), fossiliferous sandstone bodies composed of very fine to fine grained sandstone. Thin, discontinuous layers or lenses of shale or mudstone are present locally. Disarticulated and reworked, thick-shelled brachiopods and other shallow marine body fossils typically are present in low abundance. Shale rip-up clasts and/or small quartz pebbles also are present in most occurrences, as are coarse "fossil-hash" layers composed mainly of brachiopod shells. Various traces, mainly from the *Skolithos* ichnofacies, occur in varying densities.

Occurrences of this facies are composed of numerous discrete sandstone beds (Figs. 8, 11). Sand size is consistently very fine to upper fine grained (less than about 0.1 - 0.2 mm). Some discrete beds are graded in their upper portions. Individual beds generally are a few to several decimetres thick and are essentially tabular, but they have undulatory erosional bases (and truncated tops) and thus pinch and swell in thickness laterally. These bedding contacts represent surfaces of amalgamation (*sensu* Hamblin and Walker, 1979) between episodically deposited sandstone beds.

Where internal laminae are visible, sandstones exhibit metre-scale hummocky cross-stratification or flat lamination (Fig. 11). Many beds are apparently structureless but are hummocky bedded nonetheless (see comment on undulatory bedding-plane contacts above). Thin shale/mudstone lenses (or continuous, very thin shaly horizons) rarely are present at the tops of beds; argillaceous sandstone layers are more common at bed boundaries. Body fossils (mainly thick-shelled brachiopod valves and crinoid ossicles and stems) are concentrated above bedding planes or as lenses in beds. Symmetrical vortex ripples and biogenic structures (mainly *Skolithos* or *Diplocraterion*) are very rare.

Figure 11. Amalgamated hummocky cross-stratification from an occurrence of Facies 3 at Rockville (125 m level on Fig. 27). Note amalgamation surfaces, shale lens near bottom of photo, crosslaminae produced by vortex ripples at left of scale, and continuous layer of argillaceous sandstone near center of photo. Scale is 15 cm long.

Well rounded quartz pebbles (up to 1 cm diameter) are present locally. They form thin, discontinuous layers, usually only one pebble diameter in thickness, with pebbles widely dispersed along sandstone laminae. Brachiopod fragments typically are present in these layers, and shale rip-up clasts also are common. Shallow scours occur locally along some bedding planes; these typically also contain a few pebbles, shells, and/or rip-up clasts above the basal scour surface.

Many occurrences of this facies consist of amalgamated layers of sharp-based, laminated sandstone gradationally overlain by intensely bioturbated, argillaceous sandstone. The dominant sedimentary structure in the laminated portion of these layers is metre-scale hummocky cross-stratification; less commonly, flat lamination is observed. Individual layers are a few to several decimetres thick; their laminated portions range in thickness from 10 to 30 cm. The bioturbated parts of the layers typically exhibit a few recognizable forms (mainly *Teichichnus* and *Planolites*) within an intensely reworked matrix of muddy sandstone. These burrows are most easily recognized where they penetrate into underlying laminated sandstone.

Interpretation

We interpret the fossiliferous sandstone bodies of Facies 3 as storm-dominated, subtidal deposits from the sandy lower shoreface. The hummocky sands were episodically deposited by storms in shallow offshore depths immediately seaward of the fairweather mudline. This interpretation is supported by the low-density, low-diversity assemblage of trace fossils from the *Skolithos* ichnofacies present in the sandstones, indicative of high energy depositional settings (Frey and Pemberton, 1984), and by the presence of typical *Cruziana* ichnofacies traces in the more argillaceous layers included in this facies.

Facies 4: "Swaley" Sandstone Bodies

Description

Consisting predominantly of flat and nearly flat laminae of fine grained sandstone 1 m or more thick, this facies contains some of the "cleanest" looking sandstone in the Mahantango. Because the sandstone is well sorted and fine argillaceous material is absent, laminae locally are difficult to distinguish. This facies exhibits broad, very-low-angle, concave-upward scours draped by flattening-upward laminae; these "swales" typically are several metres in width (Fig. 12). Convex-upward "hummocks" are either absent or rare and are relatively smaller than the swales, and decimetre-thick intervals of flat or nearly flat, planar laminae are present locally. Thick-shelled brachiopod valves and small quartz pebbles or granules locally are concentrated above the nadirs of swales. Some occurrences contain layers or lenses of medium grained sandstone; these coarser intervals commonly exhibit trough crossbedding in isolated or grouped sets 10 to 15 cm thick. No shale rip-up clasts or biogenic structures have been observed in Facies 4.

Note that generally there is no evidence for episodic sedimentation in this facies, in contrast with Facies 3 described above. Rather, the thick sandstones of Facies 4 probably accumulated in a setting where the sandy substrate was more-or-less continuously agitated.

Occurrences of this style of lamination in thick sandstone bodies has previously been termed "swaley cross-stratification" by Duke (1980, 1985), Leckie and Walker (1982), and numerous other workers. Herein we follow the suggestion of J. Bourgeois (pers. comm.), in referring to these occurrences as "swaley facies sandstones" or "swaley sandstone bodies". These terms better reflect the definition of such occurrences, which incorporates descriptive aspects other than the dominant stratification style. Specifically, they serve to distinguish this *facies* from thinner hummocky cross-stratified sandstones (interbedded with shale) in which concave-upward swales dominate and convex-upward hummocks are absent or rare.

Figure 12. An occurrence of swaley sandstone (Facies 4) in slightly overturned strata at Rockville (95 m level on Fig. 27). Up is to the left; note prominent low-angle swale incised in flat lamination.

Interpretation

The sparsely fossiliferous sandstone bodies of Facies 4 are interpreted as storm-dominated, subtidal deposits from either the uppermost lower shoreface or the continuously agitated nearshore zone of the upper shoreface. This interpretation is supported by the absence of biogenic structures in these sandy marine deposits.

Facies 5: Thin, Tabular, Fine to Medium Grained Sandstone Beds

Description

Composed of thin, tabular, fine to medium grained sandstone beds exhibiting thin argillaceous partings at bedding-plane contacts, examples of this distinctive facies vary in thickness from about 1 to 5 m (Fig. 13). Internally, sandstone beds exhibit a wide variety of primary physical structures, and varying densities of a diverse trace-fossil assemblage derived mainly from the *Skolithos* ichnofacies. Layers of thick-shelled, disarticulated brachiopod valves are common, and a few sandstones exhibit abundant ostracodes. Coquinite beds and lenses up to 15 cm thick locally constitute a distinctive minor component of Facies 5. These coquinites are composed entirely of thick-shelled brachiopods. In some cases, virtually all brachiopod valves are articulated, suggesting little or no reworking and transport.

In contrast to Facies 4 sandstones, those of Facies 5 typically are argillaceous. Bed thickness ranges from about 1 cm

Figure 13. An occurrence of Facies 5 at Rockville (210 m level on Fig. 27). Up is to the left. Note tabular nature of sandstone beds separated by thin, continuous layers of shale and argillaceous sandstone. Scale (lower center) is 15 cm long.

to over 20 cm; average thickness is about 10 cm. Coarsening-upward trends are well developed locally, but bed thickness generally does not vary systematically. Sandstone bedding-plane contacts typically exhibit very thin partings of argillaceous sandstone or (rarely) shale or mudstone. These partings typically are only about 1 mm thick; most are continuous across an exposure, but some pinch out as a result of very shallow erosion at the base of the overlying sandstone. Shaly horizons as thick as 1 cm occur locally.

Sandstones in Facies 5 exhibit a wide range of structures (Fig. 14). The most abundant structure is flat lamination, and most beds are flat-laminated through their entire thickness. Many other beds exhibit very-low-angle lamina intersections suggestive of hummocky cross-stratification. Still other beds are apparently structureless. Trough crossbedding, in sets roughly 10 cm thick, also is present in numerous occurrences. In these cases, a single set occupies the entire thickness of an individual bed. Various forms of crosslamination are found in the upper parts of some beds. These include forms produced by both symmetrical and asymmetrical ripples, and the angle of climb varies greatly for each type. Vertical transitions from flat lamination to crosslamination are common in individual beds. Beds exhibiting these distinct lamination styles do not occur in any particular order in Facies 5, with one exception: trough crossbedded bedsets are prevalent only in the upper (coarser) part of occurrences, which commonly are dominated by this structure.

Figure 14. An occurrence of Facies 5 at Rockville (222 m level on Fig. 27). Note wavy-parallel lamination and crosslaminae in thin tabular sandstone beds, and convex-upward laminae (small-scale hummocky cross-strata?) in thicker sandstone in upper center of photo. Scale is 15 cm long.

Sandstones exhibit varying densities of a diverse trace-fossil assemblage derived mainly from the *Skolithos* ichnofacies. Most intervals are barren of biogenic structures; others exhibit many burrows. The most common forms are *Rosselia, Teichichnus*, and *Skolithos* burrows a few to several centimetres long. Also common are ornamented vertical burrows a few decimetres long and a few centimetres in diameter. *Planolites* is present on many bedding planes.

Interpretation

We tentatively associate the relatively coarse grained sandstones of Facies 5 with the upper shoreface. An open and energetic shallow marine environment is suggested by the thick-shelled brachiopods, traces from the *Skolithos* ichnofacies, and both wave- and current-formed structures. The unidentified ostracodes may reflect proximity to brackish water, consistent with a shallow nearshore setting. Salinity changes may also have been responsible for the brachiopod death assemblages.

The variable types of sandstone stratification suggest the depositing flows were not of a consistent magnitude; more importantly, the abundant shaly horizons indicate that these flows probably were intermittent. Intervals of slack water did not persist long enough to permit accumulation of sizeable thicknesses of fine material, nor were they adequate to allow the establishment of a biota that could produce abundant quiet-water trace fossils. We conclude that slack-water intervals persisted only briefly (for perhaps hours to days, but certainly not months to years). They did occur on a regular basis, however, judging from their presence at the contacts of most sandstone beds.

The notably tabular geometry of all sandstone beds in Facies 5, together with the general absence of shale rip-up clasts, suggests that the discrete sand-depositing events were not particularly erosive. This is unusual in light of the occurrence in many beds of trough crossbedding generated by large, three-dimensional ripples, which typically possess well developed scour pits. High angles of climb of asymmetrical ripples in some beds likewise suggest rapid sediment fallout. We conclude that deposition probably took place in a region of pronounced flow expansion.

The combination of rapid sand deposition followed by brief intervals of slack-water sedimentation requires that these depositional cycles (each involving one sand-depositing event and one slack-water episode) took place regularly and with a frequency of perhaps hours or days. Tidal currents are strongly indicated; reversing paleoflow data (to be presented later) strongly support this conclusion. Further support is provided by the stratigraphic context of these deposits, which also leads us to prefer an origin involving tidal channels over one involving tidal inlets (see discussion of vertical facies successions in a subsequent section).

Synthesizing these inferences, it is suggested that the deposits of Facies 5 probably accumulated in a subtidal setting within or just seaward of the funnel-shaped mouths of tidal channels incised on the uppermost shoreface. This setting was dominated by tidal currents, but the rare vortex ripples and hummocky cross-strata record the uncommon overprinting of oscillatory flows. Thus, this facies may represent channel-mouth shoals only partially exposed to the open ocean. This interpretation is supported by the presence of trace fossils derived mainly from the *Skolithos* ichnofacies. Such an assemblage suggests that the sandy substrate was more-or-less continuously agitated; however, the locally high density and variety of traces reasonably precludes deposition along a shoreline fully exposed to wave attack. Rather, a somewhat more protected environment is indicated, consistent with the channel-mouth setting inferred from other evidence.

Facies 6: Thick Bodies of Medium to Very Coarse Grained Sandstone

Description

This facies consists of relatively thick (2-10 m) sandstone bodies composed mainly of poorly sorted, medium to very coarse grained sandstone and pebbly sandstone. Layers of shale, mudstone, and fine sandstone are present locally, as are shale rip-up clasts. Thin, sandy conglomerate layers, many of which contain thick-shelled brachiopod valves, are present in many occurrences of this facies.

Sandstones display abundant crossbedding in sets 10 to 70 cm thick (Fig. 15). Cross-strata display a trough morphology in flow-transverse and oblique cross-sections. Quartz pebbles are locally concentrated near bases of crossbed sets. Some thin sandstone layers are crosslaminated, flat-laminated, bioturbated, graded, or apparently structureless, and some bedding planes exhibit vortex ripples. Some examples of this facies contain relatively thick intervals of apparently structureless, coarse, pebbly sandstone, which locally reveals faint crossbedding. Thick-shelled brachiopods are abundant locally, occurring in shelly layers as thick as 10 cm. A single crinoid ossicle was observed at one locality.

In many occurrences of Facies 6, boundaries between sets of trough crossbedding display lithological breaks and/or various minor structures. Thin shale partings are common (Figs. 15, 16, 17). Individual sets of trough crossbedding are

Figure 16. Medium-scale trough crossbedding in flow-parallel section, from an occurrence of Facies 6 at Rockville (172 m level on Fig. 27). Note thin mud layers separating sets and double mud drape on crossbeds at left of hammer.

Figure 17. Medium-scale trough crossbedding in flow-parallel section, from an occurrence of Facies 6 at Rockville (172 m level on Fig. 27). Note thin mud drapes associated with numerous reactivation surfaces, and superimposed ripple crosslaminae at right. Also note biogenic structures.

Figure 15. Medium-scale trough crossbedding in flow-parallel section, from an occurrence of Facies 6 at Rockville (172 m level on Fig. 27). Note thin mud layers separating sets and numerous reactivation surfaces associated with compound cross-strata in set above hammer.

Figure 18. Medium-scale trough crossbedding in flow-parallel section, overlain by current-ripple crosslamination. Note flow reversal in one set of the latter (arrowed). From an occurrence of Facies 6 at Rockville (178 m level on Fig. 27). Scale in centimetres and inches.

Figure 20. Lateral accretion bedding from an occurrence of Facies 6 at the Girty's Notch exposure, immediately northeast of the Watts exposure (see Fig. 1 for location). Geologist's extended arm indicates the regional bedding attitude (also indicated by the line labelled "RBA"). Several sets of laterally accreted beds are apparent; the most prominent of these is about 3 m thick and extends across the entire outcrop face.

Figure 19. Thinning-and fining-upward tidal channel-fill sequences from an occurrence of Facies 6 at the Girty's Notch exposure, immediately northeast of the Watts exposure (see Fig. 1 for location). This interval lies at a stratigraphic level just below the geologist in Figure 20. Separate channel-fill sequences are indicated by arrows; only the middle sequence is seen in its entirety. Note the conglomerate layers forming basal lags in the upper two sequences, and the reversing sets of trough crossbedding separated by shale layers (particularly obvious behind hammer). Also note mud drapes on crossbeds at left of geologist's feet.

locally capped by one or a few sets of small-scale trough crosslamination produced by asymmetrical ripples; in many instances, the paleocurrent recorded by these ripples is nearly opposite to that inferred from the underlying crossbedding (Fig. 18). Thin (1-5 cm) intervals of flat lamination also are common at set boundaries. Additionally, mud drapes (Figs. 16, 17) and reactivation surfaces (Fig. 17) occur on some crossbeds. Some of the mud drapes are paired (Fig. 16).

Fining-upward trends are common (Fig. 19), as is well developed lateral accretion bedding in erosionally-based sets 2 to 10 m thick (Fig. 20). Many of these beds display an epsilon cross-stratification. Maximum restored dip of laterally accreted beds ranges from 5 to 15°. Lateral accretion directions typically form high angles with paleocurrents from associated trough crossbeds. Reversing flow directions are recorded by adjacent sets of cross-strata (see Figs. 19, 21, and discussion below).

Pebbles in Facies 6 typically are well rounded, but many are angular. The largest exotic pebbles in the Mahantango occur in this facies; long-axis lengths up to 5 cm have been measured. Pebbles consist mainly of varicolored vein quartz, but sandstone clasts are not uncommon.

Basal contacts of Facies 6 occurrences are sharply erosional, displaying outcrop-scale relief as great as 50 cm. The typical fining-upward tendency is defined in part by an upward decrease in the abundance and size of pebbles, and minor conglomerates included in this facies generally occur just above erosional basal contacts (Fig. 22).

Large vertical, oblique, and horizontal burrows are present in this facies; locally they are highly abundant (Fig. 23). *Rosselia* and ornamented vertical burrows are the dominant forms. Other traces encountered include unidentifiable, subhorizontal, cylindrical forms roughly 1 cm in diameter and at least 10 cm long. Their fill is nearly identical to the surrounding sandstone, and no linings or ornamentation were observed.

Figure 21. Several sets of reversing trough crossbedding seen in flow-parallel section, from an occurrence of Facies 6 at the Girty's Notch exposure (immediately northeast of the Watts exposure; see Fig. 1 for location). This interval lies at a stratigraphic level just below the extended arm of the geologist in Figure 20.

Figure 22. An erosionally based, conglomeratic basal lag of a tidal channel-fill sequence from an occurrence of Facies 6 at the Midway exposure (immediately NW of the Watts exposure; see Fig. 1 for location). Up is to the left. Note quartz pebbles with long axes as long as 2 cm. The channel-floor lag is overlain by sets of trough crossbedded sandstone separated by thin shale layers (tidal channel-fill deposits, also from Facies 6). The lag fills a concave-upward scour several tens of metres wide and a few decimetres deep. The channel floor is incised into a Facies 4 sandstone body, on which the hammer rests.

Figure 23. Bioturbated sandstone from an occurrence of Facies 6 at Rockville (217 m level on Fig. 27). Scale is 15 cm long.

Unusually small, incised channel forms rarely are present along basal contacts of this facies. These are roughly 10 m wide and 1 m deep and can be remarkably regular and symmetrical in form. Their fill is largely structureless, but abundant shale rip-up clasts concentrated above the erosional bases of some occurrences define a crude, flat bedding.

Interpretation

The sparsely fossiliferous sandstones of Facies 6 are interpreted as the channelized deposits of a sandy subtidal to possibly intertidal flat, with reversing paleoflow directions representing the ebb and flood of tidal currents. Lateral accretion bedding was formed by point-bar migration within a system of sinuous channels. The thicker sets indicate that some channels were at least 10 m deep; these very likely were incised to subtidal depths. The coarseness and poor sorting of channel-fill deposits (relative to the fully marine deposits of Facies 1 to 4) suggest a connection to rivers; thus, most channels may have been either small, tide-dominated estuaries or distributaries.

Facies 7: Cycle-Bounding "Lag" Deposits

Description

This facies consists of laterally extensive, anomalously coarse "concentrates" of argillaceous, medium to very coarse grained or granular, poorly sorted, quartzose sandstones (typically pebbly) and/or argillaceous, sandy quartz-pebble conglomerates (either matrix- or clast-supported). Nearly all occurrences are sharp based, and erosion of underlying lithologies can be demonstrated in nearly all examples. The largest quartz pebbles in this facies are about 3 cm in diameter; similarly large shale rip-up clasts are also present locally. All examples of Facies 7 occur at boundaries between vertically adjacent coarsening-upward cycles (see further discussion later).

These deposits commonly are very abundantly fossiliferous, containing reworked and abraded thick-shelled brachiopods and other faunal remains. Numerous examples include rugose and tabulate corals; these have diameters as large as 12 cm (Fig. 24). Many of the sandstones (including pebbly sandstones) included in Facies 7 are intensely bioturbated.

The conglomerates typically contain varicoloured quartz grains and are cemented with silica or hematite. Certain sandstones in this facies (including the pebbly sandstones) contain iron oolites and/or abundant hematite cement; these weather a distinctive red. Many sandstone occurrences, however, are not hematitic, nor are they well cemented. These latter examples typically are very argillaceous and recessive; they weather dark grey and can easily be mistaken for similarly coloured mudstone (Fig. 8). Mahantango ironstones have been further discussed by Kaiser (1972) and Prave and Duke (1990a).

Some examples of Facies 7 are nothing but a thin, erosionally based veneer of coarse clasts, one to a few grain diameters thick. Some of these layers consist of only a few sparsely scattered pebbles lying on the basal erosional surface. Many other examples are a few metres thick and are composed of several distinct beds. All occupy stratigraphically significant positions in the succession — they only occur at or just above coarsening-upward cycle boundaries, at transitions from relatively coarse to relatively fine intervals. Many occurrences clearly represent hydrodynamic "concentrates" of the coarsest material eroded from underlying deposits (*i.e.*, they are erosional "lag deposits"). Others contain clasts much coarser than any present immediately below, and likely were transported some distance prior to deposition. Most (but not all) boundaries of both large- and small-scale coarsening-upward cycles are marked by these anomalously coarse beds.

One descriptive attribute of Facies 7 deserves special

Figure 24. A multibed transgressive lag from an occurrence of Facies 7 at Rockville (141 m level on Fig. 27). Base of lag, incised into a tidal channel sandstone of Facies 6, lies immediately above 15-cm scale. Note large coral fragments, layer of brachiopod shells below coral on right, and shale layers of varying thickness. Top of lag is not visible in this photo.

attention. A large minority of occurrences exhibit highly penetrative burrows, several decimetres to just over 1 m long, extending downward from the sharply erosional basal contact (Fig. 25). These burrows are vertical to steeply inclined, straight, non-branching, and exhibit no ornamentation. Most are about 1 to 2 cm in diameter. They may represent a form of *Skolithos*. Where present, they occur in great density, cutting through one another. Strictly speaking, these burrows are contained by deposits from various other facies *underlying* examples of Facies 7; rather than repeat their description in each of the previous summaries, however, we have described them only once, here. Clearly, these burrows are associated with conditions leading to development of Facies 7; they are *not* typical of (or depositionally related to) the various underlying facies into which they penetrate.

Cycle-bounding lag deposits from Facies 7 can be separated into four basic types: 1) single-bed lags, 2) multibed lags, 3) bioturbated lags, and 4) crossbedded lags. These are separately described below.

Single-bed lags (Fig. 25) are composed of a single erosionally based, centimetre- to decimetre-thick deposit of relatively coarse sandstone, pebbly sandstone, and/or fossil-hash debris. These coarse grains are predominantly clast-supported, but many occurrences consist of a single layer of sparsely scattered pebbles and/or fossil debris. All occurrences except one exhibit sharp, erosional bases. Disarticulated and abraded brachiopod valves are locally abundant, as are crinoid ossicles and unidentified, possibly phosphatic fossil debris. Coral fragments are present locally. Many of the thicker sandy occurrences exhibit small, nondescript burrows of variable orientation.

The multibed lags (Figs. 4, 24) are several decimetres to 2 m thick. They are composed of two or more (generally several) thin, coarse beds identical to those described above, separated by decimetre- to metre-thick intervals of nonfossiliferous shale or mudstone. These fine deposits commonly exhibit millimetre- to centimetre-thick layers of flat- or cross-laminated, fine to coarse grained sandstone. Nondescript burrows are present locally.

The bioturbated lags (Fig. 8) consist of sharp-based beds of poorly sorted argillaceous sandstone or sandy mudstone 20 to 50 cm thick. Coarse grains in individual occurrences typically span a wide range of sizes (fine to coarse grained sand), and many occurrences contain widely scattered, matrix-supported quartzose pebbles. Disarticulated and abraded brachiopod valves and other fossil remains also are

Figure 25. A cycle-capping veneer of single-bed lag material from an occurrence of Facies 7 at Watts (49.5 m level on Fig. 26). Base of lag, formed in upper medium grained sandstone, is indicated by arrowheads. This lag is only a few centimetres thick at most; it is discontinuously distributed over the erosional cycle boundary. The lag lies on fine grained sandstone of Facies 5, and is overlain by shale from Facies 1. Note long, thin, vertical burrows extending into Facies 5 from the base of the lag, and brachiopod valves and additional biogenic structures in the Facies 5 sandstone.

common. Many occurrences exhibit a few, indistinct, graded beds. All examples have been intensely bioturbated, with crosscutting burrows producing a ropey or knotted texture. Burrowing may have been responsible for mixing of the disparate grades of sediment. The tops of some occurrences grade upward to better sorted, fine grained sandstone and siltstone molded into three-dimensional, sinusoidal, symmetrical bed forms with spacings of a few decimetres; internally, these bed forms exhibit decimetre-scale hummocky cross-stratification.

Crossbedded lags are the thickest ones encountered; they range from 1 to 4 m thick and are composed of several, distinct beds. These beds consist predominantly of coarse sandstones, most of which are pebbly. Sandy conglomerates are present locally, as are medium grained sandstones and thin layers of sandy mudstone. Reworked and abraded brachiopod valves vary in abundance from bed to bed, and shale rip-up clasts are present locally. The bases of many beds exhibit local erosional relief of a few centimetres. Crude crossbedding is present in many sandstone beds. A few examples of medium-scale trough crossbedding were observed, but most crossbedding is too indistinct and/or irregular to be readily classified. On a larger scale, groups of beds in many examples exhibit low angular relationships with subjacent deposits, suggesting sedimentation over a gently inclined depositional surface. In all such examples, the restored dip direction of these depositional surfaces is westward (see further discussion later).

Interpretation

Facies 7 is interpreted as occurrences of reworked, transgressive lag deposits formed during erosional shoreface retreat associated with rises of relative sea level. Note that this interpretation is based upon both the internal characteristics of these deposits and their stratigraphic relations with other facies in the coarsening-upward cycles (see below).

Integrated Interpretation of the Mahantango Depositional System

Recognition of Depositional Cyclicity

Inspection of the measured sections presented in this paper (Figs. 26, 27) permits recognition of different types and scales of coarsening-upward cycles in these exposures. An embedded Markov-change analysis, based on observations from all ten exposures examined for this study, reveals the following preferred vertical facies succession:

$$1>2>3>4>5>6>7>1$$

The thickness of an individual facies in the preferred succession can be derived from the mean thickness of all occurrences of that facies underlying the next particular facies in the succession. Facies thicknesses thus determined produce a normative cycle about 30 m thick. Note that this value is less than the mean thickness of all large-scale, coarsening-upward cycles examined in this study (55 m). The reduced thickness of the standard facies succession reflects the non-

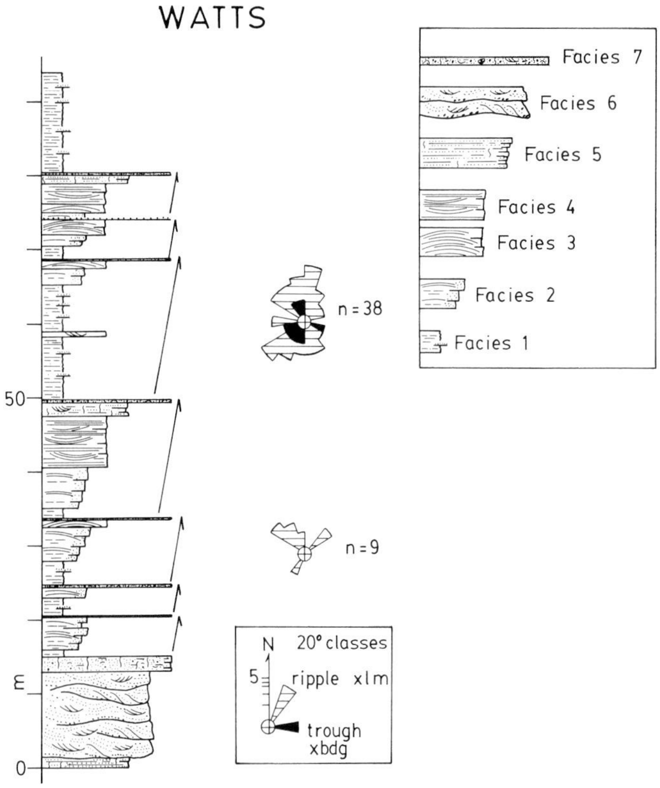

Figure 26. Measured section from the Watts Exit exposure (also known as the Amity Hall exposure). The lower and upper rose diagrams summarize paleocurrent directions from high-angle cross-strata between 15 to 50 m and 50 to 95 m, respectively. Sole marks and vortex-ripple crests from this exposure are summarized in Figure 28.

representation of superimposed small-scale cyclicity. A single, large-scale cycle closely similar to the preferred facies succession is seen at Rockville between 110 and 146 m (Fig. 27).

Interpretation of Depositional Cycles

Based on the facies interpretations presented previously, Mahantango coarsening-upward cycles are best interpreted as shoaling successions produced by the migration of a tide-dominated shoreline into a storm-dominated epicontinental sea. In a typical succession, therefore, offshore mudstones with thin "distal tempestite" sandstone beds (Facies 1) pass upward to mudstones interbedded with proximal tempestites exhibiting hummocky cross-stratification (Facies 2). These discrete storm deposits probably were emplaced in an inner shelf depositional setting. The interbedded mudstones and sandstones are succeeded by a thick lower shoreface sand body exhibiting amalgamated hummocky cross-strata (Facies 3), which in turn is overlain by swaley sandstone from the upper shoreface (Facies 4). At this point in the succession, tide-dominated deposits replace storm-dominated deposits. The lowest tidal deposits (Facies 5) represent a channel-mouth shoal or ebb-tidal delta on the upper shoreface (see further discussion below). These expanding-flow deposits are immediately overlain by channelized deposits (Facies 6) from a prograding system of fluvially

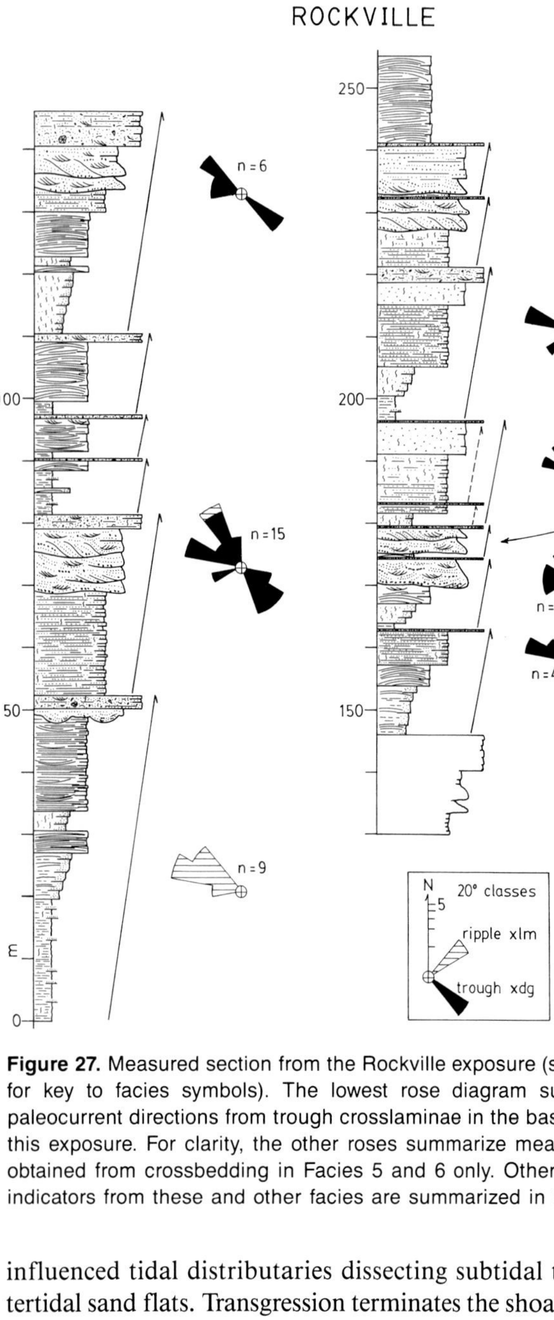

Figure 27. Measured section from the Rockville exposure (see Fig. 26 for key to facies symbols). The lowest rose diagram summarizes paleocurrent directions from trough crosslaminae in the basal 35 m of this exposure. For clarity, the other roses summarize measurements obtained from crossbedding in Facies 5 and 6 only. Other paleoflow indicators from these and other facies are summarized in Figure 28.

influenced tidal distributaries dissecting subtidal to (?) intertidal sand flats. Transgression terminates the shoaling cycle with the development of a reworked lag deposit (Facies 7). Finally, the lag is succeeded by another occurrence of distal offshore mudstone (Facies 1) associated with a new episode of progradation.

Paleoflow Data and the Paleogeography of the Mahantango Shoreline System

Paleoflow data strongly support the above interpretations. All measurements from the two sections presented herein (except those from Facies 7 lag deposits) have been summarized in Figure 28. Data from shelf storm deposits (Facies 1 and 2, Fig. 28A) indicate a shoreline oriented roughly north-northeast - south-southwest to northeast-southwest. Wave-formed vortex-ripple crests trend roughly northeast-southwest, whereas small paleocurrent indicators (tool marks and small asymmetrical ripples) are directed primarily to the west and northwest, roughly normal to the inferred shoreline trend.

The Mahantango depositional system was situated in low southern latitudes (10° S; see Table 1 of Duke, 1985). In this setting, steady bottom currents resulting from storm-generated geostrophic circulation would have been alongshore to the southwest, and the dominant storms probably were hurricanes (Duke, 1985). The roughly shore-normal orientation of small paleocurrent indicators in Facies 1 and 2 (Fig. 28A) most likely reflects the direction of peak instantaneous shear stress beneath storm-generated, oscillatory-dominant combined flows (as suggested by Duke, 1990). Such bottom flows result from superimposition of a relatively weak, geostrophically balanced steady current and the strong shore-normal wave-orbital motions that characterize storm flows on the shallow shelf (Duke, 1990). All Mahantango storm deposits, including thin distal beds from Facies 1 exhibiting Bouma sequences, probably were emplaced by similar combined flows well above storm wave base. The small amount of paleoflow data obtained from shoreface sandstones of Facies 3 and 4 (Fig. 28B) is consistent with this analysis.

Paleoflow data from Facies 5 (Fig. 28C) and Facies 6 (Fig. 28D) reflect reversing onshore-offshore currents, mainly within predominantly shore-normal tidal channels. The dispersion of data from crossbedding in Facies 6 suggests a sinuous channel geometry. Note that restored dip directions of lateral accretion bedding (produced by point-bar migration) do not indicate a preferred direction of channel migration along the Mahantango shoreline; thus, the mean wave-propagation direction probably was nearly shore-normal, and net longshore drift was low. Ebb dominance is indicated by the greater abundance of offshore-directed crossbedding in Facies 5 and 6. This dominance probably reflects the strong fluvial component of currents in the channels, rather than a tidal asymmetry *per se*.

Our reconstruction of the prograding Mahantango shoreline system in central Pennsylvania is shown in Figure 29 (upper half). As indicated in this sketch, we cannot distinguish the ebb-tidal deltas from the channel-mouth shoals from our database; perhaps both were present, as drawn in Figure 29 (upper half). Note that paleoflow data from other sections measured for this study exhibit relationships very similar to those from the exposures described herein. Also, no major *vertical* variation in the orientation of these various paleoflow indicators is apparent in the measured sections (Figs. 26, 27), suggesting our reconstruction of a straight shoreline trending approximately northeast-southwest is applicable to any individual cycle in the formation. The most similar well documented modern analogue for this ancient system is the barrierless portion of the coast of the German Bay of the North Sea, near the Weser and Elbe estuaries (Reineck and Singh, 1975, Fig. 463).

Note that the upper shoreface swaley facies (Facies 4)

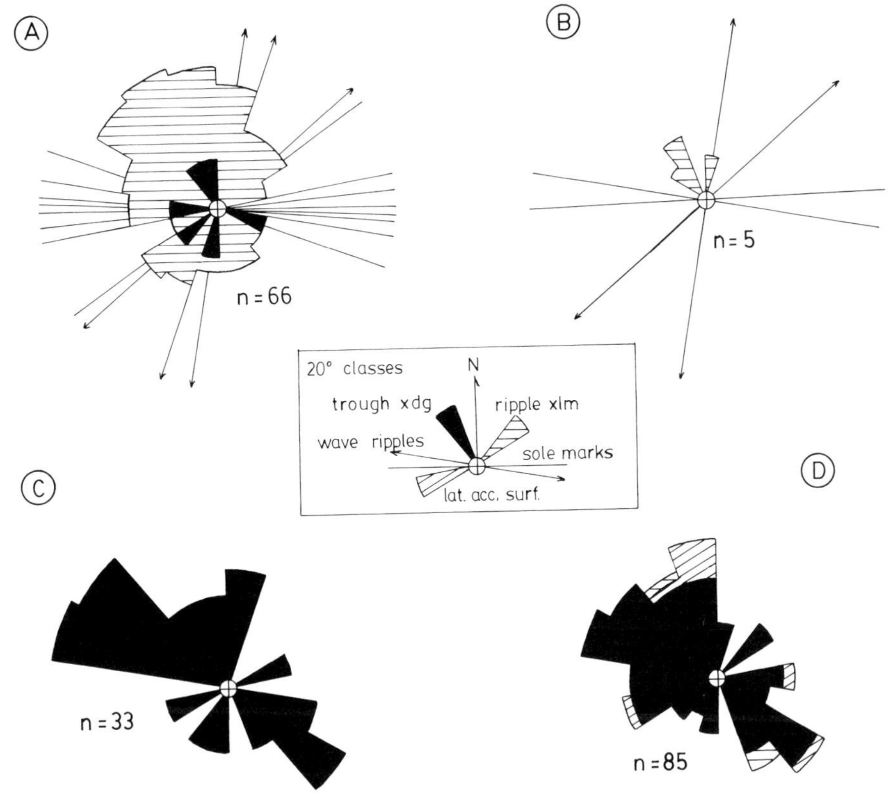

Figure 28. Summary of all paleoflow data obtained at Watts and Rockville (except those from Facies 7 transgressive lag deposits). **A.** Data from shelf storm deposits (Facies 1 and 2). **B.** Data from shoreface sandstones (Facies 3 and 4). **C.** Data from channel-mouth shoals/ebb-tidal deltas (Facies 5). **D.** Data from subtidal to (?)intertidal channels (Facies 6).

commonly is absent in Mahantango shoreface sandstones (see Figs. 26, 27); in many instances, lower shoreface storm deposits (Facies 3) are directly succeeded by tidal shoal deposits (Facies 5). This situation suggests that tidal overprinting of storm and wave effects most commonly extended well out across the upper shoreface. However, the local preservation of swaley sandstone in our measured sections indicates that this was not the case everywhere along the coast, as we have tried to represent in Figure 29 (upper half). These inferred relationships are consistent with the estimated depth of tidal channel incision. Based on the thickness of individual lateral accretion sets in Facies 6, we estimate these tidal channels were over 10 m deep locally. Thus, tidal channels probably were incised to shallow subtidal depths on the shoreface, an inference supported by the rarity of swaley sandstone. This depth of incision suggests a probable high mesotidal (possibly macrotidal) range along the Mahantango shoreline in central Pennsylvania.

The coarse, poorly sorted sediment in the tidal channel deposits clearly was supplied from a fluvial source; it could not have been derived from the much finer shoreface. We thus infer that these channels were estuaries (or deltaic distributaries) connected to rivers and maintained during progradation by strong tidal currents. The rivers of the lower coastal plain are not present in exposures examined for this study; however, the nature of the tidal channel fill indicates these rivers probably were high-gradient streams, probably braided, which debouched directly upon intertidal sand flats along the coast (Fig. 29, upper half).

Note that there is no evidence for major barrier islands (or associated inlets) in the system. On modern coasts, the absence of barriers generally coincides with a macrotidal range, or a high mesotidal range coupled with high wave energy. However, modern coasts generally are undergoing transgression; the absence of barriers in the Mahantango may simply be a consequence of progradation. We thus see no compelling need to invoke a macrotidal range for this ancient shoreline system.

Transgressive lags bounding Mahantango cycles are irregularly variable in thickness, ranging from one grain diameter to over 4 m thick. Medium-scale high-angle cross-stratification in these lags is rare, but large-scale, gently inclined crossbedding is fairly common in the thicker occurrences. Restored dip directions from these inclined beds are consistently westerly. We attribute this geometry to the seaward migration of low-amplitude transgressive shelf sand ridges formed by downwelling geostrophic circulation during storms (Fig. 29, lower half). We shall discuss the nature and origin of these transgressive deposits in greater detail in another paper.

Discussion of the Inferred Depositional System

A prograding shoreline interpretation for large-scale coarsening-upward cycles in the Mahantango has

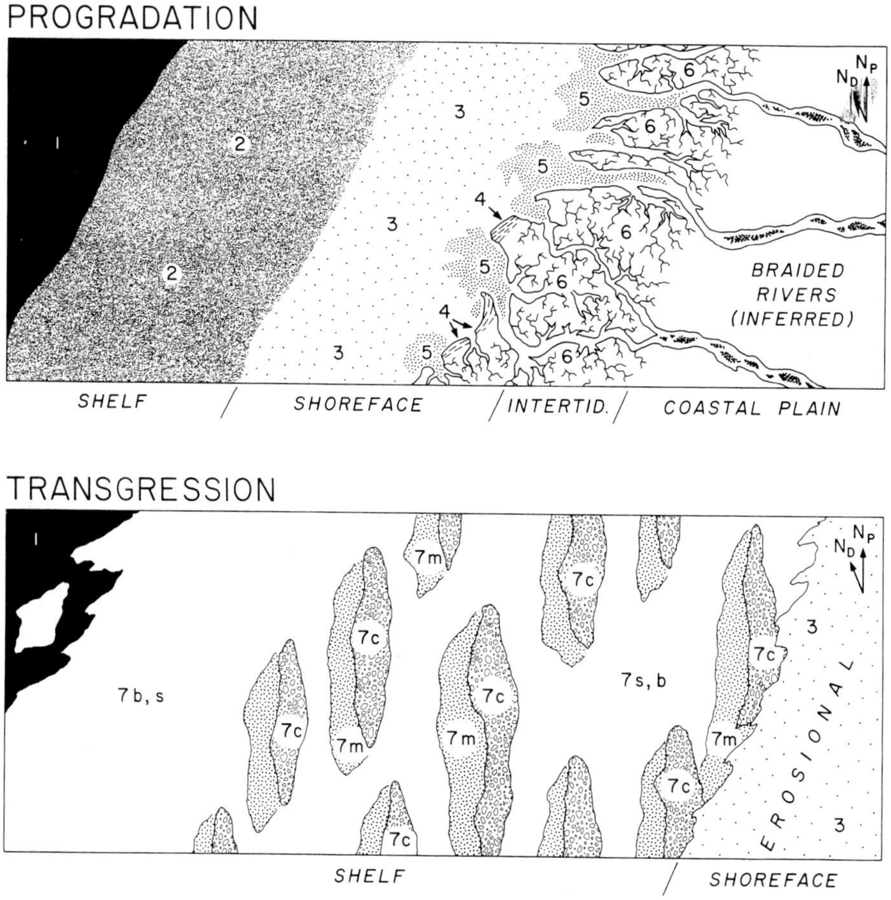

Figure 29. Reconstructions of Mahantango depositional systems in central Pennsylvania. Arrows labelled "N-sub-P" and "N-sub-D" indicate north at present and during deposition in the Middle Devonian southern hemisphere, respectively. Numbers refer to facies defined in this report. Upper panel is a reconstruction of the prograding shoreline system. Lower panel is a reconstruction of the shelf during transgression and erosional shoreline retreat, showing the inferred system of storm-generated shelf sand ridges formed in Facies 7 deposits. 7b, bioturbated lag; 7c, crossbedded lag; 7m, multibed lag; and 7s, single-bed lag. See text for discussion.

previously been advanced by Goldring and Bridges (1973), based on observations at Watts Exit (see their Fig. 8A). These workers recognized a pronounced storm influence in the two large-scale cycles fully exposed at Watts (see our Figs. 26, 30A). We largely agree with this interpretation, although our work (including regional correlations) permits us to recognize the importance of tides during deposition of the upper shoreface sandstones (see further discussion in a later section).

An essentially similar origin (involving multiple shoreline systems) is implicit in the deltaic interpretation advanced by Kaiser (1972) and Faill *et al.* (1978), although we disagree with their inference of a lobate shoreline (see further discussion below).

In contrast, the most recent sedimentological investigation of the Mahantango attributed the genesis of large-scale coarsening-upward cycles to an "autocyclic" mechanism: namely, lateral migration of storm-generated shallow marine sand ridges (Sarwar and Smoot, 1983; Sarwar, 1984). The analysis presented herein does not support this interpretation, but broadly supports the older interpretations involving shoreline migration and fluctuations in relative sea level. In our judgement, the large coarsening-upward cycles are far too thick, too uniform in thickness, and too extensive laterally to be reasonably attributed to sand-ridge migration. Likewise, the coarseness and poor sorting of pebbly sandstones near the tops of sequences strongly suggests proximity to a fluvial source. Most importantly, however, we find it hard to accept Sarwar and Smoot's interpretation of lateral accretion bedding (in our Facies 6) as representing deposition over the inclined margins of storm-generated sand ridges. The steepness of these inclined beds (Fig. 20), as well as their variable orientations and relationship to reversing trough crossbedding (Figs. 21, 28), are much better attributed to lateral channel migration in a system of sinuous tidal channels.

Our interpretation of tidal-channel deposits in Facies 6 represents, to the best of our knowledge, the first recognition of the dominance of tidal effects along the Mahantango shoreline. It should be noted, however, that Sarwar (1984) recognized reversing paleocurrent indicators at the tops of large coarsening-upward cycles, which he attributed to tidal reworking of storm-generated shelf sand ridges during episodes of eustatic sea-level lowering.

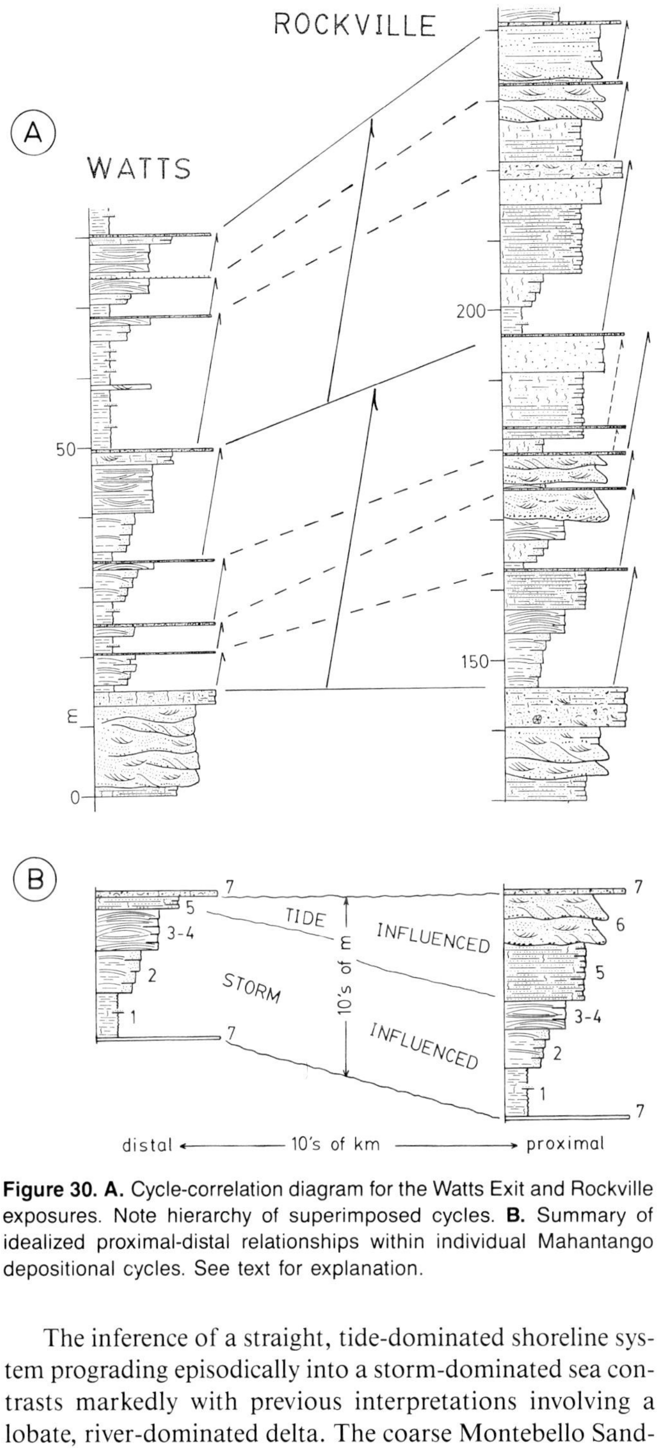

Figure 30. A. Cycle-correlation diagram for the Watts Exit and Rockville exposures. Note hierarchy of superimposed cycles. **B.** Summary of idealized proximal-distal relationships within individual Mahantango depositional cycles. See text for explanation.

The inference of a straight, tide-dominated shoreline system prograding episodically into a storm-dominated sea contrasts markedly with previous interpretations involving a lobate, river-dominated delta. The coarse Montebello Sandstone Member has long been regarded as perhaps the best example of a deltaic fan in the Appalachian Basin (Willard, 1935; Wood *et al.*, 1969; Kaiser, 1972). Thus, the paleogeography of the Mahantango shoreline deserves special consideration.

The previous deltaic interpretation is based largely upon isopach maps of the Montebello Member in surface exposures (*e.g.*, Willard, 1935; Kaiser, 1972, Fig. 6). Additionally, Montebello thickness correlates positively with the distribution of large clasts in these exposures. In contrast, our interpretation of a relatively straight shoreline is based on measured paleocurrent indicators, which reveal patterns of sedimentation that do not vary greatly either vertically or laterally. Our observations suggest that the thickening and coarsening of the Montebello in central Pennsylvania can be attributed to the combination of three effects: 1) the nonuniform distribution of exposures throughout the state has concentrated source-proximal exposures in central Pennsylvania; 2) the lithostratigraphic (rather than allostratigraphic) basis for the definition of the Montebello Member has resulted in the inclusion in this unit of a different assemblage of depositional cycles from place to place; and 3) the relative coarseness of Mahantango shoreline facies was not constant through time, but was greatest as the northeast-southwest trending shoreline prograded through central Pennsylvania. Details of these relationships are explored further by Duke and Prave (work in progress).

Cycle Relationships and Correlations

The measured sections of Figures 26 and 27 each exhibit several coarsening-upward depositional cycles of varying scale, as do the other exposures examined in this study. At each outcrop, the thicker cycles typically can be subdivided into smaller superimposed cycles. Both large-and small-scale cycles can be correlated on the basis of their internal facies successions, relative stratigraphic position and thickness, and hierarchical rank. We present a cycle-correlation diagram in Figure 30A. Note the straightforward nature of cycle correlations in these exposures.

Lateral variation in vertical facies successions within individual depositional cycles provides the basis for interpreting large-scale sedimentological relationships in Mahantango shoreline systems. One of the more interesting observations concerns proximal-distal relationships (Fig. 30B). In the source-proximal Rockville section, the vertical transition from storm-dominated shelf and shoreface deposits (Facies 1 to 4) to tide-dominated shoal and channel deposits (Facies 5 and 6) provides a fairly obvious record of the depth-controlled transition from one depositional regime to another. These relationships are much less clear in the more distal exposure at Watts Exit, because of the absence of tidal channel deposits (Facies 6) at the tops of these two coarsening-up cycles.

Had we lacked the stratigraphic context available in the more proximal exposures examined for this study, we might never have realized the hydrodynamic and environmental significance of Facies 5. At the Watts exposure in particular, where these inferred tidal shoal deposits are thin (as a result of their termination by an overlying transgression surface), they might have been mistaken for a wave-dominated nearshore sandstone. This situation prompts us to advise caution when interpreting storm-influenced stratigraphic successions — a possible strong tidal influence cannot be eliminated unless unequivocal beach and nonmarine deposits are preserved.

Origin of Depositional Cyclicity in the Mahantango

The inferred fluctuations in relative sea level at the Mahantango shoreline may possibly have been eustatic in origin; however, such a determination clearly requires evidence from many measured sections of coeval strata from at least two tectonically disconnected sedimentary basins. In contrast, there is strong circumstantial evidence suggesting that Mahantango depositional cycles are better interpreted in terms of regional tectonic events. Firstly, the poorly sorted, coarse grained sandstones and conglomerates in many cycles suggest deposition near a region of active tectonic uplift. Secondly, the Acadian Orogeny was near its zenith just southeast of the study area during Mahantango deposition (Quinlan and Beaumont, 1984). Deposition along the active margin of the well developed Acadian foreland basin is consistent with a pronounced tectonic influence on sedimentation. Thus, we tentatively attribute the large-scale coarsening-upward cycles in the Mahantango to episodes of Acadian thrusting. Note, however, that regional relationships between these cycles and under- and overlying units may perhaps be attributable to a long-term eustatic sea-level rise; such a rise is indicated in the sea-level curve of Vail *et al.* (1977), Dennison and Head (1975), Dennison (1985), and Johnson *et al.* (1985). This subject will be discussed in detail elsewhere.

The smaller, superimposed coarsening-upward cycles are much less extensive laterally; generally, we have been unable to trace them for more than a few tens of kilometres. These developments are best attributed to localized episodes of shoreline advance superimposed upon the regionally extensive progradational events. They probably resulted from relatively rapid local fluctuations of water depth induced by: 1) tectonic events of restricted temporal and spatial scale; and/or 2) local "autocyclic" systemic depositional processes, such as temporal variation in the local configuration of the advancing shoreline or "complex system response" on the lower coastal plain.

Acknowledgments

The primary support for this work was provided by a Faculty Career Initiation Grant from the Shell Companies Foundation. The Watts section was measured in 1982 as part of a project supported by a grant from the Research Fund of the American Association of Petroleum Geologists Foundation. Additional logistical support was provided by the Department of Geosciences at The Pennsylvania State University, in conjunction with graduate courses on clastic depositional environments. This support is gratefully acknowledged. Field assistance was provided by M. A. Matteliano and T. E. Narehood. The comments of R. W. C. Arnott, J. S. Bridge, and R. L. Slingerland led to substantial improvements of the text.

References

Dennison, J. M. 1985. Devonian eustatic fluctuations in Euroamerica: Discussion. Geological Society of America, Bulletin, v. 96, p. 1595-1597.

______ , and Hasson, K. O. 1976. Stratigraphic cross-section of Hamilton Group and adjacent strata along south border of Pennsylvania. American Association of Petroleum Geologists, Bulletin, v. 60, p. 278-298.

______ and Head, J. W. 1975. Sea-level variations interpreted from the Appalachian Basin Silurian and Devonian. American Journal of Science, v. 275, p. 1089-1120.

Duke, W. L. 1980. Upper Cretaceous (Turonian) Cardium Formation, southern Alberta. McMaster University, Department of Geology, Hamilton, Canada, Technical Memorandom 80-4, 59 p.

______ 1985. Hummocky cross-stratification, tropical hurricanes, and intense winter storms. Sedimentology, v. 32, p. 167-194.

______ 1990. Geostrophic circulation or shallow marine turbidity currents? The dilemma of paleoflow patterns in storm-influenced prograding shoreline systems. Journal of Sedimentary Petrology, v. 60, p. 870-883.

______ and Prave, A. R. 1989. Lateral variation within "mixed-influence" prograding shoreline sequences in the Middle Devonian Mahantango Formation in central Pennsylvania (abs.). Geological Society of America, Abstracts with Programs, v. 21, no. 6, p. A37.

Faill, R. T. and Wells, R. B. 1974. Geology and mineral resources of the Millerstown quadrangle, Perry, Juniata, and Snyder Counties, Pennsylvania. Pennsylvania Geological Survey, 4th series, Atlas 136, 276 p.

______ , Hoskins, D. M. and Wells, R. B. 1978. Middle Devonian stratigraphy in central Pennsylvania — a revision. Pennsylvania Geological Survey, 4th series, General Geology Report 70, 28 p.

Frey, R. W. and Pemberton, S. G. 1984. Trace fossil facies models. *In:* Facies Models, Second Edition, R. G. Walker (ed.). Geoscience Canada, Reprint Series 1, p. 189-207.

Goldring, R. and Bridges, P. 1973. Sublittoral sheet sandstones. Journal of Sedimentary Petrology, v. 43, p. 736-747.

Hamblin, A. P. and Walker, R. G. 1979. Storm-dominated shallow marine deposits: the Fernie-Kootenay (Jurassic) transition, southern Rocky Mountains. Canadian Journal of Earth Sciences, v. 16, p. 1673-1690.

Hoskins, D. M. 1976. Geology and mineral resources of the Millersburg 15-minute quadrangle, Dauphin, Juniata, Northumberland, Perry, and Snyder Counties, Pennsylvania. Pennsylvania Geological Survey, 4th series, Atlas 146, 38 p.

Johnson, J. G., Klapper, G. and Sandberg, C. A. 1985. Devonian eustatic fluctuations in Euroamerica. Geological Society of America, Bulletin, v. 96, p. 567-587.

Kaiser, W. R. 1972. Delta cycles in the Middle Devonian of central Pennsylvania. Unpublished Ph.D. thesis, Johns Hopkins University, Baltimore, Maryland, 183 p.

Leckie, D. A. and Walker, R. G. 1982. Storm-and tide-dominated shorelines in Cretaceous Moosebar-Lower Gates interval — outcrop equivalents of Deep Basin Gas Trap in western Canada. American Association of Petroleum Geologists, Bulletin, v. 66, p. 138-157.

Pemberton, S. G. 1987. Ichnology of the Thorold Sandstone in the vicinity of Hamilton, Ontario: a Silurian storm-influenced deposit. *In:* Sedimentology, Stratigraphy, and Ichnology of the Lower Silurian Medina Formation in New York and Ontario, W. L. Duke (ed.). Guidebook for the 1987 Annual Field Trip of the Eastern Section of the Society of Economic Paleontologists and Mineralogists, p. 66-80.

Pemberton, S. G. and Frey, R. W. 1984. Ichnology of a storm-influenced shallow marine sequence: Cardium Formation (Upper Cretaceous) at Seebe, Alberta. *In:* The Mesozoic of Middle North America, D. F. Stott and D. J. Glass (ed.). Canadian Society of Petroleum Geologists, Memoir 9, p. 281-304.

Prave, A. R. and Duke, W. L. 1989. Prograding shoreline deposits in the Middle Devonian Mahantango Formation (central Pennsylvania) reveal a vertical transition from storm-dominated sedimentation to tide-dominated sedimentation within individual shoreface sandstones (abs.). Proceedings of the Second International Research Symposium on Clastic Tidal Deposits, Calgary, Alberta, Canada, p. 77.

______ and ______ 1990a. Sequence-capping oolitic ironstones in the Mahantango Formation (Middle Devonian) of Pennsylvania: responses to basin dynamics (abs.). Geological Society of America, Abstracts with Programs, v. 22, no. 2, p. 63.

______ and ______ 1990b. Sequence stratigraphy of prograding shoreline deposits in the Acadian foreland basin: the Middle Devonian Mahantango Formation, central Pennsylvania, U.S.A. (abs.). Thirteenth International Sedimentological Congress of the International Association of Sedimentologists, Abstracts of Papers, p. 437-438.

Quinlan, G. M. and Beaumont, C. 1984. Appalachian thrusting, lithospheric flexure, and the Paleozoic stratigraphy of the eastern interior of North America. Canadian Journal of Earth Sciences, v. 21, p. 973-996.

Reineck, H.-E. and Singh, I. 1975. Depositional Sedimentary Environments. Springer-Verlag, New York, 439 p.

Sarwar, G. 1984. Depositional model for the Middle Devonian Mahantango Formation of south-central Pennsylvania. Unpublished M.S. thesis, State University of New York at Stony Brook, Stony Brook, New York, 251 p.

______ and Smoot, J. P. 1983. Depositional model for the Middle Devonian Mahantango Formation of south-central Pennsylvania. Geological Society of America, Abstracts with Programs, v. 15, p. 127.

Vail, P. R., Mitchum, R. M., Jr. and Thompson, S., III 1977. Seismic stratigraphy and global changes of sea level, part 4: global cycles of relative changes of sea level. *In:* Seismic stratigraphy - applications to hydrocarbon exploration, C. E. Payton (ed.). Tulsa, Oklahoma, American Association of Petroleum Geologists, p. 83-97.

Willard, B. 1935. Hamilton Group of central Pennsylvania. Geological Society of America, Bulletin, v. 46, p. 195-224.

Wood, G. H., Jr., Trexler, J. P. and Kehn, T. M. 1969. Geology of the west-central part of the southern Anthracite Field and adjoining areas, Pennsylvania. U.S. Geological Survey Professional Paper 602, 150 p.

KEY WORDS: PROGRADING SHORELINE SEQUENCES, MIDDLE DEVONIAN, MAHANTANGO FORMATION, SHALLOW MARINE DEPOSITS

In: Clastic Tidal Sedimentology, D. G. Smith, G. E. Reinson, B. A. Zaitlin and R. A. Rahmani (eds.). Canadian Society of Petroleum Geologists. Memoir 16, p. 371-387.

Depositional facies and controls on parasequence development in siliciclastic tidal deposits from the Lower Proterozoic, upper Mount Guide Quartzite, Mount Isa Inlier, Australia

EDWARD L. SIMPSON

Department of Physical Sciences
Kutztown University
Kutztown, Pennsylvania 19530
U.S.A.

KENNETH A. ERIKSSON

Department of Geological Sciences
Virginia Polytechnic Institute and State University
Blacksburg, Virginia 24061
U.S.A.

ABSTRACT

The upper Mount Guide Quartzite is a quartz arenite unit (2 km thick) of predominantly tidal origin. Sedimentation took place in a thermally subsiding intracratonic basin over a 20 to 25 m.y. time span between 1720 and 1800 Ma. Facies in the upper Mount Guide Quartzite consist of crossbedded arenite deposits capped by thin bedded arenites, and contain stacked, tabular parasequences ranging from 0.5 to 10.2 m thick.

Three types of crossbedded arenite subfacies are recognized: 1) tabular crossbedded sets and cosets (0.2-5.5 m thick) with common sigmoidal-shaped pause planes; 2) compound crossbed cosets (up to 9.9 m thick) characterized by three hierarchical orders of bounding surfaces; and 3) large-scale trough crossbed cosets (3.5-5.1 m thick). Internal structures within crossbed sets consist of ripple stratification and grainflows arranged in accelerating-decelerating flow cycles typical of modern and ancient tidal subaqueous dune deposits. Paleocurrent azimuths are dominantly toward the west, indicating strong asymmetrical tidal flow. The thin-bedded arenite caps contain a variety of structures: 1) current, combined-flow and wave ripples; 2) modified ripples including ladderback, round-crested, flat-topped and washed-out forms; and 3) emergence structures including aeolian ripple stratification, adhesion ripples and warts, and desiccation cracks. Vertical arrangement of facies in parasequences reflects flooding and establishment of a tidal shelf followed by shoaling to intertidal and, rarely, supratidal conditions.

Fischer plots of cumulative parasequence thickness (corrected for linear thermal subsidence) versus time show deviations in thickness from that expected for pure linear subsidence. The deviations can be attributed to changes in relative sea level rather than ''jerky'' subsidence. The Fischer plot permits recognition of systems tracts that are associated with type-2 sequences.

RÉSUMÉ

L'unité supérieure du ''Mount Guide Quartzite'' est une quartzarénite (2 km d'épaisseur) essentiellement formée par les marées. Entre 1720 et 1800 M.a., la sédimentation a pris place dans un bassin intracratonique à subsidence thermique et a duré de 20 à 25 M.a. Les faciès du ''Mount Guide Quartzite'' supérieur consistent en dépôts d'arénite à stratifications obliques, surmontés par de minces lits d'arénite; ils contiennent une succession de paraséquences tabulaires variant de 0.5 à 10.2 m d'épaisseur.

Trois types de sous-faciès à arénites à stratifications obliques ont été identifiés: 1) des ensembles à stratifications obliques tabulaires, certains ayant une même direction de courant («cosets»; 0.2-5.5 m d'épaisseur), avec des plans d'arrêt de sédimentation de forme sigmoïde; 2) un regroupement d'ensembles à stratifications obliques ayant une même direction de courant (jusqu'à 9.9 m d'épaisseur) caractérisés par des surfaces de limite hiérarchisées en trois ordres; 3) des ensembles à stratifications obliques (3.5-5.1 m d'épaisseur) avec des fosses de grandes dimensions. Les structures internes des ensembles à stratifications obliques sont constituées par des laminations onduleuses et une fabrique des grains résultant des cycles d'écoulement d'accélération-décélération, typiques des dunes subaquatiques des dépôts de marée modernes et anciens. Les directions des paléocourants pointent principalement vers l'ouest, indiquant des flots de marée asymétique forts. Les minces lits d'arénite surmontant les arénites à stratifications obliques contiennent une variété de structures: 1) des rides de courant, d'écoulement combiné et de vague; 2) des rides modifiées incluant des formes en échelon («ladderback»), des crêtes arrondies, des sommets plats et des formes d'érosion causées par le ravinement («washed-out»); 3) des structures d'émergence comprenant des rides ascendantes, des rides d'adhésion, des «nodules arrondies» («warts») et des fentes de dessication. L'arrangement vertical des faciès en paraséquences reflète l'inondation et l'établissement d'une plate-forme de marée suivi par des conditions d'eaux moins profondes de la zone intertidale, et rarement par des conditions supratidales.

Le traçage d'une courbe cumulative de Fischer — l'épaisseur des paraséquences (corrigée pour le cas d'une subsidence thermique) en fonction du temps — montre un écart avec les valeurs prévues pour le cas d'une subsidence linéaire pure. Ces écarts peuvent être attribués au changement relatif du niveau de la mer plutôt qu'à une subsidence irrégulière. Le traçage d'une courbe de Fischer a permis de reconnaître les cortèges sédimentaires ou séquences de dépôts («systems tracts») qui sont associées aux séquences de type 2.

Introduction

Parasequences have been recognized from diverse depositional environments but are most easily identified in marine and marginal-marine deposits (Van Wagoner *et al.*, 1990). Parasequences in muddy, tide-dominated shorelines are typically upward fining and thinning and are composed of subtidal sandstones conformably overlain by intertidal to supratidal sandstones and mudstones (Van Wagoner *et al.*, 1990). Subtidal sandstones are generated by migrating subaqueous dunes whereas intertidal beds are produced by smaller scale bedforms. Marine flooding surfaces separate each parasequence. Parasequences are stacked in progradational, aggradational and retrogradational sets; parasequence set types develop in response to variable accommodation space produced by a combination of changes in sea level, and subsidence (Van Wagoner *et al.*, 1987; Posamentier *et al.*, 1988; Van Wagoner *et al.*, 1988; Van Wagoner *et al.*, 1990). Parasequence stacking patterns permit recognition of systems tracks that can be grouped into sequences (Van Wagoner, 1985).

The Lower Proterozoic, upper Mount Quartzite in the Mount Isa Inlier, Australia, is an approximately 2 km-thick, quartz arenite unit. Detailed facies analysis permits recognition of two facies that make up stacked, metre-scale parasequences. It is demonstrated in this paper, through process interpretation of facies, that the parasequences consist exclusively of subtidal shelf overlain by sandy intertidal to supratidal facies and show, through the use of a Fischer plot, that systematic changes in thickness of parasequences can be used to identify different systems tracts. The approach outlined in this paper for recognizing parasequence stacking patterns and thereby relative sea level fluctuations can be applied to rock successions that are of uniform lithology, and devoid of paleontological, isotopic, paleomagnetic or lithological marker horizons.

Geological Setting, Stratigraphy and Age Constraints

Three major tectonic subdivisions are recognized in the Mount Isa Inlier: 1) the Western Fold Belt that comprises the Lawn Hill Platform, Leichhardt River Fault Trough, Ewen Block and Myally Shelf; 2) the Kalkadoon-Leichhardt Belt; and 3) the Eastern Fold Belt (Fig. 1). Basement metamorphic and granitoid rocks exposed mainly in the Kalkadoon Leichhardt Belt (Fig. 1) are greater than 1900 m.y. old (Blake, 1987) and contain Archean components (Sun and McCulloch, 1985; Page, 1985). Cover rocks attain their maximum thickness in the Leichhardt River Fault Trough where they are subdivisible into three sequences (Blake, 1987). Cover sequence 1 consists of felsic volcanic rocks dated at 1870 Ma (Page, 1983a). The Haslingden Group and overlying Quilalar Formation make up cover sequence 2, and cover sequence 3 consists of the Bigie and Surprise Creek formations and Mount Isa Group (Fig. 2). Basement and cover rocks in the Mount Isa Inlier were subjected to complex thrusting, upright folding, strike-slip faulting and regional metamorphism to greenschist and amphibolite facies between 1610 to 1510 Ma (Page and Bell, 1986).

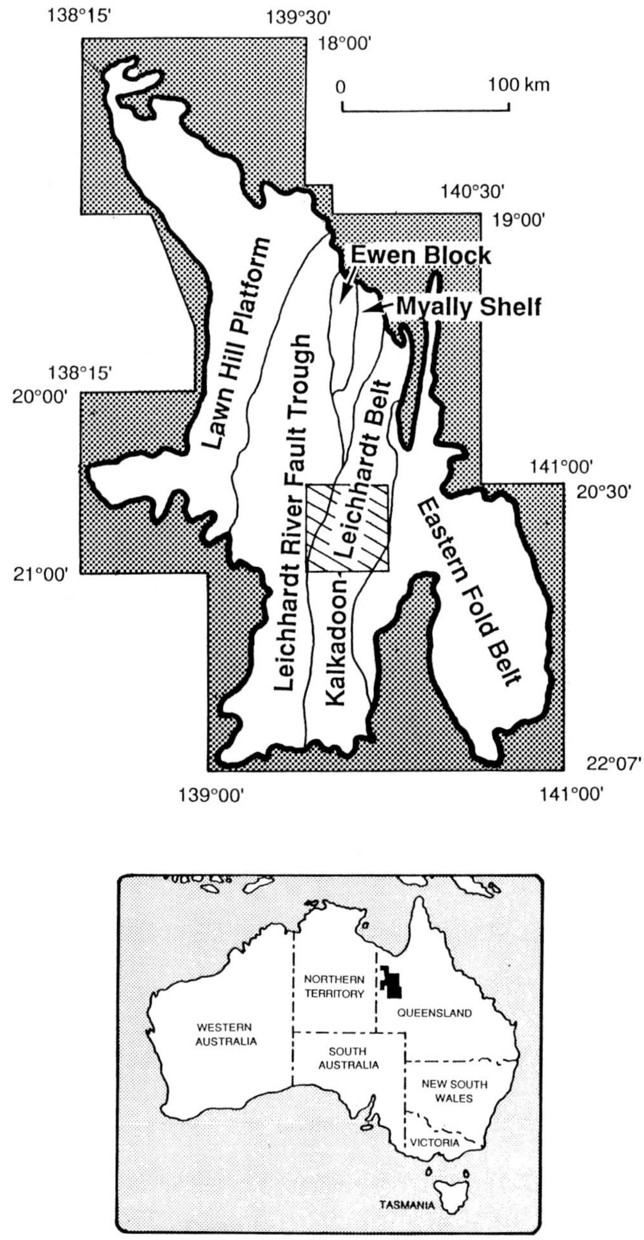

Figure 1. Location map showing tectonic framework of the Mount Isa Inlier (from Blake 1987). Crosshatched rectangle is location of Mary Kathleen 1/2° sheet.

Cover sequences 2 and 3 are interpreted as two rift-to-thermal subsidence (sag) phases of sedimentation and volcanism (Blake, 1985; Wyborn *et al.*, 1988). Bimodal volcanics, conglomerates and feldspathic sandstones in the Bottletree Formation, and conglomerates and feldspathic sandstones in the lower Mount Guide Quartzite, are compatible with a rift setting. However, the upper Mount Guide Quartzite, derived from the same source terrane as the underlying sediments, consists entirely of supermature quartz arenites and reflects a different tectonic setting. Eriksson *et al.* (1990) proposed that this unit represents an additional thermal subsidence phase (Fig. 2). Overlying mafic volcanics record initiation of the second phase of rifting that continued through the Myally Subgroup. The Quilalar

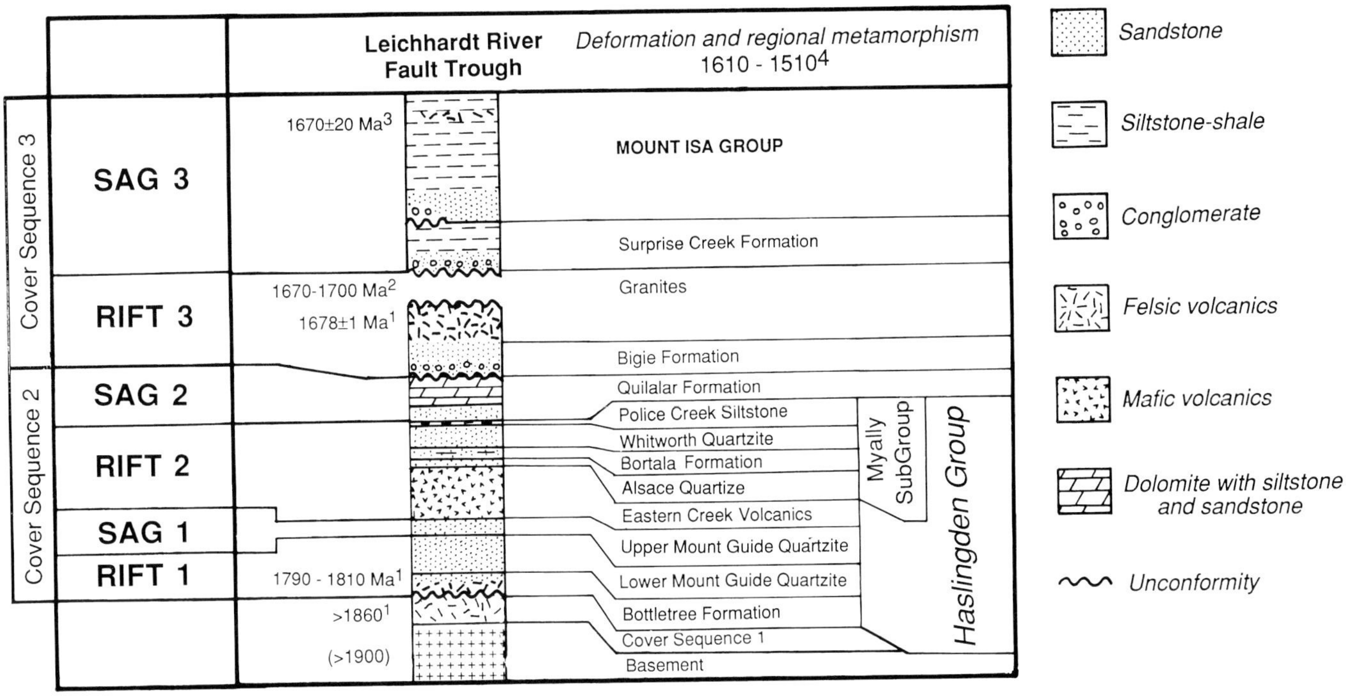

Figure 2. Generalized lithostratigraphic column showing the relationship of the upper Mount Guide Quartzite to underlying and overlying stratigraphic units (adapted from Plumb *et al.*, 1981 and Blake, 1987). Geochronological data from: Page (1983a); Page *et al.* (1981); Page (1981); Page and Bell (1986). Cover sequence nomenclature is after Blake (1987).

Formation represents the second thermal subsidence phase (Jackson *et al.*, 1990). The time interval during which the Haslingden Group and overlying Quilalar Formation developed is well constrained. Felsic volcanics in the Bottletree Formation yield U-Pb zircon ages of 1808 $^{+22}_{-17}$ and 1790 $^{+10}_{-8}$ (Page, 1983a), whereas rhyolite dykes intrusive into a correlative of the Quilalar Formation in the Eastern Fold Belt give U-Pb zircon ages of 1737 ± 15 Ma (Page, 1983b). Cover sequence 2 thus accumulated over a maximum time period of 80 m.y. and may represent as little time as 30 m.y.

The upper Mount Guide Quartzite is exposed in a number of structural blocks in the Leichhardt River Fault Trough, but primary sedimentary structures and textures are best preserved along a 10 km long strike belt on the eastern margin of the trough (Fig. 3). Observations are from gorge exposures along this outcrop belt with detailed observations including photomosaic sketches, a bed-by-bed measured section, and paleocurrent measurements from the upper 250 m. Two facies, crossbedded arenites and thin bedded arenites, dominate the measured section (Fig. 4); other outcrops along strike and throughout the upper Mount Guide Quartzite contain identical facies, but exposures are discontinuous.

Crossbedded Arenite Facies

Three subfacies are recognized on the basis of crossbed types and hierarchies of bounding surfaces. The subfacies are: 1) tabular crossbed sets and cosets (CB1); 2) compound, crossbed cosets (CB2); and 3) large-scale, trough

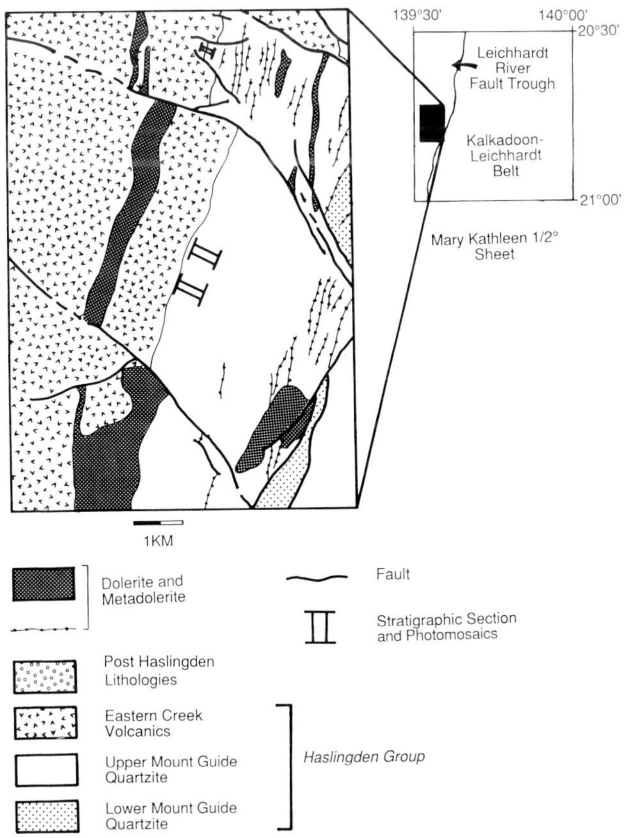

Figure 3. Location of measured section and photomosaics within the upper Mount Guide Quartzite. Map modified from Derrick *et al.* (1977). Location of Mary Kathleen 1/2° sheet is shown on Figure 1.

crossbed cosets (CB3). The distribution of the subfacies in section is shown on Figure 4. CBl makes up 88 per cent, CB2 and CB3 only 4 and 8 per cent of the crossbedded facies, respectively. Medium grained sand is the dominant grain size and coarse grained sand is subordinate. Paleocurrent data from foresets exhibit a poorly developed bimodal-bipolar pattern with a dominant southwesterly mode (Fig. 5A).

Internal structures common to CBl, CB2, and CB3 are ripple stratification and grain flows (sand-flows of Hunter, 1985; Fig. 6). Grainfall laminations may be present, but identification is problematic because of their close resemblance to ripple stratification.

Tabular Crossbed Sets and Cosets (CBl)

Description

CBl consists of tabular crossbed sets and cosets (Figs. 7-9). Thicknesses range from 0.2 to 5.5 m. Included in this facies are cosets of trough crossbeds containing similar internal structures.

Foresets vary from planar to tangential. Crossbed sets and cosets are bounded by El surfaces (*cf.* Allen, 1980) that are sharp and planar to slightly undulatory, and have little or no discernable dip when compared to regional bedding. Megaripples oriented orthogonal to underlying foresets, or cosets of troughs that downcut into underlying foresets, rarely delineate El surfaces. El surfaces truncate E2 surfaces that crosscut foresets and possess a concave-up or sigmoidal geometry (Figs. 8, 9A). E2 surfaces are analogous to reactivation surfaces.

Three types of internal stratification are developed between successive E2 surfaces. Type 1 stratification consists entirely of uniformly dipping ripple stratification. Dips are greater than that of the truncating E2 surface. Rarely, ripple stratification increases then decreases in dip between two E2 surfaces. Type 2 stratification consists of cyclic alternation of ripple stratification and grain flows (Fig. 6B). Low- or high-angle, curved E2 surfaces are mantled by ripple stratification that progressively increases in dip away from the surface. Grain flows are developed at maximum foreset

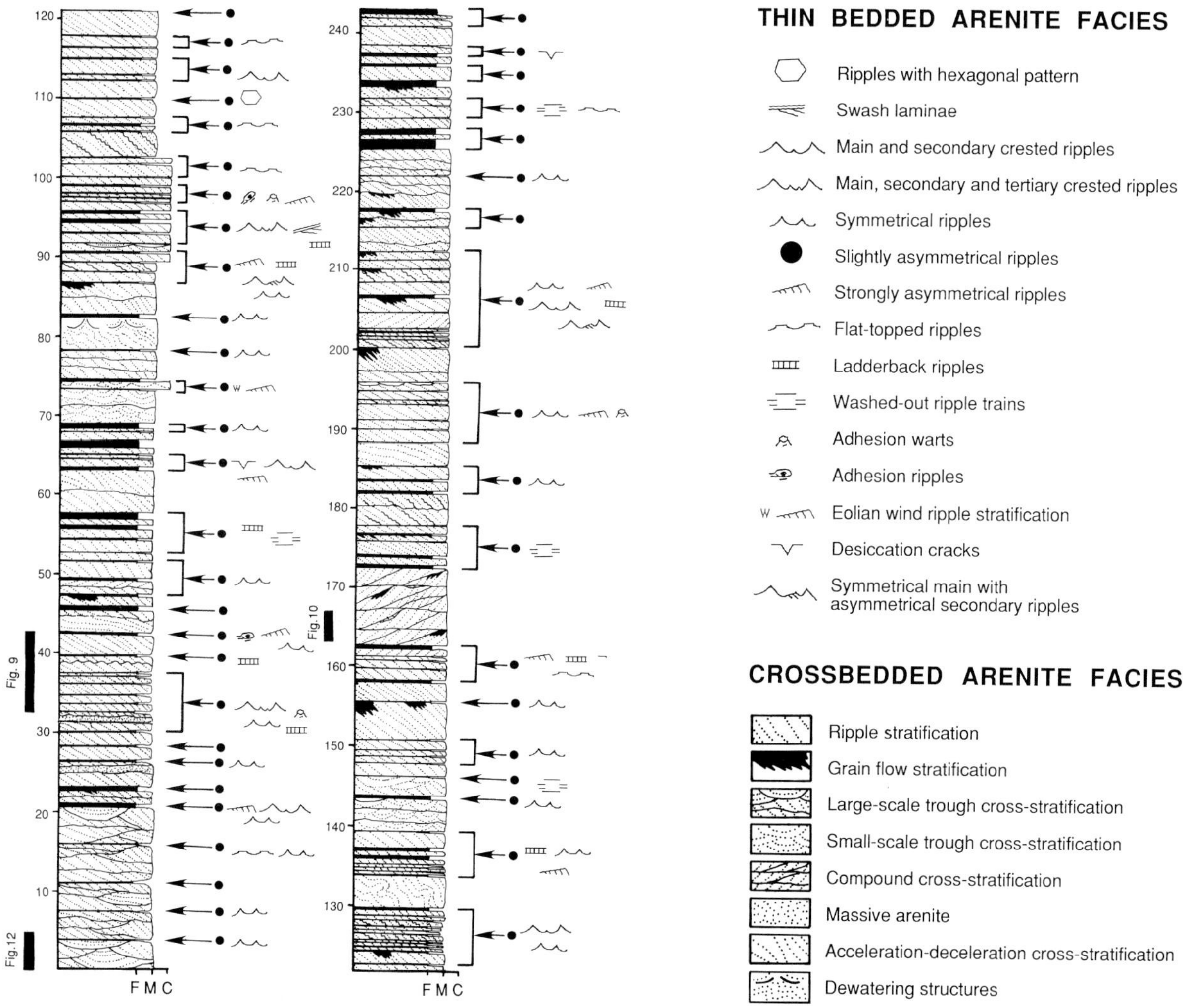

Figure 4. **(Right)** Key to stratigraphic column. **(Left)** Detailed stratigraphic column through upper 250 m of upper Mount Guide showing crossbedded arenite facies and thin bedded arenite facies. Vertical bars adjacent to the columns show locations of Figures 9, 10 and 12. F, fine grained arenite; M, medium grained arenite; C, coarse grained arenite.

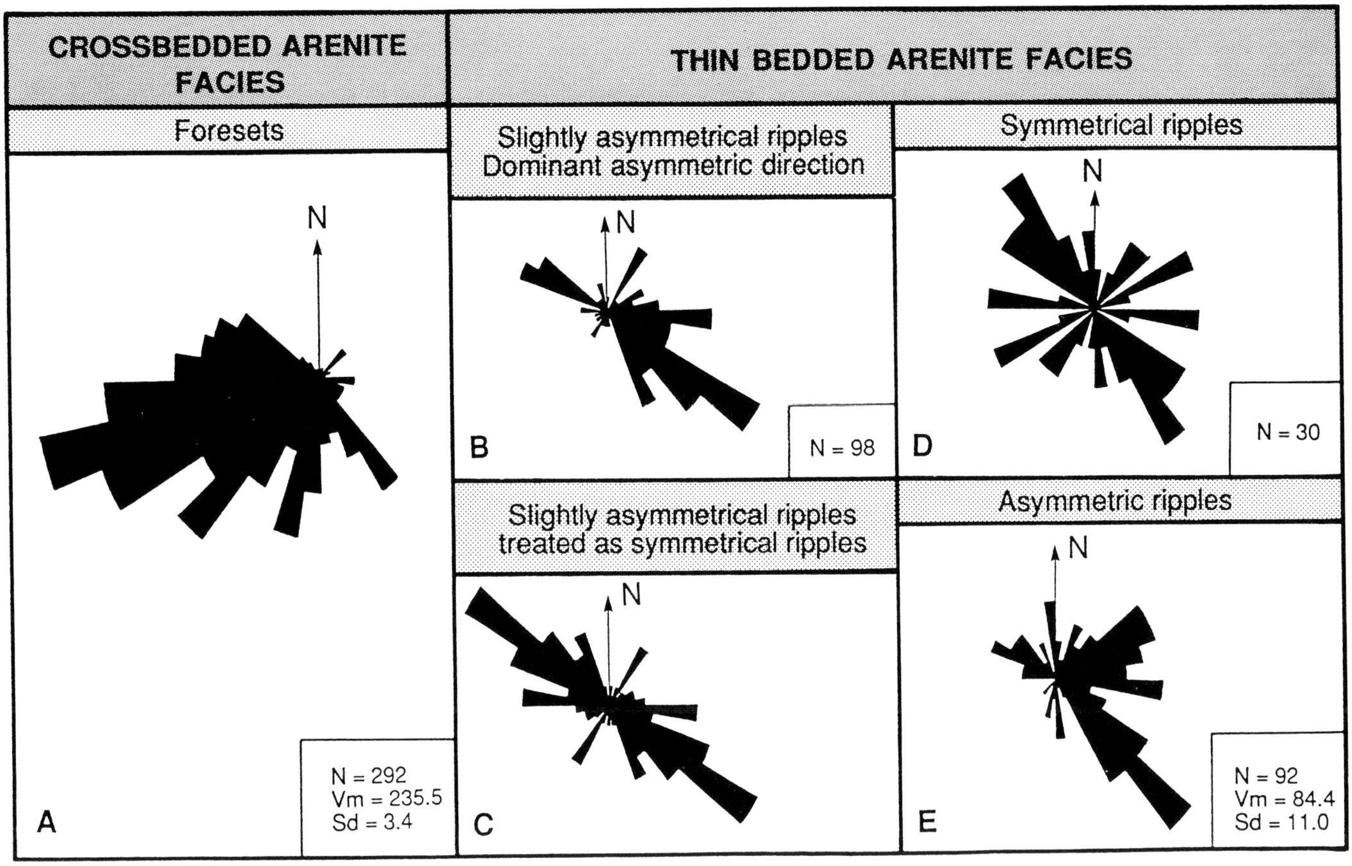

Figure 5. Paleocurrent diagram for crossbedded arenite facies and thin bedded arenite facies. N, number of observations; Vm, vector mean; Sd, standard deviation. Oscillation directions plotted for symmetrical ripples.

inclinations. Ripple stratification replaces grain flows as foreset inclinations decrease toward the next truncating E2 surface. In good exposures, E2 surfaces display a change in inclination in a downcurrent direction, and foreset bundle thicknesses are greater between low-angle than between high-angle E2 surfaces (Fig. 8A). Long toesets are associated with low-angle E2 surfaces. Type 3 stratification is developed between sigmoidal-shaped E2 surfaces and consists exclusively of ripple stratification (Fig. 8B). Away from the lower E2 surface, planar stratification displays a progressive increase in angle of dip. Topsets are preserved associated with maximum inclinations; these cap underlying foresets and are continuous into foreset stratification. The resulting foresets have a sigmoidal shape. The inflection point persists in a downcurrent direction but is accompanied by a reduction in angle between topsets and foresets related to a progressive decrease in foreset inclination. The overlying E2 surface is either parallel to or slightly truncates the foresets (Fig. 8B).

Some foreset bundles are bounded by rippled E2 surfaces. Asymmetrical ripples or small troughs are superimposed obliquely up and rarely down the E2 surface. Form-discordant crosslaminae are preserved beneath the asymmetrical ripples (Fig. 9B). Toesets contain asymmetrical, slightly asymmetrical and symmetrical ripples; slightly asymmetrical ripples have lee-sides inclined up the foreset.

Trough crossbeds are medium scale, contain E1 and E2 surfaces, and consist of grain flows and ripple stratification. Bases of troughs are defined by coarse sand lags. Parallel to the flow direction, some troughs display an increase followed by a decrease in set thickness and a corresponding decrease followed by an increase in inclination of E2 surfaces (see Fig. 8A). Grain flows are associated with greater scour depths.

Overturned and contorted foresets and dish and pillar structures are rarely present. One tabular crossbed coset contains a bed of massive arenite capped by pillar structures. Flutes are present on the base of the massive bed, which has an erosional relief of 10 to 30 cm.

Interpretation

Crossbed geometries and sequences of internal structures present within CB1 can be interpreted by comparison with documented, tide-generated, subaqueous dunefield deposits from the Holocene (Boersma and Terwindt, 1981; Terwindt, 1981; Nio *et al.*, 1983; Berné *et al.*, 1988) and the ancient (Allen and Homewood, 1984; Kreisa and Moiola, 1986). Foreset geometry and the tabular nature of most sets indicates that the majority of subaqueous dunes were straight-crested bedforms; subordinate, trough-shaped sets indicate the presence of some sinuous-crested dunes (Harms *et al.*, 1982; Rubin, 1987).

E1 surfaces within CB1 may have developed as a result

Figure 6. Photographs of ripple stratification and grain flows. (**Top**) Cross-sectional view of grain flows. (**Bottom**) Bedding plane view of alternating ripple (thin) and grain flow stratification (thick).

of bedform climb (Rubin and Hunter, 1982), and/or degradation (Smith, 1988). E2 surfaces are pause planes or reactivation surfaces that in tidal regimes are related to unsteady flow (Boersma and Terwindt, 1981), reworking of bedforms by reversed, subordinate-strength currents (Klein, 1970), or migration of superimposed bedforms (McCabe and Jones, 1977). Sigmoidal-shaped E2 surfaces indicate complete bedform preservation, whereas curved, concave-up E2 surfaces imply partial bedform preservation.

Type 1 internal stratification consisting of uniformly dipping ripple stratification between E2 surfaces indicates constant current velocities. Ripple stratification that displays an increase followed by a decrease in dip records accelerating followed by decelerating flow; the change in dip reflects adjustment of the lee face to changing equilibrium conditions. In both of the above cases, flow velocities did not attain sufficient strength to steepen the lee face and generate a slipface. Under these hydrodynamic conditions, flow expansion and bedload transport, rather than flow separation and suspended load transport, dominated on the dune lee face; current ripples migrated down or obliquely across the lee face of the dune (*cf.* Boersma and Terwindt, 1981).

Figure 7. Photograph of tabular crossbed set with megaripple top (subfacies CB1). Note lateral change in foreset inclination.

Type 2 internal stratification consisting of cyclic alternation of ripple stratification and grain flows reflects accelerating followed by decelerating tidal flow. The increase in dip of ripple stratification away from the lower E2 surface records increasing flow velocities that resulted in steepening of the dune lee face. The presence of grain flows implies that flow velocities accelerated sufficiently to generate a steep, unstable slipface. Abundant grain flow stratification reflects persistent, strong flow velocities, bedload transport and full vortex development. Ripple stratification downdip of grainflows indicates development of backflow ripples in the zone of flow separation, or migration of small ripples in response to minor deviations in the main current direction (*cf.* Boersma and Terwindt, 1981). Subsequent replacement of grainflows by ripple stratification in the downcurrent direction records decelerating flow velocities that resulted in progressive reduction in lee face steepness. Later modification of the dune lee face by a subordinate current generated the bounding E2 surface. Low-angle E2 surfaces imply that the subordinate tidal current was strong enough to significantly modify the bedform shape.

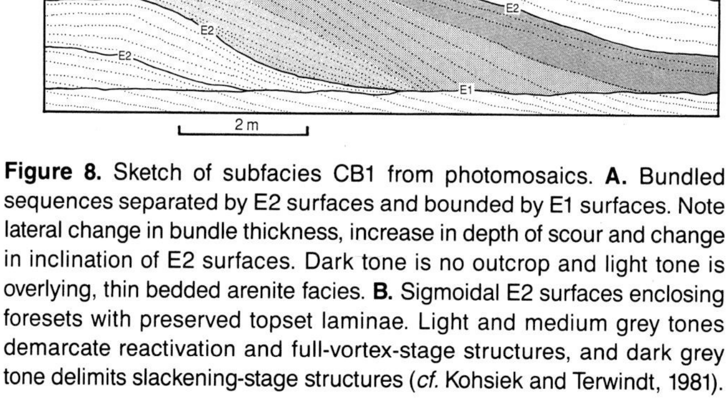

Figure 8. Sketch of subfacies CB1 from photomosaics. **A.** Bundled sequences separated by E2 surfaces and bounded by E1 surfaces. Note lateral change in bundle thickness, increase in depth of scour and change in inclination of E2 surfaces. Dark tone is no outcrop and light tone is overlying, thin bedded arenite facies. **B.** Sigmoidal E2 surfaces enclosing foresets with preserved topset laminae. Light and medium grey tones demarcate reactivation and full-vortex-stage structures, and dark grey tone delimits slackening-stage structures (*cf.* Kohsiek and Terwindt, 1981).

Tidal bundle thickness changes in Holocene and ancient tidal deposits have been used to identify period variations that are attributable to neap-spring tidal cycles (*e.g.*, Visser, 1980; Nio *et al.*, 1983; Allen and Homewood, 1984; Kreisa and Moiola, 1986). Variations in thickness of foreset bundles between E2 surfaces, variations in dip of E2 surfaces and changes in set thickness of CB1 (*e.g.*, Fig. 8A) suggests neap-spring cyclicity. Thicker foreset bundles bounded by low-angle pause planes record greater dominant as well as subordinate tidal current velocities that are best attributed to spring tides. In contrast, thinner foreset bundles bounded by high-angle E2 surfaces would be related to neap tides. Limited outcrop prevents rigorous data collection and statistical testing. The number of foreset bundles illustrated in Figure 8A is less than that characteristic of a lunar month. This discrepancy may reflect the lack of sand transport during neap tides, when velocities of the dominant tide were less than the threshold of sand transport, resulting in an incomplete record of the lunar cycle.

Type 3 internal stratification with topset preservation (Fig. 8B) also is thought to result from an accelerating-decelerating flow cycle. The increase in dip of planar stratification away from the lower bounding surface represents the reactivation stage of the accelerating flow cycle (*cf.* Kohsiek and Terwindt, 1981). Brinkpoint development (greatest inflection point angle) records greatest velocities and demarcates the full vortex stage of flow (*cf.* Boersma and Terwindt, 1981; Terwindt, 1981). As flow velocities decreased, topset and lee-face deposition took place from a concentrated suspension produced during the peak flow stage (Boersma and Terwindt, 1981). The reduction in inflection point angle and decrease in dip of foresets (Fig. 8B) records continued deceleration.

Tabular and tangential foresets bounded by rippled E2 surfaces record similar flow cyclicity to that discussed above. Preserved ripples oriented obliquely down an E2 surface record the initial stage of bedload transport by the dominant current, whereas ripples and troughs oriented obliquely up the surface probably reflect subordinate current remolding of the lee face of the bedform. Oscillatory (symmetrical) and combined-flow (slightly asymmetrical) ripples on E2 surfaces (Fig. 9B), and in toesets, may be related to either ebb lowstand or flood highstand stages; at such times, oscillatory and combined flow currents were not overwhelmed by unidirectional tidal currents.

Within trough crossbeds, the increase in scour depth of the set boundary, decrease in dip of E2 surfaces and development of grain flows reflects an increase in flow velocities to full vortex stage associated with spring tides. Reduction in depth of scour is a consequence of lesser neap tide flow velocities (*cf.* Terwindt, 1981).

The massive bed with pillar structures is interpreted as a mass-flow deposit. Eschner and Kocurek (1986) reported interbedded mass-flow and shallow-marine deposits; mass flows were generated during transgression. The presence of a tabular crossbed set below the mass-flow deposit implies

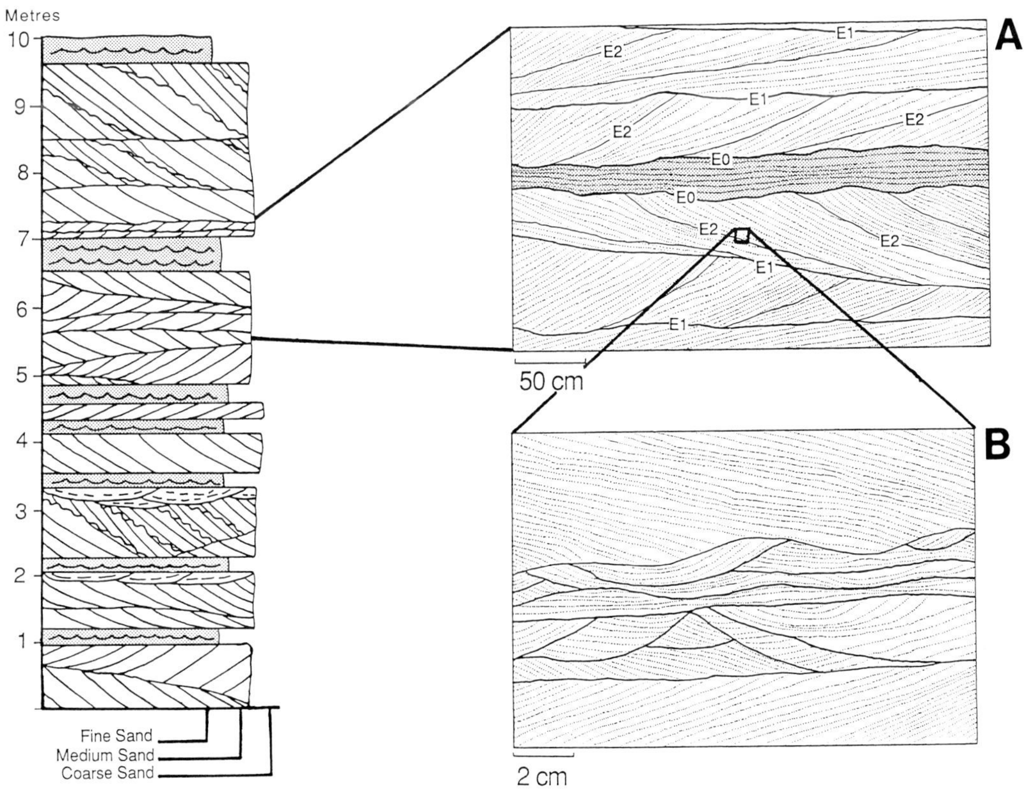

Figure 9. Detailed stratigraphic section of CB1 subfacies (unshaded) and thin bedded arenite facies (shaded). **A.** Enlargement of stratification style showing hierarchy of surfaces. **B.** Enlargement showing form-discordant ripple crosslaminae on E2 surface.

that the subaqueous dunefield was established before the mass flow was generated. Subsequent reworking by tidal processes molded the top of the mass flow into a tabular crossbed.

Compound Crossbed Cosets (CB2)

Description

CB2 consists of compound crossbed cosets that internally possess a hierarchy of bounding surfaces. These are ranked as E1, E2 and E3 based on crosscutting relationships (Fig. 10; *cf.* Allen, 1980). CB2 is 9.9 m thick. Paleocurrents from foresets display dominant southwesterly and subordinate northeasterly modes (Fig. 10).

E1 surfaces truncate lower-order surfaces and have little or no discernable angular discordance with respect to regional bedding. Rarely, symmetrical and slightly asymmetrical ripples are present on E1 surfaces. E2 surfaces originate at the upper E1 surface and dip downward to intersect the lower E1 surface. E2 surfaces display a change in inclination in the bedform migration direction. At the upper E1 surface, inclinations average 2° to 4° and steepen to as much as 22° before flattening out to ~3° at the lower E1 surface (Fig. 10). Rarely, trough sets are developed in erosional scours on the E2 surfaces (Fig. 10); trough sets are restricted to E2 surfaces with relatively low dips. Paleocurrent data from foresets within trough sets are opposed to paleocurrent trends above and below. E3 surfaces are truncated by and are steeper than E2 surfaces. E3 surfaces bounded by E2 surfaces with low dips (2° to 4°) have relatively shallow inclinations (5° to 7°; Fig. 10, top of panel A-B), whereas E3 surfaces bounded by steep E2 surfaces (22°) have higher dips (23°-25°; Fig. 10, center of panel B-C). Foresets within crossbeds bounded by E2 and E3 surfaces have higher dips than either E1,

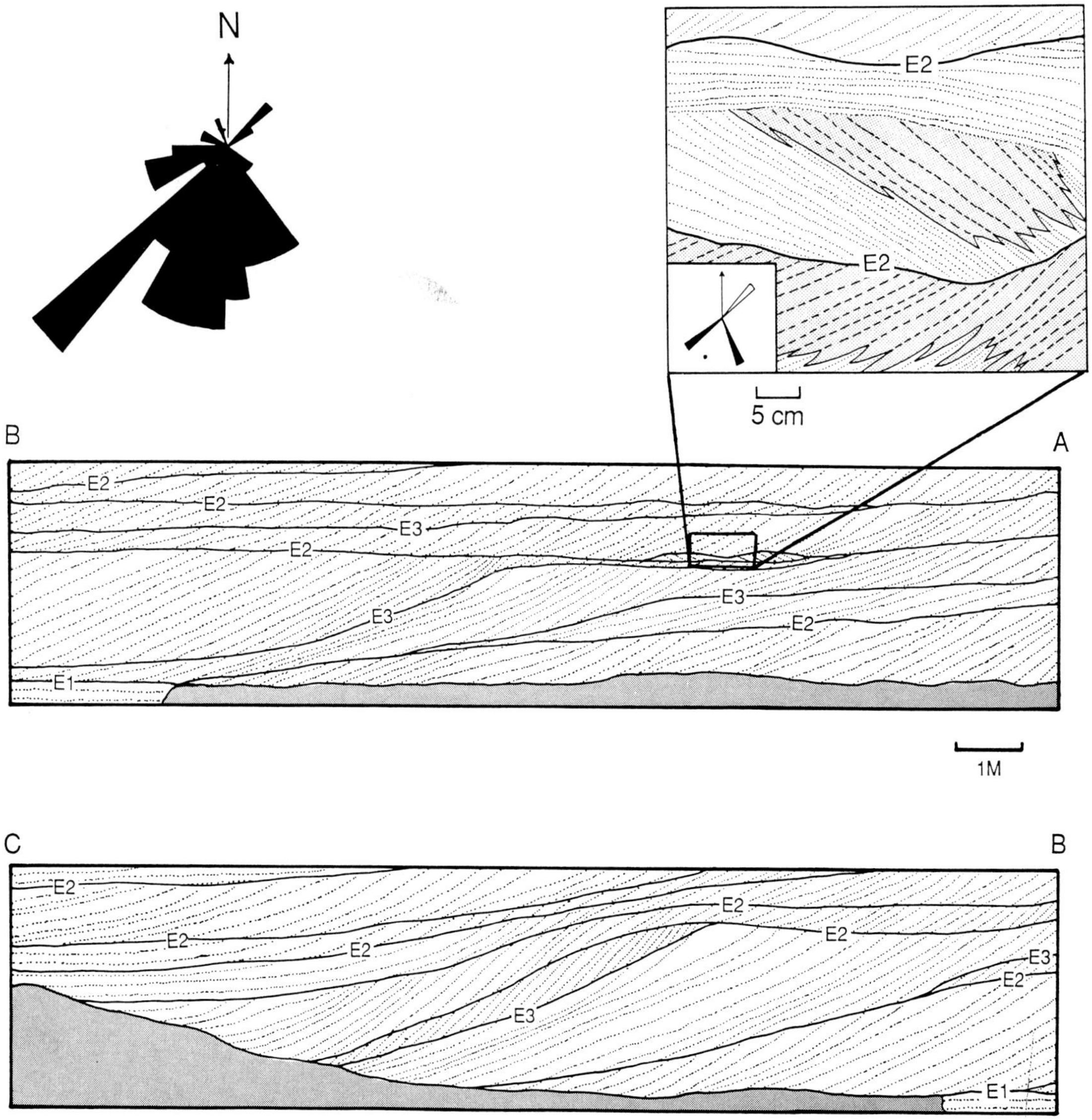

Figure 10. Detailed sketches from photomosaics of the compound crossbed cosets (subfacies CB2). Note change in dip of E2 surfaces. Shaded areas are no outcrop. Paleocurrent rose is for foresets from CB2 subfacies; these data are included in Figure 5. Enlargement is of trough scoured into E2. Foresets above and below trough are south-directed (black paleocurrent rose), whereas trough foresets are northeast-directed (unshaded paleocurrent rose). Grain flow deposits are shaded grey. Ripple stratification is unshaded.

E2 or E3 surfaces. Internal structures within these crossbeds as well as within the trough sets consist of ripple stratification and grain flows arranged in acceleration-deceleration flow cycles similar to those developed in CB1.

Interpretation

Compound crossbed cosets reflect migration of a hierarchy of tidal bedforms comparable to those from the North Sea (Houbolt, 1968), the Bay of Fundy (Dalrymple, 1984) and the Normandy coast (Berné *et al.*, 1988). E1 surfaces in CB2 are a product of the low angle of climb of the large dunes (*cf.* Rubin and Hunter, 1982). Lee side scour generated by flow separation or flow reattachment over small dunes produced E2 surfaces in response to migration of small dunes straight or obliquely down the lee sides of large dunes. Superimposed bedforms produce a geometry in which the lower-order surface (E2) originates at the upper higher-order surface (E1) and falls to meet the basal higher-order surface (Rubin and Hunter, 1982); climbing bedforms produce the converse. The large dunes were asymmetrical with a lee-side geometry that is approximated by the E2 surface. The upper portions of the lee sides of the large dunes were relatively shallow (low E2 dip); the large dunes steepened in a downcurrent direction before flattening out again. Lateral variation in E2 surface geometry may reflect the response of the large dune to different sediment transport rates and current strengths. During low transport rates of sand, the large dune morphology was subdued (low E2 dip) whereas high transport rates resulted in a better developed lee face (higher E2 dip) to the large dunes. Currents never attained sufficient strength to generate a slipface. The presence of trough crossbeds on the higher angle E2 surfaces indicates that the subordinate, opposed current was of sufficient strength to modify the lee face of the large dunes and produce sinuous-crested, small dunes. Symmetrical and slightly asymmetrical ripples (wave and combined-flow ripples) on the E1 surfaces were most likely generated during tidal stillstands.

E3 surfaces reflect reshaping of the lee side of the small dunes by subordinate currents (*cf.* Klein, 1971). Different inclinations of E3 surfaces are a consequence of the relative positions of the small dunes on the lee faces of large dunes during initiation of the opposed flow. If the small dune ceased migration near the crest of the large dune, flow compression took place as the opposed flow passed from the trough to the crestal area of the large dune. The resulting increase in current velocity caused extensive erosion of the lee side of the small dune and production of a low-angle pause plane. If the small dune ceased migration on the lee face of the large dune, no enhancement of flow velocity took place, resulting in a steep E3 surface.

The horizontal separation of successive E3 surfaces implies (Fig. 10) that small dunes migrated up to this distance between subordinate currents. The distance between E3 surfaces in CB2 is much greater than observed in CB1 or CB3. Holocene subaqueous dune complexes have been observed to migrate ~4 cm/day on the Southern Bight of the North Sea (McCave, 1971), 5 to 74 cm/day on Warts Bank (Jones *et al.*, 1965), and ~11 cm/tidal cycle in the Bay of Fundy (Dalrymple, 1984). Migration rates of 3 to 30 cm/tidal cycle (Allen and Homewood, 1984) and 15 to 90 cm/tidal cycle (Kreisa and Moiola, 1986) are inferred from bundles of foresets in the rock record. By employing echo-sounding techniques, Berné *et al.* (1988) have identified subhorizontal reflectors near the tops of subtidal dunes of the Cherbourg Peninsula in the English Channel and have attributed these features to destruction of the dunes by storm-induced currents combined with tidal and wave currents. Later studies of these and other bedforms after storms (Langhorne, 1982; Berné, 1989) showed the destruction of intermediate and small-scale bedforms on crests. During subsequent tidal cycles, bedforms recovered rapidly. The large observed migration distance in CB2 implies that the subordinate tidal currents producing the E3 surfaces were augmented periodically by storm-induced currents. The large distance between E3 surfaces thus may not reflect the migration distance of a bedform during a single tidal cycle but some longer time interval, possibly between major storms.

Large-Scale Trough Crossbed Cosets (CB3)

Description

CB3 consists of cosets of large-scale trough crossbeds (Figs. 11, 12). Coset thicknesses vary from 3.5 to 5.1 m. A hierarchy of bounding surfaces is present. E2 surfaces correspond to trough set boundaries and have maximum dips of 15° (Figs. 11, 12). E2 surfaces are mutually truncating (Figs. 11, 12) and are exposed as complete troughs (Fig. 11). E3 surfaces are mainly form-discordant and, rarely, form-concordant to E2 surfaces. E3 surfaces have maximum dips of 20°. E3 surfaces enclose packages of ripple stratification; rare grain flows are developed in conjunction with high-angle E3 surfaces. Ripple stratification varies from 1 to 3 cm in thickness and contains abundant internal crosslaminae that

Figure 11. Photograph of large-scale trough crossbed cosets (subfacies CB3) showing complete trough preservation with form-concordant E3 surfaces. Person is 1.80 m tall.

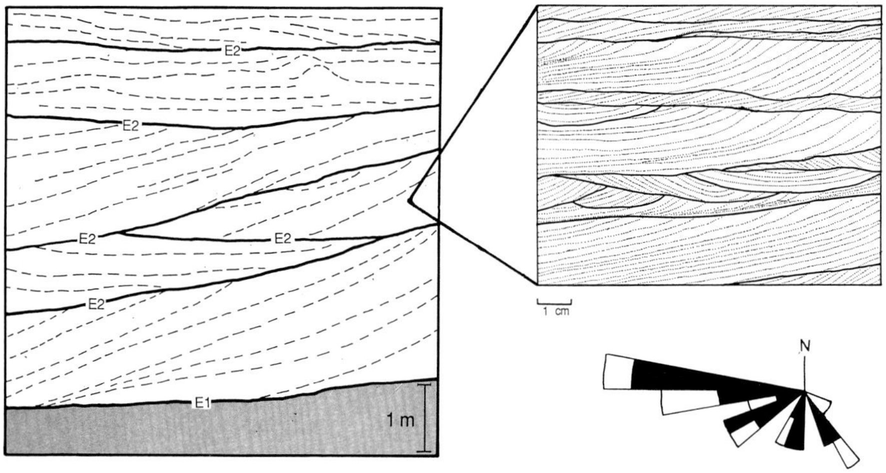

Figure 12. Detailed sketches from photomosiacs of subfacies CB3. Large sketch shows E2 surfaces (dark lines) and E3 surfaces (dashed lines). Underlying thin bedded arenite facies shaded grey. Enlargement illustrates internal structures consisting of form-concordant and form discordant crosslaminae. Paleocurrent rose: E2 and E3 surfaces (black); ripple crosslaminae (unshaded). Note apparent bimodal paleocurrent pattern.

are either form-concordant or form-discordant (Fig. 12, inset). Paleocurrent data from E2 and E3 surfaces and crosslaminae display an apparent bimodality in two-dimensional exposures (Fig. 12) with a mean flow direction to the southwest.

Interpretation

Sedimentary structures within CB3 were generated by migration of three-dimensional large dunes and ripples. Time-velocity asymmetrical flow resulted in sand transport dominantly to the southwest. The geometry of E2 surfaces closely resembles the bounding surface arrangement produced by the migration of sinuous-crested, out-of-phase dunes (*cf.* Rubin, 1987, Fig. 73). Reversals of current direction reshaped the lee sides of the large-scale dunes and sculpted the E3 surfaces. Geometry of crosslaminae reflects migration of sinuous-crested, out-of-phase ripples with a random planform geometry (form-discordant laminae) and sinuous-crested, in-phase ripples (concordant laminae; Rubin, 1987). Sedimentary structures produced by sinuous-crested, out-of-phase bedforms can mimic herringbone cross-stratification in two-dimensional exposures. The close parallelism of the E3 surfaces with the dip of the crosslaminated sets indicates that no intermediate order of bedform was present. Structures produced by an intermediate-scale bedform would truncate the crosslaminated sets. Grain flows preserved between high-angle E3 surfaces indicate periodic slipface development on the large dunes with accompanying flow separation.

Summary: Crossbedded Arenite Facies

Bounding surfaces and internal structures within the crossbedded arenite facies are analogous to tide-generated structures from the Holocene and Ancient record. Lack of emergence features is consistent with this facies having been deposited in a subtidal setting. The majority of dunes on the upper Mount Guide shelf resulted from a strong time-velocity asymmetric tidal flow reflected in the dominant southwesterly paleocurrent mode. Time-velocity asymmetrical flow, resulting in sand transport in only one direction, is common in many subtidal settings. Crossbeds generated within a regime of this type mimic unidirectional flows and commonly lack characteristic tidal bundles. Within the crossbedded arenite facies, weak subordinate tidal flows are recorded by erosive E3 surfaces, rare asymmetrical ripples oriented up E3 surfaces, and rare opposed foreset directions.

Dunes that produce structures akin to CB1, CB2 and CB3 are aligned parallel (Off, 1963) or slightly oblique (Belderson *et al.*, 1982) to local tidal flow. Dunes are less likely to equilibrate to short temporal changes in flow regime than current ripples. Paleocurrent azimuths from crossbeds (Fig. 5) thus are accurate indicators of local flow directions.

Thin-Bedded Arenite Facies

Description

The thin bedded arenite facies is composed of tabular beds dominated by medium grained arenite. A variety of sedimentary structures is developed including symmetrical, slightly asymmetrical, strongly asymmetrical and modified ripples (Table 1, Fig. 13A, B, C). and emergence structures (Table 2, Figs. 13D, 14). Ripples are the most common structures with emergence structures subordinate. Vertically within this facies, emergence structures and ripple types are interbedded randomly. Paleocurrents from symmetrical and slightly asymmetrical ripples exhibit northwest-southeast modes (Fig. 5C, D). Asymmetrical ripples have a bimodal-bipolar pattern with a dominant southeasterly mode (Fig. 5E).

Interpretation

Sedimentary structures in this facies developed in shallow-subtidal and tidal-flat environments. Wave, combined-flow and current ripples are of shallow-subtidal origin (Table 1).

Table 1. Characteristics and interpretation of ripple types

Ripple type	Description	λ (cm)	Internal structures	Interpretation
Symmetrical A	Straight-crested with trochoidal profile and rare crest bifurcation SI: 4.7-.50‡ RI: 6.5-7.2‡	1-8	1. Interwoven crosslaminae, bidirectional lenses and form-discordant drapes. 2. Climb inphase or with net translation. 3. Type A evolves into type B vertically	1. Oscillatory flow: vortex (A), rolling-grain or post vortex (B) (Clifton, 1976; DeRaaf et al., 1977; Harms et al., 1982). 2. Large sediment supply during gradual deepening with or without minor translatory component. 3. Change in maximum orbital speed.
B	Straight-crested and rounded crest Reduced crest heights. SI .48-.51* RI: 17-23*	2-10		
Slightly asymmetrical	Straight to slightly sinuous crests with some bifurcation	1-8	1. As above with preferred foreset orientation. 2. Climb inphase with translation in direction of asymmetry. 3. Translation from or to symmetrical forms	1. Combined flow; oscillatory with slight translatory component related to shoaling (Harms et al., 1982). 2. Same as above. 3. Changes in water depth
Strongly asymmetrical	Straight-crested to linguoid forms SI: 0.10-0.27* RI: 5.0-6.7*	4-15	1. Overlapping troughs with unidirectional crosslaminae. 2. Horizontal lamination traced into bedforms. 3. Horizontal lamination traced laterally into inclined lamination capped by bedforms. 4. Horizontal lamination overlain by nested troughs. 5. Decrease in amplitude upward. 6. Strongly asymmetrical overlain by symmetrical	1. Unidirectional flow: migration of out-of-phase ripples. 2. Low-angle bedform climb; preservation of lower portion of ripple (*cf.* Rubin and Hunter, 1982). 3. Increase in sand supply resulting in climb angle change (Ashley *et al.*, 1982; Rubin and Hunter, 1982). 4. Change in flow conditions during deposition from upper-to lower-flow regime (*cf.* Harms *et al.*, 1982). 5. Reduction in flow velocity (Harms *et al.*, 1982). 6. Reduction in wave period or increase in water depth resulting in change from translatory to oscillatory flow (Harms *et al.*, 1982)
Modified	Symmetrical flat-topped; straight to bifurcating crests	5-9		Shoaling (Tanner, 1960) possibly to intertidal conditions (Klein, 1977)
	Symmetrical main (M) with symmetrical secondary (S) in trough (Fig. 13)	M:2.2-4.3* S: 1.0-2.1		Shoaling(?) with progressive decrease in orbital velocity and/or wave period (Harms *et al.*, 1982)
	Asymmetrical main (M) with symmetrical secondary (S) and Tertiary (T)	M: 2.5-2.8‡ S: 1.2-1.4 T: 0.6-0.8		Same as above
	Asymmetrical or symmetrical main (M) with asymmetrical ladderbacks (L) in troughs and onto stoss at orthogonal to oblique angles (Fig. 13)	M: 3.0-6.0* L: 2.0-4.0		Runoff parallel to main ripple crest during late ebb stage (Klein, 1977)
	Symmetrical main (M) with superimposed symmetrical or slightly asymmetrical secondary (S)	M: 2.0-4.0* S: 1.0-2.0		Younger main ripple trend with superimposed combined-flow ripple that may be the result of sudden change in wave approach direction
	Symmetrical with equant hexagonal pattern	3.0-6.0*		Simultaneous formation of crests (Allen, 1982). Ripples correspond to two sets of waves or rip current disruption of wave action.
	Ripple trains with washed-out patches (Fig. 13)			Destruction of ripple trains by wind-driven small waves during reduction in water level and/or ebb runoff and emergence (Klein, 1977)
	Symmetrical ripples with wavelength of 1.5 cm (Fig. 14-2) overlying a tabular crossbed set (Fig. 14-1) and cut by small scour (Fig. 14-6). Proximal to the scour, symmetrical ripples are replaced by slightly asymmetrical ripples (Fig. 14-3). Edge of scour has parting lineations (Fig. 14-4) and center has strongly asymmetrical ripples (Fig. 14-5)			Symmetrical ripples (oscillatory flow) were generated by small waves during flood stillstand. During the initial stages of ebb flow, a unidirectional component was added to the oscillatory flow to produce slightly asymmetrical ripples (combined flow). Scour was cut during peak ebb flow. During late-stage ebb flow, remaining water drained into the channel; in shallow portions of the channel, upper flow-regime conditions existed with plane bed the equilibrium bedform; in the "deeper" channel, current ripples were in equilibrium with flow conditions. These types of drainage feature are common on modern tidal flats (Klein, 1977)

SI=Symmetry index (crest to trough length/wavelength). 0.5=symmetrical ripple.
RI=Ripple index (wave length/ripple height).
* Data from 2 bedding planes; 5 to 10 ripples from each bedding plane.
‡ Data from 1 bedding plane; 5 ripples.

Figure 13. Photographs of modified ripple types and emergence structures. (**Upper left**) Symmetrical main ripples with symmetrical secondary ripples in troughs. (**Upper right**) Asymmetrical main ripples with secondary ladderback ripples. (**Lower left**) Symmetrical flat-topped and washed-out ripples. (**Lower right**) Adhesion warts.

The bimodal-bipolar paleocurrent pattern for asymmetrical ripples (Fig. 5E) supports a tidal origin for these ripples. Symmetrical ripples with short wavelengths (~ 1 cm) record reworking by small waves with short periods in relatively shallow water with limited fetch (Komar, 1974; Boyd *et al.*, 1988). Modified ripples reflect shoaling, to possibly intertidal conditions (Table 1). Emergence features (Table 2) are indicative of supratidal settings. Interbedding of sedimentary structures generated in shallow-subtidal, intertidal and supratidal settings indicates fluctuating water levels best related to advance and retreat of tides. The dominant paleocurrent modes of the crossbedded arenite and thin bedded arenite facies are orthogonal (Fig. 5). If the asymmetrical side of the combined-flow ripples is onshore and the wave ripples are onshore-offshore, then the subaqueous dunes responsible for CB1, CB2 and CB3 were produced by shore-parallel tidal currents.

Parasequence Demarcation

The crossbedded and thin bedded arenite facies are associated in cyclic, tabular units that range in thickness from 0.4 to 10.2 m. Contacts between thin bedded and overlying crossbedded arenites are sharp and define surfaces of drowning or flooding. Subtidal dunes developed following drowning whereas the overlying, thin bedded arenites reflect shoaling to intertidal and supratidal conditions. Evidence

Table 2. Description and interpretation of emergence features

Description	Interpretation
Complete polygonal-shaped, elevated sandridges	Sand infilling of desiccated mudstone partings
Discontinuous, low-amplitude ripples with crest lengths 2 to 3 times wavelength	Adhesion ripples formed by unidirectional wind (Reineck, 1955; Kocurek and Fielder, 1982)
Isolated knobs, 1 to 3 mm in height with small protuberances (Fig. 13D)	Adhesion warts, variable wind direction (Reineck, 1955; Kocurek and Fielder, 1982)
Clusters of knob-like structures	Precursor adhesion ripples that evolve into adhesion warts as moisture content is reduced (Olsen *et al.*, 1989)
Inversely graded (fine to coarse sand) subcritically climbing translatent stratification. Rare internal crosslaminae. Stratification is horizontal to low angle	Low-angle eolian wind ripple stratification (Hunter, 1977)
Low-angle dipping foresets ($<5°$) defined by heavy mineral laminae. Low angle internal discordances	Swash laminae (Harms *et al.*, 1982)

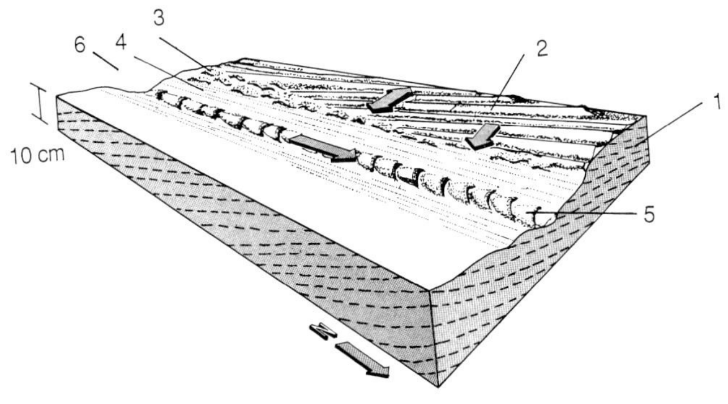

Figure 14. Exposure indicators in thin bedded arenite facies on bedding plane below subfacies CB3. 1. Southwest-directed foresets eroded by overlying bedforms. 2. Wave ripples; crests oriented approximately north-south. 3. Combined-flow ripples asymmetrical to the northeast. 4. Parting lineations oriented north-south. 5. North-directed current ripples. 6. Runoff channel oriented approximately north-south.

for eolian processes in this facies is restricted to adhesion structures and low-angle, wind-ripple stratification. The latter is characteristic of eolian sand sheets (Fryberger *et al.*, 1979). Eolian dunes require relatively long periods of time to be established whereas sand sheets require the least amount of time (Kocurek and Nielson, 1986), thus the absence of eolian dunes in the thin bedded arenite facies indicates that a short time interval is recorded in this facies.

The cyclic units in the upper Mount Guide Quartzite are comparable to parasequences. They are relatively thin in comparison to the 3 to 100 m-thick parasequences described by Van Wagoner *et al.* (1990) but in all other respects are compatible with the definition of a parasequence as "a conformable succession of genetically related strata bounded above and below by marine flooding or correlative surfaces" (Van Wagoner *et al.*, 1987, 1988, 1990). Parasequences within the upper Mount Guide Quartzite are all upward shallowing, a characteristic of almost all parasequences in the rock record (Van Wagoner *et al.*, 1990). The asymmetrical parasequences of the upper Mount Guide Quartzite are comparable to asymmetrical shoaling-upward, peritidal carbonate cycles; peritidal carbonate cycles similarly shoal from subtidal to intertidal and supratidal environments and have flooding surfaces at the base of each cycle (*e.g.*, Mazzullo and Friedman, 1975; Reinhardt and Hardie, 1976; James, 1977; Goodwin and Anderson, 1985; Grotzinger, 1986; Bova and Read, 1987; Koerschner and Read, 1989).

Three groups of parasequences are recognized in the upper Mount Guide Quartzite based on the type of crossbedded arenite subfacies (Fig. 15). Group 1 contains CB1 as the basal facies. Thicknesses of parasequences range from 0.4 to 5.7 m with a mean thickness of 1.72 m. Thin bedded arenite facies make up 2 to 70 per cent of these parasequences. CB2 makes up the basal facies of the group 2 parasequence, which is 10.2 m thick. Thin bedded arenite facies comprise 3 per cent of the parasequence. Group 3 parasequences contain CB3 as the basal facies and range in thickness from 3.8 to 5.2 m. Thin bedded arenite facies make up 2 to 10 per cent of these parasequences.

Parasequences in the upper Mount Guide Quartzite differ from those developed along muddy, tide-dominated coastlines that are characterized vertically by: 1) an upward thinning of beds, 2) an upward decrease in sandstone/mudstone ratio, 3) an upward reduction in grain size, and 4) nonmarine mudstone caps (Van Wagoner *et al.*, 1990). Parasequences in the upper Mount Guide Quartzite exhibit no apparent thinning-upward of beds within the intertidal facies and lack intertidal mudstones. The parasequences do show a reduction in grain size from the crossbedded arenite to the thin bedded arenite facies.

Controls on Parasequence Development

Various mechanisms have been proposed for the generation of parasequences, including: 1) episodic compaction and dewatering; 2) variations in sediment supply; 3) isostatic variations; 4) changes in tectonic subsidence rates; and 5) eustatic fluctuations (Spencer and Demicco, 1989). Compaction and dewatering mechanisms require less permeable units to be interbedded with highly permeable beds. Shales are

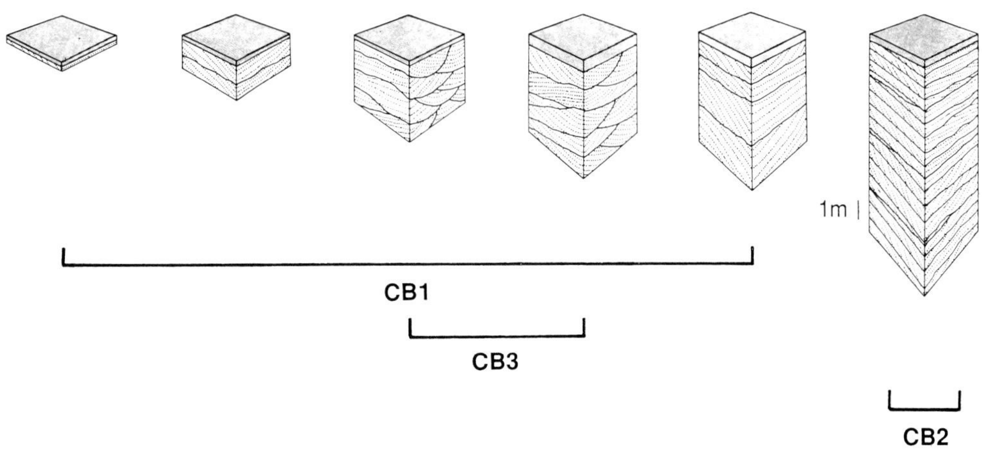

Figure 15. Parasequence thickness variations. CB1: tabular, crossbed set and coset facies. CB2: compound, crossbed coset facies. CB3: large-scale trough crossbed coset facies. Thin bedded arenite facies shaded grey.

absent from the upper Mount Guide Quartzite and in underlying stratigraphic units. Also, in well-sorted sands, compaction is minimal and takes place soon after deposition (Bond and Kominz, 1984). Dewatering structures are rare in the crossbedded arenite facies (Fig. 3). In siliciclastic sytems, cessation of sediment influx is more likely to be a consequence of rather than a precursor to flooding. Variations in siliciclastic sediment supply could be related to humid-arid climate cycles. On a land surface devoid of vegetation, more sediment would be supplied during humid than during arid periods. Flooding surfaces thus would reflect arid periods. This mechanism could possibly account for siliciclastic parasequences but does not explain the systematic variation in thickness of parasequences in the upper Mount Guide Quartzite. In addition, periodic drowning in response to isostatic adjustments or changes in rates of tectonic subsidence has not been recognized as a viable mechanism for short duration parasequences in thermally subsiding basins (Cloetingh, 1988).

Parasequence thickness is a function of accommodation space created by a combination of sea-level change and subsidence (Jervey, 1988). The influence of these two parameters on the development of parasequences of variable thickness can be investigated by use of a Fischer plot. Fischer plots are graphs of cumulative parasequence (or cycle) thickness (corrected for linear subsidence) versus time (Fischer, 1964); Fischer plots show cumulative deviations in thickness from that expected for linear thermal subsidence. The resulting graph is interpreted as being the track of relative sea level (Goldhammer *et al.*, 1987; Read and Goldhammer, 1988). If parasequence thickness is equal to the average subsidence per parasequence period, then no change in relative sea level has occurred (Fig. 16A). Where parasequence thickness is less or greater than the average subsidence per parasequence period, then the Fischer Plot shows a fall or rise in relative sea level (Fig. 16B, C). The thickness of the preserved parasequence is controlled by the rate of addition of accommodation space (Jervey, 1988). The thickest parasequence reflects the greatest increase in accommodation space for a given parasequence period, whereas the thinnest parasequence records the least increase in accommodation space. At the formation scale, deviations of stratigraphic thickness from that predicted by pure thermal subsidence mechanisms have been employed by Bond *et al.* (1988) to isolate 40 m.y. and 2 to 10 m.y. eustatic fluctuations within Cambro-Ordovician passive margins. The Fischer plots use a similar approach on a smaller scale.

The upper Mount Guide Quartzite represents a maximum time span of about 25 m.y. (see Eriksson and Simpson, 1990). A subsidence rate of 8 cm/ka and a constant period of 25 ka per parasequence were used to construct the Fischer plot (Fig. 16; Eriksson and Simpson, 1990); this subsidence rate restores the Fischer plot to the starting point at the end of the run. The underlying assumptions in the construction of Fischer plots in general, and Figure 16 in particular, are that subsidence is linear and that the same time is recorded by each parasequence. The upper Mount Guide Quartzite accumulated in a thermally subsiding basin (Eriksson *et al.*, 1990); for the time period represented by the Fischer plot, subsidence in such a basin can be approximated by a linear function. The second assumption is based on the premise that parasequence development is controlled by systematic rather than random events. Absolute constrains on the duration of parasequences and rates of subsidence are not critical to Fischer plots. Using different subsidence rates and parasequence periods results only in a change in peak heights, but the positions and relative heights of the peaks remain the same.

Estimated time durations of approximately 10^2 to 10^4 years compiled from parasequences in other siliciclastic stratigraphic units (Van Wagoner *et al.*, 1990) are comparable to the 25 ka estimate of the time duration for parasequences in the upper Mount Guide Quartzite. In basins with relatively low subsidence rates, parasequences are generated by high-frequency eustatic fluctuations (Van Wagoner *et al.*, 1990). A eustatic control on siliciclastic parasequence development in the upper Mount Guide Quartzite is implied by the Fischer plot (Fig. 16) and favored for the following reasons. Parasequences reflect abrupt drowning and establishment of a subaqueous dunefield followed by shoaling to produce the sharp contacts between the two facies. Large-scale trough and compound crossbed cosets are associated with rises in the Fischer plot (Fig. 16D, E); large-scale bedforms require greater water depths to develop than smaller-scale bedforms. Parasequences containing the greatest proportion (>20%) of the thin bedded facies are restricted to long-term falls on the Fischer plot curve; analogous, thick intertidal to supratidal caps on carbonate cycles are preferentially developed on long-duration sea-level falls (Read and Goldhammer, 1988; Read, 1989). Coarse grained faces in the upper Mount Guide Quartzite are confined to the long-term sea-level fall (Fig. 16). Read and Goldhammer (1988) and Read (1989) report that siliciclastics in carbonate cycle caps are associated with long-term sea-level falls.

Sequences and their constituent systems tracts in that portion of the upper Mount Guide Quartzite, represented by Figure 16, may be identified. Sequences consist of lowstand, shelf-margin, transgressive and highstand systems traces (Posamentier *et al.*, 1988; Van Wagoner *et al.*, 1988, 1990). Lowstand systems tracts are confined to type-1 sequences whereas shelf-margin systems tracts are developed in type-2 sequences. Transgressive and highstand systems tracts are separated by maximum flooding surfaces.

In the absence of unconformities in the upper Mount Guide Quartzite, sequence boundaries are difficult to recognize. Instead, maximum flooding surfaces are identified to coincide with the thickest parasequences. Transgressive systems tracts are located below the maximum flooding surfaces and record times of most rapid accommodation space creation in the basin. They are represented by retrogradational parasequence sets (Van Wagoner *et al.*, 1988, 1990). In the upper Mount Guide Quartzite, transgressive systems

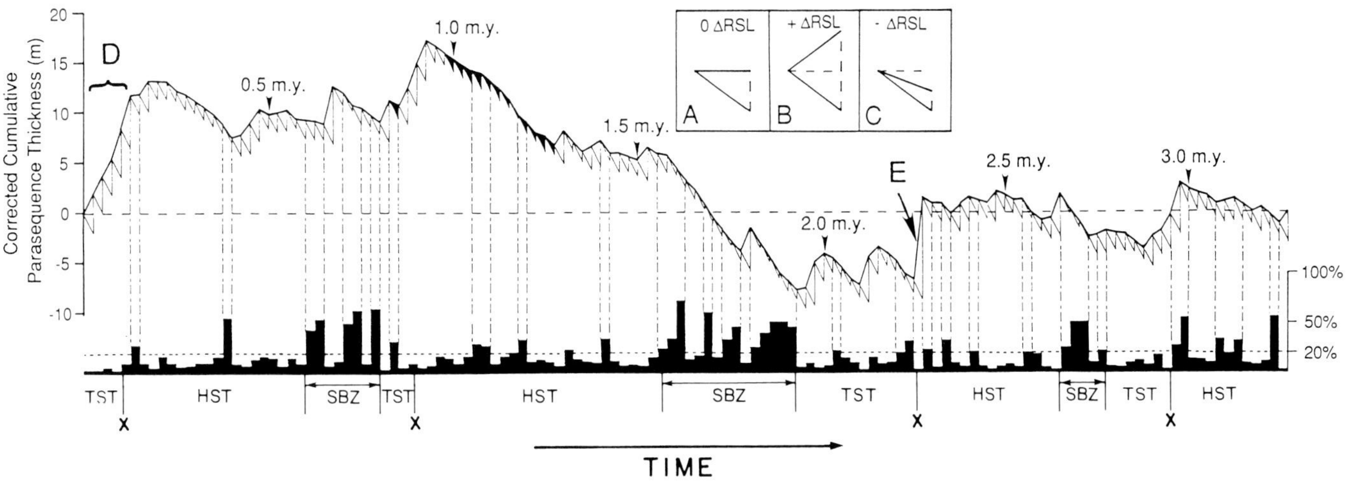

Figure 16. Fischer plot of cumulative parasequence thickness (corrected for linear subsidence) versus time for measured section through the upper 250 m of the upper Mount Guide Quartzite. Curve is a track of relative sea level. See Figure 3 for section locality. Inset **A.** Average subsidence per parasequence period equals parasequence thickness. Inset **B.** Average subsidence per parasequence period is less than parasequence thickness. Inset **C.** Average subsidence per parasequence period is greater than parasequence thickness. **D.** Position of facies CB3 in the lower part of parasequences. **E.** Location of compound crossbed cosets in the lower parts of parasequences. Bars at base of the graph show the percentage of thin bedded arenite facies in each parasequence. Parasequences composed of greater than 20 per cent thin bedded arenite facies are projected onto the Fischer plot. Solid triangles on Fischer plot indicate parasequences composed of coarse grained arenite. Note that most parasequences with large proportions of thin bedded arenite facies as well as coarse grained arenite are associated with falls on the Fischer plot. Inferred systems tracts and sequences are shown below the Fischer plot. X, maximum flooding surfaces; TST, transgressive systems tracts; HST, highstand systems tracts; SBZ, sequence boundary zones.

traces are characterized by a dominance of the subtidal, crossbedded arenite facies over the intertidal to supratidal, thin-bedded arenite facies, and by generally thickening-upward parasequence sets (Fig. 16). Highstand systems tracts reflect a decrease in rate of creation of accommodation space, and consist of progradational parasequence sets (Van Wagoner *et al.*, 1988, 1990). On Figure 16, a well-developed highstand sytems tract between 1.0 and 1.5 m.y. is characterized by an increase in the proportion of thin bedded arenite facies relative to the transgressive systems tract, and by the coarsest sediments, and is typified by a general thinning-upward of parasequences. Sequence boundaries overlie highstand systems tracts and reflect minimum space creation in the basin. The boundaries cannot be identified accurately but will be located somewhere within the zone characterized by the greatest proportion of thin bedded arenite facies (Fig. 16). The zones identified as containing the type-2 sequence boundaries will also correspond to shelf-margin systems tracts. Parasequence thickness trends are not smooth, probably due to a hierarchy of superimposed sea levels of different magnitude. Eriksson and Simpson (1990) have argued for three orders of sea level change during deposition of the upper Mount Guide Quartzite.

Conclusions

Three different types of crossbedded arenite facies are present in the upper Mount Guide Quartzite: 1) tabular, crossbed sets and cosets (CB1) produced by the migration of straight-crested, subaqueous dunes; 2) compound, crossbedded cosets (CB2) generated by the migration of subaqueous, medium-scale dunes superimposed on slower moving, larger scale dunes; and 3) large-scale trough crossbed cosets (CB3) produced by large, sinuous-crested dunes. Internal structures within the crossbedded arenite facies are analogous to modern and ancient tidal subaqueous dune deposits. Thin bedded arenites record intertidal to supratidal conditions. The crossbedded arenite overlain by thin bedded arenite facies occur in shoaling-upward parasequences bounded by flooding surfaces. These parasequences differ from typical tide-dominated shoreline parasequences in that muddy, tidal-flat facies are absent. A Fischer plot of parasequence thickness, corrected for linear thermal subsidence, versus time produces a curve that is interpreted as a track of relative sea level. Various systems tracts can be identified on the curve; these are associated in sequences.

Acknowledgments

This study was supported jointly by the National Science Foundation through Grant EAR-87-07357 to Eriksson, and the Australian Bureau of Mineral Resources Geology and Geophysics. We thank C. Harris, J. Jackson, F. Read and F. Wehr for their many discussions and W. Simpson for assisting in the field. P. Homewood and G. Reinson provided valuable comments. M. Sentelle is thanked for word processing.

References

Allen, J. R. L. 1980. Sandwaves: A model of origin and internal structure. Sedimentary Geology, v. 26, p. 281-328.

______ 1982. Sedimentary structures: their character and physical basis. Developments in Sedimentology, v. 1, Elsevier, Amsterdam, 593 p.

Allen, P. A. and Homewood, P. 1984. Evolution and mechanics of a Miocene tidal sandwave. Sedimentary Geology, v. 31, p. 63-81.

Ashley, G. M., Southard, J. B. and Boothroyd, J. C. 1982. Deposition of climbing-ripple beds: a flume simulation. Sedimentology, v. 29, p. 167-179.

______ 1987. Geology of the Mount Isa Inlier and environs, Queensland and Northern Territory. Bureau Mineral Resources Bulletin 225, 83 p.

Belderson, R. H., Johnson, M. A. and Kenyon, W. H. 1982. Bedforms. *In:* Offshore Tidal Sands, Processes and Deposits, A. H. Stride (ed.). Chapman and Hall, London, England, p. 22-57.

Berné, S. 1989. Giant subtidal dunes: a synthesis from the study of various tidal environments. Program and Abstracts from the Second International Research Symposium on Clastic Tidal Deposits, Calgary, Alberta, p. 17.

______, Auffret, J-P. and Walker, P. 1988. Internal structure of subtidal sandwaves revealed by high-resolution seismic reflection. Sedimentology, v. 35, p. 5-20.

Blake, D. H. 1985. Tectonic development of the Proterozoic Mount Isa Inlier; northwest Queensland. Abstracts of Conference on Tectonics and Geochemistry of Early to Middle Proterozoic Fold Belts. Australian Bureau of Mineral Resources, Record 1985/28.

Boersma, J. R. and Terwindt, J. H. J. 1981. Neap-spring sequences in intertidal shoal deposits in a mesotidal estuary. Sedimentology, v. 28, p. 151-170.

Bond, G. C. and Kominz, M. A. 1984. Construction of tectonic subsidence curves for the early Paleozoic miogeocline, southern Canadian Rocky Mountains: implications for subsidence mechanisms, age of breakup and crustal thinning. Geological Society of America, Bulletin, v. 95, p. 155-173.

______, Kominz, M. A. and Grotzinger, J. P. 1988. Cambro-Ordovician eustasy: evidence from geophysical modelling of subsidence in Cordilleran and Appalachian passive margins. *In:* New Perspectives in Basin Analysis. Frontiers in Sedimentary Geology, K. L. Kleinsphen and C. Paola (eds.). Springer-Verlag, New York, New York, p. 129-160.

Bova, J. A. and Read, J. F. 1987. Incipiently drowned facies within a cyclic peritidal ramp sequence. Early Ordovician Chepultepec interval, Virginia Appalachians. Geological Society of America, Bulletin, v. 98, p. 714-727.

Boyd, R., Forbes, D. L. and Heffler, D. E. 1988. Time-sequence observations on wave-formed sand ripples on an ocean shoreface. Sedimentology, v. 35, p. 449-464.

Christie-Blick, N., Grotzinger, J. P. and Van der Borch, C. C. 1988. Sequence stratigraphy in Proterozoic successions. Geology, v. 16, p. 100-104.

Clifton, H. E. 1976. Wave-formed sedimentary structures — a conceptual model. *In:* Beach and Nearshore Sedimentation, R. A. Davis and R. L. Ethington (eds.). Society of Economic Paleontologists and Mineralogists Special Publication 24, p. 126-148.

Cloetingh, S. 1988. Intraplate stresses: a new element in basin analysis. *In:* New Perspectives in Basin Analysis, K. L. Kleinspehn and C. Paola (eds.). Frontiers in Sedimentary Geology, Springer-Verlag, New York, New York, p. 205-230.

Dalrymple, R. W. 1984. Morphology and internal structure of sandwaves in the Bay of Fundy. Sedimentology, v. 31, p. 365-382.

De Raaf, J. M. F., Boersma, J. R. and Van Gelder, A. 1977. Wave-generated structures and sequences from a shallow-marine succession, Lower Carboniferous, County Cork, Ireland. Sedimentology, v. 24, p. 451-483.

Derrick, G. M., Wilson, I. H., Hill, R. M., Glikson, A. Y. and Mitchell, J. E. 1977. Geology of the Mary Kathleen 1:100,000 sheet area, northwest Queensland. Bureau of Mineral Resources, Bulletin 193, 114 p.

Eriksson, K. A. and Simpson, E. L. 1990. Recognition of high-frequency, sea-level fluctuations in Proterozoic siliciclastic tidal deposits, Mount Isa, Australia. Geology, v. 18, p. 474-477.

______, ______ and Jackson, M. J. 1990. Stratigraphic evolution of a Proterozoic rift to thermal relaxation basin: implications for lithospheric extension. Abstracts of the 13th International Sedimentological Congress, Nottingham, England, p. 155.

Eschner, T. B. and Kocurek, G. 1986. Marine destruction of eolian sand seas: origin of mass flows. Journal of Sedimentary Petrology, v. 56, p. 401-411.

Fischer, A. G. 1964. The Lofer cyclothems of the Alpine Triassic. Kansas Geological Survey, Bulletin 169, p. 107-149.

Fryberger, S. G., Ahlbrandt, T. S. and Andrews, S. 1979. Origin, sedimentary features and significance of low-angle eolian "sand sheet" deposits, Great Sand Dune National Monument and Vicinity, Colorado. Journal of Sedimentary Petrology, v. 49, p. 733-746.

Goldhammer, R. K., Dunn, P. A. and Hardie, L. A. 1987. High frequency glacio-eustatic sea level oscillations with Milankovitch characteristics recorded in the Middle Triassic platform carbonates in northern Italy. American Journal of Science, v. 287, p. 853-892.

Goodwin, P. W. and Anderson, E. J. 1985. Punctuated aggradational cycles: a general hypothesis of episodic stratigraphic accumulations. Journal of Geology, v. 98, p. 515-533.

Grotzinger, J. P. 1986. Cyclicity and paleoenvironment dynamics, Rocknest platform, northwest Canada. Geological Society of America, Bulletin, v. 97, p. 1208-1231.

Harms, J. C., Southard, J. B. and Walker, R. G. 1982. Structures and sequences in clastic rocks. Society of Economic Paleontologists and Mineralogists, Short Course no. 9, Tulsa, Oklahoma, 255 p.

Houbolt, J. J. H. C. 1968. Recent sediments in the southern bight of the North Sea. Geologie en Mijnbouw, v. 47, p. 254-273.

Hunter, R. E. 1977. Basic types of stratification in small eolian dunes. Sedimentology, v. 24, p. 361-388.

______ 1985. Subaqueous sand-flow cross-strata. Journal of Sedimentary Petrology, v. 55, p. 886-894.

Jackson, M. J., Simpson, E. L. and Eriksson, K. A. 1990. Sequence stratigraphy and relative sea-level control on sequence development in an intracratonic thermal sag basin: the Early Proterozoic, lower Quilalar Formation, Mount Isa Inlier, Australia. Sedimentology, v. 37, p. 1053-1078.

James, N. P. 1977. Facies models of shallowing-upward sequences in carbonates. Geoscience Canada, v. 4, p. 126-136.

Jervey, M. T. 1988. Quantitative geological modeling of siliciclastic rock sequences and their seismic expressions. *In:* Sea-level Changes: an Integrated Approach, C. K. Wilgus, B. S. Hastings, C. G. St. C. Kendal, H. W. Posamentier, C. A. Ross and J. C. Van Wagoner (eds.). Society of Economic Paleontologists and Mineralogists, Special Publication 42, p. 47-69.

Jones, N. S., Kain, J. M. and Stride, A. H. 1965. The movement of sandwaves on Warts Bank, Isle of Man. Marine Geology, v. 3, p. 329-336.

Klein, G. De V. 1970. Depositional and dispersal dynamics of intertidal sand bars. Journal of Sedimentary Petrology, v. 40, p. 1095-1127.

______ 1971. A sedimentary model for determining paleotidal range. Geological Society of America, Bulletin, v. 82, p. 2585-2592.

______ 1977. Clastic Tidal Facies. Continuing Education Publication Company, Champaign, Illinois, 149 p.

Kocurek, G. and Fielder, G. 1982. Adhesion structures. Journal of Sedimentary Petrology, v. 52, p. 1229-1241.

______ and Nielson, J. 1986. Conditions favorable for the formation of warm-climate aeolian sand sheets. Sedimentology, v. 33, p. 795-816.

Koerschner, W. F. and Read, J. F. 1989. Field and modelling studies of Cambrian carbonate cycles, Virginia Appalachians. Journal of Sedimentary Petrology, v. 59, p. 654-687.

Komar, P. D. 1974. Beach Processes and Sedimentation. Prentice Hall, New Jersey, 429 p.

Kohsiek, L. H. M. and Terwindt, J. H. J. 1981. Characteristics of foreset and topset bedding in megaripples related to hydrodynamic conditions on an intertidal shoal. *In:* Holocene Marine Sediments in the North Sea Basin, S. D. Nio, R. T. E. Shutenhelm and Tj. C. E. van Weering (eds.). International Association of Sedimentologists, Special Publication 5, p. 27-37.

Kreisa, R. D. and Moiola, R. J. 1986. Sigmoidal tidal bundles and other tide-generated sedimentary structures of the Curtis Formation, Utah. Geological Society of America, Bulletin, v. 97, p. 381-387.

Langhorne, D. N. 1982. A study of the dynamics of a marine sandwave. Sedimentology, v. 29, p. 571-594.

McCabe, P. J. and Jones, C. M. 1977. The formation of reactivation surfaces within superimposed deltas and bedforms. Journal of Sedimentary Petrology, v. 47, p. 707-715.

McCave, I. N. 1971. Sandwaves in the North Sea off the coast of Holland: Marine Geology, v. 10, p. 199-225.

Mazzullo, S. J. and Friedman, G. M. 1975. Conceptual model of tidal influenced deposition on margins of epeiric seas: Lower Ordovician (Canadian) of eastern New York and southwestern Vermont. American Association of Petroleum Geologists, Bulletin, v. 59, p. 2123-2141.

Nio, S. D., Siegenthaler, C. and Yang, C. S. 1983. Megaripple cross-bedding as a tool for reconstruction of paleohydraulics in a Holocene subtidal environment S.W. Netherlands. Geologie en Mijnbouw, v. 62, p. 499-510.

Off, T. 1963. Rhythmic linear sand bodies caused by tidal currents. American Association of Petroleum Geologists, Bulletin, v. 47, p. 324-341.

Olsen, H., Due, P. H. and Clemmenson, L. B. 1989. Morphology and genesis of asymmetrical adhesion warts — a new adhesion surface structure. Sedimentary Geology, v. 61, p. 277-285.

Page, R. W. 1981. Depositional age of the stratiform base metal deposits at Mount Isa and McArthur River, Australia, based on U-Pb dating of concordant tuff horizons. Economic Geology, v. 76, p. 648-658.

______ 1983a. Timing of superposed volcanism in the Proterozoic Mount Isa Inlier, Australia. Precambrian Research, v. 21, p. 223-245.

______ 1983b. Chronology of magmatism, skarn formation and uranium mineralization, Mary Kathleen, Queensland, Australia. Economic Geology, v. 85, p. 838-853.

______ 1985. Isotopic record of crustal events in northern Australia. Abstract of Conference on Tectonics and Geochemistry of Early and Middle Proterozoic Fold Belts, Bureau of Mineral Resources Record 1985/28.

______ and Bell, T. H. 1986. Isotopic and structural responses of granite to successive deformation and metamorphism. Journal of Geology, v. 94, p. 365-379.

______ , Bower, M. G., Zapasnik, T. K., Guy, D. B. and Hyett, N. C. 1981. Mount Isa Project (Geochronology Laboratory). Bureau of Mineral Resources Record 1981/46.

Plumb, K. A., Derrick, G. M., Needham, R. S. and Shaw, R. D. 1981. The Proterozoic of northern Australia. *In:* The Precambrian of the Southern Hemisphere, D. R. Hunter (ed.). Elsevier, Amsterdam, p. 205-307.

Posamentier, H. W., Jervey, M. T. and Vail, P. R. 1988. Eustatic controls on clastic deposition. I — conceptual framework. *In:* Sea-level Changes: an Integrated Approach, C. K. Wilgus, B. S. Hastings, C. G. St. C. Kendal, H. W. Posamentier, C. A. Ross and J. C. Van Wagoner (eds.). Society of Economic Paleontologists and Mineralogists, Special Publication 42, p. 109-124.

Read, J. F. 1989. Controls on evolution of Cambrian-Ordovician passive margin, U.S. Appalachians. *In:* Controls on Carbonate Platform and Basin Development, P. D. Crevello, J. L. Wilson, J. F. Sarg and J. F. Read (eds.). Society of Economic Paleontologists and Mineralogists, Special Publication 44, p. 147-165.

______ and Goldhammer, R. K. 1988. Use of Fischer plots to define third-order sea level curves in Ordovician peritidal cyclic carbonates, Appalachians. Geology, v. 16, p. 895-899.

Reineck, H-E. 1955. Haftrippeln und Haftwarzen, ablagerungsformen von flugsand. Senckenbergiana Lethaea, v. 36, p. 347-357.

Reinhardt, J. and Hardie, L. A. 1976. Selected examples of carbonate sedimentation, Lower Paleozoic of Maryland. Maryland Geological Survey Field Trip Guidebook 5, 53 p.

Rubin, D. M. 1987. Cross-bedding, bedforms, and paleocurrents. Society of Economic Paleontogists and Mineralogists, Concepts in Sedimentology and Paleontology, v. 1, 188 p.

______ and Hunter, R. E. 1982. Bedform climbing in theory and nature. Sedimentology, v. 29, p. 121-138.

Smith, D. B. 1988. Bypassing of sand over sandwaves and through a sandwave field in the central region of the North Sea. *In:* Tide-influenced Sedimentary Environments and Facies, P. L. de Boer, A. van Gelder and S. D. Nio (eds.). D. Reidel Publishing Company, Boston, p. 39-50.

Spencer, R. J. and Demicco, R. V. 1989. Computer models of carbonate platform cycles driven by subsidence and eustacy. Geology, v. 17, p. 165-168.

Sun, S. S. and McCulloch, M. T. 1985. Chemical and Nd isotope study of late Archean to middle Proterozoic mafic volcanic rocks in northern Australia. Abstracts of Conference on Tectonics and Geochemistry of Early and Middle Proterozoic Fold Belts. Bureau Mineral Resources Record 1985/28.

Tanner, W. F. 1960. Shallow water ripple mark varieties. Journal of Sedimentary Petrology, v. 30, p. 481-485.

Terwindt, J. H. J. 1981. Origin and sequences of sedimentary structures in inshore mesotidal deposits on the North Sea. *In:* Holocene Marine Sediments in the North Sea Basin, S. D. Nio, R. T. E. Shutenhelm and Tj. C. E. van Weering (eds.). International Association of Sedimentologists, Special Publication 5, p. 4-26.

Van Wagoner, J. C. 1985. Reservoir facies distribution as controlled by sea-level change. Society of Economic Paleontologists and Mineralogists, Midyear Meeting Abstracts, v. 2, p. 91.

______ , Mitchum, R. M., Campion, K. M., Rahmanian, V. D. 1990. Siliciclastic sequence stratigraphy in well logs, cores and outcrops. American Association of Petroleum Geologists, Methods in Exploration Series no. 7, 55 p.

______ , ______ , Posamentier, H. W. and Vail, P. R. 1987. Part 2: Key definitions of sequence stratigraphy. *In:* Atlas of Seismic Stratigraphy 1, A. W. Bally (ed.). American Association of Petroleum Geologists, Studies in Geology 27, p. 11-14.

______ , Posamentier, H. W., Mitchum, R. M., Vail, P. R., Sarg, J. F., Loutit, T. S. and Hardenbol, J. 1988. An overview of the fundamentals of sequence stratigraphy and key definitions. *In:* Sea-level Changes: an Integrated Approach, C. K. Wilgus, B. S. Hastings, C. G. St. C. Kendal, H. W. Posamentier, C. A. Ross and J. C. Van Wagoner (eds.). Society of Economic Paleontologists and Mineralogists, Special Publication 42, p. 39-45.

Visser, M. J. 1980. Neap-Spring cycles reflected in Holocene subtidal large-scale bedform deposits: a preliminary note. Geology, v. 8, p. 543-546.

Wyborn, L. A. I., Page, R. W. and McCulloch, M. T. 1988. Petrology, geochronology and isotope geochemistry of the post-1820 Ma granites of the Mount Isa Inlier: Mechanisms for the generation of Proterozoic anorogenic granites. Precambrian Research, v. 40/41, p. 509-541.

KEY WORDS: PARASEQUENCE, SILICICLASTIC TIDAL DEPOSITS, LOWER PROTEROZOIC, MOUNT GUIDE QUARTZITE, MOUNT ISA INLIER